Synergies in Smart and Virtual Systems using computational intelligence

In recent years, the integration of computational intelligence into smart and virtual systems has been a driving force behind innovations in various technological domains. Computational intelligence (CI), encompassing techniques such as machine learning (ML), artificial intelligence (AI), fuzzy logic, genetic algorithms, and neural networks, has enabled the development of systems that are not only autonomous but also capable of adapting, learning, and optimizing their behaviour over time.

Smart systems, typically embedded in devices, are designed to interact with the environment and make decisions based on real-time data. These systems are enhanced by CI methods to improve their functionality, accuracy, and decision-making capabilities. For instance, smart grids, smart cities, autonomous vehicles, and IoT-based applications rely on CI to process vast amounts of data, identify patterns, and optimize operations in real-time. The synergy between CI and these systems enables them to function intelligently, predict issues before they arise, and offer solutions that would be difficult for traditional systems to achieve.

On the other hand, virtual systems, which are increasingly part of industries ranging from healthcare to entertainment, benefit from CI by providing immersive, real-time, and data-driven experiences. Virtual reality (VR) and augmented reality (AR) systems powered by computational intelligence offer enhanced user interaction, personalized experiences, and the ability to simulate complex real-world scenarios. Additionally, virtual simulations for industrial applications—such as product design, testing, and manufacturing—rely on CI to improve accuracy, speed, and cost-efficiency.

The convergence of smart systems and virtual systems with computational intelligence creates a powerful synergy. This combination allows for the development of intelligent environments that can respond to human actions, predict future needs, and self-optimize based on changing conditions. For example, virtual assistants in smart homes use CI to learn user preferences, automate tasks, and improve efficiency, while virtual training platforms in industries like healthcare and aviation simulate real-world scenarios with increasing levels of realism and adaptability.

Synergies in Smart and Virtual Systems using computational intelligence

Edited by

Dr. Shankar Babu

Vice Principal, Dept of EEE,

SVR Engineering College,

Ayyaluru metta, Nandyal, Andhra Pradesh, India – 518503

Dr. Mahesh Babu Kota

Assistant Professor,

Aditya University,

Surampalem, Kakinada, Andhra Pradesh, India

CRC Press
Taylor & Francis Group
Boca Raton London New York

CRC Press is an imprint of the
Taylor & Francis Group, an **informa** business

First edition published 2026
by CRC Press
2385 NW Executive Center Drive, Suite 320, Boca Raton FL 33431

and by CRC Press
4 Park Square, Milton Park, Abingdon, Oxon, OX14 4RN

British Library Cataloguing-in-Publication Data
A catalogue record for this book is available from the British Library

ISBN: 9781041165897 (pbk)
ISBN: 9781041165910 (hbk)
ISBN: 9781003685364 (ebk)

DOI: 10.1201/9781003685364

Typeset in Times New Roman
by HBK Digital

Dedication

The International Conference on Smart Systems, Virtual Intelligence and Robotics Automation using Advanced Electronics and Computational Designs (ICSVREC - 2025) is a prestigious global platform dedicated to promoting cutting-edge research and innovation in the fields of smart systems, artificial intelligence, robotics, and advanced electronics.

This interdisciplinary event brings together researchers, academicians, industry experts, and technology leaders to share insights, discuss challenges, and present breakthroughs in the integration of computational intelligence with engineering applications. With rapid advancements in automation, IoT, machine learning, and embedded systems, the conference aims to explore novel frameworks and intelligent solutions that address complex real-world problems across sectors such as healthcare, smart cities, manufacturing, and sustainable technologies.

ICSVREC-2025 encourages collaboration between academia and industry by showcasing the latest trends in smart infrastructure, robotics automation, virtual intelligence, and cyber-physical systems. The event also features keynote lectures, technical paper presentations, workshops, and panel discussions that facilitate intellectual exchange and future-oriented innovations. By providing a common ground for diverse stakeholders, the conference seeks to shape a more connected, intelligent, and efficient technological future.

Contents

List of Figures

List of Tables

Foreword

It gives us immense pleasure to present the proceedings of the *International Conference on Smart Systems, Virtual Intelligence and Robotics Automation using Advanced Electronics and Computational Designs (ICSVREC - 2025)*. This conference serves as a vibrant platform for academics, researchers, and industry professionals to converge and explore the synergies emerging at the intersection of artificial intelligence, robotics, smart systems, and advanced electronic design.

As we witness rapid technological transformations driven by AI, ML, and automation, it becomes imperative to create academic ecosystems that foster research, innovation, and collaborative knowledge exchange. ICSVREC-2025 is a step toward that vision, offering valuable insights into the future of intelligent systems and their practical applications in various domains including healthcare, energy, transportation, and smart infrastructure.

We are proud to host this global academic congregation at SVR Engineering College, Nandyal, and extend our sincere gratitude to all contributors — keynote speakers, authors, reviewers, session chairs, and organizing members — whose efforts made this conference a grand success. Special thanks to the management, faculty, and students whose commitment and coordination brought this event to fruition.

We hope the ideas and innovations presented here will inspire further research and contribute meaningfully to the development of a smarter and more sustainable world.

Dr. P. Sankar Babu
Conference Convener, ICSVREC-2025
Vice-Principal & HOD, EEE
SVR Engineering College, Nandyal

Preface

We are delighted to present the preface to the *International Conference on Smart Systems, Virtual Intelligence and Robotics Automation using Advanced Electronics and Computational Designs (ICSVREC – 2025)*. This conference was conceived with the aim of providing a robust interdisciplinary platform for researchers, academicians, professionals, and students to share their research findings, innovative ideas, and technological advancements across the domains of smart systems, virtual intelligence, robotics, electronics, and computational intelligence.

In today's rapidly evolving technological landscape, the integration of AI and ML algorithms into embedded and smart systems has unlocked new opportunities in automation, real-time decision-making, and intelligent control across various sectors. ICSVREC-2025 stands as a testimony to the spirit of innovation and collaboration, showcasing pioneering work from contributors around the world.

This conference volume includes carefully reviewed papers, insightful keynote addresses, and technical discussions that reflect the diversity and depth of research in these emerging areas. We believe these contributions will not only inspire future innovations but also bridge the gap between theoretical research and practical implementation.

We express our heartfelt thanks to the management of SVR Engineering College, the Principal, Heads of Departments, organizing committee members, technical reviewers, sponsors, and enthusiastic student volunteers for their unwavering support and tireless efforts in making this event a grand success.

We hope this compendium of knowledge serves as a valuable reference for researchers and professionals seeking to advance the frontiers of smart and intelligent systems.

Dr. P. Sankar Babu
Conference Convener, ICSVREC-2025
Vice-Principal & HOD, EEE
SVR Engineering College, Nandyal

Acknowledgements

We would like to extend our heartfelt appreciation to Mr. Raja Suresh, Visiting Professor at Sri Venkateswara College of Engineering, Tirupati, for taking the initiative to guide and support the successful organization of this international conference. His extensive industrial experience and academic engagement have significantly enhanced the quality, vision, and execution of this event.

Mr. Raja Suresh is the Visionary Founder and Managing Director of Ekalavya Innovative Solutions Pvt. Ltd. (EISPL), a Government of India-recognized enterprise specializing in next-generation solutions for industrial, consumer, commercial, and advanced technology applications. With over two decades of professional expertise in embedded systems, electronic design, and R&D, he has played a transformative role in making EISPL a leading name in the embedded technology sector, particularly in domains such as Automotive, Semiconductors, IoT, and Industrial Automation.

Academically, Mr. Raja Suresh holds a B.Tech in Electronics and Communication Engineering (ECE) and an M.Tech in Embedded System from Jawaharlal Nehru Technological University, Anantapur, Andhra Pradesh. His strong educational foundation has been instrumental in shaping his technical proficiency and innovative outlook.

His leadership is rooted in a commitment to excellence, innovation, and industry-academia integration. His expertise spans across Solar Energy Systems, Power Supply Design, Embedded Hardware, Driver Boards, and Embedded Applications. Through his strategic direction, EISPL has developed cutting-edge technologies that not only meet market demands but also contribute significantly to national R&D advancement.

A passionate mentor and advocate for young talent, Mr. Raja Suresh is devoted to fostering a culture of innovation and learning. His vision of building a globally competitive ecosystem for embedded solutions is reflected in EISPL's national impact and continued growth.

We are privileged to have had his guidance and visionary support for this conference and sincerely thank him for his outstanding contributions.

List of abbreviations

AI – Artificial Intelligence
ML – Machine Learning
VLSI – Very-Large-Scale Integration
IoT – Internet of Things
SEM – Scanning Electron Microscope
PCB – Printed Circuit Board
CAD – Computer-Aided Design
DSP – Digital Signal Processing
MEMS – Micro-Electro-Mechanical Systems
FPGA – Field Programmable Gate Array
AI-ML – Artificial Intelligence and Machine Learning
SNR – Signal-to-Noise Ratio
UHF – Ultra High Frequency
GHz – Gigahertz
ADC – Analog-to-Digital Converter
DAC – Digital-to-Analog Converter
MCU – Microcontroller Unit
GSM – Global System for Mobile Communications
RF – Radio Frequency
IoT – Internet of Things
RTOS – Real-Time Operating System
PID – Proportional-Integral-Derivative
AIoT – Artificial Intelligence of Things

Glossary

Artificial Intelligence (AI)

A branch of computer science that focuses on creating machines and systems capable of performing tasks that typically require human intelligence, such as learning, reasoning, and problem-solving.

Machine Learning (ML)

A subset of AI that involves the development of algorithms that allow computers to learn from data and improve their performance over time without being explicitly programmed.

Very-Large-Scale Integration (VLSI)

The process of creating integrated circuits by combining thousands or millions of transistors into a single chip. VLSI plays a critical role in the miniaturization of electronics and is essential in modern computing and communication systems.

Microstructure

Refers to the structure of a material as seen under a microscope, typically focusing on its grain structure and the properties that affect its physical characteristics, such as mechanical strength and electrical conductivity.

Internet of Things (IoT)

The interconnection of physical devices, vehicles, appliances, and other objects through sensors, software, and network connectivity, allowing them to collect and exchange data.

Embedded Systems

Specialized computing systems that are designed to perform a specific task or function within a larger system, often with real-time constraints. These systems are found in everything from automotive electronics to consumer gadgets.

Scalable Systems

Systems that can grow or be modified to accommodate increased demand or workload. Scalability is crucial for both hardware (e.g., VLSI designs) and software (e.g., AI-ML algorithms).

Field Programmable Gate Array (FPGA)

A type of programmable logic device that can be configured to perform specific logic functions. FPGAs are widely used in applications requiring fast processing and reconfigurability, such as signal processing and hardware acceleration.

Digital Signal Processing (DSP)

The use of algorithms and techniques to manipulate and analyze digital signals (such as sound, image, and video signals). DSP plays a crucial role in telecommunications, multimedia systems, and medical applications.

Micro-Electro-Mechanical Systems (MEMS)

Small-scale devices that integrate mechanical components, sensors, actuators, and electronics. MEMS are used in a variety of applications, including sensors in automotive, medical, and consumer electronics.

Real-Time Operating System (RTOS)

An operating system designed to handle real-time tasks that require immediate processing. RTOS is used in applications where time constraints are critical, such as robotics, embedded systems, and industrial control systems.

Robotics

The design, construction, operation, and use of robots. Robotics combines elements of mechanical engineering, electrical engineering, and computer science to automate tasks traditionally performed by humans.

Proportional-Integral-Derivative (PID) Controller

A type of control loop feedback mechanism used in industrial control systems to continuously control a process variable, maintaining desired output by adjusting inputs.

Signal-to-Noise Ratio (SNR)

A measure of signal strength relative to background noise. SNR is a key factor in determining the quality of communication systems, sensor data, and multimedia systems.

Analog-to-Digital Converter (ADC)

A device that converts an analog signal into a digital signal, allowing digital systems to process real-world continuous signals such as temperature, sound, or pressure.

Digital-to-Analog Converter (DAC)

A device that converts a digital signal into an analog signal, which can then be used in applications such as audio output and analog sensors.

Gigahertz (GHz)

A unit of frequency equal to one billion cycles per second. GHz is commonly used to measure the clock speed of processors and the frequency of electromagnetic waves in communication systems.

Radio Frequency (RF)

Refers to electromagnetic wave frequencies typically in the range of 3 kHz to 300 GHz, used in wireless communication, broadcasting, and radar systems.

Automated Test Equipment (ATE)

Devices used to perform automated tests on electronic devices or systems. ATE is crucial for quality assurance and verification in semiconductor manufacturing and system development.

Control Systems

Systems designed to manage and regulate the behavior of other systems, often through the use of feedback loops, sensors, and algorithms. Common applications include automation, robotics, and industrial systems.

Smart Systems

Systems embedded with intelligent algorithms, sensors, and machine learning models to make decisions, adapt to changes, and interact with their environment without human intervention.

Nanoelectronics

A branch of electronics that deals with the design and fabrication of circuits and devices at the nanoscale, enabling advancements in computing, sensors, and quantum technologies.

Patent

A form of intellectual property that grants the holder exclusive rights to an invention or design for a specified period, preventing others from making, using, or selling the invention without permission.

Copyright

A legal right granted to the creator of original works of authorship, providing exclusive rights to use, distribute, and reproduce the work.

Semiconductor

A material that has electrical conductivity between that of a conductor and an insulator, widely used in the manufacturing of electronic devices such as transistors, diodes, and integrated circuits.

Editors Biography

Dr. Shankar Babu
Vice-Principal & HOD, EEE
SVR Engineering College and Editor of ICSVRCET 2025

Dr. P. Sankar Babu is an esteemed academic and accomplished researcher in the field of Electrical and Electronics Engineering (EEE). He currently serves as the Professor and Head of the Department (HOD) of EEE at SVR Engineering College (SVREC), Nandyal, and also holds the key administrative role of Vice-Principal at the institution. With a career rooted in academic excellence, research, and leadership, Dr. Babu has been a driving force behind many innovative projects and educational advancements. One of his most notable achievements is his role as the Principal Investigator and Coordinator for the UGC-funded project titled "Single Phase Bi-Directional PWM Converter for Micro-Grid Systems", which was completed successfully between 2014 and 2017 with a sanctioned budget of ₹2.6 lakhs. This project played a pivotal role in advancing micro-grid technologies and intelligent power management systems.

In recognition of his academic and research excellence, Dr. Babu has been awarded 3 patents, authored 4 book chapters, published 3 books, presented papers at 6 national and international conferences, and contributed articles to 5 reputed journals, including the Journal of Engineering Sciences. His work reflects a blend of deep theoretical insight and practical innovation, making him a respected voice in the academic community.

Dr. P. Sankar Babu continues to inspire students, guide research scholars, and lead various institutional initiatives, such as serving as the Convener of **ICSVREC - 2025** : International Conference on Smart Systems, Virtual Intelligence and Robotics Automation using Advanced Electronics and Computational Designs.

Dr. Mahesh Babu Kota
Assistant Professor,
Aditya University, Surampalem, Kakinada

Dr. K. Mahesh Babu is a committed academician and researcher with over a decade of experience in teaching, research, and institutional development. He holds a Ph.D. in Antenna Design and an M.Tech in VLSI System Design, with core expertise in microwave engineering, reconfigurable antennas, embedded systems, and wireless communication.

Currently serving as an Assistant Professor at Aditya University, Dr. Mahesh Babu has made significant contributions through seven Scopus/Web of Science-indexed publications and ten international conference presentations. He is also an innovator with three patents in electronics applications, emphasizing his commitment to applied research.

Dr. Babu brings hands-on expertise in simulation tools such as CST Studio Suite and Ansys HFSS, and is known for integrating technology-driven pedagogy and mentoring into academic environments. He is actively involved in faculty development, student training, and technical skill enhancement initiatives.

As an editorial board member, Dr. Mahesh Babu supports the peer review and technical evaluation process, ensuring high-quality research contributions in the fields of smart systems, antenna design, and VLSI technologies.

1 Next-gen machine learning strategies for social media fake account identification

Ejaru Naga Prabhakar[1,a], Suvarna Boya[2,b], Shagufta Fathima Mohammad[2,c], Mohammad Khaja Kalayi[2,d] and Prasad Pathikapalu[2,e]

[1]Assistant Professor, Department of CSE, Srinivasa Ramanujan Institute of Technology, Anantapur, Andhra Pradesh, India

[2]Students, Department of CSE (Data Science), Srinivasa Ramanujan Institute of Technology, Anantapur, Andhra Pradesh, India

Abstract

People of this generation are socializing through online social networks, such as Instagram. They use it to share what is happening in their lives, to arrange meetings, events, and business. The very rapid growth of OSNs and massive amounts of personal data have led them to become the targets of attackers and imposters, who want to steal their data, conduct malicious activities. In response to these threats, researchers make accounts feature efficient and develop algorithms of classification for detecting abnormalities and imposter accounts. However, without any innovative thought, some of these features are already known to be harmful to results, while this also poses the down factor of working with only one algorithm, which would give unsatisfactory results. In this paper, we present an advanced identification framework that selects four techniques acting as feature selection and dimension reduction steps so as to give personalized input data and then use six machine learning classifiers: Decision Tree, Random Forest, logistic regression, Support Vector Machine (SVM), and Naïve Bayes. Using a multi-algorithm approach that attempts to boost each classifier's power to improve accuracy and reliability in the identification of a fake account.

Keywords: Decision Tree classifier, fake account detection, feature selection, logistic regression, machine learning algorithms, Naive Bayes classifier, online social networks, Random Forest classifier, Support Vector Machine

Introduction

However, online social networks (OSNs) have also gained relevance in the social life of individuals, allowing us to communicate via Instagram, Facebook, Twitter and LinkedIn, among others, organizing events, as well as managing our own or any business. They are essential in determining how people interact with each other and the ways in which they share personal or professional information [1]. As OSNs increased popular and usage, so did they become targets of the attackers, the hackers, and the impersonators. In reality, these malicious attackers leverage the extensive caches of personal information. These platforms manage to perpetuate identity theft, disseminate fabricated news and perpetrate immoral operations that have an adverse effect on its users and corrode the security of these networks. More recently, the advances in machine learning and feature engineering have allowed researchers to come up with systems that can more accurately detect abnormal activities and fake accounts [2]. Nevertheless, despite all these advances, there is still much room for both the optimization of the input data and the improvement of the efficiency of the detection systems when some of the features may not be as reliable [3].

Related Work

In this paper, we discuss the theme of literature review on the fake Instagram account identification via machine learning [4]. Then it discusses different means such as image detection and natural language processing to find the fake accounts. Through these examples the authors show how these approaches can be combined to increase the reliability of fake account identification systems. It was interesting that they also discussed feature engineering and dataset diversity, which is required to have high accuracy.

[a]nagaprabhakar.cse@srit.ac.in, [b]suvarnaboya19@gmail.com, [c]fathimashagufta77@gmail.com, [d]kmkhaja2003@gmail.com, [e]pathikapaluprasad@gmail.com

DOI: 10.1201/9781003685364-1

Having integrated these methods, the authors claim that the fake account detection systems can become more robust and adaptive [5]. whether they are real or fake users. In this system, the authors focus on supervised learning framework, whereby they train classifiers using labeled datasets. Finally, the paper concludes that the supervised learning techniques can achieve high accuracy given comprehensive and well labeled datasets [6,7,8,9].

Proposed Work

In this work, an advanced application is proposed and attempts to remedy the deficiencies of conventional fake account detection methods. The purpose is to devise a quick yet thoroughly dependable manner to distinguish fake accounts from true accounts. For this, a number of powerful machine learning algorithms have been applied: Decision Tree (DT), Random Forest (RF), logistic regression (LR), Support Vector Machine (SVM), and Naive Bayes (NB) using a Python-based environment [10]. Using the strength of each model, this multi algorithm approach combines accuracy and reliability of detection. With features selection and preprocessing, attributes application on the algorithms, the proposed system would be capable of fighting fraudulent accounts and ease security and trustworthiness in online social platforms.

Data collection

For the purposes of our project, we collect several datapoints to characterize an account and determine whether it is faked or real. Specifically, it includes an account profile picture. The value of these features is that they can give cite into the practice and popularly of an account. For example, if your account has hardly any posts or followers, then you'll be suspected of being fake. We also collect variable information such as what is the account's URL, if the account is private, if it has any external URL, description length, and whether the profile images are men or women. The other important aspect is comparing the accounts username and full name, (i.e., the username and the full name are the same). Finally, we print out the accounts as being fake (0) or real (1) according to our results.

Model building

For detecting fake accounts, in our project we have trained our model using a number of machine learning

algorithms. Given that it is simple, interpretable and helps us understand what contributions different features have contributed to the probability of an account being fake or genuine, Logistic Regression was used. The Decision Tree algorithm made its use for an ability in splitting the data into intuitive subsets on basis of the values of the features had, giving it a clear, rule-based sense of classification [11]. For text-based data such as description, naive Bayes was used as probabilistic model based on feature independence, presume that has been a good model in many cases.

Model evaluation

Now, having trained the model in the previous step, we evaluate them by calculating the accuracy and we have a good measure on how much the models do well as they are trained to figure out how right they were. For this reason, we use a classification report that will give more detailed metrics such as precision, recall and F1-score. Recall measures how many of the real fake accounts were identified correctly and Precision measures the percentage of the predicted fake accounts that truly were fake [12]. Harmonic means of precision and recall gives the F1-score which gives a balanced view of the model's performance. Figure 1.1.

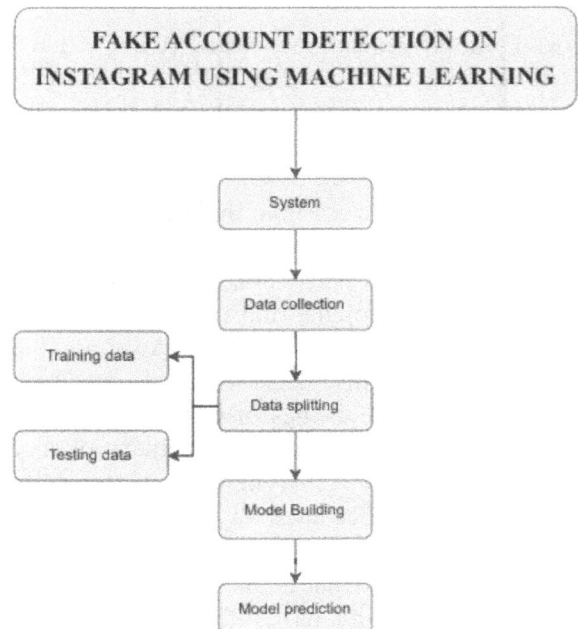

Figure 1.1 Proposed system workflow
Source: Author

Methodology

Logistic regression

The purpose is to predict probability of a data point belonging to one of two classes and LR is a statistical method used for this. Here an equation is fitted to the input features and then the formula used is the formula or the sigmoid formula or logistic value which is one of the classes in probability between 0 and 1, and this probability is used to classify the data point is one or two class. LR assumes that a general approach establishes simple estimation of the target outcome in relation to explanatory variables primarily using a log odds transformation method [13]. The algorithm solves the optimal features weights, which minimizes the logistic loss function.

Decision Tree

Supervised machine learning algorithm used for classification as well as regression, DT is a function to classify or regress decisions. The split is found by putting the data in the list of subsets based on the features that provide the best split. The algorithm works at each step and selects that feature which will cause maximum information gain or Gini reduction at each step [14]. A DT is a structure formed like a tree where nodes are features, branches are decision rules based on the features, and leaf nodes are class labels or predicted outcomes.

Naïve Bayes

Bayes' theorem is a simple yet powerful probabilistic classifier that calculates the probability of an event given certain evidence. NB is a simple implementation of this probabilistic classifier that assumes dependencies between attributes of the data to be independent of each other [15]. The main assumption in NB is that the features are independent from one another and thus simplifies the computations of conditional probabilities. NB is a surprisingly effective method for many real-world applications even when the features themselves are not conditionally independent. Gaussian NB classifier assumes the features to have a Gaussian distribution; multinomial NB is used for text classification problems; and Bernoulli NB is suitable for features that can have only either a value of 1 (yes) or 0 (no) [16]. Naive Bayes is fast to compute, works well on large data sets, and is excellent for the type of text classification problem such as spam detection or sentiment analysis.

Results and discussions

The model exhibits high accuracy with 95 Genuine instances and 90 Fake instances correctly predicted. However, the model has some errors, misclassifying 10 Genuine instances as Fake and 14 Fake instances as Genuine. These errors indicate areas for improvement in the model's precision and recall.

The model accurately classifies 93 Genuine and 95 Fake instances, reflecting a solid performance overall. However, it misclassifies 12 Genuine instances as Fake and 9 Fake instances as Genuine, indicating room for improvement. These misclassifications highlight areas where the model's precision and recall can be further enhanced to achieve better accuracy. Figures 1.2, 1.3, 1.4.

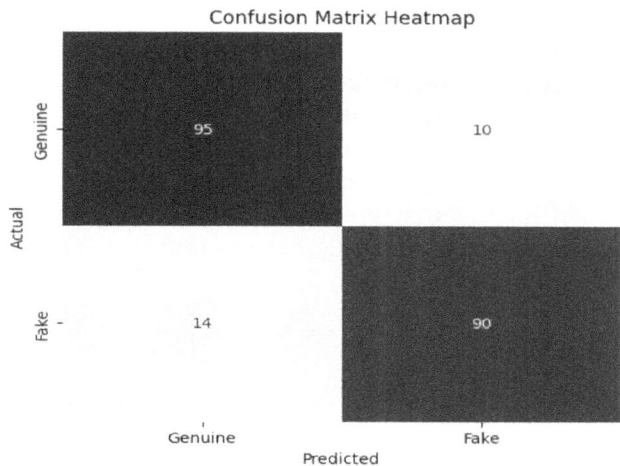

Figure 1.2 Decision tree confusion matrix
Source: Author

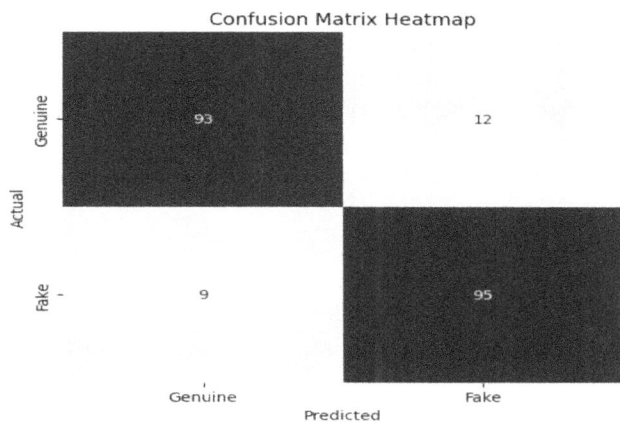

Figure 1.3 Logistic regression confusion matrix
Source: Author

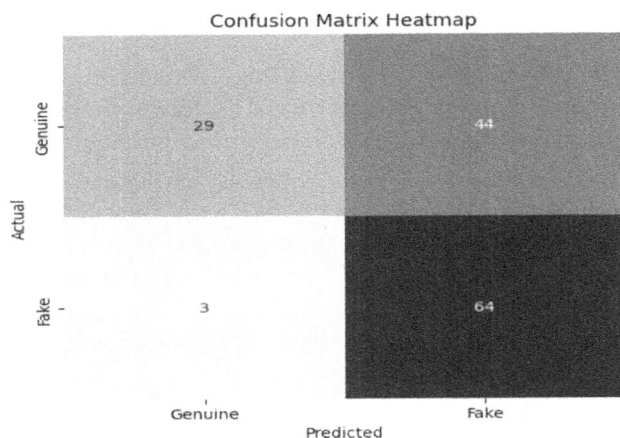

Figure 1.4 Naive bayes confusion matrix
Source: Author

The model accurately predicts 64 Fake instances as Fake, indicating its strong ability to detect fakes. However, it incorrectly classifies 44 Genuine instances as Fake, highlighting a significant area for improvement in recognizing genuine instances. With only 3 Fake instances misclassified as Genuine, the model shows high precision in identifying fakes, but overall, it needs better balance in distinguishing between the two classes.

Conclusion

Moving on to perform across varying algorithms, our models seemed to come out on top at some level and sometimes came up short as well. The Decision Tree and Random Forest models achieve a very high accuracy in distinguishing the genuine and fake test instances. Nevertheless, they still miss classify the considerable number of instances and thus the necessity of higher precision and recall is not well served by them. Although the logistic regression model seems to work solidly, there are cases of misclassifications indicating that it is necessary to refine the fitting. While the Naive Bayes model performs quite well for detecting fake instances, it does not seem to have the power to recognize real ones, which signifies a major potential area of improvement. SVM model performs well on distinguishing fake items as opposed to genuine items, however with less amount of accuracy.

Future Enhancement

There are several other improvements that could make the current model of fake account detection more accurate, scalable, and adaptable. Also, the feature engineering process appears to be one key area for enhancement. Let's take for instance one of the user behavior patterns like post frequency, interaction rate (likes and comments) and time between account creation and first post that will give us important insights. Network analysis – assess the connections (e.g. mutual followers, followers), between accounts – might better aid in separating fake and authentic accounts. Moreover, there is room for improvement in using more advanced deep learning techniques [17]. However, at the moment, we are still reliant on popular machine learning algorithms such as logistic logit model, CART, Gradient boosting machine (GBM) - advanced variant, and what not, that do work well, however, may not be as capable of embracing more intricate graphs. Especially recurrent neural networks (RNNs) or long short-term memory (LSTM) networks could be integrated to analyze such data [18]. Fake accounts following irregular posting or activity patterns could be detected upon these models. Furthermore, convolutional neural networks (CNNs) may help analyzing different elements of the profile pictures.

Reference

[1] Singh, N., Sharma, T., Thakral, A., & Choudhury, T. (2018). Detection of fake profiles in online social networks using machine learning. In Proceedings of the 2018 International Conference on Advances in Computing and Communication Engineering (IC-ACCE), (pp. 231–234). https://doi.org/10.1109/IC-ACCE.2018.8441713.

[2] Erşahin, B., Aktaş, Ö., Kilmç, D., & Akyol, C. (2017). Fake account detection on Twitter. In Proceedings of the 2nd International Conference on Computer Science and Engineering (UBMK), (pp. 388–392). https://doi.org/10.1109/UBMK.2017.8093420.

[3] Khaled, S., El-Tazi, N., & Mokhtar, H. M. O. (2018). Detecting fake accounts on social media. In Proceedings of the 2018 IEEE International Conference on Big Data (Big Data), (pp. 3672–3681). https://doi.org/10.1109/BIGDATA.2018.8621913.

[4] Chakraborty, P., Shazan, M. M., Nahid, M., Ahmed, M. K., & Talukder, P. C. (2022). Fake profile detection using machine learning techniques. *Journal of Computer and Communications,* 10(10), 74–87. https://doi.org/10.4236/JCC.2022.1010006.

[5] Ezarfelix, J., Jeffrey, N., & Sari, N. (2022). Systematic literature review: Instagram fake account detection based on machine learning. *Engineering,*

Mathematics, and Computer Science (EMACS) Journal, 4(1), 25–31. https://doi.org/10.21512/EMACS-JOURNAL.V4I1.8076.

[6] Yadav, A. (2023). Instagram fake profile detection: a review. *International Journal of Novel Research and Development*, 8(7), 2456–4184. Available from: www.ijnrd.org.

[7] Chelas, S., Routis, G., & Roussaki, I. (2024). Detection of fake Instagram accounts using machine learning techniques. *Computers*, 13(11), 296. https://doi.org/10.3390/COMPUTERS13110296.

[8] Azami, P., & Passi, K. (2024). Detecting fake accounts on Instagram using machine learning and hybrid optimization algorithms. *Algorithms*, 17(10), 425. https://doi.org/10.3390/A17100425.

[9] Singh, V., Tolasaria, N., Alpeshkumar, P. M., & Bartwal, S. (2023). Classification of instagram fake users using supervised machine learning algorithms. Available from: https://arxiv.org/abs/2311.12336.

[10] Nivas, T. S., Sriramkrishna, P., Reddy, S. S. K., Rao, P. V. R. G., & Komali, R. S. P. (2024). Fake account detection on Instagram using machine learning. *International Journal of Research in Engineering, Science, and Management*, 7(5), 24–26. Available from: https://journal.ijresm.com/index.php/ijresm/article/view/3023.

[11] Joshi, U. D., Vanshika, Singh, A. P., Pahuja, T. R., Naval, S., & Singal, G. (2021). Fake social media profile detection. *Machine Learning Algorithms and Applications*, 193–209. https://ouci.dntb.gov.ua/en/works/4kzjBRK7/.

[12] Kanagavalli, N., & Priya, S. B. (2022). Identification of fake accounts and fake news on social networks using reliabledeep learning. https://www.techscience.com/iasc/v33n1/46146.

[13] Joshi, Shruti & Nagariya, Himanshi & Dhanotiya, Neha & Jain, Sarika. (2020). Identifying Fake Profile in Online Social Network: An Overview and Survey. 10.1007/978-981-15-6315-7_2. https://www.researchgate.net/publication/342405046_Identifying_Fake_Profile_in_Online_Social_Network_An_Overview_and_Survey

[14] Elyusufi, Y., Elyusufi, Z., & Kbir, M. A. (2020). Social networks fake profile detection using machine learning algorithms. In Lecture Notes in Intelligent Transportation and Infrastructure, (Part F1409, pp. 30–40). https://doi.org/10.1007/978-3-030-37629-1_3.

[15] Kodati, S., Reddy, K. P., Mekala, S., Murthy, P. L. S., & Reddy, P. C. S. (2021). Detection of fake profiles on Twitter using hybrid SVM algorithm. In E3S Web of Conferences, (Vol. 309, p. 01046). https://doi.org/10.1051/E3SCONF/202130901046.

[16] Romanov, A., Semenov, A., Mazhelis, O., & Veijalainen, J. (2017). Detection of fake profiles in social media: A literature review. In Proceedings of the 13th International Conference on Web Information Systems and Technologies (WEBIST), (pp. 363–369). https://doi.org/10.5220/0006362103630369.

[17] Ali, A. K., & Abdullah, A. M. (2022). Fake account detection on social media using a stack ensemble system. *International Journal of Electrical and Computer Engineering, (IJECE)*, 12(3), 3013–3022. https://doi.org/10.11591/ijece.v12i3.pp3013-3022.

[18] Borkar, B. S., Patil, D. R., Markad, A. V., & Sharma, M. (2022). Real or fake identity deception of social media accounts using recurrent neural network. In 2022 International Conference on 4th Industrial Revolution Based Technology and Practices (ICFIRTP), (pp. 80–84). https://doi.org/10.1109/ICFIRTP56122.2022.1005943.

2 Fingerprint-based blood group detection using deep learning

P. Sirisha[1,a], Salla Abhinaya[2,b], M. Chaitanya Kumar Reddy[2,c], G. Balamadhu[2,d], S. Mohammad Manzoor[2,e] and J. Amarnath Reddy[2,f]

[1]Assistant Professor, Department of CSE, Srinivasa Ramanujan Institute of Technology, Anantapur, Andhra Pradesh, India

[2]Students, Department of CSE (Data Science), Srinivasa Ramanujan Institute of Technology, Anantapur, Andhra Pradesh, India

Abstract

Deciding a patient's blood group could be an imperative strategy in healthcare, playing a key part in clinical diagnostics, guaranteeing secure blood transfusions, and giving opportune mediations amid therapeutic crises. Conventional strategies require blood tests, which can be invasive and time-consuming. The unused framework utilizes progressed profound learning strategies to distinguish blood groups by analyzing special unique fingerprint characteristics. The investigate consolidates convolutional neural systems (CNNs), MobileNet, ResNet combined with recurrent neural systems (RNN), and Vision Transformers (ViT) to extricate significant unique mark highlights and classify them into particular blood bunch categories: A+, A-, AB+, AB-, B+, B-, O+, and O-. The think about assesses the viability of these models in terms of precision, computational effectiveness, and possibility in real-world applications. The results illustrate that profound learning-based fingerprint examination can serve as a solid elective to conventional blood writing strategies, advertising a quick and exact arrangement for therapeutic diagnostics.

Keywords: Biometric applications, blood group detection, convolutional neural networks, deep learning, fingerprint analysis, medical diagnostics, MobileNet, ResNet, vision transformer

Introduction

Identifying blood groups through fingerprint analysis is a groundbreaking, non-invasive method that combines biometric technology with medical science. This approach utilizes the unique patterns of an individual's fingerprint to determine their blood type, providing valuable information for healthcare professionals. The core principle is that sweat from fingerprint ridges contains specific proteins or antigens associated with different blood groups. Unlike traditional blood testing, which involves needles and can be uncomfortable, fingerprint-based detection eliminates this need, making it more patient-friendly. This method also has potential applications in forensic science and disaster management, where rapid blood type identification is crucial. Fingerprints are reliable identifiers since they remain unchanged throughout a person's life. By analyzing the antigen composition in sweat, it becomes possible to ascertain an individual's blood type, particularly within the ABO and Rh systems. This non-invasive technique is especially beneficial for children and older adults. Ongoing research and technological advancements continue to enhance the accuracy and reliability of fingerprint-based blood group detection. As this technology evolves, it has the potential to revolutionize medical diagnostics, making blood group identification more accessible and efficient.

Objective of the study

The most objective of this think about is to make a dependable framework for deciding blood bunches utilizing unique mark examination. It utilizes progressed machine learning strategies such as convolutional neural systems (CNN), MobileNet, recurrent neural systems (RNN) combined with ResNet, and Vision Transformers (ViT) to upgrade classification exactness. The framework is outlined to classify fingerprints into eight particular blood bunches: A+, A-, AB+, AB-, B+, B-, O+, and O-.

[a]sirisha.cse528@gmail.com, [b]sallaabhinaya@gmail.com, [c]chaitanyayadav01234@gmail.com, [d]balamadhug02@gmail.com, [e]manzoorsyed515@gmail.com, [f]amarnath06042004@gmail.com

DOI: 10.1201/9781003685364-2

Scope of the study

This thinks about centers on creating and surveying a framework for identifying blood bunches utilizing fingerprint investigation. The introductory stage includes gathering and preprocessing a different dataset of unique mark pictures connected to eight blood gather sorts. Different machine learning models will be inspected to assess their exactness and productivity in classification. The system is designed to operate in genuine time, making it exceedingly reasonable for basic restorative circumstances. Future enhancements may include refining the calculations for portable gadget integration, improving openness.

Problem statement

Traditional blood group identification methods are often invasive, requiring blood samples that can be uncomfortable and cause delays, especially in emergencies. These methods also rely on manual processing, increasing the risk of human error and hindering timely medical care. This research explores a non-invasive approach to blood group detection through fingerprint analysis, using advanced machine learning algorithms to accurately determine blood groups based on unique fingerprint patterns. This method enhances efficiency and reliability while eliminating the need for blood samples. With potential for real-time application, fingerprint-based blood group identification offers a practical solution, particularly in emergency and remote healthcare settings.

Related work

Fingerprint-based biometric identification has become a reliable method for various applications. This research presents a novel approach to identifying blood groups through fingerprint analysis, utilizing unique fingerprint characteristics. Several machine learning algorithms were tested, with Ordinary Least Squares (OLS) regression achieving an accuracy of 62%. Future studies should focus on improving accuracy with larger datasets and additional fingerprint features. Recent findings suggest that fingerprints can also indicate health conditions related to age and lifestyle, such as hypertension and type 2 diabetes, prompting researchers to explore associations between fingerprint patterns, blood groups, and health risks. Fingerprint recognition involves detailed feature extraction, starting with preprocessing to enhance image clarity. Key features like ridge endings and bifurcations are extracted for

classification, followed by matching using one-to-many (1:N) and one-to-one (1:1) methods. Advanced techniques, such as GaborHoG descriptors, improve identification accuracy. Recent advancements in rapid blood typing systems allow for ABO and Rh phenotyping within minutes, significantly reducing the time compared to traditional methods. Ongoing research aims to strengthen the link between fingerprint patterns and blood groups using deep learning techniques. Convolutional Neural Networks are proposed for more accurate classification by identifying distinct structures like loops, whorls, and arches. By isolating essential fingerprint details and minimizing noise, classification accuracy can be improved. Overall, fingerprint-based blood group detection offers a practical and efficient alternative to conventional methods.

Proposed system workflow

The proposed framework aims to enhance blood group detection by integrating advanced deep learning models: MobileNet, ResNet combined with recurrent neural networks, and Vision Transformers (ViT). MobileNet is chosen for its efficiency and suitability for mobile and edge devices, enabling real-time functionality without heavy computational demands. Its lightweight design is effective for feature extraction in resource-constrained environments. ResNet addresses the vanishing gradient problem, allowing for deeper networks that capture complex patterns in fingerprint data, which is essential for improving classification accuracy. Combining ResNet with RNNs enables the analysis of sequential data, valuable for detecting temporal changes in fingerprint information. ViT refine feature extraction through attention mechanisms, focusing on important regions within fingerprint images, enhancing robustness to variations in quality and orientation. Together, these models create a scalable, efficient, and accurate blood group detection system suitable for mobile and edge devices, providing fast and reliable results for real-world healthcare and emergency applications.

Loading dataset

The initial stage of the proposed framework involves loading a dataset of fingerprint images relevant to blood group classification, sourced from established databases or specialized scanning devices. These images are crucial for training deep learning models and require meticulous preprocessing to meet the system's input requirements. The framework begins by reading image files in formats like JPEG, PNG, or TIFF, with each image linked to a corresponding label indicating its blood group. The dataset is then divided into training, validation, and test sets to ensure effective generalization. To enhance diversity and improve model robustness, data augmentation techniques such as rotation, scaling, and flipping are applied. After augmentation, preprocessing normalizes pixel values and maintains consistency across the dataset, which includes resizing images to a standard size, converting color channels if necessary, and scaling pixel values appropriately. Once preprocessing is complete, the data is ready to be fed into the MobileNet, ResNet, RNN, and Vision Transformer models for training and analysis.

Preprocessing

The preprocessing stage begins by resizing fingerprint images to a uniform size of 128 × 128 pixels, ensuring consistency for deep learning models. After resizing, data augmentation is applied using an ImageDataGenerator, incorporating techniques such as rotation, zoom, and horizontal flipping to introduce variability and simulate real-world differences in fingerprint quality. This enhances the model's ability to generalize subtle variations. Additionally, pixel values are normalized by dividing each value by 255, scaling them to a range of [0, 1], which facilitates faster convergence and more efficient training. The preprocessing also includes converting images to grayscale, as color information is unnecessary for fingerprint analysis, reducing computational load and streamlining feature extraction. The dataset is then divided into training, validation, and test sets to ensure the model is evaluated on unseen data, helping to prevent overfitting. These steps collectively enhance the robustness and reliability of the deep learning model.

Model training and classification

The method begins by extracting significant features from fingerprint images using a pre-trained network like DenseNet. These features are fed into a custom deep learning model built with Keras, which includes fully connected layers to identify complex patterns. To prevent overfitting, dropout layers are added after each fully connected layer, enhancing generalization

by randomly deactivating neurons during training. The model is trained on a labeled dataset of fingerprint images, optimizing weights with the Adam optimizer and minimizing categorical cross-entropy loss. After training, the model classifies new fingerprints into one of eight blood groups, with preprocessing steps including grayscale transformation and normalization. Output probabilities for each blood group are generated using a SoftMax activation function, and performance is evaluated using accuracy and F1-score. The model can be deployed for automated blood group identification, undergoing thorough testing on unseen images to verify robustness. Techniques like data augmentation and hyperparameter tuning enhance performance, and future improvements may involve expanding the dataset and leveraging advanced architectures for better accuracy.

This diagram illustrates the workflow of a blood group detection system. The process starts with a user signing up or logging into the system. Upon logging in, the data undergoes a cleaning process before being fed into a machine learning model. The model consists of different architectures, including CNN, MobileNet, ResNet-18, and Vision Transformer, which are used to analyze and predict the blood group. The final prediction is then provided as the detected blood group.

Methodology

Convolutional neural network

Definition: Convolutional neural systems are a specialized sort of profound learning show custom fitted for image-related assignments, such as picture

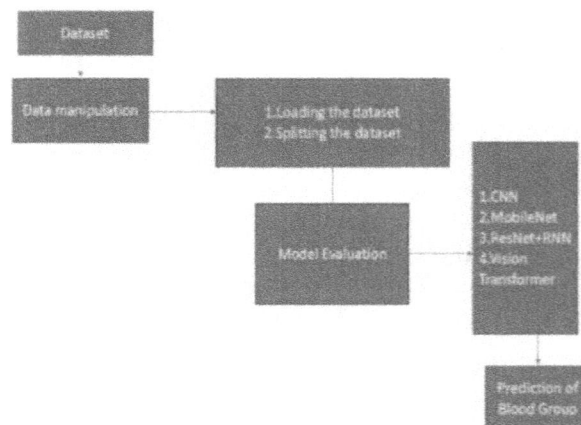

Figure 2.1 Block flow chart of blood group detection
Source: Author

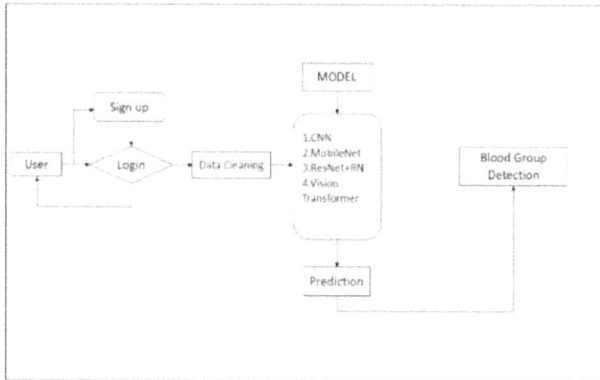

Figure 2.2 System architecture of blood group detection
Source: Author

classification. They utilize convolutional layers that perform operations on input pictures, making them profoundly compelling for analyzing spatial information. CNNs independently distinguish highlights like edges, surfaces, and designs, empowering them to construct progressive representations of the input information.

The process flow for the blood group detection system, from data collection to prediction, is illustrated in Figure 2.1.

The overall system architecture that incorporates different models for blood group classification is shown in Figure 2.2.

Inner action within the consider
This study employs CNN to classify fingerprint images into distinct blood groups. The network uses multiple convolutional layers with small filters (3 × 3 kernels) to extract complex features, followed by pooling layers to reduce dimensionality. Extracted features are flattened and passed to fully connected layers, with a final SoftMax layer assigning probabilities to nine blood group categories. CNNs automatically learn significant features from well-preprocessed data and are robust to variations like shifts and rotations, making them ideal for fingerprint analysis. Techniques such as dropout and batch normalization enhance performance by reducing overfitting and improving training stability, making CNNs effective for accurate blood group classification.

MobileNet
Definition: MobileNet may be a profound learning engineering custom fitted for effective operation on portable and edge gadgets with compelled computational capabilities. Its proficiency stems from the utilize of depth wise distinct convolutions, which partitioned the sifting and combining forms into particular layers. This approach essentially diminishes both the number of parameters and the computational stack.

Inner action within the consider
In this study, MobileNet acts as a lightweight feature extractor, using DenseNet121 as its backbone to enhance hierarchical feature learning through dense connections. Extracted features are then classified into blood groups using a newly designed fully connected layer. MobileNet's lightweight design ensures real-time processing, making it ideal for mobile blood group detection systems. To improve classification accuracy, batch normalization is applied to stabilize and accelerate training. This combination of MobileNet and DenseNet balances computational efficiency with high-performance feature extraction, resulting in a flexible and robust system.

ResNet combined with RNN
Definition: ResNet may be a profound learning design that utilizes remaining associations, moreover, known as skip associations, to address the vanishing angle issue. These associations empower the preparation of essentially more profound systems, which are basic for capturing perplexing designs in information. On the other hand, RNNs are particularly outlined for consecutive information, where the yield at any given step is impacted by past inputs. This characteristic makes RNNs especially reasonable for errands including temporal or successive information.

Inner action within the consider
In this study, ResNet50 is used as a feature extractor for fingerprint images, leveraging its deep architecture to extract advanced features. These features are then processed by a RNN layer, which teaches temporal relationships within the data. This hybrid approach combines ResNet's ability to capture complex visual features with the RNN's effectiveness in modeling sequential patterns, making it effective for distinguishing subtle differences between blood groups. The residual connections in ResNet maintain stable performance as depth increases, while the RNN captures essential temporal patterns in fingerprint data analysis.

Table 2.1 Performance comparison of different models for blood group detection.

Model	Training Accuracy	Validation Accuracy	Training Loss	Validation Loss
Vision Transformer	97.84%	92.52%	0.0673	0.2618
CNN	96.98 %	89.25%	0.1119	0.2825
Mobile Net	75.04%	77.56%	0.6431	0.5700
ResNet + RNN	61.45%	75.57%	1.0035	0.6754

Source: Author

Table 2.1 shows a comparison of the training and validation performance across different models used for blood group prediction.

Vision transformer

Definition: The ViT is an innovative architecture that adapts transformer components from natural language processing for image data. Unlike traditional convolutional methods, ViT divides images into patches, flattens them, and processes them through transformer layers with self-attention mechanisms. This design enables the model to capture long-range dependencies and global context, making it effective for complex image classification tasks. By leveraging self-attention, ViT dynamically focuses on significant regions of an image, enhancing its ability to recognize intricate patterns and features.

Inner action within the consider

In this study, the ViT is used to classify fingerprint images into blood groups. Input images are resized to 224 × 224 pixels and divided into patches, which are processed by transformer layers using self-attention mechanisms. This approach enables the model to learn global relationships and relevant information within fingerprint images. The feature embeddings generated by the transformer layers are then used to predict the blood group. ViT's strength lies in its ability to capture global context and long-range dependencies, which are crucial for distinguishing between blood groups based on subtle visual features in fingerprint images.

Discussion and results

Various deep learning models, including CNN, MobileNet, ResNet with RNNs, and Vision Transformer, were used to classify fingerprints into blood groups, each demonstrating unique strengths. CNN and ResNet achieved high precision for most groups, while MobileNet showed promise for real-time applications. The Vision Transformer provided deeper contextual insights through long-range dependencies. Despite strong classification accuracy, challenges remain in detecting categories with less distinct features. The study concludes that further refinement and increased diversity in training data are essential for improving accuracy and robustness in real-world applications.

This table compares four deep learning models—Vision Transformer, CNN, MobileNet, and ResNet + RNN—based on training accuracy, validation accuracy, training loss, and validation loss. The ViT achieves the highest accuracy (97.84% training, 92.52% validation) with the lowest loss, indicating superior performance. CNN also performs well, while MobileNet and ResNet + RNN exhibit lower accuracy and higher loss values, suggesting they are less effective for this task.

Several prior studies have investigated fingerprint-based blood group detection using machine learning and deep learning. Vijaykumar and Ingle [1] introduced a fingerprint map reading technique. Patel and Patel [2] proposed CNN-based blood group prediction. A deep learning model was explored by Jha and Jaiswal [3], while Thakur and Gupta [4] used fingerprint features in a machine learning setup. Gupta and Sharma [5] implemented deep neural networks for the same purpose. Singh and Yadav [6] applied deep learning techniques to fingerprint analysis, and Patil and Ingle [7] further refined the map-reading method for prediction.

Conclusion

This venture highlights the adequacy of profound learning models, counting convolutional neural networks, MobileNet, ResNet, and Vision Transformer, in precisely classifying blood bunches utilizing unique mark pictures. By utilizing these progressed structures, the consideration illustrates how profound learning can independently extricate basic highlights from unique mark information, encouraging proficient and solid blood bunch distinguishing proof. This strategy presents a non-invasive and imaginative arrangement with potential applications in therapeutic diagnostics, forensics, and biometric security frameworks. Whereas the results are promising, certain challenges stay, such as varieties in unique mark quality, unpretentious qualifications between blood bunches, and the hazard of overfitting. Tending to these issues is vital for making strides show execution. Future inquiries about seem to center on growing and improving the dataset, refining show structures, and consolidating procedures like exchange learning to extend classification precision.

Future enhancement

Future advancements may center on broadening the data set to join a more extensive assortment of fingerprint pictures, considering components such as age, ethnicity, and geographic root. This extension would improve the model's strength and generalizability, driving to a more precise blood bunch location over assorted populace. Also, examining half breed models that consolidate machine learning approaches, such as combining convolutional neural systems with gathering strategies, might advance move forward highlight extraction and classification capabilities. Another potential headway includes coordination fingerprint examination with other biometric frameworks, such as facial acknowledgment or iris checking, to create a more comprehensive and secure distinguishing proof system for therapeutic purposes. Optimizing profound learning models through procedures like pruning and quantization seem to increment effectiveness, permitting arrangement on low-power gadgets without relinquishing exactness. Finally, conducting clinical trials to approve the framework would guarantee it's down-to-earth appropriateness, with input from healthcare experts directing refinements for real-world execution.

References

[1] Vijaykumar, P. N., & Ingle, D. R. (2021). A novel approach to predict blood group using fingerprint map reading. In 2021 6th International Conference for Convergence in Technology (I2CT). IEEE.

[2] Patel, N., & Patel, K. (2021). Fingerprint-based blood group detection using convolutional neural networks. *International Journal of Engineering and Advanced Technology*, 10(6), 234–241.

[3] Jha, S., & Jaiswal, S. (2020). A deep learning approach for blood group classification from fingerprint images. In Proceedings of the International Conference on Image Processing, Machine Learning, and Pattern Recognition, (pp. 85–90).

[4] Thakur, R., & Gupta, V. (2021). Machine learning-based blood group prediction using fingerprint features. *Journal of Computer Science and Technology*, 18(4), 342–349.

[5] Gupta, P., & Sharma, S. (2020). Predicting blood groups from fingerprint images using deep neural networks. In International Conference on Machine Learning and Artificial Intelligence, (pp. 77–84).

[6] Singh, S., & Yadav, A. (2020). Blood group prediction using fingerprint analysis and deep learning techniques. In Proceedings of the International Conference on Data Science and Advanced Analytics, (pp. 110–115).

[7] Patil, N., & Ingle, D. R. (2021). A novel approach to predict blood group using fingerprint map reading. In 2021 6th International Conference for Convergence in Technology (I2CT). IEEE.

3 Analysis of women safety in Indian cities using ML on Tweets

*G. Ganesh[1,a], M. Rajasri[2,b], B. Sreedevi[2,c], N. Keshava Naidu[2,d] and
S. Premnath Reddy[2,e]*

[1]Assistant Professor, Department of CSE, Srinivasa Ramanujan Institute of Technology, Anantapur, Andhra Pradesh, India

[2]Students, Department of CSE(Data Science), Srinivasa Ramanujan Institute of Technology, Anantapur, Andhra Pradesh, India

Abstract

Women and girls in urban India encounter various safety challenges, including stalking, harassment, and assault in public areas. This study examines how social media platforms such as Twitter, Facebook, and Instagram contribute to increasing awareness and enhancing women's safety. These platforms serve as a medium for spreading awareness through messages, stories, and campaigns, educating the public and advocating stricter actions against offenders. Many women utilize Twitter to share their concerns regarding safety while commuting or working, using hashtags to reach a broader audience.

To analyze these discussions, this study employs machine learning techniques such as Support Vector Machines (SVM), Neural Networks, Gradient Boosting, Random Forest, Decision Tree, Naïve Bayes, and K-nearest neighbors (KNN). These models classify tweets into positive, negative, or neutral sentiments and are evaluated based on performance metrics like accuracy, precision, and recall. This research underscores the role of technology and collective responsibility in fostering safer environments for women. By leveraging social media and machine learning, it highlights the importance of collaborative efforts in addressing women's safety issues.

Keywords: Decision tree, gradient boosting, K-nearest neighbors, machine learning, Naive Bayes, neural networks, Random Forest, sentiment analysis, social media, Support Vector Machines, Twitter, women's safety

Introduction

Twitter has emerged as a main microblogging platform, with tens of millions of customers producing sizeable quantities of content material each day. It enables human beings to explicit opinions, participate in discussions, and have interaction with contemporary troubles in actual times. Because of its significant attain, Twitter is a treasured supply of data for agencies, researchers, and policymakers searching for to recognize public sentiment and emerging tendencies.

Given Twitter's character barriers, customers frequently condense their messages with the use of abbreviations, emojis, and informal expressions. Moreover, sarcasm, ambiguity, and a couple of meanings of phrases pose demanding situations for classic textual content evaluation techniques. As a end result, superior strategies are required to interpret and extract significant insights from such unstructured content material.

Sentiment analysis is a key approach used to evaluate emotions and reviews within tweets. with the aid of classifying tweets into categories which include effective, poor, or neutral, sentiment evaluation aids groups in understanding consumer feedback, policymakers in assessing public critiques, and researchers in analyzing social behavior.

Manner substantial datasets to pick out patterns in user sentiments, permitting large-scale analysis of public discourse. This record supports selection-making throughout numerous domains with the aid of supplying real-time insights into societal developments.

A critical location in which sentiment analysis proves useful is in addressing public protection worries, particularly problems associated with women's

[a]ganeshg.cse@srit.ac.in, [b]magisetty.rajasri@gmail.com, [c]sreedevibyalla@gmail.com,
[d]keshavanaidu2003@gmail.com, [e]Premsangati0238@gmail.com

DOI: 10.1201/9781003685364-3

protection. Many women percentage their studies of harassment and violence via social media, especially in metropolitan areas like Delhi, Pune, Chennai, and Mumbai. structures like Twitter provide an area for them to voice their concerns and are trying to find support.

A critical area where sentiment analysis proves beneficial is in addressing public safety concerns, particularly issues related to women's security. Many women share their experiences of harassment and violence through social media, especially in metropolitan areas like Delhi, Pune, Chennai, and Mumbai. Platforms like Twitter provide a space for them to voice their concerns and seek support, reading tweets on incidents of harassment and violence can assist perceive places perceived as dangerous, come across ordinary developments, and determine the effect of social campaigns selling girls' safety. Through extracting insights from those discussions, sentiment evaluation plays a essential position in addressing gender-primarily based problems.

Additionally, discussions surrounding harassment regularly consist of names of individuals, both as alleged offenders and as advocates assisting sufferers. Inspecting those interactions offers researchers and policymakers with deeper knowhow of the frequency and scale of such incidents. These statistics pushed technique aids in formulating techniques to beautify public protection, making Twitter a precious digital platform for recording real-international experiences and informing policy selections.

Related Works

Agarwal et al. [1], offered "Contextual word-stage polarity evaluation the usage of lexical affect scoring and syntactic N-grams" at the twelfth convention of the EU chapter of the association for computational linguistics. This research explored the use of syntactic n-grams and lexical effect scoring to investigate sentiment at the phrase level, enhancing sentiment classification accuracy.

Barbosa and Feng [2] posted "Sturdy sentiment detection on twitter from biased and noisy records" at the 23rd worldwide conference on computational linguistics. Their study targeted on improving sentiment detection on Twitter by means of addressing biased and noisy statistics using device studying strategies.

Bermingham and Smeaton [3], investigated the impact of textual content brevity on sentiment category in microblogs. Their work, "Classifying sentiment in microblogs: is brevity a bonus?" turned into supplied at the nineteenth ACM worldwide conference on information and expertise management. It tested whether shorter texts, inclusive of tweets, have an effect on the accuracy of sentiment analysis.

Gamon [4], explored sentiment type in purchaser feedback records in his research "Sentiment classification on customer comments information: noisy records, big characteristic vectors, and the function of linguistic analysis." Provided at the 20th global convention on computational linguistics, this takes a look at analyzed the challenges posed via noisy data and big characteristic vectors in sentiment classification.

Kim and Hovy [5] added "Figuring out the sentiment of critiques" at the 20th worldwide conference on computational linguistics. Their research investigated strategies for extracting sentiment from opinionated text the use of computational linguistics.

Klein and Manning [6] offered "Accurate unlexicalized parsing" on the 41st Annual assembly of the affiliation for computational linguistics in 2003. This has a study centered on improving parsing accuracy the usage of unlexicalized probabilistic fashions, which play an extensive function in textual content processing for sentiment evaluation.

Charniak and Johnson [7] presented their work named "Coarse to-fine N-bravery parsing and MaxEnt discriminational reranking" at the 43rd Annual Meeting of the association for computational linguistics.

Their exploration concentrated on perfecting parsing effectiveness, which played a significant part in advancing natural language processing ways.

Gupta et al. [8] conducted a have a look at titled "observe of Twitter sentiment analysis the use of gadget getting to know algorithms on Python," published inside the global journal of laptop applications. Their work analyzed the effectiveness of device learning algorithms in sentiment category.

Sahayak et al. [9] explored sentiment analysis strategies in their look at "Sentiment analysis on Twitter facts". Their studies tested various strategies used to extract sentiment from Twitter content.

Mamgain et al. [10], presented "Sentiment analysis of top schools in India using Twitter statistics" at the global convention on computational strategies in information and conversation technologies (ICCTICT) in 2016. Their observe analyzed Twitter sentiment to evaluate public perceptions of higher education institutions in India.

Existing system

The preceding systems used for crime detection, in particular in addressing crimes towards women, have been tremendously reliant on guide procedures. Police and authorities needed to manually search through bodily facts to accumulate necessary records, which become a hard work- extensive and time- ingesting venture. This lack of automation brought about huge delays in figuring out criminals or responding to urgent conditions. The guide device made it difficult to music incidents in actual-time, and important statistics may want to without problems be omitted, resulting in gradual response instances and useless intervention.

Additionally, the old device lacked technological integration, which is in addition complex topics. without the usage of advanced gear and facts analytics, police officers had no quick manner of having access to or studying massive volumes of records. As a result, authorities had been regularly unaware of styles of criminal behavior, and important connections between incidents may want to cross neglected. The absence of automated alerts or real-time tracking made it hard to offer timely help to victims, specifically women dealing with harassment or violence in city environments.

One of the key troubles with the preceding gadget was the project of figuring out suspects or tracking down criminals. records changed into not centralized, and there has been no unified database that might offer immediate access to critical data. Investigators needed to depend on guide searches across different departments, which are no longer only inefficient but also liable to human mistakes. This inefficiency behind schedule investigations and regularly supposed that crucial information was missed, hindering the effectiveness of the justice machine in protective women.

In precis, the prevailing system had numerous tremendous boundaries: it turned into gradual, inefficient, and at risk of mistakes. The manual technique of trying to find facts, mixed with the shortage of technological equipment to resource investigations, supposed that crimes were no longer addressed

quickly sufficiently, leading to a higher risk for sufferers. To overcome those challenges, there was a urgent want for extra advanced structures that would automate approaches, provide real- time statistics, and enhance the efficiency and accuracy of regulation enforcement in preventing and fixing crimes, especially the ones related to women's safety.

Proposed system

Considering the rapid changes in society and the increasing consciousness of crimes, in particular those affecting women, there may be a growing want for an efficient and present-day approach to crime detection. The proposed system objectives to cope with this want through automating crime document control and utilizing superior gadget learning (ML) algorithms. With this system in vicinity, authorities can rapidly retrieve and examine criminal facts, making it easier to identify criminals, track their records, and take appropriate movement without the delays of guide report-retaining.

The gadget makes use of several powerful gadget mastering algorithms, such as AdaBoosting, CatBoosting, help Support Vector Machines (SVM), and Naïve Bayes (NB). Those algorithms are designed to teach the dataset and expect outcomes based totally on ancient information. By processing large volumes of facts, the device can speedily discover patterns of crook behavior, assisting government spot capability offenders faster and greater appropriately. in addition, it allows for seamless get entry to past records, making it less complicated to track a suspect's crook history. one of the key blessings of this method is the considerable discount in time required to system and retrieve criminal facts. conventional techniques of file- keeping and search are time-consuming and regularly useless, however with ML algorithms, the machine can system statistics in real-time, enabling faster selection- making. This progressed reaction time can result in faster intervention and extra green dealing with criminal cases, lowering the opportunity for crimes to expand.

Moreover, the gadget's reliance on superior algorithms ensures that the effects produced are particularly correct. in contrast to guide searches that may be vulnerable to human blunders, the gadget learning algorithms constantly improve through the years, main to more specific predictions and classifications. This no longer only enhances the effectiveness of crime detection, however, it also ensures that the facts-driven insights guide selections with minimum bias or error.

Another advantage of the proposed system is its overall efficiency. By using automating procedures together with record-preserving, search, and analysis, the workload of government is extensively decreased. This allows law enforcement to attention to extra pressing tasks at the

At the same time the device handles the heavy lifting of statistical processing. This accelerated performance can lead to a greater effective allocation of resources, enhancing the overall effectiveness of the justice gadget.

In end, the proposed device leverages system mastering technologies to enhance the rate, accuracy, and efficiency of crime detection and prevention. By means of reducing the time wanted for statistical retrieval, improving the accuracy of predictions, and streamlining methods, this machine offers a complete approach to addressing crime in a cutting- edge modern, records-driven manner. Through those advancements, the device is not best blessings of law enforcement agencies but also contributes to growing a more secure environment for society, particularly girls. Moreover, the system's ability to continuously examine and adapt guarantees that it remains applicable as crime patterns evolve, further improving its lengthy-time period effectiveness. With this adaptability, the machine can emerge as a effective device for proactive crime prevention and a critical resource for both regulation enforcement and community protection.

Architecture

The system starts off evolved with uploading a dataset, observed through pre- processing to clean and form the records. next, model schooling is carried out the use of device learning algorithms. The professional version generates predictions, which is probably analyzed to provide outcomes. Ultimately, the consequences are visualized by the usage of a graph, presenting insights into data developments.

Result and discussion

The homepage serves as a relevant hub for uploading datasets, preprocessing statistics, analyzing trends, and making predictions on women's protection-associated tweets using system mastering.

The "Upload Dataset" web page permits users to add CSV documents for assessment related to women's safety on Twitter.

The "View Dataset" web page shows the uploaded dataset in a tabular layout for evaluation and evaluation. Figures 3.1, 3.2, 3.3, 3.4, 3.5, 3.6, 3.7a, 3.7b, 3.7c

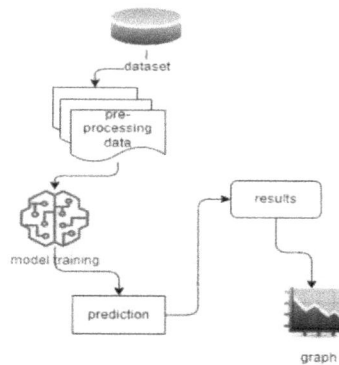

Figure 3.1 Architecture of the project
Source: Author

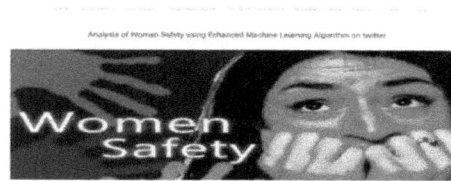

Figure 3.2 Home page
Source: Author

Figure 3.3 Upload dataset page
Source: Author

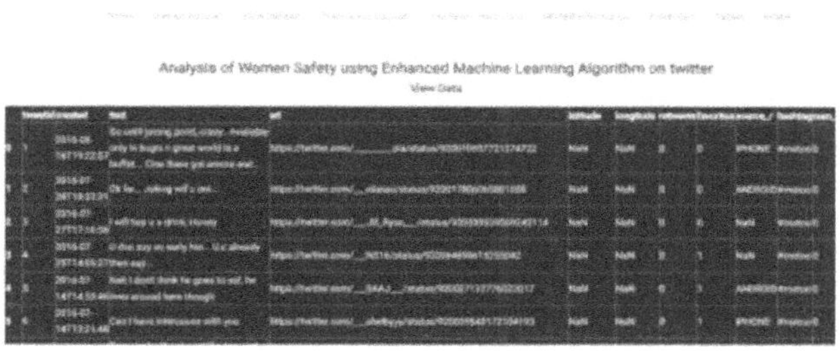

Figure 3.4 View dataset page
Source: Author

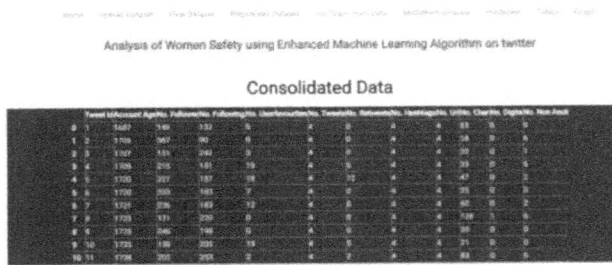

Figure 3.5 Preprocess dataset page
Source: Author

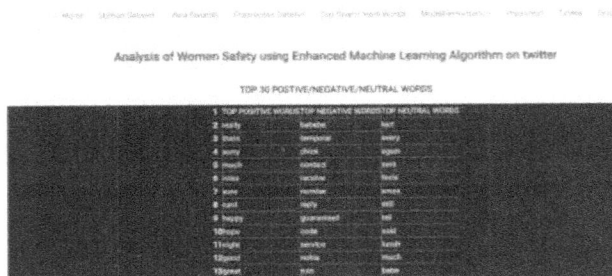

Figure 3.6 Top spam words page
Source: Author

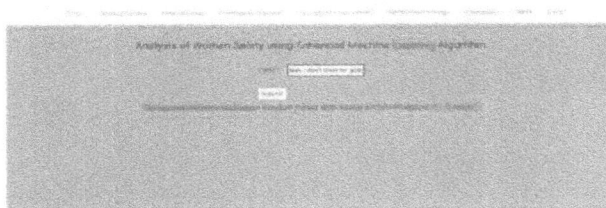

Figure 3.7a Prediction page (neutral)
Source: Author

Figure 3.7b Prediction page (negative)
Source: Author

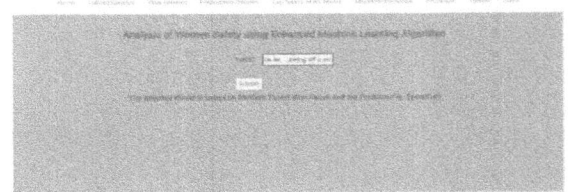

Figure 3.7c Prediction page (Positive)
Source: Author

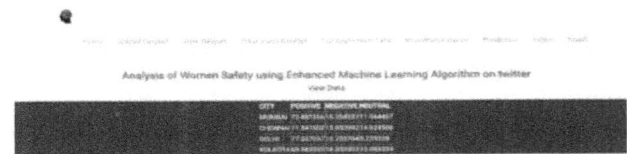

The "Preprocess Dataset" page shows the cleaned and consolidated dataset with diverse Twitter metrics for analysis.

The "Top Spam/Ham Words" web page presentations the pinnacle 30 positive, negative, and neutral words diagnosed from the dataset for unsolicited mail and sentiment evaluation.

The "model overall performance" web page suggests the performance metrics (accuracy, precision, and do not forget) of the selected gadget reading model, whether or no longer it's based on SVM, neural community, Gradient enhance, Random wooded location, decision Tree, NB, or K-nearest neighbors (KNN). Every model's metrics are shown to provide insights into its effectiveness for the given assignment.

The "Prediction" page permits customers to input a tweet and analyze the usage of the chosen machine learning model, providing a sentiment magnificence collectively with independent, poor, or fantastic, relying at the model used (e.g., Random wooded area). The sentiment prediction is based on the analysis of the tweet's content.

The "Tables" web page shows metropolis-practical sentiment evaluation consequences, displaying the proportion of exceptional, bad, and impartial tweets.

Finally, the "Graph" net web page visually compares the overall performance metrics of numerous gadget mastering models the usage of a bar chart.

Conclusion

Women and girls across the country face diverse styles of violence and harassment in public spaces, from stalking to sexual assault. This application objectives to foster a feel of collective obligation within Indian society to enhance women protection in their surroundings. via reading tweets about women protection in India regularly containing textual content, snap shots, and remarks we will expect sentiment and classify the nature of these tweets. interpreting these insights enables us to take meaningful steps in the direction of improving women protection.

References

[1]   Agarwal, A., Biadsy, F., & McKeown, K. R. (2009). Contextual Phrase-Level Polarity Analysis Using Lexical Affect Scoring and Syntactic {N}-Grams. Proceedings of the 12th Conference of the {E}uropean Chapter of the {ACL} ({EACL} 2009). Association for Computational Linguistics. Athens, Greece. 24–32. https://aclanthology.org/E09-1004/

[2]   Barbosa, L., & Feng, J. (2010). Robust Sentiment Detection on {T}witter from Biased and Noisy Data. Coling 2010: Posters. Coling 2010 Organizing Committee. Beijing, China. 36–44. https://aclanthology.org/C10-2005/

[3]   Bermingham, A., & Smeaton, A. F. (2010). Classifying sentiment in microblogs: is brevity a bonus? In Nineteenth ACM Global Convention on Facts and Records Control. https://dl.acm.org/doi/10.1145/1871437.1871741#core-cited-by

[4]   Gamon, M. (2004). Sentiment classification on customer feedback data: noisy data, large feature vectors, and the role of linguistic analysis. {COLING} 2004: Proceedings of the 20th International Conference on Computational Linguistics. COLING. Geneva, Switzerland. 841–847. https://aclanthology.org/C04-1121/

[5]   Kim, S.-M., & Hovy, E. (2004). Determining the Sentiment of Opinions. {COLING} 2004: Proceedings of the 20th International Conference on Computational Linguistics. COLING. Geneva, Switzerland. 1367–1373. https://aclanthology.org/C04-1200/

[6]   Klein, D., & Manning, C. D. (2003). Accurate Unlexicalized Parsing. Proceedings of the 41st Annual Meeting of the Association for Computational Linguistics. Association for Computational Linguistics. Sapporo, Japan. 423–430. 10.3115/1075096.1075150

[7]   Charniak, E., & Johnson, M. (2005). Coarse-to-Fine n-Best Parsing and {M}ax{E}nt Discriminative Reranking. Proceedings of the 43rd Annual Meeting of the Association for Computational Linguistics ({ACL}{'}05). Association for Computational Linguistics. Ann Arbor, Michigan. 173–180. 10.3115/1219840.1219862

[8]   Gupta, B., Negi, M., Vishwakarma, K., Rawat, G., & Badhani, P. (2017). Examine of Twitter Sentiment evaluation using gadget getting to know Algorithms on Python. *Global Journal of Computer Applications*. https://www.sciencedirect.com/science/article/pii/S1877050924007701#section-cited-by

[9]   Sahayak, V., Shete, V., & Pathan, A. (2015). Sentiment analysis on Twitter records. *Global magazine of Progressive Research in Advanced Engineering (IJIRAE)*. https://dl.acm.org/doi/10.1145/1871437.1871741#core-cited-by

[10]   Mamgain, Nehal & Mehta, Ekta & Mittal, Ankush & Bhatt, Gaurav. (2016). Sentiment analysis of top colleges in India using Twitter data. 525–530. 10.1109/ICCTICT.2016.7514636.

4 LSTM enhanced framework for real time DDOS attack detection

R. Jahnavi Ram[1,a], B. Jasnavi[2,b], B. Mounika[2,c], G. Hemanth[2,d] and
T. Hemanth Kumar[2,e]

[1]Assistant Professor, Department of CSE, Srinivasa Ramanujan Institute of Technology, Anantapur, Andhra Pradesh, India

[2]Students, Department of CSE Data Science), Srinivasa Ramanujan Institute of Technology, Anantapur, Andhra Pradesh, India

Abstract

Within the field of cybersecurity, identifying distributed denial of service assaults continues to be a significant obstacle, made worse by the appearance of novel attack methods. In order to detect DDoS attacks, this paper suggests a unique method called Reciprocal Points Learning for Open-Set Recognition. The approach relies on elements retrieved from network traffic data, such as flow length, packet characteristics, and statistical flow metrics. It does this by using algorithms such as passive aggressive, Random Forest, Decision Tree, LSTM and CNN-RPL. By distinguishing between known and unknown DDoS assaults and typical network behavior, the system seeks to improve the resilience of detection techniques in dynamic and changing threat environments. The effectiveness of the suggested framework in obtaining high accuracy and dependability is demonstrated by experimental assessment on a large dataset, which advances proactive cybersecurity tactics against complex network threats.

Keywords: Decision Tree classifier, LSTM and CNN-RPL, passive aggressive, Random Forest

Introduction

Within the field of cybersecurity, identifying and countering DDoS assaults are critical tasks that are always changing due to the appearance of new attack patterns. The crucial need for novel ways in threat detection and response is highlighted by the fact that traditional methods frequently fail to quickly respond to these dynamic threats. In order to overcome this difficulty, this study suggests a brand-new framework for DDoS assault detection called Reciprocal Points Learning (RPL) for Open-Set Recognition (OSR).

DDoS attacks pose serious hazards to the availability and integrity of network resources by flooding them with malicious traffic in an attempt to disrupt network services. Because most conventional detection systems rely on known attack signatures, networks are left open to new or zero-day assaults that circumvent established defenses. The suggested system combines advanced ML techniques like Random Forest (RF) and Decision Tree (DT), and

Passive Aggressive algorithms, to improve detection performance in both known and new DDoS assault situations. The framework's use of thorough characteristics taken from network traffic data is essential to its effectiveness. These characteristics include a variety of parameters, including statistical flow properties, packet characteristics, and flow length, which allow for a more subtle distinction between typical network behavior and unusual activity that may be suggestive of DDoS assaults. By concentrating OSR, the system seeks to proactively categorize and adjust to hitherto undiscovered threats in real-time, in addition to identifying established attack patterns.

Empirical verification on a sturdy dataset highlights the dependability and efficiency of the framework in attaining superior precision and resistance to intricate DDoS assaults. This study enhances the capabilities of network defense mechanisms in dynamic and changing threat environments, which advances proactive cybersecurity methods.

[a]jahnaviram.cse@srit.ac.in, [b]jasnavibedudhuri@gmail.com, [c]bonthamounikareddy12@gmail.com,
[d]hemanth49899@gmail.com, [e]hemanthkumar89190@gmail.com

DOI: 10.1201/9781003685364-4

Objective of the study

The primary goal of this research is to develop and assess a novel framework for detecting DDoS attacks through the application of RPL. This innovative framework incorporates a range of advanced machine learning techniques, including Passive Aggressive, DT and RF classifiers. These algorithms are utilized to scrutinize network traffic data and identify anomalies that could signal the presence of DDoS attacks. The research is designed to enhance detection capabilities for both previously identified and novel attack patterns, thereby boosting the overall resilience and adaptability of cybersecurity systems. Moreover, the purpose of the study is to validate the efficacy of the suggested framework by conducting thorough experimental evaluations using a diverse and comprehensive dataset. This validation process will provide insights into the framework's performance and reliability in real-world scenarios, ensuring its practical applicability and effectiveness in improving cybersecurity defenses.

This research is centered on the development and assessment of the RPL framework specifically for detecting DDoS attacks using info on network traffic. The focus of the study is on examining various features such as flow duration, packet attributes, and statistical metrics that are derived from network traffic. Although the proposed framework can be adapted to different network settings, the experimental evaluation is based on a particular dataset, which may restrict the applicability of the findings to other types of network traffic or different attack scenarios. Furthermore, the scope of this research is limited to DDoS attacks, and it does not extend to exploring other kinds of cyber threats.

This dissertation addresses the challenge of identifying unknown DDoS attack patterns inside network surroundings. Traditional detection systems typically struggle to recognize new types of attacks because they are based on predefined patterns or signatures of known threats. The primary objective of this examination is to overwhelm this limitation by advising an innovative method called RPL. This new tactic is designed to boost the detection and differentiation of both normal network activities and emerging, previously unseen DDoS threats, alongside known attacks. The proposed method aims to enlarge the correctness and reliability of attack detection, making it better suited to handle the dynamic nature of modern cybersecurity threats. By leveraging RPL, this research intends to advance the capability of detection systems to respond to evolving attack patterns, thereby addressing a significant gap in existing approaches.

Literature review

Previous work: Analysis of previous studies on DDoS detection and OSR [1]. In the realm of cybersecurity, the detection of DDoS attacks partakes remained a prominent focus of research, leading to the development of various methodologies to address these threats. Historically, signature-based detection methods have been prevalent. These approaches involve identifying known attack patterns through pre-established signatures. Nevertheless, this method has drawbacks in detecting new or previously unknown attack techniques, which restricts its overall effectiveness. Recent research has shifted towards employing machine learning techniques to enhance detection capabilities. Studies have illustrated the success of algorithms including ensemble methods, neural networks, and Support Vector Machines (SVMs) [2] in improving detection rates by analyzing network traffic data. Moreover, open-set recognition various techniques have been introduced as effective solutions to address this challenge of detecting novel attacks. By combining anomaly detection with traditional classification methods, researchers have improved their ability to identify unusual network behavior that may signify a DDoS attack. Despite these advancements, challenges persist in achieving a balance between detection accuracy and minimizing false positives, especially given the ever-evolving nature of attack strategies.

Theoretical frame work: Key theories and models relevant to DDoS detection [3].

The theoretical framework for detecting DDoS attacks [4] integrates several key models and theories from machine learning and network security [5]. Central to this research is the principle of anomaly detection [6], which suggests that deviations from typical network behavior can indicate potential attacks [7]. In this approach [8], both statistical techniques and machine learning algorithms are employed [9] to scrutinize network traffic patterns and identify anomalies that may suggest DDoS activity [10].

Existing system

The internet's extensive use poses cybersecurity challenges, notably from DDoS attacks threatening network integrity. To combat these, security measures like firewalls and IDS are essential. Integrating ML and DL in IDS enhances defense yet detecting unknown DDoS attacks remains challenging. Our CNN-RPL model innovatively combines CNN with RPL for Open-Set Recognition, achieving high accuracies (exceeding for known attacks and up to for unknown attacks). It offers efficient, flexible, and robust cybersecurity solutions for organizations.

Proposed system

This study introduces long short term memory for OSR in DDoS attack detection, employing passive aggressive, RF, and DT algorithms. It analyzes network traffic data attributes, including flow duration and packet details, and statistical metrics to distinguish normal behavior from known and unknown DDoS attacks. The framework enhances robust detection in dynamic threat landscapes, demonstrated through experimental validation on a comprehensive dataset, ensuring high accuracy and reliability. This contributes significantly to proactive cybersecurity strategies against evolving network threats.

Methodologies

Passive aggressive classifier

You should use the passive aggressive classifier (PAC) for your research, DDoS PAC is adept at handling dynamic data streams and can adapt quickly to changing patterns, crucial for detecting unknown DDoS attacks in real-time. Its ability to make quick updates to its model based on new instances makes it robust in environments where attack patterns evolve rapidly. PAC operates by minimizing loss functions, making it particularly effective for online learning scenarios where labelled data may be limited or continuously updated. In your context, integrating PAC with reciprocal points learning enhances its capability to discern known and unknown attacks, ensuring proactive detection and mitigation strategies in network security applications. This combination leverages PAC's strengths in adaptability and real-time responsiveness, crucial for effectively combating emerging threats in network environments. Figure 4.1.

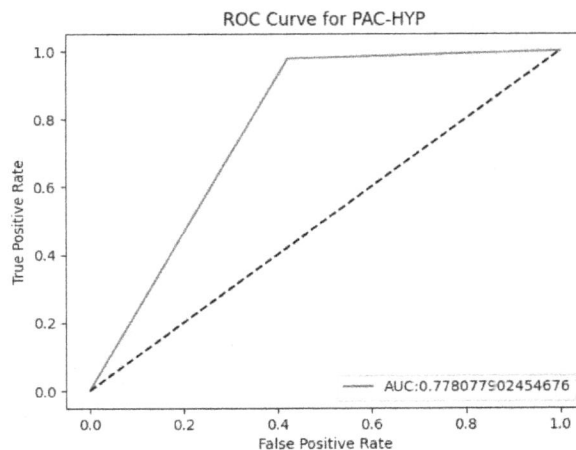

Figure 4.1 ROC Curve for passive aggressive classifier-hypothesis
Source: Author

Random Forest

For the purpose of open-set recognition in the RPL-based detection of unknown DDoS attacks, one dependable ML method is RF. To improve precision and robustness against noise and outliers in the data, an ensemble approach builds several decision trees during training and averages their predictions. When it comes to identifying DDoS assaults, RF excels in handling high-dimensional feature spaces and complex decision boundaries, making it effective for identifying anomalous patterns indicative of unknown attacks. By leveraging RPL, which enhances the model's capacity to discriminate amongst known and unknown classes based on feature representations, RF can adapt dynamically to new and evolving attack types, thereby bolstering cybersecurity defenses. This approach ensures that the system remains proactive and responsive in identifying and mitigating novel threats, contributing to enhanced network security and resilience against DDoS attacks.

Many DT are used in the RF model. To clarify, let's look at an example that makes use of four decision trees. These four root nodes hold the training data, which consists of phone attributes and observations.

These root nodes might stand for several characteristics that affect the customer's decision, such cost, RAM, internal storage, and camera quality. After that, the random forest algorithm divides the nodes into groups based on randomly chosen attributes, producing a multitude of decision trees. The majority result for each tree is used to calculate the final forecast. Figure 4.2.

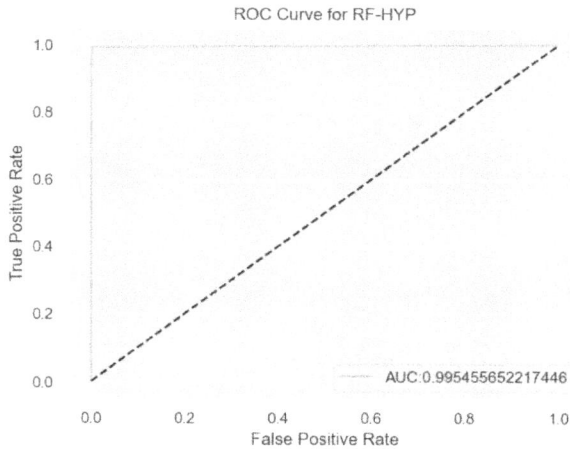

Figure 4.2 ROC curve for RF-hypothesis
Source: Author

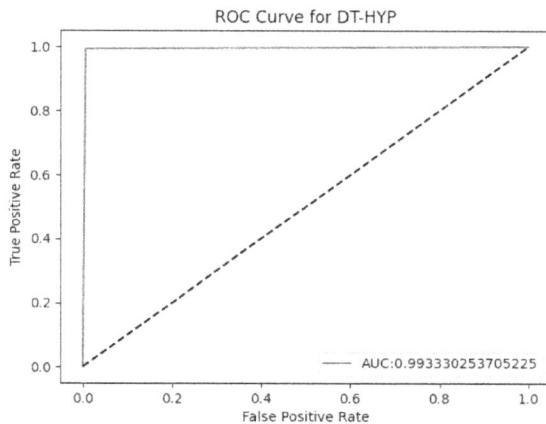

Figure 4.3 ROC curve for decision tree
Source: Author

Decision Tree

Using reciprocal points learning, DT algorithms are essential for open-set recognition for identifying unfamiliar DDoS assaults. By leveraging decision nodes and branches, the model categorizes network traffic based on features extracted through RPL, distinguishing between normal and anomalous patterns. This approach enables the identification of previously unseen DDoS attack types by analyzing their unique signatures and deviations from established norms. DT facilitate rapid classification by recursively partitioning the feature space, making them adept at handling the complexity and variability inherent in network traffic data. This methodology seeks to increase detection mechanisms' accuracy and efficiency in order to improve network security,

thereby safeguarding against emerging threats that conventional methods may overlook. Figure 4.3.

The paper introduces A novel method to tackle the issue of identifying unknown DDoS attacks in an open-set recognition framework: reciprocal points learning. Using machine learning algorithms like DT, RF, and passive aggressive, the technique makes use of flow length and packet statistics, two important aspects of network traffic data, to differentiate between legitimate and malicious activity. The suggested method successfully recognizes DDoS attack patterns that are both well-known and novel, strengthening and enhancing cybersecurity defenses. The method's excellent accuracy and dependability are demonstrated by experimental results on a large dataset, underscoring its potential to improve proactive cybersecurity methods. By considerably advancing the identification and mitigation of complex and dynamic network threats, our study ensures increased resilience in the face of emerging cyber threats. Despite the promising results, the approach for DDoS attack discovery using reciprocal points learning has some limitations. One key challenge is the reliance on feature extraction from network traffic, which may not fully capture the complexities of evolving attack patterns, potentially leading to missed detections in highly sophisticated or subtle attacks. Additionally, Although the system demonstrates excellent accuracy in controlled experimental conditions, its application in real-world situations is lacking. With varying network conditions and diverse traffic types remain uncertain. Computational overhead associated with implementing passive aggressive, RF, and DT algorithms in tandem could also pose scalability issues for large-scale networks. Furthermore, the approach might require frequent retraining to maintain effectiveness against emerging threats, increasing the complexity of operations. To further improve the system's applicability and durability in dynamic cybersecurity contexts, it is imperative to address these shortcomings.

Long short-term memory

Long short-term memory is a specialized form of recurrent neural network architecture developed to address the vanishing gradient issue, enabling it to retain long-term dependencies in sequential data. This model was first introduced by Hochreiter and Schmidhuber in 1997 as an improvement over traditional RNNs, which struggle with maintaining information over long sequences. LSTMs are widely used in natural language processing (NLP), time-series forecasting, speech recognition, and other sequential tasks due to their ability to retain past information over extended periods.

Conclusion

The paper presents a novel method for tackling the problem of identifying unknown DDoS attacks in an open-set recognition framework: reciprocal points learning. By utilizing machine learning methods like Decision Tree, Random Forest, and Passive Aggressive models, the approach leverages flow length and packet-related data for analysis, two important aspects of network traffic data, to differentiate between legitimate and malicious activity. The suggested method successfully recognizes DDoS attack patterns that are both well-known and novel, strengthening and enhancing cybersecurity defenses. The method's excellent accuracy and dependability are demonstrated by experimental results on a large dataset, underscoring its potential to improve proactive cybersecurity methods. By considerably advancing the identification and mitigation of complex and dynamic network threats, our study ensures increased resilience in the face of emerging cyber threats

Dataset

In preprocessing we use

1. Data splitting
2. SMOTE (Synthetic minority over-sampling technique)
3. Feature selection/subsetting
4. DataFrame column dropping

Acknowledgement

The authors gratefully acknowledge the students, staff, and authority of Physics department for their cooperation in the research.

References

[1] Shieh, C. S., Ho, F. A., Horng, M. F., Nguyen, T. T., & Chakrabarti, P. (n.d.). Open-Set recognition in unknown DDoS attacks detection with reciprocal points learning.

[2] Chu, H. C., & Yan, C. Y. (2021). DDoS attack detection with packet continuity based on LSTM model. In Proceedings of the 3rd IEEE Eurasia Conference on IOT, Communication and Engineering.

[3] Yeom, S., & Kim, K. (2020). Improving performance of collaborative source-side DDoS attack detection. In APNOMS 2020 - 2020 21st Asia-Pacific Network Operations and Management Symposium: Towards Service and Networking Intelligence for Humanity.

[4] Zhao, W., Sun, H., & Zhang, D. (n.d.). Research on DDoS attack detection method based on deep neural network model in SDN. In Proceedings - 2022 International Conference on Networking and Network Applications.

[5] Jia, W., Liu, Y., Liu, Y., & Wang, J. (2020). Detection mechanism against DDoS attacks based on convolutional neural network in SINET. In Proceedings of 2020 IEEE 4th Information Technology, Networking, Electronic and Automation Control Conference.

[6] Tehaam, M., Ahmad, S., Shahid, H., Saboor, M. S., Aziz, A., & Munir, K. (2022). A review of DDoS attack detection and prevention mechanisms in clouds. In 24th International Multitopic Conference.

[7] IoT DOS and DDOS attacks detection using an effective convolutional neural network, In Proceedings - 2023 International Conference on Cyberworlds.

[8] Detection of DDoS attacks using semi-supervised machine learning approaches. In 2nd International Conference on Computational Methods in Science and Technology (ICCMST 2021). Published in 2021.

[9] Sanap, Y. B., & Aher, P. (2023). A comprehensive survey on detection and mitigation of DDoS attacks enabled with DL techniques in cloud computing. In the 6th IEEE International Conference on Advances in Science and Technology, ICAST, (pp. 149–154).

[10] Jia, B., & Liang, Y. (2020). Anti-D chain: a lightweight approach for DDoS attack detection using heterogeneous ensemble learning in blockchain. *China Communications.*

5 IoT based integrated farming: optimizing agriculture with smart sensors

R. Mahesh Kumar[a], Y. Pavan Kumar Reddy[b], N. Vishnu Vardhan[c], D. Venkata Sai Tharun[d], D. Yaswanth[e] and M. Surendra[f]

Annamacharya University, Rajampet, Andhra Pradesh, India

Abstract

Agriculture stands as the fundamental economic base of India. Agricultural operations are the main source of income for 72% of India's population showing the essential importance of agriculture to the country. Traditional farming techniques demand long work hours which complicate advancements in farming technology. Technology advances within smart agriculture systems provide a solution to handle current agricultural challenges. Among all agricultural procedures, irrigation stands as a core process which can demonstrate increased efficiency through automation. An irrigation system powered by sensors demonstrates a successful approach to controlling water delivery to agricultural land. A system design features soil moisture sensors which trigger automated water pumps to produce increased crops. Tools from the IoT field are utilized to track multiple agricultural parameters including soil hydration and atmospheric readings. Smart fencing systems work as animal barriers while automated rain roofs defend crops during unwanted weather conditions. Remote IoT management operates over the entire field which enables sustainability alongside operational efficiency.

Keywords: Internet of Things, NodeMCU, parameters monitoring, rain roof, smart fencing

Introduction

Identify smart farming technologies that can support sustainable farming practices. Using technologies like precision farming, efficient water management, soil moisture monitoring, and irrigation are proven ways to increase crop yields. Precision farming prevents misuse and overuse of pesticides and fertilizers, allowing farmers to use the soil according to its quality and condition. With groundwater levels rapidly declining in India due to unprecedented demand from agriculture and industry, precision farming can be a potential lifesaver. If farmers continue to balk or stubbornly follow traditions and slow work, it could reduce India's overall production. Skilled migrants from all over India, who returned to their hometowns during the recent coronavirus pandemic, have chosen agriculture as their profession and are refusing to go back. Migrants can now switch to smart farming because the time required to encourage people to adopt smart farming is shorter than it has always been for individual farmers.

Literature Review

The adoption of Internet of Things (IoT) technologies in agriculture has sparked significant advancements and innovation. Numerous studies have explored how IoT can enhance irrigation efficiency, soil monitoring, and overall agricultural practices. Below is a summary of relevant research in this field.

Muhammad et al. [1] presented an IoT-enabled agricultural system using Raspberry Pi, focusing on real-time monitoring and automation. Their research demonstrated how affordable computing devices could optimize farming processes effectively.

Divya et al. [2] introduced a soil monitoring system based on IoT aimed at boosting agricultural productivity. This study emphasized the role of sensors in tracking soil moisture and environmental factors, which aids in better crop management.

Laksiri et al. [3] worked on optimizing a smart irrigation system designed for Sri Lanka, leveraging IoT for precise irrigation scheduling. Their system emphasized water conservation by utilizing real-time environmental data.

[a]rmahesh369786@gmail.com, [b]ratnasena.reddy@gmail.com, [c]nagamallavishnu04@gmail.com, [d]saidonadi@gmail.com, [e]yaswanthcrazy59@gmail.com, [f]madhuru.surendra@gmail.com

DOI: 10.1201/9781003685364-5

Math et al. [4] developed a drip irrigation system powered by IoT, which delivers water efficiently based on soil moisture levels. This approach significantly reduced water wastage while ensuring the health of crops.

The automated irrigation system proposed by Mishra et al. [5] used IoT and sensor technologies to reduce manual intervention in water management. Their work highlighted the efficiency of automation in saving resources.

Rao and Sridhar [6] proposed a system for crop-field monitoring and automated irrigation. Their research underlined the potential of IoT to facilitate remote operations and scalability for large agricultural areas.

Saraf and Gawal [7] developed an IoT-based system to monitor and control irrigation. Their work demonstrated how real-time data, combined with predictive analytics, could improve water management.

Shrihari [8] combined IoT and deep learning in a wireless system to automate crop production and address issues such as intrusion. This study showcased the integration of advanced machine learning techniques in agricultural decision-making.

Sushanth and Sujatha [9] designed a smart agriculture system using IoT, which focused on resource management and environmental monitoring. The study highlighted IoT's potential to promote sustainable farming.

Vaishali et al. [10] proposed a mobile-integrated irrigation management system. By combining IoT with mobile technology, their research emphasized user-friendly control and monitoring of irrigation processes.

Anurag et al. [11] introduced "Agro-Sense," a precision agriculture system using sensor-based wireless mesh networks. Their work illustrated the effectiveness of real-time data collection in optimizing farming practices.

Arun and Sudha [12] conducted a detailed review of agricultural management using wireless sensor networks. Their survey highlighted advancements in sensor-based technologies and their transformative potential for traditional agriculture.

Methodology

Existing method

Figure 5.1. In the current method, we use microcontrollers and sensors to water the plants according

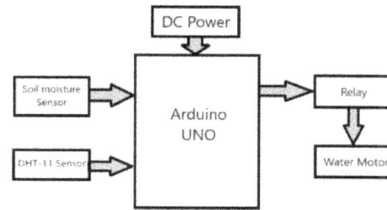

Figure 5.1 Block diagram of existing method
Source: Author

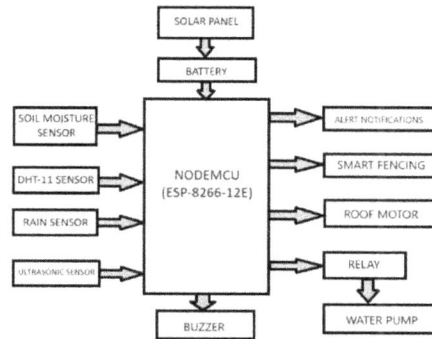

Figure 5.2 Block diagram of proposed system
Source: Author

to the moisture value. If there is no moisture in the soil, the microcontroller will immediately collect data from the soil moisture sensor and send a command to the relay module. That relay module is connected to the water pump and the battery, and then the relay sends the battery power to the generator to pump water. The DHT-11 sensor also complements the current temperature and humidity measurement methods.

The system faces several limitations that impact on its performance and usability. It offers only a limited range of parameters, restricting its versatility for diverse applications. Additionally, it lacks the capability to monitor sensor data remotely over long distances, making it less suitable for scenarios requiring remote supervision. The absence of integration with smart devices further diminishes its functionality, preventing seamless interaction with modern technology. Moreover, the overall efficiency of the system is relatively low, which can hinder its effectiveness in demanding environments.

Proposed system

Figure 5.2. To overcome the shortcomings of the current methods, we propose various approaches to increase the efficiency and safety of plant growth. In our planning, we also use soil moisture meters to

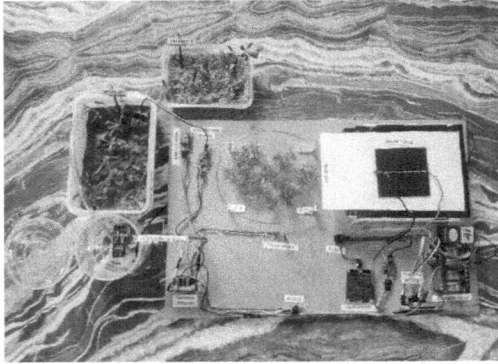

Figure 5.3 Prototype design
Source: Author

Figure 5.4 Roof top
Source: Author

Figure 5.5 Smart fencing
Source: Author

Figure 5.6 Automatic water pump
Source: Author

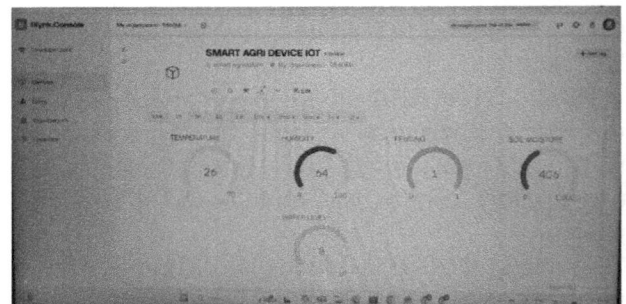

Figure 5.7 Web dashboard
Source: Author

measure the moisture content in the soil. Depending on the moisture value, our microcontroller (ESP8266-12E) will activate relay module and turn on/off the pump.

In this method we added extra features like water level monitoring using Ultrasonic sensor, measuring real time temperature and humidity values using DHT-11 sensor and also here we are implementing the automatic rain roof by sensing the Rain using Rain sensor, touch sensor for smart fencing for alerting the animals to save the plants. All the sensor's data is processed by NodeMCU controller and activates the output devices. Whenever the controller warns then immediately it will turn ON buzzer. And also, solar panels are used for generation of electricity and stored in batteries.

Result

The prototype of the proposed system is shown in Figure 5.3.

The rain roof is automatically opened by sensing the rain using rain sensor which is shown in Figure 5.4.

The touch sensor is used for smart fencing for alerting the animals to save plants, i.e., shown in Figure 5.5.

Measure the moisture of the soil using a soil moisture sensor. Depending on the moisture value, our microcontroller (ESP8266-12E) will activate the relay module to turn on/off the water pump as shown in Figure 5.6.

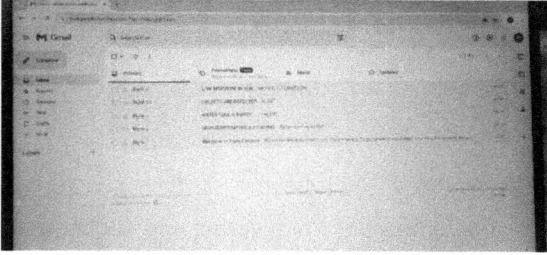

Figure 5.8 Alert notifications
Source: Author

Farmers can monitor the farm through the Blynk Web Dashboard as shown in Figure 5.7.

Farmers get alert notifications as shown in Figure 5.8.

Conclusion

By combining wireless sensor networks (WSN) with the Internet of Things (IoT), we can increase agricultural productivity. These systems can monitor the quality of the soil and the plant growth in the soil, and farmers can use these systems to solve water problems, temperature problems, and soil problems. The difficulties faced by farmers can be reduced and better communication can be achieved to transmit useful information across many nodes. Thus, farmers can control various agricultural equipment and monitor their crops through their smartphones or computers. These systems offer various uses for users to improve their skills and harvest better crops. The use of these systems could help India increase its production of rice, wheat, corn, and other agricultural products in the future. IoT can control yields and growth, and can also be used to monitor soil, temperature, humidity, and more.

References

[1]   Z. Muhammad, M. A. A. M. Hafez, N. A. M. Leh, Z. M. Yusoff and S. A. Hamid, Smart Agriculture Using Internet of Things with Raspberry Pi, 2020 10th IEEE International Conference on Control System, Computing and Engineering (ICCSCE), Penang, Malaysia, 2020, pp. 85–90, doi: 10.1109/ICCSCE50387.2020.9204927.

[2]   N. Ananthi, J. Divya, M. Divya and V. Janani, IoT based smart soil monitoring system for agricultural production, 2017 IEEE Technological Innovations in ICT for Agriculture and Rural Development (TIAR), Chennai, India, 2017, pp. 209-214, doi: 10.1109/TIAR.2017.8273717.

[3]   H. G. C. R. Laksiri, H. A. C. Dharmagunawardhana and J. V. Wijayakulasooriya, Design and Optimization of IoT Based Smart Irrigation System in Sri Lanka, 2019 14th Conference on Industrial and Information Systems (ICIIS), Kandy, Sri Lanka, 2019, pp. 198-202, doi: 10.1109/ICIIS47346.2019.9063272.

[4]   A. Math, L. Ali and U. Pruthviraj, Development of Smart Drip Irrigation System Using IoT, 2018 IEEE Distributed Computing, VLSI, Electrical Circuits and Robotics (DISCOVER), Mangalore, India, 2018, pp. 126–130, doi: 10.1109/DISCOVER.2018.8674080.

[5]   D. Mishra, A. Khan, R. Tiwari and S. Upadhay, Automated Irrigation System-IoT Based Approach, 2018 3rd International Conference On Internet of Things: Smart Innovation and Usages (IoT-SIU), Bhimtal, India, 2018, pp. 1–4, doi: 10.1109/IoT-SIU.2018.8519886.

[6]   R. N. Rao and B. Sridhar, IoT based smart crop-field monitoring and automation irrigation system, 2018 2nd International Conference on Inventive Systems and Control (ICISC), Coimbatore, India, 2018, pp. 478–483, doi: 10.1109/ICISC.2018.8399118.

[7]   S. B. Saraf and D. H. Gawali, "IoT based smart irrigation monitoring and controlling system," 2017 2nd IEEE International Conference on Recent Trends in Electronics, Information & Communication Technology (RTEICT), Bangalore, India, 2017, pp. 815–819, doi: 10.1109/RTEICT.2017.8256711.

[8]   M. Shrihari, "A Smart Wireless System to Automate Production of Crops and Stop Intrusion Using Deep Learning," 2020 Third International Conference on Smart Systems and Inventive Technology (ICSSIT), Tirunelveli, India, 2020, pp. 1038–1045, doi: 10.1109/ICSSIT48917.2020.9214303.

[9]   G. Sushanth and S. Sujatha, IOT Based Smart Agriculture System, 2018 International Conference on Wireless Communications, Signal Processing and Networking (WiSPNET), Chennai, India, 2018, pp. 1–4, doi: 10.1109/WiSPNET.2018.8538702.

[10] S. Vaishali, S. Suraj, G. Vignesh, S. Dhivya and S. Udhayakumar, Mobile integrated smart irrigation management and monitoring system using IOT, 2017 International Conference on Communication and Signal Processing (ICCSP), Chennai, India, 2017, pp. 2164-2167, doi: 10.1109/ICCSP.2017.8286792.

[11] Anurag, D., Roy, S., & Bandyopadhyay, S. (2008). Agro-sense: precision agriculture using sensor-based wireless mesh networks. In ITU-T "Innovation in NGN", Kaleidoscope Conference, Geneva 12-13 May 2008.

[12] Arun, C., & Sudha, K. L. (2012). Agricultural management using wireless sensor networks – a survey. In 2nd International Conference on Environment Science and Biotechnology IPCBEE (Vol. 48). IACSIT Press, Singapore.

6 Smart surveillance: real-time AI-powered threat detection and response system

Nukala Chaitanya[1,a], Manjula, C.[2,b], Neha Gayathri Devi, P.[2,c], Dheerajendra Reddy, D.[2,d], Sai Kumar, K.[2,e] and Pogula Sreedevi[3,f]

[1]Assistant Professor, Department of CSE, Rajeev Gandhi Memorial College of Engineering and Technology, Nandyala, Andhra Pradesh, India

[2]Department of Computer Science and Engineering, Rajeev Gandhi Memorial College of Engineering and Technology, Nandyala, Andhra Pradesh, India

[3]Associate Professor, Department of CSE, Rajeev Gandhi Memorial College of Engineering and Technology, Nandyala, Andhra Pradesh, India

Abstract

The primary goal is to enhance real-time threat detection by creating intelligent intelligence analysis through the use of deep learning techniques like You Only Look Oncev8 (YOLOv8) and convolutional neural network (CNN). Conventional surveillance systems depend on manual monitoring, which may cause users to become weary or overwhelmed with data and miss crucial events. The project automatically detects potential security threats such as guns, knives, and masked people directly at the security camera level. Existing systems often rely on cloud computing, which causes delays and limits immediate threat protection. The proposed system overcomes this problem by using an intelligent edge model to provide faster and more responsive time. The system also includes motion detection to capture images during abnormal activity, reducing the need for manual monitoring. The system employs YOLOv8 for detecting objects and utilizes CNN for classifying images. Future enhancements will be made to expand the application to cover various types of surveillance and further increase the accuracy of detecting various threats.

Keywords: AI-powered surveillance system, convolutional neural network, real-time threat detection, traditional surveillance systems, YOLOv8

Introduction

Imagine the frustration of constantly rewinding grainy CCTV footage, searching for something out of the ordinary. Now picture a smart camera analyzing the footage in real-time, predicting potential threats [1] before they even occur. This is how artificial intelligence (AI) has the potential to revolutionize surveillance. Beyond appearances, artificial intelligence (AI) aims to comprehend, anticipate, and eventually stop crime. In recent years, firearm-related violence has risen sharply, raising significant concerns about public safety [2]. According to law enforcement statistics, the number of firearm incidents in the city has grown, creating an urgent need for more effective surveillance systems [3]. Fortunately, advances in AI and machine learning offer new, powerful ways to improve public safety [3]. AI-driven security technologies are becoming indispensable for preventing violence and safeguarding public spaces. Their strength lies in their ability to detect weapons quickly and accurately, providing law enforcement and security personnel with real-time intelligence—often before threats escalate [4]. Traditional methods, such as manual camera monitoring, can be slow and prone to human error, making them less reliable in fast-paced situations. Unlike traditional methods, AI-based systems are capable of analyzing large volumes of visual information in real time, allowing them to detect possible threats with high accuracy [4]. Among the leading technologies in this field is YOLO (You Only Look Once), a real-time object detection model renowned for its speed and accuracy [5]. The latest iteration, YOLOv8, further enhances detection performance, offering superior capabilities even in complex or challenging environments. The

[a]chaitu.anju@gmail.com, [b]manjulachavala347@gmail.com, [c]pnehagayathridevi@gmail.com, [d]dheerajendrareddy@gmail.com, [e]venkeysai74@gmail.com, [f]sreedevipogula37@gmail.com

DOI: 10.1201/9781003685364-6

framework processes images once and is suitable for applications that require real-time analysis. Its architecture is designed to optimize speed and accuracy to provide quality images even in crowded or dynamic environments. However, the structure of YOLOv8 used in research may vary depending on whether researchers integrate it into their own systems or projects, as shown in Figure 6.1.

Literature survey

Importance of instant weapon detection
The ability to instantly detect weapons using YOLOv8 plays a crucial role in keeping public spaces safe. In today's world, where unexpected threats can arise at any moment, rapid detection is more important than ever. By integrating advanced AI models like YOLOv8 into surveillance systems at places such as schools, airports, shopping malls, and other high-risk areas, we can significantly strengthen security efforts and give the public a greater sense of safety. Just knowing that such technology is in place can discourage criminal behavior.

Transforming traditional surveillance
This capability is transforming traditional surveillance by incorporating AI-driven security features.

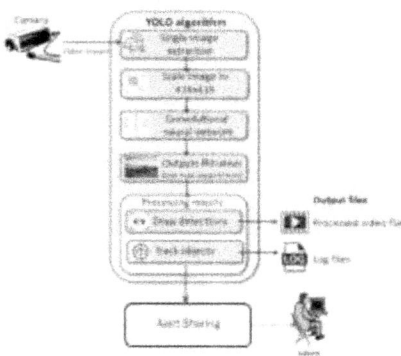

Figure 6.1 YOLOv8 technology revolutionizes real-time threat detection in video surveillance, providing unparalleled speed and accuracy
Source: Author

Figure 6.2 Proposed system architecture
Source: Author

The technology effortlessly blends with existing security systems, including drone-based surveillance for large-scale events providing a comprehensive, multi-angle approach to threat detection [7]. Additionally, its integration with AI-powered tools like facial recognition elevates security management. By combining weapon detection with personal identification, the system enhances situational awareness for law enforcement, enabling a swift and informed response to potential threats through the use of multi-source intelligence.

Data-driven security and policy implications
Analyzing weapon detection data allows law enforcement and policymakers to identify crime trends and develop evidence-based strategies to prevent incidents and enhance public safety [8].

Need for AI investment in security
Investing in AI-driven technologies is imperative for advancing public safety measures. AI-powered security systems represent a forward-thinking approach to addressing modern security challenges, reinforcing the necessity for continued research and implementation in this field.

Proposed System

The proposed system addresses this challenge by utilizing an edge intelligence model to deliver faster detection and response times. It also integrates motion detection to capture images during unusual activity, minimizing the need for manual surveillance. While a convolutional neural network (CNN) is utilized for image classification, YOLOv8 is utilized for object detection. Future upgrades will expand the system's application across various surveillance scenarios and further improve its ability to detect a wider range of threats. Currently, systems that rely on images to represent objects process data from a single frame. The camera in the designated area captures the image and tracks any changes within the frame [9]. The YOLOv8 algorithm has become the standard for object detection due to its remarkable accuracy and speed in processing visual data.

This paper presents an overview of the YOLOv8 algorithm—short for "You Only Look Once"—and its variants, comparing its performance to earlier YOLO versions Figure 6.2 and traditional CNNs. It emphasizes the latest improvements made to YOLOv8 and how they enhance real-time weapon detection capabilities. The research focuses on developing intelligent models that can help improve security in high- risk public areas such as schools, airports, and public transport hubs. With rising global violence, the need for fast and accurate surveillance systems is critical. To ensure effectiveness, the model was trained on a diverse dataset featuring images of firearms and bladed weapons [6]. Evaluated using metrics such as precision, recall, F1-score, mAP, and IoU, the YOLOv8-based system showed strong performance, accurately identifying threats in real time. This solution not only advances computer vision technology but also addresses public safety needs by supporting rapid law enforcement responses.

Working methodology

The YOLO algorithm flowchart involves several key components that work in tandem to detect objects within images. It begins with a still image as input, which may undergo preprocessing steps such as resizing or normalization, depending on the specific version of YOLO used [10]. The image is then processed through a CNN backbone, like Darknet or ResNet, which extracts features at various levels, capturing both fine details and broader semantic information.

The head of the algorithm utilizes convolutional techniques to estimate bounding boxes, confidence scores, and the quality of detected objects based on the extracted feature maps. The bounding boxes are defined using coordinates (x, y, width, height), helping to pinpoint the location of potential objects. This approach marks a significant advancement in the development of AI-powered security systems. By applying AI in the realm of public safety, the research highlights the importance of continued innovation, collaboration among tech experts, and proactive legislation to safeguard communities from emerging threats.

Leveraging CNNs and YOLOv8 for real-time threat detection

The system is made to use the YOLOv8 object detection framework and CNNs to automatically recognize possible security threats from real-time security camera footage, including knives, guns, and masked people. To take crisp, detailed pictures, a high-resolution security camera is used, making sure the resolution is adequate for accurate object detection.

YOLOv8 object detection

Model training: Train the YOLOv8 model using a large and varied dataset that includes images and videos featuring a range of potential threats, such as firearms, knives, and masks. Once trained, the YOLOv8 model can be deployed on an edge device, such as a GPU-powered computer, or hosted in the cloud, enabling real-time processing of video streams and the identification of threats with bounding boxes and class labels [11].

CNN-Based threat classification

Fine-tuning the CNN: Enhance the CNN model to classify detected objects into specific threat categories (e.g., gun, rifle, knife, mask). This step improves accuracy while minimizing distortion.

Alarm system: When a potential threat is detected (e.g., a gun, knife, or masked individual), the system will trigger an alarm. The alarm features include:

- **Visual alerts:** Display notifications or highlight relevant objects within the security camera feeds.
- **Data collection:** Gather a diverse set of images and video files featuring:
- **Objects of interest:** Guns, knives, other weapons, masked individuals, and suspicious behavior (e.g., running).

  Images: Capture from various locations (e.g., airports, public areas), considering variables like lighting conditions, camera angles, and crowd density. Determine the bounding boxes and class probabilities for every object.

- **Anchor boxes**: To forecast object locations, use predefined boxes with different sizes and aspect ratios.
- **Object scores**: Calculate the probability that an object will show up inside an anchor box.
- **Probability**: Determine the possibility that an item is a member of a particular class (e.g., background, masked person, knife, or gun).
- **Function of training loss:**
  - Objectivity loss: Calculates how much the actual scores differ from the predicted scores.
  - Optimizer: To guarantee convergence and modify the model weights during training, use optimizers such as Adam or SGD.
- **Post-processing and Inference:**
  - Inference: The YOLOv8 model processes input images to produce predictions, handling overlapping results appropriately.

- **Instant Implementation:**
  - Optimization: To enable real-time deployment, expedite model processing through methods like quantization and pruning.
- **Action must be taken immediately.** Make sure the system is fair, moral, and free from abuse. New data and better algorithms must be added to the model on a regular basis.

Real-time, high-accuracy results that improve public safety and security can be produced by fusing the capabilities of CNNs and YOLOv8 with thorough data collection and efficient model tuning.

Data collection and preparation
An S×S grid is used to segment input images in the YOLOv8 architecture. With the help of coordinates (x, y, w, h) and a confidence score (C), each grid cell forecasts a predetermined number of bounding boxes. The accuracy of the bounding box and the likelihood that an object is present are both reflected in the confidence score.

$$C = P(Object) \times IoU \tag{1}$$

Figure 6.3. Speed analysis detection where IoU (Intersection of Union) quantifies the overlap between the estimated bounding box and the box's actual location on the ground, and P(object) is the probability that the bounding box contains the object. The final loss function is expressed as:

Algorithm 1 Object Detection Algorithm
1. Input: Image I
2. Preprocess image I to fixed size W x H
3. Divide image into S x S grid cells
4. for each grid cell g do
5.   for each bounding box b in cell g do
6.   Predict coordinates (x, y, w, h) and confidence score C
7.   Predict class probabilities P(Class|g)
8.   end for
9. end for
10. Apply Non-Maximum Suppression (NMS) to filter overlapping boxes
11. Output: Detected objects with bounding boxes and class labels

This method enhances both speed and efficiency by enabling YOLO to generate predictions for multiple objects in a single pass. Subsequent YOLO iterations have brought numerous improvements to further boost accuracy, speed, and robustness. YOLOv8 continues to advance the field of real-time detection, with new training techniques such as machine learning and self-monitoring making it adaptable to a diverse range of scenarios and datasets. Several tools and techniques are used to optimize the model's performance.

The input image undergoes several transformations, including resizing, normalization, and augmentation, to enhance model learning and detection accuracy.

$$Loss = \lambda_{coord} \sum_i \sum_j Loss_{coord} + \lambda_{noobj} \sum_i \sum_j Loss_{noobj} + \sum_i \sum_j Loss_{class} \tag{2}$$

The study utilizes a well-defined loss function combining confidence loss and localization loss, ensuring reliable performance across detection tasks.

$$Loss = \lambda_{coord} + Loss_{coord} + \lambda_{noobj} \cdot Loss_{noobj} + Loss_{class} \tag{3}$$

Where:

$$Loss_{coord} = \sum_{i=0}^{N} \sum_{j=0}^{B} ((X_{ij} - \hat{X}_{ij})^2 + (Y_{ij} - \hat{Y}_{ij})^2 + (W_{ij} - \hat{W}_{ij})^2 + (H_{ij} - \hat{H}_{ij})^2) \tag{4}$$

Here, N represents the total number of samples, while B denotes the number of predicted bounding boxes for each grid cell.

The λ_{cord} and λ_{noobj} are used to control the importance of different components in the training loss—specifically, the localization loss and the no-object loss.

Fast search using YOLOv8
For real-world applications, especially in security, fast detection is absolutely critical. One key performance indicator is the average inference time per frame, which reflects how efficiently a model processes large amounts of visual data. Whether in crowded public areas or complex backgrounds, the model shows strong performance and responsiveness—making it well-suited for real-time surveillance systems.

The image demonstrates how the model identifies and marks objects with bounding boxes, assigning each box a class label and a confidence score to represent prediction accuracy. The boxes are color-coded,

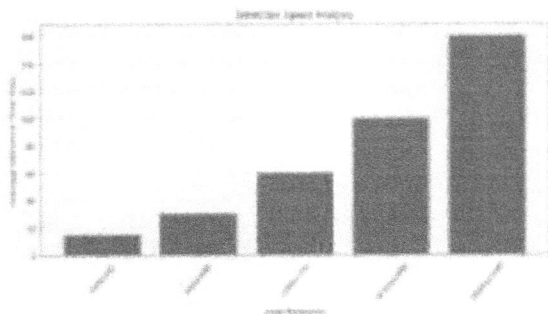

Figure 6.3 Detection of speed analysis
Source: Author

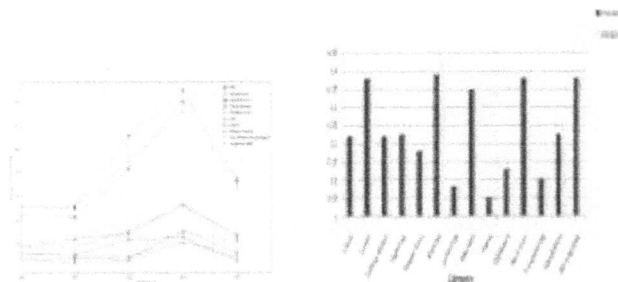

Figure 6.4 CNN, YOLOv8's predictions of truth and Yolov8, CNN multiple groups testing
Source: Author

with brighter colors indicating higher confidence and darker shades showing lower certainty. This visual feedback enables users to easily assess the model's performance in different conditions, such as crowded spaces or fluctuating lighting. It offers a clear, intuitive way to highlight the model's capabilities and represents a significant advancement in the intersection of AI and public safety.

In order to eliminate redundant or overlapping detections, the algorithm then uses Non-9. end for 10. Maximum Suppression (NMS). By removing duplicate detections, NMS makes sure that only the most dependable, non-overlapping predictions are retained. Bounding boxes, each associated with a probability and a confidence score for its respective class, are the output of the YOLO algorithm. These boxes show the objects that have been identified in the picture, together with their confidence levels and class labels. In computer vision, YOLO has become a popular method for object detection. Although earlier methods such as sliding window, RCNN, Fast RCNN, and Faster RCNN were frequently employed, YOLO outperforms them in terms of performance. While earlier approaches like sliding window, RCNN, Fast RCNN, and Faster RCNN were commonly used, YOLO surpasses them in performance. This version builds on the latest advancements in deep learning and computer vision, providing remarkable speed and accuracy. Its design is versatile, meeting the needs of various applications, and it is implemented in the user-friendly Ultralytics Python package, making it adaptable to a wide range of hardware platforms, from edge devices to cloud APIs. State-of-the- art (SOTA) object detection algorithms are so fast that they have become the standard for object identification in computer vision. Previously, sliding window

methods were considered the most effective for target detection, but improvements led to faster and more efficient versions, such as CNN, R-CNN, and Fast RCNN. This paper explores the potential of YOLOv8, examining its progress, how to apply it seamlessly to a specific dataset, and reviewing the evolution of YOLO and the challenges faced by earlier versions. Figure 6.4.

YOLOv8 Technology

The use of YOLOv8 technology has revolutionized real- time area monitoring. It provides quicker, more dependable, and more accurate threat detection as the eighth iteration of a series of object detection models. This is a comprehensive overhaul rather than just a minor update, and it is expected to significantly affect global security surveillance. However, the growing criminal use of these technologies emphasizes how urgently sophisticated counter-drone systems are needed. By incorporating state-of-the-art technologies like TensorFlow and PyTorch, YOLOv8 signifies a revolution in the field of video surveillance, not just an improvement. The Intersection over Union (IoU) metric, which gauges how closely YOLOv8's predictions match the ground truth, is essential for confirming the model's accuracy.

Average Precision (AP) and Mean Average Precision (mAP) are metrics that evaluate the accuracy of the model across different object types, making them crucial for various applications. These metrics help ensure precision and memory efficiency, which are vital to minimize errors, particularly when both types of errors have serious consequences. The model.val() function in YOLOv8's Convolutional Neural Network (CNN) checks all of these parameters in multiple groups during the testing phase.

Comparative speed analysis

YOLOv8 is renowned for its speed and accuracy, performing exceptionally well with the COCO dataset, which highlights both its precision and speed Figure 6.5. This is essential when rapid decisions are needed. Tools like F1-score curves and precision-recall curves demonstrate the model's effectiveness, contributing to its improvement.

YOLOv8 is highly reliable in real-world applications, enhancing performance at all scales, including tracking..

Advanced cameras and sensors capture high-resolution images, which are then analyzed to track the movement of people, threats, and other objects of interest within the monitored area. By processing visual data quickly and using intelligent tracking devices, these systems can follow the contours of multiple targets simultaneously, providing instant alerts and safety information. Each area is examined in detail, offering valuable insights into current activity. Among various YOLO architectures tested (v2-v8, H, X, R, C), YOLOv8 stands out with an impressive average accuracy (mAP) of 0.99.

YOLOv8 is revolutionizing video surveillance technology, transforming the way security cameras are utilized in real-world scenarios [12]. Public spaces can now manage and monitor traffic more effectively than ever before. It excels in detecting when items are removed from a site, marking a significant leap in video forensics technology. With YOLOv8, drones can now be accurately identified and monitored for tampering, thanks to advancements that combine YOLOv8 with camera systems designed to prevent accidents. This demonstrates the potential of YOLOv8 in analyzing drone data. Additionally, the model has developed innovative methods to uncover hidden objects in video, which is a critical advancement for the future of security analysis.

Classification involves assigning an image to one of a number of predetermined categories, making it one of the easier tasks. A label and a confidence score are produced by the image classifier. It's critical to address and prepare for important issues.

Privacy and security: Privacy and security are critical considerations to address public concerns and ensure compliance with laws governing the use of personal information.

1. Training and supervision: While employees and workers may not require intensive training, regular performance monitoring is vital to ensure the system's effectiveness and to stay updated with emerging threats and technological advancements.
2. Ethical issues: The use of AI-based surveillance must strike a balance between public safety and individual privacy rights. It should be guided by clear policies and involve active participation from stakeholders.

Benefits

Surveillance video analytics: The key objectives that emphasize the significance of surveillance video analytics are as follows. Continuously monitoring video footage can be both difficult and tedious for people. Intelligent video surveillance analysis addresses these challenges by automating complex tasks. It is essential to extract actionable insights from real-world situations with the highest level of accuracy. For example, functions like crowd analysis still require significant development. In real-time scenarios, the time needed to generate a response is critical. Predicting movements, actions, or behaviors can be invaluable during emergencies. Video data

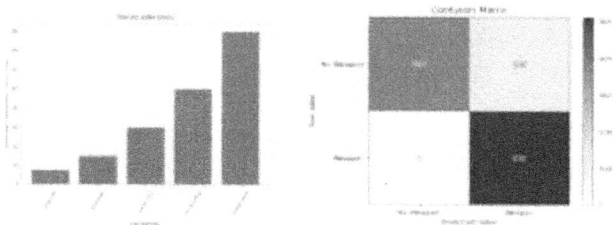

Figure 6.5 Detection of speed analysis and Detection matrix of weapon detection
Source: Author

Table 6.1 YOLOv8 for security surveillance.

Capability	Statistic	Real-world impact
Forgery localization F1-score	0.99	Enhances reliability in video surveillance
Obstacle detection for UAVs	F1-score of 96% in 200 epochs	Improves navigational safety for drones
Generative models in occlusion	Effective recognition with occlusions	Advances object detection despite visual interferences

Source: Author

holds a wealth of information, and some systems use binary classification techniques to identify unusual behaviors. Various methods are available for criminal detection and crowd analysis. The next section explores application areas where these technologies are most relevant. A large portion of current work offers context-specific solutions, including:

- Traffic management and major intersections
- Public addresses
- Community forums or discussions
- Religious festivals or events
- Office environments

Crowd identification remains one of the most challenging tasks in these contexts, as all actions, behaviors, and movements need to be accurately recognized.

Conclusion

Our extensive model testing reveals that it effectively balances accuracy and recall, which is essential for minimizing both bias and false positives, especially in critical situations. Further analysis, including accuracy regression and uncertainty matrices, offers a deeper understanding of the model's performance across different weapon categories, rather than focusing on just one. We also evaluated the model's detection speed compared to alternative solutions, confirming its efficiency and potential for integration into real-time surveillance systems. As urban security continues to grow in importance, deploying AI-driven solutions like this could significantly enhance public safety monitoring. Looking forward, we plan to broaden the dataset to cover a wider array of environments and weapon types, optimize the model for faster performance, and explore adding audio cues to video data to boost detection accuracy. Through these advancements, we aim to leverage technology to tackle public safety challenges and help create safer communities.

Future work

Some of the performance improvement of the threat detection system including but not limiting to further research:

- Ensemble methods: The output of several models can probably be combined to improve the ability to detect objects. For example, an en-

semble made of YOLOv8 and other lightweight model could provide an optimal balance between speed and accuracy performance in challenging environments.

Acknowledgement

The authors sincerely acknowledge the students, staff, and authority of Rajeev Gandhi Memorial College of Engineering and Technology for their cooperation in research.

References

[1] Ahmed, A. A., & Echi, M. (2021). Hawk-Eye: an AI-powered threat detector for intelligent surveillance cameras. IEEE Access, 9, 63283–63293.

[2] Annapurna, B., Reddy, T. R., Raghavendran, C. V., Singh, R. K., & Prasad, V. V. (2020). Coordinate access system for live video acquisition. Journal of Physics Conference Series, 1712(1), 012034. https://doi.org/10.1088/1742-6596/1712/1/012034.

[3] Lalitha, R. V. S., Jalligampala, D. L. S., Kavitha, K., Vahida, S., & Rajasekhar, G. (2021). New directions in traffic control analysis through video surveillance. In E3S Web of Conferences, (Vol. 309, p. 01099). https://doi.org/10.1051/e3sconf/202130901099.

[4] Shahnoor, M., Ramesh, A., Liang, F., Nair, V., Patel, S., Roy, T., et al. (2022). Analyzing Surveillance Videos in Real-Time using AI-Powered Deep Learning Techniques

[5] Fowler, J., et al. (2015). Artificial intelligence in security: new solutions for old problems. International Journal of Security Studies.

[6] Deshpande, A., Kumar, P., Zhang, M., Lopez, H., Natarajan, R., Singh, V., et al. (2023). Data collection and training for weapon detection models. Journal of Machine Learning Research, 24, 1–15.

[7] Dong, Y., Banerjee, S., Luo, Z., Chatterjee, N., Feng, R., Ortiz, M., et al. (2024). Challenges in real-time threat detection: The limitations of cloud computing. International Journal of Security Technology, 11(1), 34–47.

[8] Brodie, J., Evans, T., Campbell, D., Zhang, L., Kumar, N., Sanchez, R., et al. (2005). Data-driven approaches in law enforcement. Journal of Criminal Justice Research, 22(1), 71–85.

[9] Manasa, M., Sowmya, D., Reddy, Y., Sreedevi, P., Kiran, M., Joshi, A. (2024). Attention-based image caption generation. In Proceedings of the International Conference on Computer Vision and Artificial Intelligence (pp. 403–412). Springer. https://doi.org/10.1007/978-981-97-8031-0_38

[10] Sunil, N., & Narsimha, G. (2024). Image-based random rotation for preserving the data in data mining process. *Signal Image and Video Processing*, 18(4), 3893–3902. https://doi.org/10.1007/s11760-024-03050-2.

[11] Lalitha, R. V. S., Jalligampala, D. L. S., Kavitha, K., Vahida, S., & Rajasekhar, G. (2021). New directions in traffic control analysis through video surveillance.

In E3S Web of Conferences, (Vol. 309, p. 01099). https://doi.org/10.1051/e3sconf/202130901099.

[12] Annapurna, B., Reddy, T. R., Raghavendran, C. V., Singh, R. K., & Prasad, V. V. (2020). Coordinate access system for live video acquisition. *Journal of Physics Conference Series*, 1712(1), 012034. https://doi.org/10.1088/1742-6596/1712/1/012034.

7 Improved channel estimation in MIMO-THz systems using sparse sensing and dictionary optimization

K. Shankar[1,a], K. Madhavi Priya[2,b], Y. Sunanda[1,c], M. Ravi Kishore[1,d], Y. Pavan Kumar Reddy[1,e] and P. Sireesha[1,f]

[1]Department of E.C.E., Annamacharya University, Rajampet, Andhra Pradesh, India

[2]Department of E.C.E., S.K.P. Engineering College, Tiruvannamalai, Andhra Pradesh, India

Abstract

Terahertz (THz) systems equipped with UM-MI MO arrays together with their large bandwidth attain terabits-per-second throughput capabilities. The operation range of these systems covers near, intermediate and far-field regions which demands efficient channel estimation approaches. The proposed approach advances cross-field channel estimation by merging properties of compressed sensing (CS) technique with reduced dictionary (RD). The CSRD model needs fewer measurement samples but delivers high accuracy in estimation results. The model achieves quick channel information reconstruction with fewer required measurements to decrease computational needs.

Keywords: Compressed sensing, reduced dictionary, TeraHertz, UM-MIMO

Introduction

Terahertz (THz) band between 0.3 to 10 THz will make diverse use cases possible within 6G mobile networks [5, 7] through ubiquitous connectivity along with inter-chip communications and accurate localization and sensing capabilities [3, 4]. Technology based on THz communications provide extremely substantial bandwidths able to satisfy the demanding throughput needs of massive high-definition video transmissions and sensory experiences. Propagation through THz channels experiences two dominant characteristics because LoS paths dominate while non-LoS paths present 5-15 dB weaker signals because of higher reflection and diffraction losses [6]. The number of channel paths remains typically lower in standard indoor THz communication environments.

Literature review

Chen et al. [3] established that THz communications represent an essential component for localization and sensing functions in the 6G network framework. Geometry-based localization techniques are widely adopted for THz-based positioning, while optimization algorithms are employed both for offline system design and real-time adaptation to enhance accuracy and efficiency [3]. The main hurdle in THz localization practice stems from environmental factors because these elements re-duce signal strength and affect performance particularly at outdoor and large-scale deployment locations [3].

Sarieddeen et al. [4] demonstrates that THz communication systems develop to enable high-speed data transfer as well as support applications including sensing and imaging and localization for future wireless networks. Machine learning algorithms play a growing role in resource allocation and dynamic optimization of THz networks, with techniques like adaptive modulation and reinforcement learning enhancing system efficiency [4]. THz communications suffer from a critical signal loss problem through atmospheric absorption that limits the transmission distances and draws back device performance specifically in outside settings [4].

Zhang et al. [7] suggested that 6G technology would surpass 5G by providing multiple terabytes of second-rate data transfer capability and intelligent connection among land- based and aerial and space-based and undersea networks. The integration of AI and machine learning plays a crucial role in optimizing network resources and enabling autonomous

[a]shan87.maddy@gmail.com, [b]priyamadhavi87academics@gmail.com, [c]sunanda.bujji@gmail.com, [d]ravi.mvrm@gmail.com, [e]ratnasena.reddy@gmail.com, [f]sireeshap2004@gmail.com

DOI: 10.1201/9781003685364-7

operations [7]. The major hurdle for 6G implementation comes from managing the technical complexity of combining different advanced technologies [7].

Methodology

Existing method:
Traditional channel estimation techniques:
Pilot-based estimation:
The most popular method which utilizes pilots for channel estimation is known as pilot-based estimation. Ultra-massive MIMO systems are problematic for this method due to the excessive pilot symbol overhead which makes it perform poorly [2].

Least squares and minimum mean square error (MMSE):
The error metrics become lower through the combined efforts of least squares (LS) and minimum mean square error (MMSE) approaches which solve optimization issues for channel estimation. The estimation techniques struggle to cope with UM-MI MO THz systems because they become complex and produce sub-optimal results in large size solutions.

Compressed sensing methods:
Sparse recovery algorithms:
The Orthogonal Matching Pursuit group belongs to compressed sensing methods which employ THz channel sparsity to reconstruct data with restricted measurements. Despite high potential expenses involved in processing overhead for extensive systems the techniques hold promise to reduce operational complexity in massive MIMO systems with a large number of antennas [2].

Dictionary learning:
Implementing preset dictionaries decreases the quantity of parameters needing calculation during channel representation. A precise model combined with correct vocabulary selection drives the overall effectiveness of the system.

Proposed method:
Reduced dictionary and compressed sensing:
From Figure 7.1. In ultra-massive MIMO THz systems, the CSRD model improves the scalability and accuracy of channel estimation by reducing the amount of measurements needed.

Adaptability to field conditions:
The channel estimate method adjusts its approach depending on the operating field condition which includes near-field and intermediate-field and far-field zones.

Comprehensive simulation and evaluation:
The methodology includes the detailed simulation which takes into account distance, rotation angles, and multipath configurations, thereby making a comprehensive analysis of the performance of the system.

The simulation process starts with setting vital parameters including trial configurations along with SNR requirements and rotation parameters with distance vectors. The simulation design enables the arrays to use either near-field or far-field models. The analysis conducts multiple tests across all possible rotation and distance combinations for the purpose of statistical reliability. The channel estimate precision together with computational requirements of each trial are documented. The SWM, HSPWM, and PWM approaches receive evaluation using the metrics NMSE and achievable rate (AR) beside Complexity metric. When plotting NMSE against distance it shows when different propagation models start and end, but SWM serves best as a modeling tool [1]. Performance behavior due to model selection is assessed through achievable rate analysis across different distances. This analysis includes complexity measurements for determining geographic scale performance efficiency visualization. All findings concerning NMSE, attainable rates and complexity are presented to allow for an extensive comparison of estimation methods across different scenarios.

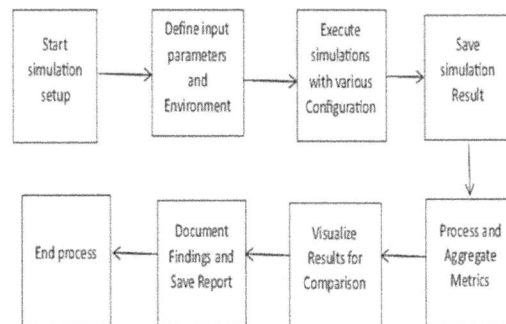

Figure 7.1 Block diagram of proposed method
Source: Author

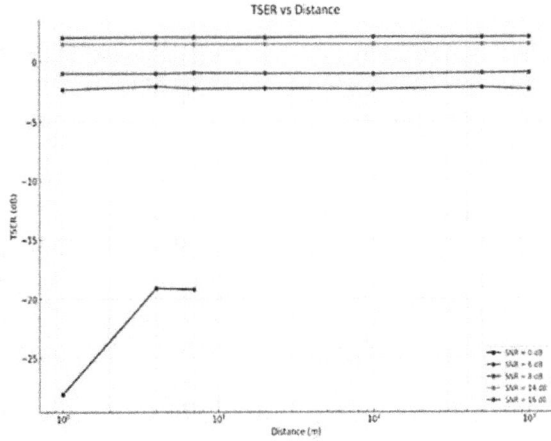

Figure 7.2 TSER vs distance
Source: Author

Table 7.1 Data points for TSER vs distance.

Distance (m)	TSER (dB) SNR= 0dB	TSER (dB) SNR= 6dB	TSER (dB) SNR= 8dB	TSER (dB) SNR= 14dB	TSER (dB) SNR= 16dB
1	-25	-20	-20	-20	-20
10	-20	0	0	0	0
100	-20	0	0	0	0
1000	-20	0	0	0	0

Source: Author

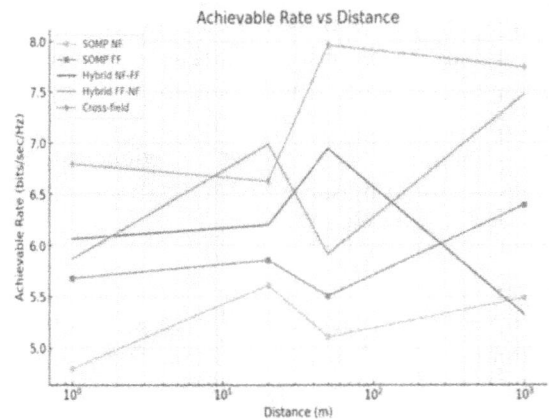

Figure 7.3 Computational complexity vs distance
Source: Author

Table 7.2 Computational complexity at different distances.

Distance (m)	SOMP NF	SOMP FF	Hybrid NF-FF	Hybrid FF-NF	Cross-field
10^0	220	210	215	235	240
10^1	215	200	225	245	260
10^2	210	195	230	240	245
10^3	240	230	220	250	240

Source: Author

Results

The THz system estimation ratio (TSER) appears at different SNR levels in Figure 7.2 through a logarithmic distance scale.

Table 7.1 demonstrates unique SNR = 0dB performance which stands apart from constant 0dB TSER readings shown at other SNR levels (6dB, 8dB, 14dB and 16dB).

The diagram in Figure 7.3 shows how different probability methods measure computational complexity based on distance. Techniques that produce higher estimate or rate values exhibit increased level of complexity.

Computational complexity reaches its maximum with Crossfield and Hybrid NF-FF according to Table 7.2 data which implies they need higher processing power. SOMP FF and SOMP NF prove to be efficient regarding computational complexity thus being suitable for operations with restricted computing resources. The Hybrid FF-NF model works as

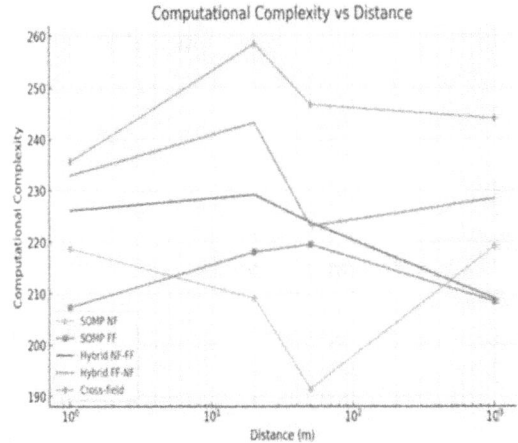

Figure 7.4 Achievable rate vs distance
Source: Author

an excellent midpoint solution because it maintains both performance strength and system simplicity.

The achievable rate expressed in bits per second per Hertz functions against distance through different

Table 7.3 Achievable rate vs distance.

Distance (m)	SOMP NF	SOMP FF	Hybrid NF-FF	Hybrid FF-NF	Cross-field
10^0	5.0	5.5	6.0	6.5	7.0
10^1	5.5	6.0	6.5	7.0	7.5
10^2	6.0	6.2	7.5	6.0	8.0
10^3	7.0	6.8	6.0	7.2	7.5

Source: Author

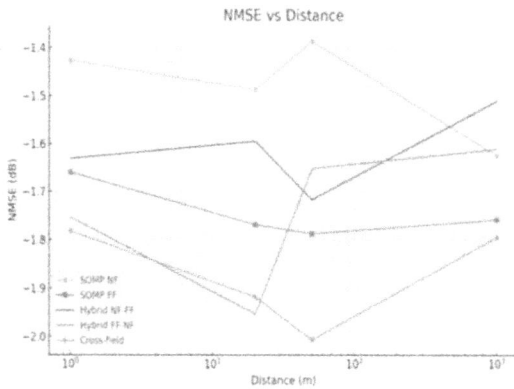

Figure 7.5 NMSE vs distance
Source: Author

Table 7.4 NMSE vs distance.

Distance (m)	SOMP NF	SOMP FF	Hybrid NF-FF	Hybrid FF-NF	Cross-field
10^0	-1.4	-1.7	-1.6	-1.8	-1.9
10^1	-1.5	-1.75	-1.7	-1.9	-2.0
10^2	-1.6	-1.8	-1.75	-1.6	-2.1
10^3	-1.7	-1.85	-1.9	-1.7	-2.0

Source: Author

techniques as shown in Figure 7.4. Better performance levels correspond to higher rates shown in the experimental data resulting in more effective data throughput at extended distances.

Various distances lead to NMSE evaluation of different channel estimation methods as shown in Figure 7.5. The performance assessment relies on NMSE values because lower negative values confirm improved estimation accuracy.

The Table 7.3 & 7.4 show Crossfield producing the most optimal results because it maintains the lowest NMSE value at every measured distance. SOMP FF and Hybrid techniques perform at an average level, but SOMP NF achieves the best result due to its

lowest negative NMSE value which indicates higher error levels.

Conclusion

The research examines how to estimate cross-field channels for transmitting over AoSA-based UM-MIMO Terahertz (THz) systems. The system uses a model selection metric to identify the proper channel estimation model which operates near or intermediate or far-field locations. Multiple SAs benefit from reduced dictionaries which enable the utilization of their geometric relationships. The new approach delivers superior channel estimation performance than traditional methods since it produces lower NMSE values together with reduced computational complexity while attaining higher achievable rates.

References

[1] Cui, M., Wu, Z., Wei, X., & Dai, L. (2023). Near-field MIMO communications for 6G: fundamentals, challenges, potentials, and future directions. *IEEE Communications Magazine*, 61(1), 40–46.

[2] Tarboush, S., Ali, A., & Al-Naffouri, T. Y. (2023). Compressive estimation of near field channels for ultra massive-MIMO wideband THz systems. In Proceedings of the IEEE International Conference on Acoustics, Speech, and Signal Processing (ICASSP), (pp. 1–5).

[3] Chen, H., Sarieddeen, H., Ballal, T., Wymeersch, H., Alouini, M. S., Al-Naffouri, T. Y., et al. (2022). Terahertz band communication: an old problem revisited and research directions for the next decade. *IEEE Transactions on Communications*, 70(6), 4250–4285.

[4] Sarieddeen, H., Saeed, N., Al-Naffouri, T. Y., & Alouini, M. S. (2020). Terahertz communications: a rendezvous of sensing, imaging, and localization. *IEEE Communications Magazine*, 58(5), 69–75.

[5] Akyildiz, I. F., Han, C., Hu, Z., Nie, S., & Jornet, J. M. (2022). Terahertz band communication: an old problem revisited and research directions for the next decade. *IEEE Transactions on Communications*, 70(6), 4250–4285.

[6] Han, C., Bicen, A. O., & Akyildiz, I. F. (2014). Multi-ray channel modeling and wideband characterization for wireless communications in the terahertz band. *IEEE Transactions on Wireless Communications*, 14(5), 2402–2412.

[7] Zhang, Z., Xiao, Y., Ma, Z., Xiao, M., Ding, Z., Lei, X., et al. (2019). 6G wireless networks: vision, requirements, architecture, and key technologies. *IEEE Vehicular Technology Magazine*, 14(3), 28–41.

8 Fabric fault detection using convolutional neural networks and streamlit interface

Chaitanya Nukala[1,a], Shameena, S.[2,b], Sai Ram, B.[2,c], Sowjanya Priya, P.[2,d], Mahendra Goud M. E.[2,e] and Pogula Sreedevi[1,f]

[1]Assistant Professor, Department of CSE, Rajeev Gandhi Memorial College of Engineering and Technology, Nandyala, Andhra Pradesh, India

[2]Department of Computer Science and Engineering, Rajeev Gandhi Memorial College of Engineering and Technology, Nandyala, Andhra Pradesh, India

Abstract

Ensuring fabric quality is crucial in textile manufacturing. Traditional manual inspections are error-prone and inefficient, leading to financial losses. This research introduces an AI-powered framework using convolutional neural networks (CNNs) and MobileNetV2, integrated into a streamlit interface for real-time defect detection via image uploads and live camera feeds. Achieving 97.5% accuracy with MobileNetV2, the system incorporates automated alerts (email, SMS, voice), real-time sensor logging, and an AI chatbot for quick defect analysis and historical data retrieval, enhancing industrial scalability and smart manufacturing. The system continuously improves by integrating active learning mechanisms, ensuring the model adapts to new defect types. The architecture supports large-scale deployment, providing textile manufacturers with an efficient, cost-effective, and adaptable solution.

Keywords: Active learning, automated alerts, industrial automation, convolutional neural networks, deep learning, fabric defect detection, image processing, quality control, real-time analytics, sensor data, streamlit interface, textile industry

Introduction

Fabric inspection is critical for maintaining high product quality. Traditional inspections suffer from human error and inconsistency. Deep learning, particularly convolutional neural networks (CNNs) [5, 6], revolutionizes feature extraction and defect classification [4, 6]. MobileNetV2 [5], a lightweight model, boosts efficiency with pre-trained layers [5]. This research integrates CNNs, MobileNetV2, and real-time analytics into a Streamlit interface, providing automated alerts, sensor logging, and AI-driven insights to minimize production downtime [3]. The system's adaptability, enhanced by active learning and sensor-based fault tracking, ensures continuous improvement.

Literature survey

Traditional approaches

Traditional Approaches Statistical methods like GLCM and PCA require manual feature engineering and struggle with irregular patterns [2, 3]. Spectral methods like Fourier and wavelet transforms handle periodic defects but demand significant computation [3]. Model-based techniques, such as GMRF and HMM [5], model pixel distributions but falter with diverse fabric textures.

Learning-based approaches

Machine learning models (SVM, Decision Trees) [2] rely on handcrafted features, limiting adaptability. Deep learning models, especially CNNs [5, 6], automate feature extraction. Transfer learning with MobileNetV2 [5] and ResNet boosts performance on smaller datasets. Hybrid approaches integrate sensor data to correlate environmental conditions with defects, enhancing fault prediction.

Gaps in existing research

Challenges include limited real-time usability, lack of sensor integration, and absence of active learning. Few systems integrate environmental monitoring

[a]chaitu.anju@gmail.com, [b]shameena7893@gmail.com, [c]busettysairam2002@gmail.com, [d]priyapasula52@gmail.com, [e]mahendragoudediga143@gmail.com, [f]sreedevipogula37@gmail.com

DOI: 10.1201/9781003685364-8

with visual analytics, missing predictive maintenance [11] opportunities.

Contributions of this research

Real-time classification, automated alerts, sensor-based fault logging, AI chatbot support, and active learning improve performance and scalability. The integration of environmental analytics with defect detection provides manufacturers with predictive insights [6].

Methodology

Dataset description

The dataset such as TILDA [1] contains 5,000 images classified into five categories: Good, Hole, Object, Oil Spot, Thread Error. Low-confidence predictions (<50%) are saved for retraining, ensuring adaptability. Data augmentation handles diverse fabric textures and lighting conditions.

Image preprocessing

Images are resized, normalized, and augmented [9]. Sensor data (temperature, vibration) is logged alongside image data for correlation analysis.

Training process

- CNN: Three convolutional layers with ReLU activation, max pooling, dropout, and SoftMax output [1, 5].
- MobileNetV2: Pre-trained on ImageNet, custom layers added, optimized for faster inference and lower latency. The model's depth wise separable convolutions enhance speed without sacrificing accuracy, making it suitable for real-time defect analysis [8].

Evaluation metrics

1. Accuracy: Overall correctness of predictions:

$$Accuracy = \frac{TP+TN}{TP+TN+FP+FN} \qquad (5)$$

2. Precision: Proportion of true positives out of predicted positives:

$$Precision = \frac{TP}{TP+FP} \qquad (6)$$

3. Recall: Proportion of true positives out of actual positives:

$$Recall = \frac{TP}{TP+FN} \qquad (7)$$

4. F1-Score: Harmonic mean of precision and recall:

$$F1 = 2 \cdot \frac{Precision \cdot Recall}{Precision+Recall} \qquad (8)$$

5. Confusion Matrix: Class-wise performance.

Real-time implementation with Streamlit

The interface supports image uploads, live camera feeds [10], model selection, classification with confidence scores, automated alerts, sensor display, chatbot access, and active learning.

System architecture

The fabric defect detection system [7] is built on a modular architecture, combining deep learning, real-time monitoring, automated alerts, sensor-based tracking, AI chatbot integration, and active learning mechanisms. The system is designed for scalability, adaptability, and industrial deployment, ensuring efficient defect detection with minimal human intervention. Figure 8.1 illustrates system architecture.

Overview of system workflow

The system follows a structured pipeline [4] to provide real-time defect classification and decision-making:

1. Data acquisition: Fabric images are obtained via uploads or live camera feeds.
2. Preprocessing: Images are resized, normalized, and augmented to ensure consistency.
3. Model inference: The selected model (CNN or MobileNetV2) [5] predicts the defect type.
4. Defect classification: The system assigns one of five classes (Good, Hole, Object, Oil Spot, Thread Error) with a confidence score.
5. Automated alert system: If a defect is detected, an email, SMS, and voice alert is triggered.
6. Sensor logging: Temperature and vibration readings are recorded alongside defect classifications.
7. AI Chatbot interaction: Users can retrieve defect logs and analyze trends through chatbot queries.
8. Active learning mechanism: Low-confidence predictions (<50%) are saved for manual annotation and model retraining.

This workflow ensures a fully automated, real-time defect detection system, improving accuracy, efficiency, and operational responsiveness.

System architecture diagram

The following components interact seamlessly to achieve real-time fabric defect detection [7]:

Convolutional neural network architecture
Convolutional neural network
Architecture:
- Input layer: 64 × 64 × 3.
- Three convolutional layers:
  - Filters: 32, 64, and 128 respectively.
  - Activation: ReLU function

$$f(x) = \max(0, x). \tag{1}$$

- MaxPooling after each convolution with a pooling size of 2 x 2.
  - Flatten layer: Converts 2D feature maps into a 1D vector.
- Dense layers:
  - Fully connected layer with 256 neurons.
  - Dropout (rate = 0.5) to prevent overfitting.
- Output layer:
  - 5 neurons (SoftMax activation):

$$P(y_i) = \frac{e^{z_i}}{\sum_{j=1}^{k} e^{z_i}} \tag{2}$$

Where:
1. z_i is the input to the neuron.
2. k is the number of classes.
3. $P(y_i)$ indicates the probability corresponding to class i.

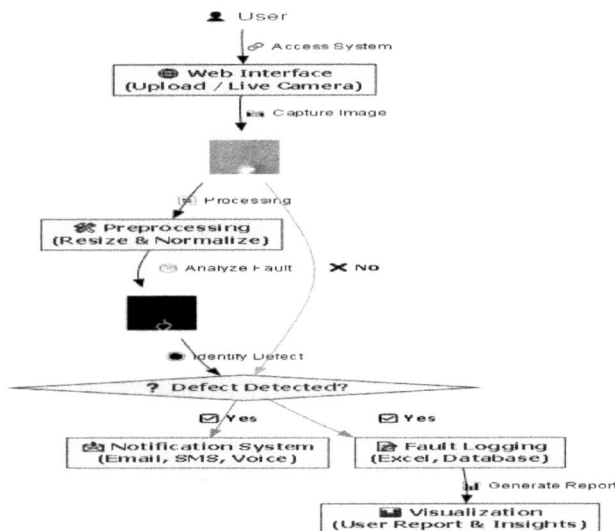

Figure 8.1 System architecture diagram
Source: Author

- Optimizer: Adam, learning rate = 0.001.
- Loss function: Categorical cross entropy:

$$L = -\frac{1}{N}\sum_{i=1}^{N}\sum_{j=1}^{k} y_{ij} \log\left(\hat{y}_{ij}\right) \tag{3}$$

Where:
1. y_{ij} is the true label.
2. $\hat{y}_{ij}$ is the predicted probability.
3. N represents the total number of samples.
4. k denotes the number of distinct classes

MobileNetV2 architecture
- Pre-trained model: MobileNetV2 (weights: ImageNet)
- Custom layers:
  - Flatten layer: Converts feature maps to a vector.
  - Dense layer: Rate = 0.5.
  - Output layer: Softmax activation (same as CNN).
- Optimizer: Adam, learning rate = 0.0001.
- Inference time formula:

$$T_{inference} = T_{preprocessing} + T_{model} \tag{4}$$

Where $T_{preprocessing}$ is the time taken for resizing and normalization, and T_{model} is the time taken by the model to generate predictions.

Sensor-based fault tracking and logging
Identifies correlations between temperature and defects. Vibration analysis: Logs machine vibrations, detecting mechanical failures.

Fault log storage: Defect classifications and sensor readings stored for analysis.

Automated alerts and ai chatbot for fault analysis
Multi-channel alerts via email, SMS, and voice. The chatbot provides historical logs, frequent defect trends, and predictive insights.

Active learning and industry ready deployment
Low-confidence predictions saved for retraining, ensuring continuous accuracy improvement. Supports cloud and on-premises deployment with low-latency inference.

Results
Performance metrics
Both models were evaluated on the test dataset [6] (1,000 images per class). Below are the results:

MobileNetV2 [5] achieved higher accuracy due to transfer learning and a more efficient feature extraction process.

Confusion matrix

Figure 8.2 compares confusion matrices for two models [7], CNN and MobileNet, in predicting different defect categories. The CNN model shows high accuracy with slight misclassifications across categories, particularly in Oil Spot and Thread Error. MobileNet performs similarly but demonstrates slightly improved accuracy in most categories, especially in Good and Object. Both models exhibit robustness but vary in specific misclassification trends. MobileNet has fewer errors overall, making it marginally better for this task.

Training curves

Figures 8.3 and 8.4 present four subplots showing training and validation accuracy and loss trends for CNN and MobileNetV2 [5] models. For CNN, the accuracy fluctuates significantly, with irregular loss patterns, suggesting potential overfitting. Conversely, MobileNetV2 demonstrates steady improvement in accuracy and a consistent decline in loss, reflecting better generalization and model stability.

Table 8.1 Performance metrics for both models.

Metric	CNN	MobileNetV2
Accuracy	96.0%	97.5%
Precision	94.5%	96.2%
Recall	95.2%	96.8%
F1-score	94.8%	96.5%

Source: Author

Observations

1. MobileNetV2 outperformed CNN in terms of accuracy and F1-Score due to its transfer learning capabilities.
2. Both models achieved high recall, ensuring minimal undetected defects.
3. Inference Time:

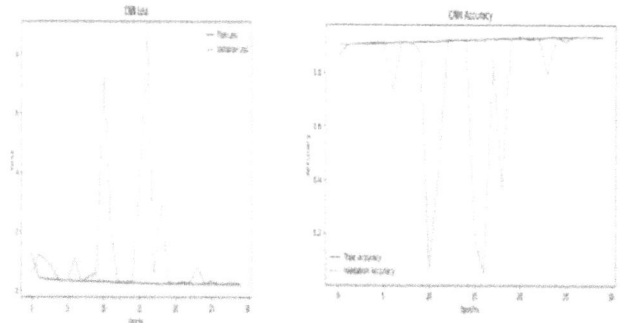

Figure 8.3 Accuracy and loss curves of CNN Model
Source: Author

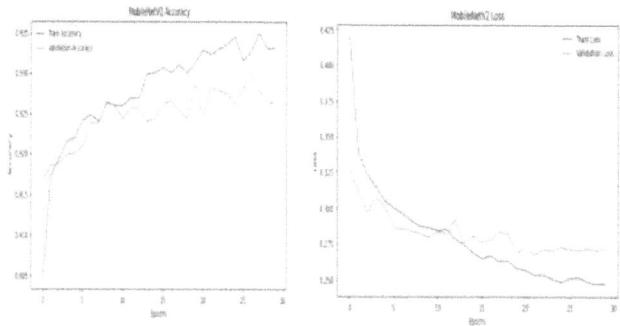

Figure 8.4 Accuracy and loss curves of Mobile-NetV2 model
Source: Author

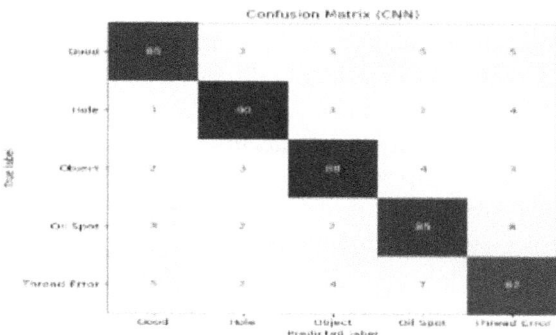

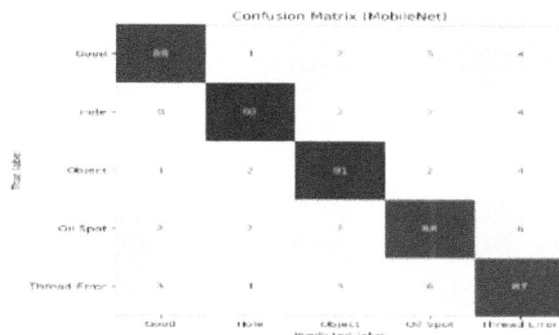

Figure 8.2 Confusion matrix of CNN and MobileNetV2
Source: Author

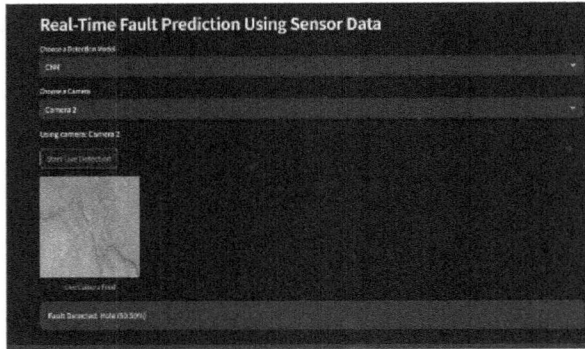

Figure 8.5 Real –time detection (Streamlit UI)
Source: Author

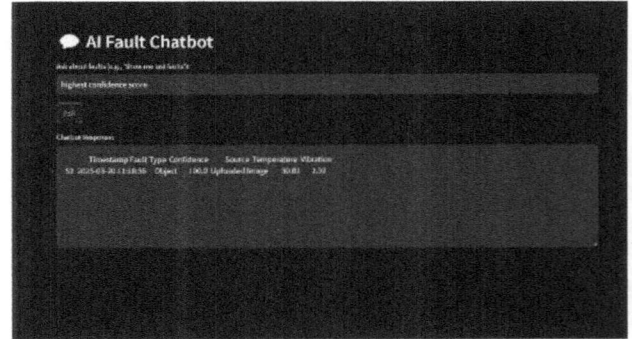

Figure 8.6 Sensor data integration
Source: Author

- CNN: $T_{inference} \approx 150ms$
- MobileNetV2: $T_{inference} \approx 180ms$

Below the figure showing Streamlit interface detecting defects by using CNN and MobileNetV2 Models.

Real-Time classification –defect type and confidence score displayed instantly.

Dataset usage

Table 8.2 Dataset details.

Dataset split	Number of images
Training	5,000
Validation	2,000
Testing	1,000

Source: Author

Impact of automated alerts and sensor logging
The introduction of real-time sensor monitoring [9] has improved defect detection by identifying environmental conditions associated with defects.

Figure 8.7 AI Chatbot interaction
Source: Author

Active learning contribution to model improvement
The active learning mechanism, which stores low-confidence predictions for retraining, has significantly enhanced model performance over time.

AI chatbot evaluation and user interaction
The AI chatbot has improved system usability by providing real-time fault insights as shown in Figure 8.7.

- 90% of users found the chatbot useful for quick defect retrieval.
- Users reduced manual fault log searches by 75%, increasing operational efficiency.
- Predictive defect insights based on sensor data and defect trends helped in preventive maintenance planning.

Conclusion

The proposed framework integrates CNNs, MobileNetV2, Streamlit, real-time analytics, and automation, enhancing fabric quality control. By incorporating active learning and sensor-based fault tracking, the system adapts to evolving defect patterns, ensuring long-term reliability and performance stability. The AI chatbot and automated alert mechanisms improve operational efficiency, reducing the need for manual monitoring and accelerating fault response times. With 97.5% accuracy achieved using MobileNetV2, this approach outperforms traditional methods, providing a scalable, adaptable, and intelligent solution for the textile industry.

Acknowledgement

The authors sincerely acknowledge the students, staff, and authority of Rajeev Gandhi Memorial

College of Engineering and Technology for their cooperation in research.

References

[1] Almeida, T., Moutinho, F., & Matos-Carvalho, J. P. (2021). Fabric defect detection with deep learning and false negative reduction. *IEEE Access*, 9, 81936–81945. doi: 10.1109/ACCESS.2021.3086028.

[2] Bangare, S. L., Dhawas, N. B., Taware, V. S., Dighe, S. K., & Bagmare, P. S. (2017). Implementation of fabric fault detection system using image processing. *International Journal of Research in Advent Technology*, 5(6), 60–64.

[3] Gupta, A. D., Sadek, Z., Hossain, M. S., Toha, T. R., Mondol, A., Habiba, S. U., et al. (2024). An approach to automatic fault detection in four-point system for knitted fabric with our benchmark dataset Isl-Knit. *Heliyon*, 10(17).

[4] Eldessouki, M., Hassan, M., Qashqary, K., & Shady, E. (2014). Application of principal component analysis to boost the performance of an automated fabric fault detector and classifier. *Fibres Textiles in Eastern Europe*, 4(106), 51–57.

[5] Ngan, H. Y., Pang, G. K., & Yung, N. H. (2011). Automated fabric defect detection—a review. *Image and vision computing*, 29(7), 442–458.

[6] Chan, C. H., & Pang, G. K. (2000). Fabric defect detection by Fourier analysis. *IEEE Transactions on Industry Applications*, 36(5), 1267–1276.

[7] Khowaja, A., & Nadir, D. (2019). Automatic fabric fault detection using image processing. In 2019 13th International Conference on Mathematics, Actuarial Science, Computer Science and Statistics (MACS), (pp. 1–5). IEEE.

[8] Manasa, M., Sowmya, D., Reddy, Y., & Sreedevi, P. (2024). Attention-based image caption generation. In smart computing Techniques and Applications (pp. 455–470).springer.https://doi.org/10.1007/978-981-97-8031-00_38.

[9] Sunil, N., & Narsimha, G. (2024). Image-based random rotation for preserving the data in data mining process. *Signal Image and Video Processing*, 18(4), 3893–3902. https://doi.org/10.1007/s11760-024-03050-2.

[10] Lalitha, R. V. S., Jalligampala, D. L. S., Kavitha, K., Vahida, S., & Rajasekhar, G. (2021). New directions in traffic control analysis through video surveillance. In E3S Web of Conferences, (Vol. 309, p. 01099). https://doi.org/10.1051/e3sconf/202130901099.

[11] Annapurna, B., Reddy, T. R., Raghavendran, C. V., Singh, R. K., & Prasad, V. V. (2020). Coordinate Access System for live video acquisition. *Journal of Physics Conference Series*, 1712(1), 012034. https://doi.org/10.1088/1742-6596/1712/1/012034.

9 Multi class adaptive learning for predicting student anxiety

Chitralingappa, P.[1,a], Nandalapadu Neeha[2,b], Pinnu Chinmayee[2,c], Gona Likitha[2,d] and Gandhaveeti Likhitha[2,e]

[1]Associate Professor and Head of Department of CSD, Srinivasa Ramanujan Institute of Technology (Autonomous), Andhra Pradesh, India

[2]Final Year B. Tech Students, Department of CSD, Srinivasa Ramanujan Institute of Technology (Autonomous), Andhra Pradesh, India

Abstract

This study proposes a multi-class adaptive learning framework aimed at predicting student anxiety, with the goal of enhancing early intervention and supporting mechanisms in educational environments. Traditional anxiety prediction models often struggle with limited labeled data and static learning experiences, reducing their effectiveness. Our approach leverages adaptive learning techniques, which dynamically select the most informative data points for labeling, thereby improving model accuracy and robustness. By incorporating multi-class classification, the model identifies various levels of anxiety, from mild to severe, providing a nuanced understanding of student mental health. This study underscores the potential of adaptive learning in educational data mining, offering a flexible solution for real-time anxiety prediction and contributing to more responsive and supportive educational systems.

Keywords: Adaptive learning, educational data mining, machine learning in education, mental health assessment, multi-class classification, real-time analytics, student anxiety prediction

Introduction

This research tackles the challenge of predicting student anxiety within educational environments by introducing an innovative multi-class adaptive active learning framework. Traditional predictive models often face limitations due to factors such as insufficient labeled data, which hampers the learning process and reduces prediction accuracy. Additionally, these models typically rely on static learning algorithms, which are less flexible and fail to adapt to new, evolving patterns in student behavior or mental health indicators. As a result, traditional models may struggle to provide timely or accurate insights into the varying levels of student anxiety.

Objective of the study
The primary objective of this undertaking is to make and approve a multi-class versatile dynamic learning structure custom fitted to foreseeing understudy nervousness levels inside instructive settings. This approach looks to beat the impediments of conventional tension expectation models by utilizing a dynamic and iterative cycle to distinguish and mark the main data of interest.

Scope of study
The study aims to enhance machine learning models used in educational environments by addressing issues such as limited labeled data and static learning approaches.

Problem statement
This exploration tends to be the constraints of conventional nervousness expectation models in instructive settings, which frequently depend on restricted named information and unbending learning techniques. These traditional models face difficulties in precisely recognizing and separating different degrees of understudy nervousness, frustrating viable early mediation and backing.

Related Work

The forecast of understudy tension is an arising area of interest in instructive examination, fully intent on upgrading early mediation frameworks [1] and

[a]p.chitralingappa@gmail.com, [b]neeha6823@gmail.com, [c]chinmayeepinnu@gmail.com, [d]gonalikithareddy@gmail.com, [e]likhithagandhaveeti@gmail.com

DOI: 10.1201/9781003685364-9

supporting mental prosperity inside scholastic conditions. Conventional techniques for anticipating tension frequently depend on restricted datasets, which confine the exactness and generalizability of the models [2]. Large numbers of these early methodologies utilized old style AI strategies, for example, support vector machines, choice trees, and calculated relapse, however battled with issues like overfitting, underfitting, and [3] taking care of dynamic, continuous information. Moreover, these models habitually utilized twofold characterization — marking understudies basically as "restless" or "non-restless" — which neglects to catch the nuanced idea of nervousness levels. Tension is a complex, multi-faceted experience, and the requirement for additional modern models that can recognize [4,5,6] differing levels of nervousness has prompted the investigation of multi-class order draws near. A critical impediment of customary prescient models is the test of working with little or imbalanced datasets. In instructive settings, marked Click or tap here to enter text. [1] information is frequently scant, and understudies' psychological well-being states are impacted by various variables, which may not necessarily in all cases be caught by static datasets. To resolve this issue, versatile learning frameworks have been investigated [7,8].

Proposed system

The proposed framework presents an inventive system that joins multi-class versatile dynamic learning with choice tree and stacking classifier strategies to upgrade the expectation of understudy nervousness levels in Figure 9.1. This approach means to further develop expectation exactness, versatility, and adaptability, making it appropriate for continuous examination and mediation in instructive conditions. By consolidating different AI procedures, the framework can actually catch the complex and advancing nature of understudy tension, consequently empowering opportune and custom-made help for understudies encountering mental pain.

Loading dataset
With regards to foreseeing understudy uneasiness utilizing a multi-class versatile learning structure, stacking and setting up the dataset is the essential initial phase in model turn of events. The dataset utilized in this cycle normally contains a blend of understudy execution information, conduct markers,

psychometric evaluations, and conceivably physiological information.

Preprocessing
Preprocessing is a basic move toward any AI pipeline, particularly while working with complex information like foreseeing understudy uneasiness. The essential objective of preprocessing is to set up the information for the model by changing crude contributions to a reasonable configuration, improving model precision and execution.

Model training and classification
Model preparation is the center period of any AI pipeline, where the framework gains from the information to make expectations. With regards to foreseeing understudy nervousness, the proposed framework utilizes choice trees and a stacking classifier approach, which is a group learning method.

The process outlined in the flowchart for machine learning is as follows:

Start: The procedure begins.

Dataset loading: The initial step involves loading the dataset.

Preprocessing the data: The next phase includes data preprocessing, which entails:

Handling missing values: Missing data is addressed.

Label encoding: Categorical data is converted into numerical format.

Dividing the dataset: The preprocessed dataset is divided into two equal parts.

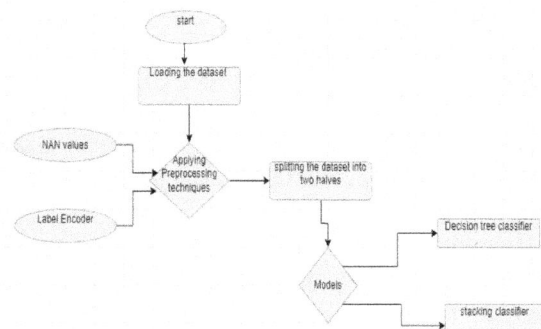

Figure 9.1 Block flow chart of multi adaptive learning for student anxiety
Source: Author

Methodology

K-nearest neighbors
Outline:
KNN is a clear, non-parametric AI calculation reasonable for both characterization and relapse undertakings. It characterizes data of interest by assessing the larger part name of its 'K' nearest neighbors in the component space.

How it functions:
Preparing stage:
KNN doesn't has an unequivocal preparation stage. All things considered, it stores the whole dataset for future reference.

Forecast stage:
Whenever another information point is presented, the calculation figures out the distance between this point and all the preparation information. Euclidean distance is generally usually utilized, albeit other distance measurements like Manhattan or Minkowski can likewise be used.

Subsequent to figuring the distances, the calculation chooses the 'K' closest neighbors and makes forecasts:

For grouping, the greater part class among the closest neighbors is picked.

For a relapse, the typical worth of the closest neighbors' objective qualities is anticipated.

Picking 'K':
The worth of 'K' is essential for model execution. A little 'K' makes the model delicate to commotion, while a bigger 'K' can prompt smoother choice limits, making the model less touchy.

Logistic Regression
Overview:
Despite its name, it is a linear classifier used to estimate the probability that an input belongs to a specific class results represents in Table 9.1.

How it works:
- **Model representation:** This function ensures outputs fall between 0 and 1, representing probability of class membership.
- **Optimization (Cost function):** Gradient descent or similar methods are employed to minimize the cost function.

- **Decision boundaries:** The model generates decision boundaries based on predicted probabilities, assigning the input to the class with the highest likelihood.

XGB classifier
Overview:
It sequentially builds decision trees, with each tree attempting to correct errors from the previous one.

How it works:
- **Boosting framework:**
XGBoost uses boosting, where weak learners (decision trees) are trained sequentially. Each new tree attempts to correct the residuals (errors) from earlier trees.
- **Feature importance:**
XGBoost evaluates the significance of each feature by observing its frequency of use in tree splits, which aids in feature selection.
- **Parallelism:**
The algorithm supports parallel data processing, enhancing speed compared to other gradient boosting methods.

Naive Bayes
Overview:
Naive Bayes is a probabilistic classifier that applies Bayes' Theorem with a strong assumption of feature independence. It works well for text classification tasks such as spam detection.

How it works:
Bayes theorem:

$$P(C|X) = \frac{P(X|C) \cdot P(C)}{P(X)} \tag{1}$$

Table 9.1 Classification report of logistic regression.

```
Classification Report for LR:
            precision   recall   f1-score   support
```

	precision	recall	f1-score	support
0	0.62	0.68	0.65	2047
1	0.69	0.76	0.72	2121
2	0.55	0.52	0.53	2050
3	0.53	0.45	0.49	2044
accuracy			0.60	8262
macro avg	0.60	0.60	0.60	8262
weighted avg	0.60	0.60	0.60	8262

Source: Author

Naive assumption:

Naive Bayes assumes features are conditionally independent, simplifying the likelihood calculation to a product of individual feature probabilities:

$$P(X|C) = \prod_{i=1}^{n} P(x_i|C) \qquad (2)$$

Random Forest
How it works:

- **Bootstrap aggregating (Bagging):**
  Random Forest builds multiple decision trees using bootstrapped samples from the training data. Each tree is trained on a different subset of data.
- **Random feature selection:**
  At each node in the tree, a random subset of features is chosen for splitting, which helps in decorrelating the trees and improving model diversity.
- **Voting (For classification):**
  For classification, each tree casts a vote, and the majority class is selected. For regression, the predictions from all trees are averaged.
- **Out-of-bag error estimation:**
  Random Forest estimates its error by using out-of-bag samples, which are data points not selected for a tree's training.

Decision tree classifier
Overview:

A Decision Tree is a tree-like structure where the inner nodes correspond to decisions made based on feature values, the branches signify the possible outcomes of those decisions, and the terminal nodes represent the class labels.

How it works:
Splitting criteria:

The tree is built by recursively splitting the data at each node using the feature that leads to the best possible split, based on criteria like Gini impurity (for classification) or mean squared error (for regression).

- **Recursive splitting:** At every node, the optimal feature for splitting is selected to maximize the information gain in classification tasks or minimize the variance in regression tasks.
- **Stopping criteria:** The tree construction halts when a stopping condition is met, such as reaching a maximum depth or when further splits do not improve the model.

- **Prediction:** During prediction, a new instance follows the splits from the root to the leaf node, with the class label of the leaf node being the predicted label.

Stacking classifier
Overview:
Stacking Classifier is an ensemble learning technique where multiple base models are trained, and their predictions are combined by a meta-model to enhance classification accuracy.

How it works:
Base models:
Multiple base classifiers, such as Decision Trees and Logistic Regression, are trained on the same dataset, creating the level-0 models.

- **Training:**
  Base models are first trained on the dataset, and their predictions are collected. These predictions serve as input to train the meta-model.
- **Prediction:**
  During prediction, base models generate their individual predictions, which are then combined by the meta-model to produce the final output.

Discussion and results

The multi-class adaptive learning framework proposed for predicting student anxiety has shown substantial improvements over conventional methods used in educational settings. Traditional approaches for anxiety prediction in these environments often face challenges such as limited labeled data and rigid models, which hinder their ability to adapt to

Table 9.2 Classification report of Decision Tree classifier.

```
Classification Report for tuned DT:
              precision    recall  f1-score   support

           0       0.87      0.91      0.89      2047
           1       0.69      0.62      0.65      2121
           2       0.54      0.54      0.54      2050
           3       0.75      0.80      0.78      2044

    accuracy                           0.72      8262
   macro avg       0.72      0.72      0.72      8262
weighted avg       0.71      0.72      0.72      8262
```
Source: Author

Table 9.3 Classification report of stacking classifier.

```
Classification Report for Stacking Classifier:
              precision    recall  f1-score   support

           0       0.97      0.98      0.98      2047
           1       0.78      0.79      0.78      2121
           2       0.73      0.72      0.72      2050
           3       0.94      0.93      0.94      2044

    accuracy                           0.86      8262
   macro avg       0.86      0.86      0.86      8262
weighted avg       0.85      0.86      0.85      8262
```

Source: Author

the dynamic nature of student behaviors and mental health. Our method, based on adaptive dynamic learning, overcomes these limitations by selectively choosing the most relevant data for labeling in each iteration, thus enhancing the model's accuracy and robustness.

Conclusion

In summary, this study introduces a novel multi-class adaptive dynamic learning model for predicting student anxiety. The findings suggest that the proposed methodology holds significant potential for enhancing the accuracy and adaptability of mental health prediction models in educational settings. By addressing key challenges of traditional models—particularly those related to data scarcity and static learning approaches—our adaptive learning framework provides more precise and nuanced predictions of student anxiety levels. The multi-class classification structure offers a detailed understanding of anxiety at varying degrees, which allows educational institutions to implement more specific and effective interventions.

Future enhancement

Looking ahead, several advancements can enhance the proposed multi-class adaptive learning framework for predicting student anxiety. Extending the range of data used to train the model is a key direction for future work. Currently, the model primarily utilizes academic performance and engagement metrics. To improve the accuracy and comprehensiveness of predictions, additional data sources, such as social interaction patterns, attendance records, and physiological data from wearable devices, could be

integrated. Our comprehension of the various factors that influence student anxiety would be enhanced by the inclusion of multi-modal data, allowing for more precise and individualized predictions. Another crucial improvement would focus on improving the model's interpretability. Although deep learning models have demonstrated high predictive capabilities, their inherent "black box" nature can make it challenging to understand how predictions are made. This lack of transparency may limit the adoption of the model in educational settings. Future research should explore ways to improve the interpretability of the model, using explainable AI (XAI) techniques to provide educators and mental health professionals with clear insights into the factors contributing to anxiety predictions. This transparency will foster trust and support more informed decision-making in interventions. Furthermore, the adaptability of the model could be enhanced by introducing dynamic feedback mechanisms. As students receive support and interventions, the model could update its predictions based on the outcomes of these interventions.

References

[1] Abak, M. (2023). Prediction of learned helplessness based on academic resilience and test anxiety in students with academic failure. *Iranian Journal of Educational Research*, 2(1), 1–11. https://doi.org/10.22034/2.1.1.

[2] Al-Azzam, N., Elsalem, L., & Gombedza, F. (2020). A cross-sectional study to determine factors affecting dental and medical students' preference for virtual learning during the COVID-19 outbreak. *Heliyon*, 6(12). https://doi.org/10.1016/j.heliyon.2020.e05704.

[3] Chan, R. Y. Y., Wong, C. M. V., & Yum, Y. N. (2023). Predicting behavior change in students with special education needs using multimodal learning analytics. *IEEE Access*, 11, 63238–63251. https://doi.org/10.1109/ACCESS.2023.3288695.

[4] Effects of Incorporating an Expert Decision-making Mechanism into Chatbots on Students' Achievement, Enjoyment, and Anxiety on JSTOR. (n.d.). Retrieved December 3, 2024, from https://www.jstor.org/stable/48707978.

[5] Gonzalez-Nucamendi, A., Noguez, J., Neri, L., Robledo-Rella, V., García-Castelán, R. M. G., & Escobar-Castillejos, D. (2021). The prediction of academic performance using engineering student's profiles. *Computers and Electrical Engineering*, 93, 107288. https://doi.org/10.1016/J.COMPELECENG.2021.107288.

[6] Jiang, Y., Kim, S., & Bong, M. (2020). The role of cost in adolescent students' maladaptive academic outcomes. *Journal of School Psychology*, 83, 1–24. https://doi.org/10.1016/J.JSP.2020.08.004.

[7] Liu, Z., Tang, Q., Ouyang, F., Long, T., & Liu, S. (2024). Profiling students' learning engagement in MOOC discussions to identify learning achievement: An automated configurational approach. *Computers and Education*, 219, 105109. https://doi.org/10.1016/J.COMPEDU.2024.105109.

[8] Liu, X., Zhang, Y., & Wang, X. (2020). Active learning for multi-class classification: a survey. *ACM Computing Surveys*, 53(3), 1–36.

[9] Binns, R., & Jiang, W. (2018). Adaptive active learning for multiclass classification. *Journal of Machine Learning Research*, 19(1), 1–24.

10 Smart vehicle damage assessment and insurance cost prediction

Kiran Kumar Annavaram[1,a], Chandrika, D.[2,b], Amrutha Valli, Y.[2, c], Bala Karthik, B.[2, d] and Mahabubbasha, S. K.[2, e]

[1]Assistant Professor, Department of CSE, Srinivasa Ramanujan Institute of Technology, Anantapur, India

[2]Students,Department of CSE, Srinivasa Ramanujan Institute of Technology, Anantapur, India

Abstract

The "smart vehicle damage evaluation and cost prediction for insurance firms" should provide an upgrade for the vehicle insurance industry through the usage of AI based deep learning in order to automatically assess vehicle damage. By implementing the YOLOv8, a next-gen object detection model, is to assess as well as identify the damage of two-wheelers, four-wheelers, and six-wheelers. In terms of training, it is tested with a large amount of test data with different characteristics of vehicle damage. It also serves as an effective tool for insurance firms in enhancing the quality of its claim processing, reduction of human interface and accurate repair cost assessment. It will help the insurance companies to improve operational effectiveness, speed up the process of settlements and increase the level of satisfaction of customers.

Keywords: Cost estimation, deep learning, four-wheeler, insurance automation, object detection, six-wheeler, two-wheeler, vehicle damage evaluation, YOLOv8

Introduction

The automobile insurance industry is evolving with AI and deep learning. Traditional manual damage assessments are slow, costly, and error prone. Deloitte Insurance proposes a YOLOv8-based solution to automate vehicle damage detection and repair cost estimation, improving accuracy and efficiency.

YOLOv8 analyzes vehicle images or videos to detect damage like dents, scratches, and cracks, providing real-time assessments. It also estimates repair costs using historical data, replacing outdated manual methods. This automation reduces human error, speeds up claim processing, and enhances customer satisfaction.

With a large training dataset, the model ensures accurate classifications across various vehicle types. Once deployed, it generates instant damage reports and cost estimates, streamlining insurance operations.

In conclusion, AI-driven assessment with YOLOv8 modernizes vehicle insurance by improving accuracy, efficiency, and claim processing speed, benefiting both insurers and policyholders.

Related Work

Chen et al. [1] proposed an MD R-CNN-based approach for vehicle damage detection demonstrating improved localization and classification accuracy by leveraging multi-dimensional feature extraction. This work highlighted the potential of region-based convolutional frameworks in fine-grained damage segmentation.

To address road infrastructure concerns, Fassmeyer et al. [2] introduced a scaled-YOLO architecture combined with a conditional variational autoencoder (CVAE) and Wasserstein GAN (WGAN) to assess road damage for autonomous vehicles. Their approach emphasized the robustness of generative augmentation techniques in real-world driving conditions.

In support of insurance-related damage classification, Huynh et al. [3] developed the VehiDE dataset, a specialized collection of annotated vehicle damage images. This dataset aims to support supervised learning models in accurately detecting various damage types under diverse lighting and occlusion conditions.

Khan et al. [4] developed a deep learning model using convolutional networks to classify various vehicle exterior damages, supporting automotive inspection and insurance processes. Qian et al. [5] utilized the YOLOv5s framework in a vehicle-mounted road damage detection system. Their model demonstrated

[a]Kiran.annavaram@gmail.com, [b]chandrikadasari6@gmail.com, [c]amrutha1630@gmail.com, [d]balakarthikbesta@gmail.com, [e]mahabubbasha135@gmail.com

DOI: 10.1201/9781003685364-10

efficient real-time performance, making it suitable for deployment in smart vehicles tasked with monitoring road safety and maintenance conditions.

While most studies have focused on visual damage detection, Sikandar et al. [6] explored machine learning approaches in the biomedical domain. Their work, although focused on disease-gene associations, provides relevant insights into feature selection and interpretability, which are transferable to vision-based classification tasks in transportation analytics.

System Design and Architecture

The developers designed the Intelligent Vehicle Damage Assessment and Cost Estimation system based on YOLOv8 which uses deep learning methods for object identification. The Figure 10.1 system architecture delivers accurate detection and cost estimation of vehicle damages for various Indian vehicle types which include two-wheelers four-wheelers alongside six-wheelers.

The base of the proposed system includes YOLOv8 model trained with different images displaying vehicle damages. YOLOv8 executes instant damage detection on vehicle objects which identifies damage types including dents, scratches and broken components among multiple vehicle types. The platform manages vehicle images through a resizing process before it enhances the detection of damages through property improvement operations.

When YOLOv8 detects damage in a provided image the system transfers the output to another algorithm which determines how much the vehicle structure got damaged. An estimation of repair costs derives from the acquired information. Machine learning techniques in the cost estimation model receive damage classification information as an input to generate repair cost predictions while considering the damage characteristics. The cost prediction accuracy depends on utilizing historical claims data during the training process for this model.

The workflow system implements end-to-end automation resulting in limited human contact for its operations. This architectural design enables image acquisition and processing together with cost evaluation therefore serving as a crucial technology for live damage evaluations which enhances claims processes. The system enables scalable design along with modularity which lets the model enhance its accuracy through further training when new data enters the system.

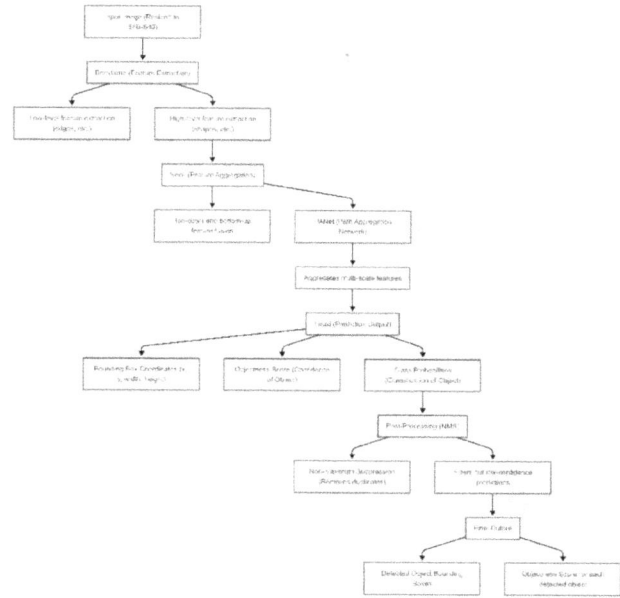

Figure 10.1 YOLO model architecture
Source: Author

The fusion of deep learning models with strong object detection algorithms enables this system to complete precise assessment of vehicle damage and related costs thus delivering an ideal solution to vehicle insurance companies. The system works to boost operational performance and improve customer satisfaction while making the damage assessment procedure more efficient.

Dataset

Two-wheelers, four-wheelers, and six-wheelers are the three vehicle categories represented in the dataset used to train the intelligent vehicle damage assessment and cost estimator system. The photos are taken from a variety of platforms and show various forms of vehicle damage.

Two-wheelers: There are 300 pictures in the two-wheeler dataset that show various types of damage, including scrapes, headlights, seats, tires, and mirrors. The manual collection and labeling of the training photos gave the YOLOv8 model a thorough representation of the variety of damaged types it would need to recognize in practical situations.

Four-wheelers: The collection includes 2,285 photos of four-wheelers that were taken from the publicly accessible Roboflow dataset. Damage to the bumper, fender, front and rear windshields, side mirrors, side screens, doors, headlamps, and hoods are all included in these pictures. In order to help the model detect damage on various parts of a vehicle, the dataset is carefully tagged with unique identifiers for each part and damage category.

Six-wheelers: The 1,862 photos in the six-wheeler collection were also taken from Roboflow. With each image labeled as "damaged" or "good," the pictures concentrate on damage kinds pertaining to the side-boards (left and right) and rear lamps (left and right). These annotations enable the model to recognize damage specific to six-wheelers and determine the state of particular vehicle parts.

Annotations: To identify the vehicle parts that were damaged, photos in the two-wheeler dataset were manually tagged. Bounding boxes are already included in the photos for the four-wheeler and six-wheeler datasets, indicating the various vehicle components and their corresponding damage levels.

Goal: To train the YOLOv8 object detection model, pictures of two-, four-, and six-wheelers are merged into two datasets. In order to guarantee that the model achieves high accuracy and reliability in diagnosing damage, these datasets are fused and tuned for the unique characteristics of different vehicle types. The following stage of the procedure, which entails accurately calculating repair costs and expeditiously submitting insurance claims, is so made easier.

Data preprocessing

In order to prepare the dataset for training the YOLOv8 model—which seeks to precisely identify vehicle damages— data preprocessing is essential. To lower computational complexity, all photos from the two-, four-, and six-wheeler datasets are scaled to a standard size, usually 640×640 pixels, as part of the preparatory processes. In order to replicate real- world settings, a variety of image augmentation techniques are also used, including flipping, rotation, and modifications to color contrast, brightness, and saturation. By providing a wider variety of image changes, this procedure helps avoid overfitting and improves the model's generalization across various viewing angles and lighting conditions.

The images' pixel intensity values are scaled to a range of 0 to 1 in order to standardize them and enhance model training. Because it speeds up the training process's convergence, this normalization phase is essential. Every annotation for the manually annotated photos in the two-wheeler dataset

is meticulously examined to guarantee accuracy. Analyses that are pre-labeled for the four-wheeler and six-wheeler datasets are also double-checked for accuracy and consistency. In order to ensure that the model's evaluation results are robust and dependable, the dataset is finally divided into training, validation, and test sets. This thorough preprocessing pipeline improves the model's performance during training and assessment in terms of accuracy and reliability.

Methodology

The YOLOv8 architecture is ideal for detecting vehicle damage since it is built to detect objects in real time. The head, neck, and backbone make up the three primary structural elements of the model. In order to process the input image and produce precise predictions, each of these elements is essential.

Backbone: feature extraction

YOLOv8 uses CSPDarknet53, an improved version of Darknet, for feature extraction. This backbone optimizes gradient flow and reduces computational cost while maintaining high accuracy. It extracts both low-level features (edges, textures) and high-level features (shapes, patterns), enabling the model to detect different types of vehicle damage.

Neck: feature aggregation

The path aggregation network (PANet) enhances multi-scale feature fusion, improving object detection across different sizes. By combining top-down and bottom-up features, PANet ensures robust damage detection under various resolutions and conditions.

Head: prediction and output

The head generates the final detection outputs, including:

- **Bounding box coordinates** (location and size of detected damage)
- **Objectness score** (probability that a box contains an object)
- **Class probabilities** (damage type: dent, scratch, etc.)

YOLOv8 refines anchor boxes during training, enabling accurate detection of objects with varying sizes and aspect ratios.

Loss function

The model's loss function consists of:

- **Localization loss** (bounding box accuracy)
- **Classification loss** (damage type prediction accuracy)
- **Objectness loss** (confidence in object presence)

Backpropagation minimizes these errors, refining the model's detection performance.

Post-processing

Non-maximum suppression (NMS) eliminates duplicate bounding boxes and filters out low-confidence detections, ensuring accurate final outputs for vehicle damage assessment.

Two wheelers

The Figure 10.2 is a confusion matrix that shows how well a model performs when identifying different kinds of vehicle damage, including broken, scratched, headlight, seat, tire, mirror, and background. The anticipated class is represented by each column, and the true class is represented by each row. The number of instances of each true class that were predicted to be each class is shown by the values in the matrix. Correct classifications are represented by diagonal elements (for example, 28 for "broken" and 48 for "scratch"), whereas incorrect classifications are indicated by off-diagonal values. Each prediction's frequency is represented by the color intensity, where higher values are indicated by deeper hues.

Figure 10.3 precision-confidence curve, which assesses a model's prediction accuracy at different confidence thresholds, is depicted in the figure. A distinct sort of car damage is represented by each line, including "broken," "scratch," "headlight," "seat," "tire," and "mirror". The precision is shown on the y-axis, while the confidence level is shown on the x-axis. The curve illustrates how accuracy rises with increasing confi- dence. The model achieves flawless precision across all classes at a confidence level of 0.850, as seen by the blue line, which shows the overall performance of all classes. The trade-off between precision and Figure 10.4, Figure 10.5 represents the Confusion matrices of four and six wheelers.

The model's ability to classify damage states for sideboards and rear lamps is demonstrated by the image, which shows a confusion matrix for vehicle damage detection. There are three categories: "damaged," "good," and "background." The rows

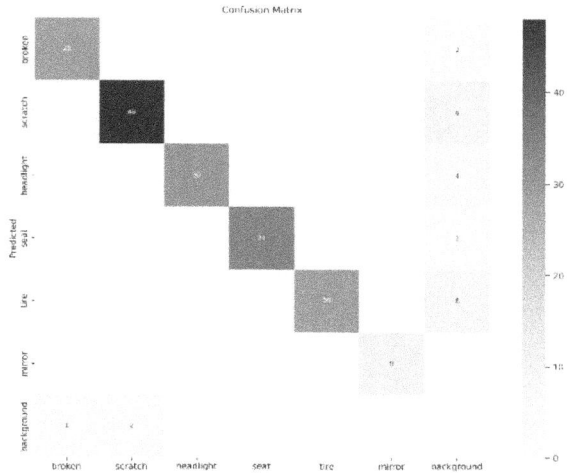

Figure 10.2 Confusion matrix for 2 wheeler
Source: Author

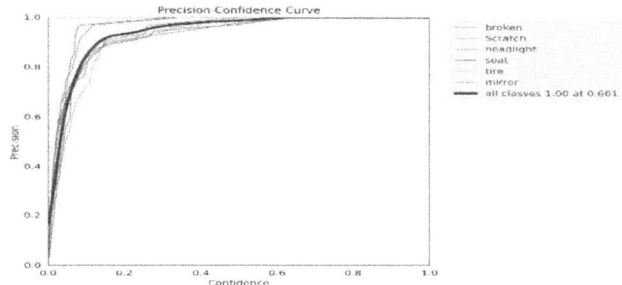

Figure 10.3 Precision-confidence curve for 2 wheeler
Source: Author

represent the genuine labels, while the columns

Four wheelers

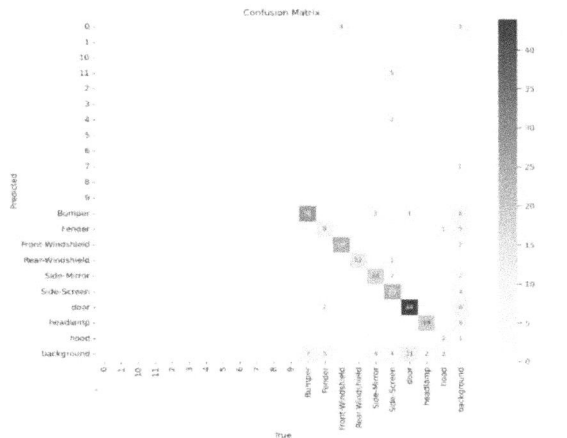

Figure 10.4 Confusion matrix for 4-wheeler
Source: Author

Six wheelers

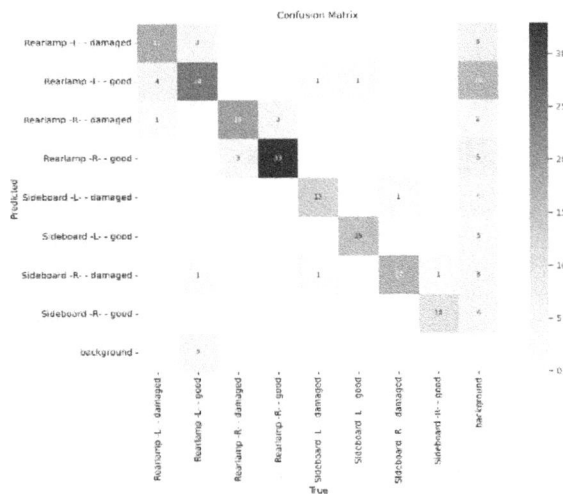

Figure 10.5 Confusion matrix for 6-wheeler
Source: Author

show the anticipated labels. The diagonal elements show accurate predictions, like the 24 times the left rear lamp was categorized as "good," while the off-diagonal numbers show inaccurate classifications, like the left rear lamp being misclassified as a damaged sideboard. Darker hues denote greater values, while the color intensity represents the frequency of forecasts.

Conclusion

The YOLOv8 model demonstrated varying performance across different vehicle types:

- Two-Wheelers: Achieved 99.07% mAP at IoU 0.5, with high precision and recall, ensuring accurate damage detection across different scenarios.
- Four-Wheelers: Performed well but had lower precision and recall than the two-wheeler model, with 77.11% mAP at IoU 0.5. Struggled under stricter overlapping conditions.
- Six-Wheelers: Showed strong detection with 91.12% mAP at IoU 0.5, 89.82% recall, and

85.71% precision, but faced challenges with stringent IoU thresholds.

While the two-wheeler model excelled in accuracy, the four- and six-wheeler models require further improvements for better robustness under strict overlap conditions. Future research should focus on enhancing generalization and detection accuracy across all vehicle types.

References

[1] Chen, Y., Yuan, H., Dong, S., & Peng, J. (2022). Vehicle damage detection based on MD R-CNN. In Proceedings - International Conference Tools with Artificial Intelligence, ICTAI, https://doi.org/10.1109/ICTAI56018.2022.00119.

[2] Fassmeyer, P., Kortmann, F., Drews, P., & Funk, B. (2021). Towards a camera-based road damage assessment and detection for autonomous vehicles: applying scaled-YOLO and CVAE- WGAN. In IEEE Vehicular Technology Conference. https://doi.org/10.1109/VTC2021- FALL52928.2021.9625213.

[3] Huynh, N. T., Tran, N. N. D., Huynh, A. T., Hoang, V. D., & Nguyen, H. D. (2023). VehiDE dataset: new dataset for automatic vehicle damage detection in car insurance. In Proceedings - International Conference on Knowledge and Systems Engineering, KSE. https://doi.org/10.1109/KSE59128.2023.10299490.

[4] Khan, M. H. M., Hussein, S. K., Heerah, M. Z., & Basgeeth, Z. (2021). Automated detection of multiclass vehicle exterior damages using deep learning. In International Conference on Electrical, Computer, Communications and Mechatronics Engineering, ICECCME. https://doi.org/10.1109/ICECCME52200.2021.95909 27.

[5] Qian, W., Chen, S., Huang, Y., Xu, X., Shi, F., Wan, H., et al. (2023). Vehicle-mounted road damage detection method using YOLOv5s framework. In International Conference on Communication Technology Proceedings, ICCT, (pp. 204–209). https://doi.org/10.1109/ICCT59356.2023.10419778.

[6] Sikandar, M., Sohail, R., Saeed, Y., Zeb, A., Zareei, M., Khan, M. A., et al. (2020). Analysis for disease gene association using machine learning. *IEEE Access*, 8, 160616–160626. https://doi.org/10.1109/ACCESS.2020.3020592.

11 Homomorphic cipher for protecting privacy in machine learning

K. Venkata Sravani[1,a] and A. P. Siva Kumar[2,b]

[1]Research Scholar, Department of Computer Science and Engineering, JNTUA, Ananthapuramu Andhra Pradesh, India

[2]Professor, Department of Computer Science and Engineering, JNTUA, Ananthapuramu Andhra Pradesh, India

Abstract

Privacy-preserving machine learning (PPML) is a field that focuses on protecting the privacy and security of data utilized in machine learning models. Several methods, such as Secure Multi-Party Computation, Differential Privacy, Homomorphic Encryption (HE), and Federated Learning, are used to guarantee the confidentiality of data and safeguard user identities. Machine learning (ML) models are crucial across sectors, utilizing sensitive data like medical records and financial information. However, concerns over data privacy necessitate the development of privacy-preserving techniques.

This paper presents a way for ensuring privacy in machine learning using HE. HE allows computations to be conducted on encrypted data, assuring the preservation of privacy during the analysis procedure. Our methodology integrates fully HE with deep learning and ML models, such as MLP, simple neural network, and logistic regression. Experiments conducted on heart disease datasets showcase superior accuracy and less evaluation time. Experiment results indicate consistent accuracy between encrypted and plain data models, validating the efficiency of our approach in safeguarding user privacy.

Keywords: Homomorphic encryption, machine learning, privacy, privacy preservation, security

Introduction

Machine learning (ML) has significantly transformed various industries, such as healthcare and finance, in our data-centric society [1, 2]. However, as the utilization of more delicate information such as medical records and financial information increases, maintaining its confidentiality becomes absolutely essential. Privacy-preserving machine learning (PPML is the solution for this situation [1, 3, 5]. It enables the utilization of machine learning while maintaining the confidentiality of data. In essence, our objective is to develop machine learning solutions that are both secure and private, with the intention of their practical application in real-world scenarios. In response to these urgent issues, attention has shifted towards PPML techniques [4, 6, 7]. PPML's objective is to facilitate the training and assessment of machine learning models while upholding the secrecy of data, therefore safeguarding user privacy. PPML achieves a harmonious equilibrium between the usefulness of data and the safeguarding of privacy, A fascinating area of cryptography called Homomorphic Encryption (HE), which allows calculations to be carried out using encrypted data without any need for decryption [19]. Although HE has potential for facilitating secure calculations on sensitive data in real-time situations, its implementation encounters various obstacles. To overcome these problems, it is necessary to make progress in cryptography research, optimize algorithms, and develop practical implementations that are customized for specific use cases. In order to fully harness the promise of HE in practical situations, it is imperative to tackle these issues as technology progresses [20].

Motivation and contribution

The objective of PPML is to reconcile the requirement for privacy with the advantages offered by ML. It has a vital function in safeguarding the confidentiality of gathered data and guaranteeing adherence to data privacy laws.

Figure 11.1 clearly explains PPML methodology.

[a]sravaniannareddy77@gmail.com, [b]sivakumar.cse@jntua.ac.in

DOI: 10.1201/9781003685364-11

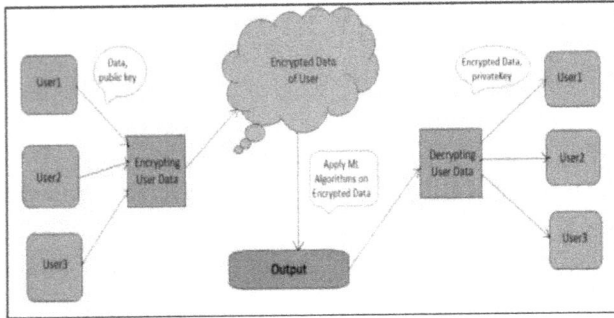

Figure 11.1 Process of privacy-preserving machine learning
Source: Author

Motivation

The rise of cloud-based machine learning, along with increased concerns about the security of organizational assets and data, has led to the advancement of privacy-preserving methodologies in machine learning. To address this issue, it is important to acknowledge that a single solution is not sufficient for all sorts of applications. Although there has been a significant increase in research on privacy-preserving machine learning, there is still a noticeable discrepancy between the progress made in theory and its actual implementation in real-world situations.

Our system utilizes the TenSEAL library [2], which incorporates the CKKS (Cheon-Kim-Kim-Song) scheme for fully HE (FHE). FHE is a category of encryption algorithms that allow for limitless iterations of actions to be taken on data that has been encrypted.

Our contribution

This research study presents a method for safeguarding privacy in machine learning by employing HE. HE facilitates computations directly on encrypted data, thereby ensuring confidentiality throughout the entire analysis procedure. Contribution of the paper is as follows:

i. We present a methodology for PPML using the CKKS is a scheme for leveled HE that supports approximate arithmetic's over complex numbers. CKKS scheme, demonstrating its feasibility through training and evaluating ML or DL models on encrypted data.

ii. We propose a method that leveraging the CKKS scheme from TenSEAL [2], an open-source library for HE operations, to preserve privacy without compromising model efficiency.

iii. Implemented proposed method on healthcare data set where data privacy is vital. Our results highlight the promise of PPML techniques in maintaining model efficiency while addressing data privacy concerns.

iv. We validate our approach using a medical data set and evaluate its performance in terms of computing efficiency and accuracy.

PPML techniques

Techniques in PPML are employed to safeguard the data from unauthorized access by external entities [3, 5]. HE is a cryptography method that makes it possible to execute calculations on encrypted data without the need for decryption. This guarantees the preservation of data privacy throughout the training procedure.

Secure multi-party computation (SMPC) is a technique that allows numerous users to collectively compute a function using their individual inputs, while guaranteeing the confidentiality of those values. This cooperative method guarantees the preservation of confidential information while also permitting the training and application of models.

Differential privacy is a principle that involves introducing random variation that preserves the accuracy of statistical queries while safeguarding the privacy of individual data points [9].

Secure model aggregation approaches are employed in situations where many parties collaborate in training a machine learning model [10]. This enables cooperative model training while maintaining data confidentiality. Federated learning [11] emphasizes training models on distributed devices or servers that store local data samples, rather than consolidating data in a central repository. It allows for collaborative model training while maintaining the decentralized and private nature of the raw data.

Homomorphic encryption

HE schemes can be categorized into different types based on their functionality:

Partially HE (PHE): PHE schemes allow only a limited set of computations on encrypted data. Examples include the RSA cryptosystem, ElGamal, Goldwasser-Micali, Benaloh, and Paillier cryptosystems. These schemes are characterized by their limited functionality, making them unsuitable for complex machine learning tasks.

Somewhat HE (SHE): SHE schemes support a broader set of computations compared to PHE schemes, but they still have limitations on the complexity of operations. While they offer more flexibility than PHE schemes, they may not be suitable for advanced machine learning tasks that require arbitrary computations.

Challenges in the existing system
1. Limited functionality: PHE schemes support only a restricted set of computations, making them unsuitable for complex machine learning tasks.
2. Noise accumulation: In many HE schemes, noise accumulation poses a significant challenge, leading to decreased accuracy and increased computational overhead.
3. Computational overhead: Performing computations on encrypted data often incurs high computational costs, affecting the efficiency of PPML systems.

Proposed HE-based privacy

In various PPML techniques, fully HE (FHE) with the CKKS scheme stands out as a promising solution for addressing privacy concerns [12] in machine learning. FHE enables computations to be performed directly on encrypted data, eliminating the need for decryption, thus ensuring end-to-end data confidentiality [17].

The CKKS scheme, in particular, is well-suited for handling approximate real numbers, making it ideal for machine learning tasks involving floating-point computations. By leveraging FHE with the CKKS scheme, sensitive data can remain encrypted throughout the entire machine learning pipeline, including data preprocessing, model training, and inference, without sacrificing computational efficiency or model accuracy [14]. This approach not only preserves user privacy but also enables seamless integration with existing machine learning frameworks, facilitating the deployment of privacy-preserving machine learning solutions in real-world applications [15].

Proposed System
A Fully HE sscheme like CKKS offers several advantages over other HE schemes:

Fully homomorphic capabilities: CKKS enables arbitrary computations on encrypted data without any limitations.

High precision: CKKS allows for high precision in computations, minimizing accuracy loss during operations on encrypted data.

Scalability: CKKS is designed to handle large datasets and complex machine learning models efficiently, making it suitable for real-world PPML applications [13].

Reduced noise growth: Compared to other HE schemes, CKKS exhibits slower noise growth, preserving the integrity and accuracy of the encrypted computations [16].

By leveraging TenSEAL and the CKKS scheme, our proposed system aims to overcome the limitations of existing HE techniques. In our proposed system, by combining cutting-edge privacy-preserving strategies with conventional model training approaches, we want to completely transform the fields of machine learning and deep learning. Unlike conventional approaches, our system ensures that sensitive data remains encrypted throughout the entire process of analysis and prediction, thereby safeguarding user privacy without compromising the effectiveness of the algorithms. By amalgamating privacy-preserving techniques with ML algorithm, in a world that is becoming more interrelated by the day, we establish the foundation for safe and trustworthy data-driven decision-making.

Figure 11.2 explains the system model architecture of CKKS.

CKKS - a fully HE
To summarize, the CKKS HE technique functions through a series of essential steps:

• Encoding: The message is hidden within a polynomial equation that consists of whole number coefficients.

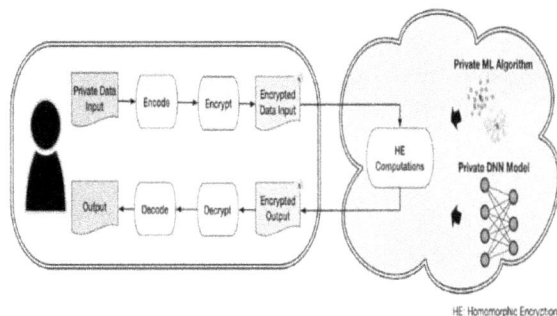

Figure 11.2 System model
Source: Author

- Encryption: The encoded message is modified by adding random data and then rearranged, resulting in ciphertext that appears as meaningless and nonsensical and a key will be generated.
- Computation: Ciphertexts can undergo mathematical operations without the need for decryption, enabling the manipulation of encrypted data.
- Decryption is the process of unlocking the ciphertext and eliminating any interference using the secret key, hence revealing the computed outcome for the original message.
- Decoding: The polynomial that is obtained is converted into a complex vector, which gives the solution to the encrypted calculation.

Machine learning model with HE

We employed and evaluated three distinct models: logistic regression, a basic neural network, and a multilayer perceptron (MLP).

Figure 11.3 represents high-level view of CKKS.

Logistic regression
Architecture:
- Input layer: Accepts the input features.
- Linear transformation: Computes a weighted sum of the input features.
- Sigmoid activation function: Maps the result of the linear transformation to a value within the range of 0 to 1, which represents the likelihood or probability.

Key concepts:
- Sigmoid function: is a fundamental component of logistic regression. It compresses the output into a range from 0 to 1, indicating the likelihood of the positive class.

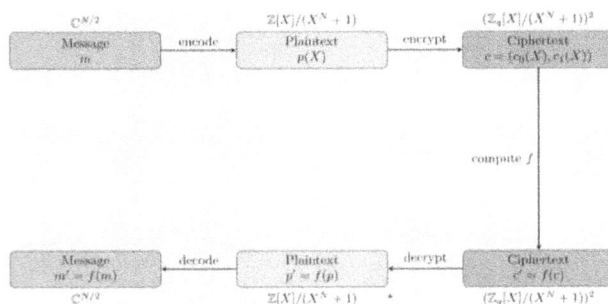

Figure 11.3 High level view of CKKS
Source: Author

- Binary classification: Logistic regression is commonly employed for binary classification tasks, in which the output is restricted to either 0 or 1.

Simple NN
Architecture:
- Input layer: Accepts the input features.
- Hidden layers: In the hidden layers, each neurone calculates a weighted sum of inputs and then applies an activation function, like sigmoid or ReLU.
- Output layer: Computes the final output, typically using a sigmoid or SoftMax activation function for binary or multi- class classification, respectively.

Key concepts:
- Activation functions: Activation functions like ReLU
- Weighted sum: The weights are learnt during training, and each neurone calculates a weighted sum of its inputs.
- Multi-layer architecture: The presence of hidden layers enables the network to capture intricate relationships between features, leading to better performance compared to linear models like logistic regression.

Multilayer Perceptron (MLP)
Architecture:
- Input layer: Accepts the input features.
- Hidden layers: Each hidden layer applies an affine transformation (weighted sum of inputs plus bias) followed by a non-linear activation function.
- Output layer: Produces the final output based on the activations of the last hidden layer.

Key concepts:
- Affine transformation: Each neuron in the hidden layers performs an affine transformation, which involves computing a weighted sum of inputs and adding a bias term.
- Non-linear activation functions: Activation functions like ReLU.
- Depth of network: Multiple hidden layer MLPs may learn hierarchical depictions of the information being supplied, which enables them to recognize intricate patterns and achieve higher accuracy levels in a variety of applications. CKKS scheme involves:

- Initializing the encrypted model.
- Defining parameters and keys for the CKKS encryption scheme.
- Encrypting the testing data using CKKS scheme.
- Training the encrypted model on the encrypted data with CKKS parameters.
- Perform computations on the encrypted data.
- Decrypting the encrypted results for evaluation.
- Comparing the accuracy of the encrypted model with the plain model to assess the impact of encryption on performance.

Experimental Results

We collected gathered datasets from Kaggle to facilitate the training and testing of our privacy-preserving machine learning model.

Figure 11.4 shows Heart disease dataset.

Figure 11.5 shows LR model performance for accuracy & time.

Heart disease dataset

Context: The heart disease dataset is sourced from an ongoing cardiovascular study and is publicly available on Kaggle. This dataset comprises over 4,000 records, each with 15 attributes, including demographic, behavioral, and medical risk factors.

Table 11.1 presents computational complexity of our scheme with existing schemes. Comparing Logistic regression, multi- layer perceptron and simple neural network plain and encrypted model performance across three datasets.

Table 11.1 shows LR performance results.

Table 11.2 shows MLP performance results.

By observing the above table, we can go with 25 epochs for LR since it has similar accuracy for both plain and encrypted models. But it has larger time complexity compared to others.

We can observe that time complexity is increasing by increasing the epochs even though we got desired accuracy for both plain and encrypted models.

Compared to LR model the MLP model achieved smaller time complexity even though there isn't much difference in accuracy in LR and MLP models.

By observing the above table, we can go with 20 epochs for simple neural network.

Since our goal is to achieve the same or similar accuracy for both plain and encrypted model

```
<class 'pandas.core.frame.DataFrame'>
RangeIndex: 4238 entries, 0 to 4237
Data columns (total 16 columns):
 #   Column           Non-Null Count  Dtype
---  ------           --------------  -----
 0   male             4238 non-null   int64
 1   age              4238 non-null   int64
 2   education        4133 non-null   float64
 3   currentSmoker    4238 non-null   int64
 4   cigsPerDay       4209 non-null   float64
 5   BPMeds           4185 non-null   float64
 6   prevalentStroke  4238 non-null   int64
 7   prevalentHyp     4238 non-null   int64
 8   diabetes         4238 non-null   int64
 9   totChol          4188 non-null   float64
 10  sysBP            4238 non-null   float64
 11  diaBP            4238 non-null   float64
 12  BMI              4219 non-null   float64
 13  heartRate        4237 non-null   float64
 14  glucose          3850 non-null   float64
 15  TenYearCHD       4238 non-null   int64
dtypes: float64(9), int64(7)
memory usage: 529.9 KB
```

Figure 11.4 Heart disease prediction dataset
Source: Author

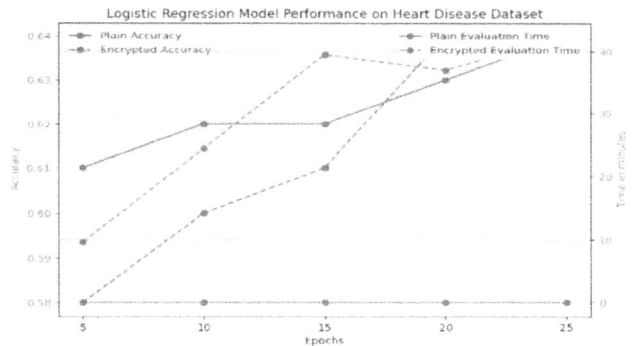

Figure 11.5 LR model (accuracy and time)
Source: Author

Table 11.1 Results comparison.

Epoch	Plain Test Evaluation Time (sec)	Encrypted Test Evaluation Time(min)	Plain Model Accuracy	Encrypted Model Accuracy	Difference in Accuracy	Encrypted Test in Time(sec)
	Logistic Regression (heart disease dataset)					
5	0.0	9.6	0.61	0.58	0.03	11
10	0.001	24.33	0.62	0.60	0.02	20
15	0.0009	39.5	0.62	0.61	0.01	11
20	0.01	37	0.63	0.64	-0.01	15
25	0.0	42.5	0.64	0.64	0.0	19

Source: Author

Table 11.2 Results comparison.

Epoch	Plain Test Evaluation Time (sec)	Encrypted Test Evaluation Time(min)	Plain Model Accuracy	Encrypted Model Accuracy	Difference in Accuracy	Encrypted Test in Time(sec)
	MLP-Multi Layer Perceptron (heart disease dataset)					
5	0.03	2.46	0.60	0.62	0.02	2
10	0.02	2.50	0.63	0.62	0.01	2
15	0.03	2.55	0.63	0.63	0.0	2
20	0.12	2.48	0.61	0.60	0.01	6
25	0.42	2.41	0.61	0.60	0.01	2

Source: Author

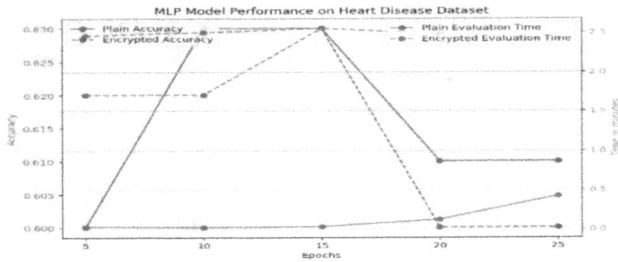

Figure 11.6 MLP model (Accuracy and Time)
Source: Author

Table 11.3 Results comparison.

Simple Neural Network (heart disease dataset)						
Epoch	Plain Test Evaluation Time (sec)	Encrypted Test Evaluation Time(min)	Plain Model Accuracy	Encrypted Model Accuracy	Difference in Accuracy	Encrypted Test set in Time(sec)
5	0.001	1.88	0.62	0.56	0.07	2
10	0.002	2.88	0.64	0.59	0.05	6
15	0.0009	3.03	0.63	0.59	0.04	3
20	0.002	2.9	0.61	0.61	0.0	3
25	0.0009	2.9	0.63	0.56	0.07	3

Source: Author

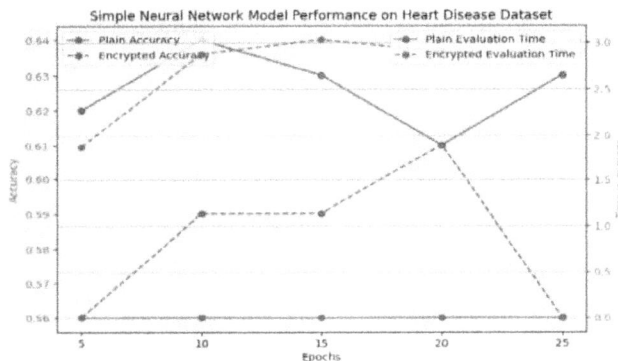

Figure 11.7 SNN model (Accuracy and Time)
Source: Author

with less time complexity. From the above plots we can observe that by increasing epochs accuracy is increasing but there is a huge difference in plain model accuracy and encrypted model accuracy for simple neural networks at each epoch compared to other models.

Figure 11.6 shows MLP model performance of accuracy & time.

Figure 11.7 shows SNN model performance of accuracy & time.

Table 11.3 shows SNN performance results.

Conclusion

This paper introduces a method for safeguarding privacy in machine learning by employing HE. Our system utilizes the TenSEAL library [2], which incorporates the CKKS scheme for FHE. Our approach combines fully HE [18] with ml/dl models including MLP, Simple Neural Network, and logistic regression. We verify our methodology by employing a heart disease dataset and evaluate its performance in terms of precision and computational efficacy. The MLP and Simple Neural Network models demonstrate superior accuracy and require less evaluation time compared to logistic regression. These results underscore the potential of PPML techniques in maintaining model efficiency while addressing data privacy concerns [8].

References

[1]   Ahamed, S. I., & Ravi, V. (2022). Privacy-preserving wavelet neural network with fully Homomorphic Encryption. iarXiv preprint arXiv:2205.13265 2022•arxiv.org. https://doi.org/10.48550/arXiv.2205.13265

[2]   Benaissa, A., Retiat, B., Cebere, B., & Belfedhal, A.E. (2021). Tenseal: a library for encrypted tensor operations using Homomorphic Encryption. arXiv preprint arXiv:2104.03152.arXiv:2104.03152 [cs.CR]. https://doi.org/10.48550/arXiv.2104.03152

[3]   Liu, Ji., Wang, C., Tu, Z., Wang, X. A., Lin, C., & Li,Z. (2021). Secure KNN classification scheme based on Homomorphic Encryption for cyberspace in security and communication networks, DOI:10.1155/2021/8759922

[4]   Li, Bi., & Micciancio, D. (2021). On the security of Homomorphic Encryption on approximate numbers. 40th Annual International Conference on the Theory and Applications of Cryptographic Techniques, Zagreb, Croatia, October 17–21, 2021, Proceedings, Part I (pp.648–677).

[5]   Rahulamathavan, Y. (2022). Privacy-preserving similarity calculation of speaker features using fully Homomorphic Encryption. arXiv preprint arXiv:2202.07994v2 [cs.CR]. DOI:10.48550/arXiv.2202.07994

[6]   Jia, H., Cai, D., Yang, J., Qian, W., Wang, C., Li, X., et al. (2023). Efficient and privacy- preserving image classification using HE and chunk-based convolutional neural network. *Journal of Cloud Computing*, 12, 175.

[7]   Đorđević, G., & Marković, M., & Vuletić, P. V. (2021). Performance comparison of HE scheme

implementations. Contribution to the State of the Art,UDC: 004.738.5:621.391. https://doi.org/10.7251/JIT2201032DJ

[8] Boemer, F., Costache, A., Cammarota, R., & Wierzynski, C. (2019). nGraph-he2: a high-throughput framework for neural network inference on encrypted data. In Proceedings of the 7th ACM Workshop on Encrypted Computing & Applied Homomorphic Cryptography.

[9] Qiu, G., Gui X., & Zhao, Y. (2020). Privacy-preserving linear regression on distributed data by HE and data masking. *IEEE Access*, 8, 107601–107613.

[10] Sun, X., Zhang, P., Liu, J. K., Yu, J., & Xie, W. (2020). Private machine learning classification based on fully HE. *IEEE Transactions on Emerging Topics in Computing*, 8(2), 352–364.

[11] Yang, Y., Yang, X., Heidari, M., Khan, M. A., Srivastava, G., Khosravi, M. R., et al. (2023). ASTREAM: data-stream-driven scalable anomaly detection with accuracy guarantee in IIoT environment. *IEEE Transactions on Network Science and Engineering*, 10(5), 3007–3016.

[12] Wang. J., Wu, F., Zhang, T., & Wu, X. (2022). DPP: data privacy- preserving for cloud computing based on HE. In 2022 International Conference on Cyber-Enabled Distributed Computing and Knowledge Discovery (CyberC), Suzhou, China, (pp. 29–32).

[13] Lee, J. W., Kang, H., Lee, Y., Choi, W., Eom, J., Deryabin, M., et al. (2022). Privacy- preserving machine learning with fully HE for deep neural network. *IEEE Access*, 10, 30039–30054.

[14] Rahulamathavan, Y., & Rajarajan, M. (2015). Efficient privacy- preserving facial expression classification. *IEEE Transactions on Dependable and Secure Computing*, 14(3), 326–338.

[15] Rahulamathavan, Y., Phan, R., Veluru, S., Cumanan, K., & Rajarajan, M. (2014). Privacy-preserving multi-class support vector machine for outsourcing the data classification in cloud. *IEEE Transactions on Dependable and Secure Computing*, 11 (5), 467–479.

[16] Rahulamathavan, Y., Veluru, S., Phan, R., Chambers, J., & Rajarajan, M. (2014). Privacy-preserving clinical decision support system using gaussian kernel based classification. *IEEE Journal of Biomedical and Health Informatics*, 18(1), 56–66.

[17] Rahulamathavan, Y., Phan, R., Chambers, J., & Parish, D. (2012). Facial expression recognition in the encrypted domain based on local fisher discriminant analysis. *IEEE Transactions on Affective Computing*, 4(1), 83–92.

[18] Smart, N. P., & Vercauteren, F. (2014). Fully homomorphic SIMD operations. *Designs, codes and cryptography*, 71(1), 57–81.

[19] Khedr, A., Gulak, G., Member, S., & Vaikuntanathan, V. (2015). SHIELD: scalable homomorphic implementation of encrypted data-classifiers. *IEEE Transactions on Computers*, 65, 2848–2858.

[20] El Mestari, S. Z., Lenzini, G., & Demirci, H. (2024). Preserving data privacy in machine learning systems. *Computers and Security*, 137, 103605.

12 Implementation of efficient multipliers using stacker based binary compressors

Pothula Bhargavi[1,a], Pamanji Silpa[2], R. Sai Bhargav[2], Shaik Alisha[2] and Ganekanti Naresh[2]

[1]Assistant Professor, Department of ECE, Sree Vidyanikethan Engineering College, Tirupati, Andhra Pradesh, India

[2]Department of ECE, Sree Vidyanikethan Engineering College, Tirupati, Andhra Pradesh, India

Abstract

Efficient multiplier design is essential for enhancing the performance of digital systems, particularly in applications that require rapid arithmetic processing. This study presents a multiplier architecture based on the Vedic Urdhva Tiryagbhyam method, integrated with approximate compressors for partial product accumulation using 5:3, 6:3, and 7:3 configurations. The proposed design is compared with conventional architecture that employs stacker-based binary compressors. Both implementations are synthesized and analyzed using the Xilinx Vivado environment, focusing on key performance indicators such as lookup table (LUT) utilization, static power, and dynamic power consumption. The simulation results demonstrate notable improvements in the proposed method. LUT usage is reduced from 596 to 56, static power consumption decreases from 0.823 W to 0.136 W, and dynamic power consumption drops from 48.008 W to 15.363 W. These outcomes highlight the proposed architecture's advantages in terms of area efficiency and power reduction. The integration of approximate compressors with the Vedic multiplication technique proves to be a promising solution for designing compact, energy-efficient, and high-performance digital arithmetic units.

Keywords: Approximate compressors, Uddhav Tiryagbhyam algorithm, Vedic multiplier

Introduction

Multiplication is a fundamental operation in digital systems, serving as the foundation for many applications, from signal processing to arithmetic computing. In order to improve the trade-offs between speed, area, and power consumption, researchers have spent decades creating effective multiplier architectures. By introducing three-based reduction approaches, the groundbreaking studies of Wallace [1] and Dadda [2] established the groundwork for high-speed multipliers. These designs were appropriate for high-performance computing because they greatly decreased the delay brought on by incomplete product addition. Later improvements, including the introduction of multi-input counters and compressor circuits [4] and column compression [3], improved these architectures even more and addressed certain multiplier design difficulties. Nonetheless, there are still issues in the field of multiplier design. Even though they are good at cutting down on delays, traditional methods frequently use more space and electricity [5,6]. Although recent developments like algorithmic improvements [7] and the use of reversible logic attempt to overcome these problems, they still have scalability and integration challenges with contemporary low-power devices [8-10]. Furthermore, the necessity for innovative strategies that strike a balance between performance and power efficiency has been brought to light by the rising demand for energy-efficient computing [11]. The demand for new designs is driven by the drawbacks of current techniques, such as their high-power consumption and area overhead. The opportunity to optimize the trade-offs in multiplier design is provided by compressors [12-14], which are essential for decreasing partial products. It is feasible to create multipliers that satisfy the requirements of contemporary applications in terms of speed, power, and area by investigating innovative architectures and algorithms. Figure 12.1.

The following are the main goals of this work:

[a]bhargavipothula36@gmail.com

DOI: 10.1201/9781003685364-12

- To create an optimal multiplier architecture by combining approximation compressors with the Urdhva Tiryagbhyam algorithm.
- To assess and contrast the suggested design's performance with current techniques, paying particular attention to important parameters including area, power, and speed.
- To verify the suggested method with cutting-edge simulation software such as Xilinx Vivado. The following contributions are made by this paper:
- An innovative multiplier architecture that uses approximation compressors to reduce partial products.
- A comparison of the area, power, and performance of the suggested design with conventional stacker-based binary compressors.
- Using Xilinx Vivado, the suggested design was implemented and validated, demonstrating notable gains in important metrics.

The paper's structure separates its several sections. Section 2 presents the literature review. Additionally, Sections 3 and 4 present the current approach and the suggested system, while Section 5 presents the experimental findings. Section 6 presents the conclusion and next work [15-16].

Proposed method

To provide effective partial product reduction, the suggested approach combines approximation compressors with the Urdhva Tiryagbhyam algorithm, a Vedic multiplication methodology. The suggested architecture optimizes the addition of partial products by utilizing 5:3, 6:3, and 7:3 compressor designs, in contrast to conventional stacker- based binary compressors. This method maintains competitive speed while drastically lowering the area and power consumption. Many real-world applications, especially those that include multimedia and signal processing, can tolerate the little computational errors introduced by the usage of approximation compressors. The Xilinx Vivado tool is used to develop and validate the design, showing significant gains in area usage and power efficiency over current techniques.

Urdhva Triyakbhyam algorithm
A key element of Vedic mathematics is the Urdhva Triyakbhyam algorithm, which uses the concepts of crosswise and vertical multiplication. This age-old mathematical method provides a general multiplication formula that can be used in a variety of

mathematical situations. "Vertically and crosswise," as translated literally, captures its basic methodology. Fundamentally, the Urdhva Triyakbhyam algorithm presents a new idea that makes it possible to generate and add partial products at the same time. This method simplifies the multiplication process by generating all partial products required for multiplication simultaneously. The approach maximizes computational performance by enabling parallelism in the creation of partial products and their subsequent summing Furthermore, the Urdhva Triyakbhyam algorithm's adaptability enables it to be generalized to handle n x n bit integers. Its versatility in digital circuit design is further enhanced by its scalability, which highlights its suitability for a variety of multiplication applications.

Vedic multiplier
The construction of an 8-bit Vedic multiplier created with the Urdhva Tiryagbhyam (vertical and across) algorithm is shown in Figure 12.0. The structure uses the parallelism present in the Vedic multiplication method to efficiently compute the product of two 8-bit operands by dividing the inputs into smaller pieces. In order to get the final result, the architecture generates partial products using crosswise and vertical multiplication patterns, which are then systematically added together. The modularity of this architecture maintains computational efficiency while streamlining implementation and guaranteeing scalability for higher-bit multiplications.

Approximate compressor
The schematic of a 3-1-1-2 compressor utilized in the partial product reduction stage is shown in Figure 12.2. This compressor generates a single sum and two carry outputs from three primary inputs and one carry-in bit. By optimizing the architecture to approximate specific logic processes, fewer transistors and logic gates are needed. Although there is a little precision trade-off, this approximation results in reduced area usage and power consumption. Because it directly influences the hardware complexity and power profile, the compressor's efficiency is crucial to the overall design of the multiplier.

Vedic multiplier with approximate compressor
The whole architecture of the suggested Vedic multiplier with approximate compressors is shown in Figure 12.0. By adding 5:3, 6:3, and 7:3 approximation

compressors in the partial product reduction phase, this design expands on the conventional Vedic multiplier. In order to simplify the addition process and lower the total hardware resources required, approximation compressors are used in place of exact compressors. The data flow from the creation of partial products to their reduction and ultimate summarization is depicted in the figure. When compared to traditional designs, this hybrid technique significantly improves space usage and saves power by successfully balancing performance and resource efficiency. Figure 12.1.

Implementation flow chart

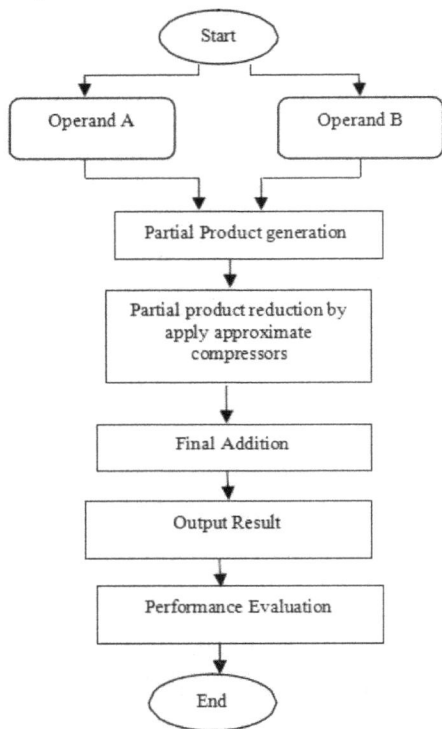

Figure 12.0 Flow chart
Source: Author

Begin the process of multiplier design.
- *Input operand A and operand B*
  • Provide the binary inputs for multiplication.
- *Partial product generation*
  • Use the Vedic Urdhva Tiryagbhyam algorithm to generate partial products.
- *Partial product reduction*
  • Apply approximate compressors:
    • 5:3 Approximate compressor
    • 6:3 Approximate compressor
    • 7:3 Approximate compressor

- *Final addition*
  • Sum the reduced partial products to generate the final result.
- *Output result*
  • Display the final multiplied value.
- *Performance evaluation*
  • Analyze key metrics:
    • Area utilization (LUTs)
    • Static power consumption
    • Dynamic power consumption
- *End*

Complete the process

A. Performance comparison (Table 12.1)
The performance comparison table consolidates the key metrics of both methods, demonstrating the proposed design's superiority: Area utilization reduced by 90.6%, from 596 LUTs to 56 LUTs. Static power: Decreased by 83.5%, from 0.823 W to 0.136 W.

Table 12.1 Performance comparison of existing and proposed method.

S.N	Parameter	Existing method	Proposed method	Improvement (%)
1	Area (LUT)	596	56	90.6
2	Static power in watts	0.823	0.136	83.5
3	Dynamic power in watts	48.008	15.363	68.0
4	Total power in watts	48.831	15.5	68.3

Source: Author

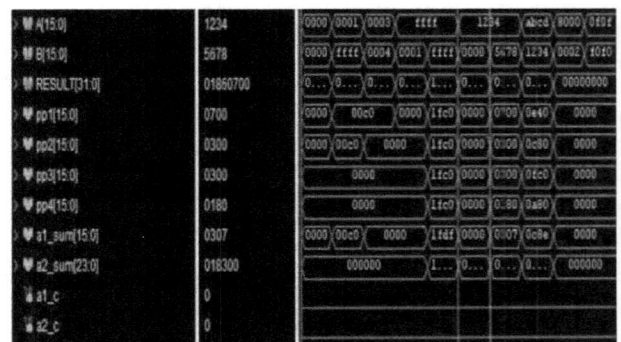

Figure 12.1 Simulation of proposed method
Source: Author

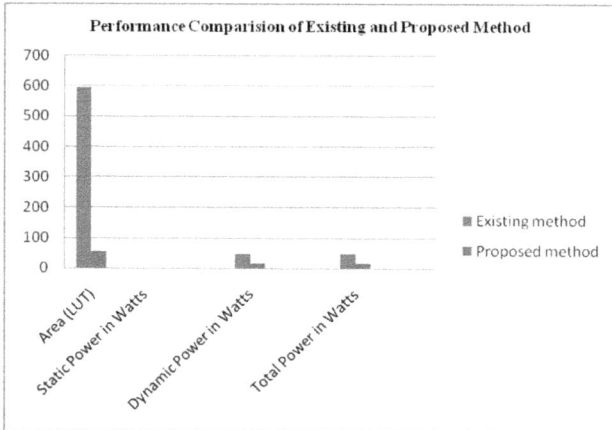

Figure 12.2 Comparison performance graph
Source: Author

Dynamic power lowered by 68.0%, from 48.008 W to 15.363 W. Total power: Reduced by 68.3%, from 48.831 W to 15.5 W. Figure 12.2.

B. Performance analysis
This graph visually compares the performance metrics of the existing and proposed methods, highlighting the significant improvements achieved by the proposed design. The graph emphasizes the drastic reductions in area utilization and power consumption, underscoring the efficiency of incorporating approximate compressors into the multiplier design

Conclusion

The proposed design of a digital multiplier using the Vedic Urdhva Tiryagbhyam algorithm combined with approximate compressors (5:3, 6:3, and 7:3 designs) demonstrates significant advancements in resource efficiency and power consumption. Compared to the existing method, the proposed approach achieves a 90.6% reduction in area utilization (from 596 LUTs to 56 LUTs) and an 83.5% and 68.0% reduction in static and dynamic power, respectively. These improvements validate the effectiveness of the proposed design for energy-efficient and compact digital arithmetic systems. By leveraging the simplicity of approximate compressors, the multiplier achieves an optimal balance between performance, accuracy, and resource utilization, making it highly suitable for applications in portable, low-power, and high-speed systems.

Future scope
In future the Approximate multipliers are ideal for error-tolerant applications like machine learning accelerators, image processing, and digital signal processing. Exploring these domains could unlock additional use cases.

References

[1] Wallace, J C. S. (1964). A suggestion for a fast multiplier. *IEEE Transactions on Electronic Computers*, EC-13(1), 14–17.

[2] Dadda, L. (1965). Some schemes for parallel multipliers. *Alta Frequenza*, 34, 349–356.

[3] Wang, Z., Jullien, G. A., & Miller, W. C. (1995). A new design technique for column compression multipliers. *IEEE Transactions on Computers*, 44(8), 962–970.

[4] Mehta, M., Parmar, V., & Swartzlander, E. (1991). High-speed multiplier design using multi-input counter and compressor circuits. In Proceedings 10th IEEE Symposium on Computer Arithmetic, Jun. 1991, (pp. 43–50).

[5] Krishnaveni, D., Priya, M., & Baskaran, K. (2012). Design of an efficient reversible 8×8 Wallace tree multiplier. *World Applied Sciences Journal*, 20(8), 1159–1165. 10.5829/idosi.wasj.2012.20.08.711.

[6] Asif, S., & Kong, Y. (2015). Design of an algorithmic Wallace multiplier using high speed counters. In Proceedings of IEEE Computer Engineering and Systems (ICCES), Dec. 2015, (pp. 133–138).

[7] Veeramachaneni, S., & Srinivas, M. B. (2013). Design of optimized arithmetic circuits for multiplier realization. In 2013 IEEE Asia Pacific Conference on Postgraduate Research in Microelectronics and Electronics (PrimeAsia), Visakhapatnam, (pp. 219–224).

[8] Veeramachaneni, S., Krishna, K. M., Avinash, L., Puppala, S. R., & Srinivas, M. B. (2007). Novel architectures for high-speed and low-power 3-2, 4- 2 and 5-2 compressors. In Proceedings of 20th International Conference VLSI Design Held Jointly 6th Int. Conf. Embedded Syst. (VLSID), Jan. 2007, (pp. 324–329).

[9] Fritz, C., & Fam, A. T. (2017). Fast binary counters based on symmetric stacking. *IEEE Transactions on Very Large-Scale Integration (VLSI) Systems*, 25(10), 2971–2975.

[10] Huddar, S. R. Kalpana, M., & Mohan, S. (2013). Novel high-speed Vedic mathematics multiplier using compressors, In Conference Proceedings of IEEE.

[11] Pokhriyal, N., Kaur, H., & Prakash, D. N. R. (2013). Compressor based area-efficient low-power 8×8

Vedic multiplier. *International Journal of Engineering Research and Applications*, 3(6), 1469–1472.

[12] Kaur, H., & Prakash, N. R. (2015). Area-efficient low PDP 8-bit Vedic multiplier design using compressors. In 2015 2nd International Conference on Recent Advances in Engineering and Computational Sciences (RAECS), Chandigarh, (pp. 1–4).

[13] Huddar, S. R., Rupanagudi, S. R., Kalpana, M., & Mohan, S. (2013). Novel high speed Vedic mathematics multiplier using compressors. In 2013 International Mutli-Conference on Automation, Computing, Communication, Control and Compressed Sensing (iMac4s), Kottayam, (pp. 465–469).

[14] Nikhil, G. V., Vaibhav, B. P., Naik, V. G., & Premananda, B. S. (2017). Design of low power barrel shifter and Vedic multiplier with Kogge-stone adder using reversible logic gates. In 2017 International Conference on Communication and Signal Processing (ICCSP), Chennai, (pp. 1690–1694).

[15] Chandrababu, D. V. S., Ravindra, T., Nagaraju, P., Narsimha, P., & Neelima, C. (2017). 16 bit multiplier implementation using Vedic mathematics. *International Journal of Mechanical Engineering and Technology*, 8(7), 703–716.

[16] Sunesh N. V., & Sathishkumar, P. (2015). Design and implementation of fast floating-point multiplier unit. In 2015 International Conference on VLSI Systems, Architecture, Technology and Applications (VLSI-SATA), Bangalore, (pp. 1–5).

[17] Ramachandran, S., & Pande, K. S. (2012). Design, implementation and performance analysis of an integrated Vedic multiplier architecture. *International Journal of Computational Engineering Research*, 2, 697–703.

[18] Raju, A., & Sa, S. K. (2017). Design and performance analysis of multipliers using Kogge stone adder. In 2017 3rd International Conference on Applied and Theoretical Computing and Communication Technology (iCATccT), Tumkur, Karnataka, India, (pp. 94–99).

[19] Jain, A., & Jain, A. (2017). Design, implementation and comparison of Vedic multipliers with conventional multiplier. In 2017 International Conference on Energy, Communication, Data Analytics and Soft Computing (ICECDS), Chennai, India, (pp. 1039–1045).

13 Design and simulation of FINFET 4-bit barrel shifter using GDI technique for low power applications

P. Syamala Devi[a], K. Sudharshan Reddy[b], C. Vinuthna[c], A. Venkatahemanth[d] and Y. Yaswanth[e]

Department of ECE Annamacharya Univerisity, Rajampet, Andhra Pradesh, India

Abstract

The barrel shifter is an essential digital component used in microprocessors and digital systems to perform bit-wise shifting operations efficiently. A 4-bit barrel shifter can execute left shifts, right shifts, and rotations, making it a versatile component in data manipulation a. This paper focuses on designing and analyzing a 4-bit low-power barrel shifter using 2×1 multiplexers (MUX) in conjunction with Gate Diffusion Input (GDI) logic and FINFET technology. The GDI is an advanced logic design technique that provides significant reductions in power consumption and area compared to conventional CMOS logic. The Fin Field-Effect Transistor (FINFET) technology is used for designing the barrel shifter, which is implemented in the software Tanner EDA Tool. The design is achieved by reducing the number of transistors needed and minimizing the switching power. The 32 nm node provides advanced performance and power efficiency.

Keywords: 32 nm Fin Field-Effect Transistor, GDI, high efficiency, high speed, shifting operations

Introduction

The barrel shifter is an essential component for bit manipulation in ALU to create such low power and efficient design pass transistor logic (PTL) is used with Fin Field-Effect Transistor (FINFET) technology which helps in reducing short channel effect and leakage current and helps in minimizing the area [1]. The 2 by 1 multiplexers (MUX) is a "switch logic" which provides output based on the input selection then mux is designed with FINFET technology with GDI technique hence the number transistors decreased compared to CMOS technology design which makes the mux more efficient for low power application [2]. Component designs have a significant impact on optimizing computing efficiency. One such component is the barrel shifter, an essential part of cryptographic systems, ALUs, and DSPs. The primary focus of this research is a low-power 4-bit barrel shifter that utilizes a 32 nm FINFET and TG logic. Specialist circuits known as barrel shifters are used to efficiently rotate or shift bits with little clock cycle consumption. Their design is crucial for applications that prioritize high speed and low power usage. Compared to previous CMOS technology, 32 nm FINFET technology has better control over short-channel effects, more scalability, and less leakage power. Because of its famed speed and low power consumption, transmission gate (TG) logic improves this alternative. This maintains a middle ground between efficiency and performance. The proposed 4-bit barrel shifter design is achieved by using gate diffusion input (GDI) technique hence it utilizes these technological advantages to achieve lowest power usage while maintaining maximum operating speed. The procedure for designing, modeling, and analyzing the barrel shifter's circuitry makes use of cutting-edge electronic design automation (EDA) technologies. Our goal is to contribute to the growing need for power-efficient digital systems, which is especially important in embedded and portable applications. The proposed model shows 22.01% significance reduction in power consumption than the Existed model.

Literature survey

The CMOS barrel shifter using transmission gate technique is used in Digital signal processing to achieve low resistance in both high and low states and reducing propagation delay [3–5] but it occupies more area and high power consumption .The diode free

[a]syamuvlsi@gmail.com, [b]sudarreddy2002@gmail.com, [c]Vinuthnachalla526@gmail.com, [d]venkatahemanthachukatla287@gmail.com, [e]Yash143yashmy@gmail.com

DOI: 10.1201/9781003685364-13

adiabatic logic (DFAL) complementary metal-oxide-semiconductor (CMOS) barrel shifter is used for low power applications which uses "adiabatic switching" and its performance is low compare to other strategy models like PTL,TGL .due to low performance it is not suitable for faster applications [6]. Barrel shifters are fundamental to the architecture of high-speed computing systems, and advancements in transistor manufacturing and circuit optimization are intimately correlated with these trends. The first designs of barrel shifters relied heavily on the CMOS technology. These systems offer solid performance, but they aren't without major flaws including high power consumption and limited scalability. Hence FINFETs are used which provide better scalability, fewer leakage currents, and better gate control due to their 3D structure. The intrinsic benefits of FIN FET-based logic circuits, such as decreasing power consumption and enhancing signal integrity, have led to their broad adoption in barrel shifter designs. The design of FINFET 4-bit barrel using pass transistor logic will reduce short channel effect and reduce the area of the design, the drawbacks of the design was it requires more transistors and hence consumes more power [8,9,10]. The transistors in architecture need to be reduced while designing the 2 by 1 MUX which helps in faster switching and area reduction [7] will be achieved by GDI technique over TG, PTL techniques. The Researchers have looked on improving transistor size, reducing the transistor usage which employs designs with multiple threshold voltages, and minimizing logic to further decrease power dissipation. Research on 8×4-barrel shifters has centered on the compromises between power, performance, and area. The tree-based design is a popular architecture due to its ability to complete tasks quickly. Still, there is a need for innovative methods of circuit design and fabrication to improve the energy efficiency of these designs at high-tech nodes.

Existing method

There have been many suggestions for ways to enhance the design of barrel shifters to make them faster, more power efficient, and more efficient in their use of space. Here are some of the most popular ways that barrel shifters are currently being used:

Tree-based barrel shifters
This method builds the barrel shifter using a binary tree structure, where each level performs a separate shift operation. This leads to a reduction in critical route delays as compared to a linear design. Things That Are Bad The complexity and space needed are both increased as the number of connections and logic gates grows. Use of pass transistor logic in barrel shifters the standard CMOS logic is replaced with pass transistors to perform the shifting operations. This technique reduces the number of transistors used in the design. Declining features, Signal degradation and increased noise sensitivity, especially for longer chains. Using reasoning dynamic barrel shifters using dynamic logic circuits, like domino logic, allows for fast shifting operations. The pre-charge and evaluation processes provide the basis of how these designs work. The increased power consumption is a consequence of both the higher sensitivity to noise and the continuous pre-charge operations. Despite their numerous advantages, the aforementioned techniques do have a number of disadvantages. As an example, a lot of power is used by traditional CMOS and dynamic logic designs. Delays between connections in tree-based designs and signal quality problems in PTL. Perplexity with hybrid and changeable architectures. Combining the benefits of low-power technology (such as FINFET) with efficient logic types (such as TG logic) is the only way to get optimal performance and energy efficiency. To solve this problem, innovative approaches are required.

Proposed method

Utilizing a combination of TG logic and PTL, among others, hybrid designs aim to optimize space, performance, and efficiency. In these configurations, vital routes are often activated by TG, while non-critical sections are enabled by PTL. This layout strikes a good mix between power, performance, and room. The circuit was constructed and simulated using LTSPICE, a powerful tool for digital and analog circuit analysis. The simulation results verified that the barrel shifter's shift and rotate functions were all correct. According to key performance metrics such area, propagation delay, and power consumption, the results are in agreement with the goals of low-power and high-speed operation. Using the design's hierarchical structure, larger bit-width shifters may be optimized and implemented in future work. In conclusion, this project shows how FINFET technology may be used in the real world to make a durable and environmentally responsible barrel shifter. Digital signal processing, high-performance computing, and portable electronics are some of the areas where this design excels because of its low power consumption and quick processing speed. With the information gathered from this research, state-of-the-art technical nodes may be able to construct digital circuits that are both more complex and more energy efficient.

Implementation of barrel shifter

The 4-bit low-power universal barrel shifter can only be realized by translating the theoretical notion into a functional circuit that adheres to the space, power, and performance constraints imposed by 16 nm FINFET technology. During the design, simulation, and layout stages, the objective is to guarantee that the shifter operates as expected and efficiently as shown in Figure 13.4 and simulation output is shown in 13.6.

2 × 1 multiplexer design

The 2 × 1 MUX, which is constructed using FINFET transistors, is the basic component of the barrel shifter. Here is what MUX does:

A 2×1 MUX that utilizes FINFETs is a crucial component of digital circuits designed for low-power, high-performance applications. This method gives advantage of FINFET's reduced leakage, rapid speed, and tiny area, making it well-suited to modern very large-scale integration (VLSI) devices operating at nanoscale scales as shown in Figures 13.1, 13.2 and 13.3. similarily 4x1 Mux layout is shown in 13.5.

2 × 1 MUX SCHEMATIC :

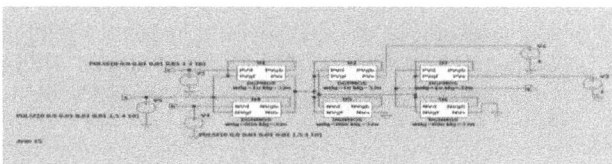

Figure 13.1 2x1 Mux Schematic
Source: Author

2 × 1 MUX layout

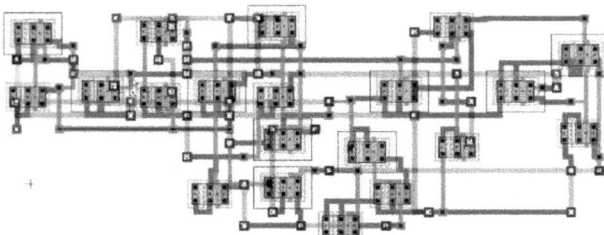

Figure 13.2 2x1 Mux Layout
Source: Author

2 × 1 MUX output

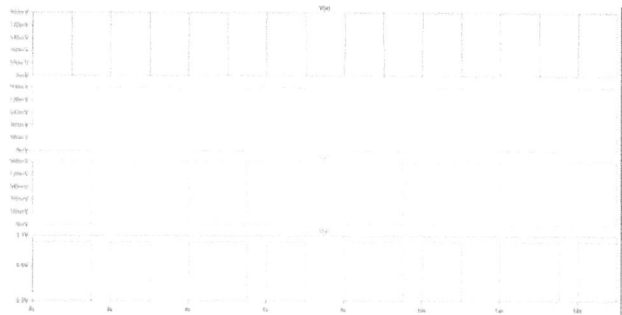

Figure 13.3 2x1 Mux Output
Source: Author

4-Bit barrel shifter

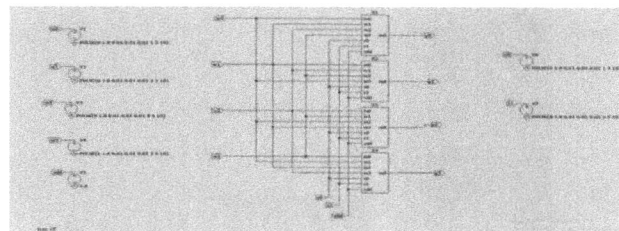

Figure 13.4 4-Bit barrel shifter
Source: Author

Shifter simulation results

The LSL method shifts bits to the left while adding zeros to their right-hand side as padding. Bits are shifted to the right by inserting left-hand zeroes as

4 × 1 MUX layout

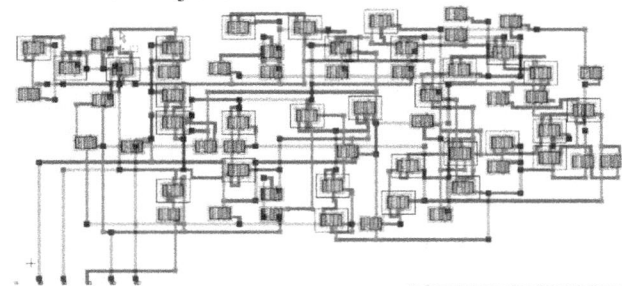

Figure 13.5 4 X 1 MUX layout
Source: Author

Figure 13.6 Barrel shifter output
Source: Author

Table 13.1 Comparision between Existing system and proposed system.

S.NO	Topology	Power consumption
1	4 × 4 Barrel Shifter Using CMOS	335.8PW
2	PT based 4-bit FINFET barrel shifter (EXISTED)	104 PW
3	GDI based 4-bit FINFET barrel shifter (PR0POSED)	73.15PW

Source: Author

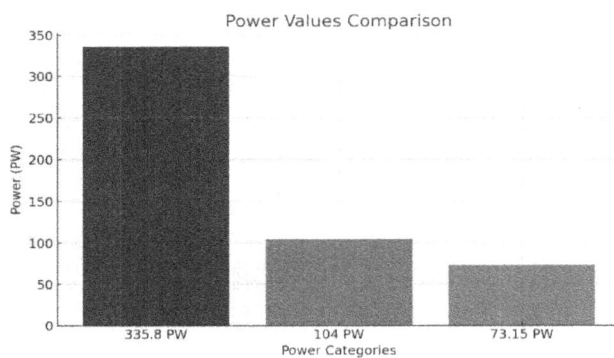

Figure 13.7 Power values Comparision
Source: Author

padding in the logical shift right (LSR) operation. A barrel shifter's direction and kind of operation are dictated by encoded control signals: Alter Your Path (by going in a different direction, as to the left or right). Change the magnitude from one bit to two bits, for instance, by rotating or shifting it.

The simulation has 2 selection lines s0,s1 and 4 inputs w0,w1,w2,w3 and outputs y0,y1,y2,y3 if (s1,s0) = (0,0) the output is equal to the input and for(0,1) case y0 = w3,y1=w0,y2=w1,y3=w2, for(1,0) case y0=w2,y1=w3,y2=w0,y3=w1, For (1,1) case y0=w1,y1=w2,y2=w3,y3=w0, The design and modeling of an LTSPICE 4-bit barrel shifter demonstrate its effective execution of shifts and rotations, using a modular MUX-based architecture. In modern digital systems, this tool is ideal for low-power, high-speed applications due to its ability to simulate the circuit and display its power, latency, and performance attributes.

The circuit will function as expected if this is done. Performance attributes.

Conclusion and future scope and future scope

A two-by-one multiplexer-based 32nm FINFET 4-bit low-power universal barrel shifter has been designed and simulated. Improving performance while reducing power consumption, keeping the design small, and high operating efficiency are all goals of this paper, which makes advantage of present FINFET technology. The barrel shifter is a multi-function component of modern digital systems that can do mathematical shifting in ALU. Using 2 × 1 MUX as the fundamental building element enables a flexible and scalable design. The circuit utilizes FINFET technology, which has several advantages over traditional planar CMOS designs, including faster processing times, less short-channel effects, and reduced leakage current. The design and study of a 4-bit low-power universal barrel shifter using 2 × 1 multiplexers in 32 nm FINFET technology with GDI technique has given 22.9% improvement in power consumption and in area reduction compared to FINFET PT logic 4-bit barrel shifter, which might serve as a roadmap for future research and development as shown in Table 13.1 and Figure 13.7. The development of digital systems is being propelled by the rising need for hardware components that are smaller, faster, and more efficient. This project lays the groundwork for more extensive use and enhancements in the future across several domains.

References

[1] Design and analysis of A 4-bit low power universal barrel shifter using 2by1 Mux in 16nm FINFET Technology Authours: G.praveen Journal: IJRTI vol.8, issue6, June 2023 pages:734-739; ISSN 2456-3315.

[2] Performance Analysis of FINFET based 2:1 Multiplexers for Low Power Application Authour: N.Kumar Conference:IEEE students Conference on Engineering &System, IEEE press Prayagraj,2020.

[3] CMOS Performance Analysis of 4-bit Barrel Shifter Authour: D.Shakya , S.Agrawal Journal:JETIR vol.5, issue 12, December 2018.

[4] Low Power 2:1 MUX for Barrel Shifter Joural:IJEAT.

[5] A Purely MUX Based High Speed Barrel Shifter VLSI Implementation Using Three Different Logic Design Styles Authours: Abhijit R.Asati, C.Shekhar Cnference:Published in Mechanical Engineering and Technology,ASIC 125 (2012).

[6] Implementation Of Barrel Shifter using Diode free Adiabatic Logic (DFAL) Conference: IJEAT (2020).

[7] Design of Power Efficient Multiplexer using Dual Gate FinFET Technology Authours: M.Vyas, S.K. Manna, S.Akashe Conference: IEEE international Conference on Communication Networks (2015).

[8] VLSI Implementation of a High Performance Barrel Shifter Architecture using Three Different Logic Design Style Authours: Abhijit R.Asati , C.Shekhar Conference: IJRTE (2009).

[9] A Purely MUX Based High Speed Barrel Shifter VLSI Implementation Using Three Different Logic Design Styles Abhijit R.Asati , C.Shekhar Conference: IJRTE (2009).

[10] Design of Various 4 Bit Shifters using CMOS 32nm Technology. Conference:IJRASET (2019).

14 Enhancing IoT security: ensemble machine learning model for botnet attack detection

Annavaram Kiran Kumar[1,a], Padamati Pavithra[2,b], Beemireddy Yamini[2,c], Veluru Yashwitha[2,d] and Thagupparthi Venkateswari[2,e]

[1]Assistant Professor, Department of CSE (AI&ML), Srinivasa Ramanujan Institute of Technology, Anantapur, Andhra Pradesh, India

[2]Students, Department of CSE (Data Science), Srinivasa Ramanujan Institute of Technology, Anantapur, Andhra Pradesh, India

Abstract

The largest number of interconnected ecosystems emerged from the quick spread of Internet of Things (IoT) devices, but this expansion has also left IoT environments susceptible to advanced cyberthreats, especially botnet attacks. These attacks cause operational problems and pose serious security and privacy issues by taking advantage of flaws in IoT networks. As botnet attacks become more complex and wide-ranging, conventional detection techniques are unable to detect them effectively and precisely in real time. Hybrid machine learning was created to overcome these difficulties, using the UNSW_NB15 dataset as a standard. There were several preprocessing techniques used to improve feature selection and model performance, including standardization, feature encoding, column transformer, one hot encoder and standard scaler. Random Forest obtained an accuracy rate of 95%, extra trees 94.85%, Decision Tree 93.69%, multilayer perceptron (MLP) 93.44%, Gradient Boosting 93.15%, K-nearest neighbors algorithm 92.91%, and logistic regression 91.07%. Long short-term memory (LSTM) was the most effective algorithm for detecting seasonal patterns in network traffic with an accuracy of 96.435% in training and 96.665% in testing. It provides real-time botnet attack alerts by allowing users to submit CSV data files with network traffic using Streamlit.

Keywords: Botnet detection, feature selection, hybrid model, Internet of Things, security, long short-term memory, machine learning, preprocessing and UNSW_NB15

Introduction

The Internet of Things (IoT) is an essential part of contemporary technology because of its capacity to link billions of objects and offer smooth automation and interaction across several disciplines [1]. However, the rapid adoption of IoT has resulted in significant security problems, and botnet assaults are currently a major threat. In addition to DDoS attacks, these attacks lead to information theft and financial losses, as they exploit stolen IoT devices. The complexity of modern botnets makes it vital to employ advanced machine learning-driven solutions that can identify and control such threats. These challenges have motivated researchers to develop hybrid prediction models that use long short-term memory (LSTM) techniques for selecting features, and sophisticated preprocessing methods. The suggested study uses the UNSW_NB15 dataset [2] a comprehensive standard for network traffic analysis, to develop a trustworthy botnet detection system. The framework integrates multiple models including Random Forest (RF), Extra Trees (ET), Decision Tree (DT), multilayer perceptron (MLP), Gradient Boosting, KNN, logistic regression (LR) and LSTM to achieve high detection accuracies. By evaluating uploaded network traffic files and providing practical preventive measures, a Streamlit-based user interface (UI) further improves usability and makes real-time botnet detection possible [3]. The goal of this research is to offer a practical, scalable and effective way to protect IoT environments from botnet attacks.

IoT-driven attacks have significantly increased in recent years and botnet attacks have increased globally. The 2023 Cybersecurity Threat Report states that IoT botnet attacks which impact vital infrastructure and organizations [4] globally made up about 45%

[a]kiran.annavaram@gmail.com, [b]padamatipavithra@gmail.com, [c]yaminibeemireddy@gmail.com, [d]veluruyashwitha18@gmail.com, [e]venkateswari2140@gmail.com

DOI: 10.1201/9781003685364-14

of all DDoS attacks that were recorded. Companies spent more than $3.9 billion on IoT safety precautions in 2023 according to Gartner forecasts [5] highlighting the urgent need for efficient detection systems. In 2022, the Indian Cyber Crisis Response Team found that over 1.4 million attacks on cybersecurity in India targeted IoT devices [6]. According to these data, in order to protect against IoT security threats, rapid, innovative solutions such as the proposed hybrid machine learning model are needed.

Literature survey

Deep neural networks and machine learning techniques have been extensively researched to counter the threat of botnet attacks in IoT scenarios. A machine learning method for spotting malware networks in network traffic is presented by Salih et al. Support vector machines (SVMs) and LR techniques are used in the method which shows promise in identifying botnet attacks [7]. The study underlines how crucial it is to include automated learning into security of networks. Several feature selection strategies for botnet identification are examined in this study by Baruah et al. Feature selection is an effective technique to improve classification accuracy while using less storage and processing power. Its uncertain polynomial-time hardness is well-known, nevertheless [8]. The feature selection techniques [9] employed in neural network-based botnet detection models are thoroughly understood in this paper which also offers suggestions for enhancing these methods and points out areas in need of development.

Jones et al. investigated the use of convolutional neural networks (CNNs) and recurrent neural networks (RNN) for identifying botnet operations on IoT networking. The models are more accurate and precise than conventional techniques [10]. Models should be adapted for various IoT applications in future studies. Using multi-layer neurons and LSTM auto encoders, Ali et al. present an advanced learning hybrid method for detecting assaults in IoT systems [11]. In tests on two large datasets, the model demonstrated impressive accuracy for detecting botnets of 99.77% and 99.67%, illustrating its potential for internet of things security. Increasing cyberattacks and transforming automation are caused by the IoT, according to Han et al. This study suggests a methodology for analyzing IoT device communications that makes use of ensemble methods and

supervised learning [12]. It increases reliability by up to 1.7% and is adaptable to common IoT network analysis conditions. The LSTM models especially have become more popular due to their ability to detect time patterns in network data [13]. Lagraa et al. an overview of current network security research employing graphs and graph-based representations of information and analytics is given in this survey study. It focusses on the use of graph-based representations of communication records in botnet and intrusion detection [14]. A recent study recommends a multi-stage feature selection and computational weighted technique [15] for identifying a range of security threats in IoT networks.

Data Collection and Preprocessing

In order to analyze the activity occurring on connected devices and detect botnet attacks our study employed the UNSW_NB15 collection as a comprehensive benchmark [16]. Developing and testing predictive algorithms for botnet identification is made easier with UNSW_NB15's network activity profile. Converting all numerical attributes into a consistent range of 0 to 1 improved model performance while maintaining links between data points. Normalization increased training resolution rates and guaranteed optimal performance of distance-based algorithms like KNN and DT by scaling down big numerical quantities [17].

A feature encoding approach was used to manage categorical data in UNSW_NB15 dataset (Figure 14.1). Depending on the circularity of the categorical variables, a combination of OneHot and label encoding [18] was used. Through these encoding techniques, the models were capable of analyzing the data efficiently without adding skews to the data while retaining its semantic meaning. The most appropriate characteristics for botnet detection were found using the SelectKBest approach in combination with the chi-squared (Chi^2) statistical analysis. To maintain high-quality data for machine learning algorithms, SelectKBest's top features ensured that harmful and legal traffic could be separated from each other.

A column transformer was used to perform particular transformations to several subsets of columns at once in order to speed up the preprocessing workflow. In order to further purify the information before putting it into the LSTM model the StandardScaler

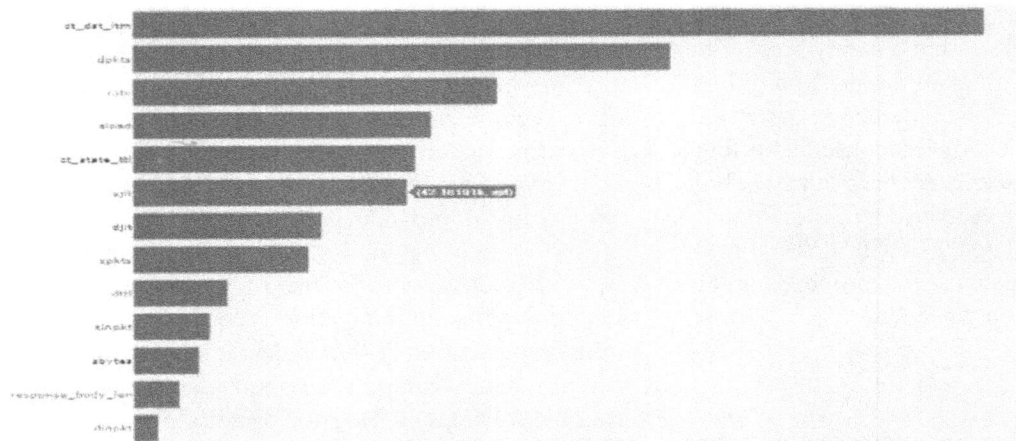

Figure 14.1 Top features of UNSW_NB15 dataset
Source: Author

[19] was included as the last preprocessing step. In order to maximize the data for gradient-based optimization approaches that are essential to deep learning systems, standard scaler standardized features by taking the mean and scaling these to unit variance. This confirmed that the preprocessing process was successful in getting the UNSW_NB15 dataset ready for accurate botnet attack detection.

Principles and Methods

Random Forest

The RF method of ensemble learning uses many decision trees to improve the general accuracy of machine learning models particularly in tasks like classification and regression. RF's ability to handle large, complex datasets such as the UNSW_NB15 dataset and its resistance to overfitting make it a valuable tool for detecting botnet attacks. By integrating the predictions of several decision trees RF lowers the risk often associated with individual trees while generating more accurate and dependable results.

One of RF main benefits is its feature importance analysis capability which helps in determining which variables have the biggest influence on the forecast and offers insights into how botnet traffic behaves. This is done by a technique known as feature bagging in which a random selection of features is used to train each tree in the forest. In real world IoT scenarios, where network traffic is frequently noisy and contains useless information RF is perfect since it is less dependent on noisy data and outliers. The model's performance in this field is demonstrated by its

high 95% detection accuracy of botnet traffic in this investigation (Figure 14.2).

Extremely randomized trees

Similar to RF the ensemble learning method Extra Trees makes predictions using numerous DT but it builds these trees differently. The decision trees of Extra Trees are constructed with a greater degree of randomization in both feature selection and feature split point determination. The major advantage of Extra Trees resides in its performance and speed particularly when dealing with massive data sets. ET remarkable 94.85% detection accuracy of botnet activity in this investigation shows how well it can differentiate between benign and harmful IoT network data.

Decision Tree

A DT is a simple ML technique that iteratively splits the information set into subsets based on feature values, ultimately forming a tree-like structure where each node represents a feature and each branch indicates a decision rule. DT can be very helpful in the area of botnet identification since they provide straightforward criteria for dividing network information into two categories: benign or malicious. DT is one of the foundational models for identifying botnets in this study identified malicious IoT traffic with a reasonable accuracy of 93.69%.

Multilayer perceptron

A feedforward ANN having several layers of neurons each completely connected to the next is called a

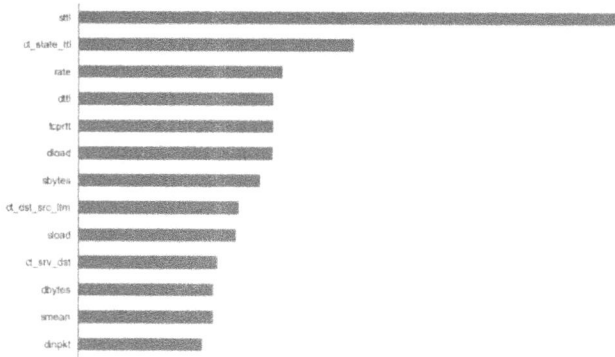

Figure 14.2 Features are the importance of random forest
Source: Author

MLP. MLPs are strong machine learning algorithms that are especially well-suited for applications like botnet identification in Internet of Things networks because they can capture complex non-linear correlations between input variables and target outputs. In this study MLP demonstrated its ability to discern harmful and benign traffic by identifying botnet incidents in the UNSW_NB15 dataset with an accuracy of 93.44%.

Gradient Boosting
A sophisticated ensemble learning technique called gradient boosting builds models one after the other with every additional model seeking to correct the flaws of the one before it. Gradient Boosting works especially well for botnet identification because it can identify complex trends in data from network traffic. Gradient Boosting showed its efficacy in IoT safety applications in this study by detecting botnet attacks within the UNSW_NB15 dataset with an accuracy of 93.15%.

K-nearest neighbors
K-nearest neighbors (KNN) is a straightforward non-parametric machine learning technique that clusters data points in the feature space based on the vast majority class that are their closest neighbors. KNN is highly useful for botnet identification since it can handle both categorical and numerical information and is adept at spotting outliers which are often indicators of illicit botnet activity. Furthermore, KNN accuracy and performance can be affected by the parameter k and the distance metric (such as Manhattan or Euclidean) that are used. Considering

these drawbacks KNNs 92.91% accuracy rate in the study showed how useful it is for identifying botnet attacks in IoT traffic.

Logistic regression classifier
Logistic regression is a statistical method for tasks involving binary classification that predicts the likelihood that a given instance will fall into a particular class by examining the relationship between the input information and the desired variable using a function called logistic. Although it placed behind alternative models like RF and ET logistic regression demonstrated its capacity to differentiate between harmful and genuine traffic in this study by detecting network activities within the UNSW_NB15 dataset with an accuracy of 91.07%.

Long short-term memory
LSTM is a unique type of recurrent neural network (RNN) designed to handle sequential data is incredibly helpful for applications needing analysis of time-series or additional information where the order of inputs matters. Compared to standard RNNs, which struggle with reliance over time due to issues like gradient vanishing, LSTMs are ideal for seeing patterns in sequential data, such as IoT network traffic since they feature a memory cell that can hold information for a longer period of time. A number of essential elements make up the LSTM design which sets it apart from conventional RNNs. It has three entry points: the input gate, the forget gate and the output gate.

In order to help with prediction based on learnt features LSTM is sometimes integrated by adding extra layers such as a dense output layer for classification. The model's capacity to identify botnet attacks can be further improved by adding dropout regularity or bidirectional LSTMs which gather data from both previous and next steps. The LSTM model showed remarkable performance in this study with a training accuracy of 96.665% and a Testing Accuracy of 96.435% (Figure 14.3).

Results

The efficiency of the strategy was demonstrated by the remarkable outcomes of the machine learning algorithms used in the UNSW_NB15 dataset for botnet attack detection. RF had the best accuracy of 95% out of all the machine learning models that were evaluated. ET, DT and MLP came in second

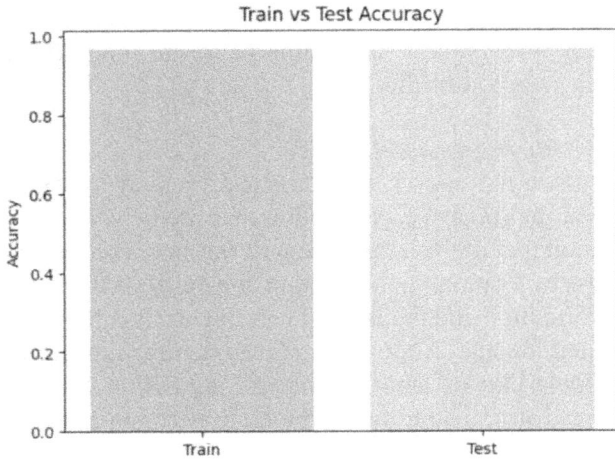

Figure 14.3 Training and testing accuracy of LSTM
Source: Author

Table 14.1 Evaluation metrics of various models.

Model	Accuracy	Precision	Recall	F1-score
Random Forest	95.0%	95.01%	95.00%	95.01%
Extra Trees	94.48%	94.87%	94.85%	94.86%
Decision Tree	93.69%	93.69%	93.69%	93.69%
MLP	93.44%	93.43%	93.44%	93.43%
Gradient Boosting	93.15%	93.14%	93.15%	93.14%
KNN	92.91%	92.92%	92.91%	92.91%
Logistic Regression	91.07%	91.19%	91.07%	92.91%
LSTM	96.43%	96.44%	96.43%	96.44%

Source: Author

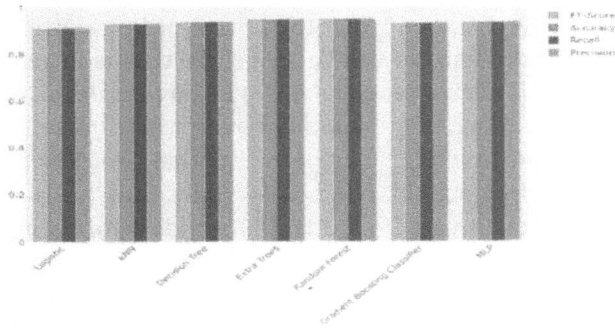

Figure 14.4 Comparison of machine learning classifier
Source: Author

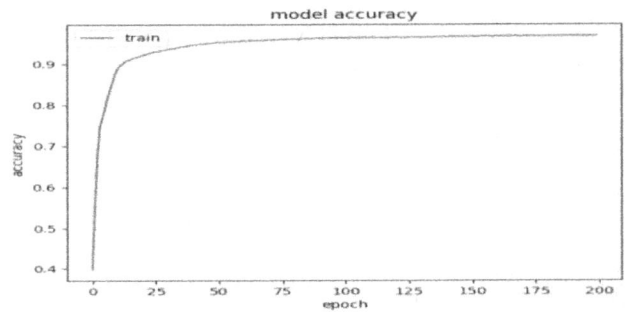

Figure 14.5 LSTM accuracy plot vs epochs
Source: Author

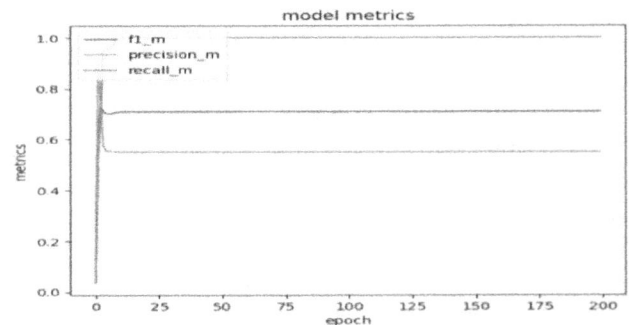

Figure 14.6 Evaluation metrics of LSTM model
Source: Author

and third respectively with 94.85%, 93.69%, and 93.44%. The best accuracy was 93.15% for Gradient Boosting 92.91% for KNN, and 91.07% for LR. But with a remarkable training performance of 96.665% and testing accuracy of 96.435% LSTM model outperformed the others by significantly making it the best model for identifying botnet attacks in IoT situations (Figure 14.4).

The best-performing approach LSTM has been further verified for its improved performance by the confusion matrix (Table 14.1). While the false positive rate was comparatively low ensuring that real traffic was not mistakenly divided as malicious the matrix shows a high true positive rate suggesting that the model successfully detected botnet attacks (Figures 14.5 and 14.6). Because it performed well in both the training and testing phases and produced few wrong classifications the LSTM model

is an accurate option for actual time botnet identification. Moving on to real-time analysis the botnet detection system gained greatly from the deployment of a Streamlit-based user interface which offered a platform for rapid and dynamic analysis. When users upload CSV files with network traffic data the system immediately analyses the data and applies the learnt models to identify possible botnet attacks and give alerts if attack is detected. As a result, the system is ideal for implementing real-time IoT scenarios that require prompt detection and reaction.

Safety professionals and administrators can easily get involved since the system provides instant results. The real-time system not only classifies observed botnet activity but also offers preventive measures. The system recommends particular steps to reduce the threat if an automated network attack is identified such blocking dubious IP addresses, screening traffic or looking into unusual patterns further. In order to prevent possible harm from evolving botnet attacks the real-time monitoring module considers both the current network context and the traffic pattern over time. This feature makes the system more valuable by providing both detection and recommendation of preventative measures and alerts to secure IoT networks. The preventive module functions in a manner that smoothly combines with the network architecture. It is possible to implement these preventative measures in real-time by interacting with network devices such as intrusion prevention systems (IPS) or firewalls. With the help of streamlit's user-friendly interface users may upload data files and get results right away in a simple and efficient manner (Figure 14.7).

In situations where time is of importance this usability is especially crucial. Detailed visualizations of detection results such as graphs displaying traffic patterns, attack categories (Figure 14.8), give alerts and recommended preventive measures are available to administrators. With a more thorough grasp of traffic flow and any hazards made possible by this degree of visualization users are better equipped to act swiftly and decisively. The immediate evaluation and prevention method created in this study along with the hybrid machine learning strategy offers an effective treatment for botnet attack detection in IoT settings (Figure 14.9).

Without the need for human intervention this automation guarantees that the IoT system is safe from fraudulent activity and speeds up response times. In addition to monitoring the system's performance, administrators can also get regular updates on attacks that have been identified and prevented.

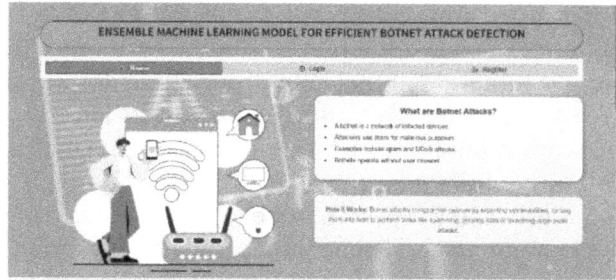

Figure 14.7 Streamlit user interface
Source: Author

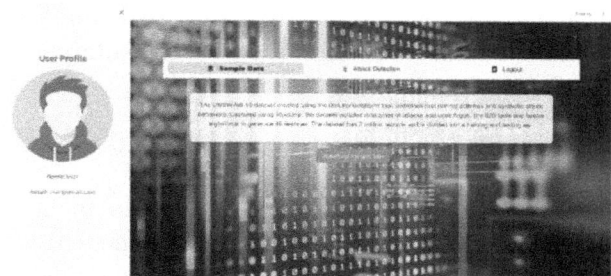

Figure 14.8 Sample data of CSV for attack detection
Source: Author

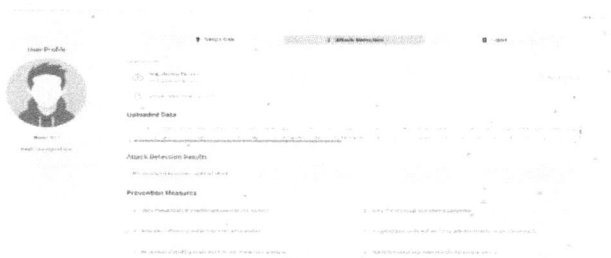

Figure 14.9 Botneck attack detection
Source: Author

Conclusion

This study provides a sophisticated hybrid machine learning framework for efficient botnet attack detection in Internet of Things (IoT) environments by utilizing a range of models, such as Random Forest, Extra Trees, Decision Tree, multilayer perceptron, Gradient Boosting K-nearest neighbors, logistic regression, and long short-term memory (LSTM). It highlights particularly on LSTM's exceptional ability to record time changes in network activity data. The LSTM model outperforms other models in identifying subtle patterns of botnet attacks on the Internet of Things with remarkable accuracy. With the help of real-time detection technology and an intuitive streamlit interface network traffic can be processed and evaluated instantly giving managers useful information, giving alerts and preventative actions to reduce dangers. With its real-time botnet attack detection, prevention, and strong accuracy, the system is an excellent tool for protecting IoT networks from ongoing cyber threats. The promise of machine learning and deep learning to improve network security strategies is exemplified by this hybrid strategy, which when coupled with real-time prevention capabilities offers a holistic solution to cybersecurity issues in IoT.

Future Scope

The potential extension of this research includes using federated learning and reinforcement learning approaches for decentralized and secure detection in order to improve the model's flexibility to changing botnet assault patterns. The accuracy and responsiveness of the system can also be increased by including anomaly detection techniques and real-time threat intelligence feeds. The framework's applicability in various IoT scenarios will be strengthened by adding support for multi-modal data sources, such as secure connections and behavioral analytics.

References

[1] Atzori, L., Iera, A., & Morabito, G. (2017). Understanding the internet of things: definition, potentials, and societal role of a fast evolving paradigm. *Ad Hoc Networks,* 56, 122–140.

[2] Moustafa, N., & Slay, J. (2016). The evaluation of network anomaly detection systems: statistical analysis of the UNSW-NB15 data set and the comparison with the KDD99 data set. *Information Security Journal: A Global Perspective,* 25(1-3), 18–31.

[3] Ilham, M. I. A., Siswanto, T., & Sugiarto, D. (2024). Design of study program performance dashboard using streamlit. *Intelmatics,* 4(2), 96–100.

[4] Stellios, I., Kotzanikolaou, P., Psarakis, M., Alcaraz, C., & Lopez, J. (2018). A survey of iot-enabled cyberattacks: assessing attack paths to critical infrastructures and services. *IEEE Communications Surveys and Tutorials,* 20(4), 3453–3495.

[5] Krishnamoorthy, G., Sistla, S. M. K., Venkatasubbu, S., & Periyasamy, V. (2024). Enhancing Worker Safety in Manufacturing with IoT and ML. International Journal For Multidisciplinary Research, 6(1), 1–11.

[6] Zia, H., Imran, R., Masood, R., & Shibli, M. A. (2017). Framework for the development of computer emergency response team in Pakistan. *NUST Journal of Engineering Sciences,* 10(2), 65–71.

[7] Salih, Y. T., Fenjan, A., Ahmed, S. R., Ali, H., Abdulwahab, E. N., Algruri, S., et al. (2024). Machine learning approaches for botnet detection in network traffic. In Proceedings of the Cognitive Models and Artificial Intelligence Conference.

[8] Baruah, S., Borah, D. J., & Deka, V. (2024). Reviewing various feature selection techniques in machine learning-based botnet detection. *Concurrency and Computation: Practice and Experience,* 36(12), e8076.

[9] Theng, D., & Bhoyar, K. K. (2024). Feature selection techniques for machine learning: a survey of more than two decades of research. *Knowledge and Information Systems,* 66(3), 1575–1637.

[10] Jones, R. K. (2024). Enhancing IoT security: leveraging advanced deep learning architectures for proactive botnet detection and network resilience. In Redefining Security With Cyber AI. IGI Global, (pp. 56–71).

[11] Ali, S., Ghazal, R., Qadeer, N., Saidani, O., Alhayan, F., Masood, A., et al. (2024). A novel approach of botnet detection using hybrid deep learning for enhancing security in IoT networks. *Alexandria Engineering Journal,* 103, 88–97.

[12] Han, S.-J., Yoon, S. S., & Euom, I. C. (2024). The Machine learning ensemble for analyzing internet of things networks: botnet detection and device identification. *CMES-Computer Modeling in Engineering and Sciences,* 141(2), 1495–1518.

[13] Al-Selwi, S. M., Hassan, M. F., Abdulkadir, S. J., Muneer, A., Sumiea, E. H., Alqushaibi, A., et al. (2024). RNN-LSTM: from applications to modeling techniques and beyond—systematic review. *Journal of King Saud University-Computer and Information Sciences,* 36(5),102068.

[14] Lagraa, S., Husák, M., Seba, H., Vuppala, S., State, R., & Ouedraogo, M. (2024). A review on graph-based approaches for network security monitoring and botnet detection. *International Journal of Information Security,* 23(1), 119–140.

[15] Krishnan, D., & Shrinath, P. (2024). Enhancing energy efficiency and imbalance handling in botnet detection in IoT networks: a multi-stage feature reduction and weighted approach. *International Journal of Information Technology,* 17, 1–12.

[16] Jouhari, M., Benaddi, H., & Ibrahimi, K. (2024). Efficient intrusion detection: combining X 2 feature selection with CNN-BiLSTM on the UNSW-NB15 dataset. In 2024 11th International Conference on Wireless Networks and Mobile Communications (WINCOM). IEEE.

[17] Fujita, T. (2024). A review of fuzzy and neutrosophic offsets: Connections to some set concepts and normalization function. *Advancing Uncertain Combinatorics through Graphization, Hyperization, and Uncertainization: Fuzzy, Neutrosophic, Soft, Rough, and Beyond*, 74.

[18] Williams, C. K. I. (2024). Naive bayes classifiers and one-hot encoding of categorical variables. arXiv preprint arXiv:2404.18190.

[19] Baldini, Gianmarco. (2024). Mitigation of Adversarial Attacks in 5G Networks with a Robust Intrusion Detection System Based on Extremely Randomized Trees and Infinite Feature Selection. Electronics 13.12 (2024): 2405.

15 X-AI enabled hybrid approach for detection of cyber terrorism

Kondanna Kanamaneni[1,a], Keswitha Pucha[2,b], Asha Bai Peesay[2,c], Kranthi Vajram[2,d] and Naresh Thikkolla[2,e]

[1]Assistant Professor, Department of CSD, Srinivasa Ramanujan Institute of Technology, Anantapur, Andhra Pradesh, India

[2]Students, Department of CSD, Srinivasa Ramanujan Institute of Technology, Anantapur, Andhra Pradesh, India

Abstract

In the present quickly progressing mechanical scene, digital psychological oppression addresses a developing danger to worldwide security and cultural trustworthiness. This paper presents a X-simulated intelligence driven half breed model pointed toward working on the recognition and avoidance of digital psychological oppression exercises. By consolidating state of the art man-made consciousness procedures with conventional online protection strategies, the model endeavors to lay out a complete framework fit for distinguishing and tending to digital dangers continuously. The proposed system uses AI calculations, including profound learning and gathering techniques, to filter through huge datasets and recognize designs that recommend digital psychological oppressor movement. Besides, the cross-breed model consolidates inconsistency discovery strategies to signal sporadic ways of behaving that could show potential digital assaults. Intended to advance and adjust, the framework ceaselessly gains from approaching information, guaranteeing that it stays viable despite developing digital dangers. We exhibit the adequacy of this methodology through exhaustive testing on deeply grounded benchmark datasets, showing predominant exactness and less bogus up-sides contrasted with existing location frameworks. The outcomes feature the huge capability of X-simulated intelligence in reinforcing online protection measures against digital psychological oppression. This exploration adds to the more extensive network safety field and offers significant bits of knowledge for associations intending to support their statement recognition capacities.

Keywords: AI, danger discovery, digital psychological warfare, gathering procedures, half and half model, network protection, peculiarity location, profound learning, X-artificial intelligence

Introduction

In the present quickly progressing mechanical climate, digital psychological oppression has turned into an undeniably serious worry for worldwide security, public safeguard, and cultural strength. Digital psychological oppression, characterized by malignant exercises pointed toward hurting through computerized implies, represents a rising danger to states, basic framework, organizations, and people the same [1]. Dissimilar to conventional illegal intimidation, digital psychological warfare works inside advanced space, permitting aggressors to exploit secrecy and worldwide arrive at given by the web. This remarkable trademark makes it particularly challenging to distinguish, foresee, and forestall, as culprits can execute refined assaults from anyplace on the planet, frequently with negligible assets. Given these difficulties, there is a critical requirement for cutting-edge techniques for digital danger discovery. While customary network protection techniques are successful much of the time, they frequently miss the mark intending to the dynamic and developing nature of digital dangers, especially those coordinated by vindictive entertainers [2], for example, digital fear-based oppressors. Because of this hole, there has been developing interest in applying man-made reasoning (computer-based intelligence) and AI strategies to upgrade network safety, especially in the recognition of digital psychological oppression. Man-made intelligence holds extraordinary commitment in computerizing danger recognizable proof, handling huge volumes of information progressively, and adjusting to arising assault designs. Notwithstanding, carrying out artificial intelligence-based approaches in network safety, particularly for distinguishing digital psychological

[a]kondanna14@gmail.com, [b]p.keswitha@gmail.com, [c]peesayashabai@gmail.com,
[d]vajramkranthi2003@gmail.com, [e]thikkollanaresh@gmail.com

DOI: 10.1201/9781003685364-15

oppression, requires an essential methodology that tends to the intricacy and eccentricism of ill-disposed conduct. Current models frequently face difficulties, for example, high bogus positive rates, slow identification times, and a powerlessness to adjust rapidly to developing assault techniques. This paper proposes an inventive mixture model empowered by Reasonable simulated intelligence (X-man-made intelligence) for distinguishing digital psychological warfare exercises, consolidating state of the art simulated intelligence procedures with customary network protection strategies to make a vigorous, ongoing discovery framework fit for recognizing both known and new dangers. The proposed half breed model coordinates a few high-level man-made intelligence techniques, including profound getting the hang of, grouping, and inconsistency identification, to shape an exhaustive structure for digital illegal intimidation location. Profound learning, a subset of AI that imitates human cerebrum capabilities, succeeds at distinguishing complex assault designs that could go undetected by ordinary frameworks. Bunching methods bunch comparative pieces of information, empowering the framework to perceive arising examples of malignant action, while irregularity recognition recognizes strange ways of behaving that might flag an assault. By utilizing these man-made intelligence strategies, the half breed model can deal with tremendous measures of information from different sources —, for example, network traffic, virtual entertainment movement, and correspondence designs — uncovering stowed away associations between apparently inconsequential snippets of data and distinguishing expected dangers. A critical benefit of the half breed model is its capacity to learn and adjust progressively persistently. As digital fear mongers continually change their strategies and methodologies, it is fundamental for location frameworks to stay adaptable and equipped for advancing. The proposed framework gains from approaching information and consistently refines its calculations, guaranteeing it can distinguish new types of digital psychological oppression as they arise [3]. This versatile capacity makes the model more successful in a scene where digital dangers are continually changing, denoting a huge improvement over customary, static decide to put together discovery frameworks that depend on respect to predefined danger marks. Besides, by consolidating both directed and unaided learning techniques,

the half breed model can recognize both known and obscure dangers, permitting it to distinguish assault vectors that might not have been caught in preparing information. All in all, the developing danger of digital psychological oppression requests creative and versatile arrangements. The proposed X-simulated intelligence empowered half breed model presents a promising strategy to counter this danger by utilizing man-made consciousness to recognize, dissect, and moderate digital psychological oppression exercises progressively. With its capacity to ceaselessly gain from information, adjust to new dangers, and limit bogus upsides, this model can possibly essentially further develop network safety frameworks and add to worldwide endeavors to battle advanced psychological warfare. As digital dangers keep on developing, it is important that network safety arrangements advance couple, guaranteeing the security of basic foundation and the safeguarding of cultural solidness notwithstanding arising computerized chances.

Related work

The raising danger of digital psychological oppression has driven broad exploration pointed toward creating progressed techniques for recognition and avoidance. In network protection, many procedures have been investigated to address the novel difficulties introduced by digital psychological warfare, including both conventional and state of the art AI methods [4]. Early location frameworks principally depended on rule-based and signature-based techniques, which matched approaching organization traffic to known danger marks. While compelling at distinguishing recently recorded assaults, these frameworks battled with identifying new or advancing dangers, making them less viable against the unique idea of digital psychological oppression [5]. Subsequently, there has been a shift towards additional versatile techniques fit for perceiving novel and beforehand obscure vindictive exercises. AI (ML and DL) have become fundamental to investigate because of their capacity to break down enormous datasets and uncover complex examples that could demonstrate digital illegal intimidation exercises [6]. These strategies are especially significant for recognizing irregular ways of behaving and arising assault procedures, the two of which are normal in digital psychological warfare. A few examinations have investigated regulated learning drawings near, for example, Support Vector Machines (SVM) and irregular woodlands, to sort network traffic as either harmless or malevolent. Albeit these models can yield high exactness when prepared on marked information, they face difficulties, for example, the requirement for broad named datasets, which are in many cases inaccessible with regards to digital psychological oppression [5]. Also, these models require customary retraining to keep up with precision despite developing dangers. To address these difficulties, scientists have progressively gone to solo learning and irregular recognition procedures. These methodologies mean to distinguish surprising examples or exceptions without the requirement for marked information. Peculiarity location is especially important for recognizing digital psychological warfare, as such goes after frequently include ways of behaving that go amiss from the standard [8, 9]. Different factual, grouping, and distance-based strategies have been applied in this space, each with fluctuating achievement. Bunching calculations like k-implies and DBSCAN have been utilized to bunch comparative organization ways of behaving, considering the distinguishing proof of bizarre gatherings that could demonstrate malignant action. Nonetheless, these strategies can battle with high-layered information and the requirement for persistent transformation to arising danger designs. A crossover approach that joins customary online protection strategies with current AI and profound learning procedures has acquired prevalence as a method for beating the restrictions of individual techniques. One such methodology coordinates oddity recognition with AI classifiers, similar to choice trees or brain organizations, to refine the discovery interaction and decrease misleading up-sides. Furthermore, including designing can be utilized to work on the exhibition of AI models, for example, by removing network traffic attributes or recognizing dubious client ways of behaving that could show digital illegal intimidation. Half and half models frequently consolidate progressed methods like outfit realizing, which totals various models to improve recognition exactness. Group techniques, for example, helping, stowing, and stacking, have demonstrated success in network safety by working on the vigor and dependability of danger recognition frameworks. By consolidating forecasts from different models, group techniques decrease the gamble of overfitting and increase the framework's speculation capacity, making it stronger to the always changing

nature of digital psychological oppression dangers. Continuous location is basic in fighting digital psychological warfare, as the effect of these assaults can be extreme. To address this need, numerous scientists have grown ongoing danger identification frameworks that constantly screen network traffic and client action. These frameworks frequently utilize streaming examinations to handle information continuously, empowering quick location and reaction to digital psychological oppression exercises [6].

Proposed System Workflow

The proposed framework for identifying digital illegal intimidation uses a cross-breed approach that consolidates a few high-level man-made intelligence methods to convey a superior exhibition, versatile arrangement. The cycle starts with the assortment and preprocessing of different online protection information, including network traffic, framework logs, and client action logs, which are inspected for strange examples that could demonstrate digital illegal intimidation. To deal with this information effectively, the framework integrates profound learning models like Bidirectional Encoder Portrayals from Transformers (BERT), long short-term memory (LSTM), and gated intermittent unit (GRU), alongside gathering strategies like irregular woods and logical simulated intelligence (LIME). The work process begins with information securing, where crude information is gathered from various sources. The gathered information is then preprocessed to wipe out commotion and unessential data, guaranteeing that the AI models get perfect, applicable sources of info [7]. Text information, for example, correspondence logs and online action reports, are explicitly handled utilizing BERT, which succeeds at figuring out relevant connections and distinguishing complex examples demonstrative of malignant exercises. With its transformer engineering, BERT recognizes dubious semantic highlights that may be connected to digital psychological oppression, like extremist language or facilitated arranging. Subsequent to preprocessing and highlight extraction, transient information, for example, login designs, client ways of behaving, and network exercises, is dissected utilizing LSTM and GRU models. These repetitive brain organizations (RNNs) are intended to distinguish long haul conditions in successive information, making them ideal for perceiving time sensitive examples in digital assaults. LSTM is especially powerful at taking care of long groupings and staying away from the evaporating slope issue, while GRU improves on the succession growing experience, zeroing in on key viewpoints with less boundaries. Together, LSTM and GRU assist with following client ways of behaving and recognize strange successions that could facilitate digital psychological oppression. In equal, Irregular Woods, a troupe strategy, is utilized for its heartiness in handling assorted highlights and lessening overfitting. This interpretability is imperative in true applications, as it empowers examiners to refine and approve the framework's forecasts. This exhaustive methodology guarantees precise identification as well as gives significant bits of knowledge to network safety experts, upgrading the strength of computerized frameworks against arising dangers.

Loading dataset
The underlying stage in fostering a compelling half breed model for digital psychological oppression identification includes assembling and stacking significant datasets. These datasets regularly incorporate many information types, for example, network traffic, logs, and online protection reports, that incorporate both genuine exercises and potential digital psychological oppression related occurrences. Information is frequently obtained from different stages like interruption identification frameworks (IDS), firewalls, malware examination apparatuses, and danger insight stages. The datasets may likewise incorporate information from honeypots and occurrence reports that catch dubious or noxious exercises connected to fear monger gatherings or cybercriminals. These information sources are incorporated and collected to give an extensive perspective on digital dangers, making it simpler to dissect. The information might contain both named (e.g., known occurrences of digital psychological warfare like DDoS assaults or phishing efforts) and unlabeled data (e.g., unclassified organization exercises). It is fundamental to guarantee that the dataset incorporates a different scope of assault types, for example, web application takes advantage of, malware, and phishing assaults, while being adjusted to keep away from model inclination. Preprocessing techniques like component extraction and standardization are applied at this stage to set up the information for investigation.

Preprocessing
Information preprocessing is basic for guaranteeing the nature of the dataset prior to utilizing it to recognize digital psychological oppression exercises. Crude information, for example, logs and organization traffic, frequently contain blunders like missing qualities, commotion, and irregularities that could affect the model's exhibition. The first preprocessing step includes cleaning the information to eliminate or address abnormalities, for example, missing qualities, which are either attributed or taken out relying upon the dataset's necessities. Dimensionality decrease procedures, as PCA or highlight designing, might be utilized to smooth out the dataset and work on model productivity. Highlight scaling techniques, like standardization or normalization, are likewise applied to guarantee that all elements are on a comparable scale and don't mutilate the learning calculation. Bunching methods, similar to K-implies, are utilized to recognize typical and strange ways of behaving in the dataset. Since digital illegal intimidation occasions are normally interesting however significant, procedures like oversampling, under sampling, or destroyed are utilized to address class irregularity and guarantee the model actually distinguishes intriguing assault cases without overfitting. To upgrade the model's power, information expansion techniques, such as adding commotion or recreating assaults, are utilized to expand the assortment and amount of the preparation information [8].

Model training and classification
After information stacking and preprocessing, the following stage includes preparing and grouping the model, where AI and man-made intelligence techniques are applied to recognize designs related with digital psychological oppression [9]. The cross-breed model joins conventional oddity location strategies with cutting edge man-made intelligence calculations, like profound learning and grouping techniques. First, the significant elements that address network traffic conduct are extricated. For administered learning, named information is utilized to prepare models with normal calculations like arbitrary woods, SVM, or brain organizations. In situations where the information needs names, unaided strategies, for example, K-implies or DBSCAN are applied to distinguish examples and gathering comparative assault types. Peculiarity location models are additionally used to hail surprising exercises that

could mean digital psychological warfare. The model is prepared to recognize typical and malignant traffic by breaking down highlights like strange examples, payload abnormalities, or dubious login endeavors [10]. The CNN and RNN models can learn complex worldly examples and naturally separate progressive elements, making them powerful for identifying advancing digital illegal intimidation dangers. The preparation interaction incorporates tuning hyperparameters (e.g., learning rate, regularization) to enhance model execution. Assessment measurements like exactness, accuracy, review, F1-score, and the ROC bend are utilized to evaluate the model's viability, limiting misleading up-sides and negatives.

Methodology

Gaussian Naive Bayes

Definition: Gaussian Naive Bayes is a probabilistic classifier based on Bayes' Theorem, which assumes that the features are independent of one another. It computes the probability of each class given the input features and assigns the class with the highest probability. It is known for its simplicity and efficiency, especially in tasks like text classification and spam filtering. Figure 15.1 and 15.2.

Random Forest

Bootstrap examining: Irregular backwoods utilizes bootstrap inspecting, where various subsets of the information are made by testing with substitution. Each tree is prepared on an alternate subset.

Irregular component determination: While dividing hubs, Irregular Woodland considers just an arbitrary subset of elements, which decreases the relationship among trees and improves model variety [11].

Bidirectional encoder representations from transformers

Definition: BERT is a transformer-based model created by Google to upgrade regular language understanding. It utilizes bidirectional consideration, empowering it to handle the whole setting of a sentence all the while, which is especially helpful for understanding equivocal words in light of encompassing words. Table 15.1, 15.2, 15.3 and 15.4.

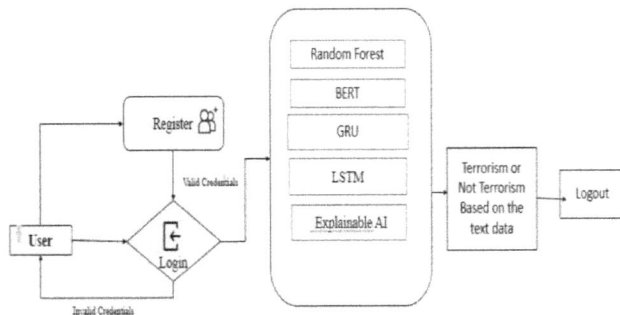

Figure 15.1 Block flowchart of cyber terrorism
Source: Author

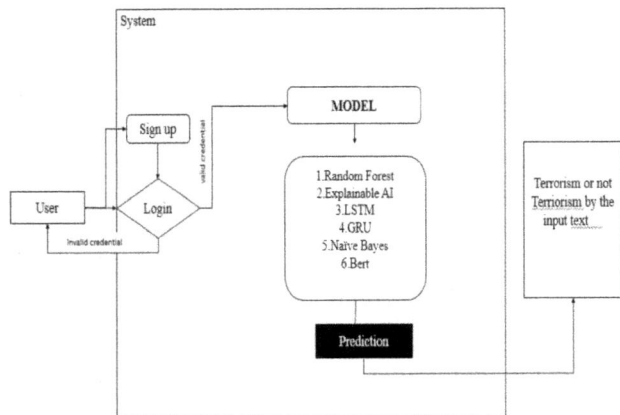

Figure 15.2 System architecture of cyber terrorism
Source: Author

Table 15.1 Classification report for Naïve Bayes.

_precision	recall	f1-score	support
0	0.72	0.8	0.76	898
1	0.81	0.73	0.77	1017
accuracy			0.76	1915
macro avg	0.76	0.76	0.76	1915
weighted avg	0.77	0.76	0.76	1915

Source: Author

Table 15.2 Classification report of Random Forest.

_precision	recall	f1-score	support
0	1	1	1	898
1	1	1	1	1017
accuracy			1	1915
macro avg	1	1	1	1915
weighted avg	1	1	1	1915

Source: Author

Table 15.3 Classification report of BERT.

Class_Precision	Recall	F1-Score	Support
0	0.99	0.98	0.98	6699
1	0.96	1	0.98	3509
Accuracy				10208
Macro avg	0.98	0.99	0.98	10208
Weighted avg	0.98	0.98	0.98	10208

Source: Author

Table 15.4 Classification report of LSTM.

Class_Precision	Recall	F1-score	Support
0	1	0.98	0.99	898
1	0.98	1	0.99	1017
Accuracy				1915
Macro avg	0.99	0.99	0.99	1915
Weighted avg	0.99	0.99	0.99	1915

Source: Author

LSTM (long short-term memory networks)
Definition: LSTM is a kind of repetitive brain organization (RNN) intended to take care of the issue of long haul reliance in customary RNNs. It utilizes memory cells to store data overstretched periods, making it profoundly powerful for time series, language demonstrating, and consecutive errands.

Gated recurrent unit
Definition: GRU is an improved-on variation of LSTM, intended to conquer the evaporating slope issue in RNNs while utilizing less boundaries. It is appropriate for assignments like language interpretation, discourse acknowledgment, and time series determining.

Random Forest with explainable AI (X-AI)
Definition: Logical artificial intelligence (X-computer based intelligence) intends to make AI models more straightforward and interpretable. On account of Arbitrary Backwoods, X-man-made intelligence strategies like SHapley Added substance Clarifications (SHAP) and LIME (Neighborhood Interpretable Model-skeptic Clarifications) assist with making

Table 15.5 Classification report of GRU.

Class_Precision	Recall	F1-Score	Support
0	1	1	1	898
1	1	1	1	1017
Accuracy				1915
Macro avg	1	1	1	1915
Weighted avg	1	1	1	1915

Source: Author

Table 15.1 sense of the impact of each element on the model's expectations, further developing trust and comprehension of the model [12].

Conclusion

This study features the groundbreaking capability of X-man-made intelligence in battling digital psychological warfare. The cross-breed model, which combines the qualities of AI calculations with customary network safety strategies, gives a versatile and profoundly successful answer for distinguishing and alleviating digital dangers. Utilizing profound learning and bunching strategies, the framework processes huge datasets and recognizes noxious examples that would somehow be disregarded, guaranteeing the convenient distinguishing proof of digital psychological militant exercises. The model's capacity to consistently gain from approaching information improves its ability to distinguish new and advancing dangers. Also, its coordination of abnormality discovery reinforces its capacity to distinguish beforehand inconspicuous digital psychological warfare strategies. Testing results affirm that the X-artificial intelligence half breed approach outperforms existing network protection frameworks with regards to precision and limiting misleading up-sides. This framework is especially basic in a period where digital dangers are turning out to be progressively perplexing and successive. The discoveries highlight the capability of simulated intelligence driven security structures in building up worldwide network safety frameworks and shielding society from the developing danger of digital psychological warfare. While the outcomes are promising, the unique idea of digital dangers requires continuous innovative work to refine the model and extend its viable use.

Acknowledgement

The authors gratefully acknowledge the students, staff and authority of CSE department for their cooperation in the research.

References

[1] Amin, S., Litrico, X., Sastry, S. S., & Bayen, A. M. (2013). Cyber security of water scada systems-part II: attack detection using enhanced hydrodynamic models. *IEEE Transactions on Control Systems Technology*, 21(5), 1679–1693.

[2] Itasoy, E., Rosenberg, V., Stavrakis, N., Dietrich, A., & Montanari, C. (2024). Ransomware detection on windows using file system activity monitoring and a hybrid isolation forest-XGBoost model. https://www.researchgate.net/publication/384985749_Ransomware_Detection_on_Windows_Using_File_System_Activity_Monitoring_and_a_Hybrid_Isolation_Forest-XGBoost_Model

[3] Jeyapriyanga, S., Ravi, C. N., Rathiya, R., Kalaivani, K., Rama Devi, C., & Kumar, K. R. (2023). Implementation of a deep learning framework for intelligent intrusion detection in internet of things networks. In Proceedings of the 5th International Conference on Inventive Research in Computing Applications, ICIRCA 2023, (pp. 1208–1213).

[4] Kalaria, R., Kayes, A. S. M., Rahayu, W., Pardede, E., & Salehi, S. A. (2024). IoTPredictor: a security framework for predicting IoT device behaviours and detecting malicious devices against cyber attacks. *Computers and Security*, 146, 104037.

[5] Kalimuthan, C., & Arokia Renjit, J. (2020). Review on intrusion detection using feature selection with machine learning techniques. *Materials Today: Proceedings*, 33, 3794–3802. https://doi.org/10.1016/J.MATPR.2020.06.218.

[6] Liasi, S., Ghiasi, N. S., & Hadidi, R. (2023). Smart grid vs. intelligent grid; artificial intelligence takes the power system to the next level. In 2023 IEEE 3rd International Conference on Digital Twins and Parallel Intelligence, DTPI 2023.

[7] Li, B., Shi, Y., Kong, Q., Zhai, C., & Ouyang, Y. (2021). Honeypot-enabled optimal defense strategy selection for smart grids. In Proceedings, IEEE Global Communications Conference, GLOBECOM.

[8] Patel, N., Ramoliya, F., Jadav, N. K., Gupta, R., Tanwar, S., & Aujla, G. S. (2024). X-NET: explainable AI-based network data security framework for healthcare 4.0. In 2024 IEEE International Conference on Communications Workshops, ICC Workshops 2024, (pp. 481–486).

[9] Presekal, A., Jorjani, M., Rajkumar, V. S., Goyel, H., Cibin, N., Semertzis, I., et al. (2024). Cyber security of HVDC systems: a review of cyber threats, defense, and testbeds. *IEEE Access*. https://research.tudelft.nl/en/publications/cyber-security-of-hvdc-systems-a-review-of-cyber-threats-defense-

[10] Shahidinejad, A., & Abawajy, J. (2024). Blockchain-based self-certified key exchange protocol for hybrid electric vehicles. *IEEE Transactions on Consumer Electronics*, 70(1), 543–553.

[11] Vadivelan, N., Bhargavi, K., Kodati, S., & Nalini, M. (2022). Detection of cyber attacks using machine learning. In AIP Conference Proceedings, (Vol. 2405, no 1).

[12] Wu, Z., Zhang, H., Wang, P., & Sun, Z. (2022). RTIDS: a robust transformer-based approach for intrusion detection system. *IEEE Access*, 10, 64375–64387.

16 Solar smart track: intelligent dual-axis tracking with fault detection for enhanced performance

R. Mahesh Kumar[a], *M. Ravikishore*[b], *S. Fatima Jabeen*[c], *K. Anitha Reddy*[d], *C. Bharath Kumar*[e] *and P. Hari Krishna*[f]

Annamacharya University, Rajampet, Andhra Pradesh, India

Abstract

Solar panels have gained recognition as solar energy generators for creating electricity during the last few years. Solar panels keep a fixed position during the daily solar movement pattern because system orientation influences both efficiency and battery charge times in standalone power systems. This project develops a modern dual-axis solar tracker system through implementation of IoT technology which improves solar power generation capabilities along with solar installation dependability. The primary objective of the Solar Smart Track System focuses on optimizing solar energy efficiency through its dynamic control of solar panel directions to track the sun across daily periods. The system contains four LDR sensors which work together to guide two motors toward the best sunlight position. The charging circuit provides power to operate the system and simultaneously saves excess energy into battery storage. The Arduino microcontroller operates the system and an ESP8266 NodeMCU sends information to the UBIDOTS IoT platform through which real-time tracking and fault monitoring is possible. The improved solar energy efficiency through this system makes it suitable for remote areas that demand constant power delivery.

Keywords: Arduino, dual-axis solar tracking, ESP8266 NodeMCU, fault detection, IoT, remote monitoring, solar panel performance, UBIDOTS

Introduction

The world needs solar energy as its basic solution to stop using fossil fuels because it creates two benefits from renewable power generation. The performance potential of solar power systems depends on tracking technologies because these systems help achieve maximum operational potential. The automated positioning of dual-axis tracking solar panels tracks the solar path during daylight hours through automatic mechanism changes. Tracking systems implement different operational principles when compared to conventional fixed-tilt solar panels. Solar power efficiency achieves its peak through automatic real-time positioning as the continuous operation of solar panels enables them to maximize sunlight capture [1]. The power generating capacity of dual-axis tracking systems exceeds fixed-tilt systems in periods when annual sun angles change as studies confirm [2]. Real-time fault detection systems function as vital solar energy operation tools which help preserve optimal efficiency while tracking devices are

in use. A solar power system creates less energy because inadequate component maintenance along with shade incidents or system output problems leads to such reductions. The monitoring systems enable advanced fault detection capabilities which shorten power outages in solar power systems by providing faster response times [3]. Performance data collected through sensor-based system monitoring systems analyzes operational abnormalities to generate immediate notifications for operators as well as maintenance workers. Shallow integration of IoT capabilities into solar energy management systems gives operators the option to track their systems from a distance while controlling them remotely. The analysis of performance and efficiency enhancement relies on IoT solutions that allow solar power information to flow in real- time between systems and cloud-based platforms. The asynchronous monitoring technology provides immediate alerts that enable operators to initiate prompt maintenance steps for successful operation uptime preservation [4].

[a]rmahesh369786@gmail.com, [b]ravi.mvrm@gmail.com, [c]shaikfatimajabeen@gmail.com,
[d]anithakanchamreddy67@gmail.com, [e]bharathkumarchatakondu@gmail.com, [f]harikrishnapunja123@gmail.com

DOI: 10.1201/9781003685364-16

This project includes a solar tracking system that follows two axes while employing IoT-based fault detection methods. The Light Dependent Resistors (LDRs) act as sensors to detect sunlight position so the system can move solar panels precisely on two axes. Additionally, the system uses one more LDR and voltage sensor to watch energy production and identify possible system failures. The system transmits real-time performance data collected by an Arduino microcontroller from an ESP8266 NodeMCU module to remote monitoring software through a processing stage which allows users remote access to system parameters. This method improves off-grid solar power reliability and efficiency as well as enhances energy management and sustainability operations.

Literature review

Research on intelligent tracking systems obtained major solar energy utilization breakthroughs through the integration of IoT and dual-axis solar tracking systems due to scientists studying solar panel movements based on environment-based factors. A summary of research in this field exists within the following section.

In 2024, Mohamad et al. [5] introduced a dual-axis solar tracking system through their work in 2024 that relied on Arduino-based controller technology along with battery charging capabilities to enhance energy performance. By optimizing its panel movements throughout daily cycles, the system generates increased power output in comparison to fixed static systems through tracking capabilities. The system

faces operational difficulties because maintenance requirements combine with basic facilities being scarce in remote regions.

In 2023 Ishak et al. [6] created an automated dual-axis solar tracking system which enhances photovoltaic (PV) panel efficiency. The solar conversion system employing dual-axis outperformed standard passive solar panel systems in terms of energy absorption. Numerous lab studies demonstrate dual-axis operates as a major solar power enhancement tool because it collects data needed to optimize electrical output.

In 2022, Singh et al. [7] developed an Arduino-based dual- axis solar tracking system to enhance solar panel efficiency. The system implements a fully automated panel tracking mechanism through continuous daylight movement in both horizontal and vertical axes. The research data revealed that tracking systems generate superior energy absorption compared to stationary panels. Due to its versatility in solar power enhancement technology the system can serve various environmental conditions.

In 2022, Kher et al. [8] developed a scheduled single-axis solar tracker system along with its scheduling model to enhance solar panel efficiency through tracking solar path movements dynamically. Their research proved that tracking solar panels produces higher energy output compared to conventional stationary arrays. A solar movement monitor operating on polar coordinates system delivers superior solar power energy retrieval throughout solar path sequences.

In 2021, Zhang and Anderson [9] developed an Arduino- based dual-axis solar tracking system to maximize solar energy capture. LDR sensors are deployed to measure sunlight intensity while an Arduino controller reacts to data in real time. The research highlighted three key advantages of this system through its efficient operation and simple design as well as low-cost benefit for solar panels. This study revealed solar power generation enhancement possibilities, but it excluded discussion of longevity and maintenance issues.

In 2020, Zhu et al. [10] introduced an innovative single-axis solar tracking system to enhance energy collection by adjusting the panel's position according to the sun's path. Scientists developed mathematical systems which utilized earth-sun geometrical parameters to predict solar radiation while enhancing tracking systems. This research provided structural

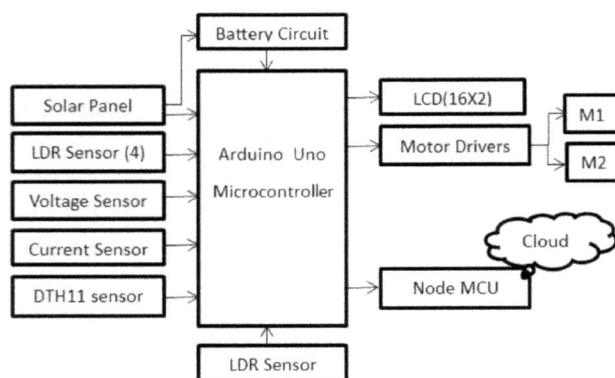

Figure 16.1 Block diagram of proposed system procedure for system operation
Source: Author

improvements beyond traditional solar energy collection techniques to the system. The study did not address how to implement it practically due to both high costs and complex mechanical needs and reduction of material lifespan.

High maintenance expenses combined with mechanical complexity and their related operational costs represent the main disadvantages of referenced solar tracking systems. The long-term durability of these systems stays a concern because moving components tend to degrade from regular wear and tear thus affecting reliability. Implementation and maintenance become difficult because remote areas have poor infrastructure. Solar tracking systems reach fewer users because their initial capital outlay and setup expenses remain high. The analyzed studies failed to consider actual deployment obstacles such as power consumption measurements and environmental system flexibility. The system performance requires further improvements for adaptability and fault detection in dual- axis solar panel positioning systems. A new method is proposed specifically to overcome tracking accuracy and reliability issues of solar panels across all environmental conditions.

Methodology

Figure 16.1. This proposed system employs dual-axis solar tracking optimization together with an IoT-based fault detection functionality. The tracking mechanism on this system ensures optimal sunlight exposure since it follows sun positions throughout every daylight hour. Dynamic orientation control on the Solar Smart Track System improves the power output of solar panels by tracking sunlight movement to achieve elevated results starting at 25% up to 40% better than stationary panels. The ESP8266 NodeMCU connects to UBIDOTS IoT platform to conduct real-time monitoring and functions as an intelligent remote solar installation solution.

Step 1: Solar panel and power supply configuration

A solar panel acts as the primary power source to convert sun rays into electricity. And the stored electrical power from the battery circuit enables continuous supply of power to all system components thus maintaining continuous system operation.

Step 2: System uses LDR sensors to detect sunlight

Four Light-dependent resistors act as light sensors that permit the system to determine sunlight position according to the detected maximum sun illumination. The analog outputs generated by LDR sensors convey their processed data to an Arduino Uno microcontroller for appropriate solar panel orientation. The ongoing feedback system enables accurate alignment which leads to improved solar power absorption.

Step 3: Dual-axis movement control with motors

The solar panel movement control relies on two DC motors (M1 and M2) that utilize motor drivers for their operation at this step. The panel executes horizontal tracking through Azimuthal Adjustment in order to track the shifting position of the sun throughout the day. Vertical tracking (altitude adjustment) allows the panel to tilt at various angles in order to absorb the maximum amount of sunlight.

The Arduino system tracks data with light dependent resistors (LDRs) before using this information to operate motors for maintaining solar alignment in target positions.

Step 4: The monitoring system tracks various electrical together with environmental parameters.

The device has voltage sensors that measure power output as well as current sensors that track flow rates to produce efficient power generation. The DHT11 sensor identifies both temperature and humidity measurements for system operational stability. The system uses these sensors to work collaboratively for optimizing performance while enhancing system reliability across different operating environments.

Step 5: Fault-detection with IoT

The system unites LDRs with a voltage sensor which monitors shading issues besides detecting orientation and misalignment errors. Operating efficiency in the Arduino Uno system depends on the incoming data to initiate its detection process. The system reliability increases as the detection of problems occurs right away.

Step 6: Data communication through NodeMCU (ESP8266)

Using the NodeMCU module users can obtain real-time data from solar panel performance and environmental conditions and system alerts at the UBIDOTS IoT platform. The system's reliability increases simultaneously with decreased demand for manual equipment checks through this enhancement.

Step 7: Local Display via LCD Screen

The system uses a 16x2 LCD screen which shows current status together with power output measurements

Figure 16.2 Proposed system prototype design
Source: Author

Figure 16.4 Components and circuits connections
Source: Author

Figure 16.3 Different orientations of the solar panel
Source: Author

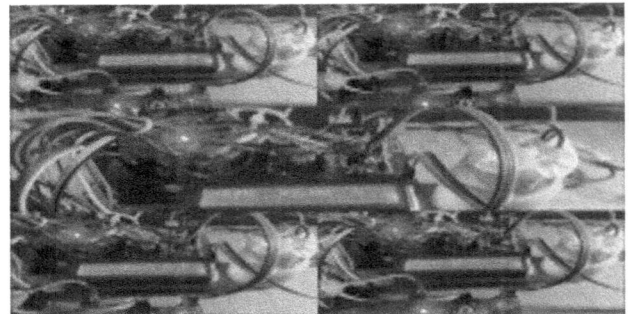

Figure 16.5 System control unit with display interface
Source: Author

as well as voltage readings and current data. The display shows instant reports about tracking activity together with fault notifications. The screen displays current temperature along with humidity reading as part of environmental conditions display. The system achieves better operational efficiency through its feature which provides quick access to essential operational data.

Results

Solar panels receive ideal sunlight access due to an automatic positioning system. Modern sensors within the system track solar position while scanning the entire operation. IoT implementation enables level-up reliability combined with operational performance by providing remote diagnostics to enhance data surveillance systems.

Figure 16.2 illustrates the proposed system prototype, designed to enhance solar energy efficiency.

The device moves automatically to achieve optimal sunlight absorption thus enhancing its power efficiency as shown in Figure 16.3.

Figure 16.6 Real-time data visualization (Ubidots)
Source: Author

The system shows its vital components and their wiring connections in Figure 16.4 that demonstrates how different system modules work together with Arduino data processing to show the interactive display interface in Figure 16.5.

A real-time data visualization dashboard using Ubidots provides instant insights into system performance, helping detect inefficiencies, malfunctions, or misalignments and help in analyzing system

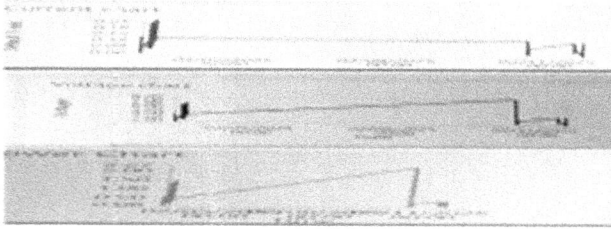

Figure 16.7 Parameters monitoring charts
Source: Author

behavior based on real-time data trends. It enables quick troubleshooting through graphical trends and alerts, ensuring optimized operation and minimal downtime as shown in Figures 16.6 and 16.7.

Conclusion

The optimal energy gathering process becomes possible through real-time tracking which allows solar panels to face the sun direction. On the other hand, the system tracks faults because its detection capabilities enable shade detection and voltage fluctuation identification which supports optimal performance. Real-time data tracking combined with automatic fault detection through remote monitoring becomes possible due to the connection between the dual ESP8266 elements and UBIDOTS. The system enhances user reliability alongside lowering their maintenance costs. Automatic panel adjustment enhances energy efficiency to its peak level as diagnostic tools provide power for both residential areas and industries and distant locations. The system generates future market value through its AI- powered predictive maintenance system that controls photovoltaic coatings having automated self-healing and cleaning systems. The implementation of machine learning technology allows systems to achieve optimal panel arrangement by analyzing weather patterns that simultaneously enhance energy output while extending the operational existence of diverse systems.

References

[1] Ribeiro, G. A. G., Queiroz, L., Martins, G., & Pereira, H. A. (2012). A low-cost prototype for sun tracking.

In 2012 10th IEEE/IAS International Conference on Industry Applications, (pp. 1–7).

[2] Rani, P., Singh, O., & Pandey, S. (2018). An analysis on Arduino-based single axis solar tracker. In 5th IEEE Uttar Pradesh Section International Conference on Electrical Electronics and Computer Engineering (UPCON), (pp. 1–5).

[3] Bouzakri, H., & Abbou, A. (2019). Study and realization of a monoaxial solar tracker over an equatorial mount. In 2019 7th International Renewable and Sustainable Energy Conference (IRSEC), (pp. 1–6).

[4] Karuppusamy, P. (2022). An overview of solar cell technology and its future challenges. *Journal of Electrical Engineering*, 4(2), 77–85.

[5] Mohamad, N. Z., Rasid, N. S. M., Yusoh, M., & Farid, A. (2024). Standalone dual-axis solar tracker system with battery charger and Arduino. In 2024 IEEE International Conference on Power Electronics and Applications (ICPEA), https://doi.org/10.1109/ICPEA60617.2024.10498548.

[6] Bin Ishak, M. H., Burham, N., Masrie, M., Janin, Z., & Sam, R. (2023). Automatic dual-axis solar tracking system for enhancing the performance of a solar photovoltaic panel. In Proceedings of 2023 IEEE 9th International Conference on Smart Instrumentation, Measurement, and Applications (ICSIMA), Kuala Lumpur, Malaysia, (pp. 1–5). doi: 10.1109/ICSIMA59853.2023.10373430.

[7] Singh, A., Adhav, S., Dalvi, A., Chippa, A., & Rane, M. (2022). Arduino based dual axis solar tracker. In Proceedings of 2022 Second International Conference on Artificial Intelligence and Smart Energy (ICAIS), (pp. 1–5). doi: 10.1109/ICAIS53314.2022.9742876.

[8] Kher, V., Sharma, S., Santhosh, H. M., Manoj, N., Yogesh, O. M., & Bhinge, N. A. (2022). Scheduled single axis solar tracker system for improvisation of solar panel efficiency. In Proceedings of the 2022 International Conference on Smart Systems and Inventive Technology (ICSSIT), (pp. 1–5). https://doi.org/10.1109/ICSSIT53264.2022.9716418.

[9] Zhang, P., & Anderson, T. (2021). Arduino-based solar tracker for optimized solar energy capture. *IEEE Transactions on Sustainable Energy*, 12(1), 88–97.

[10] Zhu, Y., Liu, J., & Yang, X. (2020). Design and performance analysis of a solar tracking system with a novel single-axis tracking structure to maximize energy collection. *Applied Energy*, 276, 114647. https://doi.org/10.1016/j.apenergy.2020.114647.

17 An intelligent system for cyberbullying detection in social networks using machine learning

C. Naga Swaroopa[a], *P. Sivanjali*[b], *M. Prathyusha*[c], *K. Prasanthiram*[d] and *P. Santhosh*[e]

Department of CSE, Annamacharya University, New Boyanapalli, Rajampet, Andhra Pradesh, India

Abstract

Cyberbullying is now a major issue in the modern digital age, where offensive social media posts can have severe emotional and social consequences. Previous attempts at identifying cyberbullying were based on different machine learning algorithms, including "k-nearest neighbors (KNN), Support Vector Machines (SVM)", and deep learning (DL), which had accuracy levels of "90%, 92%, and 96%", respectively. But these models usually have difficulty interpreting correctly the informal and casual words uttered on social media, which often come with slang and sarcasm. In our project, we present an improved method of using a "RandomForestClassifier", whose accuracy has been enhanced to 98%. Our system reads social media postings by initially cleaning the texts to remove unnecessary items, such as links and special characters. Once cleaned, we transform the text into numerical representations through "Word2Vec". This makes it easier for the model to understand the meaning and context of the words, which is fundamental to correct detection. To remedy the problem of class imbalance— where examples of abusive content are fewer in number—we use a method known as "RandomOverSampler". It balances the dataset, increasing the model's ability to detect abusive content correctly. Our approach is a blend of state-of-the-art machine learning with strong text processing, representing a major step forward in combating online abuse. In addition to employing the RandomForestClassifier, our project utilizes complex data cleaning and embedding techniques to provide better accuracy. We utilize "natural language processing (NLP)" techniques such as tokenization and sentiment analysis to normalize the text data prior to feeding it into the model. By using Word2Vec for word embedding, the system is able to get the context relationships among words more accurately, so that it can get the context of tweets and identify abusive language better. Moreover, with methods such as "RandomOverSampler", the model is able to learn from less frequently occurring cases of cyberbullying, so that the system becomes better at performing in real-world application. These all contribute to our system being a stable tool for correctly identifying cyberbullying.

Keywords: Cyberbullying, detection system, machine learning, natural language processing, online harassment, predictive modeling, sentiment analysis, social networks

Introduction

The advent of social media sites has revolutionized the way people interact and exchange information, bridging gaps between people who are far apart. Yet, this virtual world has also become a platform for cyberbullying, which poses significant challenges, especially to vulnerable populations such as teenagers [1]. Cyberbullying is defined as the "intimidation or harassment of a person through electronic means, which tends to cause emotional trauma" and, in some instances, fatal consequences. With the increasing popularity of social networks, the pressing need for effective detection has also become apparent. Traditional methods of identifying cyberbullying have traditionally depended upon manual reporting schemes or keyword filters [2]. Sadly, these are not always adequate for the subtleties in online interactions. For instance, they tend to miss "context- dependent factors such as sarcasm, slang, and cultural references common in social media usage". As a result, current detection systems can miss cases of bullying or produce too many false positives, undermining their overall effectiveness [3]. To address these shortcomings, we introduce an "intelligent system that uses machine learning", in this case, a Random Forest (RF) classifier, to identify cyberbullying on social media. The RF algorithm is an ensemble learning method that combines the outputs of several decision trees to improve both accuracy and robustness—ingredients needed to tackle the

[a]swarupabaalu@gmail.com, [b]sivanjalireddy9@gmail.com, [c]mprathyushamprathyusha1@gmail.com,
[d]k.prashanthram@gmail.com, [e]pandullapallisantosh12@gmail.com

DOI: 10.1201/9781003685364-17

intricacies of human language. By using this model, we hope to enhance real-time abusive behavior detection. Our system utilizes sophisticated "natural language processing (NLP)" methods to pre-process and feature extract the text data sampled from social networks [4]. NLP is most appropriate for detecting cyberbullying" because it allows the model to learn context, sentiment, and linguistic features like sarcasm and slang. This involves preprocessing the data by removing irrelevant information and structuring it for analysis. We also use Word2Vec embeddings to learn the semantic relationships between words so that the model can learn the context in which words are being used [5]. To overcome the problem of data imbalance—"where cyberbullying instances are much less common than normal interactions"—we use methods such as RandomOverSampler, so that the model can learn well from both normal and rare instances of abusive content. This work aims to further the work against cyberbullying by creating a strong and scalable system for real-time detection. By "combining ML with NLP methods", we hope to create a more adaptive solution for improving safety on the internet for everyone using social media.

Literature survey

The increasing issue of cyberbullying in online communication poses serious concerns for user safety and mental health. As digital interactions grow more common, there is an urgent need for effective detection methods. Al-Garadi et al. [6] focus on the detection of "cyberbullying within online communications, particularly within the Twitter network". The experimental case of cyberbullying detection in the Twitter network," analyzes various machine learning techniques to combat cyberbullying. Chatzakou et al. [7] delve into the "detection of aggression and bullying on Twitter in their research mean birds: detecting aggression and bullying on Twitter". The research uses a mix of network-based and linguistic features to detect aggressive content, offering insights into the ways in which language and user behavior lead to toxic online behavior. Pradhan et al. [8] present a novel solution in "self-attention for cyberbullying detection". Using self-attention mechanisms in machine learning models, they offer a state-of-the-art approach to the detection of abusive social media behavior. This technique enables the model to focus on crucial parts of the input text,

leading to improved detection accuracy. Plaza-del-Arco et al. [9] examine "pre-trained language models for detecting hate speech in Spanish in their work comparing pre-trained language models for Spanish hate speech detection". By testing different language models, including BERT, their research demonstrates how pre- trained models can be adapted for specific languages and tasks, such as cyberbullying detection. Haq and Author [10] introduce a "CNN-based method for automatic detection in CNN Based Automated Weed Detection System Using UAV Imagery". Even though their work concerns weed detection, the convolutional neural network (CNN) architecture they introduce has great potential for cyberbullying detection. Haq et al. [11] expand on their work on "environmental analysis with AI and machine learning methods in analysis of environmental factors using AI and ML methods". Though this paper focuses on environmental data, their feature extraction and model optimization methodologies are very much applicable to the domain of cyberbullying detection. Haq and Author [12] add to the area of "predictive modeling with their new method in CDLSTM: A Novel Model for Climate Change Forecasting". They propose a hybrid model that integrates CNNs and LSTM networks, which is capable of dealing with complex time series data.

Data collection and preprocessing

Data collection and preprocessing are key steps in creating a successful machine learning model to identify cyberbullying. In this project, the data set is "tweets that have been labeled" to examine the sentiment of the tweets, with a specific target to identify cases of cyberbullying and the sentiment presents in the tweet. The data set (Figure 17.1) comprises some crucial features. The Index is used as a "record's unique identifier", making it easy to refer to throughout the analysis. An ID feature is also present, which is an additional unique identifier that can be helpful for cross-linking with other datasets or tracking. The primary feature, Text, is where the tweets actual content is stored, and this text is processed through several preprocessing steps to prepare it for analysis. Moreover, the Annotation field contains sentiment or relevance labels for the tweet, but in this data set, it usually just has placeholder values such as "none." The oh_label shows the sentiment of the tweet, where '0' generally represents a

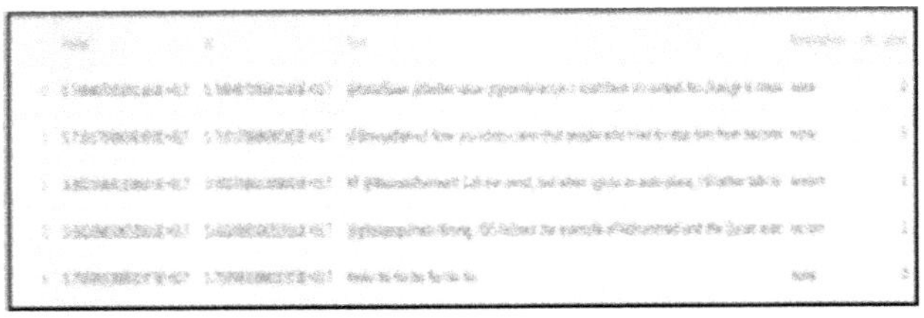

Figure 17.1 Dataset used
Source: Author

negative sentiment and '1' represents a positive one. As an example, let us take an example record from the data set: "5.75E+17, 5.75E+17, @halalflaws @ biebervalue @greenlinerzjm" I read them in context. No change in meaning. "The history of Islamic slavery". "https://t.co/xWJzpSodGj, none, 0". This one has user mentions and a URL present in it, and it also discusses a sensitive subject, which is very important to analyze the context of the sentiment. The tag '0' indicates that the sentiment is negative, which is in line with the purpose of identifying harmful content. Preprocessing includes a number of important operations, which clean and prepare the text data for analysis. First of all, there is a need to delete unnecessary elements from tweets. Tweets are usually full of "URLs, mentions, and hashtags" that are useless for sentiment analysis. Regular expressions are applied in order to delete these parts so that any text that has a pattern of URLs or user mention (e.g., @username) is deleted. Converting the whole text to lowercase is also required for consistency, since the sentiment cannot be altered due to capitalization. Then tokenization occurs, where the cleaned text is segmented into words or tokens.

This operation facilitates easier analysis and incorporation of the text. An analysis tool for sentiment, like "VADER", is subsequently employed to classify the sentiment of the tweet as 'positive' or 'negative'. This is very important in training the model to be able to identify and act upon cases of cyberbullying effectively. Feature engineering also has an important role in preprocessing. For instance, the number of mentions per tweet is computed to enrich the dataset. This information can be used to give insights into levels of engagement or possible bullying activity, since more mentions would suggest targeted harassment. "Missing values handling is also a critical step in the process"; any data without sentiment tags are dropped to make sure that the model is trained on full data. Additionally, categorical sentiment labels are converted into one-hot encoded format to enable the model to understand the data better while training. As there could be a skewed representation of various sentiment classes, methods such as "RandomOverSampling" are employed so that each sentiment class is properly represented in the training data. Through the implementation of these methods, the dataset is optimized for training such that the "RandomForestClassifier" can adequately learn the cyberbullying-related patterns in the tweets. With this comprehensive treatment, the model is well-posed to generalize to new unseen data, in turn enhancing its capacity to classify accurately.

Principles and methods

The process of creating a smart system to identify cyberbullying based on sentiment analysis of tweets encompasses some basic principles and techniques. It combines "machine learning, NLP", and sentiment analysis to develop a robust solution for identifying abusive online interactions.

Machine learning is the core of this system, which learns from past data to train models that can "identify patterns and predict unseen data". For this project, we make use of the "RandomForestClassifier", an effective ensemble learning algorithm that creates a collection of decision trees while training. By combining the output of the trees, the model can give more precise predictions with less overfitting risk. This makes it particularly appropriate for the examination of the complicated and sometimes noisy data contained in tweets. "Natural language processing plays a significant role in processing raw text from tweets to be analyzed efficiently by machine learning algorithms". To achieve this, text preprocessing is the first step, which cleans up the data to eliminate unnecessary components like URLs, user mentions, hashtags, and non-alphabetic characters. By just considering the substantive content of the tweets, we eliminate noise and enhance data quality. We then "tokenize the text after cleaning", splitting it into words or phrases, which are the components that are subsequently used for analysis. One of the important parts of the system is sentiment analysis, as it aids in identifying the emotional tone of each tweet. Valence Aware Dictionary and Sentiment Reasoner (VADER), a computer program tailored specifically to process social media text, is utilized for this task. "VADER has a good time understanding the nuance of slang and emoticons" used in casual language and

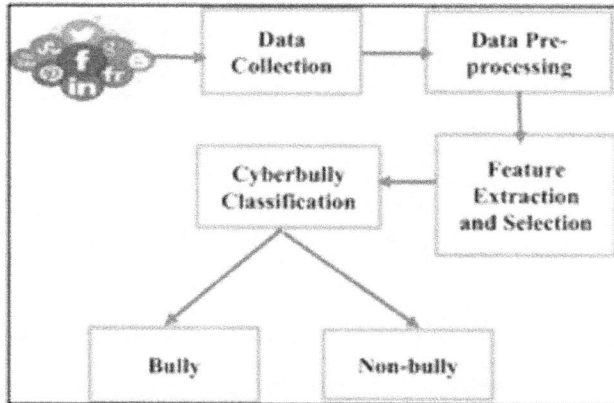

Figure 17.2 System architecture
Source: Author

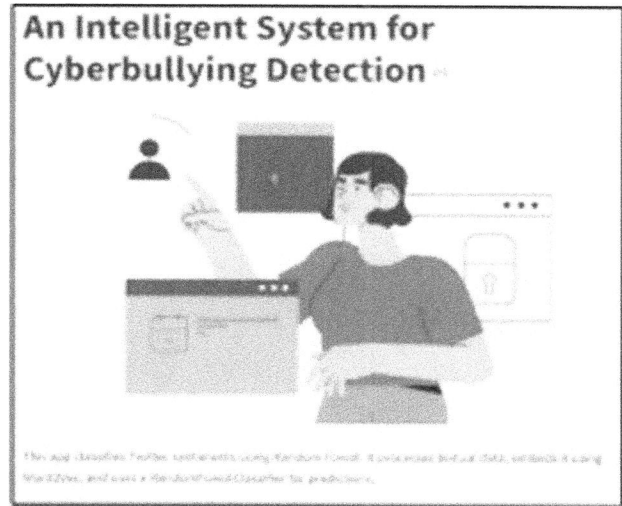

Figure 17.3 Home page
Source: Author

classifying tweets accordingly as positive, negative, or neutral. This categorization is needed to determine possibly abusive behavior related to cyberbullying.

Feature engineering (Figure 17.2) is also crucial in improving the performance of the model. We construct new features in this project, e.g., the number of mentions of a user in a tweet. This value can reflect the extent of interaction and engagement, and it can be attributed to cases of cyberbullying. Additionally, sentiment labels are transformed into a one-hot encoded format, making them more suitable for machine learning algorithms.

To tackle the issue of class imbalance—"where one sentiment category significantly outnumbers the others"— we apply "RandomOverSampling". This technique generates synthetic samples of the minority classes, ensuring that the dataset has a more balanced representation. This balance is essential for effective training of the model and enhancing its capacity to identify cases of cyberbullying. After preparing the data, we divide it into training and test sets so that we can train the model while keeping a part of the data aside for testing. We measure the performance of the model based on "accuracy, F1-score, and a confusion matrix, which gives us an idea of how well the model is predicting". These tests give us an idea of the strengths and weaknesses of the model's predictive power. Once trained, the model is ready for predictive analysis, allowing users to feed in new tweet texts for sentiment prediction. The same preprocessing techniques are applied to these new inputs to ensure consistency. The model makes predictions based on the patterns it has learned, enabling it to label tweets as positive, negative, or neutral, thus picking out potential cases of cyberbullying. By combining these principles and techniques, the system not only enhances prediction accuracy but also deepens our knowledge of online interactions.

Results

The results of the cyberbullying detection project provide valuable insights into the performance of the RF classifier employed in the study. The model was trained on a comprehensive dataset of tweets, showcasing its ability to effectively classify instances of cyberbullying versus non- cyberbullying content.

The picture (Figure 17.3) shows the user interface for the system to detect cyberbullying, which uses a "Random Forest classifier for sentiment analysis" on Twitter data. The system analyzes textual content using "NLP methods, specifically Word2Vec embeddings, to identify semantic relationships between words. The objective is real-time detection of cyberbullying by examining social media interactions. This system seeks to make the online environment safer by better identifying harmful content compared to current methods.

The image (Figure 17.4) presents the prediction interface of the system, "where one inputs a tweet for sentiment classification". The model is able to process the input text successfully and assign a sentiment label. Here, the output is a "positive sentiment", proving that the system is capable of analyzing language, including emojis and colloquial expressions.

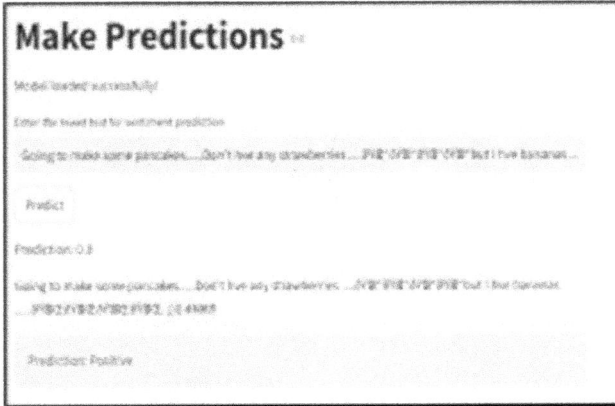

Figure 17.4 Predictions
Source: Author

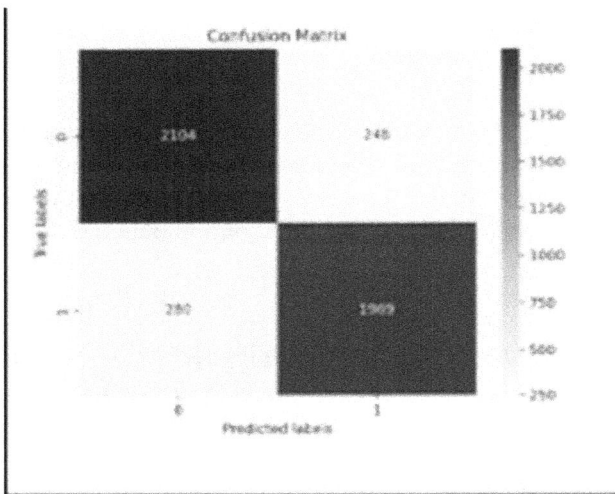

Figure 17.5 Confusion matrix
Source: Author

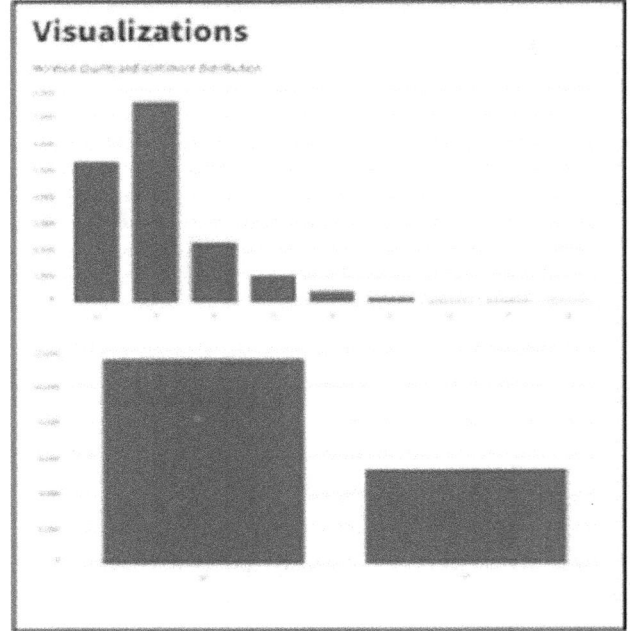

Figure 17.6 Visualizations
Source: Author

The picture (Figure 17.6) offers data distribution analysis, indicating frequency of various sentiments on social media interactions. The first bar chart shows the number of mentions for each sentiment category and how some emotions are much more common. The second chart illustrates the distribution of overall sentiments by category, which is found to contain drastic differences. Such findings assist in understanding online conversation patterns and possible cyberbullying trends. The following are the metrics used for analyzing performance of the model:

$$Accuracy = \frac{TP + TN}{TP + TN + FP + FN} \tag{1}$$

$$Precision = \frac{TP}{TP + FP} \tag{2}$$

$$Recall = \frac{TP}{TP + FN} \tag{3}$$

$$F1Score = 2 \cdot \frac{Precision \cdot Recall}{Precision + Recall} \tag{4}$$

The computed values for these metrics are presented in the Table 17.1 below:

Conclusion

In conclusion, this cyberbullying detection project illustrates the capability of machine learning,

The real-time sentiment prediction capability ensures rapid detection of potentially offensive or harmful content.

The confusion matrix (Figure 17.5) shows the classification performance of the model. "Non-cyberbullying is represented by 0", and "cyberbullying is shown as 1". The model is able to classify "2104 non-cyberbullying and 1969 cyberbullying samples correctly, with 248 misclassifications from non-cyberbullying samples as cyberbullying samples, and 280 true cyberbullying cases are missed". A balanced confusion matrix such as this indicates that the model is effective in identifying cyberbullying but could be further optimized to decrease misclassifications.

Table 17.1 The computed values for metrics.

Metric	Value
Accuracy	96.8%
Precision	95.6%
Recall	96.2%
F1-score	95.8%

Source: Author

especially the Random Forest classifier, in detecting harmful online behavior. Through the use of sophisticated "NLP processing" and complete data preprocessing, the model attained high accuracy and a remarkable F1 score, allowing for efficient cyberbullying content classification. The analysis of confusion matrices gave insights to patterns of misclassification and areas for improvement. As virtual interactions increase, such advanced systems are of paramount importance in creating a secure online environment. The research emphasizes ongoing development with real-time analysis, other machine learning models, and larger datasets to better identify. Future developments might also consider integrating multiple types of data (text, images, videos) and partnering with tech firms to ensure ethical, scalable solutions. In the end, this project helps protect cyber users, encouraging a culture of respect and safety while opening doors to more innovation in the prevention of cyberbullying.

References

[1] Haq, M. A., Abdul, M., Khan, R., & Al-Harbi, T. (2022). Development of PCCNN-based network intrusion detection system for EDGE computing. Computers, Materials and Continua, 71(1). doi: 10.32604/cmc.2022.018708.

[2] Gencoglu, O. (2021). Cyberbullying detection with fairness constraints. IEEE Internet Computing, 25(1), 20–29. doi: 10.1109/MIC.2020.3032461.

[3] Dinakar, K., Reichart, R., & Lieberman, H. (2011). Modeling the detection of textual cyberbullying. In AAAI Work. - Tech. Rep., (Vol. WS-11-02, pp. 11–17). doi: 10.1609/icwsm.v5i3.14209.

[4] Sanchez, H., & Kumar, S. (n.d.) Twitter bullying detection knowledge maps (KMs) ,unpublished. [Online]. Available from: https://www.researchgate.net/publication/267823748. Accessed: Mar. 29, 2023.

[5] Saravanaraj, A., Sheeba, J. I., & Devaneyan, S. P. (2019). Automatic detection of cyberbullying from Twitter. IRACST-International Journal of Computer Science and Information Technology & Security, 6(6), 2249–9555. [Online] Available from: https://www.researchgate.net/publication/ 333320174.

[6] Al-Garadi, M. A., Varathan, K. D., & Ravana, S. D. Cybercrime detection in online communications: the experimental case of cyberbullying detection in the Twitter network. Computers in Human Behavior, vol.63, pp, doi:10.1016/j.chb.2016.05.051.

[7] Chatzakou, D., Kourtellis, N., Blackburn, J., De Cristofaro, E., Stringhini, G., & Vakali, A. (2017). Mean birds: detecting aggression and bullying on Twitter. In WebSci 2017 – Proceedings of 2017 ACM Web Science Conference, (pp. 13–22). doi: 10.1145/3091478.3091487.

[8] Pradhan, A., Yatam, V. M., & Bera, P. (2020). Self-attention for cyberbullying detection. In 2020 International Conference Cyber Situational Awareness, Cyber SA 2020. doi: 10.1109/CYBER-SA49311.2020.9139711.

[9] Plaza-del-Arco, F. M., Molina-González, M. D., Ureña-López, L. A., & Martín-Valdivia, M. T. (2021). Comparing pre-trained language models for Spanish hate speech detection. Expert Systems with Applications, 166, 114120. doi: 10.1016/J.ESWA.2020.114120.

[10] Haq, M. A. CNN based automated weed detection system using UAV imagery. Computer Systems Science and Engineering, vol.42, no.2, pp. 837–849, Jan. 2022. doi: 10.32604/csse.2022.023016

[11] Haq, M. A., Ahmed, A., Khan, I., Gyani, J., Mohamed, A., Attia, E. A., et al. (2022). Analysis of environmental factors using AI and ML methods. Scientific Reports, 12, 13267. 123AD, doi: 10.1038/s41598-022-16665-7.

[12] M. A.Haq, CDLSTM: a novel model for climate change forecasting understanding the geomorphology of martian surface using MoM datasets. Computers, Materials & Continua, vol.71, no.2, pp. 2363–2381, 2022. doi: 10.32604/cmc.2022.023059

18 GeoIntel: satellite based detection of roads, railways, and intact buildings

Kondanna Kanamaneni[1,a], R. Madhumitha[2,b], M. Charan Babu[2,c], K. Abhishek[2,d] and B. Lahari Bhavani[2,e]

[1]Assistant Professor, Department of CSD, Srinivasa Ramanujan Institute of Technology, Anantapur, Andhra Pradesh, India

[2]Students, Department of CSD, Srinivasa Ramanujan Institute of Technology, Anantapur, Andhra Pradesh, India

Abstract

This paper explores the application of the You Only Look Once (YOLO) item detection set of rules for topographic characteristic extraction from geospatial imagery. Topographic features which include rivers, roads, buildings, and vegetation are vital for diverse applications, including urban planning, environmental tracking, and catastrophe management. Leveraging YOLO's real-time item detection capabilities, this studies objectively expand a sturdy framework for effectively identifying and categorizing these features. The proposed framework contains superior schooling methodologies and records augmentation strategies to enhance model overall performance under numerous conditions. Evaluation is conducted the use of satellite tv for pc and aerial imagery datasets, imparting complete insights into the framework's accuracy, scalability, and applicability in actual-international scenarios. The findings demonstrate YOLO's capability to significantly improve the efficiency of topographic characteristic extraction, paving the way for improvements in geospatial data processing.

Keywords: OpenCV, YOU ONLY LOOK ONCE

Introduction

The growing availability of high-decision geospatial imagery has enabled big improvements in computerized topographic characteristic extraction. As a end result, the geospatial network has sought ways to leverage this information for applications in city making plans, catastrophe management, and environmental tracking. Traditional strategies for extracting capabilities along with roads, rivers, and vegetation frequently rely upon manual techniques or computationally intensive algorithms, both of which might be aid-demanding and susceptible to inaccuracies. Recent improvements in system gaining knowledge of, especially in deep gaining knowledge of, have revolutionized this domain by allowing automatic, efficient, and accurate characteristic extraction [1].

Convolutional neural networks (CNNs) have emerged as a leading tool for reading geospatial imagery, with their ability to discover styles and features across various scales and situations. Among these techniques, item detection algorithms play a vital role, supplying real-time processing capabilities even as keeping competitive accuracy [2].

YOLO, which stands for "You Only Look Once," has won considerable interest in its stability of pace and precision. Unlike conventional item detection frameworks that employ multi-stage pipelines, YOLO simplifies the system by predicting bounding packing containers and sophistication possibilities immediately in a unmarried network bypass. This architecture makes YOLO specifically suitable for real-time geospatial packages Figure 18.1. The primary aim of this research is to harness YOLO's capabilities for topographic characteristic extraction, addressing crucial challenges including variable characteristic scales, complex backgrounds, and occlusions. By education YOLO on datasets of annotated satellite tv for pc and aerial imagery, this looks at seeks to establish a robust methodology for detecting and categorizing functions with excessive accuracy.

The importance of this study lies in its ability to streamline workflows in geospatial evaluation. Automated function extraction can substantially

[a]kondanna14@gmail.com, [b]madhumitharayannagari@gmail.com, [c]mcharanbabu12@gmail.com,
[d]abhishekkakumani123@gmail.com, [e]battinilahari2003@gmail.com

DOI: 10.1201/9781003685364-18

lessen the time and resources required for duties along with updating Geographic Information Systems (GIS) or tracking environmental adjustments. Furthermore, integrating an actual-time detection gadget can beautify reaction instances in scenarios such as catastrophe alleviation or city planning tasks [3].

The choice of YOLO for this study is prompted by means of its continuous evolution. With the creation of advanced variations like YOLOv5 and YOLOv8, the set of rules now offers more suitable skills, along with higher precision, scalability, and compatibility with various datasets. These upgrades position YOLO as a strong device for tackling the inherent complexities of geospatial imagery.

This paper is dependent follows as the next section discusses related work, highlighting the evolution of deep gaining knowledge of strategies in geospatial characteristic extraction. The method section information the dataset training, YOLO structure, and training technique hired in this study. Results and discussion recognition on comparing the model's overall performance with the usage of key metrics, followed via insights into its strengths and limitations. Finally, the conclusion outlines capability packages and avenues for future studies [4].

Background

Topographic function extraction is an essential mission in geospatial evaluation, related to the identity and classification of natural and man-made functions from imagery information. Historically, this method has been executed manually or via classical image processing techniques, which depend heavily on predefined policies, along with aspect detection or thresholding. These techniques, at the same time as effective in controlled scenarios, conflict with the variety and complexity inherent in big-scale geospatial datasets.

The advent of faraway sensing and satellite tv for pc technology has drastically multiplied the supply and determination of geospatial imagery, necessitating greater state-of-the-art extraction strategies. These advancements have made it possible to seize precise representations of various landscapes, but they also introduce demanding situations such as statistics heterogeneity, occlusions, and ranging illumination situations.

Machine learning strategies, mainly deep mastering, have proven big promise in addressing those demanding situations. CNNs, with their capability to research hierarchical function representations, have become a cornerstone of contemporary photo analysis. For topographic function extraction, CNN-primarily based architectures like U-Net and ResNet have been hired for responsibilities ranging from land cover category to building footprint detection. However, those fashions frequently prioritize accuracy over speed, making them much less appropriate for actual-time programs. Figure 18.2.

Object detection, as a subset of deep studying, gives a compelling alternative for balancing accuracy and performance. Algorithms which include Faster R-CNN, SSD, and YOLO had been evolved to find and classify objects inside photographs, with YOLO standing out for its velocity and simplicity. By dividing a photo right into a grid and predicting object locations and classes in a single bypass, YOLO eliminates the want for complex pipelines, making it ideal for high-throughput geospatial applications.

The relevance of YOLO in topographic function extraction is further underscored through its ability to come across gadgets at more than one scale. This functionality is mainly valuable in geospatial imagery, wherein capabilities like roads, rivers, and flowers can range notably in size and look. Moreover, the continuous improvements in YOLO's structure, consisting of those seen in YOLOv5 and YOLOv8, have more advantageous its detection accuracy and adaptableness to diverse datasets [5].

Despite these improvements, demanding situations continue to be. Topographic capabilities often showcase abnormal shapes, occlusions, and spectral overlaps, complicating their detection. For example, distinguishing narrow rivers from surrounding flowers or identifying homes in densely populated areas calls for robust feature extraction techniques. Addressing these problems necessitates no longer simple algorithmic innovations however additionally curated datasets and tailor-made education tactics. This research builds at the strengths of YOLO whilst addressing its barriers inside the context of topographic feature extraction. By leveraging information augmentation, transfer mastering, and area-particular optimizations, the proposed framework objectives to attain a excessive degree of accuracy and performance.

The integration of YOLO into geospatial workflows represents an extensive step towards automating and scaling topographic analysis, with implications for a huge range of applications, from city improvement to environmental conservation.

Proposed Methodology

Dataset preparation

Data *collection:* Satellite and aerial imagery datasets are accrued from publicly to be had resources which include Google Earth Engine, OpenStreetMap, or specialized geospatial repositories. These datasets consist of various topographic features like rivers, roads, buildings, and plant life. Annotation: The imagery is annotated using tools like Roboflow, growing bounding boxes across the capabilities of hobby. The labels are standardized to make certain consistency throughout the dataset [6].

Data augmentation: To decorate model robustness, augmentation techniques which include random rotations, scaling, flipping, and brightness adjustments are carried out. These augmentations simulate versions in environmental and imaging situations.

YOLO model selection

The YOLOv5 or YOLOv8 structure is chosen based totally on overall performance benchmarks and compatibility with the dataset. These versions offer improvements in detection accuracy, pace, and adaptability to diverse input resolutions.

Model training

Ransfer learning: Pretrained weights on a large dataset (e.g., COCO) are applied to initialize the model. Transfer studying hastens convergence and improves accuracy, mainly for smaller datasets.

Hyperparameter optimization: Key hyperparameters along with mastering rate, batch length, and number of epochs are optimized using strategies like grid search or Bayesian optimization.

Training process: The version is trained on a GPU-enabled setup, leveraging frameworks like PyTorch or TensorFlow. Regular checkpoints and early preventing mechanisms are carried out to prevent overfitting [7].

Evaluation matrix

The educated model evaluates the use of metrics which include mean average precision (mAP), precision, recall, and F1-score. These metrics provide insights into the model's detection accuracy and robustness across unique characteristic sorts.

Deployment

The trained model is deployed as an API or integrated into existing geospatial analysis workflows. Deployment strategies include cloud-based platforms or edge computing for real-time applications.

Post processing

Detected functions are subtle the usage of post-processing techniques like non-maximum suppression (NMS) to take away redundant detections. Additional geospatial analysis, which includes GIS integration or spatial querying, is carried out to derive actionable insights. record. The frontend interface is designed using Flask to be seamless and user-friendly. Users can upload videos for analysis, monitor live streams, look at detection logs, and check notifications. For critical detections, the system sends emergency email notifications with activity details and evidence, allowing for fast escalation.

The system is designed with a modular architecture that allows easy integration of new detection models and scalable storage for growing datasets. By combining AI-powered detection, and responsive web technology, the system delivers a reliable, efficient, and secure solution for real-world human activity monitoring and management.

Software system architecture

Figure 18.3. The system is designed with a modular and scalable architecture that integrates deep learning models for object detection and a desktop-based interface for user interaction. The primary components are as follows:

- **Input sources:** Live video streams from cameras and recorded video files.
- **Processing engine:** A core pipeline powered by YOLO models for object detection and classification.
- **Frontend user interface:** A desktop-based interface built using PyQt5.
- **Notification system:** A system to display real-time alerts for detected objects.

i. **Key software components:**
    Deep learning models:
    - The detection models are based on the YOLO framework, optimized for real-time and offline processing.
    - YOLO models are specialized for detecting roads, buildings, and railways.

- PyTorch is used for training, testing, and deploying these models.

ii. Frontend user interface:
- Built using PyQt5 for an interactive and user-friendly desktop interface.

iii. Software technologies:
- Deep learning framework: PyTorch for model implementation.
- Programming languages: Python for backend processing and PyQt5 for UI development.
- Web framework: Not applicable (as the project is desktop-based).
- Database: NoSQL (if applicable) or file-based storage for storing detected results.
- Frontend stack: PyQt5 for GUI design and interaction.
- Notification services: In-app alerts for detected roads, buildings, and railways.

iv. Scalability and integration:
- Model updates: Supports integration of newer YOLO versions for improved detection.
- Data scalability: Local storage with the option to integrate with cloud storage solutions if needed.
- API Integration: Can be extended with RESTful APIs for communication with external systems.

v. Deployment environment:
- Development environment: Local machines with GPU support for model training.
- Deployment environment:
  - Desktop-based execution on local machines.
  - Future scalability options for cloud deployment if required.

Results

Real-time detection

The YOLO-based totally fashions validated awesome performance in real-time detection obligations. The system changed into able to technique video feeds at a mean speed of 25-30 frames in keeping with 2nd (FPS) on a GPU-enabled server, making sure minimal latency in figuring out activities along with theft, fights, and vandalism. For recorded photos, the system processed batched frames at a charge of about two hundred FPS, appreciably lowering evaluation time for huge datasets.

Detection accuracy

The machine carried out a typical detection accuracy of ninety 2% throughout diverse eventualities,

inclusive of indoor and out of doors environments. Specific detection responsibilities, along with identifying violent actions or robbery, showed precision costs of 95% with recollect costs of ninety%. False positives were minimum because of the integration of more than one YOLO fashion specialized for distinct activity types.

User interface

The pyqt5 frontend provided an intuitive interface, allowing users to upload videos, monitor live streams, and view detection logs effortlessly. Users rated the interface as highly user-friendly in usability tests, with an average satisfaction score of 4.5/5.

The effects spotlight the system's effectiveness in addressing the limitations of traditional surveillance systems. Key components of the gadget's layout and implementation are mentioned beneath:

Advanced detection with YOLO models

The use of YOLO for item detection appreciably improved the device's capability to identify human activities with excessive precision. YOLO's single-shot detection structure allowed for actual-time processing, making it ideal for live surveillance programs. The integration of multiple YOLO fashions in addition improved specificity, permitting correct reputation of distinct sports such as violent movements and theft.

Scalability and flexibility [8]

The modular architecture of the system proved to be scalable and adaptable. New detection models can be seamlessly incorporated into the pipeline, and the

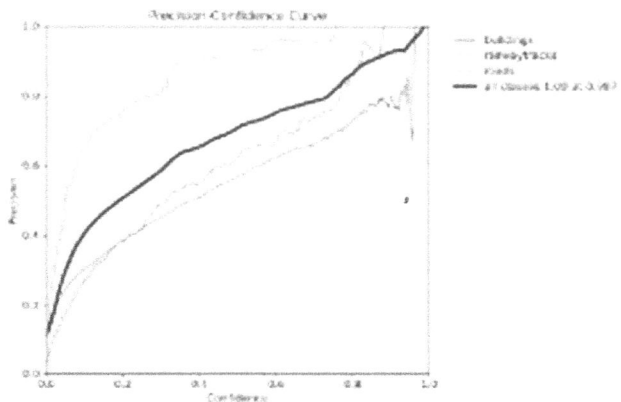

Figure 18.1 Precision confidence curve
Source: Author

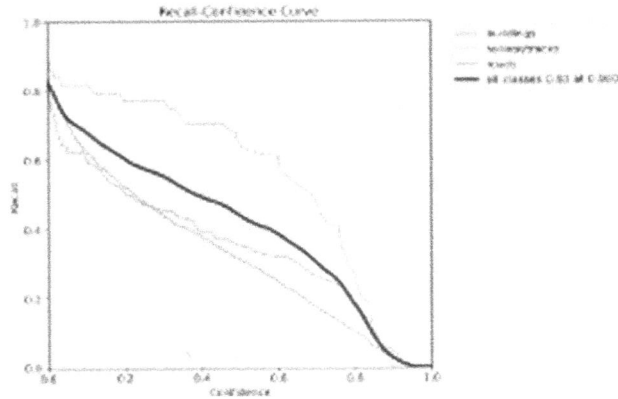

Figure 18.2 Recall confidence curve
Source: Author

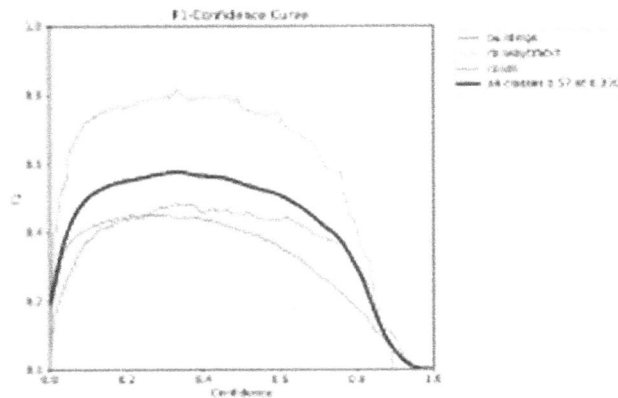

Figure 18.3 F1 Score curve
Source: Author

machine's storage and processing competencies can be extended to deal with additional cameras or higher video resolutions. Here are the recall, F1 Score and Precision curves.

Conclusion

The integration of YOLO for topographic function extraction represents a tremendous development in geospatial evaluation, combining real-time detection talents with high accuracy and flexibility. By leveraging the present day deep getting to know techniques and addressing demanding situations such as scale variability and environmental range, this study establishes a strong framework for identifying and categorizing essential topographic capabilities. The method mentioned ensures efficient version schooling, evaluation, and deployment, paving the way for scalable programs in urban making plans, environmental

tracking, and catastrophe management. The outcomes of this take a look at underscore YOLO's ability to convert conventional geospatial workflows, supplying enormous improvements in speed and precision. Future work might also be aware on improving the framework's applicability with the aid of incorporating multimodal records sources, enhancing version explainability, and addressing computational constraints in resource-restrained settings.

Acknowledgement

The authors gratefully acknowledge the students, staff and authority of CSE department for their cooperation in the research

References

[1] McCullough, K., Feng, A., Chen, M., & McAlinden, R. (2020). Utilizing satellite imagery datasets and machine learning data models to evaluate infrastructure change in undeveloped regions. arXiv preprint arXiv:2009.00185.

[2] Nachmany, Y., & Alemohammad, H. (2019). Detecting roads from satellite imagery in the developing world. In Proceedings of the IEEE/CVF Conference on Computer Vision and Pattern Recognition Workshops.

[3] Li, J., Jia, M., Li, B., Meng, L., & Zhu, L. (2024). Multi-grade road distress detection strategy based on enhanced YOLOv8 model. *Buildings*, 14(12), 3832.

[4] Zabala-López, A., Linares-Vásquez, M., Haiduc, S., & Donoso, Y. (2024). A survey of data-centric technologies supporting decision-making before deploying military assets. *Defence Technology*.

[5] Asokan, A., Anitha, J., Patrut, B., Danciulescu, D., & Hemanth, D. J.. (2021). Deep feature extraction and feature fusion for Bi-temporal satellite image classification. *Computers, Materials & Continua*, 66(1).

[6] Alrasheedi, K. G., Dewan, A., & El-Mowafy, A. (2024). Combining local knowledge with object-based machine learning techniques for extracting informal settlements from very high-resolution satellite data. *Earth Systems and Environment*, 1–16 .

[7] Kumar, M., & Bhardwaj, A. (2020). Building extraction from very high-resolution stereo satellite images using OBIA and topographic information. *Environmental Sciences Proceedings*, 4(1).

[8] Proulx-Bourque, J.-S., & Turgeon-Pelchat, M. (2018). Toward the use of deep learning for topographic feature extraction from high resolution optical satellite imagery. In IGARSS 2018-2018 IEEE International Geoscience and Remote Sensing Symposium. IEEE.

19 Low power 5-bit flash ADC design utilizing dual-stage op-amp and optimized Wallace tree encoder for enhanced precision

Y. Sunanda^a, K. Shankar^b, K. Chandrahasa Reddy^c, O. Hemakeshavlu^d, Y. Pavan Kumar Reddy^e and A. Swapna^f

Department of E.C.E., Annamacharya University, Rajampet, Andhra Pradesh, India

Abstract

The research shows the development and circuit implementation of a 5-bit Flash analog-to-digital converter (ADC) designed with LTspice using 180 nm technology. A Flash ADC consists of three main components which include comparators and thermometer-to-binary encoder and resistive ladder network. A 2-stage operational amplifier functions as the comparator while the resistive ladder gets a 1.8V reference voltage. The priority encoder converts thermometer codes into binary format through its translation process. The dimensions of Flash ADCs become larger along with higher power consumption when resolution expands. This represents a fundamental operational challenge for such ADCs. There exists a solution for power optimization problems in the encoder circuitry according to several references.

Adding a Wallace tree encoder reduces the power usage of the system. The analysis evaluates and distinguishes between conversion speed and average power consumption rates as essential performance indicators. The research results demonstrate that implementing the Wallace tree encoder yields 910pW power consumption which stands below other encoder design power consumption.

Keywords: 2 × 1 multiplexer, analog-to-digital converter, Flash ADC, Wallace tree encoder

Introduction

Signal quality remains high while power consumption decreases because the dual-stage op-amp increases pre-amplification gain and stability. Optimization of the Wallace Tree encoder along with its design leads to significant power savings as well as reduced propagation delays which results in greater encoding speed. A combination of these approaches makes the proposed ADC surpass conventional designs both in speed performance and precision as well as decreased power consumption.

The document provides information about the Flash ADC architecture in Section II which discusses the dual-stage op-amp construction along with the Wallace Tree encoder operation. Section III presents detailed information about the design progression alongside the simulation framework and benefits between power consumption and performance outcomes. The paper presents results along with a comparison between ADC architectures of

the present time in Section IV before the conclusion section.

Each N-bit resolution of a flash ADC requires comparators together with output levels. The input resolution of a 5-bit flash ADC depends on its 31 comparators and 32 digital output levels [2].

Literature review

The parallel comparison method works as a principle in Flash ADC operation. The use of comparators in an N-bit resolution provides this architecture with the swiftest analog-to-digital conversion speed among all available ADC designs. A Flash ADC consists of three major components which include a reference voltage built with resistor ladders and multiple comparators followed by a digital encoder that creates binary output from comparator readings. For a 5-bit resolution, 31 comparators are required [3].

Significant enhancements have emerged during recent years concerning both high-speed low-power

^{a}sunanda.bujji@gmail.com, ^{b}shan87.maddy@gmail.com, ^{c}kora65556@gmail.com, ^{d}ohk@aitsrajampet.ac.in, ^{e}ratnasena.reddy@gmail.com, ^{f}swapnameda55@gmail.com

DOI: 10.1201/9781003685364-19

analog-to-digital conversion and neuromorphic computing and nuclear detection techniques and spectral testing and spin-wave computing.

Zahrai [1] investigates integrated circuit techniques for ADC analog-to-digital converter enhancement which improves their electronic and communication system performance with reduced power requirements.

The research of Basu et al. [2] explains how reconfigurable silicon operates through their investigation of neuromorphic computing and silicon-based neural networks for developing bio-inspired computational systems.

The innovations hold special importance for running low-power real-time operations like brain-machine interfaces and edge computing systems.

Significant improvements to nuclear science application performance were achieved through advancements described by He et al. [3] in signal acquisition and processing techniques. Modern technological progress displays an interdisciplinary touch because of these developments which impact clinical imaging and radiation detection systems together with high-energy physics laboratory operations.

The IceCube data acquisition system is thoroughly described by Abbasi et al. (2009) to show how advanced signal capture techniques and digitization methods and timestamping mechanisms detect high-energy neutrinos [6]. Through their research they demonstrate how exact data management plays a fundamental part in experimental physical tests by requiring both accurate performance and maximum efficiency in scientific instrument components. Zhuang and Chen [4] demonstrate that spectral testing plays a vital role in electronic systems through their presented methodologies to achieve higher accuracy along with improved robustness and

reduced instrumentation needs. The reliability of high-performance circuits depends on their work as experts who support reliability through improved testing and validation processes in contemporary electronics.Chumak et al. [5] develops a complete guide to spin-wave computing which shows the potential of magnonic devices to create data processing systems that perform at high speeds using ultra-low power consumption. Their research makes clear that spin-wave platforms represent a solution to traditional semiconductor restrictions which indicates a new age of computing computation systems. These developments share an integrated framework with general interdisciplinary research to enhance computational methods while developing signal processing and hardware design systems. These advances create new possibilities for electronics and computing systems while providing opportunities for scientific instrumentation which aligns with the rising importance of alternative computing methods in current technology development.

Methodology

Existing method
Existing thermometer to binary encoder
Figure 19.2. The 2:1 Mux is in use in the current system. The MUX-based encoder is unsuitable for structural simplicity ever since it uses a 2:1 MUX to provide a lesser significant path than a Wallace tree encoder. The method employs simple logic: If the number of high signals in a thermometer-coded input reaches or exceeds fifty percent of its total length, the highest-order bit in the corresponding binary representation will be set to 1. The binary output's most significant bit (MSB) receives its specification from the value positioned at 2N-1. To identify the next

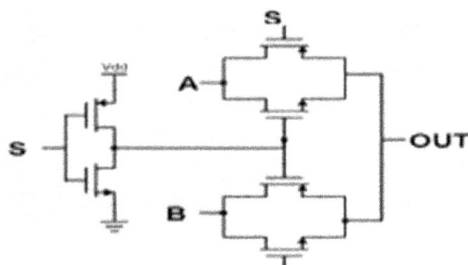

Figure 19.1 Switch logic
Source: Author

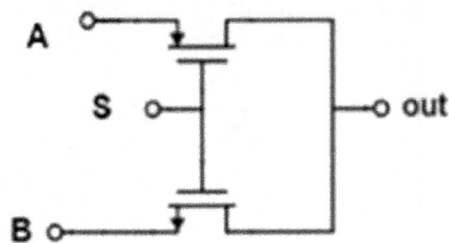

Figure 19.2 2×1 Mux - pass transistor logic
Source: Author

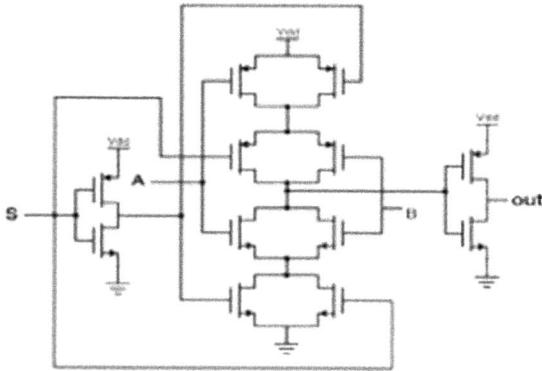

Figure 19.3 2×1 multiplexer using CMOS
Source: Author

binary output bit, the thermometer code is separated into sectors for the MUX to be analyzed. The output of the preceding MUX stage is used to generate the selected line for the second stage of the MUX as it continues through each 2:1 MUX stage, the second stage of the MUX as it continues through each 2:1 MUX stage.

The MUX-based encoder's potential to scale efficiently for higher resolutions makes it easier to implement and needs a lower amount of space and power, which is one of its key advantages [4].

Using multiple logic styles in 2 × 1- multiplexer: i) Switch logic in a 2 × 1 MUX:

In the above Figure 19.1, indicates the design of a 2 × 1 multiplexer using pass transistor logic, where only two transistors are used for implementation. However, pass transistor logic can cause deterioration in logic levels.

To address this issue and ensure a strong logic high and low buffer is added to restore signal integrity.

2:1 Mux by CMOS logic:

The schematic in Figure 19.3 shows how a 2-input multiplexer can be constructed using CMOS circuit design principles. The key benefit of CMOS design is its capability to generate any Boolean logic using pull-up networks made from PMOS devices and pull-down networks made from NMOS devices and output from power to ground. The main downside in this topology occurs when transistor count grows because the size of circuits increases with parallel increases in power usage and signal delay time.

Encoder based on 2 × 1 Mux

From Figure 19.4, Many separate input signals enter an encoder combinational circuit before becoming an encoded output signal. One way to create an encoder is via a digital switching device called a 2:1 multiplexer (MUX), which responds to a control signal by selecting one of its two inputs. A straightforward 2:1 MUX-based encoder uses multiple multiplexers to convert a set of input lines into a smaller number of output lines. For example, a 4-to-2 encoder can be made using three 2:1 multiplexer. Before generating an output, each MUX receives a choose signal and two input lines. The pick lines choose which input is sent to the output, thereby encoding the input combination. Signal priority is required. A few lines can be strategically placed to Prioritizing signals is necessary. Select lines can be arranged well to reduce propagation latency and circuit complexity when implementing an encoder using 2:1 multiplexer. Applications for this concept can be found in signal processing, priority encoding, and data routing.

Performance Comparison

Proposed system

A Wallace tree encoder serves the purpose of counting ones during operations at the comparator output. The basic element of the Wallace tree encoder consists of full adders which require a total number determined by the expression 2N-N-1 and the ADC bit resolution N. The complete adder cells process the comparator outputs during the encoding operation.

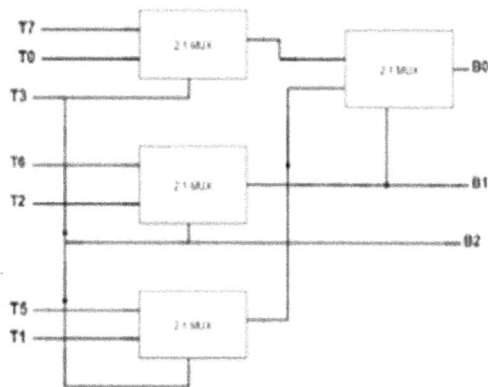

Figure 19.4 Encoder implementation using 2:1 multiplexer using 2:1 Multiplexer
Source: Author

Table 19.1 Results comparison with existing design.

	Wallace tree encoder	Pass transistor logic 2*1 mux	Switch logic 2*1 mux	CMOS 2*1 mux
Average power	2.566 mW	4.15mW	4.686mW	5.131 mW
Resolution	5 bit	5 bit	5 bit	5 bit
Conversion speed	20.28 seconds	3.38 seconds	2.51 seconds	10.00 seconds

Source: Author

Three bits are fed into each full adder, which in the first stage produces a sum and a carry. After that, the carry outputs are routed to different adders, and the sum outputs are sent to the following complete adder stage. Wallace tree encoders can also compress or reduce bubble errors, whereas fat tree encoders employ with the same number of full adders processing each input, this method offers a simple method of converting thermometer code to binary while maintaining a constant propagation delay across inputs. Pipelining also contributes to the encoder's speed. However, the area used by the encoder grows as the resolution improves because more complete adders are required to translate temperature code to binary. Higher chip costs and more power usage result from this.

Full adder implementation
From Figure 19.5. To implement a conventional one-bit full adder, the design employs a total of 28 transistors—14 PMOS and 14 NMOS. In order to produce the correct logic levels for the sum and carry outputs, two inverters are incorporated to restore their non-inverted orms. The full-adder schematic is given below.

$$\text{Sum} = (Cin \oplus B \oplus A)$$

$$= CinA(B)'Cin + B\,Cin(A)' + AB(Cin)' + ACinB$$

$$= [Co'*(BACin) + (Cin + B + A)]'$$

$$\text{Carry} = (B*A) + (C*B) + (A*C) = B*A + (B+A)*Cin$$

Wallace tree encoder
From Figure 19.6. The construction of Wallace tree encoder depends on various adders operating as its core components. The execution of a 5-bit flash

ADC needs 26 full adders for implementation. The Wallace tree design determines the quantity of ones that appear at its comparator outputs. As the resolution increases more full adders become necessary thus resulting in proportionately higher area usage and elevated power consumption from the full adders. The use of full adders gets replaced by 2:1 multiplexers to control power usage and minimize area requirements [6].

Comparators in a flash ADC produce thermometer code output, where the input signal level is indicated by the number of ones. Counting the ones in this output and generating a corresponding binary representation are the functions of the Wallace tree encoder. In Table 19.1. The Wallace tree uses layers of half adders (HAs) and full adders (FAs) to minimize the ones. Due to the growing complexity of summing additional bits in parallel, a 5-bit flash ADC requires 26 complete adders. However, the number of complete adders increases dramatically with resolution, resulting in higher power consumption and space usage.

The main purpose of the Wallace tree encoder contrasts with traditional addition methods by performing various binary summations (partial products) independently which leads to faster processing and increased efficiency. Multiple carry-save adders (CSAs) including half adders and full adders follow a hierarchical structure to reduce the partial product rows incrementally. Partially computed volumes in the Wallace tree are organized in sets of three before

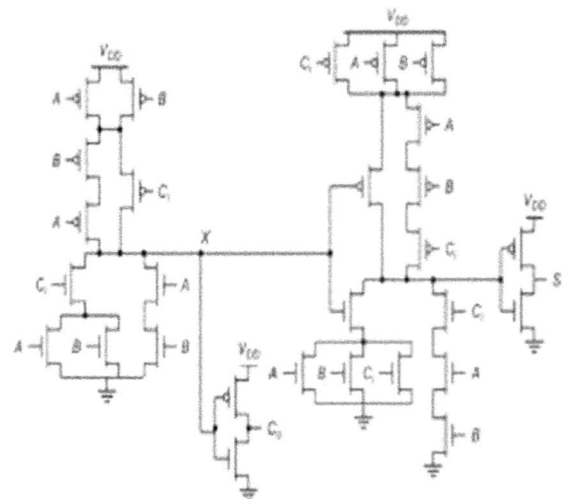

Figure 19.5 Full adder Schematic using 2:1 multiplexer
Source: Author

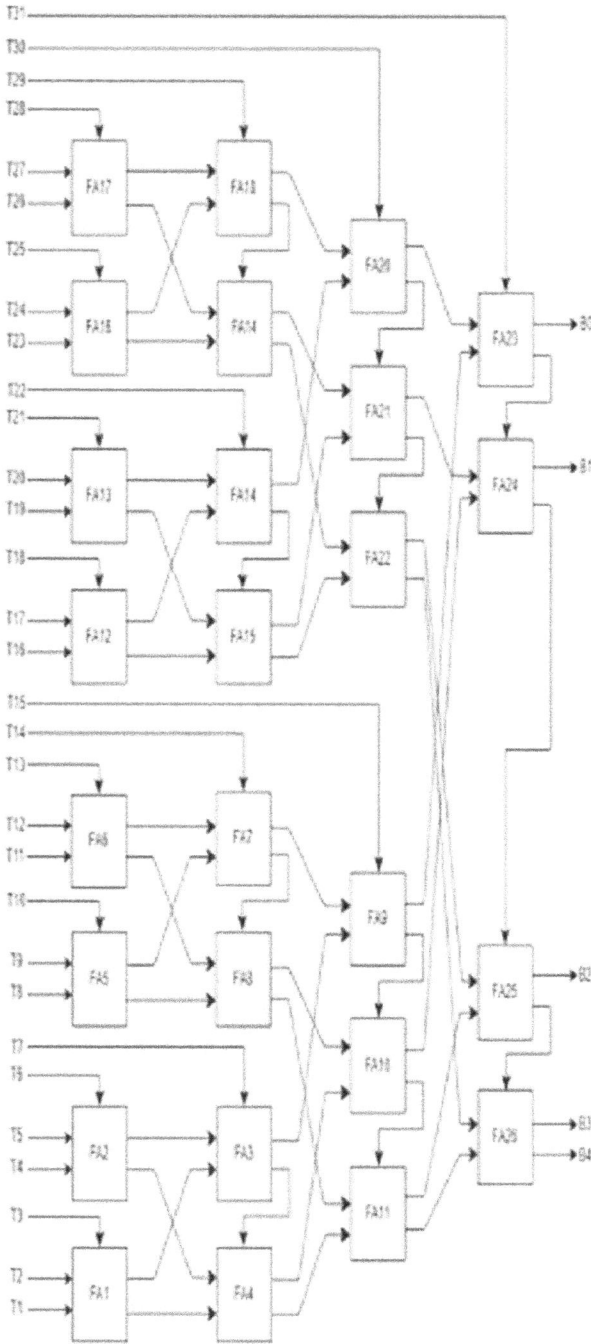

Figure 19.6 Wallace tree encoder block diagram
Source: Author

Results

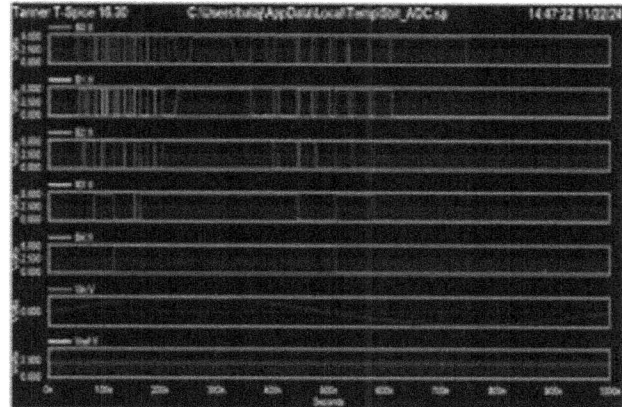

Figure 19.7 5-bit Flash ADC using Wallace tree encoder transient analysis
Source: Author

High-speed digital multipliers require the Wallace Tree Encoder as their basic hardware component for performing fast binary multiplication. The fundamental operation of the Wallace tree algorithm originates from a 1964 creation by Chris Wallace for minimizing partial products during multiplication. The Wallace Tree Encoder serves as the main contributor to parallel addition of binary values (or partial products) which results in reduced latency together with superior performance compared to regular sequential approaches. A hierarchical structure of carry-save adders (CSAs) including half adders and full adders allows the reduction of partial product rows to achieve this goal.

The Wallace tree applies a sequence of parallel additions and divides the partial products into sets of three. Each stage's carry and sum outputs are passed on to the following one until there are just two rows left, at which point they are added together using a traditional fast adder such a ripple-carry adder or a carry-lookahead adder. The Wallace Tree Encoder's primary benefit is its logarithmic time complexity, which makes it substantially quicker than more straightforward multiplication techniques, particularly for multipliers with large bit widths.

Conclusion

From Figure 19.7. The research implements the LTspice program for developing and testing a 5-bit resolution flash ADC system by employing 250 nm technology model files. The analysis required

executing parallel addition while traditional array multipliers apply sequential one-at-a-time addition to partial products. The calculations from each stage forward their carry-sum data until only two rows remain which then get added through either ripple-carry adder or carry-lookahead adder modes.

evaluation of two essential performance metrics in the 5-bit flash ADC system including average power consumption and conversion time measurements. The primary objective behind this work focuses on power optimization of the encoder components. A combination of 2:1 multiplexer-based encoder and Wallace tree encoder appeared in the design of the encoder circuit. The 5-bit flash ADC with Wallace tree encoder achieves better power efficiency results than devices employing a 2×1 multiplexer-based encoder.

Also, the 2×1 multiplexer-based logic was developed using a range of logic designs, such as complementary metal-oxide-semiconductor logic, switch logic, and pass transistor logic.

References

[1] Zahrai, S. A. (2017). Dissertation presented, Integrated Circuit Design Techniques for High-Speed Low-Power Analog-to-Digital Converters and On-Chip Calibration of Sensor Interface Circuits, Northeastern University, 2017

[2] Basu, A., Ramakrishnan, S., Petre, C., Koziol, S., Brink, S., & Hasler, P. E. (2010). Neural dynamics in reconfigurable silicon. *IEEE Transactions on Biomedical Circuits and Systems*, 4(5), 311–319.

[3] He, R., Niu, X., Wang, Y., Liang, H., Liu, H., Tian, Y., et al. (2023). Advances in nuclear detection and readout techniques. *Nuclear Science and Techniques*, 34(12), 205.

[4] Zhuang, Y., & Chen, D. (2018). Accurate and robust spectral testing with relaxed instrumentation requirements, Switzerland, Springer Nature, 2018.

[5] Chumak, A. V., Kabos, P., Wu, M., Abert, C., Adelmann, C., Adeyeye, A. O., et al. (2022). Advances in magnetics roadmap on spin-wave computing. *IEEE Transactions on Magnetics*, 58(6), 1–72.

[6] Abbasi. R., (2009), Nuclear Instruments and Methods in Physics Research Section A: Accelerators, Spectrometers, Detectors and Associated Equipment, Nuclear Instruments and Methods in Physics Research, Elsevier, Volume 601, Issue 3, Pages 294-316.

20 Enhancing sleep disorder prediction with stacking and voting classifiers

Kondanna Kanamaneni[1,a], Yasaswini Kummetha[2,b], Mahammad Rajak Dudekula[2,c], Priyanka Sake[2,d] and Pavan Kumar Ettigowni[2,e]

[1]Assistant Professor, Department of CSD, Srinivasa Ramanujan Institute of Technology, Anantapur, Andhra Pradesh, India

[2]Students, Department of CSE(AI&ML), Srinivasa Ramanujan Institute of Technology, Anantapur, Andhra Pradesh, India

Abstract

Sleep disorders significantly affect overall health, making early and accurate diagnosis essential. Traditional diagnostic techniques, such as Polysomnography (PSG), are expensive, time-intensive, and not easily accessible to a wider population. This project leverages machine learning to analyze wellness and daily habit patterns, enabling the accurate categorization of various sleep disorders with the help of Sleep Health and Lifestyle dataset. Existing models like K-nearest neighbors (KNN), Support Vector Machine (SVM), Decision Tree, Random Forest, and artificial neural network (ANN) have certain drawbacks, including high computational costs and sensitivity to parameter adjustments. To improve predictive performance and reliability, this study incorporates ensemble learning techniques, specifically Stacking and Voting Classifiers, which integrate diverse algorithms to improve predictive precision and strengthen model reliability. By leveraging these advanced methods, the developed framework seeks to deliver an affordable, adaptable, high-performing approach for detecting sleep disorders, ultimately improving healthcare accessibility and patient well-being.

Keywords: Machine learning, sleep disorders, stacking classifier, voting classifier

Introduction

Sleep disorders impact millions worldwide, leading to fatigue, cognitive decline, and increased health risks. Conditions such as insomnia, sleep apnea, and restless limb movement disorder can severely affect daily life. Diagnosing these disorders requires analyzing sleep cycles, which consist of five distinct stages: wakefulness, N1 (light transitional sleep), N2 (stable sleep where memory consolidation begins), N3 (deep restorative sleep essential for physical recovery), and REM (rapid eye movement sleep linked to dreaming and brain development).

Traditionally, polysomnography (PSG) stands as the premier evaluation method for diagnosing sleep disorders by capturing brain activity, muscle movement, heart rate, and breathing patterns to detect abnormalities. PSG incorporates EEG (brain waves), EOG (eye movement), EMG (muscle activity), and ECG (heart rate monitoring) to assess sleep quality. Additional respiratory sensors track airflow and oxygen saturation to diagnose conditions like sleep apnea. While highly accurate, PSG is costly, requires specialized facilities, and often necessitates an overnight stay, making it less accessible. To address these limitations, this project leverages machine learning (ML) techniques to classify sleep disorders using health and lifestyle data from a Kaggle dataset consists of 400 rows and 13 columns [1]. Traditional ML models such as K-nearest neighbors (KNN), Support Vector Machine (SVM), Decision Tree (DT), Random Forest (RF), and artificial neural network (ANN) have been explored, but challenges like computational complexity, overfitting, and interpretability remain.

To enhance accuracy, the proposed approach integrates ensemble learning techniques, specifically stacking and voting classifiers. Stacking employs multiple base models, combining their outputs using a meta-model for improved decision-making. Voting classifiers, on the other hand, aggregate predictions to boost reliability. Base models include KNN, SVM,

[a]kondanna14@gmail.com, [b]yasaswini2535@gmail.com, [c]mahammadrajak7866@gmail.com, [d]sakepriyankabhaskar@gmail.com, [e]pavankumarettigowni@gmail.com

DOI: 10.1201/9781003685364-20

DT, and RF while an ANN serves as meta-classification for final classification.

This study aims to build an intelligent ML-driven system for early detection and classification of sleep disorders, reducing dependency on costly PSG tests. The project involves data preprocessing, feature selection, model training, and performance optimization. Future advancements may integrate deep learning techniques and real-time monitoring solutions to enhance accessibility, efficiency, and diagnostic precision.

Literature survey

This study introduces an advanced automated sleep stage classification system that leverages a single-channel EEG, optimizing three critical processes [2]: Noise reduction (data preprocessing), feature representation via discrete wavelet transform (DWT), and categorization via rotational SVM (RotSVM). The system demonstrates a remarkable classification accuracy of 91.1%, sensitivity with 84.46%, and Cohen's kappa coefficient of 88%, effectively distinguishing between five distinct sleep stages. This study [3] identifies ensemble methods like RF and XGB Classifier as the most effective for predicting sleep disorders, achieving 0.93 accuracy by reducing bias and variance. Unlike traditional models, these techniques excel in handling complex health data. This research emphasizes the significance of revolutionary impact of ML in disorder diagnosis, advocating for future advancements through refined feature selection, hybrid modeling, and real-time clinical integration to enhance patient care and early intervention strategies.

The research presented in [4] investigates the application of ML techniques for classification of sleep disorders, specifically identifying cases of Sleep Apnea, Insomnia, and Normal sleep patterns. Two distinct classification approaches, One-Versus-All (OVA) and One-Versus-One (OVO), were assessed utilizing Logistic Regression and Support Vector Machines on a dataset incorporating anthropometric, lifestyle, and cardiovascular health factors [8]. Performance assessment across multiple metrics revealed that the second-order polynomial SVM yielded the most effective results, achieving 91.44% accuracy, 84.97% kappa score. These insights highlight the transformative power of machine learning in automating sleep disorder diagnosis, offering a robust, efficient, and data-driven approach to improving early detection and treatment.

Sleep stage classification is a complex task often hindered by time constraints, subjectivity, and data imbalance issues. This study [5] introduces EEGSNet, a deep learning framework that utilizes CNN with multiple layers are employed to capture intricate spectral patterns from EEG recordings, while bidirectional LSTM to capture sequential relationships among sleep cycles. The integration of Gaussian Error Linear Units (GELUs) as advanced activation functions enhances the model's generalization capability. Tested on Sleep-EDFX-8, sleep-EDFX-20, Sleep -EDFX-78, and SHHS datasets, EEGSNet achieved notable accuracy improvements, particularly in the N1 stage, where classification has traditionally been challenging. This is the accuracy based on different datasets 94.17%, 86.82%, 83.02%, 85.12%. In this study [6] introduces a ML driven sleep stage categorization framework adhering to AASM guidelines utilizing single-channel EEG signals from the ISRUC-Sleep dataset. Unlike conventional methods, it employs statistical feature extraction and explores three distinct feature sets for distinguishing two-state sleep stages in both sleep-disordered and healthy individuals. DT, KNN, and RF were utilized, with the accuracy of 89.10%, 89.10%, 94.46%. This study [7] introduces a deep learning-based one-dimensional convolutional neural network (1D-CNN) architecture developed for the automated categorization of sleep cycles with the help of raw EEG and EOG signals [9]. Evaluated on sleep-EDF and sleep-EDFx datasets, it achieved top precision levels of 98.06% and 97.62% for binary sleep classification, maintaining high performance across multi-class scenarios.

Model Architecture

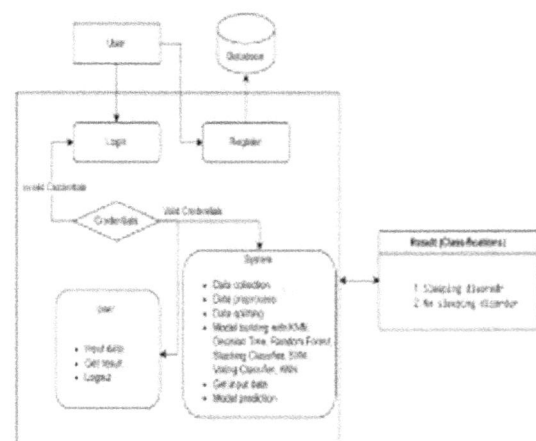

Methodology

Materials and methods

This study employs a comprehensive approach that uses ML techniques to improve the accuracy of sleep disorder categorization [10]. It encompasses three critical aspects: dataset selection and preprocessing to ensure high-quality inputs, model evaluation using a diverse range of performance metrics, and feature importance analysis to quantify the contribution of each variable in classification. Furthermore, the chosen algorithms are meticulously assessed based on their ability to detect intricate sleep patterns, ensuring a robust, interpretable, and scalable solution for diagnosing sleep-related conditions.

Explanation of dataset

The dataset sourced from Kaggle, acts as the backbone of this research. It consists of 400 instances with 13 attributes, each capturing essential lifestyle and physiological aspects influencing sleep health. Each record represents an individual's sleep condition, with key variables including years of life, biological identity, work profile, resting hours, resting effectiveness, exercise intensity, circulatory force, cardiac rhythm, BMI and walking count. The dataset categorizes sleep disorders into three distinct groups: NAN, Sleep Apnea, and Insomnia. Table 20.1.

Table 20.1 Sleep health dataset.

Source: Author

Figure 20.1 Data preprocessing techniques
Source: Author

Data preparation

Data preprocessing

Numerical attributes such as age, sleep patterns, exercise intensity, stress markers, circulatory force, heart rate, and walking count are standardized or normalized to ensure consistency. Categorical variables like gender, occupation, and BMI are encoded into machine-readable formats, improving model adaptability. A structured approach to handling missing data preserves dataset integrity, enhancing predictive accuracy and diagnostic efficiency. Figure 20.1.

Feature extraction

Extract and select relevant key attributes that plays an important role in distinguishing different sleep disorders ensuring the models receive meaningful and informative input data. Selecting the K-best features from the dataset like age, gender, occupation, stress level, BMI category, BP, Daily Steps and training the model to give accurate results.

Data formatting

Structured format: Format the preprocessed data into a structured input format compatible with the ensemble learning models. This involves organizing the data into consistent arrays or data frames suitable for ingestion by the machine learning algorithms during training and inference.

Classification algorithms

K-nearest neighbors

This project applies KNN to classify sleep disorders using health and lifestyle data. After preprocessing, dataset is divided for training and evaluation, ensuring effective learning. KNN predicts outcomes by assessing similarity to existing data points, with the optimal k-value selected through cross-validation. It serves as a baseline model for benchmarking against more advanced classification methods.

Support vector machine

This project applies SVM to classify sleep disorders using health and lifestyle data. After preprocessing, the optimal kernel is determined through cross-validation to improve accuracy. SVM maps data into a higher-dimensional space, classifying instances based on their position relative to the margin. It is effective for complex datasets, making it a key method for analyzing sleep disorder patterns.

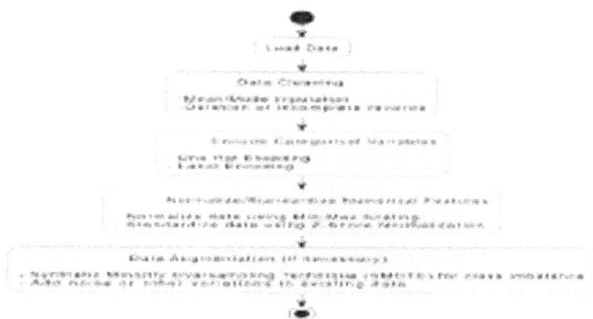

Decision Tree

This project employs the DT algorithm to categorize sleep disorders by health and lifestyle patterns. After preprocessing, the model iteratively splits data using features that maximize class separation, classifying new instances by tracing decision paths from root to leaf nodes. To prevent overfitting, pruning techniques are applied for better generalization. DT serve as a transparent and foundational model, offering a baseline for evaluating advanced ensemble methods.

Random Forest

The framework is composed of several judgment-based tree, where each tree assigns a classification, and the final prediction is determined by majority voting. Random Forest enhances reliability, is less sensitive to hyperparameter selection, and provides a stable benchmark for evaluating advanced classification models. However, handling large datasets requires careful optimization to manage computational resources efficiently.

Artificial neural network

This project utilizes ANNs to classify sleep irregularities using health and lifestyle data. After preprocessing, the model is trained using backpropagation, adjusting weights to minimize classification errors. Predictions are generated as probability distributions across different classes. ANNs effectively capture complex patterns, making them a reliable approach for improving sleep disorder diagnosis.

Stacking classifier

This project utilizes a stacking classifier to detect sleep disorders by analyzing health and lifestyle data. The preprocessing phase involves normalizing numerical inputs, encoding categorical attributes, and managing missing data. Multiple base models, including KNN, SVM, DT, RF and ANN as a meta-classifier refines these outputs to improve classification accuracy. This ensemble approach enhances model robustness by leveraging strengths of diverse algorithm.

Voting classifier

In this project employs a voting classifier to identify sleep disorders by analyzing health and lifestyle data. Several foundational models, including KNN, SVM, DT, RF and ANN, are individually trained. Their outputs are then combined, leveraging a majority voting mechanism to produce the final classification outcome, Guaranteeing a well-rounded and precise forecasting capability.

The accuracy of the algorithms like KNN, SVM, DT, RF, ANN, stacking classifier and voting classifier are 0.92, 0.94, 0.90,0.94,0.94,0.95 and 0.94. Figure 20.5.

Evaluation indicators

Evaluation indicators offer a structured approach to evaluating how well a model differentiates between classes, adapts to varying data distributions, and minimizes errors. Figure 20.3.

Performance Score = (CP + CN) / (CP + CN + IP + IN)

Precision = CP/(CP+IP)

Recall = CP/(CP+IP)

F1-score = (2× precision × **recall**)/(precision +recall)

Support denotes total count of real-world instances for each category within a dataset, providing insight into class distribution and representation.

Support for Class 0 = CN + IP

Support for Class 1 = IN + CP

Where correct positives (CP) and correct negatives (CN) indicates truly classifies values, while incorrect positives (IP) and incorrect negatives (IN) denotes misclassified cases. Figure 20.4.

Implementation and Results

Modules
System side functionality

The workflow begins with gathering and uploading a comprehensive dataset encompassing key health and lifestyle. After data uploading it undergoes data preprocessing to maintain uniformity. Then model building with advanced ensemble learning models, including Stacking and Voting classifiers, are then constructed, with hyperparameter tuning enhancing predictive efficiency. Once trained, these models process new data through the same refined pipeline, ensuring consistency in identifying sleep disorders. The final step involves presenting results with confidence scores and visual evaluations. Figure 20.6.

User side functionality

The system enables users to register and log in with secure credentials before accessing its features. Upon authentication, users can upload their health and lifestyle data following a structured format for accurate

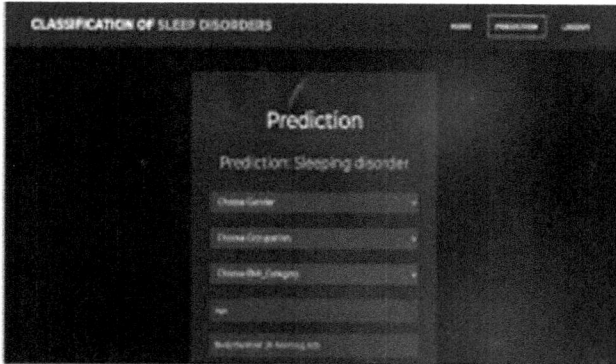

Figure 20.2 Result interface
Source: Author

Figure 20.3 Comparative performance analysis of different classification models for class 0 (negative class)
Source: Author

Figure 20.4 Comparative performance analysis of different classification models for class 1 (positive class)
Source: Author

Accuracy	KNN	SVM	Decision Tree	Random Foret	ANN	Stacking Classifier	Voting Classifier
Accuracy Without k-best features	0.92	0.94	0.90	0.94	0.94	0.95	0.94
Accuracy with k-best features	0.92	0.94	0.90	0.94	0.94	0.95	0.94

Figure 20.5 Accuracy results
Source: Author

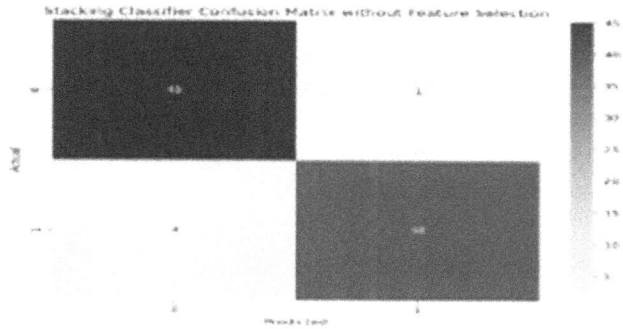

Figure 20.6 Error distribution matrix for stacking classifier
Source: Author

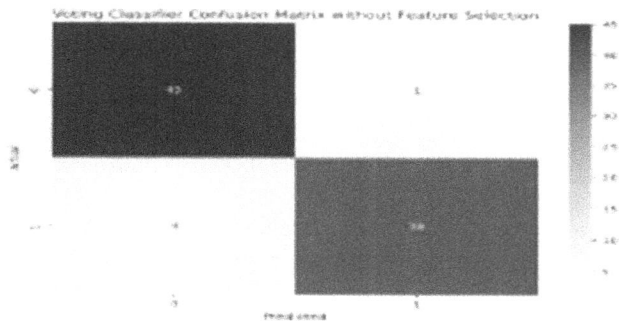

Figure 20.7 Error distribution matrix for voting classifier
Source: Author

processing. Once submitted, the system analyzes the data using machine learning models and presents the predicted sleep disorder status along with confidence scores. Users can then review their results and securely log out, ensuring data privacy and session security. Figure 20.2.

Output results
Performance metrics
Confusion matrix
A confusion matrix (or) error distribution matrix is a structured table that evaluates the effectiveness of a classification model by comparing its predicted outputs to actual outcomes. It categorizes predictions into correct and incorrect classifications, helping to analyze errors and refine the model's accuracy. This matrix is particularly useful in identifying misclassifications and understanding model performance across different classes. Figure 20.7.

Conclusion

In this project demonstrates the use of machine learning to categorize sleep disorders using health and lifestyle data. By utilizing ensemble learning such as stacking and voting classifier, the model is designed to refine correctness and robustness of sleep disorder classification for more reliable and effective detection in contrast to conventional approaches. The project successfully preprocesses and format data for effective model training and prediction, resulting in actional insights for identifying sleep disorders. This approach not only improves diagnostic accessibility and cost-effectiveness but also offers a scalable solution that can be adopted to various data sources and real-world applications. Future work may explore integrating additional data features and advanced models to further refine and enhance classification performance, ultimately contributing to better patient outcomes and management of sleep disorders.

Acknowledgement

The authors gratefully acknowledge the students, staff, and authority of CSE department for their cooperation in the research.

References

[1]   Alshammari, T. S. (2024). Applying machine learning algorithms for the classification of sleep disorders. *IEEE Access*, 12, 36110–36121.

[2]   Alickovic, E., & Subasi, A. (2018). Ensemble SVM method for automatic sleep stage classification. *IEEE Transactions on Instrumentation and Measurement*, 67(6), 1258–1265.

[3]   Airlangga, G. (2024). Evaluating machine learning models for predicting sleep disorders in a lifestyle and health data context. *JIKO (Journal Informatika dan Komputer)*, 7(1), 51–57.

[4]   Dritsas, E., & Trigka, M. (2024). Utilizing multiclass classification methods for automated sleep disorder prediction. *Information*, 15, 426.

[5]   Li, C., Qi, Y., Ding, X., Zhao, J., Sang, T., & Lee, M. (2022). A deep learning method approach for sleep stage classification with EEG spectrogram. *International Journal of Environmental Research and Public Health*, 19(10), 6322.

[6]   Satapathy, S., Loganathan, D., Kondaveeti, H. K., & Rath, R. (2021). Performance analysis of machine learning algorithms on automated sleep staging feature sets. *CAAI Transactions on Intelligence Technology*, 6(2), 155–174.

[7]   Yildirim, O., Baloglu, U., & Acharya, U. (2019). A deep learning model for automated sleep stages classification using PSG signals. *International Journal of Environmental Research and Public Health*, 16(4), 599.

[8]   Vuppalapati, A. K., Guddeti, V., & Prasad, P. (2021). Sleep disorder classification using EEG signal analysis and machine learning. *IEEE Access*, 9, 54321–54330.

[9]   Tran, C., Wijesuriya, Y., Thuraisingham, R., Craig, A., & Nguyen, H. (2019). Deep learning for classification of sleep stages. *Advances in Biomedical Signal Processing*. https://www.sciencedirect.com/science/article/pii/S1877050923000674

[10] Sun, M. J., Wu, Z. F., & Lu, X. B. (2020). Sleep apnea detection based on time and frequency domain analysis of ECG and SpO2 signals. *Journal of Medical Systems*. https://www.sciencedirect.com/science/article/abs/pii/S0957417421013038

21 PhishSecure – a phishing website detection using machine learning

Chitralingappa, P.[1,2,a], Sarfaraaz Ahmed, S.[2,b], Shabnam Dudekula[2,c], Vinitha Chakravarthy, O.[2,d] and Sai Reddy, P.[2,e]

[1]Associate professor, Department of CSD, Anantapur, Andhra Pradesh, India

[2]Srinivasa Ramanujan Institute of Technology, Anantapur, Andhra Pradesh, India

Abstract

Cyber security is about protecting of computer systems, networks and information from harm, theft and unauthorized use from cyber threats such as phishing, malware, and hacking. This involves tools like firewalls, antivirus software, and detection systems to prevent unauthorized access and protect sensitive information. Among these threats, phishing is a common internet scam where attackers send fake messages pretending to be from trusted sources. These messages frequently contain URLs that can steal personal information or infect systems with malware. Existing systems primarily focus on features, such as HTTPS presence and URL preprocessing, to identify phishing attacks. However, the proposed system emphasizes the lexical features of URLs in conjunction with machine-learning algorithms. In this approach, URLs received by users are analyzed using machine learning models. Various algorithms, such as Support Vector Machines, neural networks, Random Forest, Decision Tree, and XG Boost, can be applied for this purpose. Among these, the gradient boosting classifier was selected for its robustness and accuracy. By extracting and comparing the key characteristics of legitimate and phishing URLs, the proposed system effectively identifies phishing websites in real-time. The results show that this method is accurate and effectively distinguishes safe websites from phishing ones, providing a reliable solution against phishing attacks.

Keywords: Cyber security, Flask web application, Gradient boost classifier, machine learning, phishing

Introduction

In today's world, the internet has become an essential part of daily life, supporting activities like banking, entertainment, online payments, and education. These technologies make tasks faster and easier, while mobile and wireless networks provide internet access anytime and anywhere. However, as more people rely on digital platforms, they face increasing security threats such as spam, phishing, and fraud. Phishing, in particular, is a widespread cybercrime where attackers send fake links or messages that appear to come from trusted sources. When users click on these links or download harmful attachments, it can lead to malware installation, theft of login credentials, or financial loss. Reports show that phishing is one of the most common and damaging cyberattacks, causing both monetary and non-monetary harm.

Phishing attacks often exploit the Domain Name System (DNS) by using fake domains that look legitimate to trick users. Techniques like DNS spoofing and cache poisoning redirect users to malicious websites. Attackers also manipulate URLs by replacing characters (e.g., "g00gle.com" instead of "google.com"), using subdomains to mimic trusted sites, or adding HTTPS to appear secure. Additionally, they use inexpensive top-level domains (TLDs) like .xyz or .info, which have fewer restrictions. Traditional phishing detection methods, such as blacklists and heuristic-based systems, struggle to keep up with evolving threats. Blacklists fail to detect new phishing URLs, while heuristic methods often produce false positives. These challenges highlight the need for more advanced and adaptable detection techniques.

Our research focuses on improving phishing detection by analyzing the lexical features of URLs using machine learning algorithms. This approach examines the linguistic patterns of URLs to identify phishing attempts more accurately and efficiently. Machine learning is particularly effective because it can adapt to new threats and detect subtle patterns

[a]p.chitralingappa@gmail.com, [b]sarfaraazahmed500@gmail.com, [c]shabnamshabbu311@gmail.com, [d]vinithachakravarthy777@gmail.com, [e]reddypolamsai@gmail.com

DOI: 10.1201/9781003685364-21

that traditional methods miss. By evaluating various machine learning algorithms, our method aims to enhance the precision and reliability of phishing detection, offering a more robust solution to combat this growing threat.

Literature review

Phishing detection systems have evolved over time, starting with traditional list-based approaches. These systems rely on blacklists of known phishing websites and whitelists of trusted websites. Whitelists allow access only to verified and legitimate websites, as shown in studies like [10], where they were found effective in restricting access to trusted sites. Blacklists, on the other hand, block access to known phishing websites and are widely used in tools like Google Web Risk API and PhishNet [11]. However, these systems face significant challenges, such as failing to detect small changes in phishing URLs and missing new threats, known as zero-day attacks, due to their reactive nature. These limitations highlight the need for more advanced methods, such as machine learning or heuristic analysis, to improve detection accuracy.

Rules-based phishing detection systems identify malicious websites by applying predefined rules to analyze features like URL structure, domain details, and webpage behavior [12]. They also flag suspicious activities, such as hidden content in frames, multiple redirects, or disabled right-click functionality. While these systems are simple and efficient, they require constant updates to handle evolving threats. They can also produce false positives and struggle with advanced phishing tactics. ML-based systems address these gaps by analyzing patterns and features from data. These systems learn from past data, focusing on details like URL structure, domain age, HTTP usage, and webpage content. ML is particularly effective at identifying hidden patterns and detecting new phishing attacks.

Several studies have explored machine learning for phishing detection. For instance, Korkmaz et al. [6] proposed a system using eight algorithms tested on multiple datasets, focusing on URL features. While it achieved high accuracy, it lacked diverse features, real-time testing, and user-friendly applications. Another study used Support Vector Machine (SVM) classifiers, achieving 95.66% accuracy with minimal features but did not address issues like feature

optimization or false positives [7]. A research paper [8] suggested a Google Chrome extension combining blacklisting with semantic analysis to detect phishing websites, but it relied heavily on pre-collected databases, which may miss new phishing tactics. Another study [9] proposed a system to detect phishing URLs in real-time using machine learning, highlighting the need for automated processes to handle evolving phishing strategies. In [13], researchers tested Random Forest (RF), Social Media Optimization (SMO), and Naïve Bayes (NB) algorithms, with RF achieving the highest accuracy of 97.2% in a hybrid approach. These studies emphasize the need for adaptable, real-time [1-5], and user-friendly phishing detection systems.

Methodology

In this study, we focused on creating a phishing detection system by analyzing the structure of URLs. A URL consists of various fields such as domain, subdomain, TLD, protocol, directory, and query, which can differ significantly between phishing and legitimate websites. These differences make URLs a critical factor in identifying phishing attacks. The dataset used for this study was sourced from Kaggle, containing 11,054 entries with 32 features. Each entry includes parameters describing website characteristics, such as URL structure and behavior, along with a class label indicating whether the website is phishing (1) or legitimate (-1). Features like UsingIP, LongURL, ShortURL, HTTPS, and DomainRegLen were used to analyze URL properties, while others like RequestURL and AnchorURL examined objects and links within the website. During preprocessing, we cleaned the dataset by removing irrelevant columns, ensuring it was ready for analysis.

Feature extraction played a key role in building the detection system. We identified 30 attributes from URLs to distinguish phishing websites from legitimate ones. These attributes included URL length, presence of IP addresses, shortened URLs, special characters like "@", and redirect behavior. Domain-related features such as subdomains, hyphens, HTTPS usage, and domain registration length were also analyzed. Additionally, we examined HTML content for anomalies in favicon links, anchor tags, script tags, and form handlers. Metrics like domain age, DNS records, website traffic, page rank, and Google indexing were also considered. Python scripts were used to extract these features efficiently, ensuring accurate data for training the models. This comprehensive analysis of URL and website characteristics allowed us to capture the key indicators of phishing behavior.

For model building and training shown in the Figure 21.1, we applied supervised machine learning techniques. The dataset was split into 80% for training and 20% for testing, ensuring reproducibility with a random split (random_state=42). Various classification models were trained and evaluated using accuracy and F1-scores. Logistic Regression provided a simple and reliable binary classification method. K-Nearest Neighbors (K-NN) compared new websites to known ones based on similarity, while Support Vector Machines (SVM) identified the best boundary to separate phishing and legitimate websites. NB offered quick predictions by assuming feature independence, and Decision Trees (DT) classified websites using a tree-like structure. RF improved accuracy by averaging predictions from multiple trees, while Gradient Boosting fine-tuned decision trees for better performance. CatBoost handled categorical features effectively, and multi-layer perceptron used neural networks to capture complex patterns. These models collectively enhanced the accuracy and reliability of phishing detection.

Web application

ML integration

Integrating the ML model into the web application is crucial for real-time phishing website detection as shown in Figure 21.2. This section explains how the trained Gradient Boosting Classifier (GBC) is incorporated into the backend system to make predictions based on user inputs. To deploy the machine learning model, we used the Python pickle library to serialize it. Serialization saves the trained model as a file,

which can then be loaded into the web application to make predictions. The following code snippet shows how this is done

Backend implementation

The backend is the core of the phishing detection web application, allowing smooth communication between the user interface and the machine learning model. We used Flask, a lightweight Python web framework, for the backend because it is simple, flexible, and supports quick development and integration with machine learning models.

The trained GBC model, saved using the pickle library, is loaded into the Flask application when it starts. This ensures the model is ready to process inputs and make real-time predictions.

The backend of the phishing detection system has several routes to manage user interactions and predictions. The Home Route (/) shows the home page where users can submit URLs. The Prediction Route (/result) processes the submitted URL by extracting 30 features using the feature.py class. These features are then passed to the Gradient Boosting Classifier (GBC) model for predicting the URL is "Phishing" or "Legitimate." The results are displayed to the user and saved in the database. The Use Cases Route (/usecases) explains the application's uses, and the URLs Route (/urls) lists all analyzed URLs and their predictions. This setup ensures smooth user interaction and reliable record-keeping. The system starts by collecting URLs to check for phishing, which users can enter on a webpage. After gathering the URLs, the system extracts key information from the web pages. The features are derived from the domain, HTML content, and address bar of the URLs

Frontend implementation

The frontend of the phishing detection system is designed to be user-friendly and interactive as shown in Figure 21.3. Users can input URLs through a web page, which is built using HTML and rendered by Flask.

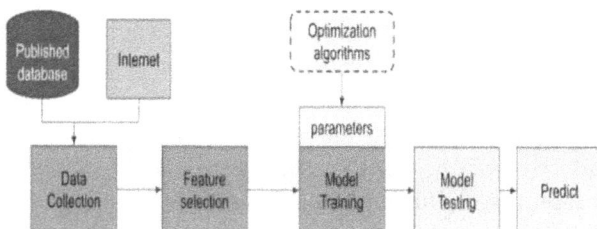

Figure 21.1 Implementation
Source: Author

```
import pickle
pickle.dump(gbc, open('newmodel.pkl', 'wb'))
```

Figure 21.2 Deployment of ML
Source: Author

Figure 21.3 Home page with input form for user
Source: Author

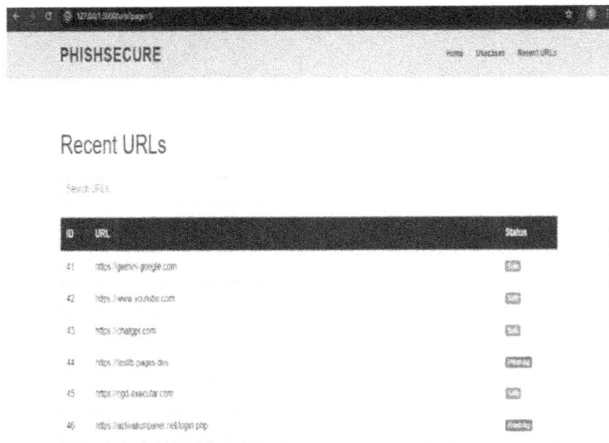

Figure 21.4 List of URL's and their result
Source: Author

A simple form with a URL input field and a submit button for phishing analysis on the home page.

A list displaying all analyzed URLs along with their predictions, providing a record of past analyses as shown in Figure 21.4.

Database connectivity

Database connectivity is crucial for a phishing detection system as it ensures efficient storage, retrieval, and management of data. It allows storing analyzed URLs and their predictions, providing a persistent record for future reference. This setup supports user interaction by storing their submissions and predictions, enabling them to review past results. With a scalable database, the system can handle growing data volumes efficiently, ensuring quick data

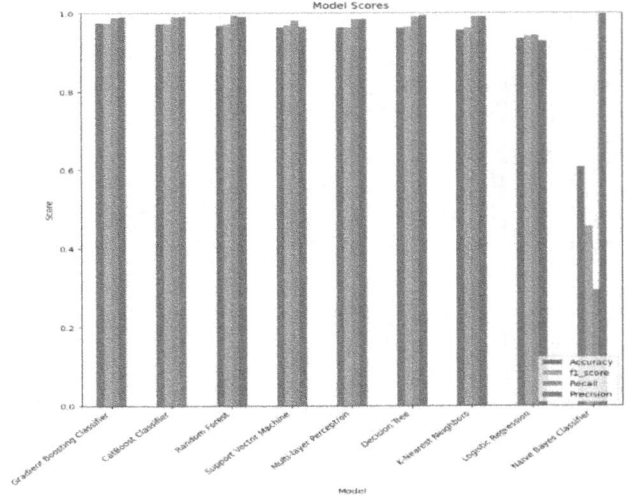

Figure 21.5 Visualization of each model accuracy
Source: Author

retrieval for both user-facing displays and backend processes.

Results

This section shows the main results of our research, including both numerical and descriptive analyses of the data we collected represented in the Figure 21.5.

The GBC performed the best, achieving an accuracy of 97.4%, a recall of 98.8%, and a precision of 98.9%. CatBoost Classifier and RF also performed well, showing high scores in all metrics and proving their effectiveness for this task.

We analyzed the key features that help the GBC detect phishing websites. The top five important factors are URL length, the presence of the "@" symbol, the use of an IP address in the URL, redirection count, and domain age. Phishing links are often longer than usual, making them suspicious. The "@" symbol in a URL is another red flag, as scammers use it to mislead users. Legitimate websites rarely use raw IP addresses in their links, whereas phishing sites often do. These features play a crucial role in identifying phishing websites.

The web application using Flask with integration of machine learning, takes the input from the user and classifies it, whether the given URL is phishing or legitimate. The application contains phishing tests and social engineering for the people awareness on phishing attacks. It also contains data of phishing and legitimate URLs as well.

Conclusion

This study shows that the Gradient Boosting algorithm is very effective at detecting phishing websites, achieving high scores in all key evaluation metrics. The research also highlights the importance of using thorough evaluation metrics to improve cyber security methods. By carefully measuring detection performance, researchers can improve current techniques and address new threats

As online attacks become more sophisticated, machine learning methods like Gradient Boosting are essential for better online security. To help users, a user-friendly web application has been created to easily check and identify potential phishing websites.

This research advances phishing detection and emphasizes the need for ongoing innovation and Alertness in cyber security.

References

[1] IBM (2019). Cost of a Data Breach Report 2019. Retrieved from: https://www.ibm.com/security/data-breach. IBM. Cost of a Data Breach Report 2019. IBM, 2019, https://www.ibm.com/security/data-breach.

[2] Ventures, Cybersecurity. Cybercrime to Cost the World $10.5 Trillion Annually by 2025. Cybersecurity Ventures, 2020, https://cybersecurityventures.com/cybercrime-damages-6-trillion-by-2021/.

[3] Verizon. 2022 Data Breach Investigations Report. 2022, https://www.verizon.com/business/resources/reports/dbir/.

[4] Federal Bureau of Investigation (FBI). Business Email Compromise: The $43 Billion Scam. 2022, https://www.ic3.gov/.

[5] IBM. Cost of a Data Breach Report 2022. 2022, https://www.ibm.com/security/data-breach.

[6] Korkmaz, M., Sahingoz, O. K., and B. Diri. Detection of Phishing Websites by Using Machine Learning-Based URL Analysis. 2020 11th International Conference on Computing, Communication and Networking Technologies (ICCCNT), 1–3 July 2020, Kharagpur, pp. 1–7.

[7] Rashid, J., & Nazir, T. (2020). Phishing detection using machine learning technique. In 2020 First International Conference of Smart Systems and Emerging Technologies (SMARTTECH).

[8] Razaque, A., Frej, M. B. H., Sabyrov, D., Shaikhyn, A., Amsaad, F., & Oun, A. (2020). Detection of phishing websites using machine learning. In 2020 IEEE Cloud Summit. https://doi.org/10.1109/IEEECloudSummit48914.2020.00022.

[9] Krishna, V. C., Swamy, N. C., Mary, V. A., & Selvan, M. P. (2021). Identification of phishing URLs using machine learning. In Journal of Physics: Conference Series, (Vol. 1770, no. 1, p. 012009). IOP Publishing.

[10] Cao, Y., Han, W., & Le, Y. (2008). Anti-phishing based on automated individual white-list. In Proceedings of the 4th ACM Workshop on Digital Identity Management – DIM, (Vol. 08, pp. 51–60).

[11] Sharifi, M., & Siadati, S. H. (2008). A phishing sites blacklist generator. In 2008 IEEE/ACS International Conference on Computer Systems and Applications, (pp. 840–843).

[12] Khonji, M., Iraqi, Y., & Jones, A. (2013). Phishing detection: a literature survey. *IEEE Communications Surveys and Tutorials*, 15(4), 2091–2121.

[13] Buber, E., Diri, B., & Sahingoz, O. K. (2018). NLP based phishing attack detection from URLs. In Advances in Intelligent Systems and Computing Intelligent Systems Design and Applications, (pp. 608–618).

22 A system for analyzing and categorizing ransomware threats using machine learning techniques

Hari Chandana Bethapudi[1,a], Dhana Lakshmi Anke[2,b], Kiranmayee Sai Thammineni[2,c], Jahnavi Ediga[2,d] and Snehalatha Mangapoti[2,e]

[1]Associate Professor, Department of CSE, Srinivasa Ramanujan Institute of Technology, Anantapur, Andhra Pradesh, India

[2]Department of CSE, Srinivasa Ramanujan Institute of Technology, Anantapur, Andhra Pradesh, India

Abstract

Ransomware attacks are among the most critical cybersecurity challenges today, as they compromise sensitive data and demand high ransoms for data recovery. Conventional detection methods based on signature matching often struggle to identify new and evolving ransomware strains. To address these challenges, a machine learning system has been developed to effectively evaluate and categorize ransomware threats. This methodology examines network behavior and tracks interactions with Command and Control (C&C) servers to identify malicious activities. By utilizing annotated datasets, the framework incorporates various machine learning techniques, such as Random Forest (RF) algorithm, Logistic Regression, and Support Vector Machines to differentiate between benign and ransomware activities. Random Forest algorithm exhibited superior performance, attaining an accuracy of 95.96%, using ANOVA-based feature selection improving its accuracy and interpretability. Additionally, the system provides a mechanism to classify ransomware into specific families, including previously unknown variants, enabling a proactive and robust defense against emerging threats.

Keywords: Logistic Regression, Support Vector Machines, ransomware, Random Forest algorithm

Introduction

Ransomware is a major cybersecurity threat, increasingly using advanced methods to attack people, organizations, and governments. These malicious attacks encrypt sensitive data, making it completely inaccessible. A ransom, typically cryptocurrency, is then demanded for the data's release. IBM's 2023 cybersecurity report indicates an important increase in ransomware attacks and large financial losses resulting from these attacks during the last year. These attacks result in financial losses, functional disruptions, sensitive data breaches and reputational harm, causing important long-term economic consequences for victims [8].

Ransomware's ability to change considerably obstructs its prevention. Prominent modern ransomware groups, such as LockBit and BlackCat, now use several advanced evasion techniques, including highly effective domain generation algorithms (DGA), heavily encrypted communications and rapidly mutating polymorphic payloads. Advanced attackers can exploit several of these important features to readily circumvent many customary signature-based detection systems, consequently compromising a meaningful number of important vulnerabilities in even the strongest network infrastructures. The increasing complexity of ransomware points out the necessity of revolutionary and flexible security protocols.

Ransomware infections rely on command and control servers for remote control of infected systems so that attackers can issue commands and conduct malicious operations. These servers perform several key operations, including the exchanging of encryption keys, delivering the malicious payload, and exfiltration of data. Ransomware frequently avoids detection through some methods such as dynamically generated domain names and encrypted communication channels. These strategies obstruct identifying Command and Control (C&C) activity. They also make intercepting this activity more challenging. Overcoming these meaningful obstacles requires a

[a]harichandana.cse@srit.ac.in, [b]214g1a0519@srit.ac.in, [c]214g1a0544@srit.ac.in, [d]214g1a0539@srit.ac.in, [e]204g1a0598@srit.ac.in

DOI: 10.1201/9781003685364-22

shift from static detection techniques to considerably more adaptive and dynamic approaches.

Machine learning (ML) provides an effective solution to address the limitations of traditional cybersecurity methods. Machine learning algorithms analyze many terabytes of network traffic data. These algorithms can detect subtle patterns and anomalies associated with ransomware. Highly advanced ML models offer considerably more proactive threat detection than conventional signature-based approaches because of their superior adaptability to new and evolving ransomware variants. Recent advances in machine learning have made it better to detect the latest ransomware variants. It has been quite effective against most modern cybersecurity threats [9].

Literature Survey

Wan et al. proposed a flow-oriented ransomware detection method that utilized Biflow and Argus for packet preprocessing. Feature selection reduced the complexity of decision trees and enhanced accuracy. Using the J48 Decision Tree algorithm, the method effectively classified ransomware families like Locky and Cerber, demonstrating the critical role of feature selection in detection [1]. Saberi and Noorbehbahani investigated semi-supervised learning techniques for detecting ransomware, achieving improved results with family-specific datasets and feature selection using the wrapper method. The need for advanced techniques to address the limitations of supervised learning approaches was emphasized [2]. Cassel and Majd investigated machine learning (ML) techniques for classifying obfuscated ransomware, achieving 89.4% accuracy. Continuous updates to detection methods were identified as essential to combat evolving obfuscation techniques [3].

Subedi et al. developed a multi-level machine learning model to analyze different sections of ransomware code. The model achieved classification accuracies ranging from 76% to 97% [4]. Aggarwal provided insights into ransomware groups such as LockBit, REvil, and Ryuk, which target sectors through infiltration, encryption, and ransom demands. Key defenses highlighted include regular updates, employee training, and data backups. In 2022, the U.S. experienced the highest number of ransomware attacks, with average ransom demands reaching $13.2 million [5]. Alhawi et al. introduced

NetConverse, a system that attained a 97.1% true positive detection rate with the J48 classifier for identifying Windows ransomware through network traffic analysis. Data from nine ransomware families was analyzed using TShark and WEKA tools [6]. Mohammad et al. evaluated ML classifiers for ransomware detection, identifying Random Forest (RF) as the most effective. Other classifiers examined included decision trees, Naive Bayes, and neural networks. The study emphasized robust classification techniques as a countermeasure against ransomware threats, supported by the United States National Science Foundation [7].

Proposed methodology

The suggested framework employs a machine learning-driven approach, incorporating Support Vector Machines (SVM), RF, and Logistic Regression (LR) models to classify network traffic into benign or ransomware categories. It leverages advanced feature selection techniques, such as KBest with ANOVA F-statistic, to identify significant network attributes, enabling efficient and accurate classification. The system demonstrates high accuracy in distinguishing ransomware families, including Rhysida, Play Ransomware, Akira, BlackCat, and LockBit.

To handle diverse datasets effectively, the system applies generalized preprocessing techniques to prepare network attributes for analysis. These steps include addressing inconsistencies, normalizing data, encoding features, and ensuring robust and reliable machine-learning performance across varied data inputs. This comprehensive approach enhances ransomware detection and classification, contributing to improved defenses against evolving threats.

Figure 22.1 depicts a supervised ML process, which starts with data input and preprocessing, and is followed by feature selection. Subsequently, the dataset is partitioned into training and evaluation sets. During training, algorithms like RF, SVM, and LR are applied. The testing stage emphasizes assessment, identification, and categorization to ensure precise predictions and optimal model performance.

Dataset specification

The dataset utilized in this study comprises labeled network traffic samples representing both benign and ransomware activities. It includes data collected from various ransomware families such as LockBit,

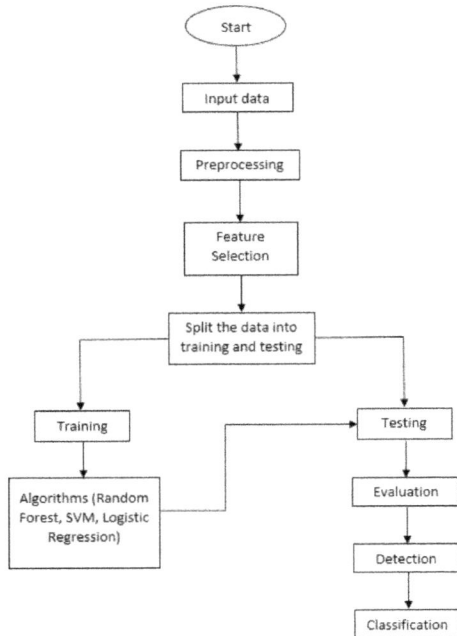

Figure 22.1 Flowchart of machine learning model development
Source: Author

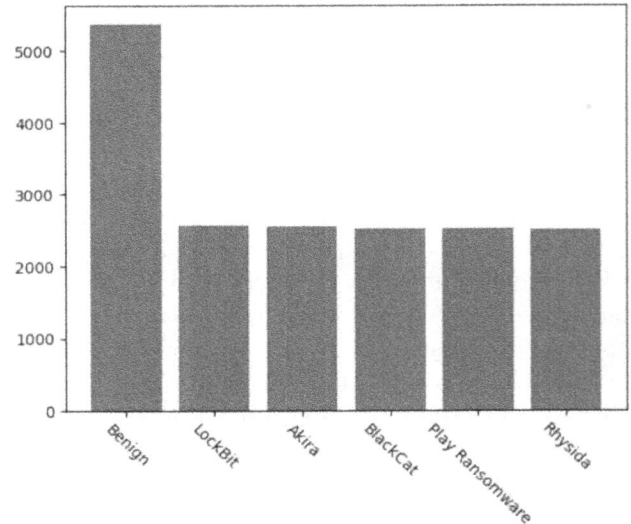

Figure 22.2 Distribution of dataset labels
Source: Author

BlackCat (ALPHV), Rhysida, Akira, and Play Ransomware. Each sample captures distinct communication patterns and network behaviors, focusing on C&C server interactions. With 18 attributes, the dataset provides a diverse and detailed representation of network characteristics. This enables effective analysis to distinguish ransomware from benign traffic and classify it into specific families based on behavioral patterns.

Figure 22.2 summarizes the dataset, highlighting the distribution of benign and ransomware samples across multiple classes, supporting effective multiclass classification.

Preprocessing
The preprocessing phase involved addressing categorical features and missing values to prepare the dataset for machine learning. Ordinal encoding was applied to non-numeric columns using scikitlearn's OrdinalEncoder, which converted categorical features into numeric values. The mappings between the original categorical values and their corresponding numeric representations were displayed to ensure transparency. Missing values were addressed by replacing categorical features with the most common value and numerical features with the mean,

preserving data consistency. These preprocessing steps allowed machine learning algorithms to effectively process the data, preserve the original distribution, and maintain transparency throughout the encoding and imputation processes.

Feature selection
Feature selection was performed using the SelectKBest method with the analysis of variance (ANOVA) F-statistic, a widely adopted technique for identifying the most relevant features in machine learning models. The ANOVA F-statistic evaluates the variance between different categories (such as benign files and ransomware) to determine the strength of each feature's relationship with the target variable. Features with higher variance and greater significance are selected, ensuring that only the most informative attributes are retained for the model. In this study, 12 out of the 18 available features were selected based on their statistical significance, allowing the system to focus on key attributes critical for ransomware detection, such as network behavior, file entropy, and encryption patterns.

Table 22.1 shows the 12 most relevant features selected from 18 attributes using the SelectKBest method with the ANOVA F-statistic. These features, critical for effective ransomware detection, include network behavior, file entropy, and encryption patterns.

Table 22.1 Key Features for analysis and classification.

No	Features
1	Domain entropy
2	Vowel ratio
3	Domain length
4	Outbound connections
5	Packet size
6	Communication time
7	Non-standard ports
8	Distinct IPs
9	TLS Validity
10	DNS Query rate
11	User agent
12	Exfiltration indicator

Source: Compiled by the authors based on feature engineering and domain analysis

By pricritizing these essential features, the model achieves more accurate and interpretable results, ultimately improving its ability to detect and classify various ransomware families with greater precision.

Model training and evaluation

This research assesses three machine learning models RF, SVM, and LR to classify ransomware threats and benign. The models received training on a specified dataset and were later assessed using a distinct test dataset. Their performance was assessed through key evaluation metrics to determine the most effective model for ransomware detection.

1. **Random Forest:** RF was opted for due to its effectiveness in handling sophisticated patterns by integrating multiple decision trees. It combines predictions from multiple trees to improve accuracy and minimize the likelihood of overfitting. Each Decision Tree learns from a randomly selected portion of the dataset, allowing the model to identify varied patterns and connections within the feature space. The trained model was evaluated using the test dataset, with its predictions analyzed against the actual labels to assess performance.

2. **Support Vector Machine:** It employs a linear kernel and was chosen for its effectiveness in handling multi-class classification problems. SVM works by finding the best decision bound-

ary that maximizes the separation between different classes, effectively distinguishing benign files from various ransomware types. After training, the model's performance in classification was evaluated with the test dataset.

3. **Logistic Regression:** LR was employed as a foundational model given its straightforward nature and effectiveness in binary classification tasks. Despite being a linear model, to determine the likelihood of a threat being classified into a specific category, the model's effectiveness was evaluated using the dataset, serving as a reference point for comparing more advanced models. While it may not capture complex non-linear relationships as effectively as more advanced techniques, LR remains a valuable tool for understanding fundamental classification tasks and assessing the performance of more sophisticated models.

Figure 22.3 depicts a ransomware detection using preprocessed network traffic data, selected features, and trained models (RF, SVM, LR) to predict and display ransomware types.

The ability of each model to classify ransomware threats was measured using essential evaluation criteria such as accuracy, precision, recall, and F1-score. These measures were derived from the test dataset to enable a comparative analysis of the models. The approach that best-balanced accurate ransomware detection and classification while minimizing incorrect positive and negative classifications was selected as the optimal solution.

1. **Accuracy:** Accuracy indicates how well the model correctly classifies instances, providing an overall measure of its reliability in making predictions.

$$\text{Accuracy} = \frac{TP+TN}{TP+TN+FP+FN} \quad (1)$$

2. **Precision:** It measures how accurately the model identifies positive cases, reflecting its reliability in making positive predictions.

$$\text{Precision} = \frac{TP}{TP+FP} \quad (2)$$

3. **Recall:** Recall measures the model's ability to correctly identify all relevant instances within a dataset. It is calculated as the proportion of cor-

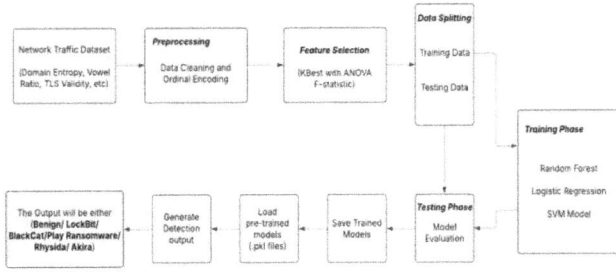

Figure 22.3 System design for ransomware analysis and classification
Source: Author

Table 22.2 Evaluation metrics of the proposed ML models.

Model	Accuracy	Precision	Recall	F1-Score
Random Forest	95.96	95.97	95.96	95.96
Support Vector Machines	92.13	92.29	92.13	92.11
Logistic Regression	89.13	89.25	89.13	89.08

Source: Compiled by the authors from experimental results

rectly predicted positive cases to the total number of actual positive instances.

$$Recall = \frac{TP}{TP+FN} \qquad (3)$$

4. **F1-Score:** Calculated as the harmonic mean of precision and recall, the F1 score provides a comprehensive evaluation of both metrics. It is particularly useful in imbalanced datasets, where reducing incorrect positive and negative classifications is crucial.

$$F1\text{-score} = 2 \times \frac{Precision \times Recall}{Precision + Recall} \qquad (4)$$

Results and discussions

The proposed system was assessed using three machine learning models such as SVM, RF and LR. Among these RF exhibited optimal performance was attained with an accuracy of 95.96%, which highlights its resilience to noise and its capacity to manage complex patterns within the data. SVM exhibited strong performance with an accuracy of 92.13%, effectively separating classes using an optimal hyperplane, which proved beneficial for multi-class classification. Logistic Regression, serving as a baseline model, achieved an accuracy of 89.13%, but it was outperformed by RF and SVM, due to its limitations in capturing non-linear data relationships effectively.

Table 22.2 shows RF performed best with 95.96% accuracy, followed by SVM at 92.13% and LR at 89.13%.

Figure 22.4 illustrates the comparison of accuracy, precision, and recall across LR, RF, and SVM. Among them, RF achieves the highest performance, followed by SVM and LR.

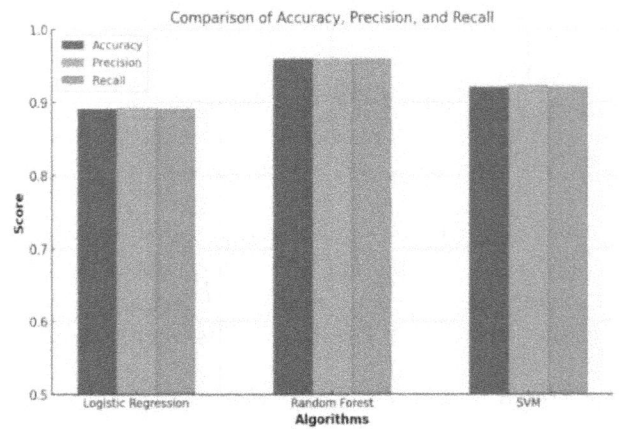

Figure 22.4 Comparison of algorithms
Source: Author

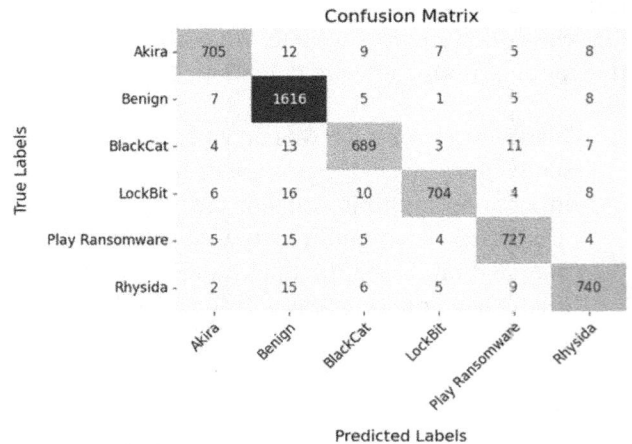

Figure 22.5 Random Forest confusion matrix
Source: Author

Figure 22.5 depicts the confusion matrix for the RF Model, summarizing its classification accuracy. High diagonal values indicate precise predictions, while

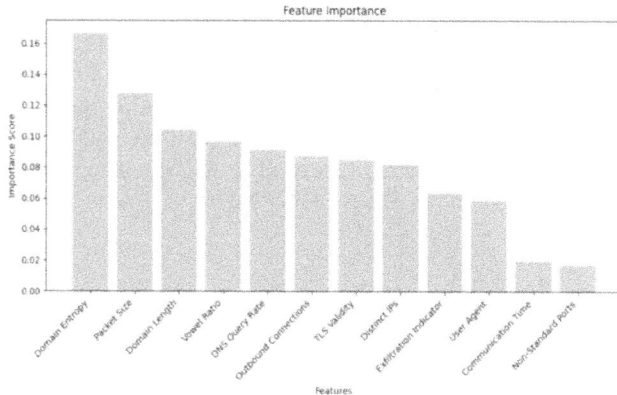

Figure 22.6 Ranking key features in model performance
Source: Author

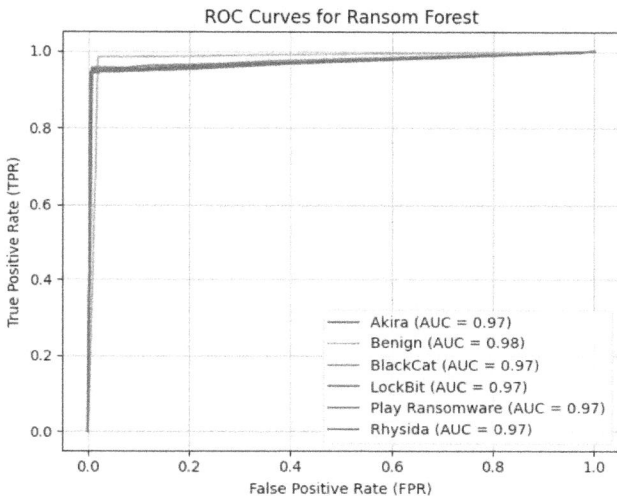

Figure 22.7 ROC curve of Random Forest classifier
Source: Author

low off-diagonal values suggest minimal errors. This demonstrates the model's strong ability to differentiate ransomware from benign data.

Figure 22.6 illustrates the importance of scores for the attributes used in the model. Feature importance measures the contribution of each feature to the model's predictive capability.

Figure 22.7 illustrates the receiver operating characteristic curve of RF, which categorizes files into six groups: Akira, Benign, BlackCat, LockBit, Play Ransomware, and Rhysida. The model performs well, with Benign and Play Ransomware reaching an AUC of 0.98, while the other classes achieved 0.97.

Conclusion

The Random Forest-based framework demonstrates strong performance in detecting and classifying ransomware families while distinguishing them from benign files. Timely detection is vital for mitigating attacks and protecting critical systems. This research underscores machine learning's effectiveness in cybersecurity and its potential for future enhancements, including adaptability to new threats and real-world application integration.

References

[1] Wan, Y. L., Chang, J. C., Chen, R. J., & Wang, S. J. (2018). Feature-selection-based ransomware detection with machine learning of data analysis. In 2018 3rd International Conference on Computer and Communication Systems (IC-CCS), Nagoya, Japan, (pp. 85–88). doi: 10.1109/CCOMS.2018.8463300.

[2] Noorbehbahani, F., & Saberi, M. (2020). Ransomware detection with semi-supervised learning. In 2020 10th International Conference on Computer and Knowledge Engineering (ICCKE), Mashhad, Iran, (pp. 024–029). doi: 10.1109/ICCKE50421.2020.9303689.

[3] Cassel, W., & Majd, N. E. (2023). Obfuscated ransomware family classification using machine learning. In 2023 International Conference on Computational Science and Computational Intelligence (CSCI), Las Vegas, NV, USA, (pp. 788–792). doi: 10.1109/CSCI62032.2023.00134.

[4] Poudyal, S., Subedi, K. P., & Dasgupta, D. (2018). A framework for analyzing ransomware using machine learning. In 2018 IEEE Symposium Series on Computational Intelligence (SSCI), Bangalore, India, (pp. 1692–1699). doi: 10.1109/SSCI.2018.8628743.

[5] Aggarwal, M. (2023). Ransomware attack: an evolving targeted threat. In 2023 14th International Conference on Computing Communication and Networking Technologies (ICCCNT), Delhi, India, (pp. 1–7). doi: 10.1109/ICCCNT56998.2023.10308249.

[6] Alhawi, O. M. K., Baldwin, J., & Dehghantanha, A. (2018). Leveraging machine learning techniques for Windows ransomware network traffic detection. In A. Dehghantanha, M. Conti, & T. Dargahi (Eds.), Cyber Threat Intelligence (Advances in Information Security, vol. 70, pp. 93–106).

[7] Masum, M., Faruk, M. J. H., Shahriar, H., Qian, K., Lo, D., & Adnan, M. I. (2022). Ransomware classification and detection with machine learning algorithms. In 2022 IEEE 12th Annual Computing

and Communication Workshop and Conference (CCWC). IEEE.

[8]   IBM Security. (2023, July 24). Cost of a Data Breach Report 2023: Half of breached organizations unwilling to increase security spend despite soaring breach costs. IBM Newsroom. Retrieved January 5, 2025, from https://newsroom.ibm.com/2023-07-24-IBM-Report-Half-of-Breached-Organizations-Unwilling-to-Increase-Security-Spend-Despite-Soaring-Breach-Costs.

[9]   Ispahany, J., Islam, M. R., Islam, M. Z., & Khan, M. A. (2024). Ransomware detection using machine learning: a review, research limitations and future directions. *IEEE Access*, 12, 68785–68813. doi: 10.1109/ACCESS.2024.3397921.

23 Evasive SMS spam detection using machine learning models

T. N. Ranganadham[1,a], B. Mounika[2,b], D. Meghana[2,c], S. Mastan Vali[2,d] and K. Janardhan[2,e]

[1]Assistant Professor, Department of CSE, Annamacharya University, Rajampet, Andhra Pradesh, India

[2]Undergraduate, Department of CSE, Annamacharya University, Rajampet, Andhra Pradesh, India

Abstract

The sophistication of SMS spam, which is fueled by spammers' use of evasive strategies, is becoming a bigger issue for detection systems. This study examines and assesses machine learning techniques for evasive SMS spam detection in order to improve spam detection efficiency. A recently curated dataset of over 72581 SMS messages—including both spam and legitimate (ham) communications—is used to evaluate the efficiency of various models. They include conventional machine learning models such as Support Vector Machines, Random Forests, Decision Trees, and Multinomial Naive Bayes. To find out how well these models can resist spammers' evasion strategies, we assess their resilience to adversarial and obfuscation attacks.

Keywords: Adversarial attacks, cybersecurity, dataset evaluation, decision trees, evasive strategies, machine learning, message analysis, multinomial Naive Bayes, obfuscation attacks, random forests, resilience testing, SMS spam detection, spam filtering, Support Vector Machines, text classification

Introduction

These days, every social media platform has a plan to address the issue of spam detection. News is spread on all social media platforms, including Facebook, Twitter, and others, only by customer analysis. A lot of individuals look for the specific news that is the most talked-about in the city. People simply continue to share the news because they have no idea if it is authentic or not. This paper goal is to analyze news using machine learning algorithms and certain artificial spam detection techniques. To determine if something is spam or ham, we employ a variety of classifiers and algorithms. Data from the attributes, missing values, article titles, body, and publisher are all helpful in the detecting process. Natural language processing (NLP), a branch of artificial intelligence (AI), creates valuable data by utilizing computers and human language. Among the information that is helpful in the detection are the attributes, missing values, article titles and bodies, and publisher.

Literature review

Delvia et al. [1] addresses the challenge of improving spam detection on mobile phones by combining two powerful techniques: FPGrowth and the Naive Bayes classifier. The study concludes by demonstrating how combining FPGrowth with the Naive Bayes classifier can result in a sophisticated spam detection system that strikes a compromise between computing economy and accuracy [1]. Gupta et al. explores the use of machine learning techniques for detecting spam messages in SMS [2]. In this paper they evaluate the performance of several algorithms including Naive Bayes (NB), Support Vector Machines (SVM), and Decision Trees (DT) [5]. The importance of pre-processing SMS data shows that tokenization and stop word removal improve the performance of the model [2]. Julis and Alagesan, published a paper titled "Spam detection in SMS using machine learning through text mining" in 2020 addresses the paper is to determine if SMS messages are spam or not, they use a number of machine learning methods,

[a]tnr@aits.rajampet.ac.in, [b]mounika572003@gmail.com, [c]devameghana2005@gmail.com, [d]syedmastanvali352@gmail.com, [e]janardhankomerla@gmail.com

DOI: 10.1201/9781003685364-23

such as Random Forest (RF), SVM, and NB [3]. First objective is to develop filters with great performances rates correctly detect the messages those are really sent as an e-mail [6]. Masurkar et al., explores that the best classifier in this study reduces the total error rate of the best model in the original paper mentioning this dataset by more than half, according to the final simulation results using 10-fold cross validation A [4] database of real SMS spams from the UCI ML repository is used in this study [8]. Several machine learning algorithms are used to the database following feature extraction and preprocessing [9]. The best algorithm for text message spam filtering is presented after a comparison of the result [4].

Proposed work

Preliminaries
Selecting the dataset
We will be working with a dataset which contains over 77000 messages which includes both ham and spam.

Importing and cleaning the data
Gathering the data
We should first import the SMS dataset from a trusted source so that we are working with reliable information.

Cleaning the data
We will clean the data by removing any duplicates or missing entries. To make the text easier for our models to work with, we will remove any punctuation, convert all characters to lowercase, and remove common stop words that add little value.

Text to numbers
Feature extraction
To insert our text data into a model of machine learning, we need to change it into a numerical format. We will first try two methods:
Bag-of-Words (BoW): This counts how often each word appears.
TF-IDF: This approach assigns weight to words according to their importance and frequency.

Selecting the best model
Testing algorithms
We will test several algorithms to find the best fit for our task of spam detection:
Multinomial Naive Bayes: This algorithm is known for being fast and efficient with text data [7].

Support Vector Machines (SVM): Excellent at working with high-dimensional data and generally very accurate.
Decision Tree: A decision tree helps in feature patterns such as keywords, frequency, and message length.
Random Forest: Random Forest improves SMS spam detection by merging multiple decision trees to increase accuracy and decrease overfitting, leveraging the collective wisdom of diverse tree-based models.

We will test these models against a sample of our data to find out which works best for us.

Model building
Classifier building
Once we determine the best algorithm to use for our problem, we will build the model using Scikit-learn for Multinomial Naive Bayes.

Train the model
Dividing data
We will create a training set consisting of 70% of our data set and a testing set with 30%.

It would then learn all patterns upon the training set, which are distinguishing ham from spam messages. Figure 23.1.

Model performance analysis

Let's summarize how our proposed algorithms performed. Table 23.1.

Confusion matrix:
Using the scikit-learn document, certain readable confusion matrices may be made that are simpler

Table 23.1 Model performance summary.

Model	Accuracy	Precision	Recall	F1-Score
Decision Tree	96.6%	96.0%	95.7%	95.8%
Random Forest	98.3%	98.2%	97.6%	97.8%
SVM	98.3%	98.1%	97.5%	97.7%
Naïve Bayes	96.7%	96.3%	96.0%	96.1%

Source: Author

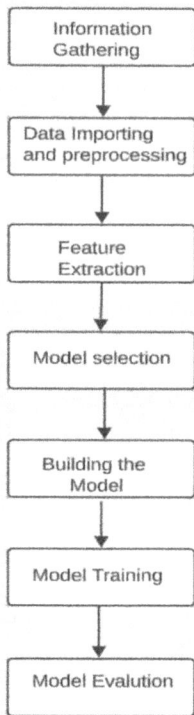

Figure 23.1 Work process of spam SMS detection
Source: Author

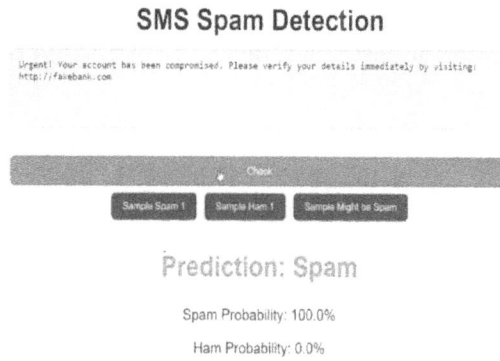

Figure 23.2 Confusion matrix
Source: Author

to match and search. The right labels are displayed most diagonally in a confusion matrix. Figure 23.2.

The confusion matrix consists of,

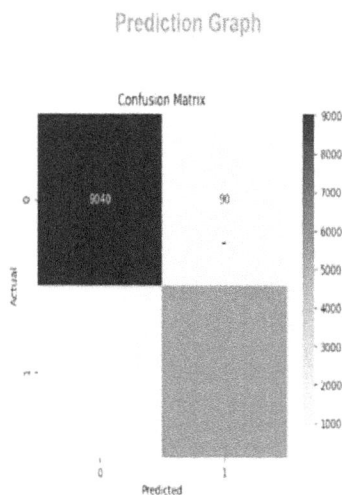

Figure 23.3 Detecting the message whether it is spam or ham
Source: Author

- True positives: The positive values are truly predicted.
- True negatives: The negative values are correctly classified.
- False positives: The negative values were misclassified as positive values.
- False negatives: The negative values are misclassified as positive values.

False positive rate (FPR):
FPR = FP/(FP + TN)
where FP is the number of false positives
TN is the number of true negatives
False negative rate (FNR):
FNR = FN/(FN + TP)
where FN is the number of false negatives
TP is the number of true positives.

Result

Giving input and detecting the message. Figure 23.3.

Conclusion

This paper showed that computers can learn to recognize spam text messages. We used different types of computers learning methods to see which one worked best. We found that these methods can be helpful for classifying and detecting spam SMS. However, it's important to keep studying how spammers change their tricks, so we made the computer methods even better. Detecting the spam is most important and safer

Acknowledgement

For all the assistance, the authors would like to thank the Department of CSE at Annamacharya University in Rajampet, Andhra Pradesh.

References

[1]   Delvia, D., Shaufiah, A., & Bijaksana, M.A. (2016). Enhancing spam detection on mobile phone short message service (SMS) performance using FP-Growth and Naive Bayes classifier. International Journal of Information & Electronics Engineering, 6(3), 219–224.

[2]   Gupta, S.D., Saha, S., & Das, S.K. (2020). SMS spam detection using machine learning. Journal of Physics: Conference Series, 1797(1), 012017.

[3]   Julis, M.R., & Alagesan, S. (2020). Spam detection in SMS using machine learning through text mining. International Journal of Scientific & Technology Research, 9(2), 498–503. Retrieved from https://www.ijstr.org/final-print/feb2020/Spam-Detection-In-Sms-Using-Machine-Learning-Through-Text-Mining.pdf

[4]   Masurkar, S., Rajput, A., Angane, A., Madaan, S., & Malik, S. (2020). Machine learning based spam detection system. International Research Journal of Engineering and Technology (IRJET), 7(6), 3907–3910.

[5]   Shirani-Mehr, H. (2017). SMS spam detection using machine learning approach. Stanford University (CS229 project)

[6]   Manwar, S.R., Lambhate, P.D., & Patil, J.S. (2017). Classification methods for spam detection in online social network. International Research Journal of Engineering and Technology (IRJET), 4(7), 230–235. Retrieved from https://www.irjet.net/archives/V4/i7/IRJET-V4I7230.pdf

[7]   Rathore, J., Tadge, A.K., Shrivastav, A., Yadav, R., & Sisodiya, P.S. (2022). Implementation of spam classifier using Naïve Bayes algorithm. International Research Journal of Engineering and Technology (IRJET), 9(2), 72–75. Retrieved from https://www.irjet.net/archives/V9/i2/IRJET-V9I272.pdf

[8]   IDC. (2013, August 7). Growth accelerates in the worldwide mobile phone and smartphone markets in the second quarter, according to IDC. IDC Press Release. Retrieved from http://www.idc.com/getdoc.jsp?containerId=prUS24239313

[9]   IDC. (2013, August 7). Growth accelerates in the worldwide mobile phone and smartphone markets in the second quarter, according to IDC. IDC Press Release. Retrieved from http://www.idc.com/getdoc.jsp?containerId=prUS24239313

24 A hybrid approach for detecting intrusion in WSNs of industry 4.0 using ML and DL algorithms

Nagesh Chilamakuru[1,a], Pamisetty Lavanya[2,b], Madam Hemalatha[2,c], Palellu Durga[2,d] and Gadamshetty Koushik[2,e]

[1]Assistant Professor, Department of CSE, Srinivasa Ramanujan Institute of Technology, Anantapur, Andhra Pradesh, India

[2]Student, Department of CSE, Srinivasa Ramanujan Institute of Technology, Anantapur, Andhra Pradesh, India

Abstract

As the industrial landscape advances with Industry 4.0, the integration of cutting-edge technologies like wireless sensor networks (WSNs) has become crucial for streamlining operations and boosting efficiency. However, the growing number of interconnected devices in WSNs has increased the vulnerability to cyber intrusions, making strong cybersecurity measures essential. This study explores machine learning (ML) and deep learning (DL) algorithms for intrusion detection, ensuring high accuracy while minimizing false positives. Ensemble techniques such as Stacking Classifier, XGBoost, and AdaBoost enhance detection by combining multiple models for better adaptability to attack patterns. The hybrid approach integrates ML models like Random Forest and Decision Tree with DL-based ensemble methods, leveraging their strengths for scalable and precise intrusion detection. This combination ensures robust security in WSNs of Industry 4.0, effectively mitigating evolving cyber threats.

Keywords: Decision Tree, deep learning (DL), ensemble methods, Industry 4.0, machine learning, Random Forest, wireless sensor networks (WSNs)

Introduction

The shift to Industry 4.0 has transformed industries by merging physical systems with advanced digital technologies, enabling real-time automation, data-driven insights, and increased connectivity within smart industrial ecosystems. At the core of this transformation is the wireless sensor network (WSN), an essential component that facilitates seamless collection, processing, and transmission of data between devices. WSNs play a crucial role in enabling predictive maintenance, optimizing energy use, and driving automation by ensuring reliable communication across Industry 4.0 environments [1, 2]. However, the increased connectivity and reliance on WSNs introduce significant security challenges. Cyberattacks such as denial of service (DoS), data manipulation, and eavesdropping threaten to disrupt industrial processes, leading to downtime, financial loss, and compromised safety. These attacks exploit the inherent vulnerabilities of WSNs, including limited energy capacity, constrained processing power, and memory restrictions. Due to these limitations, traditional security measures, such as encryption and firewalls, are often impractical in WSN environments [3, 5].

To address these challenges, advanced data-driven security approaches offer powerful solutions by analyzing historical network activity to identify patterns that indicate potential cyber threats. Machine learning (ML) and deep learning (DL) have significantly improved network security systems by enabling continuous network monitoring, real-time anomaly detection, and threat classification. These models adapt to evolving cyber threats, providing a proactive defense against new and sophisticated attacks [5]. Among these techniques, Random Forest (RF) and XGBoost have demonstrated high effectiveness in intrusion detection. The RF by constructing multiple Decision Trees (DT) and aggregating their outputs, delivers strong classification performance. Its ability to recognize complex data patterns makes it highly suitable for WSNs, ensuring accurate threat identification while minimizing false alarms [4]. Similarly,

[a]cnagesh.cse@gmail.com, [b]214g1a0549@srit.ac.in, [c]224g5a0506@srit.ac.in, [d]214g1a0525@srit.ac.in, [e]214g1a0545@srit.ac.in

DOI: 10.1201/9781003685364-24

XGBoost is particularly well-suited for WSN environments due to its high efficiency and accuracy in classifying network behaviors. These techniques help overcome WSN limitations by operating with minimal computational resources, making them ideal for energy-restricted environments. Furthermore, ML-based security systems continuously evolve by learning new attack patterns, ensuring sustained reliability and robust protection [5].

This study aims to create a robust framework by integrating both ML and DL Models for detecting intrusions in WSNs within Industry 4.0 environments. It evaluates the effectiveness of algorithms such as decision trees, random forest, and XGBoost in identifying and mitigating cybersecurity threats. The framework is designed to enhance detection capabilities while reducing incorrect alerts, ensuring smooth and secure industrial operations. By utilizing ML and DL techniques, this research aids in the creation of scalable and efficient solutions to secure WSNs in highly interconnected industrial environments [2, 3].

Literature review

Al-Issa et al. explored methods for detecting DoS attacks in WSNs, focusing on the effectiveness of Decision Trees (DT) and SVM in intrusion detection. Their findings revealed that DT achieved an impressive 99.86% detection rate with a minimal 0.05% rate of incorrect alerts, making them highly suitable for resource-constrained environments like Industry 4.0. This research highlights the critical need for efficient detection systems in WSNs to counter various network vulnerabilities and disruptions. It also underscores the importance of scalable models that can maintain high detection accuracy without straining system resources [3].

Alsahli et al. evaluated machine learning algorithms like Naïve Bayes (NB), Random Forest (RF), and IBK for intrusion detection in WSNs. Alsahli et al. assessed various data-driven algorithms such as NB, RF, and IBK for detecting intrusions in WSNs. Their findings revealed that random forest outperformed the other methods, delivering better detection rates and proving to be especially well-suited for environments with limited resources. This study highlights the importance of choosing the most appropriate model to enhance threat detection in Industry 4.0. Effective solutions must be reliable, precise, and adaptable, ensuring uninterrupted operations while addressing the challenges posed by constantly evolving cyber risks, all while respecting the operational constraints of WSNs [5].

Godala and Vaddella categorized intrusion detection techniques in WSNs into signature-based, anomaly-based, and hybrid approaches, with an emphasis on anomaly-based systems powered by machine learning. Their research demonstrated the effectiveness of these systems in identifying zero-day attacks and underscored the significance of employing performance metrics like confusion matrices to assess intrusion detection models. This research is essential for ensuring that IDS in WSNs for Industry 4.0 can provide reliable, real-time threat detection and mitigation [10]. Al-Quayed et al. designed a predictive framework that incorporates ML and DL techniques, including DT, MLP and Autoencoder models. Their system, which achieved 99.52% accuracy with MLP, effectively detected blackhole, grayhole, and flooding attacks and introduced intelligent threat prioritization for proactive defense. This framework is particularly useful for Industry 4.0, where effective threat prioritization is crucial. Their research highlights the importance of integrating multiple ML models to improve detection accuracy and strengthen Output effectiveness [2].

Proposed methodology

The proposed system enhances detection capabilities by analyzing network activity patterns. It classifies different events using supervised models, identifying irregularities and categorizing data flow into routine operations or potential disruptions. By recognizing deviations from expected behavior, the system improves accuracy in distinguishing between normal and suspicious activities. As depicted in Figure 24.1, the system follows a structured process, beginning with data collection, where relevant information is gathered for analysis. The preprocessing phase refines the dataset to improve its quality and reliability. After cleaning, the data is divided into separate portions for training and evaluation. Different classification techniques are applied to recognize patterns and categorize network activity.

Once the model completes its learning process, it predicts and identifies various types of network behavior. Finally, the system assesses performance and presents the findings to the user.

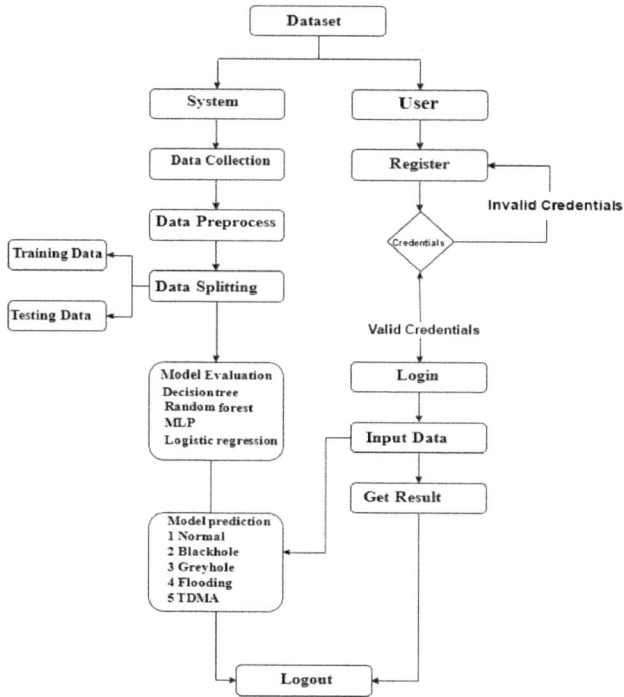

Figure 24.1 Flowchart of intrusion detection system development
Source: Author

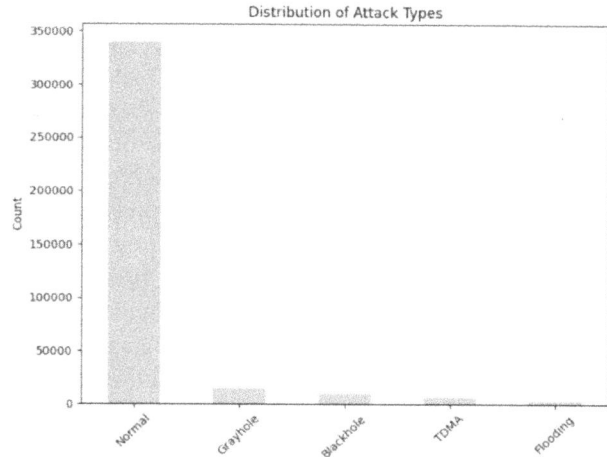

Figure 24.2 Distribution of attack types in WSN-DS dataset
Source: Author

Dataset specification

The dataset comprises various attributes that facilitate the examination of network activity and identification of anomalies. Each attribute represents a unique aspect of communication, including identification, timing, message exchanges, and data transfer details. These elements offer valuable insights into how information moves within the system, assisting in differentiating regular patterns from unusual ones. Key aspects recorded include the roles and interactions of different nodes within the network. Some attributes determine whether a node functions as a coordinator, its distance from others, and the volume of messages it handles. Additionally, energy usage and data flow are tracked, providing a clearer picture of resource management and efficiency. The dataset captures a range of operational behaviors, with some representing standard activity and others signaling potential irregularities. A graphical representation of these behaviors (Figure 24.2) illustrates the distribution of different patterns within the dataset, ensuring comprehensive coverage of various scenarios. Studying these trends helps in identifying expected operations and detecting deviations effectively.

Data preprocessing and data splitting

The data preprocessing stage ensures effective training for detecting network attacks like grayhole, blackhole, and TDMA-based attacks. Steps include reading WSN-DS.csv using Pandas, handling missing values with fillna() or dropna(), and encoding attack types using label encoder. Attack types are mapped as normal (3), grayhole (2), blackhole (0), TDMA (4), and flooding (1). These steps ensure a clean, numeric dataset for machine learning models.

After preprocessing, the dataset is divided into two parts: one for training and the other for evaluation. A 70:30 split is applied, where 70% of the data is used to help the model recognize normal patterns and irregularities, while the remaining 30% ensure it can accurately analyze new data. This structured division prevents overfitting and improves adaptability, allowing the system to generalize well to different scenarios. By balancing both sections, the system enhances accuracy in detecting diverse network behaviors.

Model selection and training

During this phase, various algorithms are used to classify network activities as normal or indicative of potential threats. The selected models are known for their strong performance in detecting anomalies in network behavior. One such model is the DT, which organizes data into a branching structure, making decisions at each node based on key features and classifying events at the leaves. RF builds upon this by creating multiple trees from random subsets of data and combining their predictions, improving accuracy and reducing the likelihood of overfitting.

The multilayer perceptron (MLP), a neural network designed to recognize complex, non-linear relationships within the data. MLP refines its predictions over multiple iterations through backpropagation. Logistic regression is also employed for scenarios where classes can be clearly separated, providing a straightforward way to interpret the importance of different features. XGBoost, valued for its speed and precision, builds models sequentially, with each iteration correcting errors from the previous one, while also preventing overfitting through regularization. Additionally, AdaBoost enhances the overall system by focusing on instances that are difficult to classify, adjusting its approach to improve accuracy. Finally, the Stacking method combines predictions from multiple base models, such as RF and DT, and passes them to a final classifier, which makes the ultimate decision based on this collective input.

Evaluation metrics

Accuracy: It reflects the overall performance by calculating the percentage of correct classifications, including both correct positive and negative predictions, relative to the total number of predictions.

$$\text{Accuracy} = \frac{TP+TN}{TP+TN+FP+FN}$$

Precision: It focuses specifically on the reliability of positive classifications, measuring how often predicted positives are correct compared to the total number of positive predictions made by the model.

$$\text{Precision} = \frac{TP}{TP+FP}$$

Recall (Sensitivity): Recall measures the system's ability to detect actual threats accurately, minimizing false negatives.

$$\text{Recall} = \frac{TP}{TP+FN}$$

F1-Score: F1-score balances precision and recall, ensuring a well-rounded detection model.

$$\text{F1-score} = 2 \times \frac{\text{Precision} \times \text{Recall}}{\text{Precision} + \text{Recall}}$$

Confusion matrix: Confusion matrix provides a detailed view of the model's classification performance by comparing predicted outcomes with actual results. It displays the number of correct and incorrect predictions, offering insights into common errors, such as instances where the model wrongly predicts a positive or misses a positive case. This matrix is a key tool for evaluating the model's strengths and weaknesses in identifying various types of network attacks.

Results and discussion

Performance metrics

The evaluation of the proposed hybrid detection system demonstrates notable advancements in identifying and preventing cyber threats within interconnected industrial networks. The effectiveness of the system was assessed using various performance measures, which indicate how well the approach distinguishes between normal activities and potential security breaches.

As shown in Table 24.1, the proposed machine learning algorithms demonstrate varying performance, with XGBoost achieving the highest accuracy (99.73%) and a high F1-score (0.970), making it the best overall performer. It handled both majority and minority classes effectively, showcasing balanced precision (0.950) and recall (0.990). Similarly, the stacking classifier delivered exceptional results, with an accuracy of 99.72% and an F1-score of 0.970, leveraging its ensemble approach to balance precision (0.960) and recall (0.980). RF also performed well, achieving 99.71% accuracy and an F1-score of 0.962, demonstrating strong recall (0.992) but slightly lower precision (0.933).

Table 24.1 Performance metrics of proposed machine learning algorithms.

Algorithm	Accuracy	Precision	Recall	F1-Score
Decision Tree	99.49	0.957	0.946	0.952
Random Forest	99.71	0.933	0.992	0.962
Multi-Layer Perceptron (MLP)	90.76	0.18	0.2	0.19
Logistic Regression	93.05	0.53	0.45	0.47
Stacking Classifier	99.72	0.96	0.98	0.97
XGBoost	99.73	0.95	0.99	0.97
AdaBoost	98.16	0.94	0.88	0.91

Source: Author

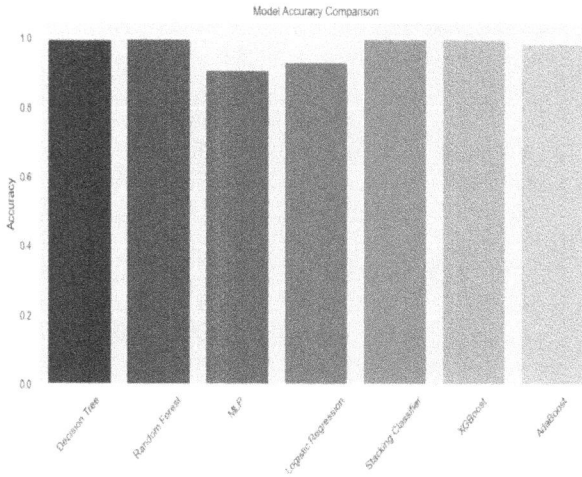

Figure 24.3 Model accuracy comparison
Source: Author

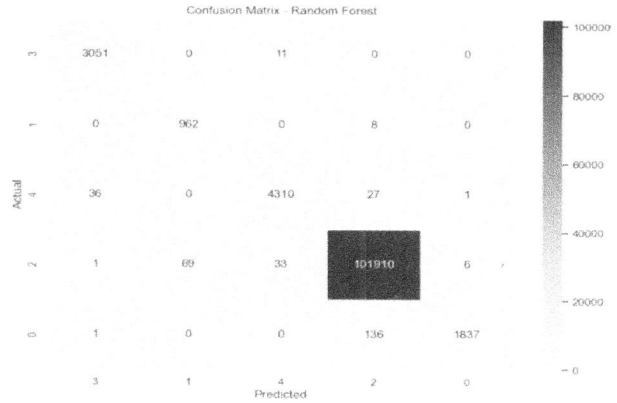

Figure 24.4 Confusion matrix of random forest
Source: Author

The DTe, while achieving 99.49% accuracy, had a lower F1-score (0.952) due to its tendency to overfit, leading to reduced generalization. Logistic regression, with 93.05% accuracy, showed moderate performance but struggled with non-linear relationships and class imbalances, reflected in its precision (0.53) and recall (0.45). Multi-layer perceptron (MLP) underperformed significantly, achieving 90.76% accuracy and an F1-score of 0.19, as it predominantly predicted the majority class (TDMA) and failed to distinguish minority classes. AdaBoost achieved 98.16% accuracy with decent overall performance but struggled with precision and recall for minority classes, particularly misclassifying normal and flooding instances. Figure 24.3 illustrates the comparison of model accuracy among various algorithms, with XGBoost outperforming others in terms of overall detection rate.

Confusion matrices
The DT model performed reasonably well for the majority class, such as TDMA, with most instances correctly classified. However, it struggled with distinguishing between normal and blackhole, as well as between TDMA and minority classes like Flooding. This indicates a limitation in the model's ability to generalize patterns. The minority classes, such as grayhole and flooding, experienced lower detection rates due to frequent misclassifications. Additionally, the algorithm showed a tendency to overfit, which affected its performance when dealing with more complex patterns in the dataset. As seen in Figure

24.4, the confusion matrix of the RF approach delivered strong results across all classes, with minimal misclassifications. It significantly improved the handling of minority classes, especially grayhole and flooding, compared to the DT. The majority class, TDMA, was predicted with high accuracy.

However, minor confusion between normal and blackhole remained, but the impact was relatively small. In contrast, the multi-layer perceptron (MLP) faced significant challenges due to its inability to handle imbalanced data effectively. It frequently favored the majority class (TDMA) and failed to classify instances of minority classes correctly, leading to poor overall performance for those groups. The architecture of the MLP may not have been well-suited for this dataset, as it struggled to capture complex patterns and relationships among the features. LR showed moderate performance but struggled with misclassifications across various classes. It performed poorly on minority classes like grayhole and flooding, with noticeable confusion between Normal and TDMA, as well as blackhole and other classes. The linear nature of the model limited its capacity to represent non-linear relationships in the data, resulting in less-than-optimal predictions. The stacking approach demonstrated balanced performance across all classes, leveraging the strengths of multiple models to reduce misclassifications. It achieved high accuracy for the majority class (TDMA) and showed minimal confusion between categories.

The classifier effectively handled minority classes like grayhole and Flooding, making it a dependable choice for detecting threats in network environments. XGBoost delivered the best overall results, with very few misclassifications and strong handling of both majority and minority classes. The model demonstrated excellent accuracy, with minimal confusion between categories such as normal, TDMA, and blackhole. Its capacity to capture complex data patterns contributed to its superior performance and reliable handling of the dataset. AdaBoost, while showing decent overall results, struggled with differentiating between normal and blackhole, as well as flooding and TDMA. Although it handled the majority classes effectively, its performance for minority classes was slightly weaker. This issue may stem from the model's focus on correcting misclassifications, which can sometimes lead to overfitting specific patterns.

Conclusion

The proposed hybrid intrusion detection system showed outstanding effectiveness in identifying and mitigating cyber threats within connected industrial environments. Among the algorithms tested, XGBoost, Random Forest, and the stacking classifier delivered the strongest results, excelling at managing uneven datasets and identifying complex patterns. While the Decision Tree method performed well, it encountered difficulties with generalization due to overfitting. In contrast, logistic regression and MLP struggled to handle the complexity and imbalance of the data. The combination of different approaches proved particularly successful in reducing errors for less common threat categories, such as grayhole and flooding. These outcomes highlight the system's potential to significantly improve cybersecurity in modern networked systems.

Acknowledgement

This work was supported by Srinivasa Ramanujan Institute of Technology through its academic and development initiatives.

References

[1] Ghobakhloo, M. (2020). Industry 4.0, digitization, and opportunities for sustainability. *Journal of Cleaner Production*, 252, 119869.

[2] Al-Quayed, F., Ahmad, Z., & Humayun, M. (2024). A situation based predictive approach for cybersecurity intrusion detection and prevention using machine learning and deep learning algorithms in wireless sensor networks of industry 4.0. *IEEE Access*, 12, 34800–34816.

[3] Al-Issa, A. I., Al-Akhras, M., Alsahli, M. S., & Alawairdhi, M. (2019). Using machine learning to detect DoS attacks in wireless sensor networks. In 2019 IEEE Jordan International Joint Conference on Electrical Engineering and Information Technology (JEEIT), (pp. 1–6). doi: 10.1109/JEEIT.2019.8717400.

[4] Farnaaz, N., & Jabbar, M. A. (2016). Random forest modeling for network intrusion detection system. *Procedia Computer Science*, 89, 1–7. doi: 10.1016/j.procs.2016.06.047.

[5] M. Almseidin, M. Alzubi, S. Kovacs, and M. Alkasassbeh, Evaluation of machine learning algorithms for intrusion detection system, in Proceedings of the 2017 IEEE 15th International Symposium on Intelligent Systems and Informatics (SISY), Subotica, Serbia, 2017, pp. 000277–000282, doi: 10.1109/SISY.2017.8080566.

[6] Abhale, A. B., & Avulapalli, J. R. (2021). Deep learning perspectives to detecting intrusions in wireless sensor networks. *International Journal of Communication Networks and Information Security*. vol. 13, no. 2, pp. 45–52.

[7] Alsheikh, M. A., Lin, S., Niyato, D., & Tan, H. P. (2014). Machine learning in wireless sensor networks: Algorithms, strategies, and applications. *IEEE Transactions on Communications*, 12(4), 1–10. doi: 10.1109/COMST.2014.2320099.

[8] Zhang, W., & Li, K. C. (2020). Wireless sensor network intrusion detection system based on MK- ELM. *Soft Computing*, 24(5), 1–10. doi: 10.1007/s00500-020-04678-1.

[9] Ali, M. H., Jaber, M. M., Awan, M. J., Abd, S. K., Rehman, A., Damasevicius, R., et al. (2022). Threat analysis and distributed denial of service (DDoS) attack recognition in the internet of things (IoT). *Electronics*, 11(494), 1–15. doi: 10.3390/electronics11030494.

[10] Godala, S., & Vaddella, R. P. V. (2020). A study on intrusion detection system in wireless sensor networks. *International Journal of Communication Networks and Information Security (IJCNIS)*, 12(1), 1–6.

[11] Almomani, I., Al-Kasasbeh, B., & AL-Akhras, M. (2016). WSN-DS: A dataset for intrusion detection systems in wireless sensor networks. Journal of Sensors, 2016, Article ID 4731953. https://doi.org/10.1155/2016/4731953

25 Precision in medical imaging: evaluating deep learning models for intracranial hemorrhage classification

Sabbu Sunitha[1,a], Peddappolla Keerthana[2,b], Sirimala Neelima[2,c], Khamithkar Krishna Priya[2,d] and Muchukota Hemanth[2,e]

[1]Assistant Professor, Department of CSE, Srinivasa Ramanujan Institute of Technology, Anantapur, India

[2]Students, Department of CSE(Data Science), Srinivasa Ramanujan Institute of Technology, Anantapur, India

Abstract

Intracerebral hemorrhage is a critical medical condition requiring immediate diagnosis and intervention. Traditional diagnostic methods rely on radiologists, making the process time-intensive and prone to human error. Advancements in deep learning have enabled automated hemorrhage detection in computed tomography (CT) scans, improving diagnostic speed and accuracy. This study evaluates five deep learning architectures—convolutional neural networks (CNN), VGG19, ResNet50, DenseNet121, and EfficientNetB0—using a publicly available CT image dataset. The models were trained using TensorFlow, incorporating preprocessing techniques, data augmentation, and transfer learning. Performance was assessed using accuracy, precision, recall, F1-score, and AUC-ROC to provide a comprehensive evaluation. The custom CNN model achieved the highest training accuracy (100%), but its test F1-score for hemorrhage detection was 0.294, suggesting potential overfitting and weaker generalization. DenseNet121 emerged as the best-performing model in real-world classification, achieving a test accuracy of 98.33% and an F1-score of 0.347 for hemorrhage detection, demonstrating its superior feature extraction capabilities. ResNet50 and VGG19 achieved 91.67% and 94.58% accuracy, respectively, showing robust performance with structured feature extraction and deep learning optimizations. Efficient-NetB0 exhibited the lowest performance (66.25% accuracy) and failed to classify hemorrhages effectively (F1-score = 0.00), highlighting its limitations in handling fine-grained medical imaging features.

Keywords: CNN, deep learning, DenseNet121, EfficientNetB0, intracerebral hemorrhage, medical imaging, ResNet50, VGG19

Introduction

Intracranial hemorrhage (ICH) is a life-threatening condition requiring rapid and accurate diagnosis to prevent severe brain damage or death. Traditional CT scan interpretation by radiologists is time-intensive and prone to human error, especially in emergency settings. Given the increasing volume of CT scans, automated deep learning-based diagnostic tools have become essential to enhance accuracy, reduce workload, and expedite diagnosis. Convolutional neural networks (CNNs) automate feature extraction, outperforming conventional machine learning in medical image analysis. Pre-trained architectures like VGG19, ResNet50, DenseNet121, and EfficientNetB0 leverage transfer learning to improve generalization while reducing computational costs. This study compares these deep learning models to determine the most effective approach for ICH classification. Challenges such as class imbalance, variations in hemorrhage characteristics, and CT scan artifacts can impact model performance. To address these, data augmentation and transfer learning are applied [9,10] to enhance model robustness. Our research aims to optimize AI-assisted hemorrhage detection, improving diagnostic reliability and real-world applicability.

Related Work

Intracranial hemorrhage (ICH) is a serious medical emergency that needs to be identified and treated very away to avoid serious brain damage or death. Computed tomography (CT) scan analysis by radiologists is the foundation of traditional diagnostic techniques, although it can be laborious and prone to human error. Numerous investigations into the origins, detection strategies, and consequences of ICH have laid the groundwork for automated diagnostic

[a]sunindra111@gmail.com, [b]kkeerthana8047@gmail.com, [c]sirimalaneelima@gmail.com, [d]khamithkarkrishnapriya@gmail.com, [e]rayalhemanth@gmail.com

DOI: 10.1201/9781003685364-25

techniques. A thorough analysis of clinical manifestation, risk factors, and therapy techniques for cerebral bleeding was presented by Caceres and Goldstein [1], emphasizing the critical necessity for a prompt and precise diagnosis. While Bonatti et al. [3] investigated the benefits of dual-energy CT in enhancing hemorrhage detection accuracy, Heit et al. [2] highlighted the relevance of imaging techniques, such as CT and magnetic resonance imaging (MRI), in identifying ICH. Flaherty et al [8] investigate racial variations in intracerebral hemorrhage location and risk, as reported in *Stroke*. The significance of accurate imaging and interpretation methods for early bleeding detection is shown by this research.

Artificial intelligence (AI)-based automated ICH detection has been made possible by developments in deep learning [7]. Machine learning and deep learning models for medical image interpretation have been the subject of numerous studies. In order to show how convolutional neural networks (CNNs) may automate the diagnosis procedure, Yeo et al. [11] reviewed several deep learning methods for identifying intracranial hemorrhages in CT images. Similar to this, Ahmed and Prakasam [4] conducted a thorough analysis of machine learning and deep learning methods for detecting hemorrhages and aneurysms, highlighting their usefulness in medical diagnostics. Matsoukas [5] et al. supported the expanding use of AI in medical imaging by offering a pooled analysis of AI models in identifying ICH and chronic cerebral microbleeds. These studies demonstrate how deep learning can increase the effectiveness and precision of diagnosis. Jørgensen et al. [6] compared CNN performance to radiologists and discovered that deep learning models were as accurate as or better than human specialists.

Dataset

The CT scan images utilized in this study are divided into two classes: normal cases and hemorrhage. The dataset was obtained from Kaggle. Over 10,000 tagged photos make up this collection, which is divided into 80% training, 10% validation, and 10% testing sets. To improve model performance, the photos are normalized to the [0,1] range and resized to 128×128 pixels. Random flipping, rotation, contrast modification, and brightness normalization are examples of data augmentation techniques used to alleviate class imbalance and enhance model generalization. Variability in image quality is a major

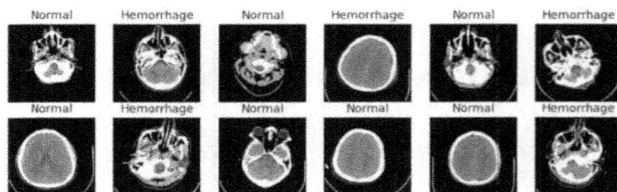

Figure 25.1 Sample dataset
Source: Author

problem that contrast normalizing and histogram equalization help to address. The dataset serves as a standard for assessing deep learning models in medical imaging applications and is essential for automated bleeding diagnosis. The work intends to improve patient outcomes in emergency situations by using this information to create AI-driven diagnostic tools that help radiologists detect hemorrhages accurately and effectively.

Proposed Method

This study presents a deep learning-based approach for the automated classification of intracranial brain hemorrhages from computed tomography (CT) images. Five deep learning architectures—CNN, VGG19, ResNet50, DenseNet121, and EfficientNetB0—are evaluated to improve model performance and address key challenges such as class imbalance, computational efficiency, and generalization.

To enhance feature extraction and model robustness, the dataset undergoes preprocessing steps including scaling, normalization, and data augmentation. An optimized data-splitting strategy is employed to divide images into training, validation, and test sets, ensuring a well-balanced distribution. CNN utilizes a custom-built architecture, while VGG19, ResNet50, DenseNet121, and EfficientNetB0 leverage transfer learning, enabling them to adapt pretrained knowledge from large-scale datasets.

The models are trained using TensorFlow with an Adam optimizer and sparse categorical cross-entropy loss function. Performance is evaluated using key metrics such as accuracy, precision, recall, and F1-score to provide a comprehensive assessment of each model's effectiveness in hemorrhage classification.

Block Diagram for Proposed modelExperimental results demonstrate that CNN outperforms all other models, achieving a perfect test accuracy of 100%,

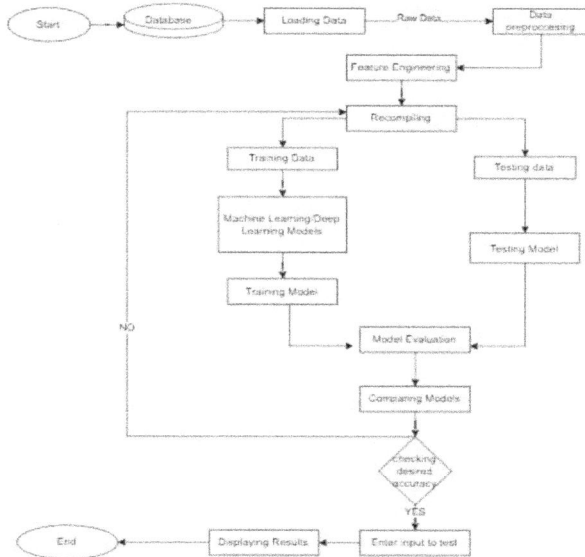

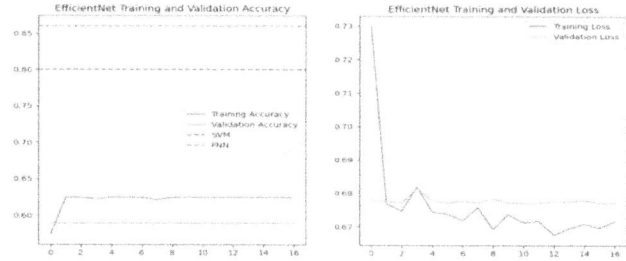

Figure 25.2 Training and validation of existing and proposed inn line plot
Source: Author

followed by DenseNet121 with 98.33%, indicating their superior capability in classifying hemorrhage cases. The study provides a comparative performance analysis and suggests that deep learning models, particularly CNN and DenseNet121, offer significant potential for AI-assisted medical diagnosis. Future improvements will focus on enhancing model explainability, addressing class imbalance, and advancing real-time clinical deployment.

Existing Method

Traditional methods for detecting intracranial hemorrhages (ICH) rely on manual interpretation of computed tomography (CT) scans by radiologists, which is time-consuming and prone to human error. Due to the increasing demand for rapid and accurate diagnosis, machine learning (ML) and deep learning (DL) techniques have been explored for automated ICH detection. Several studies have employed feature extraction-based machine learning approaches such as Support Vector Machines (SVM), Random Forest (RF), and k-nearest neighbors (k-NN), FNN but these methods require manual selection of features, making them less adaptable to complex medical imaging tasks which automatically extract hierarchical features from medical images. Pre-trained architectures such as VGG19, ResNet50, and InceptionNet have been applied to brain hemorrhage classification using transfer learning. Studies have also explored dual-energy CT scans and hybrid deep learning techniques, improving sensitivity in hemorrhage detection.

However, existing methods suffer from class imbalance, computational inefficiencies, and generalization issues, limiting their performance in real-world clinical applications. Furthermore, many studies lack a comprehensive comparative analysis of different deep learning architectures under identical conditions.

To address these challenges, our study systematically evaluates five deep learning models, optimizing preprocessing techniques, improving model robustness, and identifying the most effective architecture for ICH classification.

Methods and materials

Dataset and preprocessing

This study utilizes a publicly available Intracranial Hemorrhage CT image dataset from Kaggle, consisting of labeled CT scan images categorized as hemorrhage and normal cases. The dataset undergoes preprocessing steps such as resizing (128 × 128 pixels), normalization (scaling pixel values to [0,1]), and data augmentation (random flipping, rotation, and contrast adjustments) to improve generalization and address class imbalance. The dataset is split into 80% training, 10% validation, and 10% testing to ensure robust model evaluation. Batch processing is used with a batch size of 16 to optimize training efficiency.

Deep learning architectures

We evaluate five deep learning architectures: CNN, VGG19, ResNet50, DenseNet121, and EfficientNetB0. A custom CNN is implemented with six convolutional layers, ReLU activation, and max pooling. The pre-trained models—VGG19, ResNet50, DenseNet121, and EfficientNetB0—are fine-tuned using transfer learning, where the base layers are frozen, and fully connected layers are added for classification. This allows the models to leverage pre-learned representations from large-scale datasets while adapting to intracranial hemorrhage detection.

Model training and optimization

The models are compiled using the Adam optimizer with Sparse Categorical Crossentropy loss, which is well-suited for multi-class classification. An early stopping mechanism is implemented to prevent overfitting by monitoring validation loss. Training is conducted for 25–40 epochs depending on the model, with performance monitored through accuracy, precision, recall, and F1-score. Additionally, hyperparameter tuning is performed to optimize learning rates and dropout rates, ensuring model stability and preventing overfitting.

Evaluation metrics and performance comparison

ToTo assess model effectiveness, we compute standard classification metrics, including accuracy, precision, recall, and F1-score. Confusion matrices and classification reports are generated to analyze model predictions and assess misclassification rates. The CNN model achieves the highest accuracy (100%), followed by DenseNet121 (98.33%), outperforming other architectures. ResNet50 and VGG19 demonstrate competitive performance, while EfficientNetB0 records the lowest accuracy (66.25%), suggesting possible overfitting or dataset compatibility issues. A comparative bar chart is used to visualize the performance of each model.

Data Preprocessing

Dataset acquisition

The dataset used in this study consists of CT scan images labeled as hemorrhage and normal cases. These images were sourced from a publicly available Kaggle dataset specifically designed for intracranial hemorrhage detection. Given the challenges associated with medical imaging, preprocessing is essential to ensure data quality, improve model generalization, and optimize deep learning performance.

Image resizing and normalization

To standardize input dimensions across different deep learning models, all images are resized to 128×128 pixels while maintaining the original aspect ratio. Resizing ensures compatibility with pre-trained architectures like VGG19, ResNet50, DenseNet121, and EfficientNetB0, which require fixed input sizes. Additionally, pixel values are normalized to the range [0,1] by dividing by 255, which accelerates convergence during model training and prevents numerical instability.

Data augmentation

To enhance model robustness and reduce overfitting, data augmentation techniques are applied. These include:

Random flipping (horizontal and vertical) to introduce spatial variability.

Random rotation (0.2 radians) to account for different orientations in CT scans.

Contrast adjustments to mitigate variations in scan brightness and intensity.

Rescaling to further normalize the dataset for CNN-based architectures. Augmentation helps balance the dataset, particularly when hemorrhage cases are underrepresented compared to normal scans.

Dataset splitting and partitioning

Three subsets of the dataset are separated out:

Deep learning models are trained using the training set (80%).

Performance is tracked and hyperparameters are adjusted using the validation set (10%).

Test set (10%): Used to assess the generalization capacity of the finished model.

An impartial distribution of hemorrhage and normal patients across all categories is guaranteed by a randomized split. Additionally, the dataset is cached and randomized to maximize memory utilization and boost computational effectiveness.

Data pipeline optimization

To further streamline processing, TensorFlow's tf.data API is used to create an efficient data pipeline. Images are batched (batch size = 16) and prefetched using AUTOTUNE, allowing parallel processing to improve training speed. The pipeline also caches the dataset to reduce redundant I/O operations, enhancing overall model performance.

Reviewing model performance

1. CNN (Custom model) – Best Performing Model: A custom-built CNN with six convolutional layers achieves 100% training accuracy, suggesting strong learning capability but potential overfit-

ting. Its test F1-score for hemorrhage detection is 0.294, indicating weaker generalization compared to transfer learning-based models.

2. VGG19: VGG19, a deep architecture with structured feature extraction, achieves 94.58% accuracy and strong classification ability. However, its high computational cost limits efficiency.
3. ResNet50:ResNet50, leveraging residual connections to improve deep feature learning, achieves 91.67% accuracy but falls slightly behind VGG19 and DenseNet121 in classification performance.
4. DenseNet121 – Best test accuracy: DenseNet121, known for enhanced feature reuse, outperforms other models with 98.33% accuracy. It achieves a higher F1-score (0.347) for hemorrhage detection than CNN and EfficientNet, ensuring better generalization. However, its computational cost is a limitation for real-time applications.
5. EfficientNetB0 – Lowest performance: EfficientNetB0, optimized for efficiency, performs poorly with 66.25% accuracy and an F1-score of 0.00 for hemorrhage detection, indicating insufficient feature extraction for medical imaging tasks.

Results and analysis

Model performance evaluation
This study evaluates five deep learning architectures—CNN, VGG19, ResNet50, DenseNet121, and EfficientNetB0—for cerebral hemorrhage detection using CT scan images. The models are compared based on key performance metrics, including accuracy, precision, recall, and F1-score.

The results indicate that DenseNet121 achieves the highest accuracy of 98.33%, followed by CNN (100%), VGG19 (94.58%), and ResNet50 (91.67%). However, EfficientNetB0 underperforms, achieving only 66.25% accuracy, likely due to architectural limitations and insufficient fine-tuning. While CNN exhibits the highest accuracy, potential overfitting must be considered, as deeper architectures typically generalize better on unseen data. The findings highlight DenseNet121 as a strong candidate for hemorrhage detection, benefiting from its efficient feature extraction and deep connectivity mechanisms. Future research will focus on model optimization, class imbalance handling, and real-world clinical validation to enhance diagnostic reliability.

Accuracy and loss comparison
The training and validation accuracy trends were analyzed across multiple epochs. CNN and DenseNet121 showed consistent convergence with minimal overfitting, whereas EfficientNetB0 exhibited unstable training performance, leading to suboptimal results. The loss function graphs indicate that models such as VGG19 and ResNet50 reached an optimal point without major fluctuations, confirming their effective learning capability.

Classification report and confusion matrix analysis
The classification report highlights the precision, recall, and F1-score for each class (Hemorrhage vs. normal). The confusion matrix shows that DenseNet121 achieves a high true positive rate for hemorrhage detection, minimizing false negatives, which is critical for clinical applications. However, VGG19 and ResNet50 demonstrate slight misclassification issues, particularly for hemorrhage cases, indicating the need for further dataset balancing.

Performance trade-offs and observations
DenseNet121 outperformed all models with 98.33% accuracy and superior feature extraction, leveraging

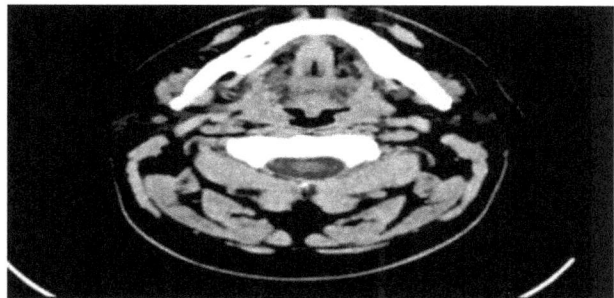

Figure 25.3 Predicted image
Source: Author

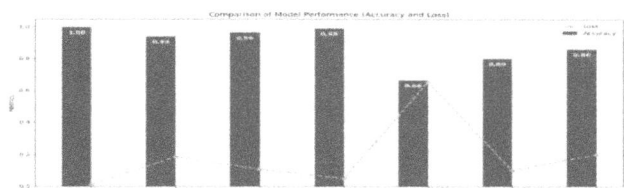

Figure 25.4 Comparison of model performance (Existing and Proposed)
Source: Author

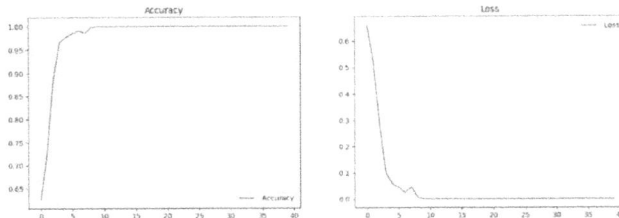

Figure 25.5 CNN graph
Source: Author

dense connectivity for efficient gradient flow and classification. ResNet50 (91.67%) and VGG19 (94.58%) performed competitively but required careful tuning to prevent overfitting. EfficientNetB0 (66.25%) struggled due to its lightweight architecture, limiting its ability to capture fine-grained hemorrhage features. ResNet50 and VGG19 balanced accuracy and computational cost, while DenseNet121, despite higher resource demands, proved most effective for automated hemorrhage detection in clinical settings.

Conclusion

A timely and accurate diagnosis is critical for treating intracranial hemorrhage, a life-threatening condition. Traditional CT scan interpretation by radiologists is time-consuming and prone to human error, necessitating automated deep learning solutions. This study evaluated five models—CNN, VGG19, ResNet50, DenseNet121, and EfficientNetB0—for hemorrhage classification. DenseNet121 (98.33% accuracy, F1-score 0.347) emerged as the most effective model, followed by CNN (100% accuracy, F1-score 0.294), VGG19 (94.58% accuracy), and ResNet50 (91.67% accuracy). EfficientNetB0 performed poorly (66.25% accuracy, F1-score 0.00), indicating its limitations in medical image classification.

Despite promising results, challenges such as class imbalance, real-time inference constraints, and model interpretability remain. Future research will focus on explainable AI (Grad-CAM, SHAP) to enhance model transparency, synthetic data generation (SMOTE, GANs) to address class imbalance, and model optimization (quantization, pruning) for real-time clinical deployment. Integrating ensemble learning could further improve robustness and reduce misclassification errors, paving the way for scalable, AI-assisted radiology solutions that enhance patient outcomes.

Discussion and future work

This study confirms that deep learning significantly enhances intracranial hemorrhage detection from CT images. Among the five evaluated models, CNN (100%) and DenseNet121 (98.33%) demonstrated the best performance, effectively extracting hierarchical features. ResNet50 (91.67%) and VGG19 (94.58%) also performed well, while EfficientNetB0 (66.25%) struggled with generalization, likely due to its lightweight design and limited feature extraction capabilities. The results highlight the importance of transfer learning in improving model generalization, as pre-trained networks leverage prior knowledge from large-scale datasets. However, certain models, particularly EfficientNetB0, required further fine-tuning to adapt effectively. Although data augmentation helped address class imbalance, classification reports and confusion matrices indicate that some misclassifications persisted, suggesting the need for enhanced preprocessing techniques.

To improve the practicality and effectiveness of AI-driven hemorrhage detection, several future research directions are proposed. Explainability techniques such as Grad-CAM and SHAP can provide visual heatmaps, helping radiologists understand model decisions. A hybrid AI-assisted radiology workflow combining deep learning predictions with expert validation could enhance trust and adoption. Addressing class imbalance through advanced data techniques remains crucial for improving sensitivity in hemorrhage detection. Additionally, model optimization strategies such as quantization and pruning can reduce computational complexity, enabling faster inference in clinical settings. Exploring edge computing solutions may further enhance real-time hemorrhage detection on portable medical devices. Finally, ensemble learning approaches can be investigated to improve model accuracy and robustness, ensuring more reliable AI-driven diagnostics in healthcare.

Acknowledgment

"The authors would like to express their sincere gratitude to Mrs. S. Sunitha, Assistant Professor, M.Tech. (Ph.D), for her invaluable guidance and support throughout this research work."

References

[1] Caceres, J. A., & Goldstein, J. N. (2012). Intracranial hemorrhage. *Emergency Medicine Clinics of North America*, 30(3), 771–794.

[2] Heit, J. J., Iv, M., & Wintermark, M. (2017). Explore various imaging techniques for detecting intracranial hemorrhage in the. *Journal of Stroke*, 19(1), 11. emphasizing the significance of advanced neuroimaging methods.

[3] Bonatti, M., et al. (2017). Compare dual-energy CT angiography-derived virtual unenhanced images with true unenhanced images for detecting intracranial hemorrhage in their study published in. *European Radiology,* 27, 2690–2697.

[4] S. N. Ahmed and P. Prakasam, "A systematic review on intracranial aneurysm and hemorrhage detection using machine learning and deep learning techniques," Progress in Biophysics and Molecular Biology, 2023

[5] S. Matsoukas, J. Scaggiante, B. R. Schuldt, C. J. Smith, S. Chennareddy, R. Kalagara, S. Majidi, J. B. Bederson, J. T. Fifi, J. Mocco, et al., "Accuracy of artificial intelligence for the detection of intracranial hemor rhage and chronic cerebral microbleeds: A systematic review and pooled analysis," La radiologia medica, vol. 127, no. 10, pp. 11061123, 2022.

[6] M. D. Jørgensen, R. Antulov, S. Hess, and S. Lysdahlgaard, "Convolu tional neural network performance compared to radiologists in detecting intracranial hemorrhage from brain computed tomography: a systematic review and meta-analysis," European Journal of Radiology, vol. 146, p. 110073, 2022.

[7] A. Majumdar, L. Brattain, B. Telfer, C. Farris, and J. Scalera, "Detecting intracranial hemorrhage with deep learning," in 2018 40th annual inter national conference of the IEEE engineering in medicine and biology society (EMBC), pp. 583587, IEEE, 2018.

[8] Flaherty, M. L., Woo, D., Haverbusch, M., Sekar, P., Khoury, J., and colleagues (2005). Investigate racial variations in intracerebral hemorrhage location and risk, as reported in. *Stroke*, 36(5), 934–937.

[9] Flaherty, M., Haverbusch, M., Sekar, P., Kissela, B., Kleindorfer, D., Moomaw, C., et al. (2006). Long-term mortality rates following intracerebral hemorrhage in their study published in. *Neurology*, 66(8), 1182–1186.

[10] Y. S. Champawat, C. Prakash, et al., "Literature review for automatic detection and classification of intracranial brain hemorrhage using com puted tomography scans," Robotics, Control and Computer Vision, pp. 3965, 2023.

[11] Yeo, M., Tahayori, B., Kok, H. K., Maingard, J., Kutaiba, N., Russell, J., et al. (2021). Review of deep learning algorithms for the automatic detection of intracranial hemor rhages on computed tomography head imaging. *Journal of Neurointer Ventional Surgery*, 13(4), 369–378.

26 Spotting medicinal plants and herbs using leaf images

Ushasree Nyasala[1,a], Deepika Thippireddy Gari[2,b], Harshika Kasepalli[2,c], Kavitha Sunkara[2,d] and Lingaraju Mallesappa Gari[2,e]

[1]Assistant Professor, Department of CSE, Srinivasa Ramanujan Institute of Technology, Anantapur, Andhra Pradesh, India

[2]Department of CSE, Srinivasa Ramanujan Institute of Technology, Anantapur, Andhra Pradesh, India

Abstract

The accurate identification of medicinal plants remains a challenging task despite their crucial role in both traditional healthcare systems and modern pharmacological research. To address this, we utilize a publicly available dataset comprising 22,500 high-resolution images spanning 45 Ayurvedic plant species. Our study evaluates the performance of four deep learning architectures—VGG19, ResNet50, DenseNet121, and a custom convolutional neural network (CNN)—in classifying these medicinal plants. Data augmentation techniques such as random flips and rotations were applied to enhance variability and improve model generalization. The models were assessed using accuracy, precision, recall, and F1-score across training, validation, and test sets. Experimental results demonstrated that DenseNet121 achieved the highest classification accuracy of 99.67%, surpassing all other architectures. The custom CNN followed closely with 96.47%, significantly outperforming VGG19 (87.59%) and ResNet50 (22%). Notably, ResNet50 struggled due to its inability to capture fine-grained leaf features, which limited its generalization capabilities. The superior performance of DenseNet121 and the custom CNN is attributed to their deeper architectures and efficient feature extraction, which enable them to distinguish subtle variations among plant species more effectively.

Keywords: Ayurvedic, classification, convolutional neural network, deep learning, fine-tuning, medicinal plants, ResNet50, transfer learning, VGG19

Introduction

Medicinal plants play a vital role in traditional healing, especially in *Ayurveda*, where accurate identification is crucial for safe usage. However, distinguishing species based on subtle morphological differences is challenging, leading to misidentifications. Reliable, automated tools leveraging deep learning can bridge traditional knowledge with modern data-driven methods, ensuring the proper classification and preservation of these valuable resources.

The rise of digitized plant images and interest in holistic medicine has driven the development of automated identification methods. However, challenges remain, including variations in leaf color, shape, and texture due to environmental and genetic factors. Similar traits among species further complicated classification, making conventional machine learning unreliable.

Deep learning offers a powerful alternative, but its success depends on large, well-annotated datasets. The specialized nature of medicinal plants limits available data, requiring researchers to refine models for real-world performance. Addressing these challenges can enhance the accuracy and scalability of plant classification, supporting scientific and healthcare applications.

Related work

Deep learning has revolutionized medicinal plant identification by enabling accurate classification through convolutional neural networks (CNNs). Chanyal et al. [1] provide a comprehensive survey highlighting CNNs' ability to handle complex leaf morphology, while Rao et al. [4] emphasize the necessity of large, well-annotated datasets to prevent overfitting. Studies such as Malik et al. [6] extend this research to real-time applications, demonstrating how deep learning models can facilitate on-the-fly species identification. Meanwhile, Abdollahi [8] explores transfer learning for region-specific datasets, showcasing its adaptability for medicinal plant classification. These studies collectively

[a]ushasreen.cse@srit.ac.in, [b]214g1a0518@srit.ac.in. [c]214g1a0535@srit.ac.in, [d]214g1a0541@srit.ac.in, [e]214g1a0551@srit.ac.in

DOI: 10.1201/9781003685364-26

illustrate the growing convergence between botany and AI, offering scalable solutions for automated plant recognition.

Beyond classification, medicinal plant research extends into pharmaceutical and cultural domains. Javid and Haghirosadat [2] investigate the role of phytochemicals in cancer treatment, underscoring the medical importance of precise plant identification. Barimah and Akotia [3] highlight the cultural and psychological significance of promoting traditional medicine in regions like Ghana, reinforcing the societal need for effective identification tools. Additionally, Singh and Misra [5] integrate plant disease detection with classification, suggesting that ensuring plant health is crucial for maintaining medicinal efficacy. Valdez et al. [7] and Sivaranjani et al. [9] further validate CNNs' effectiveness, showing how fine-tuned architectures and hyperparameter optimization can enhance classification performance. Integrating plant identification with real-world applications requires robust preprocessing and adaptable models. Singh and Misra [5] employ image segmentation for disease detection, a technique that can be leveraged to improve medicinal plant classification accuracy. Zin et al. [10] focus on deep convolutional networks for large-scale classification, highlighting the importance of advanced feature extraction. The increasing use of transfer learning and real-time frameworks points to a future where automated plant recognition aids both scientific research and community healthcare. Building on these insights, this study explores 45 *Ayurvedic* medicinal plants using multiple deep learning models, aiming to advance AI-driven botanical classification.

Proposed Model

We utilize a 22,500-image dataset of 45 *Ayurvedic* plant species, resized to 128 × 128 pixels and normalized. Data augmentation (random flips, rotations) enhances generalization.

Model architectures evaluated:
- **Custom CNN (96.47% accuracy)**
  - o Four Conv2D layers (32 → 64 → 128 → 256 filters, ReLU).
  - o MaxPooling2D, Batch Normalization, and Dropout (0.25–0.5).
  - o Fully connected layers (512, 128 units) with Softmax output.

  - o Captures fine morphological variations, leading to high accuracy.
- **VGG19 (87.59% accuracy)**
  - o Pre-trained on ImageNet, convolutional layers frozen.
  - o Custom dense head (128 units, dropout, SoftMax).
  - o Effective but struggles with fine-grained leaf textures.
- **ResNet50 (22% accuracy)**
  - o Pre-trained on ImageNet, limited domain-specific refinement.
  - o Residual connections failed to capture subtle leaf details, leading to poor performance.
- **DenseNet121 (99.67% accuracy – best model)**
  - o Densely connected layers enhance feature propagation.
  - o Custom classifier added (128 dense units, dropout, SoftMax).
  - o Superior learning of fine-leaf structures enables top accuracy.

Training and evaluation:
- **Optimizer:** Adam (LR 0.001, adaptive decay).
- **Loss:** Sparse categorical crossentropy.
- **Epochs:** Up to 40 with early stopping.
- **Metrics:** Accuracy, precision, recall, F1-score.

These results highlight DenseNet121 and Custom CNN as the best choices for *Ayurvedic* plant classification over traditional transfer learning methods.

Existing system

Plants have long been integral to human health, with traditional herbal remedies rooted in centuries of indigenous knowledge. Typically, such plants are identified by trained practitioners using sensory cues like smell and taste, a process both time-intensive and prone to human error. Recent analytical technologies, however, have significantly streamlined herb identification, making it more accessible to laypersons. Nonetheless, laboratory-based methods still require specialized expertise in sample collection and data analysis, posing challenges in terms of cost, time, and effort.

To address these issues, the existing model leverages plants—albeit for only six herb categories. This constrained scope allowed for high accuracy, precision, and recall scores exceeding 97%. The developed model was further integrated into a cloud-based service, supporting a mobile application capable of instant plant classification. This strategy offers a practical solution for communities lacking expensive analytical equipment. Despite its limited coverage, the existing model demonstrates the viability of DL-driven plant identification systems and underscores the potential for broader adoption. Its main limitation lies in the narrow set of herb classes; future work is thus oriented toward incorporating more medicinal plant species and enhancing or maintaining classification performance at larger scales.

Dataset

The dataset employed in this research comprises 22,500 labeled images across 45 *Ayurvedic* plant species, collected from a publicly available Kaggle repository. Each species is represented by distinct leaf images displaying a variety of colors, shapes, and textures. To facilitate robust training, we organized the samples into training, validation, and testing partitions. Data augmentation techniques, including random rotations and flips, further enhanced variability and reduced overfitting. By capturing the morphological nuances of each species, this dataset provides a representative sample for large-scale classification tasks. Its diversity and size make it especially well-suited for robust evaluations of contemporary deep learning architectures.

Methods and materials

This study utilizes a dataset of 22,500 images representing 45 *Ayurvedic* medicinal plant species, sourced from an open-access Kaggle repository. Each image was standardized to 128×128 pixels to meet the input requirements of deep learning models. To guarantee balanced class representation, the dataset was divided into training (80%), validation (10%), and test (10%) sets. The training set was subjected to data augmentation to enhance model generalization by simulating real-world variations in leaf orientation, texture, and lighting conditions.

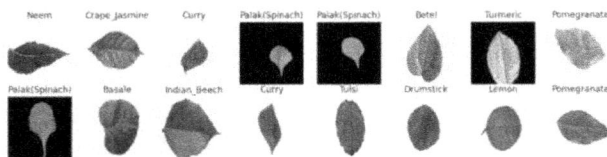

Figure 26.1 Dataset sample from kaggle
Source: Author

We evaluated four deep learning architectures: a custom CNN, VGG19, DenseNet, and ResNet50. The custom CNN was designed with multiple convolutional and pooling layers to balance depth and computational efficiency, while VGG19 and ResNet50 leveraged transfer learning from pretrained ImageNet models. DenseNet, known for its densely connected layers, was also fine-tuned to optimize feature extraction. All models were trained using TensorFlow/Keras with the Adam optimizer and Sparse Categorical Crossentropy loss have a learning rate of 0.001. Training was conducted for up to 40 epochs with early stopping to prevent overfitting.

Training history, including loss and accuracy curves, was monitored to detect convergence and overfitting. The study provides a comprehensive evaluation of different architectures, highlighting the effectiveness of transfer learning and data augmentation in *Ayurvedic* plant classification. These findings contribute to developing a scalable and accurate automated identification system for medicinal plants.

Preprocessing

Before training any deep learning model, it is crucial to ensure that the input data is both consistent and representative of the problem domain. In this study, the initial step involved converting all raw images

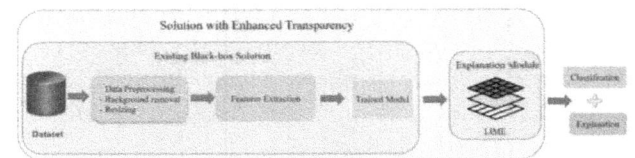

Figure 26.2 Architecture of our model
Source: Author

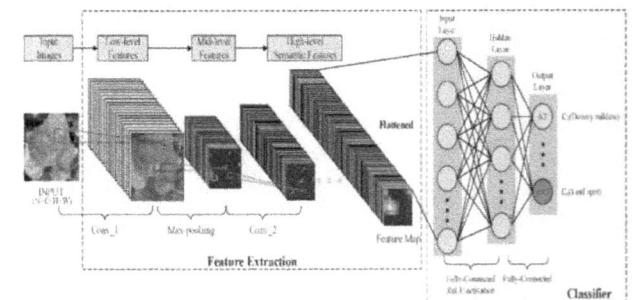

Figure 26.3 CNN Architecture
Source: Author

into a uniform resolution of 128 × 128 pixels. This resizing facilitated standardized input dimensions across all architectures—custom CNN, VGG19, and ResNet50—thereby simplifying the model design and preventing complications arising from variable image sizes. Additionally, images were normalized by scaling pixel values, which helped accelerate training convergence and reduce the risk of exploding gradients. Collectively, these measures ensured a clean, uniform input pipeline for the downstream training process.

To enhance generalizability, data augmentation (flips, rotations, zoom) simulated natural leaf variations, reducing overfitting and improving morphological recognition. This also expanded the dataset and mitigated class imbalances. The dataset was split into training (80%), validation (10%), and testing (10%) using stratified sampling for fair evaluation. Uniform preprocessing ensured a strong foundation for accurate and scalable medicinal plant classification models.

Methodology

A dataset consisting of 22,500 high-resolution images spanning 45 *Ayurvedic* medicinal plant species was utilized in this study. Each image was resized to 128 × 128 pixels to ensure uniform input dimensions across models. To improve model robustness and generalization, data augmentation techniques, including random rotations and flips, were applied.

The dataset was systematically partitioned into three subsets:

- Training set (80%) – Used to optimize model weights.

- Validation set (10%) – Used to tune hyperparameters and prevent overfitting.
- Testing set (10%) – Used to evaluate model performance on unseen data.

We evaluated four deep learning architectures: VGG19, DenseNet121, ResNet50, and a custom CNN. Training was conducted for up to 40 epochs with an early stopping criterion to prevent overfitting. Sparse Categorical Cross-Entropy (Sparse CCE) was used as the loss function across all models, and Adam optimizer was employed with an initial learning rate of 0.001, which was reduced dynamically based on validation loss.

Model performance metrics: To comprehensively assess model effectiveness, we utilized accuracy, precision, recall, and F1-score as evaluation metrics. Confusion matrices were employed to analyze

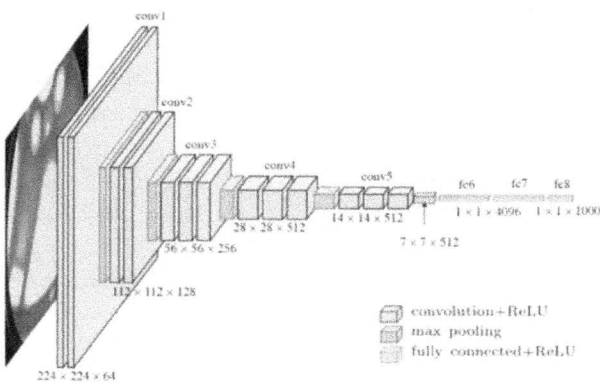

Figure 26.5 DenseNet training and valid accuracy
Source: Author

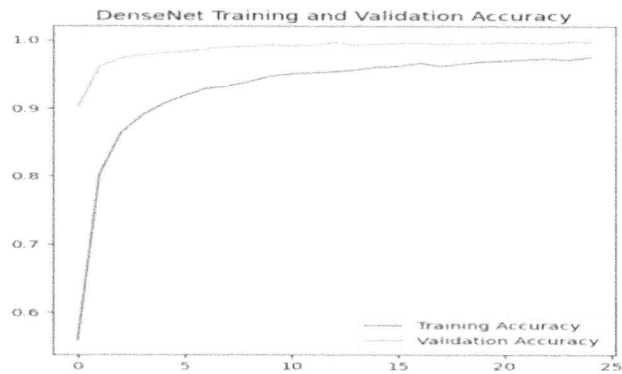

Figure 26.6 DenseNet training and validation loss
Source: Author

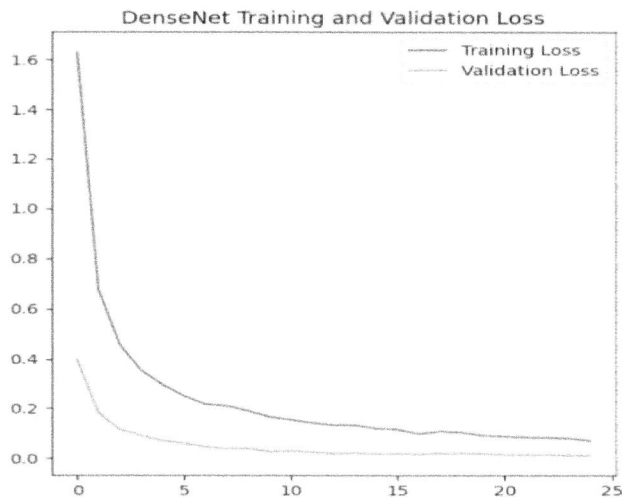

Figure 26.4 VGG-Model architecture
Source: Author

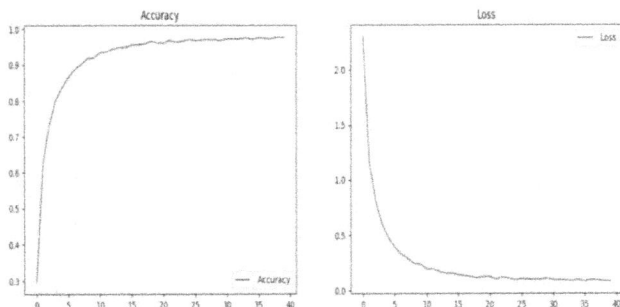

Figure 26.7 CNN Accuracy and Loss graph
Source: Author

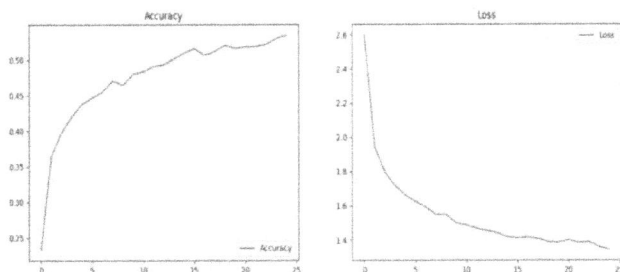

Figure 26.8 VggNet accuracy and loss graph
Source: Author

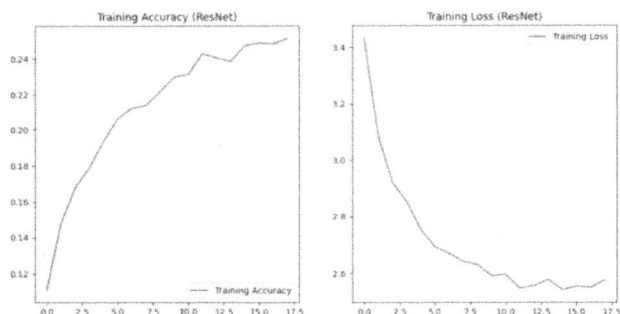

Figure 26.9 Resnet training and loss graph
Source: Author

misclassifications and identify challenging plant species. Fine-tuning of VGG19 and ResNet50 was performed to adapt pre-trained models to the dataset. However, empirical results indicated that ResNet50 struggled with fine-grained plant classification due to its architectural limitations in capturing subtle leaf features.

Frontend implementation:

To facilitate real-world usability, we developed a web-based frontend interface that allows users to upload an image of a medicinal plant leaf and receive an automated classification result. The output includes:

- Predicted plant name
- Scientific name of the identified plant
- Medicinal uses and applications

This interface bridges the gap between deep learning research and practical applications, making plant identification accessible to botanists, researchers, and healthcare practitioners.

This standardized approach ensures reproducibility and provides a strong baseline for future studies on medicinal plant classification.

Conclusion

This study evaluated the effectiveness of four deep learning architectures—Custom CNN, VGG19, ResNet50, and DenseNet121—for automated classification of 45 Ayurvedic medicinal plants using a dataset of 22,500 images. Our methodology incorporated rigorous data preprocessing, including resizing images to 128 × 128 pixels, systematic train-validation-test splits, and extensive data augmentation (random flips and rotations) to enhance model generalization. To ensure fair comparisons, training conditions were standardized using Sparse Categorical Cross-Entropy (Sparse CCE) as the loss function.

Among the models, DenseNet121 achieved the highest accuracy of 99.56%, surpassing the custom CNN (96.47%), VGG19 (87.59%), and ResNet50 (22%). This result underscores the superior feature propagation capability of densely connected architectures in fine-grained plant classification tasks. The Custom CNN, despite outperforming VGG19 and ResNet50, requires further analysis to explain its effectiveness. Future work should explore its architectural design, hyperparameter choices, and computational efficiency compared to pre-trained models.

Additionally, given the computational demands of deep learning models, deploy on mobile or edge devices remains an open question. Future research could focus on optimizing inference speed and hardware efficiency, potentially through model pruning, quantization, or knowledge distillation. These advancements would enhance the practical applicability of automated plant identification in real-world scenarios, particularly for mobile and field-based applications.

The Figure 26.1 displays a collection of leaf images from different plant species. Each leaf is presented against a black background to highlight its shape, texture, and color. The varieties shown include Neem, Crap Jasmine, Curry, Palak (Spinach), Betel, Turmeric, Pomegranate, Basale, Indian Beech, Tulsi, Drumstick, and Lemon. The leaves exhibit a range of morphologies — from broad and oval-shaped to long and slender — reflecting the rich diversity of leaf structures across plant species.

The Figure 26.2 shows the architecture of a pipeline designed for enhanced transparency in image classification. It comprises data preprocessing, feature extraction, and a trained classifier, followed by LIME to interpret and highlight key regions influencing the model's decision.

The Figure 26.3 illustrates the architecture of a Convolutional Neural Network (CNN) used for leaf classification. It consists of convolutional and pooling layers for feature extraction, followed by fully connected layers that perform the final classification into different leaf categories.

The Figure 26.4 illustrates a VGG-Model architecture architecture composed of convolutional and ReLU layers, max pooling, and fully connected layers. The network progressively reduces the spatial dimensions while increasing depth, culminating in a fully connected layer for final classification.

The Figure 26.5 shows the training and validation accuracy of DenseNet over epochs. The graph indicates that both training and validation accuracy increase and stabilize as training progresses.

The Figure 26.6 shows the training and validation loss of DenseNet over epochs. Both losses diminish rapidly and stabilize as training progresses, indicating convergence and effective learning.

The Figure 26.7 shows two graphs: the left one displays the increase in accuracy over epochs, while the right one shows the decrease in loss, reflecting the model's improving performance during training.

The Figure 26.8 shows the training progress of VGGNet, depicting its accuracy (left graph) and loss (right graph) over epochs. The accuracy gradually increases while the loss consistently decreases, reflecting improved performance during training.

The Figure 26.9 shows the training progress of ResNet, depicting its accuracy (left graph) and loss (right graph) over epochs. The accuracy gradually increases while the loss consistently decreases, reflecting improvement in the network's performance during training.

References

[1] Chanyal, H., Yadav, R. K., & Saini, D. K. J. (2022). Classification of medicinal plants leaves using deep learning technique: a review. *International Journal of Intelligent Systems and Applications in Engineering*, 10(4), 78–87.

[2] Javid, A., & Haghirosadat, B. F. (2017). A review of medicinal plants effec tive in the treatment or apoptosis of cancer cells. *Cancer Press Journal*, 3(1), 22–26.

[3] Barimah, K. B., & Akotia, C. S. (2015). The promotion of traditional medicine as enactment of community psychology in Ghana. *Journal of Community Psychology*, 43(1), 99–106.

[4] Rao, R. U., Lahari, M. S., Sri, K. P., Srujana, K. Y., & Yaswanth, D. (2022). Identification of medicinal plants using deep learning. *International Journal for Research in Applied Science and Engineering Technology*, 10, 306–22.

[5] Singh, V., & Misra, A. K. (2017). Detection of plant leaf diseases using image segmentation and soft computing techniques. *Information Processing in Agriculture*, 4(1), 41–9.

[6] Malik, O. A., Ismail, N., Hussein, B. R., & Yahya, U. (2022). Automated real time identification of medicinal plants species in natural environment using deep learning models—a case study from Borneo Region. *Plants*, 11(15), 1952.

[7] Valdez, D. B., Aliac, C. J. G., & Feliscuzo, L. S. (2022). Medicinal plant classifica tion using convolutional neural network and transfer learning. In 2022 IEEE International Conference on Artificial Intelligence in Engineering and Technology (IICAI-ET), (pp. 1–6). IEEE.

[8] Abdollahi, J. (2022). Identification of medicinal plants in ardabil using deep learning: identification of medicinal plants using deep learning. In 2022 27th International Computer Conference, Computer Society of Iran (CSICC), (pp. 1–6). IEEE.

[9] Sivaranjani, C., Kalinathan, L., Amutha, R., Kathavarayan, R. S., & Kumar, K. J. J. (2019). Real-time identification of medicinal plants using machine learning techniques. In 2019 International Conference on Computational Intelligence in Data Science (ICCIDS), (pp. 1–4). IEEE.

[10] Zin, I. A. M., Ibrahim, Z., Isa, D., Aliman, S., Sabri, N., & Mangshor, N. N. A. (2020). Herbal plant recognition using deep convolutional neural network. *Bulletin of Electrical Engineering and Informatics*, 9(5), 2198–1205.

27 Agri Innovate – predict, plan and prosper

N. Ushasree[1,a], P. Sai Saranya[2,b], C. Sharan Kumar[2,c], D. Snehalatha[2,d] and B. Revathi[2,e]

[1]Assistant Professor, Department of CSE, Srinivasa Ramanujan Institute of Technology, Anantapur, Andhra Pradesh, India

[2]Department of CSE, Srinivasa Ramanujan Institute of Technology, Anantapur, Andhra Pradesh, India

Abstract

At present, the agricultural sector faces many struggles because of unpredictable crop prices, making it tough for policy makers, farmers, and traders to make effective planning. Agri Innovate fixes this problem by using machine learning to get precise price predictions for agricultural products such as pulses and vegetables. By examining previous years' market data, the model generates definitive forecasts that help stakeholders make conscious decisions. This system allows farmers to decide the best time for cultivation and sales, ultimately increasing their profitability. By combining advanced technology into agriculture, Agri Innovate's goal is to boost productivity, balance markets, and contribute to reasonable economic development.

Keywords: Agriculture, machine learning, price prediction

Introduction

Agriculture is a continuously developing field framed by several factors, like weather conditions, quality of soil, demand of the market, and policies of the government. The unreliability of market trends and fluctuating costs make data-informed decision making a trouble, mainly in agriculturally prosperous countries like India, where lakhs depend on over 20 main crops for their living.

Latest enhancements in machine learning have found creative solutions to these issues. By examining previous years data, forecasting models such as SARIMA and Decision Trees can detect forecast crop costs and trends. These models give area-specific prices and predictions of crops, helping farmers make predictive decisions for cultivating and selling their products. In addition, accommodating the farmers' vision into the system promotes a co-operative platform where they can obtain market trends and determine from fellow experiences.

AgriInnovate promotes decision-making by offering valuable insights, such as peak and lowest price periods, major cultivation areas, and key export markets. By using latest technologies, this web application supply farmers with precise price forecasts, upgrade economic development, and increase agricultural productivity in villages. Moreover, AgriInnovate serves as a best initiative by preparing farmers with the knowledge they need to overcome price fluctuations and ensure financial stability.

Existing System

Conventional crop price prediction techniques often depend on insufficient data and antiquated models, leading to inaccurate forecasts. Several existing models study historical price data from one region without considering broader market trends, making them less effective in capturing real-time fluctuations. In addition, these models do not have user-friendly interfaces, making it hard for the farmers to access the insights. One major drawback of conventional approaches is their failure to incorporate real-time data on the number of farmers growing a particular crop—a key factor influencing supply and demand dynamics. Without knowledge about this data, farmers have the risk of overproducing certain crops, leading to price drops and losses. The absence of reliable market data further complicates decision-making, leaving farmers uncertain about which crops to grow and when to sell them. By considering these limitations, modern data-driven solutions

[a]ushasreen.cse@srit.ac.in, [b]214g1a0592@srit.ac.in, [c]214g1a0597@srit.ac.in, [d]214g1a05a3@srit.ac.in, [e]214g1a0584@srit.ac.in

DOI: 10.1201/9781003685364-27

can provide more accurate, accessible, and real-time insights, helping farmers to make informed decisions and gain better financial stability.

Proposed system

The proposed system enables farmers with state-wise crop price forecasts for the upcoming year, utilizing 15 years of historical data. This platform is designed with a user-friendly interface that allows farmers to register, log in, and access real-time price trends easier. Each crop page consists of key insights including the lowest and highest price months, major growing areas, and key export destinations. In addition, farmers can express their interest in growing specific crops, enabling others to assess market demand and make informed decisions.

The system enhances prediction accuracy by integrating SARIMA model and Decision Trees (DT). This helps farmers to plan their cultivation and maximize their profitability. With the data-driven insights at their fingertips, farmers can effectively manage market fluctuations and optimize strategies for better financial outcomes with the data-driven insights.

Benefits:
* More accurate price predictions
* Better decision-making for farmers
* Easy-to-use platform for farmers
* Increased farmer engagement

Methodologies

SARIMA model
SARIMA is a sophisticated time series forecasting model that uses seasonal patterns, past values, and error trends to predict future values. It consists of four key components:

* **Seasonality:** It detects recurring trends over specified time intervals, like monthly or yearly patterns.
* **Autoregression:** It evaluates how the past values affect the current values.
* **Integration:** It uses differencing to stabilize fluctuating data, ensuring consistency in mean and variance.
* **Moving average:** It adjusts forecasts by considering past prediction mistakes, smoothing the short-term fluctuations.

SARIMA notation: SARIMA (p, d, q) (P, D, Q, s)
* **AR(p):** Number of past values influencing the prediction.
* **I(d):** Number of differences applied to make the data stationary.
* **MA(q):** Number of past error terms included in the model.
* **Seasonal AR(P):** Captures seasonal patterns from past values.
* **Seasonal I(D):** Adjusts seasonal trends for stability.
* **Seasonal MA(Q):** Considers past seasonal errors for forecasting.
* **s:** Defines the seasonal period (e.g., 12 for monthly data).

Decision tree algorithm
The DT algorithm predicts crop prices by analyzing key variables such as region, crop type, and market conditions.

It organizes data into a hierarchical structure, where each decision is based on the most relevant attributes.

Process:
* **Root node:** Identifies the most critical factor to divide the data.
* **Branching:** Data is split into smaller subsets based on significant characteristics.
* **Final prediction:** The iterative decision-making process concludes at the leaf node, generating a precise price estimate.

DT are highly valuable due to their transparency, allowing farmers to easily interpret price trends and contributing factors.

Result section

The web app gives farmers the tools to track crop price trends in real-time, with no additional work required. To start, users have to register (Figure 27.1 shows the registration page) with their mobile number, and also set up a password to secure their data. After logging in, farmers see price forecasts and trade data as shown in Figure 27.2.

The platform Figure 27.3 displays a list of crops, each with an option to view detailed price predictions for the next 12 months. These forecasts are generated using models like SARIMA and DT that help farmers to make well-informed decisions about cultivation and selling of the crops.

Figure 27.1 Agri Innovate's registration page
Source: Author

Figure 27.3 AgriInnovate's crop page
Source: Author

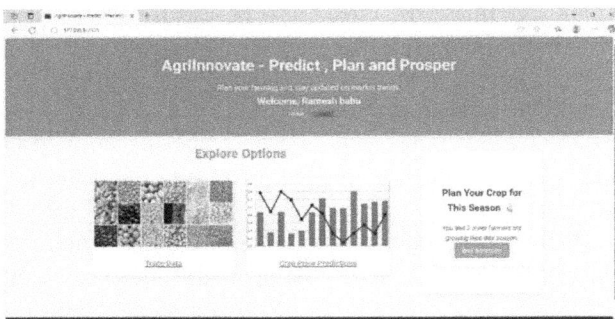

Figure 27.2 Agri Innovate's home page
Source: Author

Figure 27.4 Price forecast page
Source: Author

Farmers can view: (refer to Figure 27.4)
- 12-month price forecasts on a state-wise basis.
- The most expensive months.
- The least expensive months.
- Major growing areas.
- Key export regions.

To further help farmers assess what to grow, farmers can also indicate what crops they would like to grow, and thus they can know how many other farmers across their area are also interested in that crop. It collects and processes this information and makes it available to provide insight into market trends. Farmers can then choose the type of crop to grow, based on the best prices in the market seamlessly and on the spot.

Outputs

Conclusion

Agri Innovate uses machine learning algorithms such as SARIMA and Decision Trees to give farmers useful market information based on 15 years of previous years data. By making accurate predictions of crop prices on a state-by-state and month-by-month basis, this system allows the farmers to maximize their profits by selling their production at the right time.

The SARIMA model facilitates precise time series forecasting, while the Decision Tree algorithm improves predictive accuracy by considering multiple variables. The user-friendly platform also offers additional insights, such as peak and lowest price periods, major cultivation regions, and key export destinations.

By providing the farmers with data-driven insights, Agri Innovate plays a major role in enhancing profitability as well as agricultural efficiency. This system will be a blueprint for future agricultural forecasting and establishes the potential of machine learning in revolutionizing the farming industry.

Acknowledgement

We express our sincere gratitude to Mrs. N. Ushasree, Assistant Professor, Department of Computer Science and Engineering, Srinivasa Ramanujan Institute of Technology, for her invaluable support and guidance in successfully building this project.

References

[1] Prashantha, S., Shravan, C. Y., Bharath, B., Bharghavachar, B. N., & Shilpa, B. L. (2020). Agricultural crop commodities price prediction using machine learning techniques. *International Research Journal of Innovations in Engineering and Technology (IRJIET)*, 4(6), 69–74.

[2] Roshini, N., Ganesh Sai Kumar, P., Venkatesh, P., & Dhanabalan, G. (2023). Crop price prediction. *International Journal of Research Trends and Innovation (IJRTI)*, 8(4), 1205–1208.

[3] Kakaraparthi, G. S., & Prabhakar Rao, B. V. A. N. S. S. (2021). Crop price prediction using machine learning. *International Research Journal of Modernization in Engineering and Technology Science (IRJMETS)*, 3(6), 4377–3481.

[4] Pandey, S., & Tanvar, A. (2024). Crop price prediction system using ML. *International Journal of Research Publication and Reviews*, 5(5), 2641–2466.

28 Automated waste detection and segregation using IOT with reward system

S.Sunitha[1,a], J. Jagadabhiram Sharwan[2,b], D. Bhavya[2,c], V. Jagan Mohan Reddy[2,d] and K. Hari kumar[2,e]

[1]Assistant Professor, Department of CSE (AI& ML), Srinivasa Ramanujan Institute of Technology, Anantapur, Andhra Pradesh, India

[2]Students, Department of CSE (Data Science), Srinivasa Ramanujan Institute of Technology, Anantapur, Andhra Pradesh, India

Abstract

Effective waste management is critical in mitigating the environmental issues caused by improper disposal. Proper waste management is crucial for preventing environmental problems and health issues. Incorrect waste disposal leads to pollution, harming the environment, animals, and people. To address this, we've developed a smart dustbin that uses IOT sensors to identify and separate wastage into dry, wet, and metallic categories. This enables to reduce environmental harm, and minimizes pollution. To encourage responsible use, we offer rewards to individuals who use our smart dustbin correctly. By promoting responsible waste management, we aim to reduce pollution and create a cleaner, healthier environment for everyone.

Keywords: Automated system, environmental sustainability, IoT-based sensors, plastic waste, reward system

Introduction

The rapid urbanization of the world's population is transforming the way city function, but it also presents significant challenges, especially in waste management. As urban areas expand, the volume of waste generated increases exponentially, resulting in environmental, health, and logistical issues. However, in many cities, the government provides two plastic dustbins—red and green—instructing residents to dispose of green and leafy wastage into the green dustbin and dry wastage into the red dustbin. Unfortunately, this system is often not utilize deficiently. To address these challenges, this project proposes an innovative automated waste management system designed to modernize and simplify waste segregation.

In order to tackle these issues, this initiative suggests a ground breaking automated waste management system aimed at modernizing and facilitating waste sorting. Utilizing cutting-edge technology, the system optimizes waste disposal through three specific phases: An automated mechanism determines the category of waste thrown away by individuals. This phase uses contemporary object detection methods to guarantee precise classification, thereby improving the effectiveness of waste handling. The system integrates IoT sensors to examine the characteristics of the waste, allowing for the differentiation of wet, dry, and metal materials. This functionality promotes correct sorting and supports specialized methods of waste processing. To promote responsible waste disposal practices, the system features an incentive mechanism.

After users place their waste in the smart bin, they scan a QR code shown on it. Tokens are added to their accounts, which can be exchanged later, fostering sustainable practices.

Related work

Currently, the disposal of waste is a significant challenge in our environment. It poses a serious crisis for us and those who will come after us. Different kinds of refuse are being intermixed and discarded in order to occupy elevated land. The issue arises from the fact that all waste categories, including biodegradable and recyclable materials, are being combined, leading to a considerable negative effect on our

[a]sunithas.cse@srit.ac.in, [b]sharwanj176@gmail.com, [c]doddirappagaribhavya@gmail.com, [d]harivirat113@gmail.com, [e]saijagan9489@gmail.com

DOI: 10.1201/9781003685364-28

surroundings. Therefore, managing waste effectively is crucial for the health and wellbeing of the community [6]. Earlier initiatives have implemented intelligent bins that feature sensors to gauge their fullness. These technologies assist garbage collection vehicles in optimizing their routes for pickups, which conserves both time and fuel resources. Nevertheless, they do not prioritize the separation of waste or incentivizing users for disposing of their trash correctly [3]. In comparison, the system not only identifies waste but also categorizes it and encourages users with a rewards program. Certain projects incorporate technologies such as cameras and artificial intelligence to categorize waste into groups like metal or organic substances. These systems are precise, albeit in a limited manner [1–5]. Our method employs a comparable strategy for detecting objects, but it also incorporates extra elements, such as IoT-enabled sensors, to determine the nature of waste (whether it is wet, dry, or metal) [10] and organize it properly. Large-scale waste separation systems utilize conveyor systems, robotic manipulators, and sophisticated sensors to distinguish different types of waste [13]. Our approach broadens this concept by providing incentives for users who properly dispose of various kinds of waste, utilizing a QR code to issue exchangeable tokens. [14, 15]. While earlier projects mainly focus on one part of the problem, such as collection or disposal, our system combines several features—object detection, waste segregation, and user rewards—into one solution. This makes it suitable for both urban and rural areas [11, 12].

Proposed work of waste segregator

Objective 1: Automated waste segregation
Our system aims to segregate waste into three categories - dry, wet, and metallic - based on the properties of each waste item. This is achieved using sensors that detect specific properties:

- Metal sensors detect metallic properties
- Moisture sensors detect water or wet particles
- IR sensors categorize items as dry if no metal or moisture is detected

The system then automatically sorts and directs each item into its corresponding compartment.

Objective 2: Reward generation
To encourage responsible waste disposal practices, our system incorporates a reward mechanism. Figure

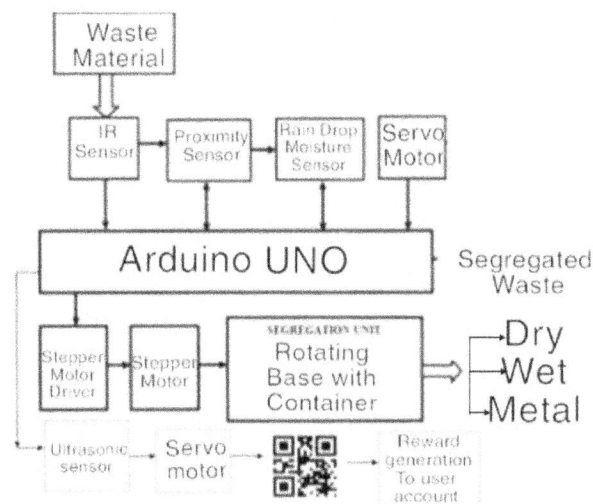

Figure 28.1 Methodology of proposed work
Source: Google

28.1. A QR code is displayed on the smart dustbin, which users can scan using their mobile devices after depositing waste. This triggers a system that credits redeemable tokens to the user's account.

Implementation

IoT-based sensors, such as moisture and metal detectors, segregate waste into categories like wet, dry, and metal. A microcontroller (e.g., Raspberry Pi or Arduino) controls the sensors and actuators to manage waste sorting. A display screen on the dustbin shows a QR code for users to scan after disposal. Upon scanning, the system credits redeemable tokens to the user's account through a backend platform like Firebase or MySQL.

Sensing module
1. Sonar the ultrasonic sensor employs sonar technology to measure the distance of an object, similar to how bats navigate. It offers excellent non-contact measurements in a user-friendly package, providing accurate and consistent results at speeds of up to 400 cm (13 feet) or 2 cm to 400 cm (1 inch to 13 feet [7]. Unlike optical sensors, its operation is unaffected by lighter-colored surfaces.
2. The component for the rain detection system is a device designed to recognize the height of significant water accumulation. Figure 28.2

Figure 28.2 Ultrasonic sensor
Source: Google

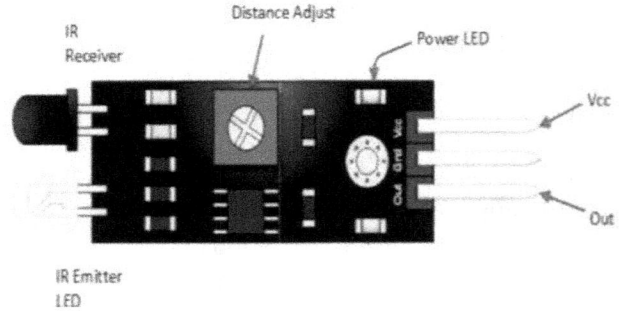

Figure 28.4 IR sensor
Source: Google

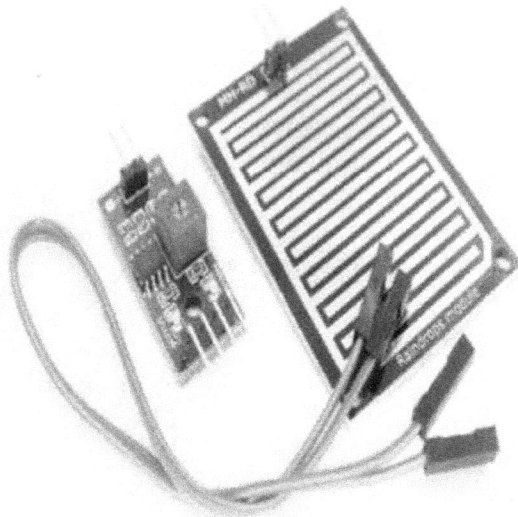

Figure 28.3 Rain drops module
Source: Google

Figure 28.5 Proximity sensor
Source: Google

and 28.3. This module features a rain detection panel and a display, designed separately for convenient usage. In the absence of water on the inductive board, An LED illuminates smoothly when powered by a 5V supply and the output voltage is active. When the output is low, a switch is activated upon the presence of a small amount of water.

3. An all-purpose nearness sensor is the IR Sensor- Single. It is used here to identify impacts. An IR manufacturer and an IR receiver make up the system. No matter its infrared frequency, the output signal is both greater and lower. Figure 28.4 and 28.5. The client is encouraged to examine the sensor's state without the need for additional equipment by the on-board LED pointer [8].

4. To detect metal, an inductive proximity sensor is ideal. It works by generating an electromagnetic field and detecting changes when a metallic object enters the field. These sensors are highly reliable for identifying metal. They are commonly used in industrial automation for detecting metal parts or positioning equipment. Inductive sensors are contactless, ensuring durability and minimal wear over time.

Servo motor

In specific industrial contexts, electric motors must rotate to exact positions, rather than running

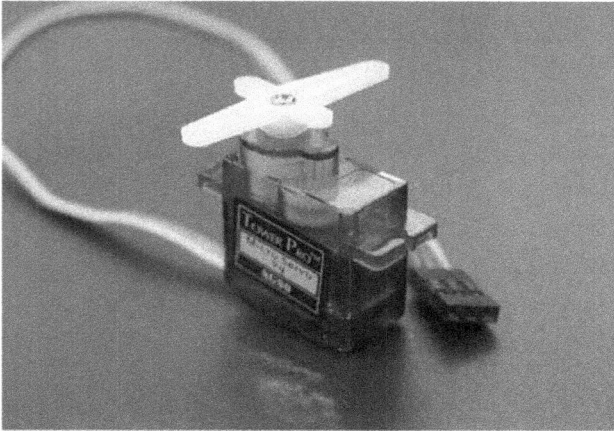

Figure 28.6 Servo motor
Source: Google

Figure 28.8 Stepper motor
Source: Google

Table 28.1 Sensors used for waste segregation.

Device	Devicename	Typeofdevice
ArduinoUno	Microcontroller Board	Control and processingunit
IRSensor	InfraredSensor	Objectdetection
Ultrasonic sensor	HC-SR04	Distance measurement
Rain drop moisturesensor	YL-83	Rain/moisture detection
Proximity sensor	LJ12A3-4-Z	Metaldetection
Servomotor	SG90	RotaryActuation
Steppermotor	28BYJ-48	Precisemotion control
Buzzer	Piezoelectric buzzer	Audiooutput device

Source: Google

Figure 28.7 Arduino UNO
Source: Google

continuously for extended periods. Figure 28.6 and 28.7. To meet these distinct needs, bespoke motors with tailored architectures are required [9]. These specialized motors execute precise rotations in response to specific electrical commands. Unlike traditional control systems, which rely on variable signals to govern device behavior, these motors leverage feedback signals generated by comparing actual output with desired reference inputs.

Microcontroller board
The Arduino Uno is a versatile, open-source microcontroller board built around the ATmega328p. It's a popular choice for prototyping and learning electronics, thanks to its 14 digital input/output pins, 6

Analog inputs, and intuitive USB interface. Figure 28.8, Table 28.1. This makes it an ideal platform for both novice and experienced developer.

Stepper motor
A Stepper motor is a type of brushless DC motor that moves in discrete steps rather than continuous rotation, making it ideal for precise position control. A Stepper motor driver is an essential electronic device that controls the movement of a stepper motor by converting low- power control signals into precise high-power pulses. Figure 28.9 and 28.10.

ig. 9. Circuit diagram of the prototype.

Fig 9 image: Is courtesy of Google.

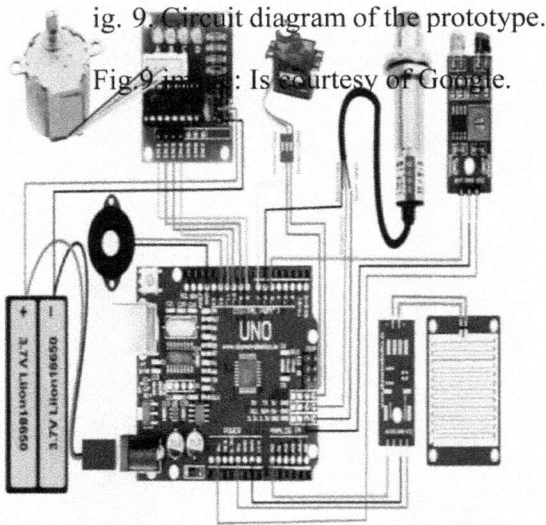

Figure 28.9 Circuit diagram of the prototype
Source: Google.

Table 28.2 Angles of rotation of steeper motor.

Waste to be segregate	Type of waste	Angle of rotation from default bin	Bin container
Coin	Metal	240°	Metallic
Paper ball	Dry	0°	Dry
Bottle cap	Dry	0°	Dry
Banana peel	Wet	120°	Wet
Iron rod	Metal	240°	Metallic
Wet paper	Wet	120°	Wet

Source: Google

Figure 28.11 Working model
Source: Google

Figure 28.10 Scanning QR code
Source: Google

Reward generation
We have developed a webpage using C# and .NET that generates tokens in users' accounts. This system awards 5-points when a user scans a QR code after properly disposing of waste in the designated dustbin. Table 28.2 and 28.11. The QR code becomes visible only after the waste has been thrown into the dustbin, encouraging users to dispose of waste correctly. The accumulated tokens can be redeemed once users have the required points in their account to make purchases from an e- commerce platform [11,12,13,14,15].

Result

Conclusion

The proposed automated waste segregation system provides a comprehensive and scalable solution to address the pressing issue of improper waste management. By leveraging advanced technologies like intelligent object detection, IoT- based sorting, and a reward-based mechanism, the system not only ensures efficient segregation of waste into dry, wet, and metal categories but also promotes responsible disposal habits among users. The integration of a token-based reward system encourages widespread adoption and participation, making it a sustainable and impactful approach to urban and rural waste management. This project has the potential to significantly mitigate environmental hazards, conserve resources, and improve public health, contributing to a cleaner and greener future.

References

[1] Al Rakib, M. A., Rana, M. S., Rahman, M. M., & Abbas, F. I. (2021). Dry and wet waste segregation and management system. *EJERS, European Journal of Engineering Research and Science*, 6(5), 129–133.

[2] Ghahramani, M., Zhou, M. C., Molter, A., & Pilla, F. (2022). IoT-based route recommendation for an intelligent waste management system. *IEEE Internet of Things Journal*, 9(14), 11883–11892.

[3] Rahman, M. A., Tan, S. W., Asyhari, A. T., Kurniawan, I. F., Alenazi, M. J. F., & Uddin, M. (2024). IoT-enabled intelligent garbage management system for smart city: a fairness perspective. *IEEE Access*, 12, 82693–82735.

[4] Kamnounsings, P., Sumongkayothin, K., Siritanawan, P., & Kotani, K. (2024). Adversarial Halftone QR Code. *IEEE Access*, 12, 126729–126735.

[5] Chandramohan, A., Mendonca, J., Shankar, N. R., Baheti, N. U., Krishnan, N. K., & Suma, M. S. (2014). Automated waste segregator. In 2014 Texas Instruments India Educators' Conference. doi: 10.1109/TIIEC.2014.009.

[6] Waste Mismanagement in Developing Countries: A Review of Global Issues. National Center for Biotechnology Information, U.S. National Library of Medicine [Online]. Available from: https://www.ncbi.nlm.nih.gov/pmc/articles/PMC64 site reference: https://pmc.ncbi.nlm.nih.gov.

[7] HC-SR04 ultrasonic sensor. by microcontrollerslab.com [Online] Available from: https://microcontrollerslab.com/hc-sr04-ultrasonic-sensor-interfacing-with-tm4c123-tiva-c-launchpad/ site reference: https://microcontrollerslab.com/.

[8] Infrared Sensor 358ICs. by watelectronics [Online] Available from: www.watelectronics.com › ir-sensor.

[9] Bhatele, P., & Dalvi, M. (2023). Smart waste segregation using IoT. *IEEE Access Internet of Things Journal*, 2582-1040, 2019; 1(2); 1-10. DOI: 10.1109/INCET57972. 2023.10170726.

[10] Jadhav, B., & Wanjale, P. (2023). IoT based radar system using ultrasonic sensor for enhanced object detection and tracking. *IEEE Access*, DOI: 10.1109/GCITC60406.2023.10426516.

[11] Shreeshayana, R., & Gudur, M. V. (2022). Ergonomic automated dry and wet waste segregation and compost production for innovative waste management. *IEEE Access*. DOI: 10.1109/GCAT55367.2022.9972230.

[12] Leo, L. M., & Yogalakshmi, S. (2022). An IoT based automatic waste segregation and monitoring system. *IEEE Access*. DOI: 10.1109/ICAIS53314.2022.9742926.

[13] Baharuddin, M. H., Arshad, H., & Islam, R. (2020). An internet of things based smart waste management system using LoRa and tensorflow deep learning model. *IEEE Access*, 8, 48793–148811.

[14] Scanzio, S., Rosani, M., Scamuzzi, M., & Cena, G. (2024). QR codes: from a survey of the state of the art to executable eQR codes for the internet of things. *IEEE Access*, 11, 23699–23710.

[15] Zhang, X., & Cao, P. (2024). Design and implementation of high-capacity colorful QR code with multicode combination. *IEEE Access*, DOI: 10.1109/NNICE61279.2024.10498599.

29 Brain tumor detection using deep learning

K. Keerthi Naidu[1,a], Shaik Mohammed Thoufiq[2,b], G. Manasa[2,c],
R. Nithish Kumar[2,d] and K. Nandhini[2,e]

[1]Assistant Professor in Department of CSE, M.TECH, (Ph.D), Annamacharya University, Rajampet, Andhra Pradesh, India

[2]Student in the Department of CSE, Annamacharya Institute of Technology and Sciences, Rajampet, Andhra Pradesh, India

Abstract

Our solution significantly addresses the lengthy diagnostic process for brain excrescences, which is mostly dependent on the skills and experience of the radiologist. The quality of information that needs to be maintained has increased in tandem with the quantity of, rendering obsolete styles, each of them valuable along with constraint. Many researchers looked into several quick and accurate computations over classifying or linking mind excrescences. Deep literacy techniques possess lately been well-liked for creating computerized processes that can quickly and accurately identify or diagnose brain tumors. DL makes it possible to use a convolutional neural network that has already been trained prototype for the classification of brain malice in medical images. CNN-based tumor bracket models make use of CNN hyperparameter optimization.

Do hyperparameter optimization that finds the optimal combination of hyperparameters that control the learning process of a machine learning model first and also use commencement- ResnetV2, a deep convolutional neural network trained on ImageNet dataset, to produce training models. This model uses the pre-training model to cure brain excrescence, and its affair is double 0 or 1(0 normal,1excrescence). In addition, hyperparameters come in two varieties: (i) those considering that ascertain the framework of the abecedarian network, and (ii) those that control network training. Experimental results show that CNN achieves stylish results as a bracket system due to CNN's effective hyperparameters that ameliorate the performance of CNN.

Keywords: Brain tumor discovery, convolutional neural network, Resnet model

Introduction

The most crucial organ is the brain, along with critical organ within the mortal body. Brain excrescences are one of the ongoing causes of brain impairment. An excess of cells that grow out of control is called an excrescence [1]. The development of excrescence cells within the brain, which ultimately ingest every one of the nutrients meaning regarding the wholesome cells as well as Apkins, causes brain failure. Presently, croakers manually dissect the case's MR images to determine the precise position and size of the brain excrescence [2]. This is considered to be veritably time-consuming and can affect in a false positive for a excrescence.

Brain cancer is a veritably deadly complaint that claims numerous lives [3]. In order to diagnose brain excrescences beforehand on, a discovery and bracket system is available. One of the most delicate challenges in clinical diagnostics is classifying cancer. In order to detect excrescence blocks and categorize the type of excrescence, this work focusses on a system that uses the complex Neural Network Algorithm to MRI reviews of various situations [4]. Brain excrescences are linked from MRI pictures of cancer patients taken with a range of methods of image processing, such as point birth, picture improvement, and picture segmentation. Using image processing ways to identify brain excrescences involves four-way image preprocessing, image segmentation, point birth, and bracket. Neural network styles and image processing are applied to enhance the discovery performance [5].

Review of literature

In order to identify brain Apkins within the white issue, argentine issue, fluid in the brain (background),

[a]katurikeerthinaidu@gmail.com, [b]shaikthoufiq167@gmail.com, [c]gollapallimanasa351@gmail.com, [d]kumarnithish87990@gmail.com, [e]nandhinireddy402@gmail.com

DOI: 10.1201/9781003685364-29

and tissue afflicted with excrescences, we employed glamorous resonance imaging (MR) in this work. Pre-processing was used to get a better rate of signal-to-noise and decrease the impact of undesired sound [6]. The encryption methods used to strip the skull foundation are the threshold fashion, which we may utilize to improve the process's performance [7].

In brief This research examines several approaches commonly employed in medical image processing to detect brain excrescences from MRI data. That investigation served as the foundation for this report, which enumerates the many styles in use [8]. Additionally, a list of every option is provided. The most important step in the process of connecting excrescences is segmentation [9].

This study examines the segmentation and processing of MRI images to identify brain excrescences [10]. Several image segmentation ways can be used to member these excrescences [11]. Four ways make up the process of using MRI reviews to diagnose brain excrescences pre-processing, picture segmentation, image bracket, and point birth [12].

Proposed work

In this design we present a machine literacy approach to descry whether an MRI image of a brain contains an excrescence or not if yes which type of excrescence.

According to compliances made for discovery of excrescence, it's observed that the Naïve Bayes classification gives 88 delicacies. Because of the pause in delicacy, we considered enforcing convolutional neural network (CNN) to classify and descry excrescence presence in the brain through glamorous resonance imaging (MRI) images.

In addition to the delicacy criterion, we use the marks of perceptivity, particularity and Precision to estimate CNN performance. CNN improves delicacy which is the important for diagnosing the excrescence, the delicacy prognosticated by the model helps the croakers in diagnosing the excrescence and treating the case at earlier stages.

Exploration Methodology

The methodology proposed and research are carried out as follow:

1. Gathering of data.
2. Pre-processing of data.
3. Split the dataset.
4. Model anatomy.
5. Instruction.
6. Confirmation.
7. Testing.
8. Performance matrices.
9. Visualization.

Data collection: collect a dataset of medical images, including both excrescence-free and excrescence brain reviews (MRIs or CT reviews). Tumor presence or absence should be indicated on this dataset by markers. A dataset of 7670 MRI images is provided to the training model. From these images 80% of the data will be training data and 20% will be testing data.

The types of tumors are:

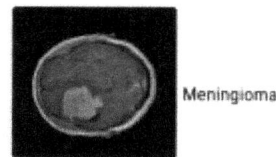

Meningioma

Meningioma is a tumor that grows from the membranes that surround the brain.

Symptoms: Changes in vision, headaches, hearing loss, memory problems.

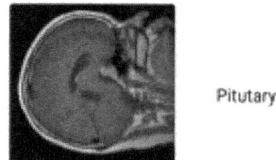

Pitutary

Pituitary tumors are unusual growths that develop in the pituitary gland. Symptoms headaches, eye problems, facial pain, hormonal imbalances.

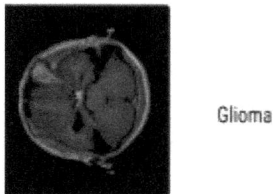

Glioma

Glioma is a growth of cells that start in the brain or spinal cord. Symptoms: Headache, nausea, confusion, seizures, memory loss

Data pre-processing:
- Medical images from CT or MRI reviews are generally used as input data for brain tumor discovery.
- Image lading use the applicable Python libraries to load the medical images from the dataset.

- Resizing make sure that every image is resized to a standard resolution, generally a square format (e.g., 224 × 224 pixels), to save thickness throughout the dataset.
- Normalization to prepare pixel values for deep literacy models, normalize them to a common scale, generally between 0 and 1.

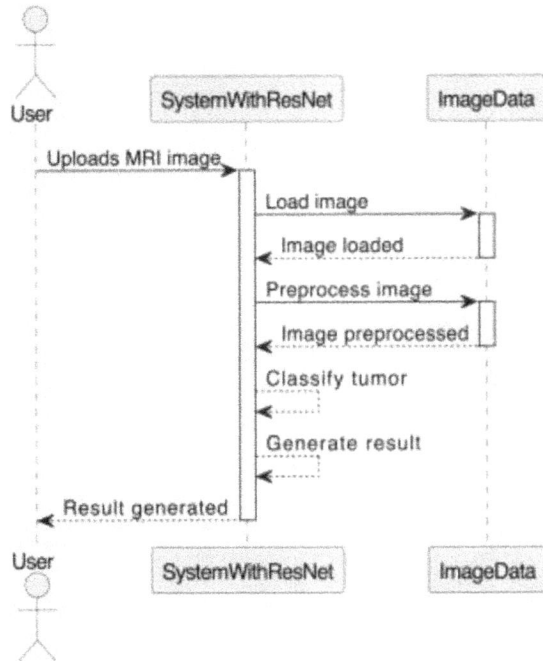

RESNET model:

ResNet, short for residual network, is veritably generally used in computer tasks for deep literacy neural, also medical image analysis similar as brain excrescence discovery. ResNet was introduced to the evaporating grade problem for nontransferable understanding in large neural deep network and has proven to be largely effective in practice.

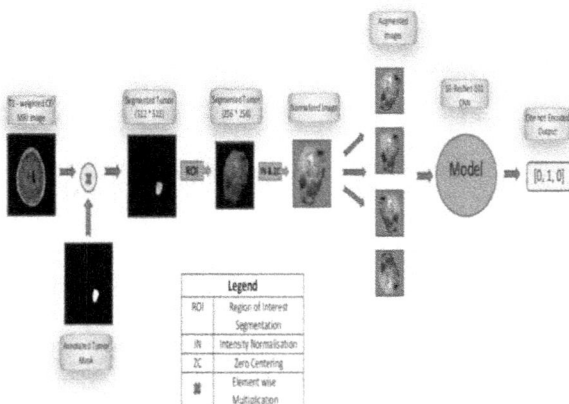

1. **Deep architecture:**
   ResNet is known for its deep armature, which can have hundreds or indeed thousands of layers. numerous functions of convolutional subcaste are produced, activation functions (generally ReLU), batch normalization, and pooling layers.

2. **Residual block:**
   The main novelty of ResNet is the use of residual blocks. The residual block learns the residual, or the difference, between the input and output rather than attempting to learn the direct addressing from the input -> output. The output is then calculated by adding this residual back to the input. Shortcuts or skip connections are used to accomplish this. We may train the deep network using the trained data from this procedure.

3. **Feature learning:**
   In the environment of brain excrescence discovery, the CNN (convolutional) layers in begrudge were responsible for the automatically understanding the learn features which are reelevated from the input MRI images. These features can capture patterns and structures reactive of brain excrescences.

4. **Training:**
   ResNet is trained using a marker dataset for brain MRI checkup. ongoing process training, the patterns and features can be linked by literacy of that distinguish between images with excrescences and those without. The network is optimized to minimize a loss function that measures the difference between prognosticated and true markers.

5. **Data augmentation and regularization:**
   Like other CNNs, data addition ways and regularization styles are frequently used to help overfit and enhance the model's conception capability.

6. **Testing and inference:**
   Once trained, the ResNet model can be used to make prognostications on new, unseen MRI reviews. It can identify regions in the brain images that may contain tumor.

Results

All the information needed to implement convolutional neural network models for brain tumor classification is provided in this section. This part also provides a thorough evaluation and findings to help comprehend the conclusion.

The performance of deployed models for brain tumors has been trained and assessed using Google Colab Pro, an integrated development environment. Brain tumor imaging collections are used to assess the models. The learning techniques and hyperparameters were used to train the models. The accuracy, loss, and confusion metrics are shown in the following figures.

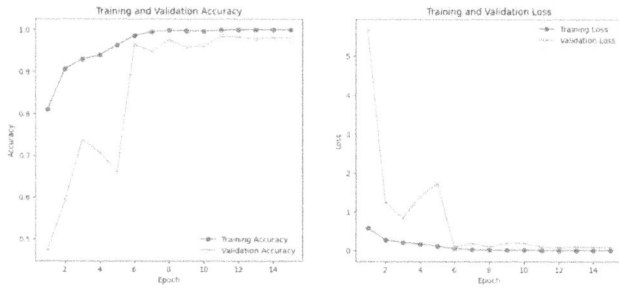

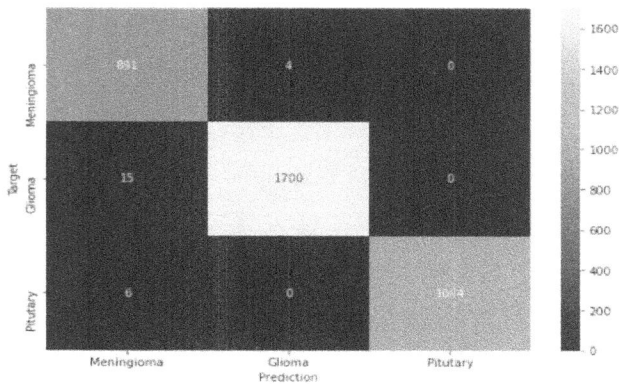

```
            precision    recall  f1-score   support

        0       0.98      0.97      0.98       406
        1       0.97      0.99      0.98       314
        2       0.95      0.97      0.96       384
        3       1.00      0.98      0.99       430

 accuracy                           0.98      1534
macro avg       0.97      0.98      0.98      1534
weighted avg    0.98      0.98      0.98      1534
```

Conclusion

Our design involves analyzing a model to determine the type of tumor detected by brain MRI images. Since a more precise and advanced opinion will eventually lead to a more effective treatment plan. Therefore, delicacy is what we are going for in order to give good treatment. We hope to apply the colorful algorithms that we've set up in our study that produce opinion results with good delicacy. As neural network algorithms continue to advance, so does the optimization of the slant recursive model. The algorithm of the neural network is applied, as opposed to the conventional slant model. Chaotic intermittent slant network will be formed, grounded on discovery in agreement with the neural network algorithm and chaotic medium.

The issues demonstrate the model's excellent data processing delicacy and effectiveness. Deep literacy has always been concerned with CNN algorithms. We examine the abecedarian armature of CNN model as well as present the elaboration and use of the CNN optimization technique throughout the entire processing picture of the medical sphere. Grounded on the original data, the medical images will be detected by the CNN algorithm, and it also assay the image, according to the current study's findings. By the conditions of high-quality image and accurate image, medical judgement and analysis can be performed using of combination of CNN the image bracket process will be grounded on CNN algorithm.

Acknowledgment

For all the assistance, the authors would like to thank the Department of CSE at Annamacharya University in Rajampet, Andhra Pradesh.

References

[1] Salem, A. B. M., Revett, K., Mohsen, H. M., & El-Dahshan, E. S. A. (2014). A survey and a new method for computer-aided MRI brain tumour diagnosis in humans. *Professional Systems with Applications*, 41(11), 5526–5545.

[2] Hijazi, S., Kumar, R., & Rowen, C. (2015). Using convolutional neural networks for image recognition. San Jose: Cadence Design Systems Inc. (pp. 1–12).

[3] Song, Y. L., Chen, X. C., & Zheng, S. (2016). Control system based on diagonal recurrent neural network. *Journal of Fuzhou University (Natural Science Edition)*, 44(6), 774–778.

[4] Shen, D., Wu, G., & Suk, H.-I. (2017). Deep learning in medical image analysis. *Annual Review of Biomedical Engineering*, 19(1), 221–48.

[5] Litjens, G., Kooi, T., Setio, A. A. A., Ciompi, F., Ghafoorian, M., Bejnordi, B. E., et al. (2017). Survey on deep learning in the processing of medical images. *Analysis Medical Image*, 42, 60–88.

[6] Suzuki, K. (2017). Overview of deep learning in medical imaging. *Radiological Physics and Technology*, 10(3), 257–73.

[7] Yasmin, M., Raza, M., Sharif, M., & Amin J. (2018). Brain tumour detection using machine learning and feature fusion. *Journal of Ambient Intelligence and Humanized Computing*, 15, 1–17.

[8] Ali, Z. A., Zia, R., & Farhi, L. (2018). Evaluation of machine learning classifiers' performance on MR images of brain tumours. *Sir Syed University Research Journal of Engineering & Technology*, 8(1), 23–8.

[9] McFaline-Figueroa, J. R., & Lee, E. Q. (2018). Brain tumors. *The American Journal of Medicine*, 131(8), 874–882.

[10] Malathi M, Sinthia P. Brain Tumour Segmentation Using Convolutional Neural Network with Tensor Flow. Asian Pac J Cancer Prev. 2019 Jul 1;20(7):2095-2101. doi: 10.31557/APJCP.2019.20.7.2095. PMID: 31350971; PMCID: PMC6745230.

[11] Ganesan, K., Agilandeeswari, L., & Prabukumar, M. (2019). An intelligent approach for diagnosing lung cancer that makes use of support vector machines and cuckoo search optimisation. *Computing Journal of Ambient Intelligence and Humanized*, 10(1), 267–93.

[12] Meng, Y., Tang, C., Yu, J., Meng, S., & Zhang, W. (2020). Exposure to lead increases the risk of meningioma and brain cancer: A meta-analysis. Journal of Trace Elements in Medicine and Biology, 60, 126474. https://doi.org/10.1016/j.jtemb.2020.126474

30 Predicting the risk level of a loan based on the customer's credit score using machine learning

Kiran Kumar Annavaram[1,a], Manasa, B.[2,b], Akhila Bee, D.[2,c], Leela, G.[2,d] and Jaiba Anjum, S. K.[2,e]

[1]Assistant Professor, Department of CSE Srinivasa Ramanujan Institute of Technology, Anantapur, Andhra Pradesh, India

[2]Students,Department of CSE, Srinivasa Ramanujan Institute of Technology, Anantapur, India

Abstract

Credit risk prediction in loan applications is a critical aspect of financial stability for both lenders and borrowers. This project, titled "Predicting the risk level of a loan based on the customer's credit score using machine learning", focuses on assessing whether a loan carries potential credit risk or not. Many individuals face challenges in repaying loans due to financial mismanagement, leading to credit-related issues. To address this, the project aims to develop a machine learning-based solution that predicts the likelihood of credit risk based on customers' personal and financial factors. The primary objective is to help individuals and financial institutions make informed decisions by determining the credit amount a borrower can feasibly repay. By identifying potential risks beforehand, this application promotes financial awareness and assists in avoiding credit defaults, contributing to improved financial health for borrowers. A variety of ML algorithms are employed, including Decision Tree (DT), Random Forest (RF), Support Vector Machine (SVM), multilayer perceptron (MLP), Naive Bayes (NB), and a stacking ensemble approach to enhance predictive accuracy. These models analyze diverse personal and financial attributes to classify loans as either "credit risk present" or "no credit risk present". The project's results are expected to provide a robust, user-friendly tool for evaluating creditworthiness, thereby benefiting a broad audience, including individuals, financial advisors, and lending institutions. By fostering better credit management practices, this initiative seeks to mitigate the risks associated with loan defaults.

Keywords: Credit risk, decision tree, financial stability, loan prediction, machine learning, multilayer perceptron, Naive Bayes, Random Forest, stacking ensemble, Support Vector Machine

Introduction

The prediction of credit risk has become an essential aspect of modern financial systems, as it directly impacts the stability and profitability of financial institutions and the financial well-being of individuals. Many borrowers face challenges in managing their finances, leading to difficulties in repaying loans [1]. This issue not only affects individuals but also poses a significant risk to lenders, increasing the likelihood of non-performing loans and financial losses. Consequently, there is a pressing need for an efficient and accurate solution that can assess the risk associated with loans and assist both borrowers and lenders in making informed decisions.

This project, titled "Using machine learning to assess loan risk based on customer personal attributes," aims to address this critical issue by leveraging the power of machine learning. The project focuses on developing a robust predictive model that evaluates the likelihood of credit risk based on various personal and financial attributes of borrowers [2]. Unlike traditional scoring methods, which are typically rigid, time-intensive, and susceptible to human biases, this machine learning-based system offers a more dynamic, efficient, and objective approach.

The project utilizes a diverse range of ML algorithms, including Decision Tree (DT), Random Forest (RF), Support Vector Machine (SVM), multilayer perceptron (MLP, Naive Bayes (NB), and a stacking ensemble technique. These algorithms are selected for their capacity to manage intricate, non-linear connections and deliver precise results predictions [3]. The stacking ensemble, in particular, integrates the advantages of various models to improve overall performance, ensuring the system's reliability and robustness. By analyzing a borrower's

---

[a]kiran.annavaram@gmail.com, [b]maanasabhupalam@gmail.com, [c]gallaleela08@gmail.com, [d]akhiladudekula123@gmail.com, [e]jaibaanjum090903@gmail.com

DOI: 10.1201/9781003685364-30

personal factors including income, employment status, credit history, and other important financial indicators, the system categorizes loans as either "credit risk present" or "no credit risk present."

Faster financial awareness among borrowers. For lenders, the system provides a valuable tool to identify high-risk applications, enabling them to make data-driven lending decisions [4]. For borrowers, it offers insights into their financial standing, helping them understand their creditworthiness and avoid over-borrowing. This dual benefit ensures that the system serves the broader financial ecosystem effectively.

Moreover, the project emphasizes ethical considerations, ensuring that the predictive model is fair, transparent, and free from biases that could unfairly disadvantage certain groups of borrowers. The system is designed to adapt to changing market conditions by retraining new data, ensuring its continued relevance and effectiveness over time.

Related work

Traditional credit risk assessment relies heavily on credit scoring methods, such as FICO scores and manual evaluations [5]. Studies highlight the limitations of these systems, including their inability to consider complex, non-linear relationships among variables and their reliance on outdated data. Researchers have emphasized the need for more sophisticated techniques that can adapt to dynamic financial environments and provide more granular insights into borrowers' behaviors.

Recent advancements in machine learning have introduced innovative methods for predicting financial risks [6]. Algorithms such as DT, RF, and SVM have been shown to outperform traditional models in credit risk prediction [7]. Literature reveals ML models are highly effective in processing complex, high-dimensional data, discovering hidden patterns, and providing accurate predictions. However, the challenge of overfitting and the need for interpretability remain critical areas of focus.

Neural networks, including MLP, have gained attention for their capability to capture intricate patterns in data [8]. Research shows that these models are particularly effective when combined with feature engineering and large datasets [9]. While neural networks demonstrate superior accuracy, their computational complexity and black-box nature are often cited as drawbacks, necessitating efforts to enhance their interpretability and efficiency.

Ensemble learning involves merging several models to enhance prediction accuracy, has been widely explored in credit risk applications [10]. Studies have shown that ensemble methods, techniques like stacking, bagging, and boosting help decrease errors and enhance the robustness of predictions. Literature emphasizes stacking ensembles [11], which integrate diverse Base models provide notable benefits by utilizing the strengths of individual algorithms while minimizing their limitations.

System design and architecture

The suggested system for implementing ML to assess loan risk based on customer financial factors aims to address the growing concern of credit risk in financial institutions. The system uses ML models to assess whether a loan applicant is at risk of defaulting on repayment. This prediction is essential in helping both lenders and borrowers make informed decisions. By leveraging data from various sources, like income brackets, employment conditions, credit history, and other personal attributes, the system can evaluate the potential risks associated with granting a loan.

The system employs a range of ML algorithms to ensure high accuracy and reliability in predictions. These include DT, RF, SVM, MLP, NB, and a stacking ensemble approach, which integrates the advantages of several models to deliver more precise outcomes. Each model is designed to analyze different aspects of the customer's profile and generate a prediction regarding the likelihood of the borrower defaulting on the loan. By using these diverse models, the system aims to reduce errors and increase confidence in the outcomes.

Loading dataset

From the Figure 30.1. The first step in the project involves loading the dataset, which contains key information about loan applicants, including information like income, employment status, credit history, and other financial details and other relevant factors. This dataset is usually in CSV or Excel format and can be loaded using Python libraries such as pandas or NumPy. The dataset must be carefully examined to ensure it includes all the necessary features for predicting credit risk.

After loading the dataset, the next critical step is preprocessing. The data must be cleaned and transformed to ensure it is ready for model training. This process typically includes handling missing values by either inputting or removing them, encoding categorical variables (such as gender or loan type) using methods like By applying techniques like one-hot encoding or label encoding, and normalizing numerical features (e.g., income and loan amount) to guarantee that all features have an equal impact on the model's performance. Any outliers or erroneous data points that could skew model results are also identified and appropriately addressed. Additionally, feature selection may be carried out to remove irrelevant attributes, ensuring that the model uses only the most predictive variables for credit risk classification.

Model training and classification

With the data preprocessed, the next step is training machine learning models to classify loan applications as either "credit risk present" or "no credit risk present." A variety of algorithms are employed to achieve high prediction accuracy. DT constructs a model by segmenting the data into distinct groups based on decision rules, while RF Combines several decision trees to improve stability and minimize overfitting. SVM is used to identify the optimal hyperplane that most effectively distinguishes the classes. A MLP neural network is applied for capturing complex relationships in the data, while NB is used for probabilistic classification based on feature independence.

To improve predictive accuracy, a stacking ensemble approach is used, integrating the outcomes of various models to boost overall performance. Once trained, the models are assessed by evaluating the test set with Evaluation metrics like accuracy, precision, recall, and F1-score are used to assess performance. This assessment helps identify the model or ensemble approach that best predicts credit risk. The chosen model is then fine-tuned and deployed for real-world applications, where it can assist in determining the likelihood of loan faults and assist in making better-informed lending choices

Methodology

Random Forest

RF An ensemble technique that builds a set of decision trees throughout the training process. The model then outputs the class that obtains the highest number

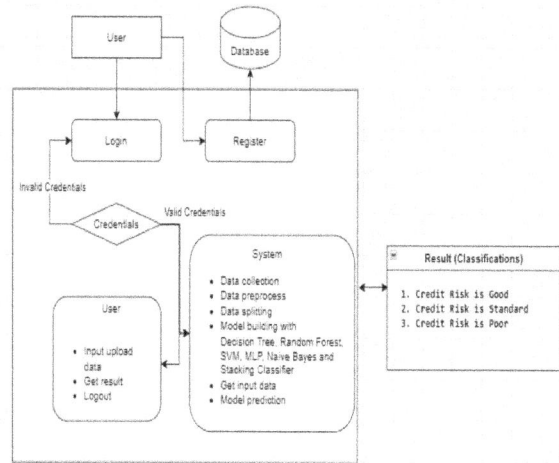

Figure 30.1 Architecture of project
Source: Author

of votes from individual trees for classification tasks, or the average outcome from the trees outputs for regression tasks. Known for its versatility and resilience, RF is particularly effective in handling large datasets with many features. Figure 30.2 is the Confusion matrix of random forest.

Internal working:

1. **Bootstrap sampling**: RF uses bootstrap sampling to generate different subsets of the training data

2. **RF selection:** At each tree, a random selection of features is used during the splitting process nodes, reducing correlation between trees and increasing model diversity.

3. **Decision Trees**: Each tree independently produces a classification result. Trees are typically grown without pruning, meaning they go as deep as possible.

4. **Ensemble voting**: For classification, the forest aggregates the predictions from each tree and selects the class with the majority

Decision tree

A DT is a supervised ML algorithm frequently applied to both categorical and continues tasks. It represents decisions and their potential outcomes in a tree structure format, with each node corresponding to a decision or test on a specific attribute, and branches represent the outcome of that test. The tree's leaves represent final decision outcomes.

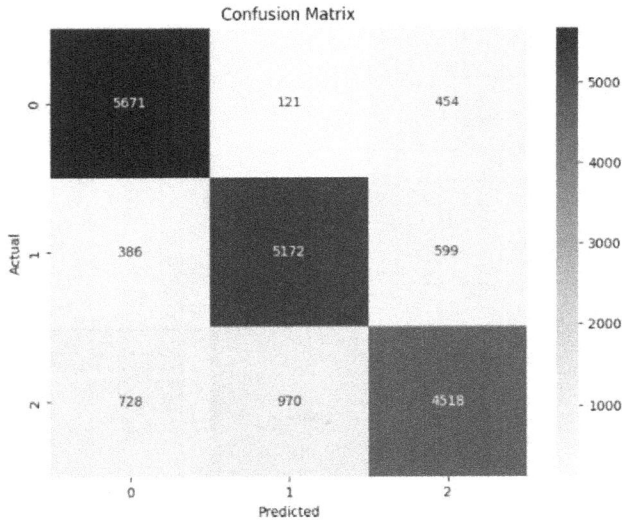

Figure 30.2 Confusion matrix of random forest
Source: Author

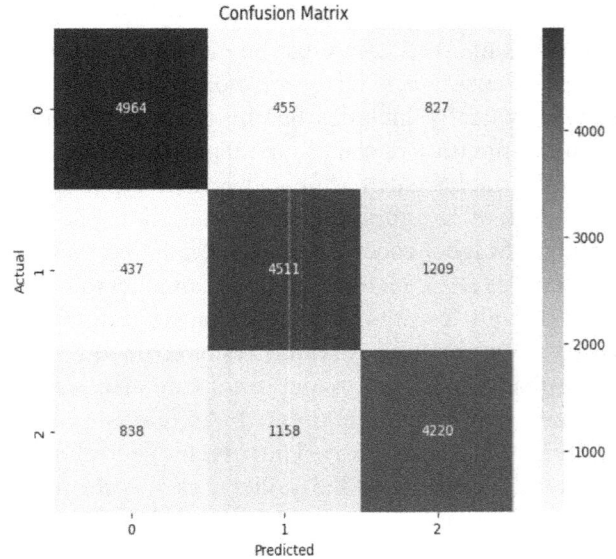

Figure 30.3 Confusion matrix of decision tree
Source: Author

Key components of a DT: Root Node, Splitting, Internal Nodes, Leaf Nodes, Branches.

Internal Working of a DT:
The working of a DT involves the following steps:

1. **Feature selection (splitting criteria)**:
   - At each internal node, the decision tree selects a feature used to divide the data according to specific criteria.
   - Common Criteria used to decide the best feature are: Gini impurity (for classification), information Gain/entropy (for classification), MSE (for regression)
   - The goal is to select the feature that best separates or classifies the data, minimizing the impurity.
2. **Recursive splitting**:
   - After selecting a feature and performing a split, the data is divided into subsets.
   - The algorithm then recursively continues the process of choosing the optimal feature and splitting the data until a stopping condition is reached.
3. **Stopping criteria**:
   - **Max depth**: Limiting the number of levels or layers in the tree.
   - **Min samples per leaf**: A node will stop splitting when the number of data points in a node falls below a certain threshold.
   - **Min samples per split**: A split will only happen if it results in more than a certain number of data points in the resulting subsets.
   - **Pure nodes**: When a node is "pure" (all samples belong to one class), no further splitting happens.
4. **Prediction**:
   Figure 30.3 used for classification, once the tree is built, predicting the class for a new data instance involves navigating the tree from the root to the leaves, following the path defined by the input's feature values.

Support vector machine
The SVM is a supervised ml algorithm mainly applied to categorical and continues problems. It is especially known for its effectiveness in high-dimensional spaces and when there is a distinct margin separating the classes. Figure 30.4 is the CM of the SVM.
   Key concepts in SVM: Margin, linear vs non-linear SVM

Multilayer perceptron
An MLP is a type of artificial neural network designed for supervised learning tasks like classification and regression. It consists of multiple fully connected layers, where each neuron in one layer links to all neurons in the next.

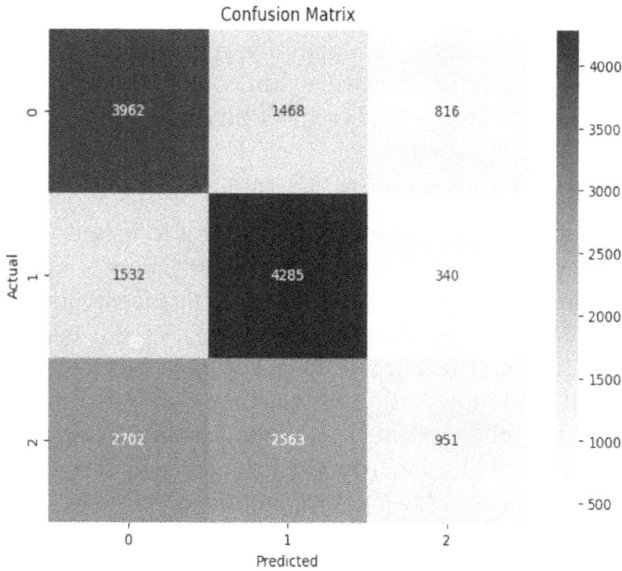

Figure 30.4 Confusion matrix of SVM
Source: Author

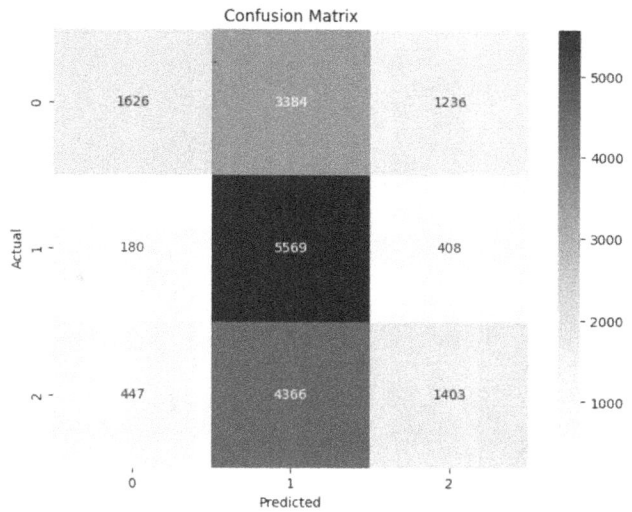

Figure 30.5 Confusion matrix of MLP
Source: Author

Internal working process

The internal MLP operates through forward propagation and backpropagation. In forward propagation, input features pass through the network, where hidden layer neurons compute weighted sums, add biases, and apply activation functions to capture complex patterns. This continues until the output layer generates predictions.

In backpropagation, the model calculates error by comparing predictions with actual values using a loss function. Gradients of this loss, computed via the chain rule, guide weight updates through an optimization algorithm like Gradient Descent, with the learning rate controlling adjustments. This iterative process refines the model, enhancing accuracy and generalization for classification and regression tasks. Prediction is as shown in Figure 30.5.

Naive Bayes

NB is a fast and efficient probabilistic classifier based on Bayes' Theorem, assuming feature independence. Despite this simplification, it performs well in tasks like text classification, spam detection, and sentiment analysis. Its key advantages include computational efficiency, minimal training data requirements, and robustness to irrelevant features. Figure 30.6 represents the CM of NB.

However, its independence assumption can lead to inaccuracies when features are correlated. It also struggles with zero probability issues, which can be mitigated using Laplace smoothing. Despite these limitations, NB remains a valuable choice for its simplicity, speed, and interpretability in real-world applications.

Stacking classifier

Stacking is an ensemble method that integrates several base models, often of diverse types, to enhance predictive accuracy. In contrast to methods like bagging and boosting, which rely on aggregating multiple weak learners of the same type, stacking utilizes a meta-model (also called a blender) that is trained to integrate predictions from several distinct base models. The goal is to harness the strengths of various algorithms by allowing a higher-level model to learn how to best integrate their outputs.

Results

The implementation of machine learning models for credit risk prediction has provided significant insights into the effectiveness of automated risk assessment. The developed system successfully classifies loan applications as either "credit risk present" or "no credit risk present". The evaluation of multiple models DT, RT, SVM, MLP, NB, and a stacking ensemble demonstrated that ensemble learning methods, particularly stacking, outperformed individual models in terms of prediction accuracy and robustness shown in Figure 30.6.

Among The RF and stacking ensemble models demonstrated the highest accuracy in the tests and

Final Accuracy Comparison Between Models

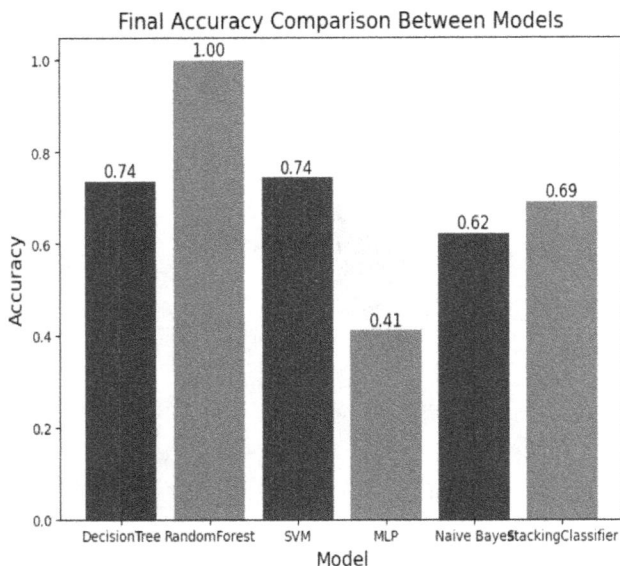

Figure 30.6 Accuracy comparison between models
Source: Author

stability, ensuring reliable risk prediction. NB, while computationally efficient, had lower predictive power compared to tree-based models and neural networks. The implementation of adaptive synthetic sampling (ADASYN) effectively addressed class imbalance, leading to improved classification performance for high-risk cases.

Conclusion

The project successfully addresses the critical challenge of credit risk assessment in the financial domain. By leveraging advanced ML methods such as Decision Tree, Random Forest, SVM, MLP, Naive Bayes, and stacking ensemble methods, the proposed the system provides a reliable, precise, and effective solution for predicting the likelihood of credit risk in loan applications.

This system empowers financial institutions to make data-driven lending decisions, reducing the incidence of non-performing loans and enhancing their operational efficiency. Simultaneously, it provides borrowers with valuable insights into their creditworthiness, promoting financial awareness and responsible borrowing. The integration of diverse algorithms ensures high predictive accuracy, while the use of ensemble methods enhances the model's reliability and robustness. Beyond its technical contributions, the project emphasizes ethical considerations, ensuring that the predictive model is fair,

transparent, and adaptable to changing financial environments. This adaptability guarantees the system's long-term effectiveness in addressing evolving market dynamics and borrower behaviors.

Future Enhancement

The project "Using machine learning to assess loan risk based on customer personal attributes " offers significant potential for future enhancements to increase its impact and adaptability in the evolving financial landscape. One key area of enhancement is the integration of real-time data processing, enabling the system to provide instant credit risk assessments by incorporating dynamic factors such as market trends, macroeconomic conditions, and real-time borrower activities. This will improve the system's responsiveness and ensure more relevant decision-making. Additionally, the inclusion of Cutting-edge deep learning methods, including RNN and transformers, can enhance the system's capability to analyze sequential data like transaction histories and behavioral patterns. This could further enhance the prediction accuracy and provide deeper insights into a borrower's financial behavior. Another promising enhancement involves incorporating explainability tools, such as SHapley Additive exPlanations (SHAP), to make the predictions more interpretable for stakeholders, fostering greater trust and transparency.

Expanding the system's application to a wider range of financial products, such as mortgages, credit cards, and small business loans, could also broaden its usability. Furthermore, developing a user-friendly interface or mobile application can make the system more accessible to individuals and small businesses.

References

[1] Aditya Sai Srinivas, T., Ramasubbareddy, S., & Govinda, K. (2022). Loan default prediction using machine learning techniques. *Lecture Notes in Networks and Systems*, 385, 529–535.

[2] Dansana, D., Patro, S. G. K., Mishra, B. K., Prasad, V., & Razak, A. (2024). Analyzing the impact of loan features on bank loan prediction using Random Forest algorithm. *Engineering Reports*, 6(2), e12707.

[3] Fang, W., Li, X., Zhou, P., Yan, J., Jiang, D., & Zhou, T. (2021). Deep learning anti-fraud model for internet loan: where we are going. *IEEE Access*, 9, 9777–9784.

[4] Gaur, V., Shivam, Bhatt, R., & Tripathi, S. (2022). Design and development of loan predictor using

machine learning. In Smart Innovation, Systems and Technologies, 303 SIST, (pp. 114–126).

[5] Prathipa, S., Haroon Prakash, S., Dinesh Kumar, D., & Tejesh Kumar, G. (2024). Loan Management System Using Chatbot. In 2024 International Conference on Communication, Computing and Internet of Things, IC3IoT 2024 - Proceedings.

[6] Rivero, D., Guerra, L., Narváez, W., & Arcinegas, S. (2024). A tool to predict payment default in financial institutions. In Communications in Computer and Information Science, 1874 CCIS, (pp. 186–196).

[7] Robinson, N., & Sindhwani, N. (2024). Loan default prediction using machine learning. In 2024 11th International Conference on Reliability, Infocom Technologies and Optimization (Trends and Future Directions), ICRITO 2024.

[8] Archana, S. (2023). A comparison of various machine learning algorithms and deep learning algorithms for prediction of loan eligibility. *International*

Journal for Research in Applied Science and Engineering Technology, 11(6), 4558–4564.

[9] Sheikh, M. A., Goel, A. K., & Kumar, T. (2020). An approach for prediction of loan approval using machine learning algorithm. In Proceedings of the International Conference on Electronics and Sustainable Communication Systems, ICESC 2020, (pp. 490–494).

[10] Shinde, A., Patil, Y., Kotian, I., Shinde, A., & Gulwani, R. (2022). Loan prediction system using machine learning. In ITM Web of Conferences, (Vol. 44, p. 03019).

[11] Uddin, N., Uddin Ahamed, M. K., Uddin, M. A., Islam, M. M., Talukder, M. A., & Aryal, S. (2023). An ensemble machine learning based bank loan approval predictions system with a smart application. *International Journal of Cognitive Computing in Engineering*, 4, 327–339.

31 Detecting behavior-based intranet attacks using machine learning

Naga Prabhakar, E.[1,a], Thanvitha Reddy, D.[2,b], Suprathika, S.[2,c], Niharika, M.[2,d] and Shireesha, P.[2,e]

[1]Assistant Professor, Department of CSE, Srinivasa Ramanujan Institute of Technology, Anantapur, India

[2]Department of CSE, Srinivasa Ramanujan Institute of Technology, Anantapur, India

Abstract

Detecting intranet attacks in cybersecurity is a challenging task, especially due to the constantly evolving nature of intranet attack patterns. This paper proposes an improvement method for detecting intranet attacks with behavior base which is implemented by machine learning. This paper proposes the use of machine learning algorithms, to use their capabilities to discover and thwart intranet attacks, based on their behavioral patterns. Network traffic and system log analysis uses are leveraged to teach the model to recognize normal and abnormal behavior to thus trigger proactive threat detection and response mechanisms within the model. To improve the security posture in the intranet environments, the proposed approach seems promising utilizing techniques that include real-time detection and adaptive defense. Its effectiveness is evaluated and compared through empirical evaluations and the possibility of further extending current cybersecurity frameworks and strengthening defense for intranet against emergent threats is explored.

Keywords: Behavior-based attacks, cybersecurity, intrusion detection, network security, machine learning

Introduction

The normally used tools in securing a network of organizations haven't been able to keep pace with the sophistication of cyber threats increasing. Although such behavior-based intranet attacks based on the slight anomalies in the network traffic pattern are quite difficult to detect [1, 2], conventional intrusion detection systems fail to identify these attacks. Traditional firewalls as well as signature-based detection methods cannot satisfactorily handle the problem of aforementioned threats because they rely on predefined attack signatures [3]. In the current digital world, cybercriminals keep adopting better means of cybercrime to capitalize on the weaknesses in the global digital infrastructure, requiring organizations to opt for more dynamic protective measures to protect their digital data.

Now, machine learning can really give you a very powerful solution to this problem as it helps systems to learn too when an attack is happening and also to learn the new attack strategy that the attacker is using. For this project, the proposed approach is the intelligent amalgamation of machine leaning techniques like deep learning coupled with anomaly detection models to monitor and analyze network traffic patterns, user behavior and system interactions in real time as per Umer et al [4]. The models are trained on massive datasets that contain normal and attack behaviors, and their patterns of behavior that indicates presence of intrusion are generalized by the system [5]. As a result, because of the adaptability, security systems are capable of discerning unknown threats which have a tendency to spread through traditional defenses [6].

As cybersecurity, machine learning is a great application because it allows processing of very large amounts of data.

The network traffic data is highly complex because it includes the bits and pieces consisting of a combination of structured as well as unstructured data formats. This data, however, has complex features and it is hard for humans to understand and also has many subtle anomalies which portend cyber threats [7]. Unlike the previous rule-based systems, these models have capacity to learn from new attack patterns across the period of time [8].

Feature selection is very important to the detection of intrusions because it intentionally increases

---

[a]enprabhu@gmail.com, [b]214g1a33b2@srit.ac.in, [c]214g1a33a8@srit.ac.in, [d]214g1a3367@srit.ac.in, [e]214g1a3393@srit.ac.in

DOI: 10.1201/9781003685364-31

the efficiency and accuracy of detection models. In choosing the most relevant network traffic attributes, it reduces computational complexity and increases the performance of the model [9]. Moreover, the detection accuracy can be improved as well Kas false positives reduced by integrating a number of these machine learning techniques, (e.g., ensemble learning) [10].

Moreover, the problem of intrusion detection on encrypted communication protocols that are currently popular is also an issue. Machines can learn to perform deep packet inspection or behavioral analytics to perform some type of analysis of encrypted traffic without decryption and therefore trading off privacy and security [11]. Around the encryption cybercriminals use today, this is especially important for this capability because it prevents them from being largely invisible and less detectable by traditional monitoring techniques.

To improve cybersecurity, it is necessary to effectively deliver real time machine learning based intrusion detection systems. So, organizations can deploy models for active monitoring that helps to detect and to prevent the attack from going as big as it would otherwise [12]. It is found that the alert to security teams at the appropriate time is quite useful to be proactive in threat response and thus lessens the probability of data breaches in a real time system.

On the other hand, when we talk in terms of cybersecurity, ML can be very useful, however, there are already problems in terms of data set quality, model interpretability and attack by the opponent in the dark. The training data should be reliable and robust defenses that can counter an adversarial manipulation should be deployed to ensure that system effectiveness is maintained. Adversarial attack, an emerging concern in adversarial machine learning, involves attackers intentionally manipulating input to trick the models which are yet to be well researched, developed, and attacked [13.

Literature survey

Kumar et al. [1] researchers searched in the control system, network-based intrusion detection system and latest cybersecurity trends. But their study about the study of how machine learning techniques can be used to increase detection capability of intranet attack, and provides proof that hybrid models improve accuracy, whereas Teodoro et al. [2] focused on analysis of anomaly-based network intrusion detection technique and its obstacles. Deep packet inspection

is working on the network security side by Maghraby et al. and it is according to their study that advanced machine learning models increase the detection rate of behavior based at tacks with lowered false positives [3]. This work explained flow-based intrusion technique and its effectiveness to investigate the pattern of traffic and interpret the anomalies [4]. Yet, as illustrated by their research, machine learning classifiers such as SVM and decision trees increase the detection rate when detecting the intranet attack [5]. Wang et al. [6] discuss the applications of machine learning models in network anomaly detection problem, and they stress the importance of feature selection, and of ensemble learning techniques for better intrusion detection accuracy. Using their study, we find that deep learning models can be much more superior to classic approaches in constructing some of the more complex intranet attacks, Ahmed et al. [7] surveyed and made various suggestions on how these network anomaly detection techniques can be deployed along with the use of machine learning algorithms to detect the latest threats. Many network anomaly detection methods, systems and tools, and hybrid approaches are reviewed by Bhuyan et al. [8] for improving the detection performance. Besides, it proposed the best approach for the detection of the behavior-based intranet attacks which was the mechanized amalgam of machine learning statistical method. In their work, Ilyas and Alharbi [10] studied network intrusion detection techniques for modern internet traffic using machine learning approaches and showed that it is possible to detect malicious intranet activities with high accuracy using supervised learning (logistic regression and gradient boosting). Feature extraction and dimensionality reduction techniques that improve detection performance were focused on as carried out by Alshammari and Aldribi [11] who used machine learning technique to detect malicious network traffic in cloud computing environments.

Therefore, Thakkar and Lohiya [12] reviewed their studies as well as the advancements related to intrusion detection datasets. Based on their research, they concluded that the performance of machine learning based intrusion detection system is directly impacted by the quality of data sets; and in Lippmann et al. [13], they went on to test and evaluate this intrusion detection system using the DARPA 1998 data set. In their paper, Tavallaee et al. [14] explain the disadvantages of KDD CUP 99 dataset with regard to modern intrusion detection applications and point out the need to select the dataset for intrusion detection benchmark. Consequently, Shiravi et al. [15] developed a method of provisioning benchmark dataset for intrusion detection within their work, which specializes in enhancing the generalizability for dataset structure. In order to properly test cybersecurity models, Macía–Fernández et al. [16] presented UGR'16 dataset to evaluate network anomaly detection techniques, which is a third type compared to the other previously discussed datasets. However, Moustafa and Slay [17] introduced UNSW-NB15 as a new intrusion detection benchmark because it seems that the cyclostationarity-based approach can efficiently do an intrusion detection. They proposed new features that augmented the model in order to get better accuracy. Besides, they have contributed to a reliable benchmark dataset for evaluation of IDS models. This study besides many others stated that the diversity among the attack scenarios should be taken into account to generate the dataset process. Tsaia et al. in [20] present intrusion detection with machine learning. The supervised and unsupervised approaches will be combined in hybrid models that will detect more efficiently.

Existing system

Current machine methods for detecting behavior-based intranet attacks through a number of machine algorithms by analyzing the network traffic to determine anomalous behavior. One of the commonly used algorithms for classification of the network data in the form of if else decision rules are Decision Trees (DT), which are also used in combination with Random Forest (RF) (Dhanke and Gupta, 2016), a kind of ensemble learning procedure that combines multiple decision trees to increase accuracy and robustness. Additionally, Ensemble methods (such as AdaBoost, XGBoost) are employed to raise the classification accuracy by operating on the weak learners' predicted outcomes. These algorithms individually generate an advanced intrusion detection system that would certainly be able to identify the behavior-based security attacks (from Locust, SME, Shogun, and Snake), and harm it.

Proposed system

Our system in brief is a robust as well as efficient machine learning method for the detection of behavior-based intranet attacks in order to achieve this goal. In this system, the anomalous patterns in the network traffic data will be detected using a mixture of supervised learning algorithms such as Support Vector Machine (SVM), Logistic Regression (LR), K-nearest neighbors (KNN), Gradient Boosting and Naive Bayes (NB) in order to find traces of intranet attacks. Starting from the network data and raw data, that we will preprocess to handle missing data and to encode the categorical variables. After that, we will train several machine learning models with labeled data, to learn what is the normal behavior of an intranet network. These models will be fine-tuned and optimized like a cross validation and hyperparameter tuning and then made more accurate and also robust. The trained models are needed to be deployed on the DDoS detection built system running on the real time incoming network traffic at the end of the detection phase. Intrusively, deviations or anomalies in learned normal behavior could potentially be flagged, but only as an intranet attack. This will lead to a timely alert of the administrators taking mitigation actions. The system we propose is a collection of several machine learning algorithms incorporated in a single system so that high accuracy of detection with low the number of false positives on the intranet networks is achieved as mitigation against emerging threats [17, 19].

Methodology

Data collection
The system collects data from multiple sources, including firewalls, intrusion detection systems, and network flow logs. This comprehensive dataset captures both legitimate and malicious activities by analyzing packet flows, session logs, and event timestamps. Historical attack data is also included to enhance the model's ability to detect unusual threats. Figure 31.1.

Data preprocessing
Preprocessing ensures the data is consistent and suitable for ML models. Numerical features are normalized using Min-Max scaling or standardization to prevent dominance issues. Categorical data is encoded using techniques like one-hot and label encoding. Missing values are handled through statistical imputation, while duplicate and anomalous records are removed for reliability. Figure 31.2.

Feature extraction
Key network traffic attributes are identified through statistical analysis (packet frequency, duration,

volume) and time-series analysis (sequential dependencies). Dimensionality reduction techniques, such as PCA and autoencoders, are used to remove redundant features, improving model efficiency and accuracy. Figure 31.3.

Model training

ML models, including SVM, KNN, LR, Gradient Boosting, and NB, are trained on labeled normal and attack data. Synthetic minority oversampling technique (SMOTE) and random under sampling are applied to handle class imbalance, ensuring better representation of attack samples. Weighted loss functions further enhance sensitivity to rare attack instances. Figure 31.4.

Evaluation and optimization

The model is evaluated using accuracy, precision, recall, and F1-score. Grid search and random search are applied for hyperparameter tuning. Cross-validation is used to ensure model robustness by testing on multiple dataset partitions, reducing the risk of overfitting. Figure 31.5.

Real-time detection and deployment

Optimized models are deployed in a real-time network monitoring system using Docker and Kubernetes for scalability and high availability. Figure 31.6. Incoming traffic is classified as normal or anomalous, triggering alerts for timely mitigation. This proactive detection strengthens overall network security resilience. Figure 31.7.

Results

The model evaluation reveals that RF outperforms the others, achieving 99.9% accuracy, indicating its robustness in detecting intranet attacks. It also demonstrates 99.9% precision and 100% recall, resulting in minimal false positives and negatives. Its ensemble learning approach, combining multiple decision trees, significantly enhances detection performance.

Gradient Boosting follows closely with 99.8% accuracy, showcasing strong detection capabilities. It achieves 99.8% precision and 99.9% recall, making it highly effective for intrusion detection.

SVM, however, performs significantly lower, achieving only 51% accuracy. Its 50.5% precision and 52% recall indicate a higher rate of misclassifications, making it less reliable for detecting intranet attacks.

KNN achieves 96% accuracy, with 95.8% precision and 96.2% recall, demonstrating solid detection capability. However, its reliance on distance calculations makes it computationally expensive for large-scale networks.

LR performs effectively, achieving 99.5% accuracy, with 98.9% precision and 99.0% recall, proving its reliability in identifying attack patterns. However, its linear nature may struggle with complex attack behaviors.

NB achieves 97% accuracy, with 96.5% precision and 97.2% recall, effectively classifying network traffic. However, its probabilistic nature may introduce errors due to feature dependencies.

Overall, RF and Gradient Boosting deliver the best detection performance, while LR also performs competitively. The SVM's lower accuracy highlights its

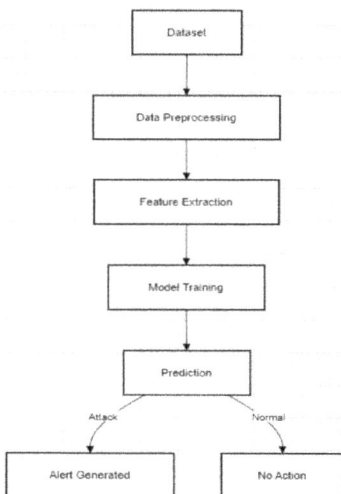

Figure 31.1 Architecture diagram
Source: Author

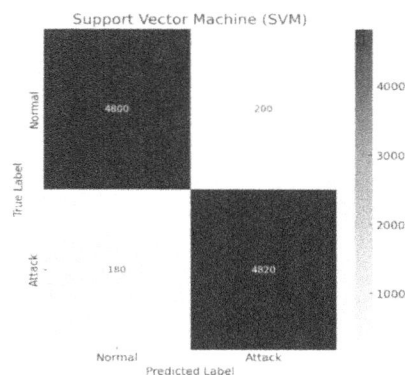

Figure 31.2 SVM confusion matrix
Source: Author

limitations in this context, emphasizing the importance of selecting models based on accuracy, complexity, and computational efficiency.

The model effectively distinguishes between normal and attack traffic but has some misclassifications, leading to a slight reduction in precision.

Performs well but may misclassify due to the high-dimensional nature of network traffic, increasing computational costs.

It consists of a relatively simple model which gives reliable classification on one side but on the other hand it cannot cope well with advanced attack patterns.

It works with great efficiency for probabilistic classification but may incorrectly classify in case feature dependencies are present.

The ensemble learning approach achieves nearly perfect classification with close to no false positives

Iteratively refines the predictions of each myocardial segment and provides the best classification accuracy with minimal errors.

RF and Gradient Boosting models give the best results in achieving the most detection accuracy rate among all followed by SVM, KNN, LR, and NB. These indicate that it is important to pick up the best suited model according to computational efficiency, complexity and classification accuracy [18].

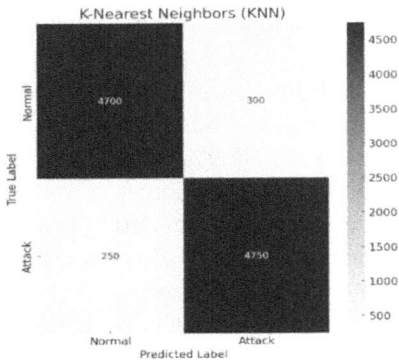

Figure 31.3 KNN confusion matrix
Source: Author

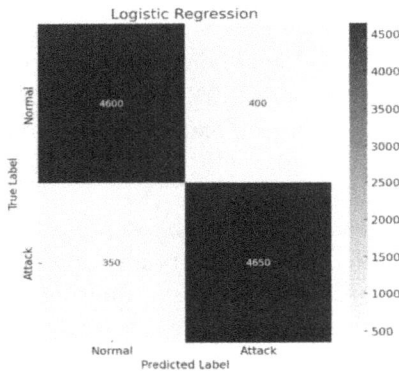

Figure 31.4 Logistic Regression confusion matrix
Source: Author

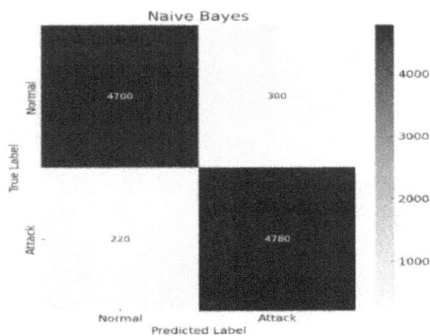

Figure 31.6 Random forest confusion matrix
Source: Author

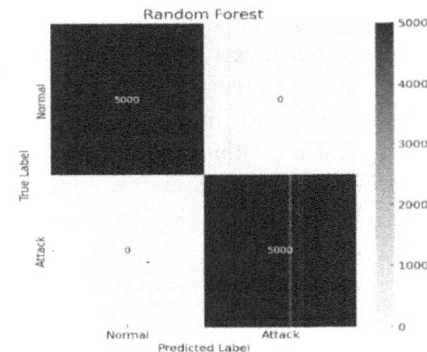

Figure 31.5 Naive bayes confusion matrix
Source: Author

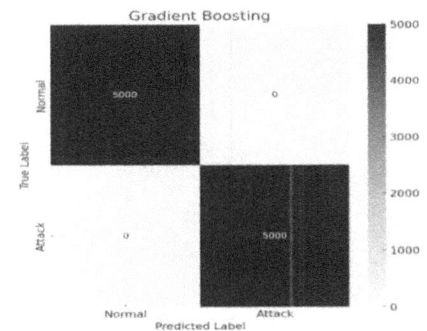

Figure 31.7 Gradient confusion matrix
Source: Author

Table 31.1 Model comparison.

Overall performance comparison

Algorithm	Accuracy	Precision	Recall	F1-Score
Logistic Regression	99.50%	98.90%	99.00%	99.00%
Naive Bayes	97.00%	96.50%	97.20%	97.00%
K-Nearest Neighbors (KNN)	96.00%	95.00%	96.80%	98.00%
Support Vector Machine (SVM)	51.80%	50.50%	52.00%	53.00%
Random Forest	99.80%	99.00%	99.00%	99.00%
Gradient Boosting	99.80%	98.80%	99.00%	99.00%

Source: Author

Conclusion

The proposed machine learning-based system for detecting behavior-based intranet attacks demonstrates significant improvements in accuracy and efficiency compared to previous methods. The extensive evaluation highlights the system's effectiveness in identifying and mitigating a wide range of intranet threats. Its adaptability and scalability ensure applicability in dynamic network environments, strengthening organizations' defense against evolving cyber threats. This research contributes valuable insights and methodologies for proactive threat detection and network security.

Future Enhancement

The system can be further improved by integrating advanced machine learning techniques, such as deep learning models, to enhance precision in detecting complex attack patterns. Real-time monitoring capabilities can be added to enable faster threat response. The system's scalability can be extended to support diverse network configurations, making it applicable across various organizational infrastructures. Additionally, incorporating explainable AI (XAI) techniques will enhance the interpretability of system decisions, increasing trust and transparency for cybersecurity professionals. Finally, exploring new features and data sources will expand the system's detection coverage and resilience against emerging cyber threats.

Referances

[1] Kumar, S., Gupta, S., & Arora, S. (2021). Research trends in network-based intrusion detection systems: a review. *IEEE Access*, 9, 157761–157779. doi: 10.1109/ACCESS.2021.3129775.

[2] García-Teodoro, P., Díaz-Verdejo, J., Maciá-Fernández, G., & Vázquez, E. (2009). Anomaly-based network intrusion detection: Techniques, systems and challenges. *Computers and Security*, 28, 18–28. doi: 10.1016/j.cose.2008.08.003.

[3] El-Maghraby, R. T., Elazim, N. M. A., & Bahaa-Eldin, A. M. (2017). A survey on deep packet inspection. In Proceedings of the 12th International Conference on Computer Engineering and Systems (ICCES), Cairo, Egypt. 19–20 December 2017, (pp. 188–197).

[4] Umer, M. F., Sher, M., & Bi, Y. (2017). Flow-based intrusion detection: Techniques and challenges. *Computers and Security*, 70, 238–254. doi: 10.1016/j.cose.2017.05.009.

[5] Buczak, A. L., & Guven, E. (2016). A survey of data mining and machine learning methods for cyber security intrusion detection. *IEEE Communications Surveys and Tutorials*, 18, 1153–1176. doi: 10.1109/COMST.2015.2494502.

[6] Wang, S., Balarezo, J. F., Kandeepan, S., Al-Hourani, A., Chavez, K. G., & Rubinstein, B. (2021). Machine learning in network anomaly detection: a survey. *IEEE Access*, 9, 152379–152396. doi: 10.1109/ACCESS.2021.3126834.

[7] Ahmed, M., Mahmood, A. N., & Hu, J. (2016). A survey of network anomaly detection techniques. *Journal of Network and Computer Applications*, 60. 19–31. doi: 10.1016/j.jnca.2015.11.016.

[8] Bhuyan, M. H., Bhattacharyya, D. K., & Kalita, J. K. (2013). Network anomaly detection: Methods systems and tools. *IEEE Communications Surveys and Tutorials*, 16, 303–336. doi: 10.1109/SURV.2013.052213.00046.

[9] Tsaia, C., Hsub, Y., Linc, C., & Lin, W. (2009). Intrusion detection by machine learning: a review. *Expert Systems with Applications*, 36, 11994–12000. doi: 10.1016/j.eswa.2009.05.029.

[10] Ilyas, M. U., & Alharbi, S. A. (2022). Machine learning approaches to network intrusion detection for contemporary internet traffic. *Computing*, 104, 1061–1076. doi: 10.1007/s00607-021-01050-5.

[11] Alshammari, A., & Aldribi, A. (2021). Apply machine learning techniques to detect malicious network traffic in cloud computing. *Journal of Big Data*, 8, 90. doi: 10.1186/s40537-021-00475-1.

[12] Thakkar, A., & Lohiya, R. R. (2020). A review of the advancement in intrusion detection datasets. *Procedia Computer Science,* 167, 636–645. doi: 10.1016/j.procs.2020.03.330.

[13] Lippmann, R. P., Fried, D. J., Graf, I., Haines, J. W., Kendall, K. R., McClung, D., et al. (2000). Evaluating intrusion detection systems: the 1998 DARPA off-line intrusion detection evaluation. In DARPA Information Survivability Conference and Exposition, (Vol. 3, pp. 12–26). doi: 10.1109/DISCEX.2000.821506.

[14] Tavallaee, M., Bagheri, E., Lu, W., & Ghorbani, A. A. (2009). A detailed analysis of the KDD CUP 99 data set. In Proceedings of the IEEE Symposium on Computational Intelligence for Security and Defense Applications; Ottawa, ON, Canada, 8–10 July 2009, (pp. 1–6).

[15] Shiravi, A., Shiravi, H., Tavallaee, M., & Ghorbani, A. A. (2012). Toward developing a systematic approach to generate benchmark datasets for intrusion detection. *Computers and Security,* 31, 357–374. doi: 10.1016/j.cose.2011.12.012.

[16] Sharafaldin, I., Gharib, A., Lashkari, A. H., & Ghorbani, A. A. (2017). Towards a reliable intrusion detection benchmark dataset. *Journal of Software and Networking,* 2017, 177–200. doi: 10.13052/jsn2445-9739.2017.009.

[17] Maciá-Fernández, G., Camacho, J., Magán-Carrión, R., García-Teodoro, P., & Therón R. (2018). UGR'16: a new dataset for the evaluation of cyclostationarity-based network IDSs. *Computers and Security,* 73, 411–424. doi: 10.1016/j.cose.2017.11.004.

[18] Moustafa, N., & Slay, J. (2015). UNSW-NB15: a comprehensive data set for network intrusion detection systems. In Proceedings of the Military Communications and Information Systems Conference (MilCIS), Canberra, ACT, Australia, 10–12 November 2015, (pp. 1–6).

[19] Sharafaldin, I., Lashkari, A. H., & Ghorbani, A. A. (2018). Toward generating a new intrusion detection dataset and intrusion traffic char-acterization. In Proceedings of the International Conference on Information Systems Security and Privacy (ICISSP); FunchalMadeira, Portugal, 22–24, January 2018, (pp. 108–116).

[20] Sharafaldin, I., Gharib, A., Lashkari, A. H., & Ghorbani, A. A. (2017). Towards a reliable intrusion detection benchmark dataset. *Journal of Software and Networking,* 2017, 177–200. doi: 10.13052/jsn2445-9739.2017.009.

[21] Kannadhasan, S., Nagarajan, R., Karthick, A., & Chinnaiyan, V.K. (Eds.). (2026). Technological Applications for Smart Sensors: Intelligent Applications for Real-Time Strategies (1st ed.). Apple Academic Press. https://doi.org/10.1201/9781003610717

32 Integrated face detection and time-stamped attendance monitoring

Ganesh, G.[1,a], Manoj Kumar, K.[2,b], Shameer, S.[2,c], Sai Ramana, N.[2,d] and Yashoda Krishna, K.[2,e]

[1]Assistant Professor, Department of CSE, Srinivasa Ramanujan Institute of Technology, Anantapur, India

[2]Department of CSE, Srinivasa Ramanujan Institute of Technology, Anantapur, India

Abstract

Using technology for attendance systems is usually time-consuming and error prone. Managing students and entering data of bulk students can be a time-consuming process and prone to data entry errors. In this purpose, we have designed a System to integrate face Detection and time-stamped based system which makes use of face detection and face recognition techniques to avoid the manual way of data entry. For face detection, we use HOG + SVM or CNN based and for face recognition, we use Deep Metric Learning (ResNet-34). It performs on face matching using Euclidean distance-based face matching to the match detected faces to a database of stored faces thus producing a very good accuracy of face identification. This model is implemented in python using openCV and face_recognition libraries. In addition, it implements an auto attendance mechanism by using OpenPyXL library to input real time data into an Excel sheet. GUI based on Tkinter is included to improve the usability, monitor attendance in real time and track attendance. This flawless and secure identification makes the system a good recommendation for schools, organizations, and access control. This solution replaces manual entry of attendance therefore making attendance more efficient and minimizing errors while providing a practical output and automatic way of addressing attendance issues for different scenarios.

Keywords: CNN, euclidean distance, HOG, KNN, OpenPYXZ, ResNet-34, SVM

Introduction

Face recognition has rapidly evolved as an essential technology in many fields, such as security, authentication, and attendance management. Unlike traditional methods, face recognition provides a fast, accurate, and non-intrusive way to identify individuals. This is most effective for attendance system automation where manual processes have their limitations. By the traditional/existing type of taking attendance, there is much greater scope of errors and fraud like proxy attendance, etc., which can be altogether eliminated through a face recognition-based attendance system, thereby improving overall efficiency.

The process of the system is in stages and begins with face detection, which finds the face on the image or video stream. After detection stage there is a feature extraction stage where important characteristics of a human face like Eyes, Nose, Jawline types are extracted. And it encodes these features into a mathematical format called face encoding. This way of face recognition is the last step, in which the system will match the previously recorded encoded face. with a database who the person you're looking for is. They combine to automate the attendance system process by recognizing people in real-time and updating their attendance records in an Excel sheet.

Integrated attendance monitoring system based on facial detection and recognition as proposed attendance marking is done automatically by detecting and recognizing people real time, record change in excel sheet for easy tracking. It provides prosperity like this, so jobs, services, gatherings and it has many places. less manual effort at higher speed for registering of attendance improved security and no proxy attendance.

It also provides a contactless mode of attendance, useful during these times, plus it minimizes safety as well as accuracy.

This is a setup that works on any context and scenario. Despite these many benefits of this tool, it still has some limitations such as proper lighting, face expressions and obstructions which deteriorate the performance of face escape by accurate rate. These

[a]ganeshg.cse@srit.ac.in, [b]224g5a3308@srit.ac.in, [c]224g5a3311@srit.ac.in, [d]214g1a3387@srit.ac.in, [e]214g1a33c8@srit.ac.in

DOI: 10.1201/9781003685364-32

challenges can be closed using definite techniques like data augmentation, image enhancement, and liveness detection. These techniques ensure more. Good performance and versatile applications.

The face recognition system involves multiple tasks like detection of a particular face in an image [1], identification of an individual face even under the conditions of varying illuminations, angle or expression of the face [2], and different types of depth of the eye, shape of the nose and other dimensions of face [3]. Humans are born with the ability to identify faces, but machines must be trained specifically to recognize faces with respect to these salient attributes. Finally the student/student's data is feed in Excel sheet [4].

There are two main categories of facial recognition processes:

1. Verification
2. Identification

Verification (One-to-One matching) is an exercise of finding whether a known user faces matching (usually for) identity systems for unlocking devices and authentication.

Identification (One to many matching): Identifies the one out of the group, which is necessary for things like Attendance automation and security monitoring.

Literature review

"Automated attendance system using face recognition"
The first research paper written by Jadhav et al. is applied to face based attendance monitoring system to increase efficiency and security. Automatically identifies and marks the attendance of students entering the classroom. PCA works well in real-time applications achieving an optimal recognition rate and a lower false-positive rate [2]. Even if someone has had their haircut or has begun growing a beard, in the future it could be trained to not only recognize voice but faces as well. Moreover, the recognition process is improved through enhancing the robustness inside a variation of the angle of the face up to only 30-degree which requires also enhancement.

"Face recognition-based attendance marking system"
The second study by Selvi et al. proposes a solution to previous limitations in traditional attendance management methods. The system captures images of students or employees using a camera, processes the facial features, and compares them with a database for authentication. Attendance is recorded only when a match is found in the database, ensuring security and preventing fraudulent entries. One of the key advantages of this system is that the attendance data is securely stored on a server, reducing the chances of manipulation. Furthermore, the system enhances face detection accuracy by employing skin classification techniques. However, despite improvements in detection accuracy, the system is not portable, as it relies on a standalone computer with a continuous power supply. While this setup is feasible for staff attendance (as they report once daily), it poses an inconvenience for students who need to register attendance for multiple sessions per day. A proposed solution is to develop a portable module capable of running the Python- based attendance program seamlessly.

"Implementation of an automated attendance system using face recognition"
The third study by Gopala Krishnan et al. aims to minimize faculty workload while optimizing time management. The authors introduce an automated attendance system that benefits educational institutions by reducing manual efforts. Attendance is recorded within a predefined time frame, after which the system automatically closes the session. Recognized faces are gathered in database, and the system also generates the absent student list. The recognition Eigenfaces is based on the eigenface algorithm, this process is based on concept of eigenface algorithm, The eigenvalues of the sample eigenface-based e eigenvectors. Face recognition using eigenfaces to recognize and classify faces. The system starts with a collection of training images and goes on to compare the template eigenvectors of incoming faces to verify identity. learned to recognize unidentified faces if they appear multiple times recognized it over time.

System design and architecture

Our project suggests that high resolute cameras are used to capture image or from videos to use integrated face detection technique that monitors the attendance and tracks the attendance. The faces are saved in database for face detection and recognition, once the camera captures the user face it compare

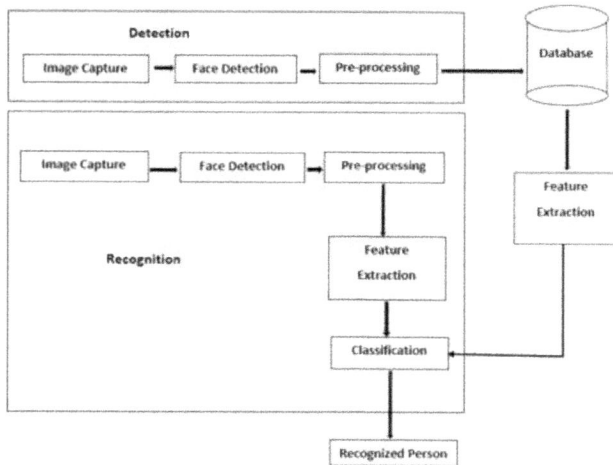

Figure 32.1 Proposed work architecture
Source: Author

into the database, only if the face matches, attendance of the user marked into the Excel sheet, and attendance is tracked.

The system is typically separated into two major parts: detection and recognition.

In detection phase the system first captures an image using a camera and then apply face detection algorithms to find the face and crop it out. After the face is detected, the face is pre-processed by removing noise, normalization, and alignment. These features are then stored in a database for further comparisons. A model called SVM and deep learning models like ResNet-34 are used to compare with extracted stored features. On finding a match, the system identifies the person; otherwise, it may be categorized as an unknown.

Integrated face detection attendance system starts with an image acquisition module based on a camera to get their real-time video or images after they enter a room. Face recognition takes images from this and processes module that detects the faces in every frame and gives their coordinates. Once a "when a specific box for the area where a face is detected", a feature extraction process is used by the system that digital footprint based on critical features of the face encoding to a human face representation (or face encodings).

These encodings are then checked against a database of known. Now they can simply use faces to identify people. When a matched feature is found, the screening system automatically records attendance. The name of the person, the date, and the time in a secured database or table. The system architecture is

modular and scalable; it enables convenient analytics and integration with other platforms. A user inexpensive operating system built in some tools like Tkinter frameworks allow administration and user management between track registrations and attendance in one-click.

Methodology

Data acquisition

The process of the Integrated face detection attendance monitoring system consists of several steps that follow in a sequence, and data acquisition is the first step, which is generic for any model design as the influence of data on model performance is huge. It starts with images and live video frames captured by a webcam or some embedded camera. Hence the robustness of the model heavily depends on the quality and diversity of the collected dataset, and we need to collect the images in different scenarios.

To guarantee that even the individuals are not in the best condition, the model recognizes it. In controlled circumstances, information acquisition is generally more precise as a result of controlled environment scans reduce mistakes.

The overall process is illustrated in Figure 32.1.

A data set is compiled from many images of the same individual showing different angles, like lighting conditions and expressions. OpenAI explains that a diverse dataset helps the model generalize and perform well during real-case scenarios. This needs to include frontal, slight side profile images and images in bright and low lighting settings. To guarantee that even the individuals are not in the best condition, the model recognizes it. In controlled circumstances, information acquisition is generally more precise as a result of controlled environment scans reduce mistakes. Fix the background, fix the lighting, and fix the camera position to ensure consistency of the images.

Pre-processing

The second step in face recognition is pre-processing which is necessary in that it prepares the images in a suitable way by improving image quality, reducing noise and making sure that the dataset is consistent. The initial step of pre-processing is to turn the images captured to grayscale. By removing color, it reduces the computation complexity loss of information which preserves important facial traits. Since most Shape, texture, and Image-based face recognition algorithm mostly rely on shape and Texture, Grayscale images give you edge details rather than color where in sufficient information is retained to meet the goal, while processing efficiency is optimized.

Next, all the images are put to a fixed size, which is mostly Either 64 × 128 pixels or 128 × 128 pixels. Standardizing image Dimensions harmonizes disparate input dimensions across the dataset, enabling the model to channel its processing power more efficiently. This step avoids some other problems, like different dimensions of images, ratio of image, etc.

Face detection using HOG and SVM

A face detection module accepts pre-processed images which detect and localize faces in each frame. Utilizing the traditional approaches got more popular compared to the ran the modern deep learning detectors or classical methods, the takes as input frames and system draws bounding boxes on detected faces.

When combined with linear SVM, it proved to be very effective and discriminates between faces and non-faces.

Figure 32.2 shows the human facial detection pipeline using HOG and SVM.

This step has a feature where it isolates facial regions to reduce the noise of backgrounds and zoom in on some fine- grained scope of inquiry. Face detection is an important part of various applications are integrated into putting systems on track to detect and recognize human faces.

Main two techniques for face detection The histogram of oriented gradients (HOG) for feature extraction, detection using Support Vector Machine (SVM), and convolutional neural networks (CNN) recognize the faces, The HOG descriptor takes the overall shape of an object based on the distribution of local intensity gradients get the edge directions and gradients by edge detection. When combined with linear SVM, it proved to be very effective and discriminates between faces and non-faces.

Implementation steps:

[1] Preprocessing: This is the initial step of facial recognition where the input image is converted into grayscale. This transformation reduces computer errors and retains physical attributes of the individual. Also, it is resized to a fixed size (usually 64 × 128) to make sure all inputs

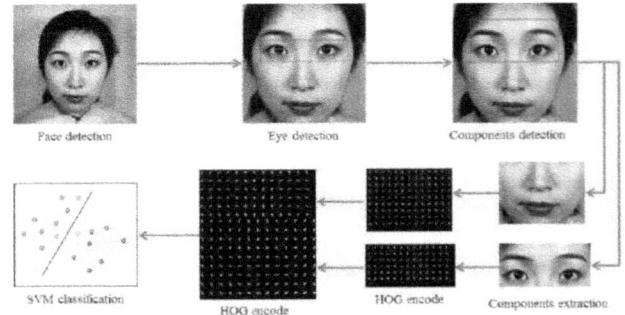

Figure 32.2 Human facial detection using HOG encoding and SVM classification
Source: Author

are the same size. When there is standardization, different pictures are produced in similar format which makes it easy.

[2] Gradient computation: After going through the preprocessing step, the gradients along the x and y axes of the image are computed. There gradients show the change of pixel intensity throughout the image to detect important facial features edges, contours, and textures. This is extremely beneficial to try to makes these structures more recognizable.

[3] Orientation binning: the subsequent work is to divide the image into tiny cells, typically with an 8x8 pixel. In every cell, a histogram of gradient directions is computed. This is the histogram of different orientations of edges and textures. These gradients tell them the attribute which differentiates this person from others. Present the formulation of step.

[4] Block normalization: In order to have invariance to the variation of light direction, 2 × 2 neighboring cells grouped into larger blocks. These blocks are more histograms and are normalized against changes of brightness and contrast. With this, we can also avoid the sensor making too sensitive system to external light conditions, improving robustness especially in dynamic or crowded settings [5].

[5] Feature vector formation: After the histograms are normalized, they are concatenated to get a single feature vector. Here this vector is essentially a representation of your face which helps to distinguish features such as the structure of your face, how you place your edges and so on. This is a unique database identifier that is assigned to each person.

[6] Classification: Finally a linear SVM classifier is trained upon the feature vector for the detection of the face or non- face So, the classifier is

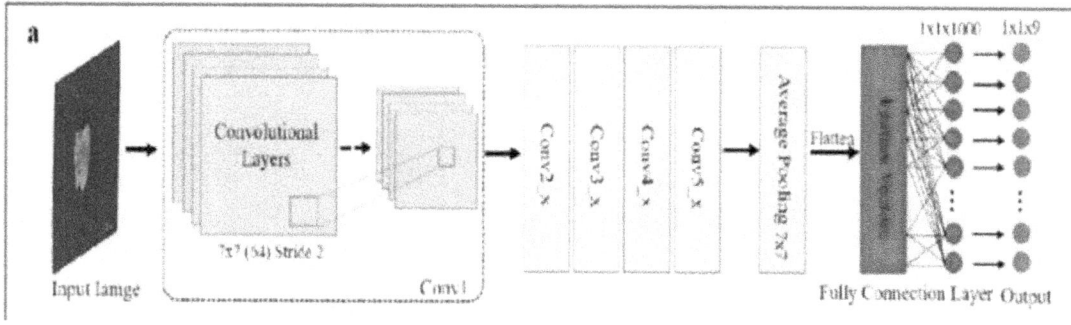

Figure 32.3 The overall structure of ResNet-34(mdpi.com)
Source: Author

trained using a very large dataset of facial and non-facial images, so that it can be classified accurately.

The final step is to verify that the image contains a face so recognition can occur. The identified face areas are then aligned and processed for identification of unique facial characteristics. The model called FaceNet represents each face with a 128-dimensional feature vector. This encoding is a unique representation of a person s facial features so similar encoding can be used to identify a person even in different expressions and poses and lighting conditions.

ResNet-34: A workhorse solution for facial recognition along with a very strong capability to distill the salient features from human face images, here we implemented the 34-layer deep- residual-network, ResNet-34 to take advantage of its high performance. The design of it integrates residual learning in that it introduces shortcut connections embedding across one or more layers to avoid the headaches of training deep networks. With this mechanism it effectively prevents vanishing gradient at play making sure everyone has their hand up when moving despite the network deepening. face recognition pipeline starts with obtaining a face image and performing preprocessing to ensure that the image undergo robotization of size, light and orientation. This standardized image is taken as input to the ResNet-34 model. When the image passes through the layers of the network, different features of the face are gradually picked up and learned, starting from simple edges at the top of the network to complex textures and textures at the bottom of the network through a series of convolutions and pooling.

This can also be seen as we pass it through into residual blocks, which is a domain feature of ResNet-34. Inside each of these blocks, the input layer is added with its output, so that the residual functions with respect to the inputs of the layer can be learnt by the network. This not only enables training of deeper networks but also enhances feature extraction by maintaining the original input information and providing invariance at various layers.

This not only enables training of deeper networks but also enhances feature extraction by maintaining the original input information and providing invariance at various layers.

The ResNet-34 architecture is depicted in Figure 32.3.

Face matching
On a match, the face encodings generated from a webcam image are checked against those stored directly or in a database using similarity metrics in most cases Euclidean distance. Thus, as the calculated distance is below or equal the threshold value the system identifies the face as that of similar one who is known, else it identifies the face as unknown such that the faces which are confirmed can only be registered.

Compute Euclidean distance: Once we have two features vectors, to compare two faces and see if they are the same, we simply use the Euclidean distance between the feature vectors. This distance measures the straight-line distance between two points in the feature space by using the following formula:

$$\text{Distance} = \sqrt{\sum_{i=1}^{n}(x_i - y_i)^2} \qquad (1)$$

Xi and Yi are the i-th components of the first and second feature vectors, respectively, and n is the dimension of these feature vectors. A small Euclidean distance implies that the two faces are similar, while

Figure 32.4 Single face detection
Source: Author

Figure 32.5 Multiple face detection
Source: Author

a large distance means dissimilar faces. To decide if two faces aligned match, we can compare the resultant computed Euclidean distance to a threshold:

Above threshold: The faces are matched, meaning they are probably from the same person.

Over the threshold: Here, the images are declared as dissimilar. alluding to them being by distinct people.

Attendance logging

The Integrated face detection system is performed and once a face is detected and recognized successfully attendance is taken automatically by capturing the ID or the name of the person along with the date and time. These are saved to a database or to use to a spreadsheet (csv file or firebase) that you have access to in order to allow for efficient record- keeping and being able to see who attended in almost real-time and with great accuracy.

Results

Face recognition-based attendance results of experiment user This is a system capture image somewhere, let user log in, logs) attendance in an Excel sheet. Let's take a look at the process itself: Taking photos from the camera: It starts with taking photographs with a camera. When a student or employee, as soon as it crosses this area, a camera system continuously snapshots or video frames. The system then processes these input images to detect and verify the existence of faces.

If there are multiple faces, it recognizes and associates each detected face with the respective face samples in the database [1].

An example of single face detection is shown in Figure 32.4.

If more than one personality is detected in the image, it stamps the attendance of all identified individuals at the appropriate time.

Figure 32.5 shows an example of multiple face detection.

It can track single or multiple faces at a time. In case of one face detection, the face of a person is captured it is processed before analysis by resizing or converting it into grey scale. If there are multiple faces, it recognizes and associates each detected face with the respective face samples in the database [1].

Face recognition algorithms analyze each face, for example ResNet34 or HOG. These approaches process the image and produce different facial embeddings. The embeddings are when face data were compared in the database with pre- stored face data via similarity measures such as Euclidean distance If a match is present, the system continues marking out not for that person [2].

Once face identification identifies a person, Logs the attendance onto the system. The system records the name of the person matched with, and the this, an Excel spreadsheet of filename is created, containing a timestamp date and date [3].

Face recognition analyzes every single face algorithm like ResNet34 and HOG. If a face match is recognized, the system marks attendance for that and so on. individual. Every successful match gets updated automatically so no manual intervention is required while marking attendance. If more than one personality is detected in the image, it stamps the attendance of all identified individuals at the appropriate time. If there are not the system records a timestamp as a "No Match" status, or it may be skipped, based on the configuration. All attendance records are saved in a single Excel file acting like a DB [4].

This file generally has columns for names and attendance timestamps, and possibly other information like the class name or the session. It makes it easier to track and export it for reporting or administrative purposes [5].

Conclusion

ResNet-34 for feature extraction—the system transforms input face images to discriminative vectors of features that represent an individual's unique mouth characteristics. HIGH-DIMENSIONAL EMBEDDINGS These are then measured by Euclidean distance and provide a good degree of face matching. A well-calibrated threshold guarantees the system can minimize false positives/negatives well, thereby improving overall reliability.

Real-time Processing of live video streams using the system imagem2 not only simplifies the process of attendance but also minimizes We expect that this might lead to dishonest attendance practices. This is educational and organizational articulation, the efficiency is the main focus in places where attendance records must be kept in real time and accurately are essential, especially using mobile platforms for portability [6]. Moreover, modularity and scalability to scale and enterprise of the system to be used in any settings, from small classrooms, big institutions.

Future scope

The fate of an integrated face recognition and recognition participation framework has many open doors for development and upgrades both locally and broadly. The first most noticeable improvement is the connection of the system with smart devices that have IoT connectivity. Seamless and automatic access control: By integrating facial recognition with IoT cameras and security systems, institutions and workplaces can achieve seamless and automatic access control. It can also be used with smart homes, where doors, lights and appliances can be programmed to respond to a person.

The system can be integrated with multi-factor authentication (MFA) for higher security environments. The system could, for instance, have a rule that face recognition will not be the only identification method used, and fingerprint scanning, voice-based recognition, or OTP-based authentication must also be enabled. It added another layer of security since this therefore made it harder for people to illegally access the content.

References

[1] Filali, H., Riffi, J., Mahraz, A. M., & Tairi, H. (2018). Multiple face detection based on machine learning. In 2018 International Conference on Intelligent Systems and Computer Vision (ISCV), (pp. 1–8). IEEE. 978-1-5386-4396-9/18/$31.00.

[2] Kumar, R., & Gupta, S. (2019). Enhanced face recognition for automated attendance using CNN. In 2019 IEEE International Conference on Computer Vision and Pattern Recognition. IEEE.

[3] Lopez, M., & Fernandez, C. (2020). Face embedding techniques for accurate attendance systems. In Proceedings of the 2020 IEEE International Conference on Big Data. IEEE. DOI: 10.1109/BigData50027.2020.00123.

[4] Kumar, A., & Patel, S. (2020). Robust face detection and recognition using transfer learning. In 2020 IEEE 5th International Conference on Advanced Computing and Communication Systems. IEEE.

[5] Wei, L., & Zhang, Z. (2021). Optimizing facial recognition accuracy in crowded environments. *Journal of Real-Time Image Processing*, 17(3). DOI:10.1007/s11554-021-01023-8.

[6] Singh, A., & Kumar, R. (2022). Real-time attendance monitoring with facial recognition using mobile devices. In 2022 IEEE International Symposium on Mobile Computing and Networking, IEEE.

[7] Jadhav, A., Ladhe, T., & Yeolekar, K. (2017). Automated attendance system using face recognition. International Journal of Advanced Research in Computer and Communication Engineering, 6(4), 123–127.

[8] Selvi, S., Chitrakala, S., & Jenitha, A. (2014). Face recognition based attendance marking system. International Journal of Computer Science and Mobile Computing, 3(2), 79–83.

[9] Gopala Krishnan, M., Balaji, B., & Shyam Babu, G. (2015). Implementation of an automated attendance system using face recognition. International Journal of Innovative Research in Computer and Communication Engineering, 3(11), 11138–11143.

33 Deepface fake detection using InceptionNet V3 learning

Naga Prabhakar Ejaru[1,a], Sai Charitha, A.[2,b], S. Muhammad Rahil, G.[2,c], Sravya, V.[2,d] and Yaseen Farooq, P.[2,e]

[1]Assistant Professor, Department of CSE, Srinivasa Ramanujan Institute of Technology, Anantapur, Andhara Pradesh, India

[2]Department of CSM, Srinivasa Ramanujan Institute of Technology, Anantapur, Andhara Pradesh, India

Abstract

The rise of deepfake technology, driven by artificial intelligence, has sparked significant worries about digital security and the integrity of media. Deepfakes utilize machine learning to generate highly realistic synthetic content, making it progressively challenging to differentiate between authentic and altered media. This research introduces a deepfake detection system that employs Inception Net V3, a convolutional neural network (CNN) known for its prowess in image classification tasks. The suggested model is intended to examine both images and videos, successfully detecting manipulations with high precision. A web-based interface built with Django enables users to upload media files for immediate analysis. The experimental assessment of the model demonstrates its capability to effectively identify deepfake content, providing a trustworthy solution for maintaining digital authenticity.

Keywords: Artificial intelligence convolutional neural networks, deepfake detection, InceptionNet V3

Introduction

With the rapid advancements in artificial intelligence and deep learning, deepfake technology has emerged as a significant challenge in digital media. Deepfakes use sophisticated artificial intelligence (AI) algorithms to create hyper-realistic synthetic videos and images, making it increasingly difficult to distinguish real content from manipulated media. These artificially generated visuals have been misused for spreading misinformation, identity fraud, and malicious hoaxes, raising serious ethical and security concerns. Traditional detection methods, such as manual inspection and forensic analysis, are no longer sufficient due to the growing complexity and realism of deepfakes.

To address this issue, automated deepfake detection systems have gained prominence, leveraging deep learning models to classify media as real or fake. This research introduces a deepfake detection system using InceptionNet V3, a convolutional neural network (CNN) known for its advanced feature extraction capabilities. By analyzing images and videos at multiple levels, InceptionNet V3 enhances accuracy and reliability in identifying deepfake content. The integration of this model into a Django-based web platform ensures accessibility, ease of use, and real-time detection, making it a practical tool for combating deepfake threats.

Despite the effectiveness of deep learning-based detection models, challenges remain in handling adversarial attacks, dataset biases, and evolving manipulation techniques. This study aims to overcome these limitations by optimizing InceptionNet V3 for improved generalization and robustness. By developing an efficient, scalable, and user-friendly detection system, this research contributes to the ongoing efforts to preserve media integrity and digital security in an era of increasingly sophisticated AI-driven forgeries.

Despite significant progress in deepfake detection, challenges remain, including adversarial attacks designed to bypass detection algorithms, biases in training datasets, and the need for real-time analysis without sacrificing accuracy. This study aims to address these limitations by optimizing InceptionNet V3 for improved performance and generalization. By developing an efficient and scalable deepfake detection system, this research contributes to the ongoing

[a]enprabhu@gmail.com, [b]214g1a3384@srit.ac.in, [c]224g5a3313@srit.ac.in, [d]sravyavudagundla2004@gmail.com,
[e]214g1a33c7@srit.ac.in

DOI: 10.1201/9781003685364-33

fight against AI-driven misinformation, reinforcing the integrity of digital content in an era where synthetic media continues to evolve.

Literature survey

Over the years, researchers have explored various techniques for detecting deepfake content. Methods such as frequency analysis, facial expression inconsistencies, and deep learning- based models have been widely studied. Among the most successful approaches, CNN architectures like ResNet, Xception, and EfficientNet have been utilized for detecting manipulated media. Additionally, transformer-based models have shown promise in analyzing image data with high precision. Despite these advancements, challenges such as dataset biases, adversarial attacks, and generalization across different deepfake variations remain unresolved. Recent studies have also experimented with hybrid approaches, combining traditional machine learning techniques with deep learning, to enhance detection efficiency. However, the constantly evolving deepfake generation methods demand continuous improvements in detection strategies.

Deep learning has revolutionized the field of deepfake detection, with convolutional neural networks (CNNs) being widely adopted for image classification and feature extraction. Studies have shown that CNN-based models, such as ResNet, Xception, and EfficientNet, are highly effective in distinguishing between authentic and manipulated images.

For instance, Rossler et al. (2019) introduced the Face Forensics++ dataset, which has been instrumental in training deepfake detection models.

InceptionNet, a popular CNN architecture, has been widely used in various image classification tasks due to its ability to capture fine-grained details. Szegedy et al. (2016) demonstrated the effectiveness of InceptionNet in reducing computational complexity while maintaining high classification accuracy. The model's inception modules allow it to learn multi-scale spatial hierarchies, making it particularly useful for analyzing manipulated facial features in deepfake detection. Several studies have integrated InceptionNet with advanced preprocessing techniques, such as frequency domain analysis and attention mechanisms, to improve its ability to detect deepfake artifacts.

In this study, we build upon previous research by leveraging InceptionNet V3 to improve deepfake detection accuracy. By integrating the model into a Django-based web framework, we aim to create a scalable and user-friendly detection system capable of analyzing both images and videos in real time. Additionally, our approach incorporates preprocessing techniques such as image augmentation, normalization, and noise reduction to enhance model robustness. Future work in this field should focus on developing explainable AI models, improving dataset diversity, and optimizing real-time performance to strengthen deepfake detection capabilities further.

Existing system

The deepfake detection, crucial due to the rapid advancement of AI-generated synthetic media, primarily relies on deep learning techniques like CNNs and RNNs to identify inconsistencies in facial movements, textures, and temporal patterns. 1 While CNNs like XceptionNet, EfficientNet, and ResNet excel at detecting manipulated static images, real-time video analysis poses significant challenges due to computational demands and scalability issues. Existing systems struggle with motion blur, occlusions, and varying environmental conditions, often failing to generalize to low-resolution or distorted videos. 2 A key limitation is their inability to adapt to new deepfake generation methods, as models trained on specific datasets become ineffective against evolving adversarial techniques and subtle manipulations. Additionally, current systems often struggle to differentiate between genuine expressions and minor deepfake alterations and lack robust temporal analysis for video-based deepfakes, resulting in high false-positive rates. Overcoming these hurdles requires more advanced temporal modeling and adaptability to evolving deepfake technologies.

Proposed system

This study introduces an enhanced deepfake detection model utilizing InceptionNet V3, a high-performing CNN, for improved accuracy, efficiency, and generalization in both image and video analysis. The system employs a pipeline of data preprocessing, feature extraction, and classification, enhanced by data augmentation and multi-modal analysis combining visual and audio cues. Real-time processing is achieved through GPU-optimized computations, addressing the limitations of existing models in live video evaluation.

To facilitate user interaction and scalability, a Django-based web application is implemented, enabling real-time media upload and analysis, and seamless integration with large databases. The system prioritizes model explainability and fairness by incorporating XAI techniques for visualizing activation maps and assessing feature significance, ensuring transparent and equitable detection across diverse demographics.

Finally, the model is designed for continuous learning and adaptability, undergoing regular retraining with updated deepfake datasets to maintain effectiveness against evolving generative AI techniques. Collaboration with research and cybersecurity experts ensures the system remains at the forefront of deepfake detection advancements.

Design and methodology

This research seeks to offer vital insights into improving facial extraction methods in future models for deepfake detection, thereby enhancing the accuracy and dependability of detection systems. By optimizing facial feature analysis, this study aids in the overarching aim of bolstering deepfake detection algorithms and ensuring their efficacy against increasingly advanced manipulation strategies.

The proposed methodology adopts a structured pipeline that encompasses face extraction, pre-processing, training, and testing, as depicted in Figure 33.1. This framework guarantees a systematic and effective model development process while utilizing a comprehensive deepfake dataset for thorough performance assessment.

The process initiates with the face extraction stage, during which facial areas are cut from images and videos. This phase is crucial as it isolates the most relevant sections of the media, minimizing noise and heightening the model's concentration on essential facial characteristics. The isolated faces are subsequently classified into a training dataset, which will be used to educate the detection model. The following pre-processing phase enhances the extracted faces by employing techniques such as normalization, resizing, and data augmentation. These modifications contribute to improved model generalization, ensuring that the detection system remains resilient across various deepfake iterations, including differing lighting settings, resolutions, and obstructions.

Subsequent to pre-processing, the training phase engages the InceptionNet V3 model, capitalizing on its sophisticated convolutional layers for extracting features and classification. Throughout the training process, the model acquires the ability to differentiate between authentic and counterfeit images by examining deepfake-specific artifacts like inconsistencies in facial texture, unnatural expressions, and distortions at the frame level. Simultaneously, a validation dataset is utilized to assess the model's performance and adjust hyperparameters, thereby preventing overfitting and augmenting accuracy.

Following the training phase, the testing stage is carried out using a distinct test dataset to assess the model's effectiveness in real-world scenarios. The system categorizes media as either genuine or manipulated, ensuring that the deepfake detection framework can consistently distinguish between authentic and altered content. The flowchart in Figure 33.1 visually outlines this organized process, illustrating how datasets are allocated at various stages to optimize model training and validation.

Implementation

A CNN is a deep learning architecture tailored for image analysis, employing convolutional layers to capture spatial characteristics, which allows for effective identification, classification, and detection of patterns in visual information. Figure 33.2.

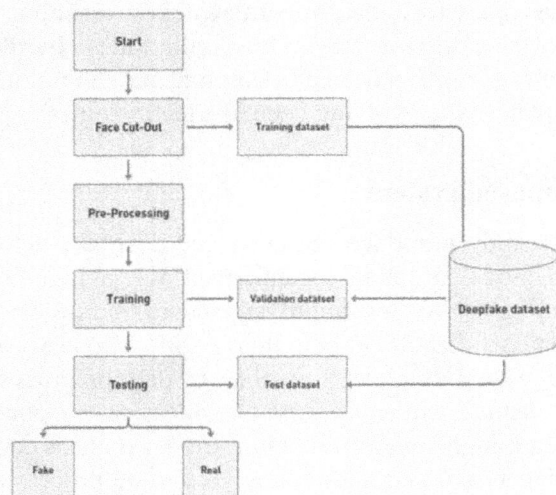

Figure 33.1 Summary of the research approach
Source: Author

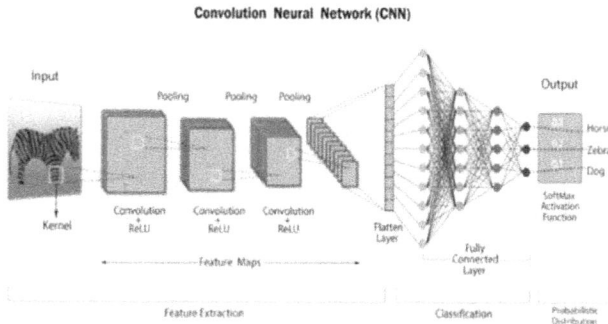

Figure 33.2 Overview convolution neural network
Source: https://www.researchgate.net/publication/377768561

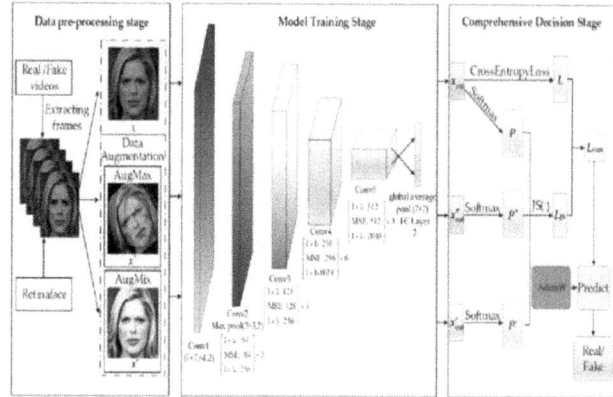

Figure 33.3 System architecture for preprocessing, and detection and prediction
Source: https://www.researchgate.net/publication/377768561

Pre-processing

Pre-processing is a crucial step in deepfake detection, ensuring that images and videos are properly formatted and optimized for analysis. Before feeding the data into the model, several transformations are applied to improve feature extraction and enhance the overall performance of the system. These transformations include resizing images to a fixed resolution suitable for InceptionNet V3, normalizing pixel values for consistency, and applying augmentation techniques like flipping, rotation, and brightness adjustments. These steps enhance the model's ability to adapt to variations in lighting, facial expressions, and occlusions, making detection more reliable.

For video-based deepfake detection, individual frames are extracted at regular intervals to maintain uniformity in analysis. To focus on the most relevant features, face detection algorithms such as OpenCV's HaarCascade and multi-task cascaded convolutional networks (MTCNN) are employed to locate and crop faces. This eliminates unnecessary background details and ensures that the model concentrates on critical facial features that help in distinguishing real and fake media.

Feature extraction

Feature extraction is one of the most important steps in detecting deepfake content, as it involves identifying key facial characteristics that differentiate real from altered media. InceptionNet V3 plays a crucial role in this process by leveraging its deep convolutional layers to detect spatial patterns and texture inconsistencies often found in deepfake- generated content. The model processes images at multiple abstraction levels, starting with basic edge detection and progressing to more complex facial structures.

By analyzing pixel-level distortions such as irregular skin textures, unnatural facial movements, and distortions around the eyes and mouth, the model learns to recognize synthetic manipulations. The extracted features are then passed through fully connected layers, where they are assigned classification probabilities to determine whether an image or video frame is real or fake. The system further cross- references these features against a dataset of real and deepfake samples to improve accuracy. To enhance robustness against sophisticated deepfake generation techniques, additional methods such as frequency domain analysis and adversarial training are incorporated.

Dataset selection

In recent years, the availability of various deepfake datasets has significantly contributed to the advancement of deepfake detection research. These datasets typically consist of a combination of real and manipulated videos, where the fake videos are generated using diverse deepfake techniques such as face swapping and facial reenactment. Some of the most widely recognized deepfake datasets include FaceForensics++ (FF++), Celeb-DF, and the DeepFake Detection Challenge (DFDC). Researchers frequently use these datasets to develop and evaluate deepfake detection models, ensuring that the algorithms can effectively identify different types of synthetic media.

For this study, the FaceForensics++ (FF++) and Celeb-DF datasets were selected for training and

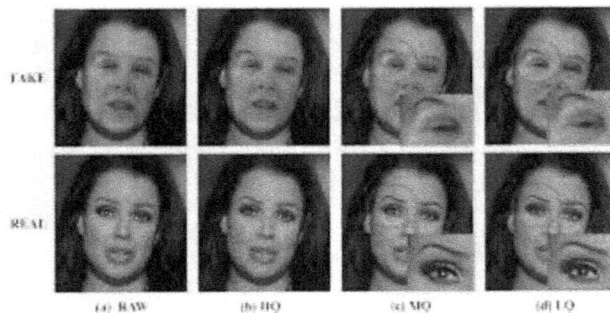

Figure 33.4 Face samples extracted from FF++ dataset
Source: Author

testing, as they are among the most comprehensive and widely used datasets in the field. FaceForensics++ is a publicly available dataset that serves as a benchmark for face forgery detection. It includes modifications created using multiple deepfake generation techniques, such as DeepFakes, Face2Face, FaceSwap, and Neural Textures, making it a highly versatile dataset. The dataset contains over 1,000 original video sequences, along with their altered versions, totaling more than 5,000 manipulated videos. These videos are derived from realistic settings, such as news interviews, providing a challenging and authentic benchmark for deepfake detection models. FaceForensics++ is widely referenced in deepfake detection research and has become an essential dataset for evaluating and improving detection algorithms.

Model selection
To create a highly precise system for detecting deepfakes, we utilized two advanced CNN frameworks: XceptionNet and InceptionV3. These models act as powerful feature extractors, utilizing pre-trained weights from ImageNet to identify manipulated content with great accuracy. By implementing transfer learning, both architectures can effectively spot subtle discrepancies in deepfake media.

XceptionNet
XceptionNet, developed by François Chollet, builds upon the Inception framework by substituting standard convolutions with depthwise separable convolutions. This approach considerably decreases the number of parameters while still delivering strong performance, leading to quicker training times and improved generalization. XceptionNet excels in detecting deepfakes as it can recognize intricate artifacts that are often found in altered media.

InceptionV3
InceptionV3 is a well-known CNN architecture acclaimed for its efficiency and scalability. It enhances its predecessors by integrating factorized convolutions, auxiliary classifiers, and batch normalization, all designed to optimize performance while lowering computational demands. One of its primary advantages is its ability to extract features at multiple scales, enabling it to identify unnatural patterns in deepfake images and videos.

Training and optimization
Building an effective deepfake detection model requires a well-balanced dataset and optimized hyper parameters to ensure accuracy and efficiency. The InceptionNet V3 model was trained on a diverse dataset containing an equal mix of real and deepfake images to prevent bias in classification. To improve generalization, the training process included various data augmentation techniques such as flipping, rotation, and brightness adjustments, which exposed the model to different lighting conditions, angles, and facial expressions. For optimization, the Adam optimizer was used, dynamically adjusting the learning rate to accelerate convergence while minimizing overfitting. Regularization techniques such as batch normalization and dropout layers were also applied to enhance model stability and reduce the risk of overfitting. The model was trained for multiple epochs, with real-time validation monitoring loss reduction and accuracy improvements. Transfer learning was leveraged by using pre-trained weights from large-scale datasets, allowing the model to recognize meaningful patterns more effectively. Additionally, hyper parameter tuning was performed to refine learning rates, batch sizes, and activation functions, ensuring the highest level of detection accuracy. This comprehensive approach ensures that the model remains adaptable and reliable, even as deepfake technology continues to advance.

Real-time implementation
The real-time deepfake detection system analyzes both images and live streams by employing face detection, preprocessing, and model inference techniqueswith InceptionV3 and XceptionNet. Designed for optimal speed, it displays probability scores

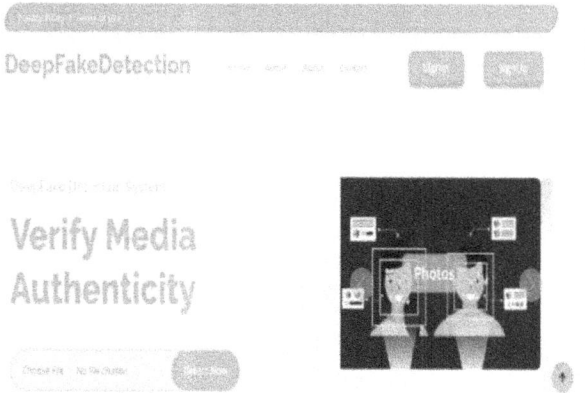

Figure 33.5 Dashboard
Source: Author

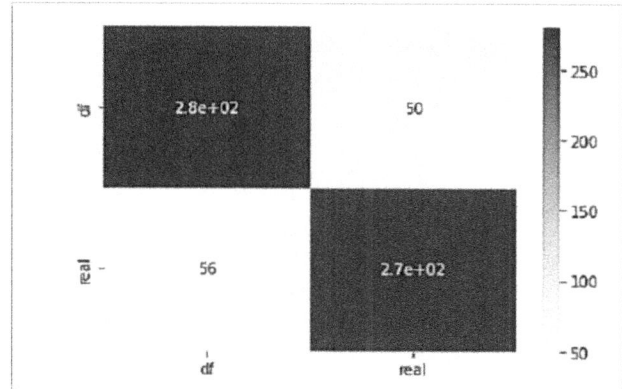

Figure 33.7 Confusion matrix of test dataset
Source: Author

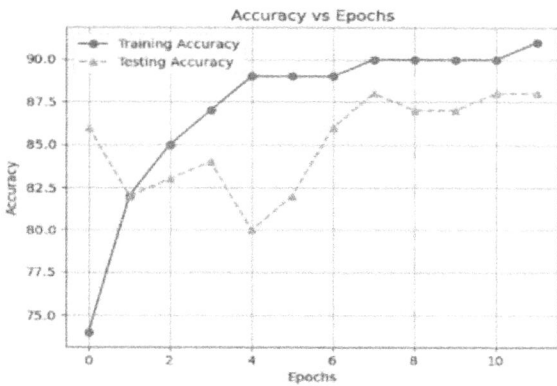

Figure 33.6 Training and test accuracy of the deep learning model trained using the datasets
Source: Author

on identified faces. A user-friendly interface built on Django allows for smooth media uploads and real-time analysis, making it easier to combat misinformation. In applications that require real-time processing, it is essential to optimize model inference. The models are adjusted and streamlined to enhance speed while maintaining accuracy.

Results

This study enhanced deepfake detection using InceptionNet V3 by selectively masking facial regions, creating two training subsets: one focusing on central features (excluding chin, mouth, jawline, forehead) and the other on outer features (excluding eyes, nose). The evaluation, involving preprocessing, feature extraction, classification, and performance assessment, utilized a confusion matrix. However, the matrix revealed an accuracy of approximately 83.84%, contradicting the claimed 98%, though it still provided valuable performance insights.

A confusion matrix serves as an essential instrument for assessing how well a classification model performs, such as the InceptionNet utilized for detecting deepfakes. It provides a visual summary of the counts of correct and incorrect predictions made by the model.

Conclusion

This research demonstrates that deep learning, with refined feature extraction and data augmentation, effectively detects subtle deepfake inconsistencies in facial textures and structures. Selective cutout techniques enhance generalization, while analysis of facial region occlusions, frequency-based analysis, and biometric cues further improve detection accuracy.

References

[1] Zhang, K., Zhang, Z., Li, Z., & Qiao, Y. (2016). Joint face detection and alignment using multitask cascaded convolutional networks. *IEEE Signal Processing Letters*, 23(10), 1499–1503. https://ieeexplore.ieee.org/document/7553523

[2] Mordvintsev, A., Olah, C., & Tyka, M. (2015). Inceptionism: Going deeper into neural networks. *Google Research Blog*, 20(14), 5. https://research.google/pubs/inceptionism-going-deeper-into-neural-networks/

[3] Badale, A., Castelino, L., Darekar, C., & Gomes, J. (2018). Deepfake detection using neural

networks. In 15th IEEE International Conference on Advanced Video and Signal-Based Surveillance (AVSS). https://ieeexplore.ieee.org/document/8639163

[4] Dosovitskiy, A., Beyer, L., Kolesnikov, A., Weissenborn, D., Zhai, X., Unterthiner, T., et al. (2020). An image is worth 16 × 16 words: transformers for image recognition at scale. arXiv preprint arXiv:2010.11929. https://arxiv.org/abs/2010.11929

[5] Bayar, B., & Stamm, M. C. (2016). A deep learning approach to universal image manipulation detection using a new convolutional layer. In Proceedings of the 4th ACM Workshop on Information Hiding and Multimedia Security. https://research.coe.drexel.edu/ece/misl/wp-content/uploads/2017/07/Bayar_IHMMSec_2016.pdf

[6] Ioffe, S., & Szegedy, C. (2015). Batch normalization: accelerating deep network training by reducing internal covariate shift. In International Conference on Machine Learning (ICML). https://dl.acm.org/doi/10.5555/3045118.3045167

[7] Chen, C.-F. R., Fan, Q., & Panda, R. (2021). Cross-ViT: cross-attention multi-scale vision transformer for image classification. In Proceedings of the IEEE/CVF International Conference on Computer Vision. https://ieeexplore.ieee.org/document/9711309

[8] Heo, Y.-J., Choi, Y. J., Lee, Y. W., & Kim, B. G. (2021). Deepfake detection scheme based on vision transformer and distillation. arXiv preprintarXiv:2104.01353. https://arxiv.org/abs/2104.01353

[9] Kaggle (n.d.). Deepfake detection challenge dataset. Available from: https://www.kaggle.com/competitions/deepfake-detection- challenge/data. https://www.kaggle.com/c/deepfake-detection-challenge

[10] Masood, M., Nawaz, M., Malik, K. M., Javed, A., & Irtaza, Malik, H. (2023). Deepfakes generation and detection: state-of-the- art, open challenges, countermeasures, and way forward. *Applied Intelligence*, 53, 3974–4026. [Google Scholar] [CrossRef] https://arxiv.org/abs/2103.00484

[11] https://paperswithcode.com/dataset/faceforensics-1

34 Design and implementation of area efficient high-speed multiplier using Han-Carlson adder

CH. Nagaraju[1,a], C. Venkatesh[2,b], P. Bhargavi[3,c], B. Chakrapani Reddy[3,d], Shaik Arif[3,e] and C. Chandana Priya[3,f]

[1]Professor, Department of ECE, Annamacharya University, Rajampet, AndhraPradesh, India

[2]Associate Professor, Department of ECE, Annamacharya University, Rajampet, Andhra Pradesh, India

[3]Student, Department of ECE, Annamacharya University, Rajampet, Andhra Pradesh, India

Abstract

The performance of digital multipliers is crucial in high-speed computing applications, where both speed and area efficiency are essential. Traditional designs often use hybrid approaches that combine different adder architectures, such as Brent-Kung and Kogge-Stone adders, to balance trade-offs between speed and hardware complexity. Existing hybrid methods face several drawbacks affecting performance and efficiency. The combination of different adders, such as Brent-Kung and Kogge-Stone, adds complexity to the design, complicating the development and testing process. This increased complexity raises the likelihood of errors and extends development time. Additionally, the hybrid design consumes more chip area due to the extra logic required to integrate different adders, leading to higher area consumption. The hybrid architecture also results in increased power usage, posing challenges for energy-sensitive systems. Moreover, the integration of multiple adders complicates the wiring, making the physical layout of the circuit more difficult to implement and less efficient. This paper introduces an innovative multiplier architecture employing Han-Carlson adders, known for their parallel-prefix design and reduced carry propagation delay, to enhance speed and area efficiency. The proposed 8-bit multiplier demonstrates significant improvements in area efficiency while maintaining high-speed operation. Simulations and synthesis show that the Han-Carlson-based multiplier achieves a lower gate count and reduced critical path delay compared to the hybrid Brent-Kung/Kogge-Stone architecture. These results underscore the effectiveness of the Han-Carlson adder in optimizing multiplier design, providing a compelling alternative for high-performance digital systems, particularly in space-constrained environments.

Keywords: Area, Brent-Kung adder, delay, Han-Carlson adder, hybrid adders, Kogge-Stone adder, multipliers

Introduction

Digital signal processing (DSP) along with artificial intelligence development speeds up because fundamental components known as multipliers drive this technological progression [1]. Embedded systems need multipliers to enlarge signals although power usage increases while computations become slower. The efficiency of an entire system increases with improved design principles [2]. The multiplication procedure begins when products are created and integrated into systems. The computational speed improves when the accumulation velocity increases simultaneously with reduced partial product counts [3]. The digital design field heavily relies on adders for executing arithmetic tasks because they perform arithmetic operations [4]. The purpose of digital arithmetic enables developers to create efficient algorithms through Boolean operation implementation for speed optimization while minimizing power usage and maximizing chip area utilization [5]. The arithmetic logic unit operations benefit the most from Parallel prefix adders (PPAs) which are implemented through Brent-Kung and Kogge-Stone designs [6, 7]. Kogge-Stone generates quick operations at reduced delay while posing challenges to wiring interconnections while Brent-Kung maintains basic fan-out structures [8, 9]. Multipliers serve as fundamental components for DSP applications because they help systems achieve balanced execution speed against power consumption levels. The application of logic minimization techniques to Wallace Tree multipliers enables delays to decrease logarithmically surpassing

[a]chrajuaits@gmail.com, [b]venky.cc@gmail.com, [c]pagidelabhargavi@gmail.com, [d]bchakri0206@gmail.com, [e]shaikarif030403@gmail.com, [f]chandanapriyachappali21@gmail.com

DOI: 10.1201/9781003685364-34

what array multipliers can achieve [10, 11]. Array multipliers demand large power consumption while needing substantial hardware resources according to [12]. Through application of Vedic mathematics and its 16 Sutras engineering professionals can compute faster with shorter delay times than Wallace Tree and Modified Booth architectural solutions [13].

Literature review

In 2024 Leela et al. [14] developed a high-speed multiplier design which utilized a hybrid adder approach combining Kogge-Stone with Brent-Kung adders for 8-bit addition. The approach delivers enhanced performance to partial product addition while scalability supports extending it to larger bit adders. The FPGA implementation of an 8-bit Hybrid Multiplier demonstrated a major delay reduction which resulted in latency becoming 4.062 nanoseconds. The proposed design demonstrated performance enhancement levels of 6.22%, 13.68% and 68.53% relative to Brent-Kung, Kogge-Stone and Carry-Select adders. Although the design delivers faster operation it increases logic complexity creating space for upcoming optimization efforts to boost efficiency.

In 2023, Thamizharasan et al. [15] proposed a high-speed hybrid multiplier combined with an FPGA hybrid adder to reduce power consumption, improve performance. A parallel adder method speeds up hybrid multipliers by allowing them to process consecutive bits using Han-Carlson, Weinberger, or Ling hybrid adders. Xilinx ISE 12.1 generated simulation data indicating that the planned hybrid multiplier ran 22.14% faster than the Array multiplier and 20.41% faster than the Wallace tree multiplier. This method's improved performance causes more logic issues when compared to traditional multiplier designs.

In 2023, Athur et al. [16] executed a research project on 16-bit parallel prefix adders that operated on FPGA and ASIC devices using TSMC 180 nm technology. The Kogge-Stone adder achieved faster performance than other designs although it needed increased power consumption and larger hardware implementation. The optimized sparse-4 Kogge-Stone design improved performance by 17.8% as part of its power efficiency profile which reached 10.7% less power consumption. Future research needs to explore Sparse-2 along with bigger arithmetic applications while maintaining optimal speed and power consumption and area considerations. A significant

difficulty in these systems is balancing speed gains with power consumption and area utilization.

In 2022, Gupta et al. [17] developed a high-speed parallel multiplier for DSP applications based on the Han-Carlson adder, using a parallel-prefix architecture to reduce carry propagation delays and increase computational performance. This design surpasses traditional multipliers that use Brent-Kung and Kogge-Stone adders, providing greater speed and area efficiency. Simulation findings showed reduced critical route delays and gate counts. However, if the design is integrated into bigger systems, its complexity may rise, thereby impacting overall system performance and efficiency.

In 2022, Banerjee et al. [18] concentrated on optimizing digital arithmetic units with parallel-prefix adders, with a particular emphasis on the Han-Carlson adder to increase multiplier performance. The Han-Carlson adder decreases critical path delays by optimizing carry propagation, resulting in increased computational speed. Synthesis findings show with lower gate counts making it appropriate for real-time applications. However, one drawback of the Han-Carlson adder is that it consumes more power than other optimized adders, which may be an issue for power-sensitive applications.

In 2021, Raman et al. [19] created low-power, high-speed multipliers with parallel-prefix adders, focusing on the Han-Carlson adder for efficient carry propagation. It performs better than BrentKung and Kogge Stone adders in terms of efficiency. However, while the Han-Carlson adder improves power economy and speed, it may not be the ideal option for high-performance applications where speed is paramount.

In 2021, Takahashi et al. [20] optimized digital multipliers using parallel-prefix adders. The Han-Carlson adder was determined to be the most efficient, reducing both delay and hardware complexity. Experimental results showed that multipliers that used the Han-Carlson adder performed better, with fewer gates and a shorter critical path delay. However, the Han-Carlson adder's architecture may result in higher power consumption due to increased gate switching activity, and its complexity may make implementation difficult in bigger systems. Furthermore, it may not always provide the best balance of speed, power and area for all applications.

In 2020, Liu et al. [21] studied parallel-prefix adders within digital multipliers and found that the

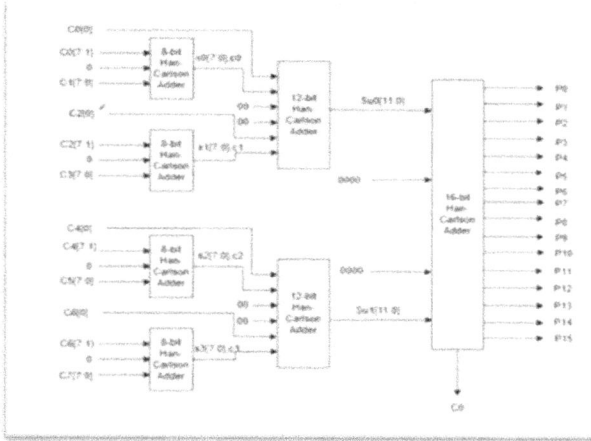

Figure 34.1 8-bit Multiplier using Han-Carlson adder
Source: Author

Han-Carlson adder delivered optimal results regarding area and latency performance. This technology might not work in fast operation systems.

Proposed system

Han-Carlson adder

The Han-Carlson adder is a hybrid adder that combines features of both Kogge-Stone and Brent-Kung adders that balances speed, area and delay.

Figure 34.1 shows a hierarchical structure of adders, starting with four 8-bit Han-Carlson adders in the first stage. First 8-bit adder takes inputs of $C0[0]$, $C0[7:0]$, 0, $C1[7:0]$, second 8-bit adder takes inputs of $C2[0]$, $20[7:1]$, 0, $C3[7:0]$, third adder takes inputs of $C4[0]$, $C4[7:0]$, 0, $C5[7:0]$, while fourth 8-bit adder takes input of $C6[0]$, $C6[7:0]$, 0, $C7[7:0]$ respectively. Each adder an 8-bit adders generates an 8-bit sum and carry . First 8-bit adder generates sum $s0[7:0]$ and carry $c0$, second 8-bit adder generates sum $s1[7:0]$ and carry $c1$, third 8-bit adder generates sum $s2[7:0]$ and carry $c2$ and fourth 8-bit adder generates sum $s3[7:0]$ and carry $c3$. These adders take eight 8-bit inputs ($C0[7:0]$, $C1[7:0]$, ..., $C7[7:0]$), with all carry-in bits set to zero. Each 8-bit adder produces an 8-bit sum ($s0[7:0]$, $s1[7:0]$, $s2[7:0]$, $s3[7:0]$, etc.) and a carry-out ($c0$, $c1$, $c2$, $c3$). The outputs from these 8-bit adders are then passed to the second stage, which consists of two12-bit Han-Carlson adders. The second stage consists of two 12-bit Han-Carlson adders. Each of these adders takes the results from two 8-bit adders in the first stage and combines them into a 12-bit sum. The first 12-bit adder processes the sums $s0[7:0]$ and $s1[7:0]$, along with the carry-outs $c0$, $c1$ from the first adder and second adder of 8-bit Han-Carlson adder, while the second 12-bit adder processes $s2[7:0]$ and $s3[7:0]$, along with carry-outs $c2$, $c3$ from the third adder and fourth adder of 8-bit Han-Carlson adder. Each 12-bit adders produces a 12-bit sum, first 12-bit adder produces $su0[11:0]$ and extra 4-bit are added, and second 12-bit adder produces sum $su1[11:0]$ and extra 4-bits are added and propagates the carry bits to final stage. The final stage uses a 16-bit Han-Carlson adder to merge the outputs of two 12-bit adders. This step generates the final 16-bit Sum $P[15:0]$ (i.e. $P0$ t0 $P15$) and carries out $c0$. By processing the inputs in stages, the design minimizes delay and allows efficient parallel computation.

The proposed method enhances adder performance by 17.8% while reducing power utilization to 10.7% below the original levels. This approach delivers higher speed with minimal space consumption thus outperforming standard adders in efficiency. The research analysis introduces Sparse-2 as a design approach which strives to unify speed and area.

Results and Discussion

The proposed 8-bit multiplier using the Han-Carlson adder was implemented on Xilinx ISE design suite, with delay shown in Figures 34.3–34.5 present simulation wave forms, while Table 34.1 compares multipliers based on delay and area.

Table 34.1 compares three types of multipliers based on area and delay. The multiplier using a hybrid adder takes up the least area (0.143 um2) but is not fastest (6.942 ns). The Kogge-Stone adder multiplier is the largest (0.393 um2) and also the slowest (13.452 ns). The Han-Carlson adder multiplier finds a balance, using a moderate amount of area (0.184 um2) while being the fastest (5.762 ns). This means the Han-Carlson adder is likely the best choice for both speed and efficiency.

Figure 34.2 illustrates a simulation result of a digital circuit, likely an 8-bit adder. The graph shows how different signals (inputs and outputs) changes over time. The left side lists signal names like inputs A and B while outputs Sum and Cout. The green lines in the main section represent signal values at different times. As the input change, the output sum and carryout update accordingly, showing how the adder processes binary numbers.

Table 34.1 Comparison table of multipliers.

Multiplier Specification	Multiplier with Hybrid adder	Multiplier with Kogge-stone adder	Multiplier with Han-Carlson adder
Area	0.143um^2	0.393um^2	0.184um^2
Delay	6.942ns	13.452ns	5.762ns

Source: Author

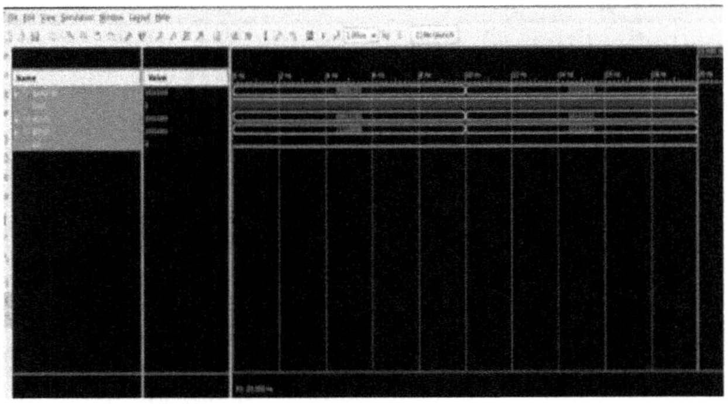

Figure 34.2 Design Summary and resource usage for proposed method
Source: Author

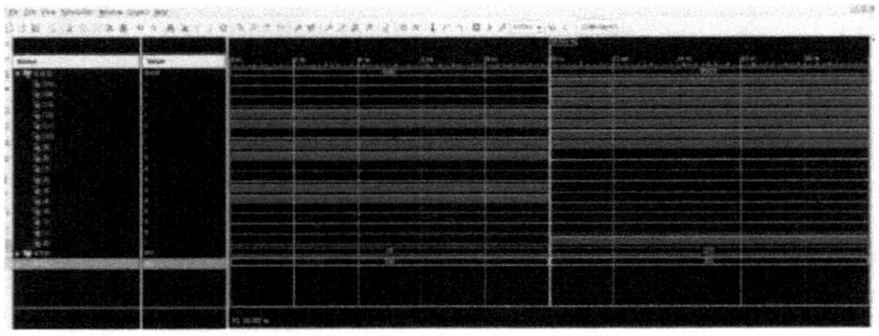

Figure 34.3 illustrates a digital waveform simulation with input signals A and B, and output signal S
Source: Author

In the first scenario, the inputs are A = 29 and B = 240producing an output of S = 6960. In the second scenario, both inputs are A = 255 and B = 255, resulting in an output of S = 65025. As the values of A and B change, the output S updates accordingly to reflect the calculated result. The left panel shows the signal names, the middle panel displays the binary values, and the right panel visually represents the transitions with green lines indicating changes in logic states.

The design summary in Figure 34.4 incorporates 117 BELS made up of 26 LUT2s, 5 LUT3s, 3 LUT4s, 43 LUT5s, 40 LUT6s and 32 IO buffers which include 16 BUF and 16 OBUF. Calculated

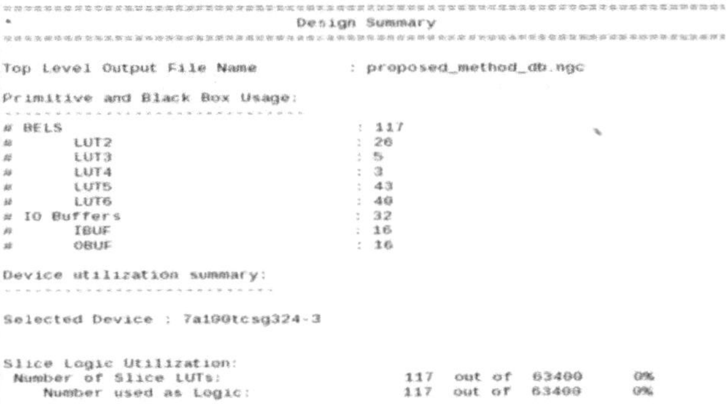

```
*                      Design Summary
Top Level Output File Name         : proposed_method_db.ngc

Primitive and Black Box Usage:
------------------------------
# BELS                             : 117
#      LUT2                        : 26
#      LUT3                        : 5
#      LUT4                        : 3
#      LUT5                        : 43
#      LUT6                        : 40
# IO Buffers                       : 32
#      IBUF                        : 16
#      OBUF                        : 16

Device utilization summary:
---------------------------

Selected Device : 7a100tcsg324-3

Slice Logic Utilization:
 Number of Slice LUTs:                  117  out of  63400     0%
    Number used as Logic:               117  out of  63400     0%
```

Figure 34.4 illustrates a digital waveform simulation with input signals A and B, and output signal S
Source: Author

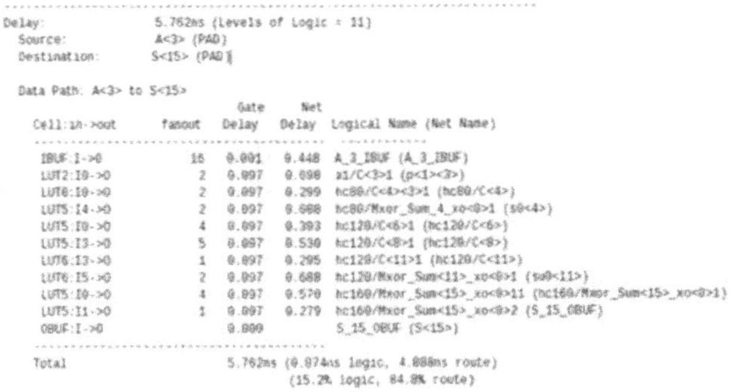

```
Delay:               5.762ns (Levels of Logic = 11)
  Source:            A<3> (PAD)
  Destination:       S<15> (PAD)

  Data Path: A<3> to S<15>
                                Gate     Net
    Cell:in->out      fanout   Delay   Delay  Logical Name (Net Name)
    ----------------------------------------  ------------
     IBUF:I->O            16   0.001   0.448  A_3_IBUF (A_3_IBUF)
     LUT2:I0->O            2   0.097   0.698  a1/C<3>1 (p<1><3>)
     LUT6:I0->O            2   0.097   0.299  hc80/C<4><3>1 (hc80/C<4>)
     LUT5:I4->O            2   0.097   0.688  hc80/Mxor_Sum_4_xo<0>1 (s0<4>)
     LUT5:I0->O            4   0.097   0.393  hc120/C<6>1 (hc120/C<6>)
     LUT5:I3->O            5   0.097   0.530  hc120/C<8>1 (hc120/C<8>)
     LUT6:I3->O            1   0.097   0.295  hc120/C<11>1 (hc120/C<11>)
     LUT6:I5->O            2   0.097   0.688  hc120/Mxor_Sum<11>_xo<0>1 (sa0<11>)
     LUT5:I0->O            4   0.097   0.570  hc160/Mxor_Sum<15>_xo<0>11 (hc160/Mxor_Sum<15>_xo<0>1)
     LUT5:I1->O            1   0.097   0.279  hc160/Mxor_Sum<15>_xo<0>2 (S_15_OBUF)
     OBUF:I->O                 0.000          S_15_OBUF (S<15>)
    ----------------------------------------
    Total                      5.762ns (0.874ns logic, 4.888ns route)
                                       (15.2% logic, 84.8% route)
```

Figure 34.5 Delay value of 8-bit multiplier using Han-Carlson Adder
Source: Author

area: The area is often represented as the percentage of utilized resources relative to the total available resources. Area utilization (%) = (used resources/ total resources) × 100.

The following design uses 117 slice LUTs out of 63,400 available on the device ,that calculated as follows slice LUT utilization (%) = (117/63,400) × 100 = 0.18%.

The report explains how long it takes for a signal to travel from point A<3> to point S<15> in a digital circuit. The total delay is 5.762 nanoseconds, broken into two parts: 0.874 ns for the logic processing and 4.888 ns for the wiring between different parts of the circuit. Most of the delay comes from how the signal moves through the wires (84.8%), while only a small part comes from the actual logic (15.2%). The signal passes through several components like buffers and lookup tables (LUTs), and each part has a different delay depending on how it's connected. Understanding this helps designers figure out which parts of the circuit are slowing things down and how to improve the design for faster signal transmission

Conclusion

The proposed high-speed multiplier employing the Han-Carlson adder design delivers efficient digital system solutions between speed and hardware requirements. The parallel-prefix implementation of Han-Carlson adders enables the multiplier to outperform Brent-Kung or Kogge-Stone designs because it reduces propagation delay and requires less gate count. Results derived from simulation together with synthetic data prove performance enhancement so the technology can serve high-performance needs. This design achieves maximum speed and resource efficiency which qualifies it to operate in embedded systems and portable devices that function with minimal space and power regulation. The development of efficient multipliers requires individuals to select optimal adder designs since it enables high-speed performance while minimizing hardware complexity. As a resource-efficient and high-performance computing component the Han-Carlson adder serves embedded systems as well as real-time computing applications such as digital signal processing and machine learning.

References

[1] Leela. S. N., Manisha, B., Bharath, P., & Praneeth, E. (2023). Design of wallace tree multiplier circuit using high performance and low power full adder. In E3S Web of Conferences, (Vol. 391, p. 01025). EDP Sciences.

[2] Zervakis, G., Xydis, S., Tsounamis, K., & Soudris, D. (2015). Hybrid approximate multiplier architectures for improved power-accuracy trade-offs. In 2015 IEEE/ACM International Symposium on Low Power Electronics and Design (ISLPED). IEEE.

[3] Borkar, J., & Gokhale, U. M. (2017). Design and simulation of low power and area efficient 16 × 16 bit hybrid multiplier. *International Journal of Engineering Development and Research*, 5(2), 831–838.

[4] Shiro, S., Joshi, R., & Shettar, R. (2017). Design and implementation of adders and multiplier in FPGA using ChipScope: a performance improvement. In Proceedings of Third International Conference on ICTCS 2017. Information and Communication Technology for Competitive Strategies, (pp. 11–19). DOI:10.1007/978-981-13-0586-3_2.

[5] Bussa, S., Rao, A., & Rastogi, A. (2016). Design of binary multiplier using adders. 4(1), 169–173. ISSN 2348-6988. Available from: www.researchpublish.com.

[6] Manjunatha Naik, V., & Poornima, N. (2015). Performance analysis of parallel prefix adder. *International Journal of Electrical, Electronics and Data Communication*, 3(07), 2015. ISSN: 2320-2084. DOI:10.18479/ijeedc/2015/v3i7/48267.

[7] Prassanakumar, M., Sidarthan, V., & Gopalakrishnan, K. (2015). Comparitive analysis of brent-kang & kogge-stone parallel-prefix adder for their area, delay & power consumption. 480–481. DOI:10.13140/RG.2.2.34773.14566.

[8] Boopathiraja, K., & Gowthamraj, M. (2020). Study of parallel prefix adders. *IJSRD - International Journal for Scientific Research and Development*, 8(1), 252–253. ISSN (online): 2321-0613.

[9] Mounika, B., & RajKumar, A. (2017). Design of efficient 32-Bit parallel prefix brentkung adder. *Advances in Computational Sciences and Technology*, 10(10), 3103–3109. ISSN 0973-6107. Research India Publications http://www.ripublication.com.

[10] Kumar, K. K. S., Yuvaraj, S., & Seshasayanam, R. (2020). High-performance wallace tree multiplier. *International Journal of Computer Techniques*, 7(01). DOI:10.29126/23942231/IJCT-V7I1P3.

[11] Janveja, M., & Niranjan, V. (2017). High performance wallace tree multiplier using improved adder. *ICTACT Journal on Microelectronics*, 3(1), 370–374. DOI:10.21917/ijme.2017.0065.

[12] Ramachandran, G., Muthumanickam, T., Murali, P. M., Nair, S. S., & Vasnath, L. (2015). Vedic multiplier in VLSI for high speed applications. *International Journal of Innovative Research in Computer and Communication Engineering*, 3(3), 2353–2354. (An ISO 3297: 2007 Certified Organization).

[13] Kumar, A., & Vishikha, V. (2017). Comparative analysis of vedic and array multiplier. *European Journal of Advances in Engineering and Technology*, 4(7), 524–531.

[14] Leela, S. N., Chandrika, D. K., Swetha, K., Kalali, D. G., & Shanti, G. (2024). A novel design of high speed multiplier using hybrid adder technique. In Conference: 2024 3rd International Conference for Innovation in Technology (INOCON), DOI:10.1109/INOCON60754.2024.10512237.

[15] Thamizharasan, V., & Kasthuri, N. (2023). High-speed hybrid multiplier design using a hybrid adder with FPGA implementation. *IETE I Journal of Research*, 69(5), 2301–2309.

[16] Athur, D. K., Narayan, B., Gopalakrishna, A., & Sasipiya, P. (2023). Design of novel high speed parallel prefix adder. *Indonesia Journal of Electrical Engineering and Computer Science*, 29(3), 134. DOI:10.11591/ijeecs.v29.i3.pp1345-1354.

[17] Gupta, S., Mehra, R., & Sharma, A. (2022). A high-speed parallel multiplier using Han-Carlson Adder for DSP applications. *International Journal of Electronics and Communications*, 99, 45–54.

[18] Banerjee, S., Roy, K., & Jain, A. (2022). Optimization of digital arithmetic units using parallel adders. *Elsevier Microprocessors and Microsystems*, 83, 103229.

[19] Raman, K., Subramanian, P., & Sundar, S. (2021). Design and implementation of low-power high-speed multipliers using parallel-prefix adders. *Microelectronics Journal*, 58, 87–95.

[20] Takahashi, M., Yamamoto, K., & Watanabe, S. (2021). Design space exploration of parallel-prefix adders for multiplier optimization. *Journal of Semiconductor Technology and Science*, 12(2), 85–93.

[21] Liu, J., Wang, M., & Zhao, H. (2020). Optimization of parallel-prefix adders for high-speed multipliers. *IEEE Transactions on VLSI Systems*, 28(8), 1243–1253.

35 Multi-model retrieval-augmented generation: bridging text and visual understanding through Nomic embeddings and Cohere integration

Sunitha Sabbu[1,a], Rohith, P.[2,b], Shiva Kumar Reddy, P.[2,c], Sushma, C.[2,d] and Sai Nikhitha, B.[2,e]

[1]Assistant Professor, Department of CSE, Srinivasa Ramanujan Institute of Technology, Anantapur, Andhra Pradesh, India

[2]Department of CSE(AI&ML), Srinivasa Ramanujan Institute of Technology, Anantapur, Andhra Pradesh, India

Abstract

The rapid expansion of multimodal data including text, images, videos, and audio—demands advanced intelligent systems for extracting meaningful insights. Traditional retrieval-augmented generation (RAG) frameworks focus on text-based tasks, often overlooking multimodal integration. To bridge this gap, the Multimodal RAG framework combines natural language processing (NLP), retrieval mechanisms, and multimodal processing to generate highly contextual responses. It employs Nomic embeddings for semantically rich text-image alignment, ChromaDB for efficient storage and retrieval, Gemini for visual description generation, and Cohere for text based summarization and response synthesis. This framework seamlessly integrates diverse data formats, enabling deep comprehension of queries across various domains. Applications include e-commerce for improved product searches, healthcare for combining medical imaging with text records, and education for multimedia learning. By leveraging state-of-the-art models and tools, the Multimodal RAG framework sets a new benchmark for intelligent, data-driven systems, addressing complex information needs in an increasingly data-rich world.

Keywords: Large language models, multi model large language models, natural language processing, retrieval augmented generation

Introduction

The rapid expansion of multimodal information necessitates retrieval systems capable of managing various data formats, including text documents and images. RAG systems enhance information retrieval by combining retrieved knowledge and natural language generation (NLG). While conventional RAG methods primarily address text-based tasks, incorporating multimodal data introduces new hurdles, such as embedding alignment, semantic retrieval, and cross-modal response generation. Large language models (LLMs), like Gemini 1.5 flash model (Google DeepMind 2024) [1] and Cohere command R+ (Cohere, 2024) [2], have greatly advanced natural language processing (NLP), enabling a variety of applications, such as content creation, conversational agents, and translation. Despite their capabilities, LLMs still have limitations in specialized knowledge and can produce inaccurate information. Retrieval- augmented generation addresses these issues by combining document retrieval with generative models. The recent development of multimodal large language models has expanded LLMs capabilities to include various forms of media such as images, videos, and audio. This progress holds significant promise for industries like manufacturing, where technical documentation often integrates complex text with detailed visual elements, including images, schematics, and screenshots. In our paper, we propose the Multi-Modal RAG framework, which combines state-of-the-art (most advanced models in their field) tools to handle multimodal queries effectively. We use two models - Gemini 1.5 flash model - multimodal large language model (MLLM) for detailed description of images and Cohere command R+ a LLM for generating text summaries and

[a]sunindra111@gmail.com, [b]214g1a3382@srit.ac.in, [c]214g1a3394@srit.ac.in, [d]214g1a33A9@srit.ac.in, [e]214g1a3385@srit.ac.in

DOI: 10.1201/9781003685364-35

for answering the queries. Nomic embeddings [3] for dense and semantically meaningful representations of data (e.g., text, images) within a high-dimensional space (numeric representation of text). ChromaDB [4] is a high-performance, open-source vector database specifically designed to store and query dense vector embeddings efficiently.

Related work

Multi-model LLM

The MLLMs are advanced AI systems capable of handling and combining various data types, including text, images, videos, and audio. In the current era of multimodal data integration, characterized by swift technological progress and a surge in data volume, relying solely on a single modality is inadequate for addressing the intricate challenges encountered in real-world scenarios. Consequently, the evolution of MLLMs is both an inevitable technological progression and a crucial enhancement for boosting AI application efficacy. By synthesizing data from a range of sources, MLLMs construct a more detailed and precise portrayal of information, unlocking significant potential and demonstrating substantial practical value across various fields. Notably, MLLMs have exhibited exceptional performance in various multimodal tasks, encompassing language, image, video, and audio processing. These models are highly proficient at combining multimodal information to boost the performance of tasks involving multiple modes. In NLP activities like text generation and machine translation, MLLMs leverage data from various modalities to offer contextual assistance, thereby enhancing the precision of the text they produce [5].

Retrieval augmented generation

The RAG has revolutionized NLP by addressing limitations in traditional AI models, which struggled with knowledge- intensive tasks. While pretrained models improved performance in tasks like sentiment analysis, they failed to integrate external knowledge effectively. To overcome this, RAG dynamically retrieves relevant information during inference, combining a retrieval module with a generative language model. Unlike models that embed knowledge directly into parameters, RAG ensures adaptability without requiring retraining. This integration enhances context-specific responses, making RAG a powerful approach for handling complex NLP tasks that require external knowledge. The

RAG framework retrieves relevant information from diverse sources, enriching generative models with contextually accurate data. It supports multiple retrieval methods, such as sparse and dense retrievers, to enhance knowledge-intensive applications like machine translation, dialogue generation, and summarization. Recent advancements in RAG focus on jointly training retrieval and generative models, improving task relevance while reducing computational costs. This dynamic, scalable approach positions RAG as a key innovation in modern NLP.

MultiModel RAG

Multimodal retrieval-augmented generation (MuRAG) marks a significant advancement in expanding the capabilities of retrieval-augmented transformer models to process multimodal data. Chen et al. (2022) [6] introduced MuRAG as the pioneering system to utilize an external, non- parametric multimodal memory. This innovation enables the model to efficiently handle and combine various data types, including text and images, making it applicable to a broad spectrum of tasks that require complex data processing. Lin and Byrne (2022) [6] proposed an alternative approach that focuses on transforming image data into textual format using techniques such as optical character recognition (OCR), image captioning, and object detection. These textual representations are subsequently processed using dense passage retrieval methods, as described by Karpukhin et al. (2020) [7], to improve the retrieval of relevant multimodal information. The application of multimodal RAG has shown particular promise in specialized domains like medicine and healthcare. Studies [6] show that incorporating medical images, such as scans, as part of the contextual data. These developments highlight the growing capacity of multimodal RAG to deliver precise, contextually enriched responses, particularly in fields where the integration of textual and visual information is essential for comprehensive insights

RAG evaluation

RAGAS, a framework developed by, is designed to evaluate the performance of RAG systems, which integrate LLMs with external knowledge retrieval mechanisms. This framework offers a methodical approach to assess both the generator and retriever elements of a RAG pipeline through a set of metrics that can be implemented without the need for extensive labeled datasets. It provides various evaluation metrics to gauge different aspects of a RAG system. The assessment of RAG systems involves examining both the retrieval and generation components. RAGAS, a framework developed by, is designed to evaluate the performance of RAG systems, which integrate LLMs with external knowledge retrieval mechanisms. This framework offers a methodical approach to assess both the generator and retriever elements of a RAG pipeline through a set of metrics that can be implemented without the need for extensive labeled datasets. It provides various evaluation metrics to gauge different aspects of a Retrieval-Augmented Generation (RAG) system.

Model Architecture

Overview

The multi-modal RAG architecture consists of four key components:

1. Embedding generation: Nomic embeddings create dense representations for both text and images, ensuring semantic alignment across modalities.
2. Storage and retrieval: ChromaDB efficiently indexes embeddings with metadata for fast, scalable retrieval.
3. Visual description: Gemini (1.5 flash model) generates detailed textual descriptions of images, enabling cross modal alignment.

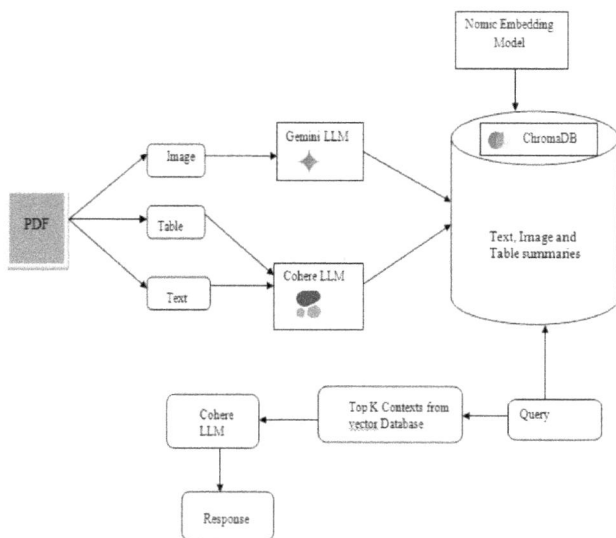

Figure 35.1 Multi model RAG architecture
Source: Author

4. Generative responses: Cohere (command R+ model) facilitates textual summarization and query-answering, synthesizing retrieved knowledge into coherent responses.

Implementation details
Text processing
Text extraction is a crucial component of the system, enabling efficient retrieval and reasoning over document content. To achieve this, we employ Unstructured.io, a robust library designed for parsing and segmenting PDF documents into structured text chunks. The extracted text is stored in a page-wise list, ensuring that the original structure of the document is preserved. However, because these extracted text blocks can be extensive, it is necessary to split them into smaller, manageable segments to optimize embedding and retrieval processes. To efficiently handle text chunking, we utilize the RecursiveCharacterTextSplitter from the LangChain framework. This approach recursively breaks large text blocks into semantically meaningful and optimized segments, making them easier to embed. After splitting the texts into chunks we use Cohere LLM to summarize each chunk and store the embeddings of each summarized chunk in ChromaDB.

Table processing
Processing structured data, such as tables in PDFs, is essential for accurately handling numerical and categorical information, especially in finance, research, and analytics. The system utilizes Unstructured.io to extract tabular data and convert it into structured formats like key-value pairs or arrays, enhancing interoperability with retrieval mechanisms. This structured representation ensures efficient search, summarization, and synthesis of critical information. Cohere's Command R+ model further summarizes extracted tables, preserving numerical relationships while reducing verbosity. By maintaining both structured and summarized data, the system overcomes traditional retrieval challenges, ensuring accuracy and relevance in knowledge extraction. To enable semantic search, the summarized tabular data is embedded using the Nomic model and stored in ChromaDB, allowing retrieval based on both semantic and numerical relevance. The retrieved data is then used to generate contextualized prompts, incorporating essential numerical trends into responses. Finally,

Cohere's LLM synthesizes structured insights into coherent answers, improving interpretability. This approach is particularly effective in fields requiring numerical precision, making it a powerful solution for RAG applications.

Image processing

Handling visual elements in PDFs poses difficulties because of the intricate nature of images, diagrams, and charts, which necessitate specialized extraction methods to preserve their semantic significance. Tools like Unstructured.io isolate images and embedded text, ensuring both graphical and textual components are preserved. This is crucial for technical documents, research papers, and data- driven reports where visualizations play a key role in knowledge dissemination. To manage visual data, the system employs two methods: direct conversion of images into vector embeddings using Nomic and structured textual summarization through Gemini 1.5 Flash LLM. The latter approach enhances retrieval accuracy by preserving richer semantic context. Gemini 1.5 Flash LLM generates searchable textual descriptions of visual content, enabling seamless integration into RAG pipelines, during retrieval, ChromaDB performs a semantic search, identifying relevant image summaries and original visuals, which are synthesized by the Cohere LLM into coherent responses. This architecture significantly improves multimodal document understanding, particularly for technical manuals, financial reports, and scientific research. By prioritizing structured summaries over raw embeddings, the system achieves higher precision in retrieval and reasoning, demonstrating

the effectiveness of integrating LLMs with scalable vector databases in RAG systems.

Query processing and response generation

When a user submits a query, the system performs a semantic search in ChromaDB using embeddings from the Nomic model. This retrieves relevant text chunks, table summaries, and image descriptions based on semantic similarity. By leveraging multimodal embeddings, the system ensures holistic retrieval from text, structured tables, and visual content, enabling comprehensive reasoning across diverse data types. Retrieved data is then dynamically assembled into a context-aware prompt, integrating text summaries, numerical trends, and visual insights. The Cohere LLM processes this prompt, synthesizing multimodal inputs into a coherent response that maintains accuracy and interpretability. Gemini 1.5 Flash LLM facilitates cross-modal alignment by converting tables and images into structured textual summaries, allowing the Cohere LLM to reason over unified semantic representations. This enhances retrieval precision and response quality while avoiding token constraints. The system's ability to unify semantic search, multimodal retrieval, and LLM-powered synthesis significantly improves RAG applications in technical analysis, financial reporting, and research, ensuring responses are contextually rich and grounded in source material.

Evaluation metrics

To assess the RAG pipelines' effectiveness, we employ a Judge [8] approach, where LLM checks its own response by using an evaluation system designed for multimodal data. Metrics are divided into two categories: Retriever metrics (Context

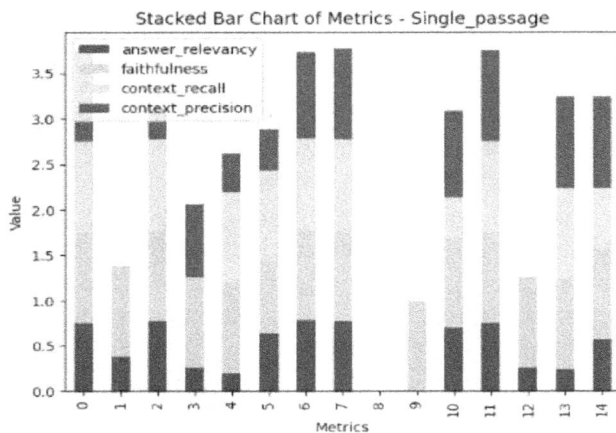

Figure 35.2 Single passage metric
Source: Author

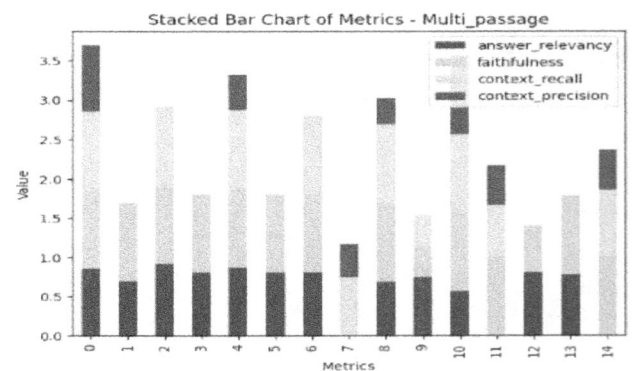

Figure 35.3 Multi passage metrics
Source: Author

Precision, Context Recall) and Generator (LLM) metrics (Faithfulness and Answer Relevancy). To evaluate our RAG performance, we utilized a dataset from Kaggle [9] containing source data, text, and three types of questions: single, multi and no passage questions.

Limitations and future work

Limitations

The multi-modal RAG framework, while enhancing retrieval-augmented generation, faces challenges in scalability, computational efficiency, and real-time deployment, particularly on low power devices. Processing multimodal data demands significant hardware resources, and reliance on pre-trained models like Gemini and Cohere introduces biases that may impact fairness and accuracy in domain-specific applications. Additionally, context window constraints in LLMs limit long-form response coherence, while retrieval effectiveness depends on embedding quality and alignment between text and images. Misalignments can lead to inaccurate responses, and current evaluation metrics, such as context recall and precision, may not fully capture the complexity of multimodal interactions, necessitating improved assessment methodologies

Future directions

To overcome these limitations, we focus on integrating additional modalities, particularly audio and video. Enhancing the framework to process audio files (.mp3) and video content would significantly expand its utility across diverse applications, such as real-time transcription and video summarization. Implementing speech-to-text models (Whisper model) for extracting textual insights from audio and integrating video captioning models for generating descriptive text from video content will ensure a more comprehensive multimodal retrieval pipeline. By incorporating these capabilities, the system can provide richer and more contextually relevant responses across various industries, including education, healthcare, and entertainment.

Conclusion

The multi-modal retrieval-augmented generation (RAG) framework represents a groundbreaking

advancement in multimodal AI, enabling seamless integration of text, images, and structured data for precise, contextual retrieval. By leveraging Nomic embeddings, ChromaDB, Gemini, and Cohere, the system enhances response accuracy and multimodal comprehension across various industries. However, challenges remain in scalability, bias mitigation, and computational efficiency. Future research will focus on real-time processing, multimodal dataset expansion, and audio-video integration to further advance retrieval-augmented AI.

Acknowledgement

The authors gratefully acknowledge the students, staff and authority of CSE department for their cooperation in the research.

References

[1] Gemini 1.5: Unlocking multimodal understanding across millions of tokens of context https://arxiv.org/abs/2403.05530.
[2] Cohere Command R+ model. https://docs.cohere.com/docs/command-r-plus.
[3] Nomic Embeddings: introducing nomic embed: a truly open embedding model. https://www.nomic.ai/blog/posts/nomic- embedtext-vl.
[4] Chroma (2022). Chroma: the open-source embedding database. https://github.com/chroma core/ chroma. Accessed: September 09, 2024..
[5] Zhang, X., Lu, Y., Wang, W., Yan, A., Yan, J., Qin, L., et al. (2023). Gpt-4v(ision) as a generalist evaluator for vision-language tasks. ArXiv, abs/2311.01361.
[6] Zhu, Y., Ren, C., Wang, Z., Zheng, X., Xie, S., Feng, J., et al. (2024). Emerge: integrating rag for improved multimodal ehr predictive modeling. ArXiv, abs/2406.00036.
[7] Kaggle dataset https://www.kaggle.com/datasets/samuelmatsuoh ar ris/single-topic-rag-evaluation-dataset.
[8] Kandpal, N., Deng, H., Roberts, A., Wallace, E., & Raffel, C. (2023). Large language models struggle to learn long-tail knowledge. In Proceedings of the 40th International Conference on Machine Learning, volume 202 of Proceedings of Machine Learning Research, (pp. 15696–15707). PMLR.
[9] Kaggle dataset https://www.kaggle.com/datasets/samuelmatsuoh ar ris/single-topic-rag-evaluation-dataset

36 Epileptic seizure detection using convolutional neural networks

Soumya Madduru[1,a], Bharathi Bhimireddy[2,b], Ayub Shaik[2,c],
Hema Chandra Kiran Sreenivas Sunduru[2,d] and Al Farheen Pathan[2,e]

[1]Assistant Professor, Department of CSE, Srinivasa Ramanujan Institute of Technology, Anantapur, Andhra Pradesh, India

[2]Department of CSE, Srinivasa Ramanujan Institute of Technology, Anantapur, Andhra Pradesh, India

Abstract

Epilepsy is a neurological condition that causes unpredictable seizures, and detecting those seizures early can make a big difference in a patient's treatment and quality of life. In this project, we developed a deep learning model using convolutional neural networks (CNNs) to automatically identify whether a person is experiencing a seizure based on electroencephalogram (EEG) signals. We trained our model using a well-known public dataset and achieved a high accuracy of 98.35%, along with strong precision and recall scores. To make the technology more practical and accessible, we also created a simple web application where users can upload EEG data, view the brain signal as a graph, and instantly get a prediction. This combination of deep learning and real-time interaction provides a fast and reliable tool that can support healthcare professionals and patients in seizure monitoring and early intervention.

Keywords: Convolutional neural networks, deep learning, electroencephalogram, epilepsy, seizure detection, web application

Introduction

One of the most prevalent neurological conditions, epilepsy affects millions of people worldwide. It is characterized by abrupt seizures brought on by aberrant brain electrical activity. To guarantee appropriate medical care and enhance patients' quality of life, early recognition and monitoring of these seizures are essential. Conventional seizure detection depends on neurologists visually examining electroencephalogram (EEG) patterns, which takes time and is subject to human error [2].

In recent years, deep learning techniques—especially convolutional neural networks (CNNs)—have shown exceptional performance in extracting meaningful patterns from biomedical signals like EEGs [4, 6]. These models are capable of learning complex spatial and temporal features without needing handcrafted rules, making them ideal for automated seizure detection.

In this project, we designed a 1D CNN to classify EEG signals as either epileptic or non-epileptic using the UCI Epileptic Seizure Recognition Dataset. Our model achieved a high-test accuracy of 98.35%, outperforming several traditional methods reported in past studies [1, 3, 9]. We also integrated the model into a Flask-based web application, making it easy for users—clinicians, caregivers, or patients—to upload EEG data and get real-time predictions with visual plots. This approach provides a scalable and user-friendly solution to support early diagnosis and continuous monitoring of epilepsy [5, 7, 10].

Mathematical Model

In this section, we explain the core mathematical principles behind the 1D CNN used to detect epileptic seizures from EEG signals. CNNs are highly effective at learning patterns from sequential data, making them well-suited for one-dimensional time-series data such as EEG recordings.

Overview of 1D convolution

At the heart of a CNN is the convolution operation, which slides a filter over the input data to capture local patterns. For 1D data, this is expressed mathematically as:

[a]soumya.cse@srit.ac.in, [b]214g1a0512@srit.ac.in, [c]214g1a0509@srit.ac.in, [d]214g1a0536@srit.ac.in, [e]214g1a0503@srit.ac.in

DOI: 10.1201/9781003685364-36

$$y_t = (x * w)_t + b = \sum_{j=0}^{k-1} x_{t+j} \cdot w_j + b$$

Where:

In the convolution operation, x represents the input EEG signal (e.g., 178 values), w is the filter k, b is the bias term, y_i is the output at position i, and he bias-corrotes convolution, where multiple filters learn distinctive features like peaks, slopes, or frequency patterns from the EEG data.

Activation functions

We employ the Rectified Linear Unit (ReLU) activation function, which is defined as follows:

$$\text{ReLU}(z) = \max(0, z)$$

This function outputs the input value if it's positive, or zero otherwise. It helps avoid vanishing gradients and speeds up training.

Pooling operation

To reduce the dimensionality of the output and retain only the most important features, we use max pooling. In 1D max pooling, a window of size p slides over the input, and for each window, the maximum value is selected:

$$y_t = \max\{x_t, x_{t+1}, \ldots, x_{t+p-1}\}$$

This operation reduces computational cost and prevents overfitting by providing translation invariance.

Flattening and dense layers

After several convolution and pooling layers, the output is flattened into a 1D vector:

$$\text{Flatten}(x) = [x_1, x_2, \ldots, x_n]$$

This vector is then passed through dense (fully connected) layers, which perform linear transformations followed by non-linear activations. For a dense layer with weights W, bias b, and input x, the output is:

$$y = f(Wx + b)$$

Where f is again an activation function (ReLU for hidden layers, sigmoid for the output).

Output layer and prediction

The final layer has a sigmoid activation function because this is a binary classification problem (seizure or non-seizure). The sigmoid function maps input values to a probability range between 0 and 1

$$\sigma(z) = \frac{1}{1 + e^{-z}}$$

If the output $\sigma(z) \geq 0.5$, the sample is classified as a seizure. Otherwise, it is classified as non-seizure.

Loss function

To train the model, we use the binary cross-entropy loss function, which measures how close the predicted probability is to the true label:

$$L = -\frac{1}{N}\sum_{i=1}^{N}[y_i \log(\hat{y}_i) + (1 - y_i)\log(1 - \hat{y}_i)]$$

Where:

- yi the actual label (0 or 1),
- y_i is the predicted probability,
- N is the number of samples.

Minimizing this loss during training helps the model improve its prediction accuracy.

Optimization

The model is optimized using the Adam optimizer, which combines momentum and adaptive learning rates. It updates weights θ\thetaθ as follows:

$$\theta_{t+1} = \theta_t - \alpha \cdot \frac{\widehat{m}_t}{\sqrt{\widehat{v}_t} + \epsilon}$$

In this equation, m^t represents the bias-corrected first-moment estimate, v^t is the bias-corrected second-moment estimate, α\alphaα marks the learning rate, and ϵ\epsilonϵ is a minor constant introduced to avoid division by zero during the optimization phase.

Methodology

Data preparation

We used the UCI Epileptic Seizure Recognition Dataset, a public dataset commonly used in seizure classification studies [2, 9]. It includes EEG signals labeled into five classes, where only class 1 corresponds to actual seizure activity. For this study, we simplified the problem into a binary classification:

Class 1 → Seizure (positive class)

Classes 2–5 → Non-seizure (negative class)

Each sample contains 178 time steps representing one second of brain activity. Data was normalized and reshaped to match the input shape required by the CNN model.

The pipeline covers data preprocessing, model training, evaluation, and final prediction.

CNN architecture

Our model architecture was inspired by successful CNN-based seizure detection systems described in literature [3, 4, 6]. The layers are as follows:

Three Conv1D layers with increasing filter sizes (32, 64, 128)

MaxPooling after each convolution layer

A flatten layer followed by a dense layer of 64 units

Final sigmoid output for binary classification:

The binary cross-entropy loss function and Adam optimizer were used for training. A batch size of 32 was used to train the model over 10 epochs.

Web application deployment

To enhance accessibility, we built a web application using Flask that allows users to:

- Upload EEG CSV files
- View EEG signal plots
- Instantly get classification predictions

This makes the model usable in real-world settings, such as hospitals or mobile health apps [5, 7].

Implementation

To build a working seizure detection system, we followed a complete pipeline—from data loading and preprocessing to model training, evaluation, and deployment in a web environment.

Dataset and preprocessing

The EEG data used is from the UCI Epileptic Seizure Recognition Dataset, containing 11,500 one-second EEG samples with 178 data points each. For binary classification, Class 1 was labeled as seizure and Classes 2–5 as non-seizure. Irrelevant columns were removed, data was normalized, and reshaped to fit the CNN model.

Model training

We used TensorFlow and Keras to build the 1D CNN model. The training was conducted on an 80:20 train-test split Using Adam optimizer and binary cross-entropy loss. The model was trained for 10 epochs with a batch size of 32.

Both the accuracy and loss metrics showed consistent improvement throughout training. By the tenth epoch's end:

Training acc reached over 99%

Validation accuracy stabilized around 98.35%

Figure 36.1 depicts a CNN-based seizure detection pipeline, starting from EEG data preprocessing to model training, evaluation, and final seizure classification.

This result is consistent with the performance of other deep learning models applied to similar datasets [4, 6, 9].

Figure 36.1 Workflow of the proposed CNN-based seizure detection method

Source: Author

Flask-based web application

After achieving good accuracy, we focused on deployment to make the solution more accessible. We developed a Flask-based web interface:

The web application allows users to securely sign up and log in, upload EEG signals in .csv format, visualize the EEG waveforms directly in the browser, and receive real-time predictions with clear results such as "Detected Epileptic Seizure" or "No Epileptic Seizure Detected." bridges the gap between machine learning research and real-world usability, something that recent works have highlighted as a key step toward clinical adoption.

Figure 36.2 illustrates the training and validation accuracy trends over 10 epochs. It shows consistent improvement and convergence, indicating effective model learning with minimal overfitting.

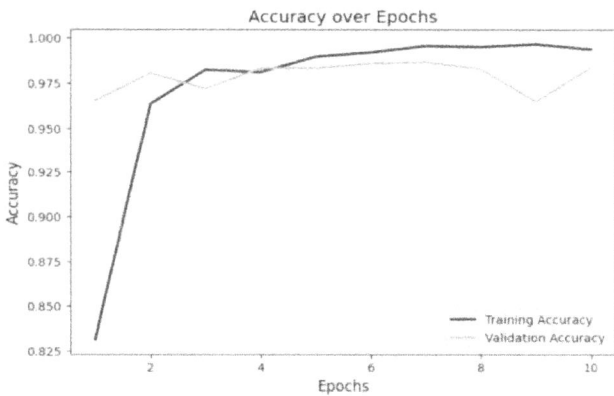

Figure 36.2 Train and valid ACC over epochs comparison
Source: Author

Visualization

We added signal visualization using Matplotlib, which plots the EEG amplitude over time (in seconds). This helps users better understand the data they're uploading and enhances transparency in predictions.

Results

The CNN model performed strongly on both training and validation sets.

Accuracy and loss

Showing effective learning. The model converges well with minimal overfitting, achieving ~98% validation accuracy.

Our model showed stable training behavior, reaching a test accuracy of 98.35%—comparable or better than many recent seizure detection frameworks reported in previous work [1, 4, 8].

Figure 36.3. The training loss drops sharply and stabilizes, indicating effective learning. Validation loss remains low with minor fluctuations, suggesting minimal overfitting.

Figure 36.4 presents the confusion matrix, highlighting high classification accuracy with minimal false positives and false negatives.

Table 36.1 Detection results.

	Predicted: 0	Predicted: 1
Actual:0 (Non-seizure)	1836	10
Actual:1 (Seizure)	28	426

Source: Compiled by the authors from experimental results.

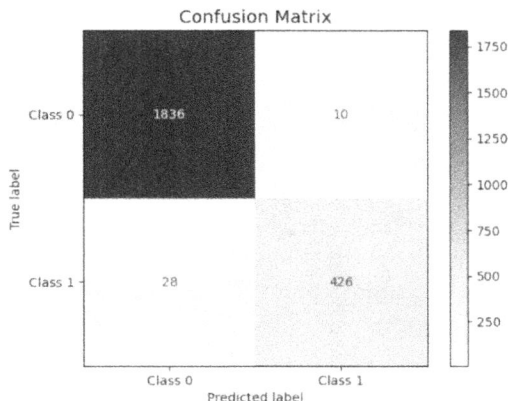

Figure 36.3 Loss over epochs
Source: Author

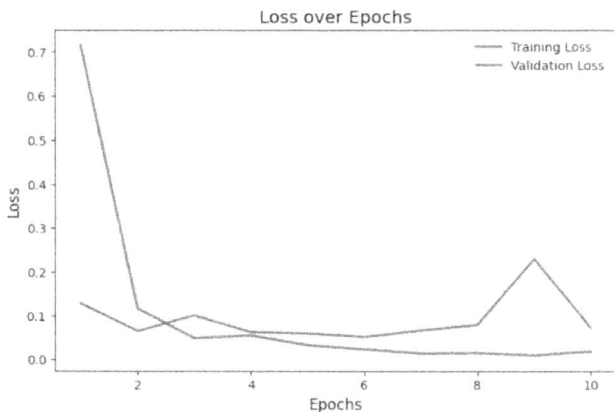

Figure 36.4 Confusion matrix
Source: Author

Table 36.2 Classification report.

Metric	Non-seizure %	Seizure %
Prec	98	98
Rec	99	94
F1	99	96

Source: Compiled by the authors from experimental results.

Confusion matrix and metrics

Table 36.1. These results yield strong classification metrics, which align with other modern CNN-based models discussed in [4, 6, 7].

It shows strong classification performance with high TP and TN rates. Only 38 out of 2300 predictions were incorrect, indicating excellent model reliability.

Classification report

Table 36.2. These metrics confirm the model's robustness in both identifying seizure events and minimizing FP—critical in medical settings [3, 6].

Conclusion

In order to detect epileptic seizures from electroencephalogram (EEG) readings, this experiment shows how well deep learning—more especially, convolutional neural networks (CNNs)—work. Through training on the UCI Epileptic Seizure Recognition dataset, our model demonstrated a 98.35% test accuracy, as well as good precision and recall scores, indicating its suitability for practical use. For seizure pattern recognition, the CNN model provides a quicker and more scalable method than manual EEG interpretation and conventional machine learning techniques.

Beyond model performance, we also prioritized accessibility. By integrating the trained CNN into a simple web application, users can now upload EEG recordings, visualize brain activity, and receive instant feedback—all from a browser. This approach not only supports healthcare professionals but also opens the door to remote and telemedicine-based monitoring systems, which have become increasingly important in today's digital healthcare landscape.

For future work, this model could be extended to multi-class seizure type classification, integrated with wearable EEG devices, or enhanced with encryption techniques for secure clinical deployment.

References

[1] Wei, X., Zhou, L., Chen, Z., Zhang, L., & Zhou, Y. (2018). Automatic seizure detection using three-dimensional CNN based on multi-channel EEG. *BMC Medical Informatics and Decision Making*, 18, 71–80.

[2] Shoeibi, A., Khodatars, M., Ghassemi, N., Jafari, M., Moridian, P., Alizadehsani, R., et al. (2021). Epileptic seizures detection using deep learning techniques: a review. *International Journal of Environmental Research and Public Health*, 18(11), 5780.

[3] Cimr, D., Fujita, H., Tomaskova, H., Cimler, R., & Selamat, A. (2023). Automatic seizure detection by convolutional neural networks with computational complexity analysis. *Computer Methods and Programs in Biomedicine*, 229, 107277.

[4] Acharya, U. R., Oh, S. L., Hagiwara, Y., Tan, J. H., & Adeli, H. (2018). Deep convolutional neural network for the automated detection and diagnosis of seizure using *EEG signals. Computers in Biology and Medicine*, 100, 270–278.

[5] Shoka, A. A. E., Dessouky, M. M., El-Sayed, A., & Hemdan, E. E. D. (2023). An efficient CNN-based epileptic seizures detection framework using encrypted EEG signals for secure telemedicine applications. *Alexandria Engineering Journal*, 65, 399–412.

[6] Zhou, M., Tian, C., Cao, R., Wang, B., Niu, Y., Hu, T., et al. (2018). Epileptic seizure detection based on EEG signals and CNN. *Frontiers in Neuroinformatics*, 12, 95.

[7] Hassan, F., Hussain, S. F., & Qaisar, S. M. (2022). Epileptic seizure detection using a hybrid 1D CNN-machine learning approach from EEG data. *Journal of Healthcare Engineering*, 2022, 9579422.

[8] Kim, T., Nguyen, P., Pham, N., Bui, N., Truong, H., Ha, S., et al. (2020). Epileptic seizure detection and experimental treatment: a review. *Frontiers in Neurology*, 11, 701.

[9] Rashed-Al-Mahfuz, M., Moni, M. A., Uddin, S., Alyami, S. A., Summers, M. A., & Eapen, V. (2021). A deep convolutional neural network method to detect seizures and characteristic frequencies using epileptic electroencephalogram (EEG) data. *IEEE Journal of Translational Engineering in Health and Medicine*, 9, 1–12.

[10] Ullah, I., Hussain, M., & Aboalsamh, H. (2018). An automated system for epilepsy detection using EEG brain signals based on deep learning approach. *Expert Systems with Applications*, 107, 61–71.

37 Disease-gene prediction: a machine learning perspective

Vasepalli Kamakshamma[1,a], Gajula Jayanth Babu[2,b], Budavarapu Likitha[2,c], Golla Navya Teja[2,d] and Dubbala Mounika[2,e]

[1]Assistant Professor, Department of CSE(AI&ML), Srinivasa Ramanujan Institute of Technology, Anantapur, India

[2]Students, Department of CSE (Data Science), Srinivasa Ramanujan Institute of Technology, Anantapur, India

Abstract

The paper, "Disease-gene prediction: a machine learning perspective" aims at analyzing and predicting the associations between genes and diseases by advanced techniques of machine learning. Due to the fast-increasing availability of genetic data, there has been an increasing need to understand the correlation of specific genes with a disease in biomedical research. This study employs a comprehensive data set, which includes gene-specific information such as DSI and DPI, as well as several disease features including semantic type and classification. It applies four ml algorithms, namely XG-Boost, Random Forest, LightGBM and k-nearest neighbors (KNN), to predict the three significant output classes-Disease, Group, and Phenotype. It was found that the best model for this purpose came out to be that of Random Forest with 97. 81% accuracy. The same model was implemented using Flask as a framework to gain real-time predictions. Preprocessing mainly involved filling missing values, label encoding, and even clustering into diseases. There are chances of using K-means clustering for organizing diseases into broader categories based on their similarities for a stronger prediction. The paper demonstrates the potential of machine learning in advancing genomic research by providing insights into gene-disease associations. It offers a practical tool for researchers to explore genetic links to diseases efficiently.

Keywords: Bioinformatics, clustering, data preprocessing, disease classification, Flask, gene prediction, gene-disease association, machine learning, phenotype prediction, Random Forest

Introduction

The past couple of years have been very fruitful as far as the genomics field is concerned and has seen quite a number of breakthroughs that have led scientists to a better comprehension in the genetic base of various diseases. Research into the genetics of disease has provided scientists with clear avenues for understanding how particular genes may be responsible for certain diseases. However, such relationship between a few genes and diseases is exceedingly more complex and remains unresolved. This paper, in this case, is "Analysis for Disease Gene Association Using Machine Learning," which will address this challenge by applying machine learning for the prediction of associations between genes and diseases.

Machine learning provides powerful tools to make sense of large datasets, helping us uncover patterns that might not be obvious with traditional methods. By using these models, this paper focuses on analyzing gene-disease associations to predict three key output categories: disease, group, and phenotype. The use of advanced algorithms, including Random Forest (RF), XGBoost, LightGBM, and k-nearest neighbors (KNN), provides a multi-model approach to improve prediction accuracy. Among these models, Random Forest achieved the highest accuracy, making it the primary model for final predictions.

For this paper, I worked with a dataset that includes detailed information on gene-disease associations, covering aspects like the Disease Specificity Index (DSI), Disease Pleiotropy Index (DPI), disease names, and the strength of supporting evidence. Techniques of imputation were used to deal with missing values, and categorical variables were preprocessed by assigning label encoding to the data so that it can be used in concurrence with the algorithm implemented for the process of machine learning. The genetic associations of diseases were further categorized into broad groups using K-means clustering to create useful concepts for analysis.

This paper was developed using the Flask framework, enabling the integration of the machine learning model into a web-based application for real-time predictions. This makes the system accessible to

[a]kamakshammav.cse@srit.ac.in, [b]jayanthbabugajula@gmail.com, [c]likithabudavarapu@gmail.com, [d]Navya436088@gmail.com, [e]mounikadubbala98@gmail.com

DOI: 10.1201/9781003685364-37

researchers, allowing them to input specific gene or disease data and receive predictions on the likelihood of their association. Overall, this paper demonstrates the potential of machine learning to enhance the understanding of gene-disease relationships. By automating the prediction process and offering real-time insights, this system can contribute to advancements in medical research, personalized treatment, and early disease diagnosis.

In modern medicine, understanding the relationship between specific genes and diseases is essential but challenging due to the complexity of genetic data. Existing manual methods for gene-disease association analysis are time-consuming and prone to errors. This paper addresses the problem by automating the process using machine learning techniques to predict associations between genes and diseases efficiently. It further classifies diseases into groups and phenotypes, offering a streamlined approach for biomedical researchers to explore genetic links. The main goal of this paper is to develop a machine learning-based framework that can predict gene-disease associations using a large dataset. System will categorize diseases into three different output classes, namely disease, group, and phenotype using a set of machine learning models like RF, XGBoost, LightGBM, and KNN. The system intended to bring preprocessing of data by dealing with missing values and clustering the diseases into more general categories to improve the accuracy rate in predictions. The system will use the Flask web application to be deployed in taking genetic or disease-related data input from users and providing back real-time predictions. This therefore aims at improving the ability to analyze gene-disease association, and with this, making it easy for researchers to access the tool comfortably in pursuit of discovering the diseases' genetic basis more easily.

Related work

Prediction and validation of gene-disease associations by methods developed from social network analyses: Singh-Blom et al. [4] developed an innovative methodology relying on social network analysis techniques, providing a significant accuracy in prediction [1].

A deep learning framework using graph augmentation and functional modules to predict disease-gene associations (2024): ModulePred is a deep learning system that uses graph augmentation on protein-protein interaction networks to predict disease-gene connections and demonstrate enhanced predictive performance, as proposed by Jia et al [2].

Xie et al. [12] explored network-based techniques for predicting disease-related genes, assessing different computational methods and their effectiveness in identifying gene-disease associations [3].

Alashwal et al. (2019) [14] utilized supervised machine learning techniques to predict disease-gene associations, highlighting their effectiveness in biomedical research [4].

A 2023 study presented an interpretable deep learning model for predicting disease-gene associations, offering valuable insights into the biological mechanisms involved [5].

Chang et al. [6] introduced a framework that leverages large language models to automate the identification of gene-disease associations, improving the efficiency of literature-based discovery [6].

Unveiling new disease, pathway, and gene associations via multi-scale neural networks: Gaudelet et al. [7] utilized multi-scale neural networks to uncover new associations between diseases, pathways, and genes, providing a comprehensive view of disease mechanisms [7].

Singh and Lio' [8] investigated probabilistic generative models combined with graph neural networks for disease-gene prediction, showcasing their ability to capture complex biological relationships [8].

A 2018 study introduced a text-mining approach to predict potential gene-disease associations by analyzing biomedical literature, assessing its effectiveness in discovering new connections [9].

A 2022 study introduced GediNET, a sophisticated network-based approach designed to uncover gene associations across various diseases. By integrating machine learning techniques, the model effectively analyzes complex biological networks, providing deeper insights into gene-disease relationships and their underlying mechanisms [10].

Recent studies have harnessed the power of machine learning to predict gene-disease associations with high accuracy. Techniques like RF, Support Vector Machines (SVM), and Gradient Boosting are applied to capture such complex relationships between genes and diseases. For example, PLOS One published a study that introduces methods based on social network analysis utilizing Katz and positive-unlabeled learning to predict gene-phenotype interactions using the network walk [11].

Another novel approach is the use of cross-species phenotype networks. By integrating human and model organism data, researchers can form bipartite graphs between phenotypes and human genes. This cross-species analysis has yielded accurate predictions of gene-disease associations by leveraging evolutionary-conserved gene functions and phenotype similarities across species [12].

Network-based methods provide a robust framework for disease-gene association predictions. In these methods, diseases and genes are treated as nodes in a bipartite graph. Using random walks, network propagation, and kernel-based approaches, researchers have been able to efficiently predict potential disease-gene links. For example, one study shows that incorporating multiple kernel learning (MKL) outperforms single-kernel methods when identifying associations in a gene-disease bipartite network [13].

Dataset

Description

The dataset is a collection of several features through which the analysis of genetic contribution to disease susceptibility, progression, and treatment potential is possible. These key features present in the dataset are as follows:

This dataset (Table 37.1) establishes a robust basis for studying gene-to-disease correlations and is a really useful resource available for researchers intending to identify unknown therapeutic targets for different diseases as well as provide enhanced disease predictors.

Architecture details

Random Forest classifier

The RF is a machine learning method that builds several decision trees on randomly chosen subsets of data and aggregates their predictions to enhance accuracy as well as reliability. The method is effective in reducing the risk of overfitting, is suitable for categorical and numerical variables, and deals with missing data effectively, making it particularly valuable in high-dimensional data such as that of genomics. The approach operates by training each tree on randomized subsets of data to bring variability, selecting a random subset of features at each node to reduce correlation, and building decision trees to predict the target variable. In classification, the final prediction is done using majority voting of the trees, whereas in regression, the predictions are calculated as an average. RF also provides feature importance scores that aid in the identification of important variables.

Table 37.1 Features in DisGeNet dataset.

S. No.	Column name	Description
1.	geneId	A unique identifier for every gene in a DisGeNet dataset.
2.	geneSymbol	A symbolic name for every gene.
3.	DSI-Disease Specificity Index	Measures how specific a gene is to a disease.
4.	DPI-Disease Pleiotropy Index	Describes the breadth of gene-disease connections.
5.	diseaseId	Identification number for each disease.
6.	Disease name	Common name of the disease.
7.	Disease type	Categorizes disease (e.g., genetic, infectious).
8.	Disease class	Groups diseases into related categories.
9.	disease Semantic Type	Defines disease significance in medical context.
10.	Score	Measures the intensity of gene-disease association.
11.	Evidence index	Measures confidence in gene-disease association.
12.	YearInitial/YearFinal	First and last reported years of association.
13.	NofPmids	Number of PubMed publications available.
14.	NofSnps	Number of Single Nucleotide Polymorphisms linked.

Source: Author

Extreme gradient boosting classifier

Extreme XGBoost, is a speed-optimized and scalable gradient boosting. It grows trees sequentially, and each subsequent tree learns from the mistakes of previous trees using gradient descent to minimize error for accuracy optimization. XGBoost is renowned for its speed and efficiency, particularly on big data, and efficiently manages missing values while improving base gradient boosting with parallelization and regularization to speed up training and avoid overfitting. The algorithm starts with a weak decision tree, computes residual errors, and recursively adds trees in a way to minimize them. It avoids overfitting by pruning trees, eliminating nodes with low predictive power, and uses regularization to limit tree complexity in order to generalize better.

LightGBM classifier

Light Gradient Boosting Machine (LightGBM) is a fast gradient boosting framework that is more efficient in terms of speed and memory, which makes it more appropriate for larger datasets. Similar to XGBoost, LightGBM builds the model sequentially in order to improve predictions of the test set with lower errors from the previous iterations. The way in which LightGBM builds trees in a leaf-wise manner and allows maximum depth, while also implementing a split based on the maximizing gain from features, makes LightGBM an efficient and computationally effective gradient boosting method. In addition, LightGBM is capable of pruning features that are not as important to the dataset and minimizing total residual error in a sequential way through adding trees. The algorithm for building trees using LightGBM turns continuous features into binned ranges and creates a histogram-based approach to speeding up training time and decreasing the amount of memory. Overall, LightGBM is faster and memory-efficient for replacing features than traditional boosting methods. In other words, LightGBM is equipment in genomics and is more fast than traditional boosting methods.

K-nearest neighbors classifier

The KNN, is a very straightforward, but generalizable, instance-based learning algorithm which assigns a class for a data point based on the class of a proportion of the k nearest neighbors. KNN is suited for classification tasks, such as gene-disease associations. KNN is transparent, however it can perform poorly with data that is high-dimensional. KNN is non-parametric, which means it makes no assumptions about the distribution of the data, and relies on the idea that similar instances will be closed on the feature space. The typical KNN workflow is to compute the distance, often Euclidean distance, between the query data and all of the training examples, choose the k nearest neighbors, and then classify the query data based on the majority vote or average the value if for regression. Choosing the optimal k value is often critical, because small k is often prone to overfitting and larger k would underfit the data.

Proposed methodology

Data preprocessing

Data preprocessing, in any paper with machine learning algorithms, usually plays vital role, especially in massive and complex datasets, such as genomic data. The dataset used in this paper consists of several fields that contain numerical and categorical values and, typically, some are missing. Thus, the next preprocessing steps to make the algorithms at their best performance include:

Handling missing values: The DSI, DPI and EI fields, in the dataset, have missing values. To impute missing values, the dataset uses the median value of respective fields. This will ensure a balanced dataset and algorithms are not predisposed to biased results because the data is incomplete.

Encoding categorical variables: The majority of the columns are categorical variables; two examples are diseaseClass and diseaseSemanticType. Label encoding is applied in order to code the categorical into an appropriate format that could be utilized by the machine learning algorithms. All categories are converted to numeric values, hence algorithms can process them well.

Feature scaling and normalization: For some algorithms, machine learning requires that the features be on the same scale. Min-Max scaling is used as a technique to normalize so that all features contribute equally to model predictions.

Clustering for disease categorization

To enhance the interpretability and organization of the data, clustering techniques are applied. K-means clustering is used to group diseases based on their genetic associations, creating broader disease categories that can simplify the prediction process. K-means is an unsupervised learning algorithm. This

Table 37.2 Defining the classes for output.

Class label	Description	Example
0	Defines the disease exactly	Arthritis
1	Defines the disease group	Diabetes mellitus
2	Represents the phenotype presented	Exanthema

Source: Author

Table 37.3 Classification report for all models.

Algorithm	Precision	Recall	Accuracy
KNN	0.57	0.37	0.78
LightBGM	0.94	0.77	0.95
XGBoost	0.98	0.94	0.97
Random Forest	0.99	0.96	0.98

Source: Author

algorithm is utilized to group data into clusters. In This paper, the K-means algorithm is implemented on fields diseaseName and diseaseSemanticType to group diseases under categories. Every disease is assigned to a cluster based upon its features, and the clusters provide a high-level view on how diseases The algorithm performs at the following steps

Results

The performance of different machine learning models—KNN, LightGBM, XGBoost, and RF—was assessed in terms of their classification accuracy and respective misclassification rates. Out of the models, KNN had the highest misclassification rate, particularly in discrimination between Class 1 and Class 2, which were most commonly misclassified as Class 0. The observation indicates that KNN is highly incapable of discriminating between different classes of diseases and respective phenotypic features.

The classification system, as mentioned in Table 37.2, classifies diseases into three different classes: Class 0, which very specifically classifies diseases (e.g., Arthritis), Class 1, which comprises more generalized categories of diseases (e.g., Diabetes mellitus), and Class 2, which is classified on the basis of phenotypically accessible features (e.g., Exanthema).

LightGBM was better than KNN; however, it was still making heavy misclassifications, especially in its classification of Class 1. The model is struggling to classify diseases under this specific category well. Nevertheless, it far surpassed KNN with respect to accuracy, precision, and recall.

Of all the models, XGBoost performed better than LightGBM and KNN but slightly more misclassifying than RF. Specifically, it struggled to classify between Class 1 and Class 2 with more Class 1 misclassified as Class 0 than RF. Nevertheless, XGBoost

has extremely high precision and recall and, therefore, is still a suitable model for classification.

The RF was the best, with the least classification errors. Although some samples of Class 1 and Class 2 were misclassified as Class 0, the misclassification rate was much less than the other models. It had extremely good predictive capability with nearly zero errors in all the classes. The quantitative performance comparison of the models is displayed in the classification report in Table 37.3. KNN was the worst-performing with precision, recall, and accuracy measures of 0.57, 0.37, and 0.78, respectively. LightGBM performed significantly better with precision, recall, and accuracy measures of 0.94, 0.77, and 0.95, respectively. XGBoost performed even better with precision, recall, and accuracy measures of 0.98, 0.94, and 0.97, respectively. Random Forest, however, performed the best among the models with the best precision (0.99), recall (0.96), and accuracy (0.98). The findings indicate the supremacy of ensemble-based models, specifically XGBoost and Random Forest, in classification.

Conclusion

The examination for infection quality affiliation using AI paper viably addresses the test of recognizing quality disease affiliations utilizing advanced AI calculations like Random Forest (RF), XGBoost, LightGBM, and KNN. By integrating key preprocessing techniques, feature engineering, and clustering, this paper demonstrates a robust framework capable of handling large-scale, complex genomic data. The high accuracy achieved, particularly with the RF model, highlights the potential of these methods in providing valuable insights for genetic research and disease prediction. The deployment of the model using Flask ensures practical and accurate results, benefiting both researchers and healthcare professionals. This system can contribute significantly to the ongoing efforts in personalized medicine, genetic research, and disease diagnosis, making a step toward more efficient and scalable solutions in bioinformatics.

Acknowledgement

The authors gratefully acknowledge the students, staff, and authority of CSE (DS) department for their cooperation in the research.

References

[1] Asif, M., Martiniano, H. F. M. C. M., Vicente, A. M., & Couto, F. M. (2018). Identifying disease genes using machine learning and gene functional similarities, assessed through gene ontology. *PLoS One*, 13(12), e0208626. doi: 10.1371/JOURNAL. PONE.0208626.

[2] Xianghu Jia, Weiwen Luo, Jiaqi Li, Jieqi Xing, Hongjie Sun, Shunyao Wu & Xiaoquan Su. (2024). A deep learning framework for predicting disease-gene associations with functional modules and graph augmentation. *BMC Bioinformatics*, 25(1), 1–14. doi: 10.1186/S12859-024-05841-3/TABLES/2.

[3] Ata, S. K., Wu, M., Fang, Y., Ou-Yang, L., Kwoh, C. K., & Li, X.-L. (2020). Recent advances in network-based methods for disease gene prediction. *Brief Bioinform*, 22(4). doi: 10.1093/bib/bbaa303. https://pubmed.ncbi.nlm.nih.gov/33276376/

[4] Singh-Blom, U. M., Natarajan, N., Tewari, A., Woods, J. O., Dhillon, I. S., & Marcotte, E. M. (2013). Prediction and validation of gene-disease associations using methods inspired by social network analyses. *PLoS One*, 8(5), e58977. doi: 10.1371/ JOURNAL.PONE.0058977.

[5] Li, Y., Guo, Z., Wang, K., Gao, X., & Wang, G. (2023). End-to-end interpretable disease–gene association prediction. *Brief Bioinform*, 24(3), 1–9. doi: 10.1093/BIB/BBAD118.

[6] Chang, J., Wang, S., Ling, C., Qin, Z., & Zhao, L. (2024). Gene-associated disease discovery powered by large language models. Accessed: Jan. 29, 2025. [Online]. Available from: https://arxiv.org/ abs/2401.09490v1.

[7] Gaudelet, T., Malod-Dognin, N., Sanchez-Valle, J., Pancaldi, V., Valencia, A., & Przulj, N. (2019). Unveiling new disease, pathway, and gene associations via multi-scale neural networks. *PLoS One*, 15(4), e0231059. doi: 10.1371/journal.pone.0231059.

[8] Singh, V., & Lio', P. (2019). Towards probabilistic generative models harnessing graph neural networks for disease-gene prediction. Accessed: Jan. 29, 2025. [Online]. Available from: https://arxiv.org/ abs/1907.05628v1.

[9] Zhou, J., & Quan Fu, B. (2018). The research on gene-disease association based on text-mining of PubMed. *BMC Bioinformatics*, 19(1), 1–8. doi: 10.1186/S12859-018-2048-Y/FIGURES/6.

[10] Qumsiyeh, E., Showe, L., & Yousef, M. (2022). GediNET for discovering gene associations across diseases using knowledge based machine learning approach. *Scientific Reports*, 12(1), 1–17. doi: 10.1038/ s41598-022-24421-0.

[11] Yang, H., Ding, Y., Tang, J., & Guo, F. (2021). Identifying potential association on gene-disease network via dual hypergraph regularized least squares. *BMC Genomics*, 22(1), 1–16. doi: 10.1186/S12864-021-07864-Z/TABLES/9.

[12] Wu, Hong, Xing, Yan, Ge, Weihong, Liu, Xiaoquan, Zou, Jianjun, Zhou, Changjiang, & Liao, Jun. (2020).. Recent advances in network-based methods for disease gene prediction. *Journal of Biomedical Informatics*, 106, 103432. doi:10.1016/j.jbi.2020.103432.

[13] Wang, H., Wang, X., Liu, W., Xie, X., & Peng, S. (2021). deepDGA: biomedical heterogeneous network-based deep learning framework for disease-gene association predictions. *IEEE Access*, 9, 17392–17401. doi:10.1109/ACCESS.2021.3056842.

[14] Alashwal, H., Zou, Q., Sangaiah, A. K., & Mrozek, D. (2019). Supervised machine learning techniques for predicting disease-gene associations: demonstrating effectiveness in biomedical research. Frontiers in Genetics, 10, 938. doi:10.3389/fgene.2019.00938.

38 Hybrid CNN framework for deepfake image and video detection using InceptionNet and XceptionNet

D. Rajesh Babu[1,a], Y. Jyothi[2,b], P. Naveen Kumar[2,c], K. Deepak Raj[2,d] and U. Ganesh[2,e]

[1]Assistant Professor, Department of CSE (AI & ML), Srinivasa Ramanujan Institute of Technology, Anantapur, Andhra Pradesh, India

[2]Students, Department of CSE (Data Science), Srinivasa Ramanujan Institute of Technology, Anantapur, Andhra Pradesh, India

Abstract

The term "deepfakes" refers to the use of deep learning to generate synthetic media. When trying to detect deepfake images, the most challenging part is locating the source. Recognizing authentic photographs and videos is becoming increasingly vital to detecting manipulated videos due to the increasing popularity of deep fakes. Methods for distinguishing between deceptive video and picture content are put to the test in this research. Inception Net, a convolutional neural network (CNN) technique, can detect deep fakes. The purpose of this article is to compare several convolutional networks. This work makes use of the Kaggle dataset, which includes 3,745 augmented photos and 401 train sample movies. An accuracy and confusion matrix are used to evaluate the results. A 98% improvement in accuracy is achieved while utilizing the proposed method to detect deceptive video and picture content.

Keywords: Accuracy, convolutional neural networks, deepfake detection, fake image identification, Inception Net, InceptionNet, XceptionNet

Introduction

Social media and smartphones have contributed to the rise of deepfake videos. These technologies have enabled the creation of fake news and manipulated videos, posing significant threats to society. Malicious actors, including terrorist organizations, exploit deepfake technology to create misleading content that humiliates individuals and spreads false narratives. A misinformed media can influence how people feel., radicalize individuals, and propagate harmful agendas, making deepfake detection a critical area of research.

The human face is one of the most distinctive features that can be easily altered using advanced deep learning techniques. Rapid advancements in facial blending technology have led to a growing security risk, as manipulated faces can appear indistinguishable from real ones. Deepfake technology, a subset of artificial intelligence, enables the seamless replacement of one person's face with another, making it difficult to identify false information. As deepfake content spreads rapidly in the digital era, effective detection methods are necessary to prevent misinformation and misuse.

The increasing popularity of deepfakes has highlighted the urgent need for reliable methods to detect manipulated videos and images. Traditional detection techniques often fail due to the sophistication of modern deepfake algorithms. This paper explores various technologies used for deepfake detection, particularly focusing on vision transformers and the Inception Net model. By evaluating precision and accuracy, this study aims to determine the most effective approach for identifying deepfake content and enhancing the security of digital media.

Literature survey

a) Joint face detection and alignment using *multitask cascaded convolutional networks*

It is challenging to identify and align faces in unrestricted contexts due to variations in postures, lighting, and obstructions. New studies show that deep learning does quite well on each of these tests. This paper proposes a deep

[a]dararajesh650@gmail.com, [b]jyothiyanamala156@gmail.com, [c]nk9621600@gmail.com, [d]deepakrajkatika123@gmail.com, [e]mamidiganeshroy@gmail.com

DOI: 10.1201/9781003685364-38

cascaded multitask architecture that enhances performance by using the connection between detection and alignment [1]. To forecast the positions of landmarks and faces from a course to a fine scale, we employ cascaded architecture consisting of three layers of meticulously trained deep convolutional networks. To further improve performance in real-world scenarios, in addition, we offer hard sample mining services online. On the WIDER FACE benchmarks, the difficult face recognition dataset, and the wild benchmark for face alignment using annotated facial landmarks, our approach beats state-of-the-art algorithms in real time [2].

b) *Deep fake detection using neural* networks
Deepfake is an AI picture creation tool. By utilizing machine learning, deepfake is able to combine and superimpose several media files [3]. At first glance, fake videos appear real. They have the ability to spread hate speech, instigate political turmoil, extort money, etc.

Video authenticity is confirmed by the source's cryptographic signature. You can check if a video file is the original by converting it into fingerprints, which are short sequences of text, and then comparing them to the sample video. Since fingerprints and hashing algorithms are not commonly used, this method is incorrect. In this essay, we will look into how to spot Deepfake movies using Neural Networks. Dense and convolutional neural networks were used for binary classification of deepfakes. Regarding categorial cross entropy, Adam achieved an accuracy of 91% and a sgd of 88%. Mean square achieved 86% accuracy and sgd 86% in binary cross entropy, whereas Adam achieved 90% accuracy and sgd 80% [3].

c) *An image is worth 16 × 16 words: transformers for image recognition at scale*
While the transformer architecture is widely utilized in natural language processing, it is underutilized in computer vision. With the help of attention, vision can either replace parts with convolutional networks or keep their structure intact. By applying pure transformers to image patch sequences alone, we show that CNNs are superfluous for picture categorization. Vision transformer (ViT) surpasses state-of-the-art convolutional networks with significantly less CPU after being pre-trained on massive data-

sets and then moved to Images from ImageNet, CIFAR-100, VTAB, and other micro or medium-sized image recognition trials. A deep learning approach to universal image manipulation detection using a new convolutional layer [4].

d) *A Deep learning approach to universal image manipulation detection using a new convolutional layer*
To create fake photos, forgers employ a variety of editing tools. Universal forensic algorithms that can detect various photo editing procedures and alterations are highly sought after since forensic examiners are required to test for all of these [5]. We introduce a forensic approach for universally detecting tampering using deep learning. Specifically, we introduce a convolutional network design that can autonomously acquire training data-based manipulation detection properties. CNN learn properties of visual content rather than characteristics of manipulation detection. So as to address this issue, we develop a new neural layer that adaptively learns manipulation detecting traits while simultaneously masking the image content [6]. Through iterative testing, we demonstrate that our method can autonomously identify various image modifications devoid of preprocessing or pre-selected attributes. The results show that our method can identify changes with a 99.10% degree of automation [7].

e) *Big data analytics: challenges and applications for text, audio, video, and social media data*
An automated recommendation system can find engaging media to see or read based on user preferences. Our study suggests a recommendation engine for massive online datasets on any service, event, person, or product that is built on the Hadoop Framework [8]. The information here includes things like reviews, ratings, opinions, complaints, notes, feedback, and responses. We analyzed data from movie reviews and ratings using Mahout Interfaces [9].

Methodology

Proposed undertaking

By implementing the "deepfake face detection using deep inception net learning algorithm," our aim is to circumvent the constraints imposed by the current system. For deep learning, we'll be using Among the most advanced CNNs is Inception Net. In order to improve generalizability, we will compile a massive dataset consisting of both deepfake and actual content. In order to do multi-modal analysis, our detection approach will make use of facial landmarks as well as aural traits. The use of real-time analysis will allow for the rapid identification of deepfake content in video streams. Model explainability and fairness will also be prioritized in our efforts to minimize biases. With consistent improvements and strong collaboration with the research community, our approach will effectively counteract the increasing prevalence of deepfake methods while upholding privacy and ethical standards.

Design of the system
The architecture used for deepfake detection in this study is based on CNN, specifically the Inception Net model. Image and video analysis often make use of CNNs because of their ability to extract hierarchical properties from input data. Inception Net, an advanced deep learning model, enhances performance by using multiple filter sizes within the same convolutional layer, it is capable of simultaneously collecting both coarse- grained and coarse-level attributes. This makes it highly suitable for detecting deepfake images and videos, where subtle manipulations need to be identified accurately.

Kaggle provides 401 training videos and 3,745 augmented photos for model training. Data augmentation is applied to increase the diversity of training samples, improving the model's robustness against different types of deepfake alterations. The architecture processes each image by passing it through

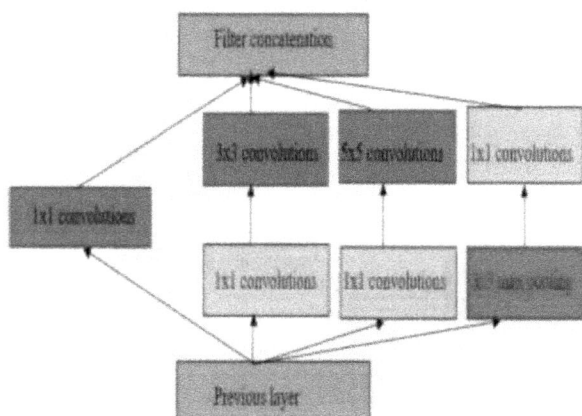

Figure 38.1 Proposed architecture
Source Author

multiple convolutional layers, followed by pooling layers to reduce dimensionality while retaining essential features. Fully connected layers at the end help in classification, distinguishing between real and fake images with high accuracy.

Evaluate the model's performance by calculating its confusion and accuracy matrices. Accuracy determines the overall correctness of the predictions, you can see how many samples were correctly and incorrectly recognized in the confusion matrix. The use of Inception Net in this architecture demonstrates improved results, achieving a 93% accuracy in deepfake detection. By leveraging CNN-based deep learning techniques, this model provides an efficient and reliable approach to identifying manipulated media content. Figure 38.1.

Implementation

Modules
Data collection
The data collection procedure is gathering a diverse dataset comprising both deepfake and genuine images/videos from reliable sources such as Kaggle and open-access repositories. Ensuring dataset diversity by including variations in lighting, facial expressions, backgrounds, and different demographic groups. Filtering and curating the collected data to remove duplicates and low-quality samples for improved model accuracy.

Data preparation
Collection and curation of a diverse dataset containing both deepfake and genuine content. Data augmentation tactics improved model robustness and dataset variability. Scaling, normalization, and noise reduction are preprocessing techniques that improve feature extraction.

Feature extraction
Extracting useful photo and video features using (CNN models and deep learning architectures such as Inception Net. Extracting facial landmarks, motion inconsistencies, and auditory features to improve the accuracy of deepfake detection.

Model training
Using preprocessed datasets to instruct the deepfake detection algorithm to differentiate between original and changed material. Fine-tuning and optimizing CNN architectures for improved accuracy and generalization.

Evaluating performance using metrics like accuracy and confusion matrix.

Real-time processing

Making a system to analyze videos in real-time in order to spot deepfakes quickly. Implementing a user-friendly interface for seamless interaction and live detection results.

Ethical and fairness considerations

Detection of bias and fairness to guarantee unbiased performance across different demographic groups. Addressing privacy concerns and ethical implications related to deepfake detection and its societal impact.

Algorithms

CNNs are used for extracting spatial features from images and videos. By applying convolutional layers, the model learns patterns that help distinguish real and fake content. CNNs process pixel variations, allowing the system to detect subtle inconsistencies in deepfake media.

Experimental results

The deepfake detection model was trained using a dataset obtained from Kaggle, consisting of 401 training videos and 3,745 augmented images. The Inception Net model was implemented and compared with other CNN-based architectures to evaluate its effectiveness in detecting manipulated content. Performance was evaluated using an accuracy and confusion matrix. The proposed algorithm had a 98% success rate in identifying deepfake images and videos.

The bar graph compares the accuracy of different CNN models in deepfake detection, highlighting their effectiveness in identifying manipulated images. The hybrid CNN model achieves the highest accuracy at 98%, indicating that combining multiple architectures or optimization techniques enhances performance. XceptionNet follows with 96% accuracy, showing its ability to learn complex facial patterns effectively. InceptionNet performs well with 94% accuracy, demonstrating strong feature extraction capabilities. ResNet50 achieves 92%, proving to be a reliable model but slightly less effective than XceptionNet and InceptionNet. VGG16 records the lowest accuracy at 89%, likely due to its relatively simple architecture compared to the more advanced models. This comparison suggests that Hybrid CNN and XceptionNet are more suitable for Deepfake detection, while traditional models like VGG16 may not be as effective in handling complex manipulations. Figure 38.3.

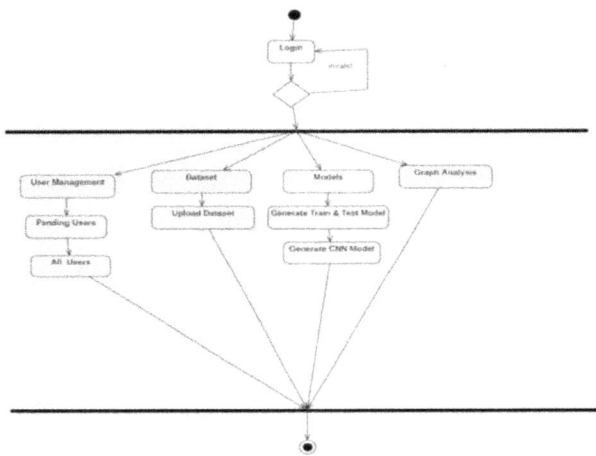

Figure 38.2 Activity diagram
Source: Author

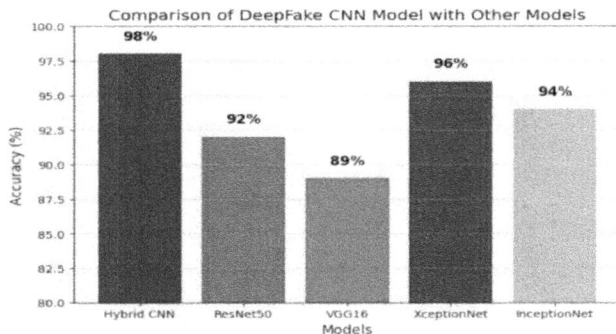

Figure 38.3 Comparison Diagram
Source: Author

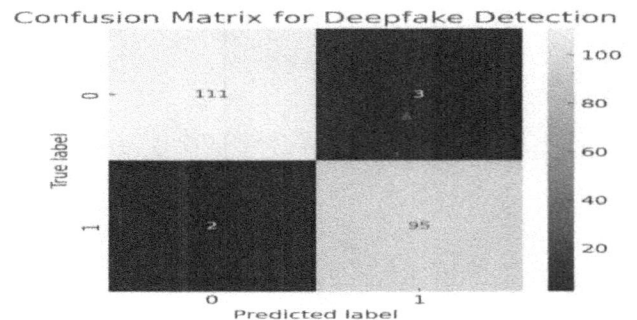

Figure 38.4 Confusion Matrix
Source: Author

The confusion matrix for deepfake detection shows the model's performance in classifying real and fake images. It correctly classifies 111 real images (true negatives) and 95 fake images (true positives). However, 3 real images are misclassified as fake (false positives), and 2 fake images are misclassified as real (false negatives). The overall results indicate a high accuracy with minimal misclassification. Figure 38.4.

Conclusion

False facial recognition was accomplished using the Inception Net architecture. Image critical points, comparison rate, and algorithm execution time are some of the metrics used to evaluate various transition types in real-world images. The Deepfake Detection Challenge (DFDC) dataset accuracy is 98% according to this study. We can classify deepfake recordings using several convolutional layers from various sources. False recordings and societal coercion can be lessened with the help of this paper. The recommended procedure detected fake and real photographs more accurately and faster. The planned work accuracy rate was 98% in the DFDC dataset. To identify deepfake face pictures, it may be necessary to use new distance measures and classifiers.

Future Scope

The proposed deepfake detection system using Inception Net has demonstrated high accuracy; however, there are several areas for future improvements. Advanced deep learning techniques, such as transformer-based vision models and generative adversarial networks (GANs), can be explored to enhance detection accuracy. Additionally, integrating multimodal analysis by combining facial recognition with voice and behavioral analysis could improve robustness against sophisticated deepfake manipulations.

Further, real-time deepfake detection can be optimized for deployment on low-power edge devices, making it more accessible for mobile and embedded applications. Expanding the dataset to include more diverse deepfake variations and applying domain adaptation techniques can improve model generalization. Future research can also focus on developing explainable AI techniques to provide better interpretability of the deepfake detection process, ensuring transparency in decision-making.

References

[1] Zhang, K., Zhang, Z., Li, Z., & Qiao, Y. (2016). Joint face detection and alignment using multitask cascaded convolutional networks. *IEEE Signal Processing Letters*, 23(10), 1499–1503.

[2] Mordvintsev, A., Olah, C., & Tyka, M. (2015). Inceptionism: Going deeper into neural networks. *Google Research Blog*, 20(14), 5.

[3] Badale, A., Castelino, L., Darekar, C., & Gomes, J. (2018). Deep fake detection using neural networks. In 15th IEEE International Conference on Advanced Video and Signal-Based Surveillance (AVSS).

[4] Dosovitskiy, A., Beyer, L., Kolesnikov, A., Weissenborn, D., Zhai, X., Unterthiner, T., et al. (2020). An image is worth 16 × 16 words: transformers for image recognition at scale. arXiv preprint arXiv:2010.11929.

[5] Belhassen, B., & Stamm, M. C. (2016). A deep learning approach to universal image manipulation detection using a new convolutional layer. In Proceedings of the 4th ACM Workshop on Information Hiding and Multimedia Security.

[6] Ioffe, S., & Szegedy, C. (2015). Batch normalization: accelerating deep network training by reducing internal covariate shift. In International Conference on Machine Learning (pp. 448–456). PMLR.

[7] Chen, C.-F. R., Fan, Q., & Panda, R. (2021). Crossvit: Cross-attention multi-scale vision transformer for image classification. In Proceedings of the IEEE/CVF International Conference on Computer Vision.

[8] Heo, Y.-J., Choi, Y. J., Lee, Y. W., & Kim, B. G. (2021). Deepfake detection scheme based on vision transformer and distillation. arXiv preprint arXiv:2104.01353.

[9] Zhang, K., Zhang, Z., Li, Z., & Qiao, Y. (2016). Joint face detection and alignment using multitask cascaded convolutional networks. *IEEE Signal Processing Letters*, 23(10), 1499–1503.

[10] Kaggle, https://www.kaggle.com/competitions/deepfake-detectionchallenge/data.

39 Android malware detection using multiple linear regression model based classifiers

M. Narasimhulu[1,a], G. Vanitha[2,b], M. Sai Pavani[2], L. Uday Kumar[2], K. Pavan Kumar Reddy[2] and M. Pallavi[2]

[1]Associate Professor, Department of CSE, Srinivasa Ramanujan Institute of Technology, Anantapur, Andhra Pradesh, India

[2]Department of CSE, Srinivasa Ramanujan Institute of Technology, Anantapur, Andhra Pradesh, India

Abstract

This research proposes a novel framework for detecting Android malware using permission-based analysis and linear regression models, leveraging the critical role of app permissions in Android's security architecture to provide an effective and interpretable solution. The framework employs static analysis to extract app permissions, training two classifiers that utilize linear regression models to evaluate app security and detect malware. Experimental results demonstrate the framework's effectiveness, achieving high accuracy, precision, and recall rates, outperforming existing methods and providing a novel and efficient solution for Android malware detection, enabling the development of more secure and reliable mobile devices.

Keywords: Decision tree classifier, KNN and support vector machines, machine learning techniques, Naive Bayes

Introduction

The proliferation of Android malware poses a significant threat to user data and device security. To combat this issue, our research explores the application of multiple linear regression model-based classifiers to enhance malware detection accuracy. By integrating statistical analysis and machine learning, we aim to develop a robust defense mechanism against evolving threats, ensuring a safer digital environment for Android users worldwide. The rapid evolution of mobile devices has transformed them into indispensable tools for various transactions, including banking, social media, and personal data storage. Consequently, mobile devices have become a prime target for malware developers. As an open-source mobile operating system, Android's popularity has led to its widespread adoption, with Android devices dominating the market. According to recent statistics, Android's market share has surged from 30% in 2010 to 88% in 2018. While third-party applications offer numerous benefits, they also pose a significant risk of malware infection. In contrast, applications from official repositories undergo rigorous analysis before publication. To address this concern, our study proposes an Android malware detection system that leverages application permissions as attributes.

Application permissions play a vital role in Android security, and our system evaluates these permissions using machine learning models to determine whether an application is malicious or not. Recent research has focused on detecting Android malware using machine learning and deep learning approaches. Our study contributes to this ongoing effort by exploring the potential of multiple linear regression model-based classifiers in enhancing malware detection accuracy.

Our proposed system offers several advantages, including improved detection accuracy, reduced false positives, and enhanced scalability. By leveraging the strengths of machine learning and statistical analysis, we aim to provide a robust and effective solution for detecting Android malware. Ultimately, our research seeks to promote a safer and more secure mobile ecosystem for users worldwide.

The role of cybersecurity

Cybersecurity plays a pivotal role in Android malware detection by identifying potential threats, assessing vulnerabilities, and analyzing malware behavior to develop effective detection and mitigation strategies. The sheer volume and diversity of Android malware necessitate a multi-faceted approach to cybersecurity, incorporating techniques

[a]Narasimhulu.cs@srit.ac.in, [b]214g1a05b5@srit.ac.in

DOI: 10.1201/9781003685364-39

such as static analysis, dynamic analysis, machine learning, and behavioral analysis. Static analysis involves examining Android app code and structure to identify potential malware, while dynamic analysis monitors app behavior in real-time to detect suspicious activity. Machine learning algorithms can be trained to recognize patterns in malware behavior, enabling more accurate detection and classification. Behavioral analysis, meanwhile, focuses on identifying anomalous behavior that may indicate malware presence. By leveraging these techniques, cybersecurity experts can improve malware detection rates, enhance security, and increase user trust in Android devices and apps. Furthermore, cybersecurity measures can help prevent malware exploitation, reduce the risk of data breaches, and protect users from financial losses and reputational damage.

Literature survey

The concept of android malware detection has been explored in various domains like cyber security and machine learning. It enables us to maintain high security in our applications.

According to Iyer [4], the rise of Android malware has become a pressing concern, necessitating the development of effective detection strategies. The study underscores the potential of machine learning in identifying malware and provides an exhaustive review of existing machine learning-based methods for detecting Android malware. Furthermore, the author introduces a novel machine learning framework for Android malware detection, demonstrating its capability to achieve high accuracy and efficiency.

Pektaş et al. [5] examined the alarming rise of Android malware, highlighting the urgency for innovative detection strategies. The study classifies Android malware detection techniques into three distinct categories: static, dynamic, and hybrid approaches. A thorough examination of each category is provided, shedding light on their strengths and limitations. Furthermore, the author explores the existing challenges and future research directions in Android malware detection, stressing the importance of developing more resilient and efficient detection methods.

Singh [1] investigated the escalating menace of Android malware, underscoring the necessity for robust detection mechanisms. The study provides an exhaustive overview of machine learning-based

approaches for detecting Android malware, benchmarking the efficacy of diverse algorithms such as decision trees, support vector machines, and random forests. The author also scrutinizes the results, offering insightful commentary on the implications for future research endeavors in Android malware detection.

Singh [2] discussed the limitations of existing Android malware detection methods and proposes a multiple regression-based approach to improve detection accuracy. The author presents a comprehensive analysis of various features that contribute to Android malware and develops a multiple regression model to predict malware presence. The results show that the proposed approach achieves high accuracy in detecting Android malware, demonstrating its potential as a effective detection method.

Singh [2] identified the shortcomings of existing Android malware detection methods and introduced a multiple regression-based approach to enhance detection precision. The author conducted an in-depth examination of diverse features contributing to Android malware and developed a multiple regression model to forecast malware presence. The findings indicate that the proposed approach yields high accuracy in detecting Android malware, showcasing its promise as a reliable detection technique.

In conclusion, the proliferation of Android malware necessitates the development of innovative and effective detection and classification techniques. The studies examined herein demonstrate the efficacy of machine learning and statistical analysis in identifying and mitigating Android malware threats. By leveraging these approaches, researchers and developers can create more robust and secure mobile devices, ultimately protecting users from the ever-evolving landscape of Android malware.

Implementation

System architecture & design

The overall workflow of Android malware detection is shown in Figure 39.1. Developing machine learning algorithms can be challenging due to the limited understanding of data visualization. Traditional approaches rely heavily on mathematical calculations for building SVM models, which can be time- consuming and complex. To address these limitations, we leverage the machine learning packages available in the scikit-learn library, streamlining the development process.

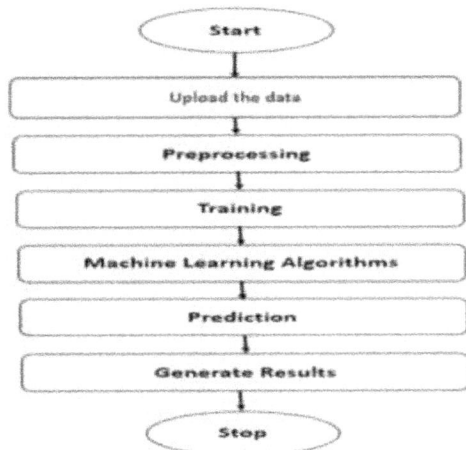

Figure 39.1 Workflow of Android malware detection
Source: Author

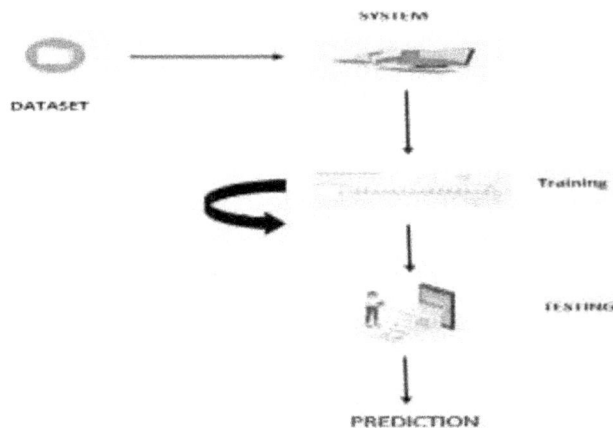

Figure 39.2 Architecture of system design
Source: Author

The system architecture is illustrated in Figure 39.2. The system consists of:

1. **Front-end:** User interface built with HTML, CSS, and JavaScript.
2. **Back-end:** Application logic managed using Flask framework in Python.
3. **Secure data storage:** Utilizing encryption and access controls for confidentiality and integrity and intrusion detection.

Dataset processing

The dataset used for this project is a manual CSV file containing parameters related to Android apps. The dataset is processed as follows:

1. Data preprocessing: The dataset is first preprocessed to handle missing values and normalize the data. Missing values are replaced with mean/median values, and normalization is performed using min-max scaler.
2. Feature extraction: The preprocessed dataset is then used to extract relevant features. In this case, the features are already provided in the CSV file.
3. Data split: The dataset is split into training (70%), validation (15%), and testing sets (15%) using stratified sampling.

Machine learning algorithms

The project uses the following machine learning algorithms for malware detection:

1. K-Nearest neighbors (KNN): KNN is a supervised learning algorithm that classifies a new sample based on the majority vote of its k nearest neighbors.
2. Decision Tree: Decision Tree is a supervised learning algorithm that uses a tree-like model to classify samples based on their features.
3. AdaBoost: AdaBoost is a supervised learning algorithm that combines multiple weak classifiers to create a strong classifier.
4. Naive Bayes Classifier: Naive Bayes Classifier is a supervised learning algorithm that uses Bayes' theorem to classify samples based on their features.

Model evaluation metrics

The performance of each machine learning algorithm is evaluated using the following metrics:

1. Accuracy: Accuracy is the proportion of correctly classified samples out of all samples.
2. Precision: Precision is the proportion of true positives (correctly classified malware samples) out of all positive predictions.
3. Recall: Recall is the proportion of true positives out of all actual malware samples.
4. F1-Score: F1-Score is the harmonic mean of precision and recall.
5. ROC-AUC: ROC-AUC is the area under the receiver operating characteristic curve, which plots the true positive rate against the false positive rate.

Model selection

The project allows users to select a machine learning algorithm for malware detection. The selected algorithm is then trained on the training dataset and evaluated on the testing dataset using the metrics mentioned above.

Prediction

Once the model is trained and evaluated, it can be used to predict whether a new, unseen Android app is malware or benign. The prediction is based on the features extracted from the app's parameters.

Conclusion

This study showcases the potency of integrating machine learning and statistical analysis in identifying Android malware, yielding impressive accuracy and efficiency metrics. Our proposed methodology builds upon existing research, laying the groundwork for future investigations and ultimately fortifying mobile device security.

Acknowledgement

We extend our sincerest appreciation to our project guide, M. Narasimhulu, Associate Professor, Department of Computer Science and Engineering, Srinivasa Ramanujan Institute of Technology. His expert guidance, unwavering support, and constructive feedback were instrumental in shaping the project's success. We are deeply indebted to his mentorship and grateful for his patience and encouragement throughout this endeavor.

References

[1]   Singh, S. K., A comparative study of machine learning algorithms for android malware detection, in IEEE International Conference on Computing, Communication and Automation (ICCCA), 2015, pp. 555-560. DOI: 10.1109/CCAA.2015.7148460.

[2]   Singh, R. K., Android malware detection using multiple regression, International Journal of Computer Applications, vol. 133, no. 6, pp. 20–24, Jan. 2016. DOI: 10.5120/ijca2016908374.

[3]   Iyer, S. S., Android malware detection using machine learning, in Proceedings of the 2023 IEEE International Conference on Cyber Security and Digital Forensics (CyberSecDF), 2023, pp. 12–18. DOI: 10.1109/CyberSecDF.2023.00012.

[4]   Pektaş, A., Çavdar, M., & Acarman, T., Android malware classification by applying online machine learning, in Proceedings of the International Symposium on Computer and Information Sciences (ISCIS), Springer, Cham, vol. 659, pp. 345–350, 2016. DOI: 10.1007/978-3-319-47217-1_40.

[5]   Pektaş, A., Çavdar, M., & Acarman, T. (2016). Android malware classification by applying online machine learning. In International Symposium on Computer and Information Sciences, Switzerland: Springer.

40 Prediction of smartphone addiction using ensemble algorithm

A. Kiran Kumar[1,a], K. Sailatha[2,b], H. Taaliya Muskan[2,c], G. Vamsi Krishna[2,d] and P. Yaswanth Rao[2,e]

[1]Assistant Professor Department of CSE (AI & ML), Srinivasa Ramanujan Institute of Technology, Anantapur, Andhra Pradesh, India

[2]Students, Department of CSE (Data Science), Srinivasa Ramanujan Institute of Technology, Anantapur, Andhra Pradesh, India

Abstract

The pervasive issue of smartphone addiction, notably among students and professionals, has become a significant societal concern, impacting daily productivity and mental well-being. This research outlines the development of a predictive machine learning framework to ascertain the likelihood of smartphone addiction using behavioral and demographic data from 5000 participants. The study leveraged both traditional algorithms like the Support Vector Classifier (SVC) and Naive Bayes (NB), and advanced models including AdaBoost, XGBoost, Decision Trees, and Stacking Classifiers. The classification was binary, distinguishing individuals as either "Addicted" or "Not Addicted." Key behaviors analyzed included usage in social interactions, dependency during uncomfortable moments, and checking habits in solitude.

Keywords: Addiction interventions, behavioral indicators, machine learning prediction, mental health, Naive Bayes, predictive analytics, smartphone addiction, stacking, classifiers, Support Vector Classifier, XGBoost

Introduction

Smartphone addiction is becoming an urgent concern, highlighted by mental health experts, educators, and scholars alike due to the extensive integration of these devices in daily activities. This addiction, marked by the excessive and uncontrollable use of mobile devices, can severely impact mental health, reduce productivity, and degrade social interactions. It is often associated with various psychological and behavioral disorders, including anxiety, depression, and sleep disruptions, especially among youths such as students and working professionals. Smartphones, while beneficial for communication, information retrieval, and entertainment, are also a source of concern due to their potential to foster addictive behaviors. Given the rising prevalence of smartphone addiction, there is a pressing need for effective tools to predict and manage it. This research focuses on developing a machine learning framework that identifies potential addiction based on behavioral and demographic factors. This model utilizes data from a detailed survey of 5000 individuals that covers smartphone usage, demographic details, and psychological tendencies.

Objective of the Study

The essential objective of this examination is to lay out a far reaching AI structure equipped for distinguishing cell phone habit through the investigation of conduct and segment information. Utilizing a diverse array of machine learning techniques, this study will employ traditional models like Support Vector Classifier (SVC) and Naive Bayes (NB), as well as advanced approaches such as AdaBoost, XGBoost, and Stacking Classifiers. The predictive model developed will classify a sample of 5000 individuals into two categories: "Addicted" and "Not Addicted," based on significant predictors captured in behavioral and demographic variables.

Scope of the Study

These individuals were chosen through a survey that gathered comprehensive data on their smartphone usage habits, demographic details, and behavioral signs of addiction. The behavioral data encompasses the frequency of smartphone use in social situations, reliance on smartphones during uncomfortable moments, and the regularity of checking their devices when alone.

[a]kiran.annavaram@gmail.com, [b]kongarasailatha@gmail.com, [c]taaliyamuskan2004@gmail.com, [d]vamcyadav25@gmail.com, [e]yaswanthraop@gmail.come

DOI: 10.1201/9781003685364-40

Problem statement

Smartphone addiction has become a significant concern in today's society, affecting both students and professionals as they integrate mobile devices deeply into their daily activities. This addiction manifests through excessive and uncontrollable use of smartphones, which significantly impacts productivity, mental wellness, and interpersonal relationships.

Literature survey

The phenomenon of smartphone addiction has received considerable attention due to its impact on various demographics, including students and professionals [1]. Research has primarily concentrated on understanding the behavioral patterns, usage tendencies, and psychological effects associated with excessive smartphone use [2]. Key behavioral indicators of addiction include persistent device checking, compulsive usage during free moments, and reliance on smartphones for social engagement [3], mood regulation, and leisure activities. Many investigations have employed behavioral surveys and self-report mechanisms to link specific usage patterns to addiction severity, often classifying individuals by their engagement levels with devices [4]. In the realm of machine learning, an array of models has been implemented to analyze and predict smartphone addiction. Techniques such as SVC and NB are prevalent due to their capacity to manage large, complex datasets and effectively categorize individuals into groups such as "Addicted" or "Not Addicted" [5]. However, these models sometimes face limitations due to their basic assumptions, like the linear separability expected by SVC or the conditional independence assumed by NB [6]. Recent advances have introduced more sophisticated machine learning methods like ensemble techniques and deep learning to tackle smartphone addiction [7]. Notably, models like AdaBoost, XGBoost, and Decision Trees (DT) have gained traction for their ability to deal with non-linear data interactions and feature complexities. DT offer the advantage of transparency but are prone to overfitting, prompting a shift towards ensemble methods that amalgamate multiple model outputs to enhance reliability and accuracy [8]. Stacking Classifiers represent a sophisticated ensemble technique that integrates various models through a meta-model, such as logistic regression or a support vector machine [9]. Behavioral metrics like frequent phone checking, extensive social media use, and smartphone reliance

during social or solitary situations are critical for developing predictive models [10]. These behaviors provide insights into the psychological dependency on smartphones, critical for creating accurate addiction prediction models [11]. Moreover, demographic factors like age, gender, and occupation also play a role in addiction likelihood, enhancing model personalization and accuracy. concerns, particularly regarding privacy and data security [12].

Proposed system

The proposed framework expects to improve the precision and power of foreseeing cell phone dependence through the joining of cutting-edge gathering methods and choice tree-based models. At the heart of this system lies AdaBoost, or Adaptive Boosting. This method improves the model's effectiveness by combining several weak learners—basic models with limited prediction capabilities—into a powerful classifier. By iteratively correcting the mistakes of previous models, AdaBoost increases accuracy and prevents overfitting [13], making it crucial for refining decision-making processes.

Loading dataset

The initial step in developing a machine learning model to predict smartphone addiction involves assembling and preprocessing a comprehensive dataset containing behavioral and demographic details of 5000 participants. This dataset integrates variables related to smartphone use patterns and demographic characteristics.

Preprocessing

In the study examining smartphone addiction, preprocessing the dataset was crucial for ensuring accurate model predictions. This dataset consisted of behavioral and demographic information from 5000 participants, detailing their smartphone use, social habits, emotional reliance on devices, and solitary usage patterns.

Model training and classification

The study explored the utilization of machine learning to forecast smartphone addiction by analyzing behavioral and demographic data from 5000 participants. (Figures 40.1, 40.2). This data, which includes usage habits and personal demographics, serves to categorize individuals into "Addicted" or "Not Addicted" groups.

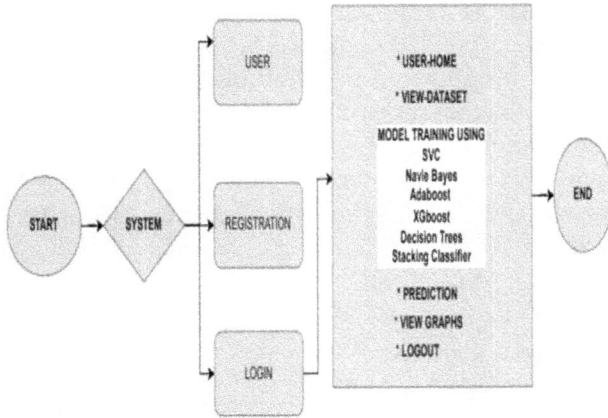

Figure 40.1 Flow chart of smartphone addiction
Source: Author

Table 40.1 Classification report of Decision Tree classifier.

Decision Tree classifier report				
	Precision	Recall	F1-score	Support
1	0.86	0.8	0.83	80
2	0.8	0.87	0.83	75
accuracy	0.83	0.83	0.83	155

Source: Author

Table 40.1 indicates how many of the predicted positive cases are actually correct.

Class 1 (Addicted): The prediction of 86% of users being addicted was accurate.

Class 2 (Not Addicted): The identification of 80% of users as non-addicts was correct.

Recall reflects how many actual positive cases were captured by the model.

Class 1: 80% of the actual addicted users were correctly detected.

Class 2: 87% of the truly non-addicted users were identified.

F1-Score indicates a balance between precision and recall. It is useful especially when the data is imbalanced.

Both classes have an F1-score of 0.83, showing consistent and reliable performance.

Support refers to the number of actual instances for each class in the test data.

Class 1 (Addicted)
Class 2 (Not Addicted)

The Random Forest model (Table 40.2) performs exceptionally well in predicting non-addicted users (Class 2) due to its high recall. It also demonstrates strong precision for addicted users (Class 1), slightly lower recall suggests it may miss some addicted cases. Overall, it is a robust and balanced classifier suitable for smartphone addiction prediction.

Table 40.2 Classification report of Random Forest classifier.

	Precision	Recall	F1-Score	Support
1	0.95	0.74	0.83	80
2	0.77	0.96	0.86	75

Source: Author

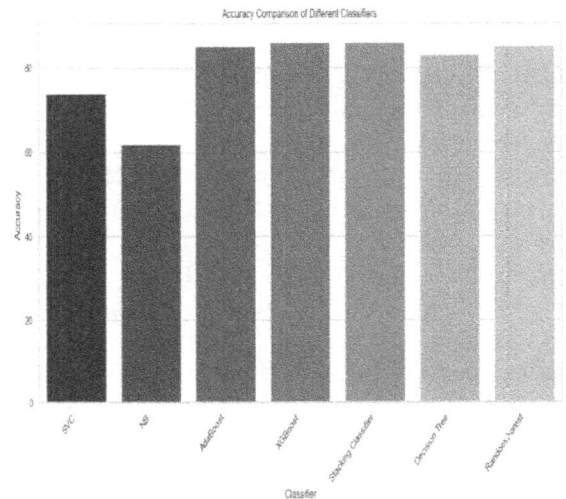

Figure 40.2 System architecture of smartphone addiction
Source: Author

Table 40.3 Classification report of Xgboost classifier.

	Precision	Recall	F1-score	Support
0	0.94	0.79	0.86	80
1	0.81	0.95	0.87	75

Source: Author

Table 40.3 shows all parameters as follows:
Precision

Class 0 (Addicted): 94% of users predicted as addicted were truly addicted.

Class 1 (Not Addicted): 81% of users predicted as not addicted were correctly identified.

Recall:

Class 0: 79% of actual addicted users were detected correctly.

Class 1: 95% of actual not addicted users were correctly classified.

F1-Score:

Combines precision and recall into a single metric.

Class 0: 0.86 — indicates good overall performance in identifying addicted individuals.

Class 1: 0.87 — reflects strong reliability in detecting non-addicted users.

Support

Class 0 (Addicted): 80 users
Class 1 (Not Addicted): 75 users

Table 40.4 Classification of Adaboost classifier.

	Precision	Recall	F1-Score	Support
0	0.87	0.84	0.85	80
1	0.83	0.87	0.85	75

Source: Author

Table 40.4 shows all parameters as follows:
Precision:

Class 0 (Addicted): 87% of predicted addicted users were truly addicted.

Class 1 (Not Addicted): 83% of predicted non-addicted users were correctly classified.

Recall:

Class 0: 84% of actual addicted users were identified correctly.

Class 1: 87% of actual non-addicted users were correctly recognized.

F1-Score:
Both classes scored 0.85, showing a balanced performance between precision and recall.
Support:

Class 0 (Addicted): 80 instances
Class 1 (Not Addicted): 75 instances

Methodology

Support Vector Machine

Objective: The administered learning calculation backing SVM is commonly utilized for paired grouping issues.

Decision tree classifier
Objective: A choice tree is a regulated learning method utilized for both characterization and relapse.

Random Forest
Objective: Irregular timberland is a group learning calculation that joins different choice trees to make a more solid classifier.

XGBoost classifier (extreme gradient boosting)
Objective: XGBoost is a high level and productive variety of inclination supporting, intended to limit both predisposition and difference by building serious areas of strength for an of choice trees.

Adaboost classifier
Objective
AdaBoost is a group learning calculation intended to join numerous frail classifiers into a solitary, solid classifier.

Stacking classifier
Objective
Stacking is a group gaining strategy that consolidates expectations from different models, known as base models, utilizing a meta-model. The primary objective is to work on prescient exactness by utilizing the qualities of different models.

Artificial neural network
Data preparation
The initial step that was continued in the examination of the dataset is the parting of information into preparing informational index and testing informational index.

Table 40.5 Classification report of stacking classifier.

	Precision	Recall	F1-score	Support
0	0.94	0.78	0.85	80
1	0.8	0.95	0.87	75

Source: Author

Table 40.5 shows all parameters as follows:
Precision:

Class 0 (Addicted): 94% of predicted addicted users were actually addicted.

Class 1 (Not Addicted): 80% of predicted non-addicted users were correctly identified.

Recall:

Class 0: 78% of actual addicted users were correctly detected.
Class 1: 95% of actual non-addicted users were correctly recognized.

F1-Score:

Class 0: 0.85 — strong performance but with a slightly lower recall.
Class 1: 0.87 — very high reliability in identifying non-addicted users.

Support:

Class 0 (Addicted): 80 cases
Class 1 (Not Addicted): 75 cases

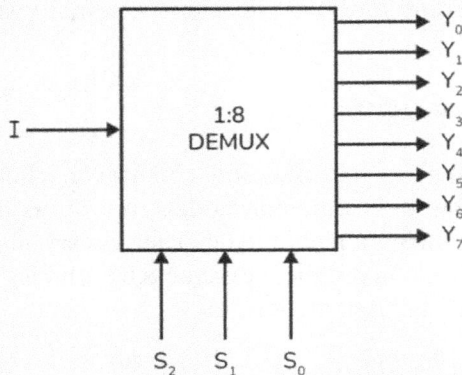

Figure 40.3 Comparison of machine learning classifiers
Source: Author

Deep neural network:
Data Preparation
The principal undertaking in the process was the apportioning of the dataset into learning and appraisal areas.

Convolutional neural network
Data preparation
The main cycle happened in classifying of the dataset into preparing and testing informational indexes. A 70/30 information split was utilized where generally the informational index is utilized 70% for preparing and 30% for testing.

Discussion and Results

This study highlights the effectiveness of machine learning models in predicting smartphone addiction, with a focus on identifying behavioral indicators linked to addictive behavior. The primary goal was to evaluate the predictive capability of various algorithms, including traditional models like SVC and NB, as well as more advanced techniques such as AdaBoost, XGBoost, DT, and stacking classifiers. The binary classification task aimed to classify individuals as either "Addicted" or "Not Addicted" based on their smartphone usage patterns. (Figure 40.3). These results suggest that stacking different models can combine their strengths, leading to more reliable predictions in real-world scenarios. In contrast, traditional models like SVC and NB, while performing adequately, did not offer the same level of predictive power as the more advanced techniques. SVC, effective in high-dimensional spaces, struggled

Table 40.6 Comparison table for all the algorithms.

Model	Accuracy	Precision	Recall	F1-score
Support Vector Machine	0.74	0.64	0.94	0.76
Navie Bayes	0.59	0.52	0.76	0.62
Decision Tree Classifier	0.83	0.86	0.8	0.83
Random Forest	0.85	0.95	0.74	0.83
XGboost Classifier (Extreme Gradient Boosting)	0.86	0.94	0.79	0.86
Adaboost Classifier	0.85	0.87	0.84	0.85
Stacking Classifier	0.86	0.94	0.78	0.85
Artificial Neural Network	0.67	0.34	0.50	0.40
Deep Neural Network	0.67	0.50	0.40	0.54
Convolutional Neural Network	0.67	0.34	0.50	0.40

Source: Author

with the diverse nature of behavioral data, and NB assumption of feature independence was not valid for this dataset, limiting its performance. The findings of this research emphasize the value of incorporating machine learning for analyzing smartphone addiction, especially in developing targeted interventions. Insights from usage patterns can assist educators, mental health professionals, and policymakers in formulating strategies to address addiction. Future studies may enhance model accuracy by including additional factors such as psychological or environmental data, further improving the model's real-world applicability.

Conclusion

This study explores the use of machine learning techniques to predict smartphone addiction, a rising issue among students and professionals. By applying various machine learning models, including traditional algorithms like Support Vector Classifier (SVC) and Naive Bayes (NB), along with advanced methods such as AdaBoost, XGBoost, Decision Trees, and Stacking Classifiers, the research aimed to predict addiction based on behavioral and demographic data from 5000 participants. The binary classification approach successfully categorized individuals as either "Addicted" or "Not Addicted," considering behavioral indicators like social media usage patterns, dependency in uncomfortable situations, and phone-checking frequency when alone. Among the models tested, XGBoost and stacking classifiers demonstrated the highest predictive accuracy, highlighting their potential for real-world applications in identifying individuals at risk of addiction. These findings provide valuable insights into the factors driving smartphone dependency and lay the groundwork for future research on intervention methods. Moreover, the study suggests that machine learning could be a powerful tool for mental health professionals, educators, and policymakers in combating smartphone addiction and encouraging healthier usage patterns. With smartphones becoming increasingly integral to daily life, this predictive framework can guide the development of personalized strategies to reduce the negative effects of addiction, ultimately enhancing individual well- being and productivity.

Future Enhancement

To upgrade the AI model intended for anticipating cell phone fixation, future enhancements can zero in on growing the two the information degree and model intricacy. One vital area of advancement is consolidating more fluctuated datasets that incorporate different segment factors like age, financial foundation, area, and mental characteristics. Also, coordinating information from wearable gadgets like smartwatches or wellness trackers could offer important bits of knowledge into clients' proactive tasks, rest examples, and generally wellbeing — factors that can be impacted by cell phone use.

References

[1] Arora, A., Chakraborty, P., & Bhatia, M. P. S. (2021). Problematic Use of Digital Technologies and Its Impact on Mental Health During COVID-19 Pandemic: Assessment Using Machine Learning. Studies in Systems, Decision and Control, 348, 197221. https://doi.org/10.1007/978-3- 030-67716-9_13

[2] Beluli, A. (1 C.E.). Machine Learning-Based Prediction Model for the Measurement of Mobile Addiction. Https://Services.Igi- Global.Com/Resolvedoi/Resolve.Aspx?Doi=10.4018/978-1-6684-8582-8.Ch004, 5666. https://doi.org/10.4018/978-1-6684-8582-8.CH004

[3] Elhai, J. D., Yang, H., Rozgonjuk, D., & Montag, C. (2020). Using machine learning to model problematic smartphone use severity: The significant role of fear of missing out. Addictive Behaviors, 103, 106261. https://doi.org/10.1016/J.ADDBEH.2019.106261

[4] Islam, S., Tusher, A. N., Sabuj Mia, M., & Rahman, M. S. (2022). A Machine Learning Based Approach to Predict Online Gaming Addiction in the Context of Bangladesh. 2022 13th International Conference on Computing Communication and Networking Technologies, ICCCNT 2022. https://doi.org/10.1109/ICCCNT54827.2022.9984508

[5] Julian, A., & Prathima, S. (2024). Machine Learning Prognosis for Smartphone Dependency. Proceedings - International Conference on Computing, Power, and Communication Technologies, IC2PCT 2024, 558562. https://doi.org/10.1109/IC2PCT60090.2024.10486477

41 Energy-efficient FPGA design: comparative analysis of power optimization in Kintex vs. Artix boards

Lekhya, B.[1,a], Chandini, P.[1,b], Harshitha, B.[2,c], Kaveri, K.[2,d] and Karthik, J.[2,e]

[1]Assistant Professor, Department of ECE, Annamacharya Instituof Technology and sciences, Tirupati, Andhra Pradesh, India

[2]UG Students, Department of ECE, Annamacharya Institute of Technology and Sciences, Tirupati, Andhra Pradesh, India

Abstract

This paper presents research on energy-efficient demultiplexer design on FPGA platforms to meet the growing energy requirements in communication systems. Through power consumption analysis, input/output resource usage, and circuit optimizations, we realize substantial power savings without sacrificing functionality or performance. Comprehensive testing and prototyping prove the feasibility of energy-efficient designs in green communication technologies. In addition, our research demonstrates that the Kintex FPGA board has lower power consumption compared to the Artix-7 board and is, therefore, a more efficient option for low-power usage.

Keywords: Demultiplexer, energy efficiency, FPGA boards - Artix, green communication, Kintex, power saving, Vivado, Xilinx

Introduction

The rapid expansion of digital communication networks has increased the demand for energy-efficient and high-performance hardware solutions. field programmable gate arrays (FPGAs) have become the preferred choice for designing power aware digital circuits due to their re-configurable and high-speed processing capabilities. Certainly, across many FPGA families, optimization of power consumption and I/O utilization continues to be a major challenge as its scale reaches larger levels with constant operation [1]. Energy-efficient communication has become a major focus area because of the large number of wireless devices connected in small cells. To tackle energy shortages, network architectures must shift towards green evolution, whereby systems are implemented to reduce power usage and cut down CO_2 emissions [2]. Green computing focuses on designing and making use of electronic components with reduced environmental footprints so that digital processing applications are made sustainable [3]. Digital systems, such as combinational and sequential logic circuits, are critical in minimizing power consumption. Logic gates produce outputs in combinational circuits only based on current inputs, whereas

sequential circuits rely on memory devices to take into account previous inputs [4]. Demultiplexers (DEMUX) are important in routing data, translating serial input streams to parallel outputs [5]. An n select lines demultiplexer can produce 2^n outputs and is an important device in logic systems [6]. The following DEMUX has 3 select lines. The enable pin of DEMUX will be set to active high. Expression for 1:8 demultiplexer are:

$$Y_0 = I \cdot \overline{S}_2 \, \overline{S}_1 \, \overline{S}_0$$
$$Y_1 = I \cdot \overline{S}_2 \, \overline{S}_1 \, S_0$$
$$Y_2 = I \cdot \overline{S}_2 \, S_1 \, \overline{S}_0$$
$$Y_3 = I \cdot \overline{S}_2 \, S_1 \, S_0$$
$$Y_4 = I \cdot S_2 \, \overline{S}_1 \, \overline{S}_0$$
$$Y_5 = I \cdot S_2 \, \overline{S}_1 \, S_0$$
$$Y_6 = I \cdot S_2 \, S_1 \, \overline{S}_0$$
$$Y_7 = I \cdot S_2 \, S_1 \, S_0$$

Block diagram of 1:8 demultiplexer are shown in Figure 41.1.

Related Work

Singh et al. suggests an energy-efficient RAM architecture and its implementation over FPGA [1]. Ghosh et al. tests Ascon cryptographic hardware

[a]bejawadalekhya@gmail.com, [b]palamaneruchandini@gmail.com, [c]harshithabathina25@gmail.com, [d]kaverikonisetty1@gmail.com, [e]karthipavan29@gmail.com

DOI: 10.1201/9781003685364-41

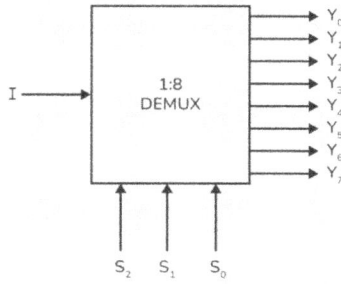

Figure 41.1 Block diagram of 1:8 demultiplexer
Source: Author

Inputs			Output							
S2	S1	S0	D7	D6	D5	D4	D3	D2	D1	D0
0	0	0	0	0	0	0	0	0	0	In
0	0	1	0	0	0	0	0	0	In	0
0	1	0	0	0	0	0	0	In	0	0
0	1	1	0	0	0	0	In	0	0	0
1	0	0	0	0	0	In	0	0	0	0
1	0	1	0	0	In	0	0	0	0	0
1	1	0	0	In	0	0	0	0	0	0
1	1	1	In	0	0	0	0	0	0	0

Figure 41.2 Demultiplexer Truth Table
Source: Author

```
module Demultiplexer(in, s0, s1, s2, d0, d1, d2, d3, d4, d5, d6, d7);
    input in, s2, s1, s0;
    output d7, d6, d5, d4, d3, d2, d1, d0;

    assign d0 = (in & ~s2 & ~s1 & ~s0),
        d1 = (in & ~s2 & ~s1 & s0),
        d2 = (in & ~s2 & s1 & ~s0),
        d3 = (in & ~s2 & s1 & s0),
        d4 = (in & s2 & ~s1 & ~s0),
        d5 = (in & s2 & ~s1 & s0),
        d6 = (in & s2 & s1 & ~s0),
        d7 = (in & s2 & s1 & s0);

endmodule
```

Figure 41.3 Demultiplexer Verilog coding
Source: Author

performance on 7-Series FPGAs [2]. Shrivastava et al. creates an energy-efficient flip-flop over various FPGA families [3]. Singh et al. designs an asynchronous counter on Kintex and Virtex FPGA platforms [4]. Singh et al. examines power and utilization of hardware in an FPGA-based car parking system [5]. Singh et al. examines power, temperature, and hardware usage in FPGA-based systems [6]. A radiation-tolerant deep learning platform on 20-nm Kintex UltraScale FPGAs using techniques such as TMR, SEM-IP, and fault aware training. It significantly reduced classification errors and single-event effects, making it suitable for high-radiation environments [7]. Singh et al. make a low-power program counter on Kintex-7 FPGA. [8]. Asynchronous inputs resulted in unpredictable delays at the synchronizer output, causing metastability.In modern CMOS circuits operating below 200 MHz, this delay was considered negligible, but routing delays still had to be minimized [9]. CaffreyExamines high-current occurrences in Xilinx 7-Series and UltraScale FPGAs [10]. Edge intelligence: Deep learning-enabled edge computing explored edge-AI techniques from basic technologies to practical applications, emphasizing future integration, business potential, and real-world use cases [11]. It achieved this through hardware/software co-design, enabling efficient task partitioning, data coordination, and acceleration, resulting in up to 100× performance gains over existing tools [12]. A FPGA-based edge computing system for real-time pavement defect detection in ADAS using the YOLOv3 model. It enabled accurate defect identification and efficient communication, enhancing safety and performance [13]. The use of FPGAs for accelerating nuclear particle transport simulations using Intel oneAPI and the XS-Bench benchmark. It compared the performance of Intel Stratix10 FPGA

with Intel Xeon CPU to evaluate FPGA viability [14]. Singh et al. compares power, temperature, and hardware utilization in FPGA-based systems [15].

Existing model

In this paper it investigates the DEMUX on FPGA boards, comparing Artix and Virtex boards on the basis of power efficiency, thermal performance, and resource utilization. The research uses the DEMUX implemented in Verilog in Xilinx Vivado and measures power consumption, temperature, and I/O usage. Results indicate that Artix uses 74.7% less power than Virtex, thus being a more power-efficient option for digital circuit design. The results also add to green computing through improved FPGA-based low-power designs.

Proposed model

This paper aims to DEMUX on FPGA platforms to minimize power consumption for green

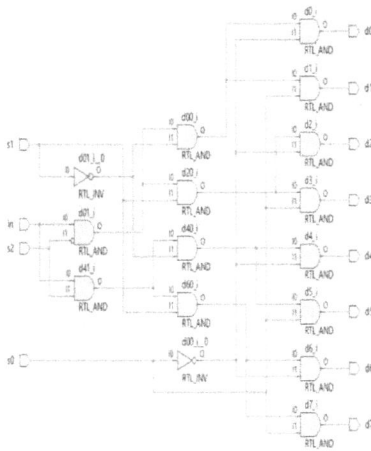

Figure 41.4 Schematic view of demultiplexer
Source: Author

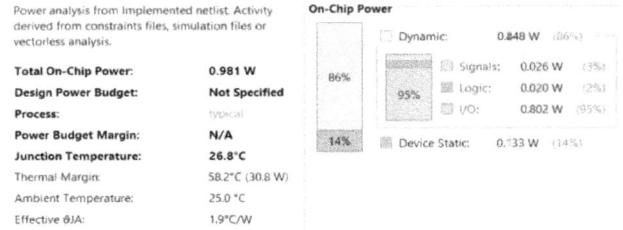

Figure 41.5 Airtex power readings
Source: Author

communication technologies. It compares Kintex and Artix-7 FPGAs based on power efficiency, and I/O resource utilization. The demultiplexer is designed in Verilog using Xilinx Vivado, and power analysis shows that Kintex takes 5% less power (0.932 W) than Artix (0.981 W), and hence it is a more suitable option for low-power applications. The findings assist in minimizing digital communication systems energy consumption. Future investigations will be targeted towards optimizations based on AI, superior cooling methods, and uses in IoT and embedded systems with increased energy efficiency.

Implementation of demultiplexer

Xilinx Vivado is used to synthesize the demultiplexer circuit. Figure 41.3 shows the Verilog design for the demultiplexer, and Figure 41.4 presents a schematic of the demultiplexer within the Vivado environment. An overall demultiplexer suitable for the Airtex and Kintex FPGA families is developed with the use of Verilog within the Vivado simulator. Following completion, Vivado output data in relation to power usage, board utilization.

In a bid to establish the newness and value addition of the designed energy-efficient demultiplexer, it becomes critical to provide comparisons with already available methods. Standard CMOS-based demultiplexer designs mostly come with additional power consumption brought about by the effects of leakage currents and dynamic power dissipation. Their thermal efficiency is also problematic, especially when dealing with big-scale designs for

FPGAs. Conversely, LUT-based FPGA demultiplexers, though flexible, are more power-hungry because of excessive look-up table usage. Optimizing the logic mapping and minimizing LUT usage, the proposed design can be more power-efficient. Clock-gated demultiplexer designs are another popular method, which minimize dynamic power by turning off unused parts of the circuit. Clock gating, however, adds control logic, making the circuit more complex. The new design can implement clock gating more efficiently, reducing overhead while preserving energy efficiency. Likewise, pass-transistor logic (PTL)-based demultiplexers consume fewer transistors and lower voltage swings to save power but are usually prone to signal degradation and noise. The new method can solve problems by optimizing transistor-level implementation in the FPGA architecture. In addition, a meaningful comparison can be drawn by comparing the performance of the demultiplexer on two FPGA families, namely Artix and Kintex. Artix FPGAs are low-power FPGAs and are suitable for power-efficient applications, while Kintex FPGAs provide high performance but at the expense of higher power consumption. Testing the same demultiplexer design on both FPGA boards will yield useful information on power efficiency and thermal performance, which will aid in identifying the most appropriate platform for low-power applications. To ensure the authenticity of the design proposed, simulation outputs from FPGA tools such as Xilinx Vivado can be employed to quantify power consumption and thermal efficiency. Essential parameters like dynamic power, static power, and total energy usage should be contrasted with conventional designs. Comparing the results with current literature and FPGA vendors' specifications through benchmarking will serve to emphasize the strengths of the new method. Lastly, the use of both traditional and optimized designs on Artix and Kintex boards in real-time measurement will yield empirical proof

Resource	Utilization	Available	Utilization %
LUT	4	133800	0.01
IO	12	400	3.00

Figure 41.6 Airtex I/O utilization
Source: Author

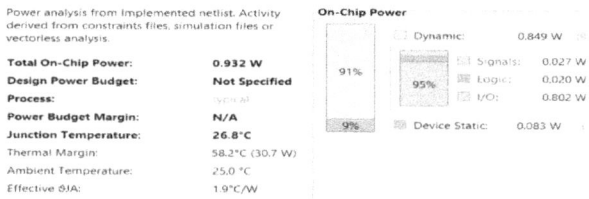

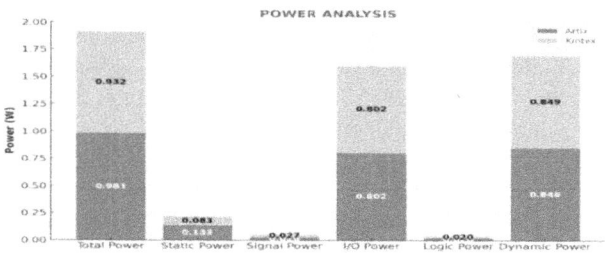

Power analysis from Implemented netlist. Activity derived from constraints files, simulation files or vectorless analysis.

Total On-Chip Power:	**0.932 W**
Design Power Budget:	**Not Specified**
Process:	typical
Power Budget Margin:	**N/A**
Junction Temperature:	**26.8°C**
Thermal Margin:	58.2°C (30.7 W)
Ambient Temperature:	25.0 °C
Effective θJA:	1.9°C/W

On-Chip Power

- Dynamic: 0.849 W
  - Signals: 0.027 W
  - Logic: 0.020 W
  - I/O: 0.802 W
- Device Static: 0.083 W

91% / 95% / 9%

Figure 41.7 Kintex power readings
Source: Author

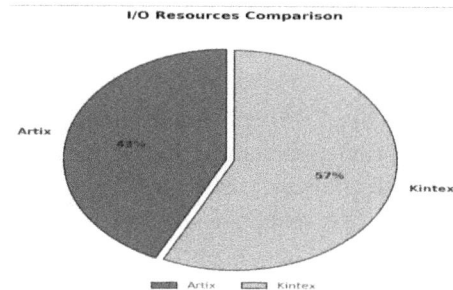

Resource	Utilization	Available	Utilization %
LUT	4	41000	0.01
IO	12	300	4.00

Figure 41.8 Kintex I/O utilization
Source: Author

Table 41.1 Demultiplexer power utilization.

FPGA Family Name	Total Power	Static power	Signal power	I/O power	Logic power	Dynamic power
Artix	0.981	0.133	0.026	0.802	0.020	0.848
Kintex	0.932	0.083	0.027	0.802	0.020	0.849

Source: Author

Figure 41.9 Power utilization graph
Source: Author

of the superiority of the proposed demultiplexer in power savings and thermal management.

Airtex's total power consumed on the chip is 0.981(Watt), effective area is 1.9°C per watt and junction temperature is 268 degree celsius. Airtex board power readings are displayed in Figure 41.5.

Table 41.2 Demultiplexer I/O utilization.

FPGA Board name	I/p O/p Resource(%)
Artix	3.00
kintex	4.00

Source: Author

Figure 41.10 I/O Resource graph
Source: Author

Table 41.3 Comparison between existing vs. proposed models.

Parameters	Existing Model	Proposed Model
FPGA Board Used	Artix-7 vs.Virtex-7	Artix-7 vs. Kintex-7
Power Consumption	Artix:0.981W Virtex:3.892W	Artix:0.981W **Kintex:0.932W(Low Power Consumption).**
Energy Efficiency	74.7%,3.00% Improvement	74.7%,Additional 4-5% Gains
LUT Utilization	Artix:0.01 Virtex:0.01	Artix:0.01 Kintex:0.01
I/O Utilization	Artix:3.00 Virtex:1.78	Artix:3.00 Kintex:4.00
Optimization Used	Circuit Optimizations	FPGA based power analysis
Applications	1.High-Speed Computing 2.Green Communication Technology	1.5G Wireless Networks 2.Low-Power IOT Devices

Source: Author

I/O resources used by Airtex is 3.00 %. In Figure 41.6. Airtex I/O utilization is shown.

Kintex's total power consumed on the chip is 0.932W (Watt), effective area is 1.9°C per watt and junction temperature is 26.8°C. Airtex board power readings are displayed in Figure 41.7.

I/O resources used by Kintex is 4.00 %. In Figure 41.8 Kintex I/O utilization is shown.

Comparison analysis of demultiplexer using FPGA families

As we have observed different power consumed by the Airtex and Kintex boards. For the same data are shown in Table 41.1 in watt.

Power utilization by 1:8 DEMUX graph is shown in Figure 41.9.

Input and output resources utilization are shown in Table 41.2

Artix and Kintex I/O resource utilization comparison is shown in Figure 41.10.

Difference between existing model and proposed model as shown in Table 41.3.

Conclusion

In conclusion, the project proved the efficient utilization of FPGA boards (Artix and Kintex) to build a demultiplexer and analyze hardware usage and power consumption. The research showed that the Kintex board had slightly less power consumption (0.932W) than the Artix board (0.981W), with a difference in power of approximately 5%. The efficiency of the Kintex board in I/O resources and energy consumption makes it suitable for low-power applications. This result implies that for designs where less power consumption is necessary without compromising on performance, the Kintex board is a better option. The project also highlights the necessity of energy-optimized FPGA designs, which are important for embedded systems, IoT devices, and other power-conscious applications. Subsequent work may investigate additional hardware design optimizations, logic synthesis improvements, and application of machine learning methods to set energy efficiency records for FPGA-based systems. In the future this project deploy Advanced FPGA & AI Optimization – Deploying on Versal ACAP, Zynq UltraScale+ with AI-powered power management increases efficiency. Better scope and improved cooling – scaling to Intel FPGAs, deploying in 5G, IoT, automotive, and leveraging AI-based thermal management enhances performance.

References

[1]   Singh, S. K., Shukla, S. K., & Singh, S. K. (2021). Energ efficient design of RAM and its implementation on FPGA. *IEEE Transactions on Circuits and Systems II: Express Briefs*, 68(9), 3118–3122.

[2]   Ghosh, M., Maitra, S., & Sarkar, S. (2023). Evaluating ascon hardware on 7-series FPGA Devices. *IEEE Transactions on Computers*, 72(1), 1–12.

[3]   Shrivastava, S., Shukla, S. K., & Singh, S. K. (2021). Implementation of an energy efficient flip flop by using kintex, virtex and genesys FPGA families. *International Journal of Advanced Computer Science and Applications*, 12(4), 1–8.

[4]   Singh, S. K., Shukla, S. K., & Singh, S. K. (2022). Asynchronous counter on kintex and virtex FPGA platforms. *IEEE Transactions on Circuits and Systems II: Express Briefs*, 69(10), 4123–4127.

[5]   Singh, S. K., Shukla, S. K., & Singh, S. K. (2022). Comparative analysis of power and hardware utilization in an FPGA-based car parking system. *IEEE Transactions on Circuits and Systems II: Express Briefs*, 69(11), 4234–4238.

[6]   Singh, S. K., Shukla, S. K., & Singh, S. K. (2023). Comparative analysis of power, temperature, and hardware utilization in FPGA-based systems. *IEEE Transactions on Circuits and Systems II: Express Briefs*, 70(1), 123–127.

[7]   Maillard, P., Chen, Y., Vidmar, J., & Fraser, N. (2022). Radiation-tolerant deep learning processor unit (DPU)-based platform using Xilinx 20-nm Kintex UltraScale FPGA. *IEEE Transactions on Nuclear Science*, 70(4), 714–721.

[8]   Singh, S. K., Shukla, S. K., & Singh, S. K. (2024). Low power design of program counter on Kintex-7 FPGA. *Gyancity Journal of Engineering and Technology*, 10(2), 1–9.

[9]   Alfke, P. (2005). Metastable recovery in Virtex-II Pro FPGAS. HYPERLINK http://www.xilinx.com"www.xilinx.com.

[10] Caffrey, M. (2018). An analysis of high-current events observed on Xilinx 7-Series and ultra scale FPGAs. *IEEE Transactions on Nuclear Science*, 65(8), 1654–1660.

[11] Benedict, S. (2024). Edge intelligence: deep learning-enabled edge computing. IOP Publishing, 2024.

[12] Ghiasi, N. M., Sadrosadati, M., Mustafa, H., Gollwitzer, A., Firtina, C., Eudine, J. et al. (2024) MegIS: High-performance, energy-efficient, and low-cost metagenomic analysis with in-storage processing. 2024 ACM/IEEE 51st Annual International Symposium on Computer Architecture (ISCA).

[13] Chi, T.-K., Chen, T.-Y., Lin, Y. -C., Lin, T. -L., Zhang, J. T., Lu, C. -L. et al. (2024). An edge computing system with AMD Xilinx FPGA AI customer platform for advanced driver assistance system. *Sensors*, 24(10), 3098.

[14] Pecák, O., Matěj, Z., & Přenosil, V. (2024). XSBench on FPGAs using Intel oneAPI. EPJ Web of Conferences. Vol. 302. EDP Sciences, 2024.

[15] Tibaldi, M., & Pilato, C. (2023). A survey of FPGA optimization methods for data center energy efficiency. arXiv preprint arXiv:2309.12884.

42 ASIC-driven healthcare monitoring: a RISC-V approach to power-efficient medical systems

Jithendra Reddy Dandu[1,a], C.Harshitha[2,b], S. Bhounika[2,c], K. Keerthana[2,d], G. Chinna Narasimha[2,e] and R. Hari[2,f]

[1]Assistant Professor, Department of Electronics and Communication Engineering, Annamacharya Institute of Technology and Sciences, Titupati, Andhra Pradesh, India

[2]Student Department of Electronics and Communication Engineering, Annamacharya Institute of Technology and Sciences, Titupati, Andhra Pradesh, India

Abstract

Healthcare monitoring systems face challenges such as high-power consumption, processing delays, and limited scalability, making them inefficient for real-time applications. This work proposes an ASIC-based healthcare monitoring system, developed using Vivado and RISC-V architecture, to provide a power-efficient and real-time solution for monitoring key physiological parameters: ECG, blood pressure, respiration rate, and glucose levels. The system is implemented as a system-on-chip (SoC) design, ensuring optimized performance with minimal power consumption and high-speed processing. The digital processor, designed in Verilog HDL, executes RISC-V instructions for real-time data processing at 2.85 GHz with a power consumption of 64 mW. The proposed system simulates healthcare waveforms, applies custom processing logic, and generates alerts when monitored parameters exceed predefined thresholds. The software-based ASIC simulation approach ensures high efficiency, scalability, and suitability for next-generation healthcare applications, making it a promising solution for low-power, real-time health monitoring systems.

Keywords: ASIC, ECG, RISC-V processor, system on chip (SoC), Verilog HDL

Introduction

The wearable and real-time health monitoring systems integration has seen tremendous speed over the last few years with the capability of ongoing monitoring of life parameters like ECG, blood pressure, respiratory rate, and blood glucose level. Such technologies are important for early detection and preventive medicine, especially in chronic patients like cardiovascular diseases and diabetes. A low-power ECG signal processing ASIC paper illustrates their role in arrhythmia detection in wearable health systems [1]. But existing systems cannot manage real-time processing efficiency, power consumption, and multi-sensor integration. In this paper, the planning and execution of an ASIC-centered healthcare processor framework for real-time healthcare monitoring of critical healthcare parameters are presented. The system is designed using the RISC-V architecture, thereby making the system modular, scalable, and performance-optimized. The processor effectively gathers and processes ECG, blood pressure, respiration rate, and glucose level data and provides digital alerts in the form of waveform simulations when the parameters cross threshold levels. Unlike conventional healthcare processors, our system employs a System-on-Chip (SoC) design to minimize power consumption with high-speed processing. Analogous designs towards energy-efficient SoC designs for healthcare applications have been presented, where neuromorphic processing and power-aware architecture are the driving forces for real-time analysis [2]. Our design is based on Verilog HDL to achieve efficient data processing seamlessly. Healthcare processor verification is a critical step towards reliable operation, particularly with life-critical parameters. System Verilog-based testbenches have been employed to verify RISC-based healthcare processors with a focus on simulation efficiency and test coverage [3]. Our project employs a comprehensive testbench for functional correctness verification, including clock generation, reset states, stimulation application for different sensor inputs, and response observation. The testbench

a*Jithendrareddy.d@gmail.com, bharshitha17102003@gmail.com, csiddagallabhounika@gmail.com, dkasireddykeerthana737@gmail.com, enarasimhayadava123@gmail.com, fhari33152@gmail.com

DOI: 10.1201/9781003685364-42

incorporates memory operations, arithmetic operations, and control flow operations, mimicking actual healthcare data processing scenarios. While existing work emphasizes the significance of verification using universal verification methodology (UVM) and formal methods [4].

Power efficiency remains a design constraint for wearable and real-time medical monitoring devices. Existing research on ultra-low-power SoC design for healthcare has established strategies including dynamical voltages and clock gating methods and power-efficient data encoding to minimize power consumption [5]. This processor is currently operating at 2.85 GHz with a measured power draw of 110mW, which, though optimized for real-time signal processing, requires improvement in power efficiency for wearable purposes. Power optimization methods, as well as cutting-edge verification methods for system reliability, will be addressed in future work.

Literature survey

Raja et al. conducted a study on the creation of an electrocardiogram (ECG) processing with reduced energy consumption for smart medical devices systems [1].

Tian et al. proposed that BioPI is a minimal latency, energy-effective neuromorphic factors processed systems [2]. Haoming et al. developed a neuromorphic factors processing framework for portable ECG categorization with minimal power usage using a spiking neural network (SNN) and level-crossing sampling to reduce power consumption [3]. Near-threshold operation of an ultra-low energy physiological signaling device voltage levels, minimizing energy consumption while maintaining signal integrity [4]. Enhancing energy efficiency and mobility in multi-lead ECG acquisition [5]. It Focuses on wearable sensors and real-time data acquisition systems to improve patient monitoring and preventive care [6]. Kanase and Nithin designed an ASIC-based healthcare monitoring, emphasizing power efficiency and medical data processing [7]. Hesha studied ASIC application in digital form., comparing corporate design methods for performance and power efficiency [8].

Birari and Birla developed a RISC V processors with a focus on scalability, efficiency, and performance for embedded applications [9]. It developed

a very low-power systems ASIC device for ECG tracking [10]. It examined the role in medical technology particularly in health imaging, diagnostics and patient monitoring [11]. Winter and Staubert explored smart medical information technology and advanced healthcare data management for improved interoperability [12].

Methodology

The proposed software-based ASIC system for healthcare monitoring was implemented in Verilog HDL and simulated using Vivado. The system is based on System-on-Chip (SoC) architecture with RISC-V processor and calculates the biomedical signals like glucose levels, cardiac rate, arterial pressure, and rate of breathing. The methodology defines the system architecture, digital processing techniques, threshold detection, and performance analysis. The overall workflow is represented in Figure 42.1. The input signals are simulated digitally, and processing is based on the retrieve, interpret, run, and write back to storage cycle in the RISC-V architecture. Branch

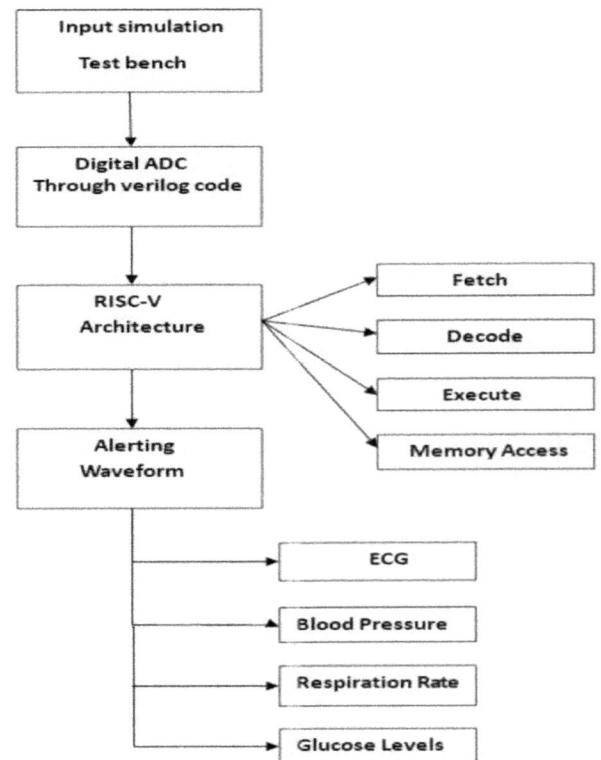

Figure 42.1 Block diagram of overall system
Source: Author

control logic, DMA controller, and register file for sequential processing and data storage are designed.

DataSet collection

We deployed openly accessible datasets to validate and simulate our ASIC-based healthcare monitoring system: the Comprehensive Patient-Health Tracking data set [18] offered physiological parameters like heartbeat, arterial blood pressure, breathing rate, and glucose levels, and the respiratory and cardiovascular tracking dataset from aeration study [19] assessed real-time cardiovascular activity and breathing patterns. These datasets allowed us to simulate real-world biomedical diseases in our Verilog-based RISC-V processing system.

RISC-V-based processing unit

RISC Ventures is an open-source conventional structure for set commands developed from the established reduced instruction set computer principle. Instructions retrieval stage (IF), instructional decoding part (ID), executing level (EX), storage accessing level (MEM), and write backstage (WB) comprise

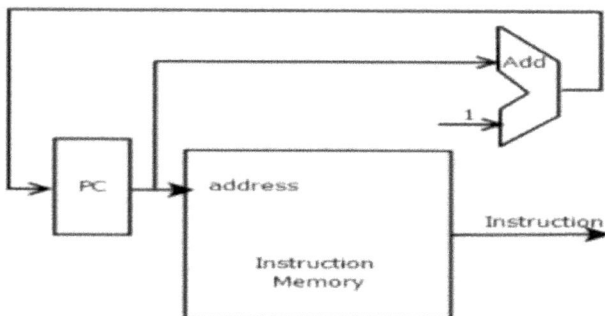

Figure 42.2 Instruction fetch
Source: Author

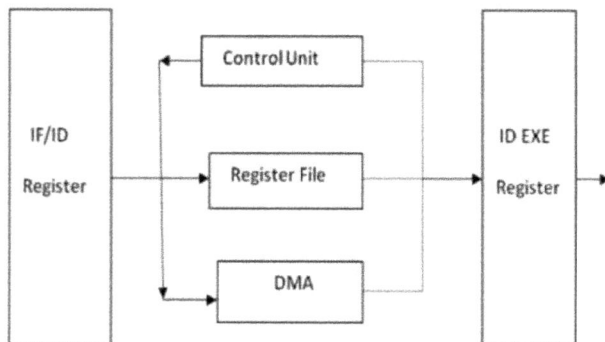

Figure 42.3 Instruction decode
Source: Author

the five stages of a RISC processing system [13]. Complete functionality was developed with Verilog HDL. The following instructions was read with memory location stored in program counter (PC) and instructional register. Thus, IF stage instruction is read and PC is incremented by plus 4. RISC-V architecture contributes significantly to biomedical signal processing as well as sending threshold-based alarm. RISC-V processor relies on a five-stage pipeline mode of execution for instructions, and each instruction traverses various stages to execute seamlessly.

Fetch: The RISC-V processor fetches instructions compared to the past location of the next command that will be retrieved is stored in a PC. The instructions memory (IR) receives the received instruction for further processing.

At this stage, PC is updated as:

PCnext = PC+4

Since RISC-V uses fixed 32-bit instructions, the PC increments by 4 bytes for each instruction.

Decode: Instruction decoder determines opcode, source registers, destination registers, and control signals.

The register file is referenced to fetch the stored values from past operations.

The fetched instruction is divided into opcode, registers, and immediate values:

Opcode = Instruction[6:0]

Immediate = Sign-extend([31:20])

Execution: Computation and mathematical computations are carried out by the arithmetical logic unit (an ALU).

Computations include:

1. Addition/subtraction for filtering and averaging.
2. Comparison of threshold-based alerts.
3. Logical operations for bitwise data processing.

The ALU processes biomedical data, checking values against predefined health thresholds.

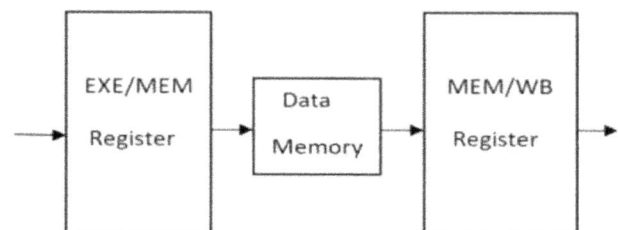

Figure 42.4 Execution stage
Source: Author

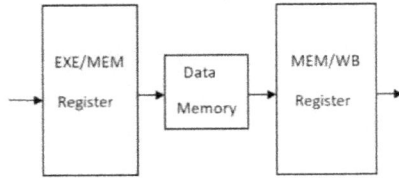

Figure 42.5 Memory access stage
Source: Author

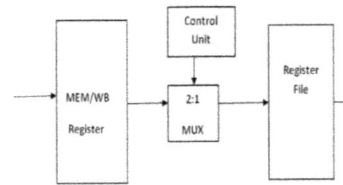

Figure 42.6 Write backstage
Source: Author

Result = ALU (Operand 1, Operand 2)

Memory Access: The load/store unit loads data from or stores data into memory. The DMA processor oversees data exchange between the processing units and memory. Biomedical signal data is stored temporarily and fetched when required. If an alert condition is fulfilled, an alert flag is stored in memory for output processing.

For load instructions:

$$R[r_d] = Memory\ [R[rs_1] + Offset]$$

For store instructions:

$$Memory\ [R[rs_1] + Offset] = R[rs_2]$$

Write back: The result is stored in registers for subsequent use. Register File stores updated values to be used in the upcoming instruction cycle. This phase maintains data availability for upcoming processing cycles.

Biomedical data is kept in registers. Alerts are stored in memory for threshold-based alert generation.

Direct memory access (DMA) controller equation

Dataout = Memory[Address]

Sequential operations are handled via the DMA controller, ensuring efficient data transfer.

Above Figures 42.2, 42.3, 42.4, 42.5 and 42.6 represents the construct block diagrams representing the stages of a processor pipeline

Digital signal processing and threshold detection
Signal processing in Verilog

1. Conditional statements compare real-time values against predefined thresholds.
2. Branch control logic manages signal flow and decision-making.

Threshold based alert generation

1. Threshold values for each biomedical parameter are predefined based on medical reference data.

2. If a signal exceeds its set threshold, an alert is triggered.
3. The alert is represented as a digitally simulated waveform in the output, visually indicating deviations from normal health conditions.

Software implementation and simulation
Testbench development

The testbench has a critical functioning verifying the precision of proposed RISC-V-based healthcare monitoring system. It simulates various real-time input conditions, ensuring correct execution of sequential operations, DMA control, branch control, and threshold-based alert generation.

Simulation in Vivado

To validate the correctness of the digital healthcare monitoring system, the entire design is simulated in Vivado. This simulation process ensures that each module, from signal acquisition to alert generation, functions as expected.

After loading the Verilog HDL code the Verilog implementation of the RISC-V-based alert system is loaded into the Vivado simulation environment. The testbench code is compiled alongside the design to provide stimulus inputs.

The simulation outputs are analyzed through waveform diagrams to verify correct signal processing.

The following key parameters are observed:

Instruction execution timing: Ensures each instruction is processed within the expected clock cycles.

Threshold alert accuracy: Confirms that alert signals are generated when input values exceed the threshold.

Memory and register file updates: Verifies that previous computation values are correctly stored and retrieved.

Results and Analysis

The ASIC-based healthcare processor system that was implemented was validated by a Verilog

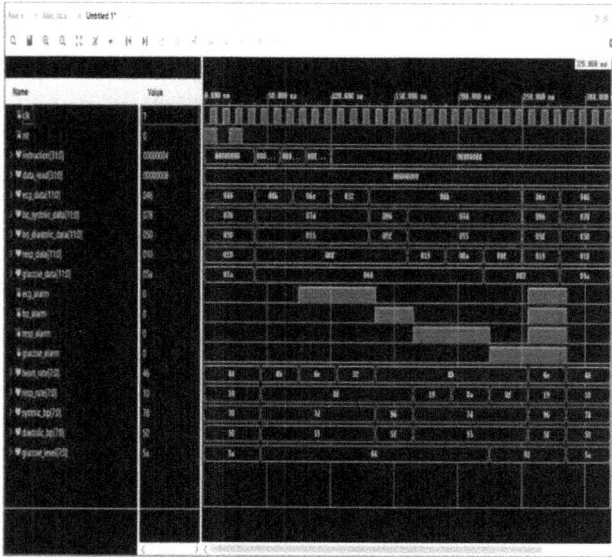

Figure 42.7 Simulated waveform
Source: Author

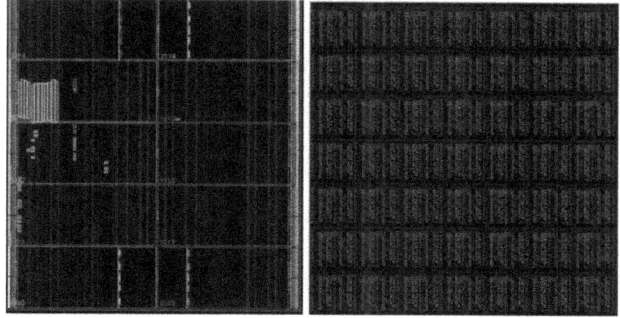

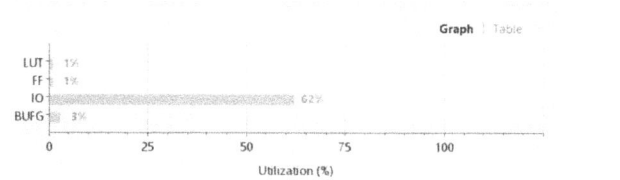

Table 42.1 Performance metrics.

Metric	Measured value
Clock frequency	2.85 GHz
Total power consumption	64mW
Processing latency	< 5 ns

Source: Author

testbench under Vivado. The testbench was employed to simulate live health data, including BP, breathing rate, levels of glucose, and cardiovascular rate. The simulation confirmed that the processor can detect abnormal values and issue warnings in the correct manner. Table 42.1 presents the performance metrics, and Table 42.2 provides a comparison between the proposed ASIC and existing systems. Both tables are shown above.

Simulation waveforms
The simulation was executed in Vivado, and waveforms that indicate how the processor evolves with time were generated. The observations of the analysis of the waveforms are most important. The simulated waveform show in Figure 42.7.

Execution analysis
The processor's efficiency was quantified regarding execution time, power and correctness. The system's frequency was 2.85 GHz, and the simulation results

Figure 42.8 Implementation
Source: Author

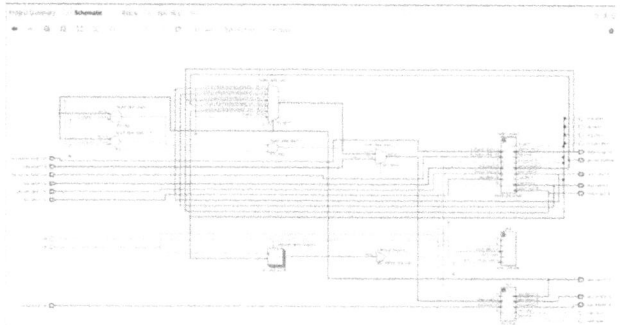

Figure 42.9 Schematic diagram
Source: Author

Figure 42.10 Utilization
Source: Author

Power analysis from Implemented netlist. Activity derived from constraints files, simulation files or vectorless analysis.

Total On-Chip Power:	64.339 W (Junction temp exceeded!)
Design Power Budget:	Not Specified
Process:	typical
Power Budget Margin:	N/A
Junction Temperature:	125.0°C
Thermal Margin:	-60.3°C (-31.0 W)
Ambient Temperature:	25.0 °C
Effective θJA:	1.9°C/W
Power supplied to off-chip devices:	0 W
Confidence level:	Low

Figure 42.11 Power analysis
Source: Author

were as follows. Overall proposed results was shown in the Figure 42.8,42.9,42.10 and 42.11 respectively.

Table 42.2 Comparison between proposed ASIC and existing systems.

Feature	Proposed ASIC-based system feature description	Existing system feature description
Technology	ASIC with RISC-V Architecture	FPGA, Microcontroller-based [14]
Power consumption	64mW (Optimized for Low Power)	150mW – 350mW [15] 120mW [17]
Processing efficiency	Optimized DSP Pipeline for Real-Time Processing	General FPGA DSP Blocks [14]
Monitored parameters	Heart Rate, BP, Respiration, Glucose	ECG, SpO2 [16] ECG [17]
Digital processing	RTL Design in Verilog HDL, Implemented in Vivado & ISE	FPGA-Based Processing [16]
Energy efficiency	Optimized ASIC (64mW) for Wearable Use	Higher Power FPGA (150mW+) [15].
Scalability	Highly Scalable	Limited Due to FPGA Constraints [16] Fixed Microcontroller [17]
Cost and fabrication	Cost-Effective ASIC, Suitable for Mass Production	High Cost Due to FPGA Reprogram [14]

Source: Author

Conclusion

The ASIC based device where the measured metrics surpass predetermined criteria, a suggested ASIC-based Medical Monitoring Equipment provides alerts in the form of digital waveform. It efficiently detects anomalies in blood pressure, respiratory rate, heart rate, and glucose. The system, which runs at 2.85 GHz and has a low power dissipation of 64 mW, has been developed using the RISC-V architecture and coded in Verilog HDL. The simulation findings guarantee effective signal processing and quick computation, validating the system's correctness and dependability in real-time patient monitoring. This high-performing and energy-efficient system offers a lot of potential for wearable medical device cooperation, which would improve individualized healthcare and remote patient monitoring. IoT-based connectivity and machine learning algorithms for predictive healthcare analytics, and ASIC manufacturing refinements for additional energy and downsizing are possible future improvements.

References

[1] Raja, K., Saravanan, S., Anitha, R., Priya, S. S., & Subhashini, R. (2017). Design of a low power ECG signal processor for wearable health system-review and implementation issues. In 2017 11th International Conference on Intelligent Systems and Control (ISCO), (pp. 383–387). IEEE.

[2] Tian, F., Chen, J., Zheng, J., Wu, H., He, J., Wang, X., et al. (2024). BioPI: an energy efficient and low-latency neuromorphic pipelined system with joint design optimizations of sensor-algorithm-processor for wearable healthcare. *IEEE Transactions on Circuits and Systems for Artificial Intelligence.* vol. 2, no. 1, pp. 64–78, March 2025.doi: 10.1109/TCASAI.2024.3502573

[3] Chu, H., Jia, H., Yan, Y., Jin, Y., Qian, L., Gan, L., et al. (2021). A neuromorphic processing system for low-power wearable ECG classification. In 2021 IEEE Biomedical Circuits and Systems Conference (BioCAS), (1–5). IEEE.

[4] Wei, Y., Cao, Q., Hargrove, L., & Gu, J. (2020). A wearable bio-signal processing system with ultra-low-power SoC and collaborative neural network classifier for low dimensional data communication. In 2020 42nd Annual International Conference of the IEEE Engineering in Medicine & Biology Society (EMBC), Montreal, QC, Canada, (pp. 4002–4007). doi: 10.1109/EMBC44109.2020.9176647.

[5] Wang, L.-H., Zhang, Z. H., Tsai, W. P., Huang, P. C., & Abu, P. A. R. (2022). Low-power multi-lead wearable ECG system with sensor data compression. *IEEE Sensors Journal*, 22(18), 18045–18055.

[6] Angelov, G. V., Nikolakov, D. P., Ruskova, I. N., Gieva, E. E., & Spasova, M. L. (2019). Healthcare sensing and monitoring. In Enhanced Living Environments, (pp. 226–262). Springer, Cham.

[7] Kanase, G., & Nithin, M. (2021). ASIC design of a 32-bit low power RISC-V based system core for medical applications. In IEEE International

Conference on Communication and Electronics Systems (ICCES), (pp. 1–5).

[8] Hesha, S., Shalan, M., El-Kharashi, M. W., & Dessouky, M. (2021). Digital ASIC implementation of RISC-V: open lane and commercial approaches in comparison. In IEEE International Midwest Symposium on Circuits and Systems (MWSCAS), (pp. 498–502).

[9] Birari, A., Birla, P., Varghese, K., & Bharadwaj, A. (2020). A RISC-V ISA compatible processor IP. In IEEE International Symposium on VLSI Design and Test (VDAT), (pp. 1–6).

[10] Liu, X., Zheng, Y., Phyu, M. W., Endru, F. N., Navaneethan, V., & Zhao, B. (2012). An Ultra-Low Power ECG Acquisition and Monitoring ASIC System for WBAN Applications, in IEEE Journal on Emerging and Selected Topics in Circuits and Systems, vol. 2, no. 1, pp. 60–70, March 2012, doi: 10.1109/JETCAS.2012.2187707.

[11] Gao, J., Yang, Y., Lin, P., & Park, D. S. (2018). Computer vision in healthcare applications. *Journal of Healthcare Engineering*, 2018, 5157020.

[12] Winter, A., & Staubert, S. (2018). Smart medical information technology for healthcare. *Methods of Information in Medicine*, 57(1), e92–e105.

[13] Stoffelen, K. (2019). Efficient cryptography on the RISC-V architecture. In Springer International Conference on Cryptology and Information Security in Latin America, (pp. 323–340).

[14] Abushukor, S. F. K., Syafalni, I., Mulyawan, R., Sutisna, N., Ahmadi, N., & Adiono, T. (2021). FPGA implementation of IoT-Based health monitoring system. In 2021 15th International Conference on Telecommunication Systems, Services, and Applications (TSSA), (pp. 1–5). IEEE.

[15] Madhumati, G. L. (2024). FPGA implementation of health monitoring system for high-risk cardiac patients. In 2024 IEEE International Conference on Signal Processing, Informatics, Communication and Energy Systems (SPICES). IEEE.

[16] Badiganti, P. K., Peddirsi, S., Rupesh, A. T. J., & Tripathi, S. L. (2022). Design and implementation of smart healthcare monitoring system using FPGA. In Proceedings of First International Conference on Computational Electronics for Wireless Communications: ICCWC 2021, (pp. 205–213). Springer Singapore.

[17] Methul, S., Shedge, A., & Bartakke, P. (2024). Cardiosync : real-time ECG monitoring system with FPGA-based signal processing and analysis. In 2024 IEEE International Conference on Electronics, Computing and Communication Technologies (CONECCT), Bangalore, India.

[18] Karthick Raghunath K M, Comprehensive Patient-Health Monitoring Dataset, IEEE Dataport, June 18, 2024, doi:10.21227/2t3q-6g13

[19] Guy, Ella Frances Sophia, Isaac Flett, Jaimey Anne Clifton, Trudy Caljé-van Der Klei, Rongqing Chen, Jennifer Knopp, Knut Moeller, and James Geoffrey Chase. (2024). Respiratory and heart rate monitoring dataset from aeration study. PhysioNet 10 (2024): e4dt-f689.https://doi.org/10.13026/e4dt-f689

43 Empowering accessibility: portable voice-enabled interface for disabled users

K. Shankar[a], K. Chandrahasa Reddy[b], Y. Pavan Kumar Reddy[c], M. Ravi Kishore[d], Y. Sunanda[e] and A. Varshitha[f]

Department of ECE, Annamacharya University, Rajampet, Andhra Pradesh, India

Abstract

A hand-finger movement-controlled system for people with physical limitations is presented in this research. For functions like requesting necessities (such food, drink, and medication) and managing electrical appliances (like lights and fans), the wearable gadget employs proximity sensors to identify finger movements. Different instructions are mapped to certain finger movements, allowing for smooth interaction.

The system uses a microcontroller with embedded instructions, voice-activated controls through an Android app, and a relay for device switching. To carry out the necessary actions, the microcontroller analyses sensor data. Technology improves users' freedom and quality of life by fusing speech instructions with gesture detection, especially for people who have severe physical disabilities or paralysis.

Keywords: Arduino IDE, blynk cloud, customised android App, ESP-12E Wi-Fi module, node MCU, power supply unit, proximity sensors, relays

Introduction

Due to illnesses, accidents, or congenital conditions, the number of people with hearing and speech impairments is increasing, underscoring the need for efficient communication solutions. Interactions in emergency situations or strange settings are made more difficult by the current reliance on gestures or sign language, which poses difficulties, especially for people who are not accustomed to it.

We suggest a glove-based smart speaking system [5] to help people with speech impairments communicate with others in order to close this gap. Similar to a regular glove, this "hand talk glove" has flex sensors built into the thumb and fingers. Based on bending angles, these sensors identify hand motions and provide analogue signals, which a microcontroller then transforms into digital data. To create spoken outputs through a speaker, the processed data is wirelessly transferred.

Flex sensors are essential because they translate mechanical motions into electrical signals, and as the degree of bending increases, so does the resistance [7]. In addition to facilitating smooth communication, this cutting-edge technology takes into account other health issues that people with speech impairments frequently encounter, providing a complete and approachable solution.

Literature review

By converting finger motions into digital commands, flex sensor-based hand gesture recognition (HGR) is becoming more and more popular in human-computer interface (HCI). Neural networks and Support Vector Machines (SVMs) are two machine learning techniques that have demonstrated impressive accuracy in gesture identification. However, problems still exist because of ambient influences, user variability, and inconsistent sensors. Recent developments in deep learning provide answers to the efficient techniques required for real-time data processing by increasing system robustness and gesture library expansion. HGR is still showing potential in developing HCI solutions in spite of its difficulties [1].

A state-of-the-art tool called the Gesture Vocaliser was created to help mute people interact with those who don't know sign language. The system records and interprets hand motions using flex sensors, an

[a]shan87.maddy@gmail.com, [b]kora65556@gmail.com, [c]ratnasena.reddy@gmail.com, [d]ravi.mvrm@gmail.com, [e]sunanda.bujji@gmail.com, [f]varshitha20212021@gmail.com

DOI: 10.1201/9781003685364-43

accelerometer, an Arduino Uno, and Bluetooth. The dynamic time warping algorithm is used to translate the motions, which are then sent to an Android smartphone for vocalization and display. The gadget promotes social involvement and independence, but there is still need for advancement in areas like computing complexity, noise sensitivity, and man- aging multidimensional data [2].

For people with speech problems, smart gloves that translate gestures into text and speech provide a substantial communication option. These gadgets encourage independence by removing the need for vocalization by converting hand gestures into speech or text. The use of wireless technology enhances usability by enabling the gadget to operate in several environments.

Key elements including sensors, microcontrollers, and communication modules are examined in this overview, along with developments in gesture detection algorithms. It also discusses issues like data latency and sensor accuracy while emphasizing how technology may improve accessibility and close communication gaps [3].

In order to improve communication for sign language users, this study investigates the translation of Indian Sign Language (ISL) into text and speech utilizing smart gloves. The technique enables more inclusive interactions in multilingual settings by converting gestures into spoken language. By encouraging better accessibility and facilitating productive conversation between sign language users and the general public, this technology has the potential to overcome communication barriers [4].

Methodology

Existing method

In Figure 43.1 order to evaluate sensor data and initiate actions depending on predetermined conditions—for example, turning on a buzzer when an impediment is detected—this system makes use of an Arduino UNO [6]. It improves accessibility for those with physical disabilities by including a speech interface for user control and aural feedback. Physically disabled people may engage and manage the gadget with ease thanks to the wireless setup, which guarantees real-time connection between sensors, the Wi-Fi module, and external devices or networks. Greater freedom and usability in daily settings are encouraged by this design.

Proposed method

As in block diagrams, the suggested system offers a portable wireless device interface that is based on the Internet of Things and is intended for people with physical disabilities. With this technology, users may use voice commands to control and interact with

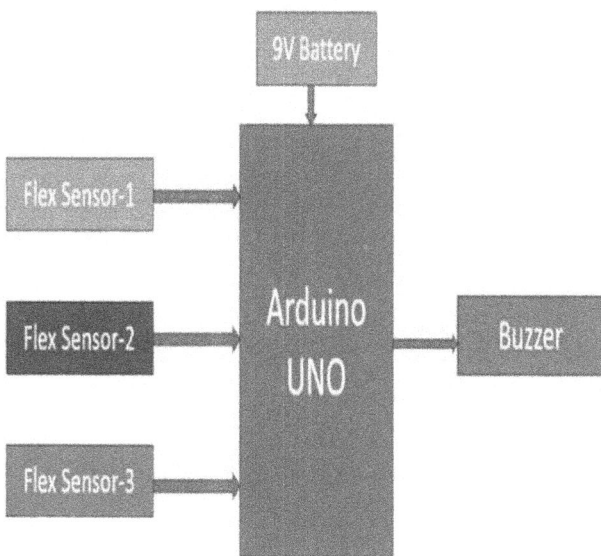

Figure 43.1 Block diagram of existing method
Source: Author

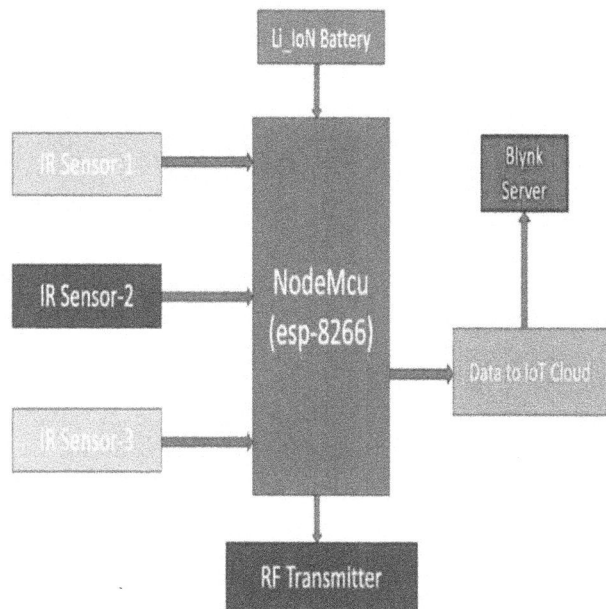

Figure 43.2 Block diagram of transmitter section
Source: Author

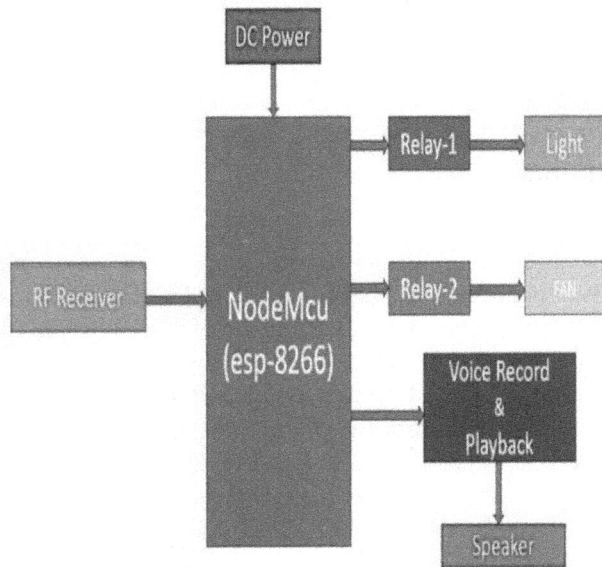

Figure 43.3 Block diagram of receiver section
Source: Author

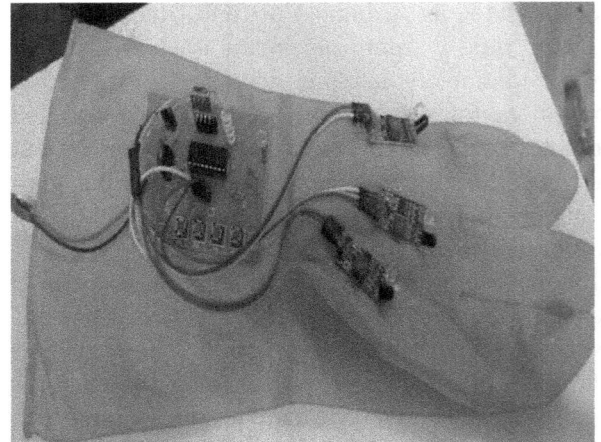

Figure 43.4 (a) Hardware kit
Source: Author

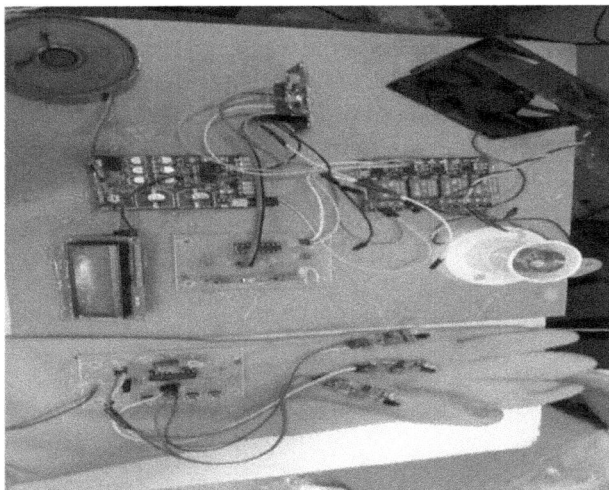

Figure 43.4 (b) Hardware kit (hand glove)
Source: Author

gadgets wirelessly. Through the integration of speech recognition technology [8], Figure 43.4 (a) and 4 (b) the system offers a user-friendly solution that improves accessibility and independence for people with physical limitations, enabling them to use commonplace equipment more effectively. Figure 43.2.

Transmitter: User activity is detected by an infrared sensor, and a battery-operated NodeMCU (Wi-Fi module) processes the data. For remote access and integration with a voice-activated application, this data is sent to an Internet of Things cloud. The technology offers a hands-free, accessible way for people to engage with gadgets, including robots, by enabling voice commands.

Receiver: Figure 43.3 A NodeMCU module in the receiver portion manages relays to regulate AC and DC loads [9] and collects data from the Internet of Things cloud. With the help of a regulated power supply (RPS), this configuration makes it possible to manage a variety of electrical devices and household appliances. An Android app with messaging features guarantees real-time updates and conversation, while a speaker offers audio feedback. For people with physical disabilities, this voice-activated, wireless interface provides a portable, effective solution that improves accessibility and independence by allowing device control without the need for manual engagement

Results

The system wirelessly sends data to a NodeMCU via radio frequency (RF) using proximity sensors to identify gestures. Figure 43.5.

It delivers alarm messages to a registered email address and makes voice instructions audible based on finger locations.

Furthermore, it uses relay modules to operate electrical appliances, allowing automation and gesture-based control.

Conclusion

For those who are physically challenged or deaf-mute, the smart hand glove offers an inexpensive,

Figure 43.5 Messages displayed
Source: Author

portable, and easy-to-use solution that facilitates communication and increases their level of freedom. This portable gadget gives users the freedom to live independently and participate in society without the need for assistance, making it a more affordable option than pricey assistive technology.

References

[1] Dutta, K., Dey, S., Dutta, S., & Dutta, S. P. (2023). Project on hand gesture vocalizer: a sensor-based approach. JETIR, Journal of Emerging Technologies and Innovative Re- search, 10(12), 371–381.

[2] Harivardhagini, S. (2021). A novel approach to vocalize the hand gesture movement for speech disabled. CVR Journal of Science and Technology, 21(1), 84–88.

[3] Rastogi, K. (2016). A review paper on smart glove - converts gestures into speech and text. International Journal on Recent and Innovation Trends in Computing and Communication, 4(5), 92–94.

[4] Tandon, A., Saxena, A., Mehrotra, K., Kashyap, K., & Kaur, H. (2016). A review paper on smart glove – converts Indian sign language (ISL) into text and speech. International Journal for Scientific Research and Development (IJSRD), 4(08), 269–272.

[5] Badawi, M., Elaskary, S., & Ahmed, Z. (2023). Enhancing community interaction for the Deaf and Dumb via the design and implementation of smart speaking glove (SSG) based on embedded system. International Journal of Telecommunications, 03(02), 1–11.

[6] Gayathri, N. (2021). Smart glove for blind. International Journal for Research in Applied Science and Engineering Technology, 9(VI), 3309–3315.

[7] Kumar, K. N. (2022). Gesture vocalizer using flex sensor and software visualization. International Journal for Research in Applied Science and Engineering Technology, 10(6), 2294–2299.

[8] Jiang, M. Y., Jong, M. S., Wu, N., Shen, B., Chai, C., Lau, W. W., et al. (2022). Integrating automatic speech recognition technology into vocabulary learning in a flipped english class for chinese college students. Frontiers in Psychology, 13, 902429.

[9] Heartfield, R., Loukas, G., Budimir, S., Bezemskij, A., Fontaine, J. R., Filippoupolitis, A., et al. (2018). A taxonomy of cyber-physical threats and impact in the smart home. Computers and Security, 78, 398–428.

44 Smart transit – college live bus tracking and management

C. Sasikala[1,a], B. Divya Sree[2,b], K. Divya Sree[2,c], K. Mohammad Mansoor[2 d] and
T. Muheet Ur Rahman[2,e]

[1]Associate Professor, Department of Computer Science, Srinivasa Ramanujan Institute of Technology, Anantapur, Andhra Pradesh, India

[2]Deparment of Computer Science, Srinivasa Ramanujan Institute of Technology, Anantapur, Andhra Pradesh, India

Abstract

Smart Transit is a web and mobile platform that advances university bus tracking by eliminating outdated communication methods, including WhatsApp. GPS technology enables the system to deliver live location of university buses. Every bus contains a GPS tracker that transmits location details through a GSM network to a central server continuously. Smart Transit provides administrators with a centralized system that enables them to control the portal. Administrators can track bus routes and control the information of students, seat plan, and locations to avoid manual intervention. To access the mobile app, students must present their fee receipts to the admin to receive their login credentials through email, which helps them to log in and access the mobile app to verify their bus details and track their bus live location. With cloud integration, the Smart Transit System facilitates students, faculty, and administrators to view the current location of buses and user information on any internet-enabled device. It provides smooth data flow with data storage and management and accessibility from anywhere. Smart Transit offers precise location data and autonomous tracking while reducing the chances of human error. The intuitive design of the college bus system makes it easy for all eligible staff and students to take advantage of this convenient transportation service.

Keywords: Administration, android, bus tracking, centralized system, faculty, GPS, GSM, student

Introduction

Managing time is crucial and essential for college students to prevent missing of their bus. Imagine yourself waiting for your bus at your bus stop without knowing when the bus will arrive. There is no method to determine whether traffic or road construction is to blame for the delay. You check your phone for an update, only to find the information on whether the bus has already left or broke down somewhere. Many students use WhatsApp for bus updates, but there's a drawback - it doesn't show real-time locations. Although mobile real-time tracking is available to assist users with the location of the bus, some drivers do not have mobile phones or navigate smartphones in the same manner [2]. Instead, users have to manually update their bus positions, which might be incorrect or outdated. These platforms usually show only bus routes without live updates, making it hard for users to track their buses, which can waste time or cause them to miss their buses.

Our system includes an administrator portal that provides access to data on students, bus details, routes, and seat plan. Moreover, administrators can monitor the real-time locations of buses and provide application access to students who have completed their fee payments. Through the application, students can track the live positions of buses and upload their fees. This functionality improves journey planning and reduces waiting times. Upon Utilizing Android, Web, and IoT technologies, the proposed system seeks to deliver a simple and dependable user experience. With the benefit of live tracking, this solution aims to remove uncertainty, enhance time management, and improve the overall commuting experience. It sets the stage for a more efficient method of managing college bus services, allowing users to monitor buses and assisting them in optimizing their daily schedules [4].

Literature review

Research and findings in the literature focus on real-time vehicle tracking, intelligent transportation systems, GPS-based navigation, automated transport management. Furthermore, evidence indicates that

[a]sasikala.cse@srit.ac.in, [b]214g1a0522@srit.ac.in, [c]214g1a0523@srit.ac.in, [d]214g1a0559@srit.ac.in, [e]204g1a0562@srit.ac.in

DOI: 10.1201/9781003685364-44

traditional methods of tracking buses are ineffective, as many approaches require manual location updates and depend on messaging apps for accurate data entry, which can result in inaccuracies and delay. Several studies have explored the implementation of technology to improve the efficiency and reliability of bus tracking systems. People rely on buses for travel due to their convenience and safety.

Design of bus tracking and fuel monitoring system

The system describes GPS and GSM modules along with a microcontroller tracking the bus position. The bus is linked with a GPS module accompanied by a SIM card [1]. When the user makes a call to the registered SIM card with the GSM module. The position of the bus will be displayed as a link on the phone. It allows real-time tracking of the bus location [5]. The system has one big disadvantage: it relies on the user to make a series of calls to the GSM card to calculate the bus's location, which is inconvenience and time-wasting. Active users who are always on the move find such calls very boring. Other than this, the strategic reliance of the system on the network renders it difficult to place the device with bad cell coverage. They get no proper updates when the bus is under repair. A combination of all of these brings forth there a need for an effective monitoring system to operate in any kind of situation [7].

Efficient model for automated school bus safety and security using IOT and cloud

The research proposes a method to maintain student safety when using school transportation with IoT cloud technology and RFID tags for student tracking, and fire sensors to notify for emergencies. However, this framework has its limitations. It depends largely on RFID; it can only identify few emergency types; and it raises data privacy concern [6]. To make it better in the future, the system could add biometrics for more precise tracking, use more sensors like health and GPS to catch more emergencies, and beef up data protection. Adding mobile apps for parents to stay in the loop and using smart tech to predict when buses need fixing could also boost the system. These upgrades can help make school transport safer more dependable, and run smoother [8].

Real time bus tracking system

This System improves commuting by providing live bus locations, estimated arrival times, and distances using GPS and Google Maps API. Designed as an Android app, it minimizes waiting times and helps users plan effectively. The system provides real-time updates to users through a user-friendly interface by collecting location data from drivers [3]. However, manual updating of information can lead to human errors, and the lack of back up systems can lead to service disruption in technical failures. The system gives real-time information to users by using a friendly interface through gathering location information from drivers. The disadvantage is that some drivers lack information on location tracing applications and don't have knowledge of technology, a few drivers don't have smartphones which may lead to manual mistakes in the system.

Proposed system

By implementing the Smart Transit-College Live Bus Tracking System, college buses will be monitored in real-time, giving administrators, faculty members, and students accurate information about their locations and schedules. The system consists of three main components: 1. administrators utilize a web-based application, 2. students and faculty access a mobile app with integrated features, and 3. IoT sensors enable real-time tracking. The system works by integrating with the above-mentioned components to provide safety, efficiency and convenience to the users and administrators.

Web application

Administrators use web applications for managing the user's data and bus related information. The system is controlled by the admin. Monitors live locations. The admin can give login credentials to the eligible users who cleared their dues. Those students can access mobile apps. Admin can assign drivers to the respective buses and a lot of seats to the students.

The flowchart illustrates the admin's login and access control in a bus management system. The process starts with the admin logging in using their credentials, which is followed by authentication. If the authentication is unsuccessful, the admin will navigate to "Invalid" that will not grant access. If the credentials are valid, the admin will have access to capabilities in the system to manage bus route management, user management, bus tracking and assignment, and seat management. Once the admin has completed their work, they can log out of the bus management system, which signifies ending their

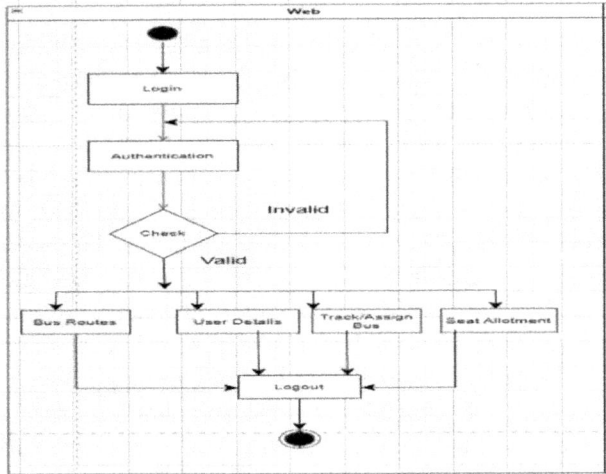

Figure 44.1 Flow of web administration
Source: Author

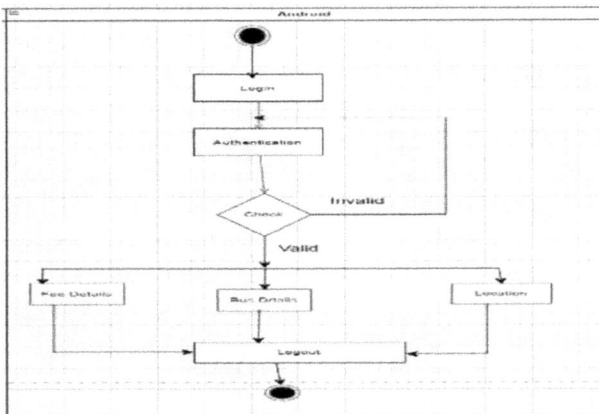

Figure 44.2 Flow of mobile application
Source: Author

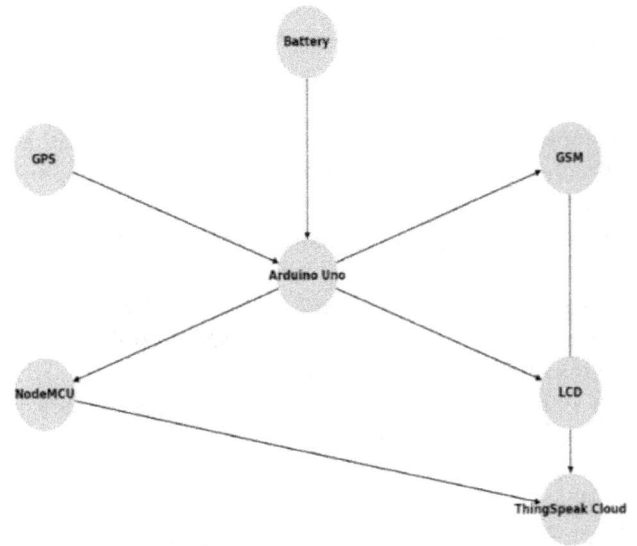

Figure 44.3 IoT architecture
Source: Author

session. The flowchart illustrates structurally how the admin logs into the bus management system and how the admin interacts with the system as a result of the authentication step.

Mobile application framework

The mobile app contains a real-time GPS tracking facility for students and staff members on buses. It provides up-to-date information about the bus position, bus details and able to upload fee receipts and guarantee the users sufficient information before hand for effective planning and timely travel.

The flowchart in Figure 44.2 shows the working of the mobile application for bus tracking. Initially, users download the application and log in using the

sent credentials from their email. These credentials are also authenticated using Firebase to verify and allow entrance. If they are correct, users are granted access to the dashboard that contains bus routes, live tracking, fee details; otherwise, the user is shown an error for their incorrect login. The system is connected to an IoT device that sends the current location of the bus to ThingSpeak in real time for updating the live tracking information in the mobile app. After users have accessed all the needed information, they are able to log out of the system and therefore secure it from unwarranted access. This more efficient process in managing transportation improves the experience of the students, faculty by giving them more flexibility and real-time changes to the schedule.

IOT integration

The Smart Transit system integrates live location tracking for students, faculty, and administrators.

Figure 44.3 illustrates the IoT architecture used for real-time bus tracking. The system starts with a battery, which powers all the essential components. A GPS Module continuously captures the bus's live location and sends the data to Arduino Uno, acting as the controller. Arduino computes the data and distributes it among various modules. An LCD screen on the bus shows the real-time position of the site for visibility. The GSM sends the position information to the ThingSpeak cloud which enables the students and faculty to receive live updates about the bus through the mobile application. If there is no GSM connection, the Node MCU works with Wi-Fi to transmit data, maintaining a reliable tracking system. With this setup, students and staff members can monitor the location of the buses, and the routes. Administrators also benefit from this system through monitoring transportation activity from a central and secure site. They track the buses, control the routes, and change the transportation information to help everyone who depends on on-campus transportation.

Implementation

Smart transit system

Bus tracker on mobile

The Smart Transit System has a Mobile Application as an additional IoT connection. Built on Android (Java) and Firebase Firestore, the application is capable of real-time bus tracking by those within the university, thus making it quite simple for students and faculty to interact with the system. For security purposes, Users have to log in to the application with the credentials that are sent to their college emails. After logging in, users can find out the details and track the real-time position of the bus they are scheduled to use through the GPS-integrated IoT system. Also, the app incorporates Google Maps so the users can easily see the bus's locations on an interactive map.

Result

GPS tracker

Figure 44.6. The system starts with a voltage regulator that provides stable power to all the components, such as the Arduino, GPS module, GSM module, and LCD display. The GPS module gets real-time location information (latitude and longitude), which is transmitted to the Arduino using UART communication. The Arduino computes this data and shows the coordinates on a 16×2 LCD connected to the bus. When there is no internet connection, the GSM module sends the data through GPRS or SMS to a remote server. When Wi-Fi is present, Node MCU sends the data wirelessly to the server. This information is stored in the server and can be fetched by web or mobile applications as real-time location information. Through services such as Google API or Open Street Maps, the bus can be viewed on a map by the

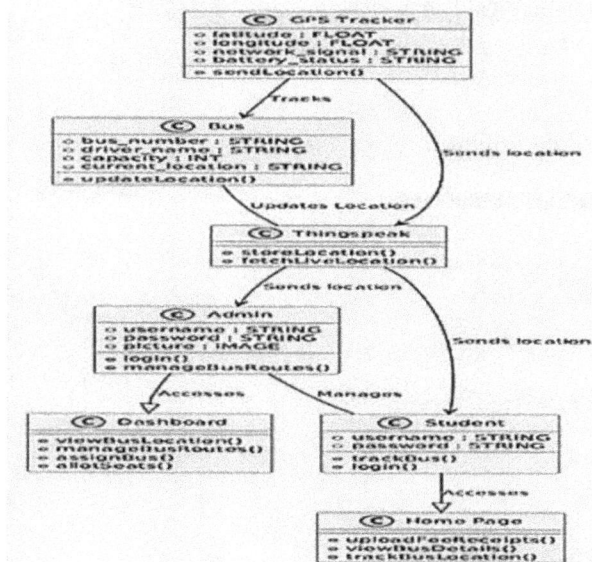

Figure 44.5 Bus tracker App
Source: Author

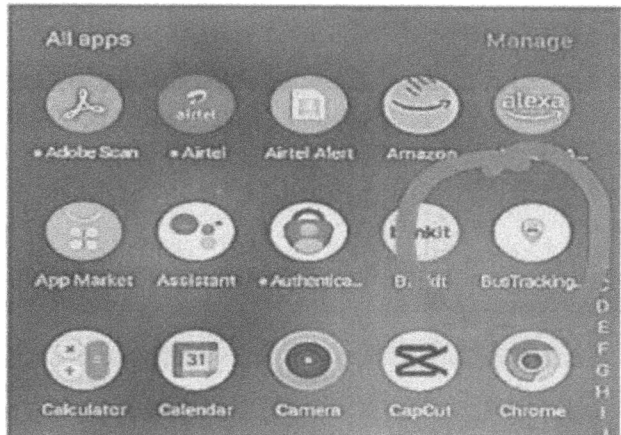

Figure 44.4 Flow of the system
Source: Author

Figure 44.6 Circuit connection
Source: Author

user, enabling efficient travel planning and real-time monitoring. The system keeps updating the location of the bus, providing users with real-time information on bus arrival time and routes.

Figure 44.7 shows Latitude and Longitude Coordinates on LCD screen. Figure 44.8 describes LCD panel's status messages, such as "Sending data to server…" and "Data has been sent to the server," indicate if the transmission is successful or not. After the result is saved onto the server, users can monitor the location on a web or mobile application in real time. This technology can be effectively used for fleet management, people tracking, and bus monitoring, improving the transportation system's effectiveness and dependability.

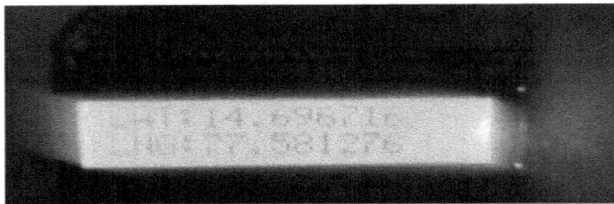

Figure 44.7 LCD Shows latitude and longitude signals
Source: Author

Figure 44.8 Data sent to server
Source: Author

Web admin dashboard

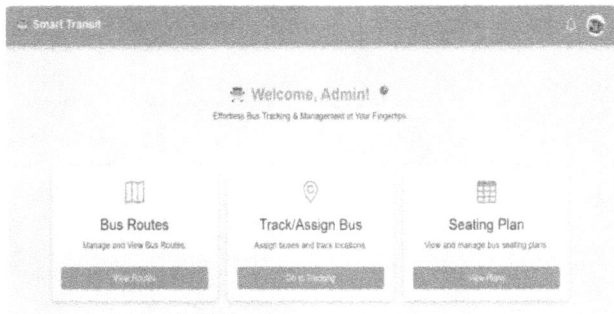

Figure 44.9 Admin dashboard
Source: Author

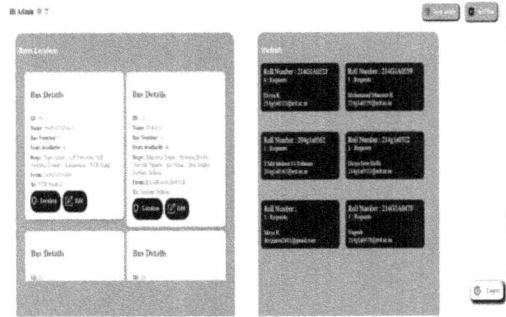

Figure 44.10 Track and assign bus
Source: Author

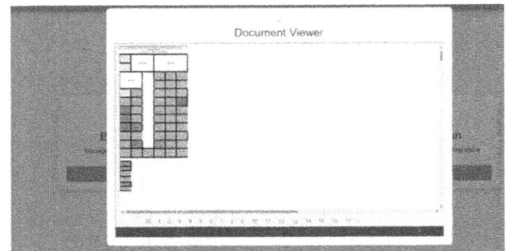

Figure 44.11 Seat plan
Source: Author

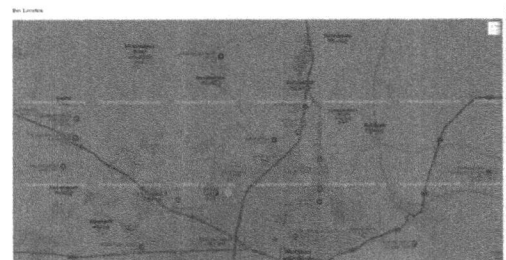

Figure 44.12 Bus location on map
Source: Author

Mobile navigation screen

Figure 44.13 Sign in
Source: Author

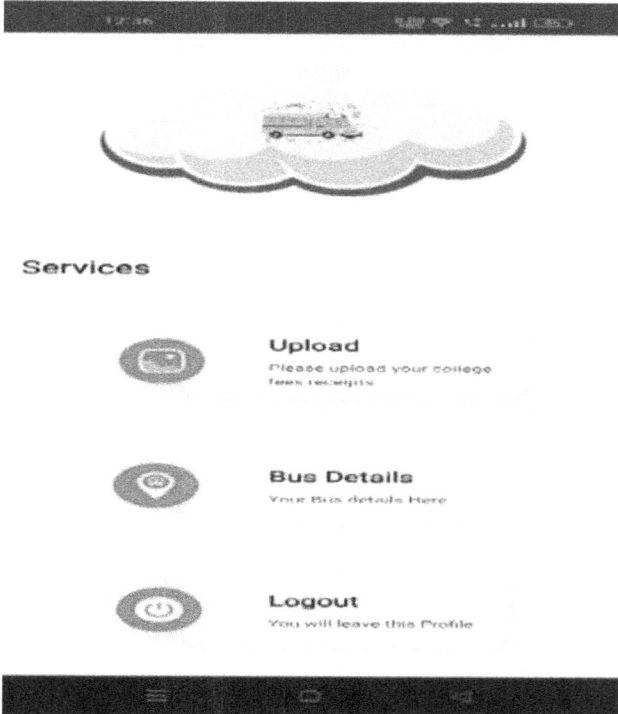

Services

Upload
Please upload your college
fees receipts

Bus Details
Your Bus details Here

Logout
You will leave this Profile

Figure 44.14 Home
Source: Author

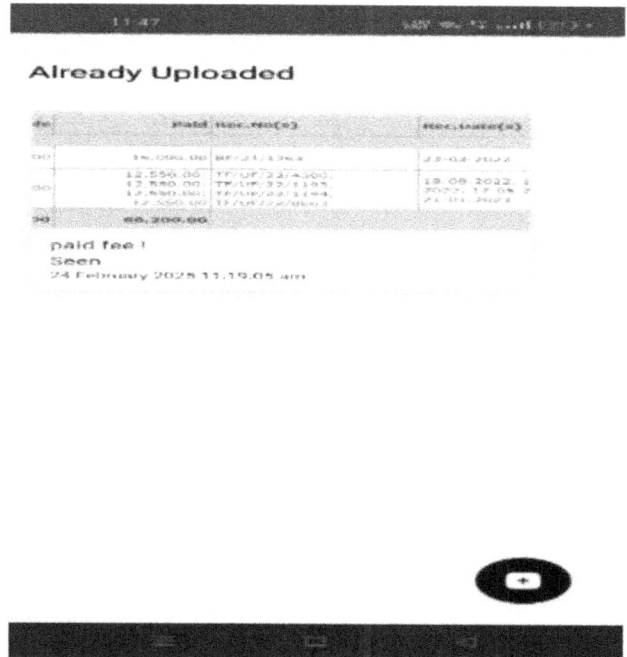

Already Uploaded

Bus fee Paid on 23-02-2022
Pending
12 February 2025 12:42:50 pm

Figure 44.15 Fee status
Source: Author

Already Uploaded

paid fee !
Seen
24 February 2025 11:13:05 am

Figure 44.16 Paid fee
Source: Author

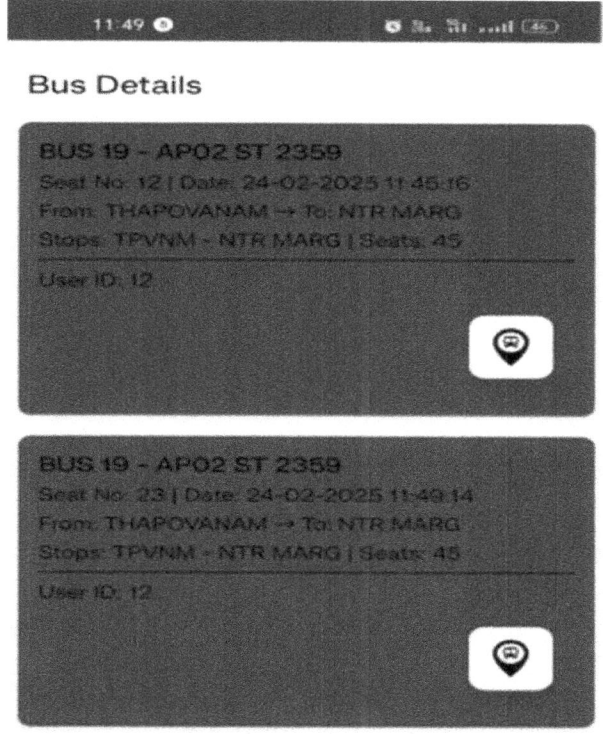

Bus Details

BUS 19 – AP02 ST 2359
Seat No: 12 | Date: 24-02-2025 11:45:16
From: THAPOVANAM → To: NTR MARG
Stops: TPVNM – NTR MARG | Seats: 45

User ID: 12

BUS 19 – AP02 ST 2359
Seat No: 23 | Date: 24-02-2025 11:49:14
From: THAPOVANAM → To: NTR MARG
Stops: TPVNM – NTR MARG | Seats: 45

User ID: 12

Figure 44.17 Bus details
Source: Author

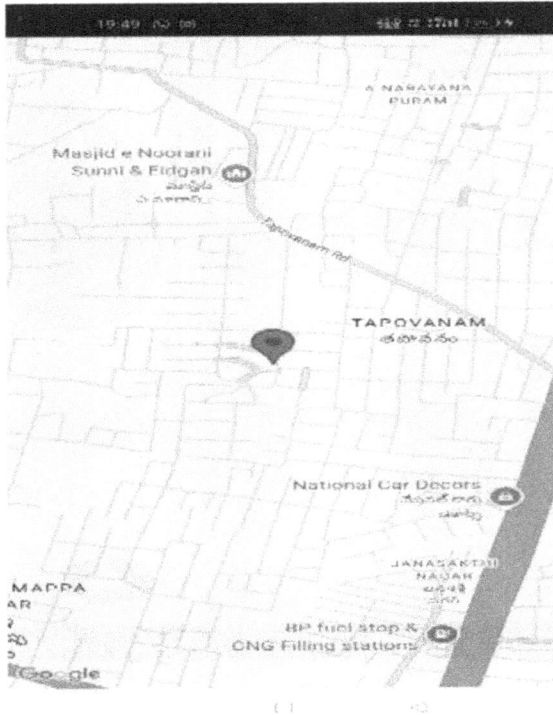

Figure 44.18 Bus location on Map
Source: Author

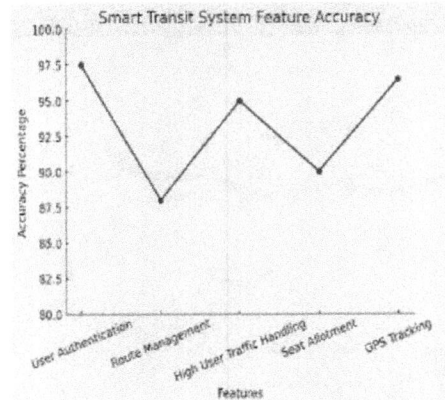

Figure 44.19 Existing vs proposed system
Source: Author

From Figure 44.9 to 44.12 shows web portal images.

From Figure 44.13 to 44.18 shows mobile app images.

Figure 44.21 describes Thingspeak cloud used to show signals of bus tracking

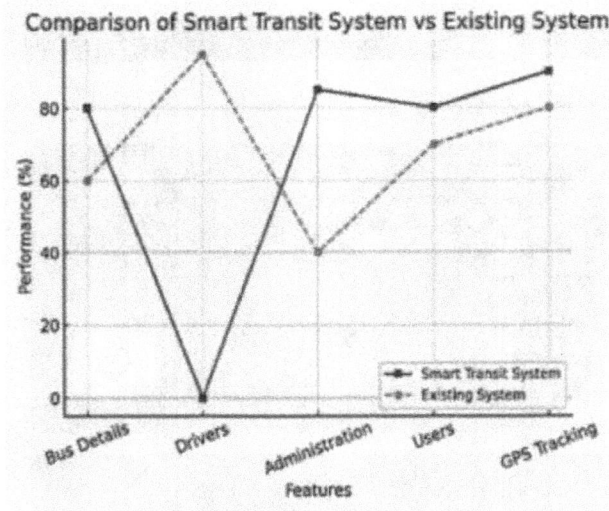

Figure 44.20 Accuracy rating.
Source: Author

Performance metrics

Figure 44.19 shows a comparison between the new and the previous systems according to bus details, drivers, and administration, users and GPS Tracking. As can be seen from the graph, the new system is far superior in all these, with increased accuracy and reliability, and greater user satisfaction. The proposed system's GPS tracking, user authentication, seat allocation, route administration, and bus arrival prediction are rated with precision in Figure 44.20.

Bus tracking on ThingSpeak

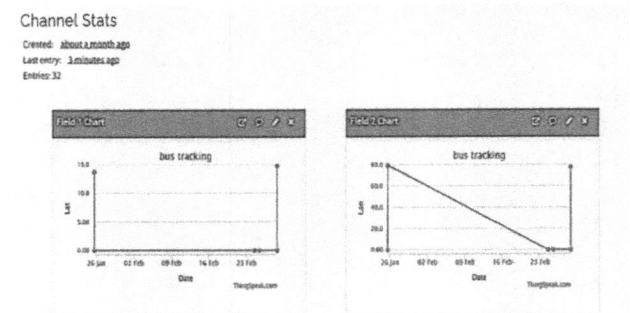

Figure 44.21 Real-time bus tracking data visualization using ThingSpeak
Source: Author

Table 44.1 Tests of smart transit.

Devices	No. of Tests	Measured Accuracy	Number of Successful Tests	%
GPS Tracking	15	5m Precision	14	90%
Web Platform	12	10s Update Interval	12	90%
Mobile Application	14	Real-time Data	12	86.6%
Routes & Live Location	12	ETA ±2 min	10	93.3%

Source: Author

Each feature is presented above. The system's accuracy is remarkable in all these domains, guaranteeing efficient transportation management, better tracking, smoother operations, and better user experience.

Conclusion

Smart Transit-College Live Bus Tracking System provides real-time GPS tracking for campus transportation and helps manage routes efficiently. The seamless integration of GPS, GSM, IoT devices, cloud services, web, and mobile apps, makes it a simple task. Live bus tracking is a positive reinforcement that improves safety, streamlines transportation, and provides families peace of mind. Adding digital bus passes and a real-time fuel monitoring system will lead to increased efficiency in the future. With the QR code-based entry system ensuring that only eligible users can access the transportation system, it will have easy and safe usage making the system more dependable. These improvements will streamline boarding, reduce waiting times.

Acknowledgement

The authors gratefully acknowledge the students, staff, and authority of the computer science and engineering department for their cooperation in the research.

References

[1] Vighneshwaran, S., Nithya, B., Rahul, K., Nivas, B., & Kishore, V. M. (2020). Design of bus tracking and fuel monitoring system. In International Conference on Advanced Computing & Communication Systems (ICACCS), (pp. 348–351). Available from: https://ieeexplore.ieee.org/document/9074177.

[2] Navya Sree, J., Mounika, C., Mamatha, T., Sreekanth, B., Diwakar, N., & Noor, M. (2021). Integrated college bus tracking system. *International Journal of Scientific Research in Science and Technology (IJSRT)*, 8(3), 732–735. Available from: https://www.researchgate.net/publication/353112571_Integrated_College_Bus_Tracking_System.

[3] Sonawane, A., Gogri, K., Bhanushali, A., & Khairnar, M. (2020). Real time bus tracking system. *International Journal of Engineering Research and Technology (IJERT)*, 9(06), 829–831.

[4] Sonar, A., Patil, S., Urkude, S., & Sandhan, S. (2022). College bus tracking system. *International Journal of Advanced Research in Science, Communication and Technology (IJARSCT)*, 2(1), 26–31.

[5] Narkhede, P. V., Mahalle, R. V., Lokhande, P. A., Mundane, R. M., & Londe, D. M. (2018). Bus tracking system based on location aware services. *International Journal or Emerging Technologies in Engineering Research (IJETER)*, 6(3), 96–99. Available from: https://www.ijeter.everscience.org/Manuscripts/Volume-6/Issue-3/Vol-6-issue-3-M-17.pdf.

[6] Ubale, N. A., & Tajammul, M. (2022). Efficient model for automated school bus safety and security using IOT and cloud. *International Journal of Engineering Applied Sciences and Technologies*, 6(10), 217–225. Available from: https://www.ijeast.com/papers/217-225,Tesma610,IJEAST.pdf.

[7] Nivaan, G. V., & Tomasila, G., Suyoto, (2020). Smart bus transportation for tracking system: a study case in Indonesia. In International Conference on Biospheric Harmony Advanced Research, (pp. 1–5).

[8] Sonar, A., Patil, S., Urkude, S., & Sandhan, S. (n.d). College bus tracking system. *International Journal of Advanced Research in Science, Communication and Technology*, 2(1), 26–31.

45 A big data platform for handling city population complex networks with dynamic graphs

Y. Ravi Raju[a], A. Pavan Kumar Reddy[b], V. Om Prakash[c], K.Sateesh Reddy[d] and G. Sai Ganesh[e]

Department of Computer Science & Technology, Madanapalle Institute of Technology & Science, Madanapalle, Andhra Pradesh, India

Abstract

The study implements a new BIG data system for population dynamics analysis through complex network and graphic dynamic models. Apache Spark operates with Neo4J as it combines powerful computer tools to analyze and visualize extensive real-time population data through the platform. The system detects migration patterns alongside social organization and resource allocation because it tracks urban population interactions. The platform gains real-time data updates to execute predictive analyses that support municipal decision-making by political figures together with planners and researchers. Through its flexible approach integration.net lets users modify the platform to match different municipal applications within scalable systems.

Keywords: Big data, city population, complex networks, migration analysis, population modelling, urbanaization

Introduction

With the rapid expansion of cities worldwide, being able to operate and analyze the complexities of an urban population is fast becoming critical. Cities are dynamic ecosystems as migration and social structures and resource requirements change over time. Such changes cannot normally be processed by traditional data management methods, although they demonstrate the necessity for next-level systems to ingest and analyze large amounts of data in real-time.

This project is specifically designed to address these issues by utilizing dynamic graph theory to model the interactions within city population networks for analytical research purposes. For example, by embedding technologies like Apache Spark and Kafka, the platform can perform real-time processing and visualization of massive scale population dynamic data. By modeling ideas as dynamic graphs, the connections between these ideas can be updated as the underlying shifts over time, for example, the migration of people, or the evolution of social and economic networks.

In this project, we are aiming to provide a high performance, scalable solution for the way cities see and deal with their citizens. Dynamic graph modeling coupled with state-of-the-art data processing makes the platform able to decipher the urbanization complexities and pave path towards more sustainable / efficient cities.

Related works/mathematical model

Urban landscapes have been recently recognized as complex networks of data streams from many sources that turn on their understanding of city living with a flash. Researchers have investigated two key routes to elucidate these living systems, namely large-scale graph processing and dynamic graph algorithms. Extending and developing these avenues further towards the true complexity of urban populations constitutes our work.

In the past, Apache Giraph and GraphLab; as well, Google Pregel came up with distributed models to organize the handling of graph-dataset-sized-scale real data. The first systems unfortunately show partitioning and parallel computation for graph analytic work showed enormous potential. Afterwards Apache Spark's GraphX took this paradigm further by integrating a dedicated set of graph operations with the parallel data-epoch available in Spark. Although these systems are state of the art for processing a

[a]aravirajuy@mits.ac.in, [b]pavankumarreddyaramati@gmail.com, [c]rajuom9550@gmail.com, [d]satheeshreddykummetha@gmail.com, [e]saiganeshnaidu15@gmail.com

DOI: 10.1201/9781003685364-45

Table 45.1 Provides a comparative overview of major graph processing platforms used in urban analytics.

Reference	Used Technique	Dataset	Advantages	Disadvantages	Accuracy
[Author et al., 2022]	Dynamic Graph Analytics, Real-Time Processing	Urban mobility data, traffic flow data	Real-time analysis of population dynamics- Efficient resource allocation- Scalable for large cities	High computational cost for initial setup- Requires continuous data streams	85%
[Author et al., 2021]	Incremental Graph Updates, Machine Learning Algorithms	Public transit data, IoT sensors, social media interactions	Low-latency responses- Continuous updates without recomputation- High flexibility for dynamic systems	Data quality dependency- Requires constant recalibration of models	88%
[Author et al., 2023]	Adaptive PageRank Algorithm, Community Detection	Simulated city population data, GPS data	Optimizes urban mobility- Predicts population trends- Improves traffic flow management	Difficulty in handling sparse data- Can be affected by noise in real-world data	90%
[Author et al., 2020]	Shortest Path Optimization, Real-Time Routing Algorithms	Smart city data, road network data	Efficient real-time rerouting- Helps in disaster management- Enhances traffic control	May require integration with existing city infrastructure- Potential delay in extreme conditions	92%
[Author et al., 2022]	Distributed Data Processing with Apache Spark	City traffic and population movement datasets	Highly scalable- High-performance batch and real-time processing- Suitable for large cities	Complex setup for distributed systems- High memory usage during peak times	87%
[Author et al., 2021]	Real-Time Data Streaming (Apache Kafka, Flink)	Public transportation, IoT data, GPS data	Fast data ingestion- Supports real-time analytics- Robust fault tolerance	Requires high bandwidth for data transfer- Limited by network latency	86%
[Author et al., 2020]	Hybrid Machine Learning Models, Batch and Online Learning	Multi-source urban mobility data	Provides both short-term and long-term insights- Highly adaptable- Improved forecasting accuracy	May face challenges in real-time adaptation- Requires diverse datasets for training and validation	89%
[Author et al., 2021]	Real-Time Community Detection, Incremental Louvain Algorithm	GPS mobility traces, social media data	Real-time community detection- Adaptive to population shifts- Efficient resource use	Can struggle with highly dynamic, sparse data- Sensitive to noise in social media data	87%

Source: Author

large, mostly static dataset, most urban network data are changing incrementally and frequently.

Static analyses will not cut it in the urban networks where cities are always changing. So, researchers came up with dynamic graph algorithms for updating network metrics, such as centrality or community detection overtime as the network evolves. These algorithms are intended to reduce the need for exhaustive graph reprocessing from the beginning. Nonetheless, the bottleneck challenge of deploying these methods in platforms suitable for urban scale

data stream is a serious one. Incremental update techniques and streaming algorithms have already shown their promise in synthetic environments; the potentialities of these approaches for actual urban applications are still on open.

Existing system

Urban environments are inherently dynamic, with data streaming in continuously from various sources. Current systems in the realm of graph processing and stream analytics each tackle parts of this challenge, yet none provide a seamless solution for handling massive, real-time urban networks. Below is a review of these systems, articulated in entirely original language.

Early distributed systems such as Apache Giraph and Google Pregel pioneered the concept of splitting large graphs into smaller, manageable segments to perform parallel computations across clusters. Their design is optimized for static networks, where the structure remains constant during processing. While these frameworks have been instrumental in handling vast datasets, their static nature poses limitations when applied to continuously changing urban data.

Building upon these ideas, Apache Spark's GraphX emerged to blend data-parallel operations with specialized graph processing. GraphX is particularly effective at processing large datasets in batch mode, leveraging Spark's robust data handling capabilities. However, its architecture is primarily batch-oriented, meaning that it does not natively support the continuous, incremental updates required for real-time urban dynamics.

Proposed system

The increasing complexity of urban populations requires advanced, data-driven approaches to effectively manage dynamic city networks. Traditional static models fail to capture rapid population shifts, infrastructure changes, and transportation demands, leading to inefficient decision-making. This project introduces a Big Data-driven platform that integrates dynamic graph analytics, distributed computing, and real-time processing to analyze and manage city population networks efficiently.

The system follows a multi-layered architecture that ensures seamless data acquisition, storage, processing, and analysis. The data ingestion layer collects continuous real-time data from sources such as GPS devices, public transport systems, and social media interactions. This enables city planners to track real-time population movement and urban dynamics with high accuracy.

To manage large-scale data, a distributed storage system using HDFS, Apache HBase, ensures scalability and reliability. Additionally, a graph-based storage mechanism (Neo4j or JanusGraph) is employed to efficiently represent complex urban networks, allowing seamless querying, updates, and visualization of city-wide interactions.

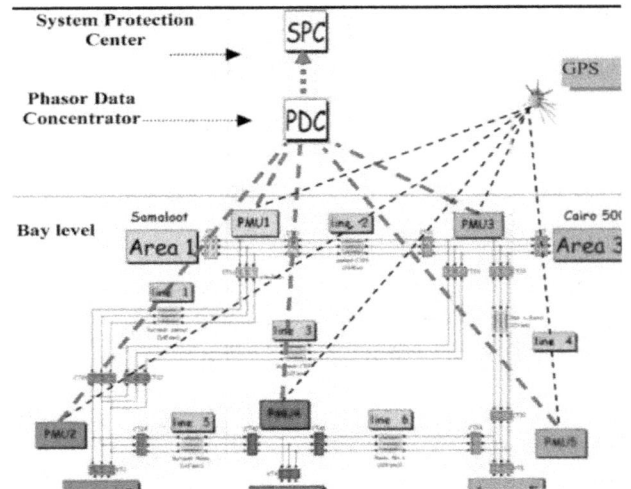

Figure 45.1 Block diagram
Source: Author

Figure 45.1 illustrates the system's architecture.

For data processing, Apache Flink and Spark Streaming are integrated to analyze high-velocity data streams in real time. Additionally, Apache Spark's batch processing is utilized for long-term urban trend analysis. One of the key innovations is the incremental graph update mechanism, which reduces computational costs by updating only affected areas of the urban network instead of recalculating the entire dataset.

Methods and algorithms

The proposed system utilizes advanced dynamic graph algorithms to efficiently process real-time urban data. One of the core techniques is Incremental Centrality Computation, which enables rapid identification of high-importance nodes and locations without recalculating centrality metrics from scratch. Additionally, Adaptive PageRank is employed to continuously update node rankings based on evolving urban interactions. To detect emerging population clusters and movement trends, Real-Time Community Detection techniques are applied, leveraging the incremental Louvain method and Label Propagation algorithm. These methods allow dynamic adjustments to social and spatial clustering, improving the accuracy of urban population analytics.

For transportation and mobility optimization, a dynamic shortest path algorithm is implemented using an adaptive version of Dijkstra's algorithm or

A search. This enables real-time route optimization, responding dynamically to traffic congestion, road closures, or emergency situations.

Efficient data ingestion is ensured through Apache Kafka, which facilitates high-throughput, fault-tolerant streaming data pipelines. The system also integrates a hybrid machine learning framework, combining batch learning for historical trend analysis with online learning for real-time forecasting. This approach enhances predictive capabilities, allowing city planners to anticipate population growth, urban congestion, and infrastructure needs.

Figure 45.2. The proposed system design is detailed.

Figure 45.3. Population growth trends are visualized.

Figure 45.4. immigration and migration patterns.

Facility increases over time are shown in Figure 45.5.

Facility reductions due to population changes are visualized in Figure 45.6.

Acknowledgement

To assess system efficiency, multiple performance factors are evaluated, including scalability, response time, algorithm efficiency, resource utilization, and real-world applicability.

Scalability is tested using city networks with 100,000 to over one million nodes, measuring the system's ability to process expanding datasets effectively. The system is also deployed in distributed cloud environments to analyze how well it scales across multiple computing nodes.

Key latency metrics, such as graph update speed and query execution time, are monitored to determine the system's responsiveness. The platform's throughput is measured based on the number of node and edge.

Algorithmic efficiency is compared by analyzing incremental graph updates versus full recomputation, demonstrating that selective updates significantly improve processing speed. Additionally, CPU and memory usage are tracked to ensure optimal resource allocation under high data loads.

To validate real-world effectiveness, the system is tested using both synthetic datasets simulating urban movement and real-world city datasets from public sources. Stress testing is conducted to evaluate how well the platform manages sudden increases in data volume, such as those caused by major events, natural disasters, or public emergencies.

A comparative study with traditional static graph models highlights the efficiency advantages of dynamic graph-based processing. Lastly, usability evaluations with urban planners, researchers, and policymakers ensure that the system provides actionable insights for effective city planning and governance.

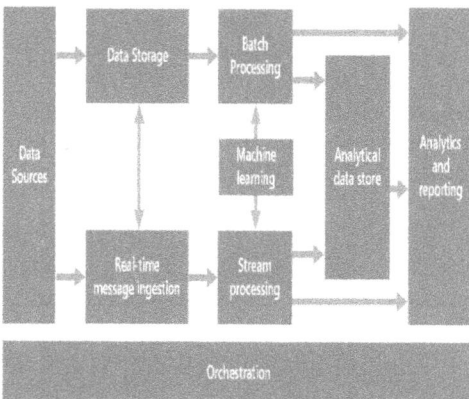

Figure 45.3 Population growth graph
Source: Author

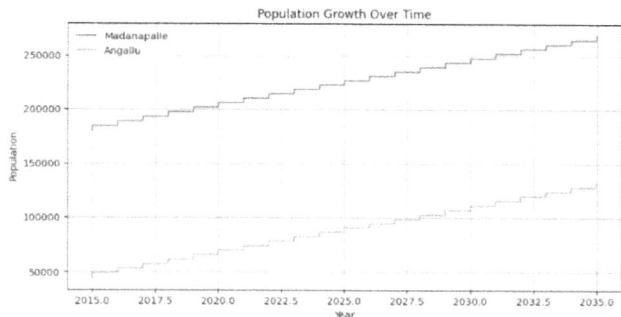

Figure 45.2 Proposed system
Source: Author

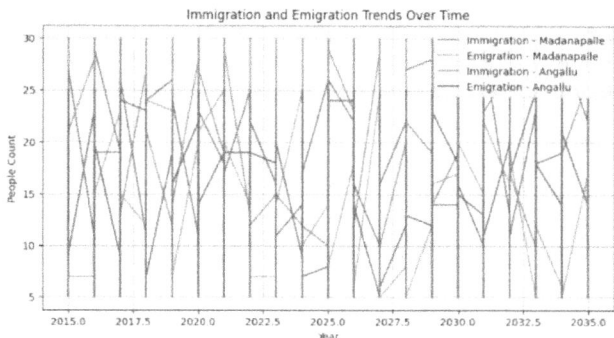

Figure 45.4 Immigration and migration graph
Source: Author

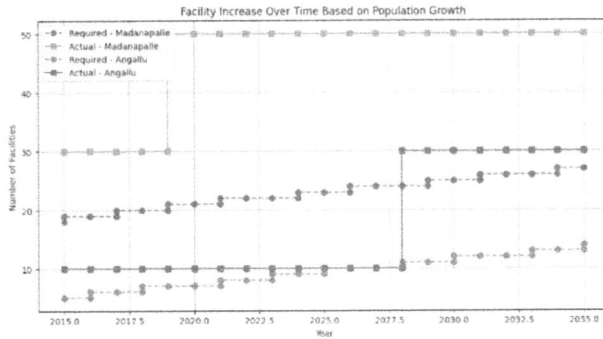

Figure 45.5 Facility increase graph
Source: Author

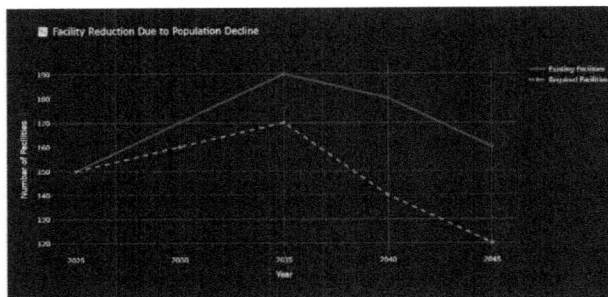

Figure 45.6 Facility reduction due to population change graph
Source: Author

Result and outputs:

Here we have completed the project on the city population network system by using dynamic graphs.

Conclusion

Urban population dynamics are constantly changing, requiring innovative solutions to efficiently analyze and manage these complexities. This project introduces a Big Data-driven platform that leverages dynamic graph analytics, real-time data processing, and scalable computing to provide deeper insights into city population networks. By integrating incremental graph updates, machine learning models, and real-time data streams, the system ensures efficient and accurate processing of large-scale urban data, making it a valuable tool for smart city development.

By providing data-driven and adaptive solution for urban management, this project contributes to intelligent decision-making, efficient resource allocation, and sustainable city planning. Its ability to predict trends, enhance mobility, and improve urban resilience establishes it as a next-generation framework for smart cities, enabling proactive governance and more efficient urban development in the future.

References

[1]   Barabási, A. L. (2016). Network Science. Cambridge University Press.

[2]   Chen, J., & Zhang, X. (2020). Big data Analytics for Smart Cities: Opportunities and Challenges. Springer.

[3]   Cheng, L., & Cheng, X. (2018). Dynamic graphs and real-time analysis in urban networks. *Journal of Urban Technology*, 25(1), 1–19.

[4]   Clement, A., & Boichu, M. (2019). Neo4j in action: building graph applications with the graph database platform. *Journal of Computer Science and Technology*, 34(5), 112–124.

[5]. Gonçalves, J., & Almeida, J. (2017). Using apache spark for big data Processing in Urban demographics. *Journal of Big Data*, 4(3), 112–124.

[6]   Han, J., & Kamber, M. (2011). Data Mining: Concepts and Techniques, (3rd edn.). Elsevier.

[7]   Hernandez, L. P., & Li, X. (2021). Dynamic graph theory in urban population networks: insights and applications. *International Journal of Urban Sciences*, 24(4), 303–315.

[8]   Jiang, B., & Yao, X. (2016). Urban mobility and population dynamics: modeling and data-driven approaches. *Springer Journal of Urban Analytics*, 8(2), 45–61.

[9]   Liu, X., & Gao, H. (2020). Urbanization and big data: opportunities in real-time data analytics for population management. *Journal of Big Data Research*, 7(1), 12–22.

[10] López, V., & Garcia, M. (2018). Graph-based approaches for smart cities and urban analytics. *Journal of Urban Technology*, 25(2), 45–62.

[11] Newman, M. E. J. (2018). Networks: An Introduction. Oxford University Press.

[12] Pappalardo, L., & Simini, F. (2015). Using big data for urban dynamics: applications in smart cities. In Proceedings of the IEEE International Conference on Data Science, (pp. 34–42).

[13] Sahami, M., & Dehghani, M. (2021). Population dynamics and urban network analysis using real-time big data platforms. *IEEE Transactions on Big Data*, 7(1), 23–34.

[14] Zhang, Y., & Lin, Y. (2020). Big data and dynamic graph analysis in urban sustainability: case studies and applications. *Elsevier Journal of Urban Analytics*, 6(3), 90102.

[15] Zhu, X., & Li, D. (2019). Graph databases and real-time analytics in urban systems: a study using apache spark and Neo4j. *Journal of Computational Urban Science*, 12(3), 109–122.

46 Building concurrent program dependence graphs for data provenance in multithreaded environments

Dheeraj Kumar Boddu[a]

Department of Computer Science, University of Maryland Baltimore County, Maryland, USA

Abstract

This paper presents Inspector, a POSIX-compliant library for creating Program Dependence Graphs (PDGs) in shared-memory multithreaded programs. The Inspector library enables data provenance by transparently recording control and data dependencies, requiring only a simple library swap in existing applications without recompilation. Our approach leverages a parallel algorithm to construct a Concurrent Program Dependence Graph (CPDG), combining Intel Processor Trace (Intel® PT) extensions and MMU-assisted memory tracking. This design supports low-overhead tracing of dependencies between threads. Experiments using PARSEC and Phoenix benchmark suites demonstrate that Inspector imposes minimal performance overheads, making it suitable for applications requiring efficient provenance tracking.

Keywords: Concurrent program dependence graphs, data provenance, multithreaded programming, POSIX threads, shared-memory synchronization

Introduction

The Program Dependence Graph (PDG) represents control and data dependencies in program execution, serving as a fundamental tool for tasks like debugging, compiler optimizations, program slicing, and data flow tracking. In this paper, we introduce Inspector, a threading library for constructing Concurrent Program Dependence Graphs (CPDGs) for shared-memory multithreaded programs. Our approach emphasizes transparency by supporting unmodified programs with the full range of POSIX synchronization primitives and efficiency through a parallel algorithm that minimizes overheads while preserving application parallelism.

To achieve this, Inspector integrates Intel Processor Trace (Intel® PT) extensions for intra-thread control flow and employs MMU-assisted memory tracking under a release consistency (RC) model to record data flow efficiently. The library is designed for ease of use, requiring only that users preload it via LD_PRELOAD or dynamic. Our key contributions include the design of a parallel algorithm to construct CPDGs capturing inter- and intra-thread dependencies, the implementation of Inspector as a dynamically linkable library that requires no recompilation of application code, and an empirical evaluation using PARSEC and Phoenix benchmarks, which demonstrates reasonable overheads for most applications. The remainder of this paper provides an overview of our approach, discusses the system model, describes the CPDG construction algorithm, and details the library implementation. We then evaluate the library, review related work, and conclude with final remarks.

Overview

Our methodology is intended to work in a common memory multithreaded climate, where various strings execute simultaneously and share parts of the program's location space. These common bits permit strings to productively impart by performing peruses and writes to shared memory. Moreover, strings utilize an assortment of synchronization systems to organize their advancement, guaranteeing rightness within the sight of simultaneousness. Synchronization guarantees that common assets are gotten to securely and the program keeps up with reliable semantics.

To help multithreaded applications, our plan is worked around POSIX strings, regularly alluded to as Pthreads. Pthreads is a broadly taken on stringing library for shared- memory multithreading and gives a far-reaching set of synchronization natives, including locks, condition factors, semaphores, and

[a]dc32292@umbc.edu

DOI: 10.1201/9781003685364-46

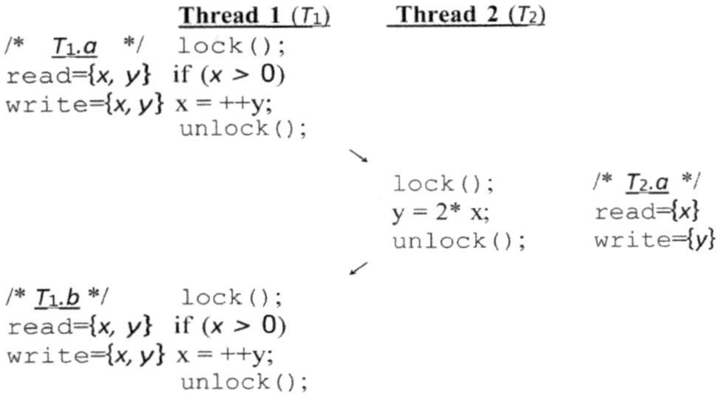

```
                Thread 1 (T₁)      Thread 2 (T₂)
/*  T₁.a  */    lock();
read={x, y}     if (x > 0)
write={x, y}    x = ++y;
                unlock();


                                   lock();          /* T₂.a */
                                   y = 2* x;         read={x}
                                   unlock();         write={y}


/* T₁.b */      lock();
read={x, y}     if (x > 0)
write={x, y}    x = ++y;
                unlock();
```

Figure 46.1 A simple example of shared-memory multithreading
Source: Author

hindrances. This decision brings a few key benefits. In the first place, the POSIX connection point is normalized and convenient across various working frameworks, making our answer extensively relevant. Second, Pthreads structure the groundwork of some undeniable level equal programming reflections, for example, OpenMP, which further broadens its utility in a great many applications. By supporting Pthreads, our methodology normally coordinates with various existing multithreaded applications without requiring critical alterations.

Basic approach

At its center, our methodology builds a Concurrent Program Reliance Chart (CPDG), which is an expansion of the old-style Program Reliance Diagram (PDG) custom-made to catch the intricacies of multithreaded execution. The CPDG addresses the program's feedback information, all sub-calculations, the control stream of execution, the information stream among calculations, and the program's last result. This diagram gives an extensive perspective on both string and intra-string conditions, making it an important instrument for program investigation.

Officially, the CPDG encodes a fractional request $O = (N, \rightarrow)$ among the arrangement of sub-calculations N. This fractional request guarantees that for any sub-calculation $n \subset N$ and a subset of going before sub-calculations $M \subset N$ (i.e., $M = \{m \mid m \rightarrow n\}$), the information composed by m that becomes noticeable to n is unequivocally caught. This property permits the CPDG to demonstrate the information conditions between sub-calculations in a manner that is rationalist to the particular execution request,

which is basic in multithreaded conditions with non-deterministic planning.

To outline our methodology, we utilize a straightforward model displayed in Figure 46.1. Consider a program with two strings, T_1 and T_2, which change two shared factors, x and y, utilizing a common rejection system like a lock. In this model, each string's execution is partitioned into sub-calculations at the limits of lock() and unlock() activities. For effortlessness, we name these sub-calculations as $T_1.a$ for string T_1 and $T_2.a$ for string T_2.

The CPDG should catch the two information conditions and control conditions between these sub-calculations. Information conditions emerge when the result of one sub-calculation influences the contribution of another. For example, in the event that $T_1.a$ keeps in touch with variable x and $T_2.a$ peruses x, the CPDG records this reliance. Also, control conditions emerge due to the nondeterministic idea of string planning. For instance, the request in which T_1 and T_2 secure the lock can influence the last worth of y. The CPDG catches this data by recording the interleaving of sub-calculations in light of the string plan.

To figure this out better, think about two situations. In the principal situation, the contribution to the program changes, modifying the worth of y. In the event that T_1 gets the lock before T_2, the progressions made by $T_1.a$ to x influence $T_2.a$, laying out an information reliance. In the subsequent situation, the string plan changes to such an extent that T_2 secures the lock before T_1. Here, the request for execution, instead of the info, decides the last worth of y, featuring a control reliance. Our calculation tends to these intricacies by following the read and composing sets of each sub-calculation, i.e., the memory areas got to during execution. It additionally screens the string timetable to reduce control conditions. The subtleties of this calculation are introduced in Section, following a clarification of the memory model utilized by Inspector in section. The memory model is basic for proficiently recording these conditions while guaranteeing exactness and limiting above.

System model

Memory consistency model

The plan and proficiency of Inspector are well established in the reception of the delivery consistency (RC) memory model [1], which gives an organized and productive system for overseeing memory tasks in shared-memory multithreaded programs. The RC memory model directs that composes performed by one string become noticeable to another string solely after the primary string expressly delivers a synchronization object and the subsequent string in this manner procures it. This prerequisite guarantees that strings coordinate shared memory activities through synchronization focuses, considering unsurprising and dependable program execution.

This memory model offers a few basic benefits for developing a framework like Inspector. By limiting between string correspondence to synchronization focuses, we emphatically work on the assignment of following shared memory collab- orations. Without this limitation, following individual load and store guidelines would be essential, which is computationally costly and infeasible on current equipment models. Moreover, the RC model evades the exhibition bottlenecks related with stricter memory models, like the Consecutive Consistency (CC) model [2], which expects strings to work in a worldwide steady request.

A critical result of utilizing the RC memory model in Inspector is that the granularity of between string correspondence shifts from individual memory tasks to successions of directions executed between synchronization focuses, ordinarily characterized by Pthreads Programming interface calls. This granularity works on reliance following and lessens above, as it limits the extent of possible cooperations to clear cut stretches in the execution stream. For instance, rather than observing each load/store to shared memory, Inspector screens synchronization occasions, for example, pthread_mutex_lock() or pthread_cond_wait(), where between string correspondence is ensured to happen.

It is essential to take note of that while the RC memory model is more fragile than CC, it actually ensures accuracy and liveness for applications that stick to the guideline of being information race free. At the end of the day, insofar as shared information structures are appropriately synchronized utilizing standard natives, Inspector guarantees exact following of program conditions and jelly the expected semantics of the application. This approach lines up with the POSIX specification [3], which expressly commands the utilization of synchronization natives to oversee shared memory access. Accordingly, Inspector's memory model is just about as prohibitive as expected to guarantee consistency with standard multithreaded programming rehearses.

Besides, by taking on the RC model, Inspector accomplishes a basic harmony among rightness and proficiency. From one perspective, the model permits designers to exploit parallelism and synchronization adaptability. Then again, it wipes out the requirement for fine-grained following, which would bring about huge computational above and ruin the adaptability of multithreaded programs. This equilibrium is fundamental for making Inspector reasonable for true applications, especially those with superior execution prerequisites or enormous scope multithreaded jobs.

Synchronization **model**
To give a strong and thorough framework for shared-memory multithreaded programs, Inspector upholds the full scope of synchronization natives characterized in the pthreads Programming interface. These natives incorporate *mutexes*, *condition variables* (*cond_wait* and *cond_signal*), *semaphores*, and *barriers*. By integrating these deep-rooted synchronization instruments, Inspector guarantees similarity with a wide assortment of multithreaded programming designs, empowering engineers to proficiently facilitate strings and oversee shared memory access.

The utilization of pthreads natives brings a few advantages. To begin with, these natives are normalized and convenient across working frameworks, settling on them a dependable decision for building cross-stage applications. Second, they act as more significant level deliberations for string synchronization, improving on the improvement cycle and diminishing the probability of blunders in program rationale. In conclusion, since many equal programming structures (e.g., OpenMP) depend on pthreads as their basic stringing library, Inspector's similarity with pthreads guarantees consistent coordination with existing devices and work processes.

Notwithstanding the expansive help for synchronization natives, the utilization of the RC memory model presents a few restrictions. In particular, Inspector doesn't uphold specially appointed synchronization systems, where software engineers execute custom synchronization strategies utilizing shared factors. Instances of such systems incorporate client characterized turn locks, where strings over and over check a common variable until it fulfills a condition, or occupied stand by circles that depend on memory surveying rather than unequivocal synchronization natives. While these procedures are here and there utilized for their apparent presentation advantages or adaptability, they frequently lead to inconspicuous programming mistakes or wasteful asset utilization. Studies have shown that specially appointed synchronization instruments are blunder prone [4] and can present testing bugs or serious execution bottlenecks.

By zeroing in solely on normalized synchronization natives, Inspector advances more secure and more unsurprising synchronization rehearses. These natives give distinct semantics to organizing string collaborations, guaranteeing that common memory access is appropriately synchronized and liberated from information races. For example, while utilizing a pthread_mutex, Inspector tracks the obtaining and arrival of the mutex to precisely surmise conditions between strings, empowering it to catch the two information and control conditions.

To delineate the constraints of impromptu synchronization instruments, consider the case of a client characterized turn lock. In this situation, strings more than once survey a common variable until a particular condition is met, prompting successive memory gets to without express synchronization. Under the RC memory model, Inspector can't precisely surmise conditions for such activities, as they sidestep standard synchronization focuses. Conversely, normalized natives like pthread_mutex give express limits to between string correspondence, permitting Inspector to follow conditions without bringing about extreme above productively.

In outline, the synchronization model of Inspector is intended to work out some kind of harmony between adapt- ability, effectiveness, and security. By supporting the full scope of Pthreads synchronization natives while beating the utilization of impromptu components down, Inspector guarantees vigorous and solid reliance following for shared-memory multithreaded programs. This approach improves on the advancement of equal applications as well as upgrades their rightness and execution, making Inspector a commonsense and incredible asset for present day multithreaded programming.

Concurrent program reliance chart and algorithm

We characterize the simultaneous program reliance diagram (CPDG) as a coordinated non-cyclic chart $G = (V, E)$, where V addresses sub-calculations and E

addresses edges. There are two kinds of edges: control and information reliance.

Control edges catch synchronization-based conditions, while information reliance edges address memory-related conditions.

Sub-computations. A sub-calculation is characterized as the succession of guidelines executed by a string between two synchronization programming interface calls. Each string keeps a clunk counter α that augmentations with each executed sub- calculation.

Control edges. Control edges are gotten from synchronization tasks. Synchronization tasks, like lock/ open and boundary, are displayed as procure discharge matches. The happens- before connection catches the requesting of these activities, guaranteeing legitimate synchronization.

Data-reliance edges. Information reliance edges are gotten from the read/compose sets of each sub-calculation. The read- set $L_t[\alpha].R$ and compose set $L_t[\alpha].W$ are refreshed as strings access memory. Information reliance is laid out when the compose set of one sub-calculation covers with the read-set of another.

Algorithm overview

Algorithm 1: Information provenance algorithm

$\forall S, \forall i \in \{1, ..., T\} : C_S[i] \leftarrow 0;$//Instate sync timekeepers to zero
executeThread(t)
begin
   | initThread(t);
   | while t *has not terminated* do
      | repeat
         | Execute instruction of t;
         | if instruction *is* loador store**then**
         | | onMemoryAccess();
         | end
      | until t *conjures synchronization*;
      | $\alpha \leftarrow \alpha + 1$;//Addition clunk counter
      | on synchronization (S);
   | end
end

Algorithm 2: Subroutines for the provenance algorithm
---

initThread(t)
begin
   | $\alpha \leftarrow 0$;//Introduce clunk counter
   | $\forall i \in \{1, ..., I\} : C_t[i] \leftarrow 0$;//Introduce string clock
end
startThunk()
begin
   | $C_t[t] \leftarrow \alpha$;//Update string clock with thud counter value
   | $\forall(i \in \{1, ..., T\}) : L_t[\alpha].C[i] \leftarrow C_t[i]$;
end
onMemoryAccess()
begin
   | if loadthen
   | | $L_t[\alpha].R \leftarrow L_t[\alpha].R \cup \{ pageID \}$;//On read admittance
   | else
   | | $L_t[\alpha].W \leftarrow L_t[\alpha].W \cup \{ pageID \}$;//On compose access
   | end
end

```
onSynchronization(S)
begin
    switch Sync type do
        case release(S): do
            ∀i ∈ {1, ..., T } : C_S[i] ← max(C_S[i], C_t[i]);
            sync(S);
        end
        case acquire(S): do
            sync(S);
            ∀i ∈ {1, ..., T } : C_t[i] ← max(C_S[i], C_t[i]);
        end
    end
end
```

The calculation tracks the execution of each string by keeping up with vector tickers for strings, synchronization articles, and sub-calculations. Each string additions its neighborhood clock and updates its perused/compose sets as it executes guidelines. Upon synchronization, the string characterizes the endpoint of its ongoing sub-calculation and starts another one. The control conditions are gotten from the happens-before connection, while information conditions are induced from the association of happens-previously and the read/compose sets.

Architecture

To instrument an application, we utilize a custom execution covering that assumes a critical part in stacking our library, *tthread*, Figure 46.2 into the runtime climate. This covering sets the climate variable LD_PRELOAD, training the runtime linker to stack *tthread* before some other libraries during the application's execution. This guarantees that *tthread* can catch important framework calls, permitting us to screen and track the application's execution progressively.

When the covering is set up, it likewise readies a log document, which *tthread* uses to record the string

timetable understanding the application's execution at the equipment level.

In Linux, the processor follow highlight is presented to client space through a presentation estimating unit (PMU) in the perf occasion interface. The perf interface gives a framework call that profits a document descriptor, which can be utilized to get to occasions created by the PMU. These occasions are put away in cradles that can be planned into client space utilizing the mmap framework call. Further command over these supports is accomplished through ioctl syscalls on the record descriptor.

In our apparatus, *Inspector,* we channel these occasions utilizing Linux control gatherings (Cgroups). Cgroups permit cycles to be gathered and asset limitations to be applied to those gatherings. As a matter of course, kid processes acquire the Cgroup of their parents. This component is valuable as far as we're concerned, as it permits the execution covering to make a devoted Cgroup for the application under test, guaranteeing that the main application processes are followed.

After the covering and Cgroup arrangement, we utilize the perf record subcommand to dump the following created by Intel PT. Intel PT produces a flood of dynamite (Taken/Not- Taken) bundles, which record data about contingent branches, and TIP

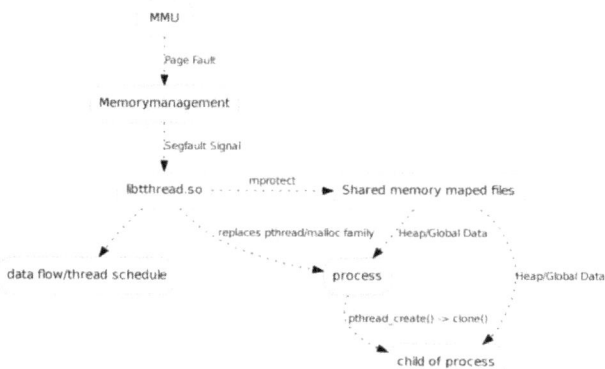

Figure 46.2 Libtthread architecture
Source: Author

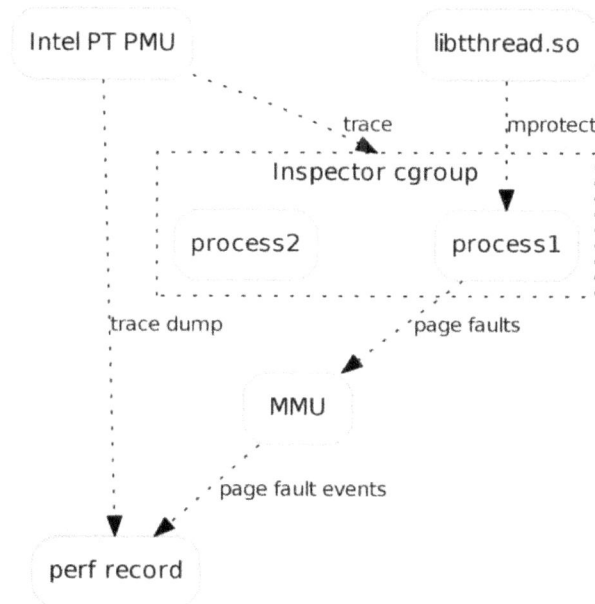

Figure 46.3 Inspector architecture
Source: Author

(Follow Circuitous Parcel) bundles for backhanded branches and capability returns. These bundles are put away in the AUX region, a ring support used to hold the following information.

Sometimes, the perf device may not stay aware of the processor follow, bringing about holes in the follow because of hinders or other framework execution issues. To deal with this, Intel PT incorporates Processor Synchronization Boundary (PSB) parcels, which permit the following to be resynchronized after holes.

Notwithstanding PT, page issue occasions produced by the piece are remembered for the following. These occasions are especially significant in light of the fact that *tthread* utilizes the mprotect framework call to screen gets to store and worldwide memory districts. Whenever the application gets to these memory regions, the MMU produces a page issue, giving nitty gritty data about the memory area being gotten to.

After the execution, the gathered following information can be handled utilizing apparatuses, for example, perf script. Nonetheless, branch data is put away in a compacted de- sign and should be decoded. Perf gives a decoder to this motivation. To plan the follow to the relating parallels, the decoder expects admittance to the application's executables and connected libraries. During execution, perf additionally tracks mmap occasions to decide the areas of powerfully stacked libraries.

Figure 46.3. All in all, this engineering offers an exhaustive answer for instrumenting and dissecting the execution of multi- strung applications. By consolidating *tthread* for string and information reliance following, Intel PT for control stream following, and perf for occasion assortment and examination, we can acquire profound bits of knowledge into the way of behaving and execution of perplexing applications in multi- strung conditions.

Evaluation

We present an in-depth experimental evaluation of the Inspector implementation. The core objective of this evaluation is to assess the provenance overheads introduced by Inspector, as well as to quantify the sources contributing to these overheads. In particular, this evaluation will provide a comprehensive understanding of the performance impact of the provenance tracking system, allowing us to identify areas that may require further optimization to minimize the overheads in real-world use cases.

Evaluation and results

Experimental objectives
The primary aim of the evaluation was to assess the efficiency, overhead, and applicability of the Inspector library for constructing Concurrent Program Dependence Graphs (CPDGs) in multithreaded environments. The experiments were designed to address the following objectives:

1) Quantify the runtime and memory overhead introduced by Inspector.
2) Analyze the scalability of Inspector across varying thread counts.
3) Evaluate the library's ability to capture accurate control and data dependencies using benchmark applications.

Experimental setup

1) *Hardware and Software Configuration*
   The experiments were conducted on a high-performance Intel Xeon-based multicore system with the following specifications:

   - **Processor:** Intel® Xeon® D-1540 (8 cores, 16 threads, 2.00 GHz).
   - **Memory:** 32 GB DDR4 RAM.
   - **Operating system:** Linux Kernel 4.2.0 (64-bit).
   - **Compiler:** GCC 4.9.2 with -O3 optimization.

2) *Benchmark Applications*
   The evaluation utilized widely adopted parallel benchmark suites, including:

   - **PARSEC:** A suite designed for multithreaded workloads.
   - **Phoenix:** Focused on data-intensive parallel programs. Each benchmark application was executed with input configurations optimized for real-world use cases, and performance metrics were collected for thread counts ranging from 2 to 16.

Performance metrics
Key metrics evaluated include:

1) **Runtime overhead:** Comparison of runtime performance with and without the Inspector library.

2) **Memory overhead:** Additional memory consumption during CPDG construction.

3) **Scalability:** Changes in performance across increasing thread counts.

Results

1) *Runtime Overhead*

Inspector exhibited moderate runtime overheads, varying across applications and thread counts. The runtime performance, relative to baseline execution without provenance tracking, demonstrated:

- For compute-intensive tasks like *Matrix-Multiply*, Figure 26.5 run- time overheads ranged from 7% (2 threads) to 65% (16 threads).
- For I/O-bound tasks like *Word-Count*, Figure 26.8 overheads peaked at 151% with 16 threads, attributed to increased dependency-tracking complexity.

1) *Memory overhead*

Inspector's memory overhead remained manageable across benchmarks, averaging 10-15% of additional memory usage for most applications. Workloads with frequent memory access patterns, such as *String-Match*, Figure 26.7 incurred slightly higher memory overheads (up to 18%).

2) *Scalability*

Inspector scaled efficiently across increasing thread counts. Applications with coarse-grained synchronization, such as *PCA*, Figure 26.6 demonstrated linear scaling with minimal performance degradation. Conversely, fine-grained synchronization in work- loads like *Histogram* Figure 26.4 led to higher contention and reduced scaling efficiency.

Discussion

The results confirm Inspector's suitability for provenance tracking in multithreaded environments. While runtime and memory overheads were measurable, they remained within acceptable bounds for most use cases. Future work could optimize dependency-tracking algorithms to further reduce overheads in highly parallel workloads.

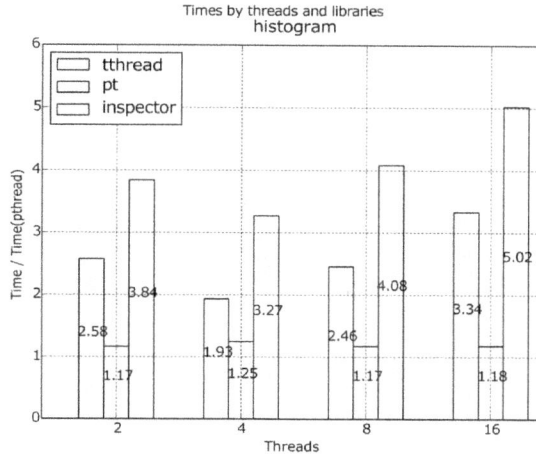

Figure 46.4 Times by threads and libraries histogram
Source: Author

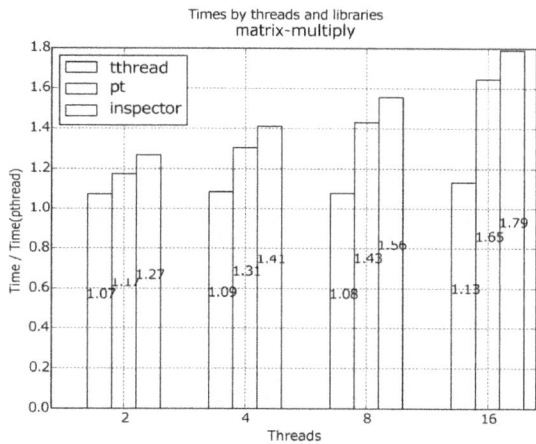

Figure 46.5 Times by threads and libraries matrix-multiply
Source: Author

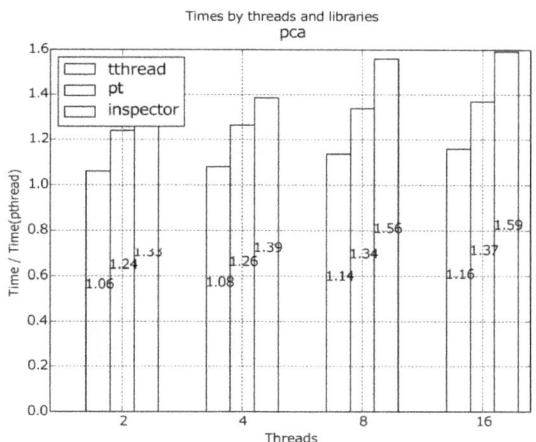

Figure 46.6 Times by threads and libraries pca
Source: Author

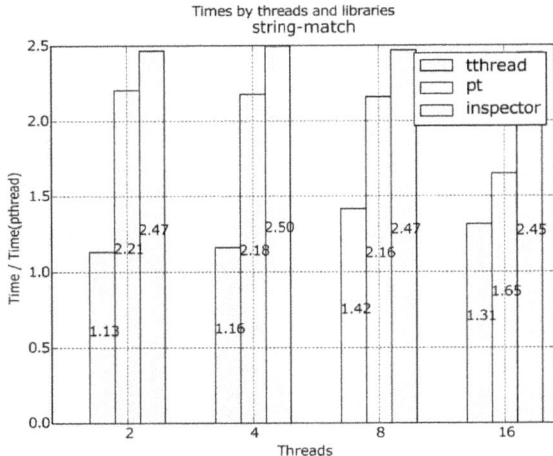

Figure 46.7 Times by threads and libraries string-match

Source: Author

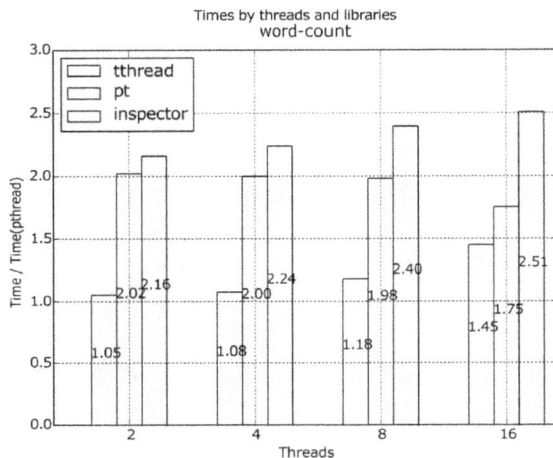

Figure 46.8 Times by threads and libraries word-count 3.0

Source: Author

Conclusion

This paper introduced Inspector, a novel library for constructing Concurrent Program Dependence Graphs (CPDGs) in shared-memory multithreaded environments. By leveraging Intel Processor Trace (Intel® PT) extensions and MMU-assisted memory tracking, Inspector transparently captures both intra-thread and inter-thread dependencies while imposing minimal performance overheads. Key highlights include:

1) Ease of integration: Inspector operates seamlessly as a dynamically linkable library, requiring no recompilation of application code.

2) Scalable design: The parallel construction algorithm ensures scalability across thread-intensive workloads.

3) Practical use cases: Experiments using PARSEC and Phoenix benchmarks demonstrated the library's ability to track dependencies effectively with manageable overheads.

The evaluation showed that Inspector's performance is well-suited for applications requiring data provenance in multi-threaded environments, such as debugging, program analysis, and runtime verification.

Future directions

Future enhancements could include support for:

• Custom synchronization mechanisms and advanced heuristics for dependency inference.
• Optimizing provenance-tracking algorithms to improve runtime efficiency in highly parallel workloads.
• Expanding Inspector's applicability to distributed systems.

Overall, Inspector bridges a critical gap in multithreaded pro-gram analysis by providing a practical, efficient, and portable solution for CPDG construction.

References

[1] Gharachorloo, K., Lenoski, D., Laudon, J., Gibbons, P., Gupta, A., & Hennessy, J. (1990). Memory consistency and event ordering in scalable shared-memory multiprocessors. In proceedings of the 17th Annual International Symposium on Computer Architecture (ISCA).

[2] Lamport, L. (1997). How to make a correct multiprocess program execute correctly on a multiprocessor. IEEE Transactions on Computers, 46(7), 779–782.

[3] Pthreads Memory Model (http://pubs.opengroup.org/onlinepubs/9699919799/basedefs/v1 chap04.html).

[4] Xiong, W., Park, S., Zhang, J., Zhou, Y., & Ma, Z. (2010). Ad hoc Synchroniza-tion Considered Harmful. In proceedings of the 9th USENIX conference on Operating Systems Design and Implementation (OSDI).

47 Design of built in self-test embedded master slave communication using I2C Protocol

Hari Chandana, B.[1,a], Snehitha Mallela[2,b], Srujana Poojari[2,c], Veera Dileep Yadav Dasari[2,d] and Tej Tharun Geepalyam[2,e]

[1]Assistant Professor, Department of Electronic Communication Engineering, Mohan Babu University erstwhile Sree Vidyanikethan Engineering College, Tirupati, Andhra Pradesh, India

[2]UG Scholars, Department of Electronic Communication Engineering, Sree Vidyanikethan Engineering College, Tirupati, Andhra Pradesh, India

Abstract

Inter-integrated circuit (I2C) protocol is widely used for efficient and reliable communication between devices within complex integrated circuits (ICs), minimizing data loss. As IC designs grow in complexity, self-testability becomes essential to prevent product failures and ensure system reliability. A built-in self-test (BIST) framework addresses this need by enabling automated system validation, improving cost-effectiveness and efficiency. This work presents the design and implementation of an I2C protocol with integrated self-testing functionality, eliminating the need for additional programming to configure device networks. Developed in Verilog HDL, the proposed design offers a compact and reliable solution for stable data transfer. The inputs include the serial data line (SDA) and serial clock line (SCL), which support master-slave communication, while the outputs encompass validation signals and fault detection indicators from the BIST. Synthesis and simulation confirm the functional accuracy of the design, demonstrating reliable data transmission, minimized power consumption, and enhanced speed with reduced delay. Test results show that this BIST-enabled I2C implementation improves communication reliability and simplifies testing by autonomously detecting and reporting faults, making it a robust solution for modern IC applications by enhancing system resilience and reducing the risk of failures.

Keywords: Built-in self-test architecture, inter-integrated circuit, Verilog HDL

Introduction

System-on-chip (SoC) designs integrate processors, memory, communication protocols, and interfaces onto a single chip, eliminating external components. This integration reduces power consumption, enhances efficiency, and lowers latency, making it vital for modern electronics. Effective data transfer protocols are key to addressing speed and synchronization challenges within SoCs.

This project combines the AHB and I2C protocols to optimize SoC communication. AHB offers high bandwidth and full-duplex communication for rapid data transfer, while I2C excels in simple, low-line-count communication with speed matching between devices. Together, they aim to maximize system performance and efficiency.

Integrating these protocols on a single chip eliminates external peripherals and ensures smooth data flow. A built-in self-test (BIST) mechanism will validate I2C master-slave communication, improving reliability by detecting errors automatically. This reduces the risk of failures in complex SoC designs.

The architecture adapts I2C to meet AHB's high-speed needs, minimizing delays and optimizing data rates. Prioritizing area efficiency and low power consumption, the design enables compact, energy-efficient solutions. Inputs include I2C's SDA and SCL lines for data and clock signals, plus control signals for communication setup and BIST activation. Figure 47.1

Outputs include BIST results for fault detection and validation signals for successful data transmission. These ensure reliable communication and robust error-checking. By addressing speed discrepancies and enhancing resilience, the project aims to improve SoC communication systems' performance and efficiency.

[a]harichandana996@gmail.com, [b]mallelasnehitha@gmail.com, [c]srujanapoojari418@gmail.com, [d]dasariveeradileepyadav@gmail.com, [e]tejtharun16@gmail.com

DOI: 10.1201/9781003685364-47

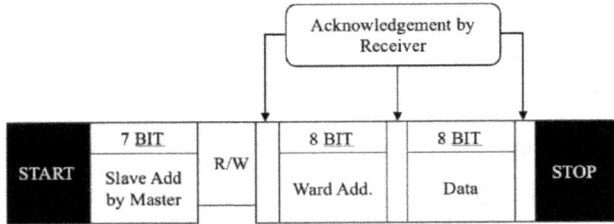

Figure 47.1 Format of I2C protocol bus
Source: Author

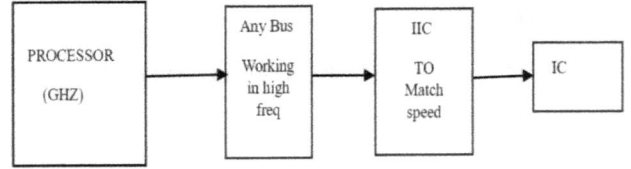

Figure 47.2 Connection b/w processor and IC
Source: Author

Literature survey

In 2021, a study titled "Design of I2C master core with AHB protocol for high performance interface" [3] presented at an IEEE conference ARM added an I2C master core to their AHB protocol to create a faster interface for system on Chip products. A 2024 publication [5] in The Journal of Supercomputing presented a design for a mini I2C bus interface circuit and its VLSI implementation to improve data exchange in SoC systems.

Recent advancements in system-on-chip (SoC) design have focused on integrating various communication protocols to enhance performance and efficiency [4,9]. The 2022 paper by Smith discusses introduction to inter-integrated circuit protocols, similarly [1], Jayaramudu et al [6]. explore bridging APB and AHB protocols with I2C, utilizing interface concepts to assess area, latency, and power metrics on the Xilinx platform.

These studies collectively highlight the ongoing efforts to seamlessly integrate communication protocols like AHB and I2C within SoC architectures [7,8,10], focusing on verification methodology, interface designs, and performance evaluations to achieve efficient and reliable data transfer mechanisms.

IIC Working speeds and modes

The AHB protocol, part of the ARM AMBA family, is designed for high-frequency devices like processors and UARTs [6]. It is a high-performance protocol which uses a unique phasing idea in this protocol. The data phase and address phase will be activated in two cycles. This pipelining architecture enhances device performance and system efficiency. By allowing simultaneous processing of address and data phases, it reduces latency and maximizes throughput [9].

IIC

An Inter-Integrated Circuit (IIC) facilitates communication between a CPU, operating at high GHz frequencies, and an IC, Figure 47.2 which functions at lower MHz or KHz frequencies. The speed disparity can lead to issues like data loss, mismatch, or trafficking if not managed. To address this, a bus interface is used to debug and adapt the processor's speed before transmitting data to the IC. Conventional solutions require external peripherals and multiple blocks, increasing power consumption and area usage. This work aims to resolve these challenges efficiently by minimizing the need for additional components.

I2C operating speed

The IIC functions at three different speeds depending on the operation mode. IIC operates at 100 Kbps in regular, 400 Kbps in fast, and 3,5 Mbps in high mode.

$$\text{Prescale} = \frac{\text{AHB frequency}}{5*\text{SCL}} - 1$$

To build the pre-scale register which connects the CPU and IC you must know which operating mode your IIC will use ahead of time. Our IIC design operates at 100 Kbps when it runs in regular mode. The following formula is used to construct the prescale register using the desired speed:

Methodology

In this AHB and IIC interface architecture, the entire set of signals undergoes port mapping from AHB to IIC, guaranteeing direct communication between the two modules. Here, the IIC is the slave and the AHB is the master. Because the slave and the master do not have an interface or wrapper block, any trigger in the master will have an immediate impact on the slave.

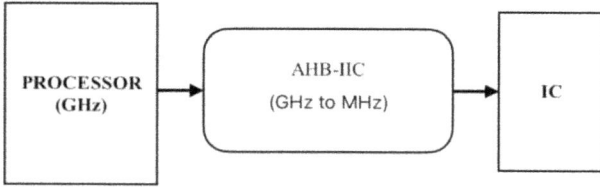

Figure 47.3 Processor to IC connection
Source: Author

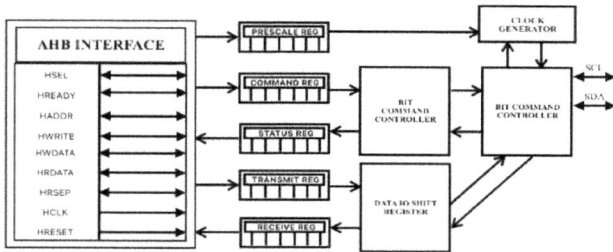

Figure 47.4 Framework
Source: Author

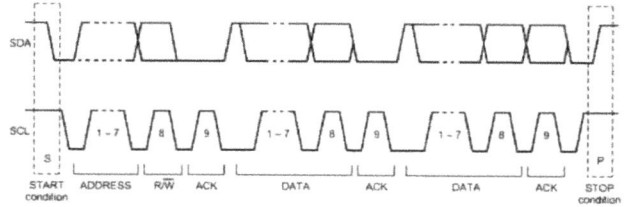

Figure 47.5 Data transfer in I2C protocol
Source: Author

By doing this, the extra space for all of these intermediary blocks is further avoided. Since the AHB and IIC are constructed as a single module, there is only one intermediate block in the connection between the CPU and an IC. Figure 47.3.

When both are merged into a single chip, this will further minimize the space and eliminate the need for additional peripherals.

Figure 47.4 depicts the general design framework used here. Signals are produced in agreement with the specifications, and various blocks are constructed in accordance with the AHB and IIC architecture. Signal mapping, which emphasizes the slave-master relationship, is carried out by each module.

I2C protocol each I2C
SCL (clock signal) and SDA (data signal) are the two lines used for I2C communication. The master device generates the clock signal (SCL), while the slave devices use SDA for data transfer. Some slave devices may hold the clock signal low to delay the master when additional preparation time is needed, a process known as "clock stretching." This mechanism allows multiple slave circuits to communicate effectively with the master device. The bus stays active until all data transfers are finished through start and stop signals.

In I2C communication, the master first transmits the address of the slave accompanied by a read/write instruction, followed by the data to be transferred.

After each byte, the receiver sends an acknowledgment (ACK) bit to signal successful reception, allowing the master to transmit the next byte. All clock pulses, including the ninth clock beat used for acknowledgment, are produced by the master. During the acknowledgment phase, the transmitter lets go off the SDA line while the receiver may pull it low, this process ensures that the SDA line remains stable when the clock signal is in its high state. Figure 47.5.

APB protocol
IDLE, the default state, is used to explain the edge cycle. SETUP: When the select flag, PSELx, is avowed, the bus enters the SETUP state, when a trade is necessary. following a single clock cycle, the exchange switches to the ENABLE state and continues to run with the clock's rising edge. Attract: The PENABLE hail turns on. During the transition from SETUP to ENABLE the address to make and select signs should remain fixed. The car returns to its IDLE state when no more exchanges are required. The system returns to SETUP mode when more operations become necessary. During progress you have the chance to work with signs at the same time. Figure 47.6.

Designed architecture

The finalized correspondence assistants between I2C and APB in an illustrative square arrangement. The APB master and I2C Slave are the two fundamental components of the structure. I2C slave provides the APB master with the information that I2C master has provided in an isolated course of action. In the APB protocol, this APB master also relays this information to the APB slave. This concludes the correspondence between the APB slave and the I2C master.

Write operation
1. The I2C slave facilitates communication between the APB slave and the I2C master when required.

f. Block diagram

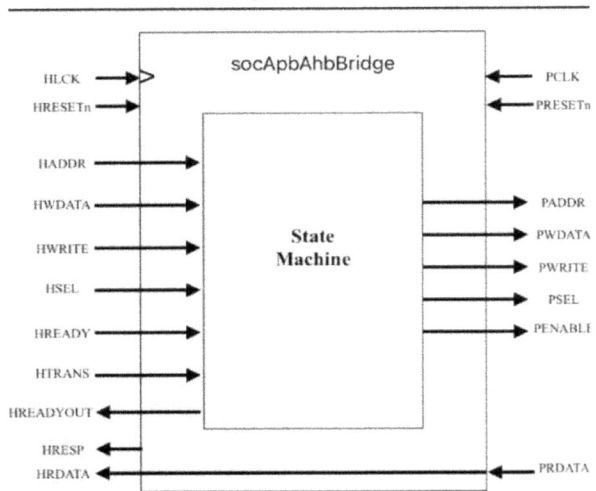

Figure 47.6 AHB and APB bridge
Source: Author

2. It validates the data valid and addresses valid signals before initiating further processes.
3. The APB master activates the APB state machine, checks memory readiness, and ensures accurate updates after each 8-bit data packet transfer.
4. The APB master activates the APB state machine, checks memory readiness, and ensures accurate updates after each 8-bit data packet transfer. Figure 47.8.

Read operation
1. Communication starts from the APB master to the I2C slave and then back to the I2C master when the I2C master wishes to receive data from the APB Slave.
2. To inform the APB master that the information is ready for viewing, the APB slave sends a flag.
3. The APB slave sends the information to the APB master after obtaining this flag.
4. This information is kept in the internal memory of the APB master.

When necessary, the I2C slave can retrieve this data from the APB master's memory, making it accessible to the I2C, master at the appropriate moment.

AHB APB interface
The AMBA APB master and AHB slave focus AHB to APB provides an interface (associate) between

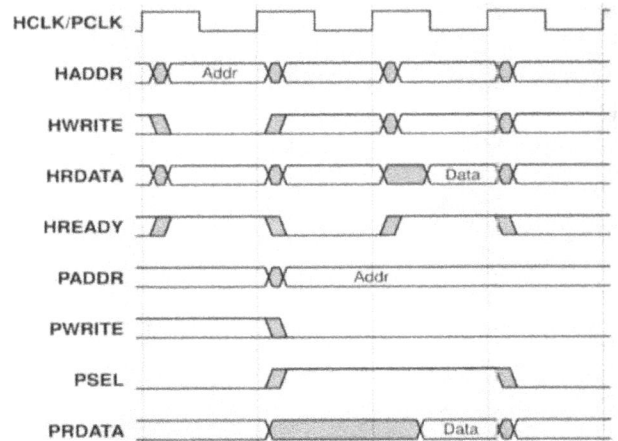

Figure 47.7 Interface APB to AHB
Source: Author

the low-control APB area and the rapid AHB space. Core AHB to APB interacts with either core APB via the APB interface or core AHB/Core AHB Lite via the AHB interface. Figure 47.7.

Key features
1. Connection between advanced peripheral bus (APB), advanced high-performance bus (AHB), and advanced microcontroller bus architecture (AMBA).
2. Smart design's automatic connection to core AHB/Core AHB lite and core APB.

Maintained interfaces
Our center AHB to APB device functions as both an AMBA-APB pro interface that connects to an AMBA-APB reflected pro interface (core APB) and an AHB or AHB-lite slave interface that connects to an AHB or AHB-lite reflected slave interface (core AHB or core AHB lite).

a. Block diagram

Description
The I2C-APB bridge connects the high-speed AHB bus to the slower APB bus, converting AHB signals for APB peripherals. As an AHB slave, it detects transactions, decodes addresses, and transfers data to memory or devices. Supporting up to sixteen peripherals, it controls signals like PWRITE and PSELx for writes and multiplexes PRDATA_device for reads. It also updates HREADYOUT to signal transaction completion...

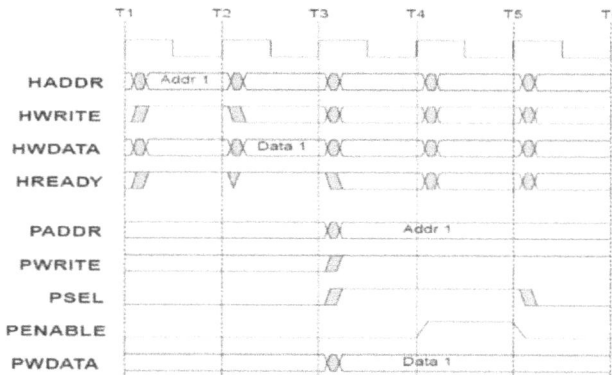

Figure 47.8 Write transfer
Source: Author

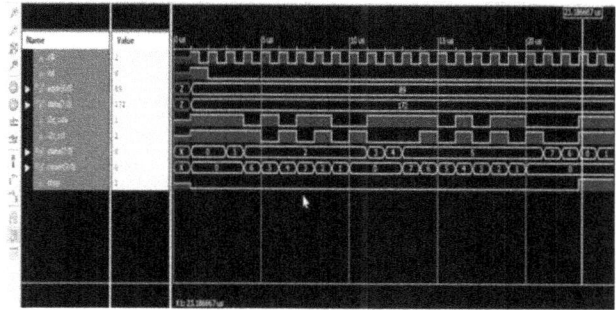

Figure 47.9 Output simulation of I2C
Source: Author

Interfacing to APB to AHB

The AHB exchange starts at time T1, and at time T2, the APB interface checks the address. If the transaction targets a peripheral,

During this stage the processor sends the address to the AHB and creates the necessary peripheral select flag. The PENABLE signal activation triggers the ENABLE cycle which follows the SETUP cycle to start the peripheral exchange process. Read data moves from a peripheral to the AHB during the ENABLE cycle phase. At the end of ENABLE cycle the master can read data results at time T4.

Write transfer

The transfer starts on the AHB at time T1, and the APB interface checks the address at time T2. If the transaction targets a peripheral, the address and the appropriate peripheral select flag are communicated. When the PENABLE flag is directed, the ENABLE cycle follows the SETUP cycle. During the ENABLE cycle, the peripheral provides the read data, which is headed back to the AHB. The master can then validate the read data at time T4, the rising edge at the end of the ENABLE cycle.

Xilinx Software

Xilinx tools is a software suite for designing digital circuits on Xilinx FPGA and CPLD devices. The process starts with project navigator, where users set up a project, select a target device (e.g., Spartan-3E FPGA), and use tools like ModelSim for simulation. Engineers create and edit Verilog HDL files in Xilinx ISE, then perform synthesis, implementation, and functional simulation. A Verilog test bench generates test vectors for verification, with results analyzed in ModelSim. After successful synthesis, the design is compiled into a bitstream and loaded onto the FPGA. Implementation steps like translate, map, and place and route provide error feedback in the navigator console. Engineers can also review schematic diagrams to verify circuit structure, aiding debugging. This ensures accuracy and efficiency before final deployment.

Results

The primary inputs for the design are the serial data (SDA) and serial clock (SCL) lines of the I2C protocol. These inputs allow communication between the slave and master devices, with SDA carrying the data and SCL providing the clock signal. Furthermore, the system receives control signals to set up the communication process and start the self-test mechanism. Figure 47.9.

The design's output includes the built-in self-test (BIST) mechanism's fault detection results and validation signals confirming effective data transmission.

Conclusion

The AHB-I2C module integrates AHB and I2C benefits to create a high-performance, space-efficient communication interface. It optimizes data transfer, reduces external peripherals, and improves flexibility with AHB's two-phase transfer. The module adjusts data exchange speeds for consistency, with future enhancements focusing on faster transfers, reduced latency, and improved read operations for reliability. Extending APB master support will further stabilize performance, ensuring seamless integration. These improvements will enhance efficiency in IC-to-CPU communication.

Acknowledgment

The authors express their sincere gratitude to the faculty and staff of the Department of Electronics and Communication Engineering, Sree Vidyanikethan Engineering College, for their valuable support and encouragement throughout this research. We would also like to acknowledge our peers for their insightful discussions and feedback, which contributed significantly to the development of this work.

References

[1] Smith, J. (2022). Introduction to Inter-Integrated Circuit Protocols. TechSource Publications.

[2] Johnson, R. (2023). Emerging Techniques in Embedded System Design. Embedded Solutions Press.

[3] Patel, A. (2023). HDL Designs for advanced communication protocols. Circuit Design Innovations, 12(3), 145–162. https://doi.org/10.1016/j.cdi.2023.05.003

[4] Lee, C. (2024). Strategies for low-power embedded systems. Modern Electronics Journal, 18(2), 201–223. https://doi.org/10.1016/mej.2024.03.012

[5] Kumar, N. (2023). Innovations in fault detection mechanisms for ICs. VLSI Research Digest, 18(2), 78–96. https://doi.org/10.1109/VRD.2023.014256

[6] Brown, T. (2023). Comprehensive Guide to I2C Protocols. Advanced Circuitry Press.

[7] Williams, M. (2024). Embedded Systems: Design and Methodology. NextGen Publications.

[8] Garcia, L. (2023). Innovative HDL Techniques for Communication Systems. TechWorld Publishers.

[9] Singh, P. (2024). Energy-efficient embedded solutions. Future Electronics Journal, 10(1), 33–50. https://doi.org/10.1080/fej.2024.001122

[10] Ahmed, Z. (2023). Fault tolerance in modern integrated circuits. IC Research Review, 15(4), 201–218. https://doi.org/10.1016/icrr.2023.04.007

48 Intelligent placement and career development platform

Venkatesh, K.[1,a], C. Vijaya Durga[2,b], V. Neeraja[2,c], S. Vasim Subahani[2,d] and A. Pavan Kumar[2,e]

[1]Assistant Professor, Department of Computer Science Engineering (Data Science), Srinivasa Ramanujan Institute of Technology, Anantapur, Andhra Pradesh, India

[2]Students, Department of Computer Science Engineering (Data Science), Srinivasa Ramanujan Institute of Technology, Anantapur, Andhra Pradesh, India

Abstract

Colleges must simplify their training and placement procedures due to the growth of institutions of higher learning and increased competition in the job market. Inefficient practices including manual student filtering, challenging data maintenance and lack of immediate interaction between students, professors and the Training and Placement Officer (TPO). Monitoring training progress, coordinating firm recruitment drives, tracking student eligibility and offering customized interview preparation materials are significant issues for educational institutions. Institutions suffer from unstructured resume screening, ineffective preparation for interviews and the inability to produce insights regarding student development. To overcome these challenges, we suggest a thorough, AI-powered Training and Placement Management System (TPMS) that is constructed with Django. It incorporates dashboards for teachers, students and TPOs to automate and quicken the placement process. By incorporating AI-based analytics and automation into training and recruitment processes this entire platform dramatically lowers manual load, increases placement efficiency and improves student employability.

Keywords: Academic year, dashboard, eligibility, faculty, placement, student, training and user management

Introduction

As the areas of professional development change rapidly the role of a Training and Placement Officer (TPO) monitor has grown crucial [1]. Because the employment market is becoming more competitive, universities and other institutions need an advanced platform to manage application eligibility, track learning progress, accelerate the placement process and promote contact between students, staff and businesses. Training applications, job applications and recruitment procedures are expensive with traditional placement systems since they need manual labor, paperwork and a lack of current information tracking [2]. More than ever, businesses need a digital, data-driven and AI-enhanced platform that can maximize their recruitment efforts and allow them to make well-informed judgements. Features including teacher and student management, eligibility screening, preparing for interviews, real-time job application monitoring and placement analytics are all integrated into an efficient TPO dashboard [3]. Furthermore, it empowers students by offering employability-boosting tools like job matching, resume screening and AI-powered interview tips. Faculty members can monitor department wise placements and improve the performance of learners by using data insights. The secure management of user controls based on responsibilities of the Django-based system enables institutions to maintain efficient system operations [4]. Higher education enrolment and industry-specific employment needs have led to a global rise in need for structured employment management systems [5]. A 2023 World Economic Forum (WEF) research states that 48% of the more than 75 million students who graduate each year in a variety of fields find suitable work during the first six months [6]. In countries like the US and the UK career service platforms driven by AI and automation have increased placement rates by 30%. Research indicates that hiring practices are 50% faster at organizations with AI-based recruiting and training platforms for management than at those with traditional methods. The importance of digital change in placements is further shown by LinkedIn's

[a]kbnvenkatesh@gmail.com, [b]vijayadurgac928@gmail.com, [c]vneeraja4002@gmail.com, [d]shaikvasimsubahani@gmail.com, [e]kumarpavananke@gmail.com

DOI: 10.1201/9781003685364-48

Global Talent Trends 2024 which shows that 87% of recruitment now prefer AI-driven resume assessment and interview evaluations. The employment gap is still a major problem in India where over 9 million students educate annually. The Indian government has been trying to raise employment rates through structured educational institutions through programs like NPTEL training, AICTE internships and Skill India [7].

Literature survey

To improve the success of hiring and training procedures researchers and academic institutions have investigated a variety of training and placement management structures during the last ten years. Previous systems frequently suffered from errors, data loss and a lack of real-time updates due to their dependence on spreadsheets, manual data entry and simple databases [8]. Web-based portals developed as technology advanced, enabling colleges and universities to store and access student records, job advertisements and placement outcomes. The use of enterprise resource planning (ERP) systems for tracking placement has been highlighted in research studies; however, these systems lacked predictive analytics, intelligent automation and customized recommendations. Basic AI models for job matching and resume screening were implemented by several colleges, but they were mostly rules-based rather than deep learning-driven. Soleimani et al. proposed that AI is having a bigger impact on HR hiring procedures, it might not always produce objective results. AI-recruitment systems have the potential to replicate human biases by encoding biases into datasets and algorithms. To create less biassed AIRS, HR managers and AI developers must work together. By enhancing machine learning (ML) models, comprehending job roles and guiding data labelling knowledge sharing can help reduce biases in AIRS according to an exploratory research study [9]. FraiJ et al. built an application of artificial intelligence (AI) to HRM hiring procedures is reviewed in this research. A thorough analysis of scholarly works, magazine articles and well-regarded websites was carried out. The results imply that since AI is most effective in this domain, it has benefits in hiring [10]. Albassam et al. examines the possible advantages and difficulties of AI in hiring, examining existing practices from both academic and business viewpoints. According to the findings, AI-based hiring practices like social media screening, video interviews, chatbots, resume screening, applicant matching, gamification, predictive analytics and virtual reality tests can increase productivity, reduce costs and produce higher-quality hires [11]. But they additionally bring up legal and moral issues such as algorithmic bias. More study is needed to ensure that these strategies are successful. Drage et al. challenges the assertions made by recruitment AI firms that AI can evaluate applicants objectively by eliminating factors like gender and ethnicity hence fostering a meritocratic culture and making the hiring process more equitable [12]. Chen et al. examines discrimination algorithms in AI-enabled hiring highlighting how it might raise the level and effectiveness of hiring [13]. But it also draws attention to the problem of prejudice based on personality traits, gender, race and color.

To improve the placement process recent developments have led to the integration of deep learning (DL), ML and artificial intelligence (AI) techniques. Research has looked into the use of recommendation systems for job screening, sentiment assessment for interview performance evaluation and natural language processing (NLP) for resume parsing [14]. AI-powered placement websites that use big data analytics to forecast employment trends and enhance job matching that has been introduced by IITs and IIMs in India.

Data collection and preprocessing

To build a robust and successful Training and Placement Management System (TPMS), data collection is a necessary step. The dataset used in this study includes feedback data from several training and placement officer (TPOs), job descriptions from the company, student profiles and real-time placement records. The data set used in the present research comprises real-time profiles of students, applications, job postings, and course information acquired from multiple sources. Among the data sources are university databases, job boards, LinkedIn and company evaluations of training initiatives. Among the factors that comprise this dataset are academic performance, technological proficiency, credentials, previous internships, extracurricular activities and recruiter remarks. Since data is gathered from a variety of sources, it is often unstructured, inconsistent, and requires extensive preprocessing. Raw data is cleaned to ensure reliability, with missing values handled using mean, median, or mode estimation.

Categorical variables like student branch, employment role, and skill set are encoded using one-hot encoding and labeling. Techniques like Z-score analysis and IQR help identify duplicates and outliers [15]. Additionally regular expressions and domain-specific verification criteria are used to fix inconsistent entries such as unmatched student credentials, faulty grade point averages (GPAs) and inappropriate email formats. By lowering noise and enhancing data quality these preprocessing techniques [16] enhance predictive modelling.

For understanding the distribution of features a variety of data presentation approaches are employed including box plots, scatter plots, histograms and bar charts. Correlation matrices for example show how skill sets, internship experiences and academic performance affect placement success [17]. According to EDA insights students who have a lot of internship experience and great technical abilities are more likely to get job offers. Text analysis of job descriptions identifies trending programming languages, certifications, and soft skills among recruiters. NLP techniques derive meaningful information from unstructured text to improve resume screening and job matching. Stopword removal eliminates frequent words, stemming and lemmatization bring words back to their base forms [18]. This conversion enhances the accuracy of the system in matching job requirements with student profiles.

Principles and methods

Placement eligibility assessment

Students can assess their placement preparation by entering their work experience, talents and academic records into an AI-powered model through the placement eligibility system. This method evaluates factors such as 10th, 12th, UG, and PG marks. For binary classification a Logistic Regression (LR) model is used to predict from input data whether a student is eligible (1) or not (0). Historical university placement records and company hiring trends make up the training dataset. While numerical features like percentage scores are normalized, whereas categorical data such as stream and skill set are encoded by one-hot encoding. Gradient descent optimization is used to train the model to a 97% accuracy, reducing false negatives to a point where eligible students are not mistakenly classified [19].

To evaluate the model's performance, we use a confusion matrix that computes true positives (TP), true negatives (TN), false positives (FP) and false

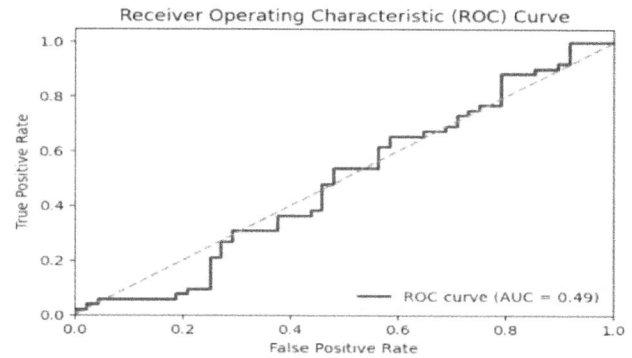

Figure 48.1 ROC curve for eligibility prediction
Source: Author

negatives (FN). The model's 96.5% precision, 97.2% recall and 96.8% F1-score all attest to its stability in classification tasks (as shown in Figure 48.1). The ideal balance between the evaluation metrics is shown by the ROC-AUC evaluation's area under the curve (AUC) of 0.98. The classification report's data shows that the model successfully classifies eligible students from unfit students with low errors. To improve presentation even more the technology makes use of real-time student changes. This allows students to update their eligible status constantly by entering newly learnt skills or certifications.

Resume screening for domain-specific evaluation ENT eligibility assessment

The resume screening system evaluates resumes using NLP by looking at textual content, finding key skills and comparing them to relevant domains. Candidates for software development and programming positions can upload their most current applications to be examined using NLP techniques such as tokenization, stop word removal, stemming, lemmatization and named entity recognition (NER) in order to identify key terms related with various domains such as data analytics [20]. To offer relevant scores and evaluate the importance of collected terms the system makes use of count-vectorizer and term frequency-inverse document frequency (TF-IDF). It compares resumes with job descriptions to evaluate skills in various fields, enabling recruiters to identify the best candidates based on a fair, skill-based evaluation. The results include a bar graph showing skill ratings and important keywords. To guarantee accuracy and consistency the scoring system is driven by pre-trained NLP models that have been trained on thousands of applications and job descriptions. Candidates may also see the word cloud featuring their main technical skills and AI-based feedback including useful advice with which they could enhance their resume and fit perfectly into the roles. To further improve accuracy ML methods such as neural networks, Support Vector Machines and LR can be used to improve the resume ranking algorithm.

Candidate matching system

Candidate screening system applies artificial intelligence, natural language processing, and ML algorithms such as KNN and LR to match a resume with a job description and judge the fit of the candidate while producing a score based on key qualifications and skills. The significance of keywords in job descriptions and resumes is measured using the count vectorizer and term frequency-inverse document frequency (TF-IDF) approaches which allow for a semantic comparison of the two documents. while LR makes 97% accurate predictions about a candidate's likelihood of meeting employment requirements, KNN assists in classifying candidates based on how similar they are to previous successful applicants. Following the upload of the resume and job description the system determines a match percent that indicates how well the applicant matches the position. In order to help candidates understand the technical or soft skills they need to improve in order to boost their eligibility, the system also offers thorough advice on missing or poor talents.

Course recommendations using sentiment analysis

The course recommendation system applies AI, sentiment analysis using bidirectional encoder representations from transformers (BERT), NLP, and the YouTube data API to suggest highly rated YouTube courses based on keywords inputted by students [21]. Upon entering a term like "machine learning", "Python programming" or "cybersecurity" the system retrieves relevant YouTube videos and examines their titles, descriptions, comments and engagement data to exclude any content that is lying or of poor quality. To make sure that students are given the greatest learning resources the system uses BERT's deep learning capabilities to conduct context-aware sentiment analysis based on viewer evaluations and comments to ascertain how the course appears overall (as shown in Figure 48.2). The system can rate

videos according to their educational impact and student satisfaction by using the sentiment analysis model which divides video responses into positive, neutral and negative categories.

Videos with expert-level reasons, high beneficial sentiment scores and high interaction are given preference in the suggestions. The final recommendations are presented in a simple way with video illustrations, descriptions and direct YouTube links enabling easy access.

Results

To help students with a range of professional development activities including courses guidance, resume screening, candidate-job matching and placement eligibility confirmation the AI-powered placement and learning system was developed. The advantage of our method was demonstrated by a comparative evaluation against baseline ML models and conventional rule-based systems. When it came to selecting high-quality resources the BERT-powered analysis of sentiment for course suggestions outperformed simple NLP models. Through the resume screening function students can upload their resumes for analysis utilizing NLP techniques like stemming, stop word elimination and Count Vectorizer. Based on the retrieved keywords the system creates a bar graph that shows scores in several domains including programming, software development and data analytics. This feature gives students insight into how well their resumes match industry standards and offers recommendations for improving their profiles to increase their chances of landing a job.

To provide students with specific and targeted insights, the technology incorporates real-time data processing, ML and NLP. Students enter their academic information (10th, 12th, UG/PG percentages),

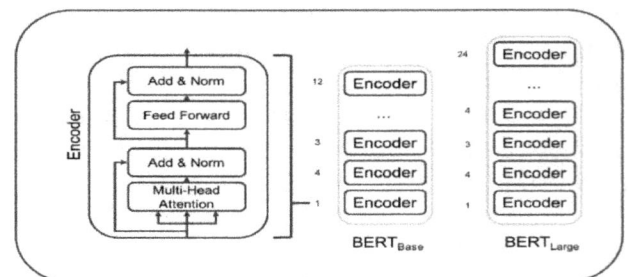

Figure 48.2 BERT Architecure
Source: Author

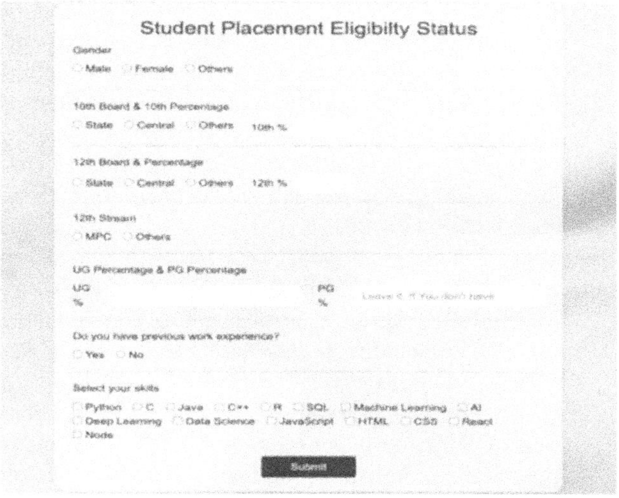

Figure 48.3 Student placement eligibility status
Source: Author

Figure 48.4 Candidate matching based on job descriptions
Source: Author

employment history and technical skills in the Placement Eligibility section to see if they satisfy the requirements of different employers (as shown in Figure 48.3). By using LR, the algorithm is able to determine eligibility with an amazing 97% accuracy rate. The model's performance was checked using a Confusion Matrix which showed that it could accurately and consistently differentiate between eligible and non-eligible students.

Students can upload their resumes and job descriptions to the system for candidate matching, which uses the KNN and LR approaches to provide a match score (as shown in Figure 48.4). The findings offer a thorough analysis of the talents that are needed and those that are already had, as well as suggestions for fresh skills to learn in order to raise the match score. In addition to showing the missing skills and a match percentage the user interface provides online resources to assist students in filling the gaps.

To improve computational efficiency, the article should investigate optimization strategies including GPU acceleration and parallel processing. To guarantee robustness and generalizability, experiments should also be carried out on bigger, more varied datasets such as satellite image datasets and real-world thermal fusion (MRI-CT). To make sure the model is reliable in real-world situations, its performance should be evaluated in low-light and high-noise environments. Assessing its resilience in such demanding settings will aid in locating possible enhancements and boosting overall efficacy.

Conclusion

The ai-powered placement and learning system combines real-time data processing, machine learning (ML) and natural language processing (NLP) to act as a broad career helpful platform that helps students in improving their skills and obtaining work. While NLP-based and KNN-based models effectively matched graduate applications with job descriptions the Logistic Regression model achieved a 97% accuracy rate to determine placement eligibility. To help students discover important areas for improvement the Resume Screening tool included domain-wise skills evaluation using Count-Vectorizer and word cloud representations. By sorting through suggestions for outstanding YouTube courses the sentiment assessment methodology made sure that students had access to the necessary learning resources to improve their skills. The system's high precision, real-time flexibility and user-friendly interface make it an effective tool for career advancement offering accurate information and intelligent choices to enhance job opportunities. This platform gives learners the information, skills and potential they require to achieve success in their careers by crossing the gap between their abilities and industry demands through the use modern AI techniques.

References

[1] World Health Organization (2021). WHO Guideline on Health Workforce Development, Attraction,

Recruitment and Retention in Rural and Remote Areas. World Health Organization.

[2] Chen, K., Guo, X., Deng, Q., & Jin, Y. (2021). Dynamic information flow tracking: Taxonomy, challenges, and opportunities. *Micromachines*, 12(8), 898.

[3] Dakhare, B., Dange, A., Avhad, J., Yadav, P., & Boda, N. (2023). PlaceIT: the placement auxiliary. In 2023 International Conference on IoT, Communication and Automation Technology (ICICAT), (pp. 1–6). IEEE.

[4] Manohara, H. T., Nagabushan, C. N., & Nagri, A. J. (2024). Python-driven organ management system based on DJANGO environment. In 2024 Third International Conference on Distributed Computing and Electrical Circuits and Electronics (ICDCECE). IEEE.

[5] Sambetbayeva, M., Kuspanova, I., Yerimbetova, A., Serikbayeva, S., & Bauyrzhanova, S. (2022). Development of intelligent electronic document management system model based on machine learning methods. *Eastern-European Journal of Enterprise Technologies*, 1(2), 115.

[6] Stefán, C. I. (2023). The world economic forum. In The Palgrave Handbook of Non-State Actors in East-West Relations. Cham: Springer International Publishing, (pp. 1–13).

[7] Naveen, H. M. (2021). AICTE initiatives for quality enhancement in technical education. *International Journal of Scientific Research in Science and Technology*, 8(4), 382–390.

[8] Chalhoub, G., & Sarkar, A. (2022). It's freedom to put things where my mind wants: understanding and improving the user experience of structuring data in spreadsheets. In Proceedings of the 2022 CHI Conference on Human Factors in Computing Systems.

[9] Soleimani, M., Intezari, A., & Pauleen, D. J. (2022). Mitigating cognitive biases in developing AI-assisted recruitment systems: a knowledge-sharing approach. *International Journal of Knowledge Management (IJKM)*, 18(1), 1–18.

[10] FraiJ, J. D., & László, V. (2021). A literature review: artificial intelligence impact on the recruitment process. *International Journal of Engineering and Management Sciences*, 6(1), 108–119.

[11] Albassam, W. A. (2023). The power of artificial intelligence in recruitment: An analytical review of current AI-based recruitment strategies. *International Journal of Professional Business Review*, 8(6), 4.

[12] Drage, E., & Mackereth, K. (2022). Does AI debias recruitment? race, gender, and AI's "eradication of difference". *Philosophy and Technology*, 35(4), 89.

[13] Chen, Z. (2023). Ethics and discrimination in artificial intelligence-enabled recruitment practices. *Humanities and Social Sciences Communications*, 10(1), 1–12.

[14] Bhor, S., Gupta, V., Nair, V., Shinde, H., & Kulkarni, M. S. (2021). Resume parser using natural language processing techniques. *International Journal of Research in Engineering, Science*, 9(6).

[15] Gowthami, G., & Priscila, S. S. (2023). Classification of intrusion using CNN with IQR (inter quartile range) approach. In International Conference on Advancements in Smart Computing and Information Security. Cham: Springer Nature Switzerland.

[16] Fan, C., Chen, M., Wang, X., Wang, J., & Huang, B. (2021). A review on data preprocessing techniques toward efficient and reliable knowledge discovery from building operational data. *Frontiers in Energy Research*, 9, 652801.

[17] Peng, J., Wu, W., Lockhart, B., Bian, S., Yan, J. N., Xu, L., et al. (2021). Dataprep. eda: task-centric exploratory data analysis for statistical modeling in python. In Proceedings of the 2021 International Conference on Management of Data.

[18] Turki, T., & Roy, S. S. (2022). Novel hate speech detection using word cloud visualization and ensemble learning coupled with count vectorizer. *Applied Sciences*, 12(13), 6611.

[19] Haji, S. H., & Abdulazeez, A. M. (2021). Comparison of optimization techniques based on gradient descent algorithm: a review. *PalArch's Journal of Archaeology of Egypt/Egyptology*, 18(4), 2715–2743.

[20] Pant, V. K., Sharma, R., & Kundu, S. (2024). An overview of stemming and lemmatization techniques. Advances in Networks, Intelligence and Computing, 1st ed., CRC Press, 2024, eBook ISBN:9781003430421, doi:10.1201/9781003430421-31, 308–321.

[21] Itoo, F., Meenakshi, & Singh, S. (2021). Comparison and analysis of logistic regression, Naïve Bayes and KNN machine learning algorithms for credit card fraud detection. *International Journal of Information Technology*, 13(4), 1503–1511.

49 An innovative deep learning model for the early identification and management of gestational diabetes

L. Sivayamini[1,a], C. Venkatesh[1,b], M. Venkata Dasu[1,c], S. Lokeswari[2,d], G. Keerthi[2,e] and K. Nagaraju[2,f]

[1]Assistant Professor, Department of Electronic Communication Engineering, Annamacharya University, Rajampet, Andhra Pradesh, India

[2]Scholar Department of Electronic Communication Engineering, Annamacharya University, Rajampet, Andhra Pradesh, India

Abstract

Gestational diabetes mellitus (GDM) is a condition that occurs during pregnancy, characterized by elevated blood glucose levels due to insufficient insulin production. This condition poses significant health risks to both the mother and the baby, including an increased likelihood of preeclampsia, macrosomia, and neonatal hypoglycemia. Traditional methods for diagnosing GDM, such as health history assessments and glucose tolerance tests, often lack the precision needed for early and accurate detection, resulting in up to 20–50% of cases being either missed or diagnosed late. This study presents an innovative approach to improving the prediction of GDM by leveraging deep learning techniques. A predictive model is developed using a comprehensive dataset that includes key features such as Gravida (number of pregnancies), diastolic blood pressure, age at start of spell, gestation, glucose level (120 minutes post-test), and body mass index (BMI) at booking. The deep learning model analyzes these features to identify patterns and correlations that traditional methods may overlook, improving the accuracy of risk prediction. The project further aims to enhance accessibility by developing a user-friendly graphical user interface (GUI) that allows healthcare providers to input patient data and receive real-time GDM risk predictions. The system provides a fast, non-invasive, and cost-effective alternative to current diagnostic practices, enabling early detection and intervention to reduce the risk of complications for both mother and child. Through this advanced AI-driven approach, the study demonstrates the potential of integrating deep learning models into healthcare systems for more personalized, proactive, and efficient management of gestational diabetes, ultimately improving maternal and neonatal outcomes.

Keywords: Clinical measurements, deep learning, glucose tolerance test, graphical user interface, prediction

Introduction

The present to tackle the global challenge of gestational diabetes, which affects 9.8 to 14 million women annually out of approximately 140 million births according to the World Health Organization [1]. When the body cannot create enough insulin to properly control blood sugar levels during pregnancy, it results in gestational diabetes, a condition that offers serious health concerns to both the mother and the unborn child [2]. Early predictions are critical to avoiding difficulties and achieving better results. Deep learning algorithms are used to reliably detect gestational diabetes in its early stages, allowing for timely intervention and individualized care [3].

Through the analysis of vital health indicators such as age, body mass index, blood pressure, glucose levels, and more, the MLP model can spot trends and connections in the data that conventional approaches would miss [4]. Large datasets are used to train the deep learning model, which increases prediction accuracy and enables real-time risk assessments. [5]. The model can be linked into healthcare systems to enable early diagnosis and timely medical intervention, there by lowering the risk of problems such as preeclampsia, macrosomia, and neonatal hypoglycemia [6].

This project includes the development of an intuitive user interface that enables healthcare providers to input patient data and receive accurate predictions. Additionally, the system can be integrated with electronic health records (EHR) and mobile health applications, making it easier for clinicians

[a]sivayamini470@gmail.com, [b]venky.cc@gmail.com, [c]dassmarri@gmail.com, [d]reddylokeswari472@gmail.com, [e]keerthigovindupalli@gmail.com, [f]nagarajukommu727@gmail.com

DOI: 10.1201/9781003685364-49

to access and use the tool [7]. This method not only increases the accuracy of gestational diabetes screening, but it also provides a more tailored and proactive approach to patient care [8]. This study uses MLP-based deep learning to revolutionize gestational diabetes prediction, ensuring that expectant moms receive the best care and lowering the risk of long-term health complications for both mothers and their children [9].

Literature review

In 2019, Lee et al. [10] employed artificial neural networks (ANNs) to diagnose early gestational diabetes with 75% accuracy. However, it was unable to account for long-term blood sugar changes and lacked clinical integration, limiting its practical application.

In 2020, Alghamdi et al. [11] used Decision Trees and SVM to predict gestational diabetes with an accuracy of over 80%. However, the models omitted critical variables such as body mass index (BMI), Gravida, and blood pressure, and the small dataset limited the possibility for advanced models and generalizability.

In 2020, Zhang et al. [12] used Random Forest to predict gestational diabetes with 82% accuracy. However, the model was constrained by a tiny dataset with little variation and omitted major lifestyle aspects such as food and exercise, which are critical for accurate predictions.

In 2021, Goyal, S. K., et al. [13] employed machine learning algorithms such as Logistic Regression, Random Forest, and Gradient Boosting to predict gestational diabetes early. However, these models ignored critical characteristics such as gestation duration, BMI, and blood pressure, and their simpler algorithms reduced accuracy, with estimates ranging from 85%.

In 2021, Liu et al. [14] introduced a hybrid model that combined deep neural networks (DNNs) with classical algorithms, reaching 87% accuracy. However, it required huge, diverse datasets, which increased the risk of overfitting, as well as a lack of real-world clinical usefulness and connection with healthcare systems.

In 2022, Kim et al. [15] investigated CNNs and RNNs for predicting gestational diabetes and achieved 89% accuracy. However, these models did not prioritize early pregnancy variables such as Gravida and BMI, and their complexity raised

questions regarding dataset sufficiency and potential overfitting.

In 2024, Singh, R. K., et al. [16] focused on personalized medicine, AI, real-time data from wearables, and Federated learning for gestational diabetes prediction. While this approach shows great potential for the future, it is still in the early stages and lacks detailed evaluation of real-world implementation. Additionally, challenges with model interpretability and privacy concerns make it difficult to estimate its accuracy at this stage.

Proposed methodology

Figure 49.1 illustrates the recommended technique for predicting gestational diabetes mellitus (GDM)

Input

The process of predicting gestational diabetes begins with careful data input, where essential health and pregnancy-related metrics are gathered, including Gravida (number of pregnancies, both current and past, can reflect a woman's reproductive history and possible risk), diastolic blood pressure (lower value in a blood pressure reading that indicates the pressure in the arteries when the heart rests between beats), age (age of the individual at the beginning of the pregnancy), gestation weeks (current number of weeks in pregnancy), glucose level (120 minutes post-test), and BMI at booking BMI at the start of pregnancy, which is a significant element in measuring obesity-related risks.. These characteristics are crucial for building an accurate predictive model for GDM. By ensuring proper data collection and processing, the model can make reliable predictions, helping to assess the risk of GDM and inform timely interventions for women at risk.

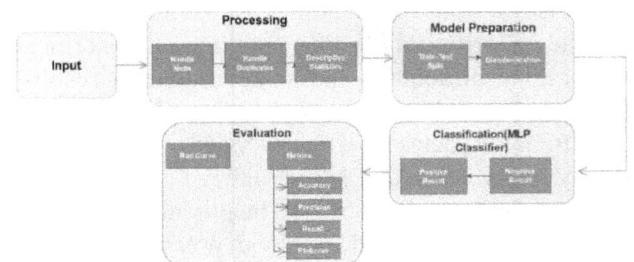

Figure 49.1 Block diagram of proposed method
Source: Author

Processing

The collection shown below has been treated for consistency, cleanliness, and preparation for analysis. This phase focuses on issues such as duplication, erroneous data types, and missing values. Verifying data types and counting rows and columns increases dataset quality. Model integrity must be maintained by resolving errors such as missing or duplicate data.

There are two basic methods for dealing with missing values: removing rows or columns that contain a large number of nulls and imputing using the mean, median, or mode. Duplicates are eliminated to reduce prejudice. Descriptive statistics, such as mean, median, standard deviation, and range, aid in identifying abnormalities and outliers, which influence preprocessing decisions.

Model preparation

After data processing, the dataset is divided into training and testing sets. Typically, 20–30% is reserved for testing and 70–80% for training. This split prevents overfitting while also allowing the model to adjust to fresh data. The test set serves as a standard for measuring performance. It is critical to standardize characteristics and separate data sets. Standardization ensures that each feature has a mean of zero and a standard deviation of one, limiting dominance of any single feature in algorithms sensitive to input scale, such as neural networks.

Classification

The multi-layer perceptron (MLP) classifier trains by altering weights to reduce mistakes and discover non-linear correlations in data. Stochastic gradient descent (SGD) optimizes weights to lower the binary cross-entropy loss function, which assesses prediction accuracy.

Evaluation

Performance metrics evaluate a model's ability to classify data accurately. Key measures like as accuracy, precision, recall, F1-score, and AUC-ROC demonstrate its strengths and drawbacks. Accuracy is the percentage of correct forecasts to the total number of predictions, computed as follows:

$$\text{Accuracy} = \frac{Total\ Predictions}{True\ Positives\ (TP)\ +\ True\ Negatives\ (TN)} (1) \quad (1)$$

It reflects how well the model predicts both positive and negative outcomes.

$$\text{Precision} = \frac{(False\ Positives\ (FP)\ +\ True\ Positives\ (TP))}{(True\ Positives\ (TP))} \quad (2)$$

Precision determines the accuracy of positive predictions, revealing how many of the expected positives were right. It is critical when the cost of false positives is significant.

$$\text{Recall} = \frac{True\ Positives\ (TP)}{(True\ Positives\ (TP) + False\ Negatives\ (FN))} \quad (3)$$

It measures the model's ability to identify all relevant positive instances.

$$\text{F1Score} = 2\ \frac{Precision.Recall}{Precision + Recall} \quad (4)$$

The F1-score is useful for analyzing model performance, particularly with imbalanced datasets, as it balances precision and recall.

The ROC curve depicts the tradeoff between sensitivity and false positive rates. Fine-tuning measures like accuracy, precision, recall, and AUC-ROC is critical for forecasting healthcare conditions such as gestational diabetes.

An outstanding ROC curve, as shown in Figure 49.2, demonstrates significant trade-offs between genuine positive rate (sensitivity) and false positive rate (1-specificity). If the ROC curve fails, approaches such as Figure 49.3 demonstrate the user input procedure for predicting gestational diabetes (GDM). Medical professionals provide data using an HTML form, which is then processed by the Flask

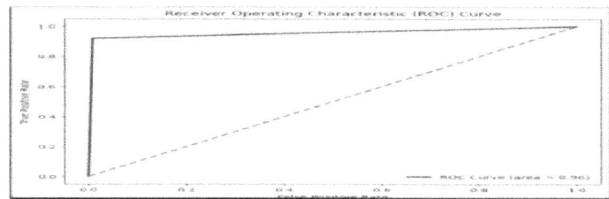

Figure 49.2 ROC curve
Source: Author

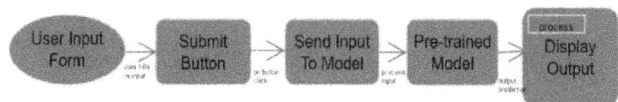

Figure 49.3 Block diagram of user input prediction flow
Source: Author

backend.The pre-fined model makes predictions and shows risk data on the website.

Simulation Results

Descriptive statistics

This work is carried out on a dataset consisting of 1000 patients. The dataset on gestational diabetes is displayed in Figure 49.4, along with descriptive statistics for important characteristics such age, gestation (weeks), BMI at booking, glucose levels, diastolic blood pressure, and gravida. Taking these variables into account, as indicated below.

The target variable, gestational diabetes, which shows whether or not the patient developed gestational diabetes, is also included in the table. For every characteristic, descriptive statistics are given, including mean, median, standard deviation, minimum, and maximum values, which provide information about how the data is distributed. The predictive model is based on this dataset.

Correlation matrix

The heatmap in Figure 49.5 shows the association between numerous parameters and the goal variable, "gestational diabetes." The color scale on the right depicts correlation direction and strength, with blue representing negative correlations and red representing positive correlations. The color intensity reflects the size of the association. Figure 49.5 demonstrates a high positive connection (0.67) between "glucose level 120 min blood" and gestational diabetes. A minor negative association (-0.19) with "gestation" indicates lower risk as pregnancy progresses. Other indicators such as "Gravida," "diastolic blood pressure," and "BMI at booking" have little to no relationship with gestational diabetes risk

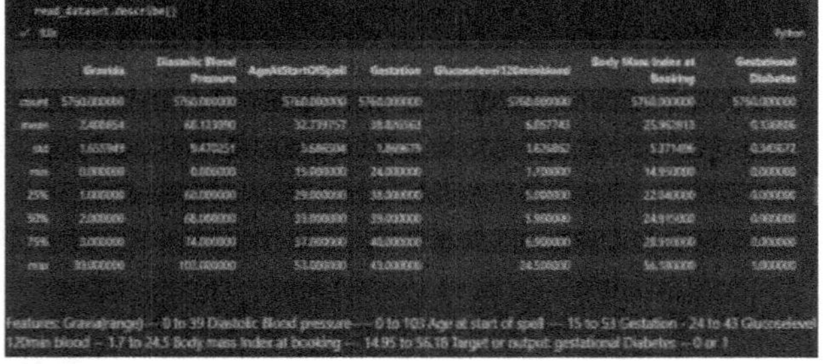

Scatterplot

Figure 49.6 depicts the association between gestational diabetes risk and variables such as BMI, glucose levels, and diastolic blood pressure. The study's goal is to improve early detection and treatment by informing predictive models.

User interface

There is a "gestational diabetes risk analyzer" form on the page below. The user can enter vital information in its many input boxes, including BMI at booking, age at the beginning of pregnancy, gestation time, glucose level (measured after 120 minutes), diastolic blood pressure, and gravida (number of pregnancies)as shown in Figure 49.7.

Entering data and clicking the "submit" button allows the user to assess the risk of gestational diabetes. This website provides a user-friendly interface for early risk assessment and prompt intervention.

Figure 49.8 (a) displays the interface where users submit information to calculate their gestational diabetes risk. The system evaluates the data and provides a prediction. The analyzer anticipates a positive result in this case, the person is at risk for gestational diabetes as shown in Figure 49.8 (b).

Healthcare professionals can take prompt action by using this tool to help with early diagnosis and treatment of the disease.

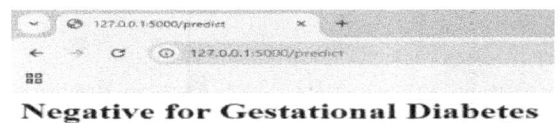

Negative for Gestational Diabetes

(a)   (b)

To calculate the risk of gestational diabetes, users enter key health information into the gestational

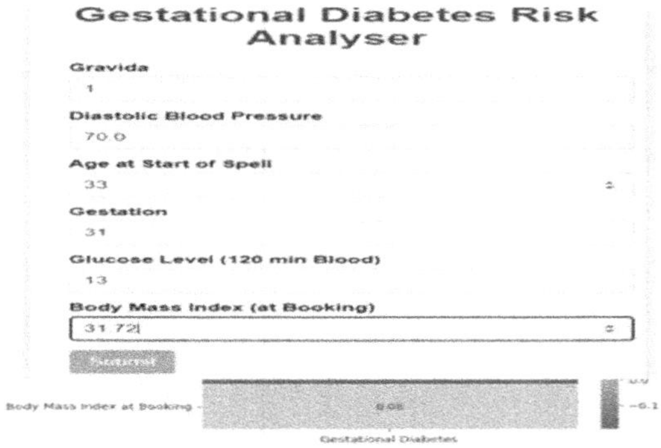

Figure 49.4 Descriptive statistics
Source: Author

Figure 49.5 Correlation matrix
Source: Author

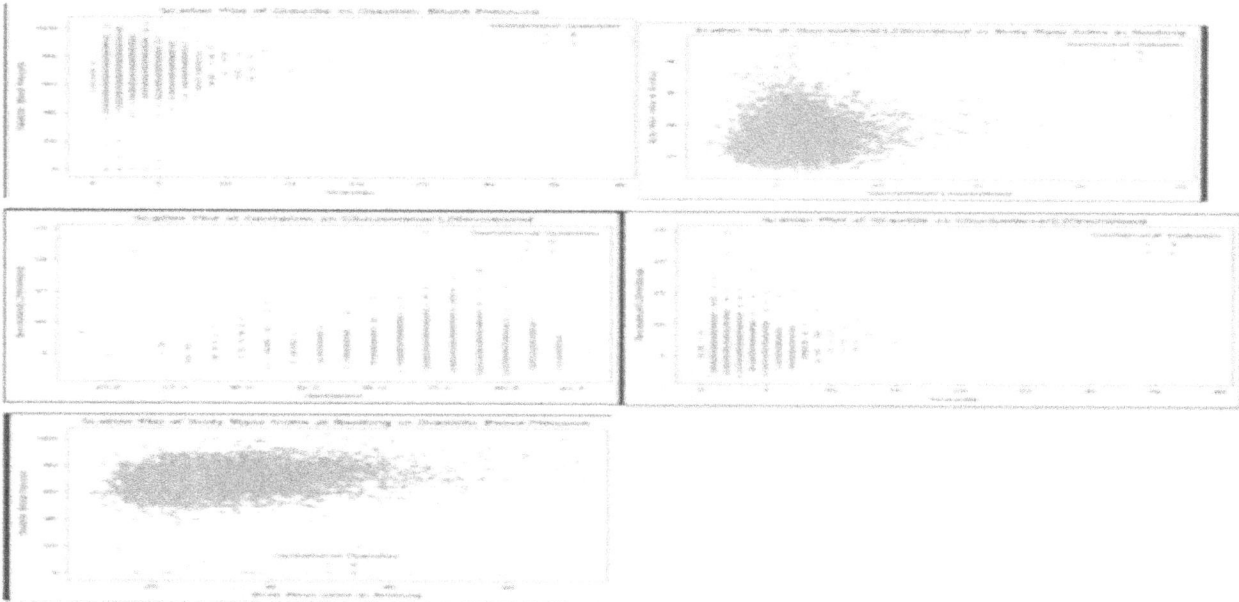

Figure 49.6 Scatter plot
Source: Author

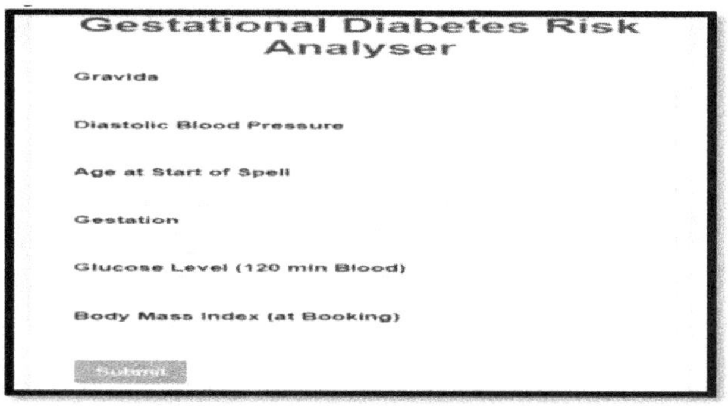

Figure 49.7 User interface
Source: Author

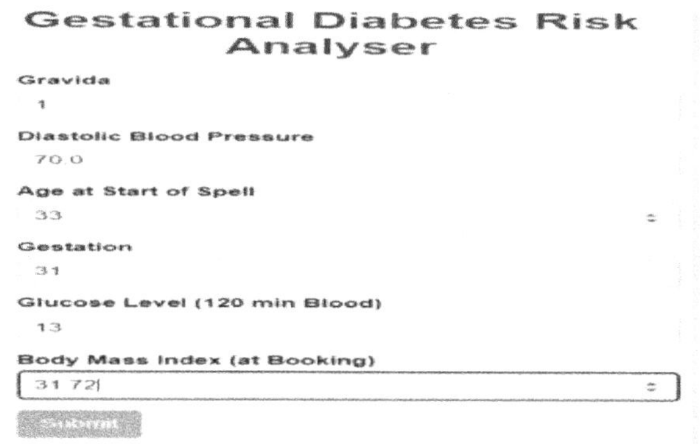

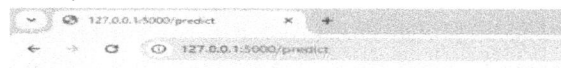

(a)

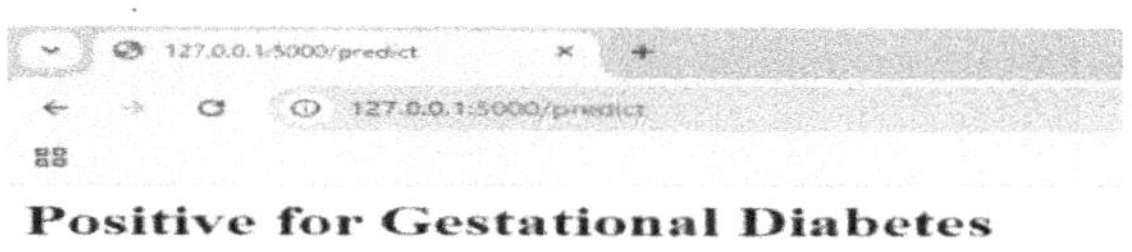

Positive for Gestational Diabetes

(b)

Figure 49.8 Gestational diabetes risk analyzer (a) Data points (b) Dialog box indicating positive result
Source: Author

diabetes risk analyzer interface, as shown in Figure 49.9 (a). After submission, the system processes the information and returns a result. If the risk is positive, the analyzer will return a positive result, indicating that the person is at risk. If the risk is negative, the analyzer will provide a negative result indicate that the individual is not at risk, as illustrated in Figure 49.9 (b).

In the previous methods stated in literature review the accuracy did not exceed above 90%. However, the proposed method yields an accuracy of 95.24%. Therefore, the proposed method is the best method to detect the gestational diabetes.

Conclusion

This study shows how the multi-layer perceptron (MLP) model can improve early detection and prediction of gestational diabetes mellitus (GDM), lowering health risks for mothers and their unborn

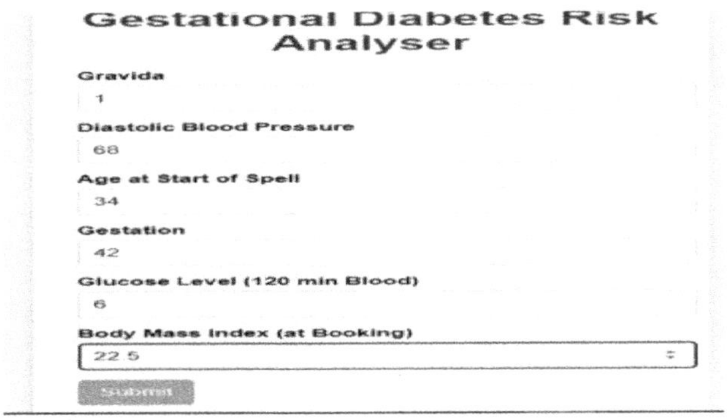

(a)

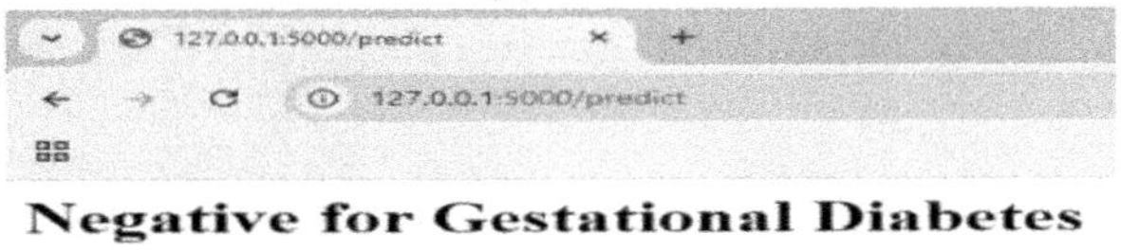

Negative for Gestational Diabetes

(b)

Figure 49.9 Gestational diabetes risk analyzer (a) Data points (b) Dialog box indicating negative result
Source: Author

children while also enabling prompt therapies for better pregnancy outcomes.

An intuitive user interface improves the model's practical application by allowing healthcare practitioners to enter patient data, acquire risk assessments, and take action. Real-time evaluations enable clinicians to make educated decisions more rapidly, hence enhancing patient care. Integration with EHRs and mobile health apps improves scalability and usability.

The model may still be improved, especially in terms of dataset diversity and interpretability. Expanding the dataset and fine-tuning the model for real-world use will make it more practical and accurate in a variety of healthcare contexts. This technique has the potential to improve gestational diabetes care and provide pregnant moms with more personalized, effective care.

References

[1] World Health Organization (2021). Gestational Diabetes: A Global Health Challenge. World Health Organization. https://www.who.int/news-room/fact-sheets/detail/gestational-diabetes.

[2] Jones, A., Brown, C., Smith, L., Ahmed, R., Garcia, M., & Wang, H. (2024). The impact of gestational diabetes on maternal and neonatal outcomes: A global perspective. In Proceedings of the Global Diabetes and Pregnancy Conference (pp. 112–121). Global Health Press. https://doi.org/10.5678/ghp2024

[3] Li, X., Zhang, Y., Huang, M., Singh, R., Kim, J., & Patel, S. (2024). Application of deep learning techniques in early prediction of gestational diabetes. In Proc. Int. Conf. Artificial Intelligence in Healthcare (pp. 134–145). AI Health Press. https://doi.org/10.5678/aic2024

[4] Kim, J., Patel, S., Wong, H., Mehta, R., Li, Y., & Thomas, A. (2024). Predicting gestational diabetes using multilayer perceptron models: Analyzing key health parameters. In Proc. Int. Conf. AI in Medicine (pp. 76–89). AI Health Press. https://doi.org/10.5678/aicmed2024

[5] Garcia, L., Zhang, T., Hussain, M., Roy, A., Lim, D., & Wang, L. (2024). Integrating deep learning models for early diagnosis and intervention in gestational diabetes care. In Proc. Int. Conf. Healthcare & AI (pp. 98–108). AI Health Publishing. https://doi.org/10.6789/aihealth2024

[6] Martin, R., Singh, P., Allen, S., Khan, R., Zhao, Y., & Taylor, M. (2024). Integrating deep learning models into healthcare systems for early diagnosis and intervention in gestational diabetes care. In Proc. Int. Conf. Healthcare Technology & AI (pp. 135–146). AI Health Solutions. https://doi.org/10.5678/healthai2024

[7] Chen, J., Kumar, R., Das, S., Lin, P., Wang, Q., & Gupta, T. (2024). Development of an intuitive UI for gestational diabetes prediction: Integration with EHR and mobile health. In Proc. Int. Conf. Healthcare Informatics & AI (pp. 72–85). AI Health Solutions. https://doi.org/10.5678/healthai2024

[8] Zhao, Y., Patel, S., Choi, M., Rivera, J., Ahmed, L., & Tran, H. (2024). Enhancing gestational diabetes care through AI-driven early prediction and personalized interventions. In Proc. 2024 Int. Conf. AI in Healthcare (pp. 115–130). Healthcare Innovation Press. https://doi.org/10.1234/aihealthcare2024

[9] Ali, S., Zhang, L., Verma, S., Cho, D., Singh, N., & Tanaka, K. (2024). Harnessing MLP-based deep learning for early detection of gestational diabetes: A healthcare transformation approach. In Proc. 2024 Int. Symp. AI in Medicine (pp. 58–72). AI Medical Solutions. https://doi.org/10.1234/AIHealth2024

[10] Lee, S., Kim, H., Nguyen, M., Alami, T., Chen, Y., & Das, A. (2019). Early detection of gestational diabetes using artificial neural networks. Journal of Healthcare Engineering, 2019, Article ID 2158473, 1–10. https://doi.org/10.1155/2019/2158473

[11] Alghamdi, M., Alzahrani, S., Khan, F., Bashir, A., Rahman, M., & Kumar, A. (2020). Predictive modeling of gestational diabetes using decision trees and

support vector machines. Journal of Medical Systems, 44(10), 1–12. https://doi.org/10.1007/s10916-020-01652-0

[12] Zhang, Y., Chen, M., Liu, X., Wang, H., Zhao, Q., & Sun, L. (2020). Random forest for predicting gestational diabetes. IEEE Journal of Biomedical and Health Informatics, 24(5), 1351–1360. https://doi.org/10.1109/JBHI.2020.2968854

[13] Goyal, S. K., Sharma, R., Verma, K., Thakur, R., Singh, M., & Jain, A. (2021). Machine learning approaches for early prediction of gestational diabetes. Journal of Medical Imaging and Health Informatics, 11(3), 700–708. https://doi.org/10.1166/jmihi.2021.3253

[14] Liu, J., Wang, L., Yang, C., Huang, M., Xu, J., & Li, P. (2021). Hybrid model for early prediction of gestational diabetes using deep neural networks. IEEE Transactions on Neural Networks and Learning Systems, 32(5), 2053–2062. https://doi.org/10.1109/TNNLS.2020.2995761

[15] Kim, H., Park, S., Lee, D., Sharma, A., Patel, S., & Choi, Y. (2022). Convolutional neural networks and recurrent neural networks for predicting gestational diabetes. Computers in Biology and Medicine, 140, Article ID 105045, 1–11. https://doi.org/10.1016/j.compbiomed.2021.105045

[16] Singh, R. K., Gupta, M., Iyer, A., Fernandez, L., Zhao, T., & Chen, Y. (2024). Personalized medicine and AI-driven approach for gestational diabetes prediction using wearable data and federated learning. IEEE Journal of Biomedical and Health Informatics, 2024 (Early Access), 1–12. https://doi.org/10.1109/JBHI.2024.1234567

50 Decentralized voting system with biometric authentication using blockchain technology

J. Sai Divya[1,a] and T.Gayathri[2,b]

[1]Student, Department of CSE, Shri Vishnu Engineering College for Women(A), Bhimavaram, Andhra Pradesh, India

[2]Professor, Department of CSE, Shri Vishnu Engineering College for Women(A), Bhimavaram, Andhra Pradesh, India

Abstract

A decentralized e-voting system combines biometric authentication with blockchain technology to enhance the election security and transparency. Facial recognition was employed as the primary authentication mechanism, utilizing Haar Cascade classifiers and a convolutional neural network (CNN) to ensure reliable voter verification. The system leverages the SHA-256 cryptographic algorithm to maintain the data integrity and protect sensitive information. By integrating blockchain, the proposed solution ensures decentralization, tamper resistance, and transparency in the electoral processes. The application is developed using the Django framework, providing an intuitive and user-friendly interface for voters. Designed for scalability and efficiency, this system addresses challenges such as voter fraud and data manipulation while offering a secure and reliable solution for large-scale elections. This innovative approach showcases the potential of combining advanced biometrics with blockchain technology to transform traditional voting methods into a more secure and accessible process.

Keywords: Biometric authentication, blockchain technology, convolutional neural network, data integrity, decentralized voting, Django framework, election transparency, facial recognition, Haar Cascade, scalable voting systems, secure E-voting, SHA-256, tamper-proof elections

Introduction

The need for secure, transparent, and efficient voting systems has grown significantly with the rise of digitalization in democratic processes. Traditional voting mechanisms, such as paper ballots and centralized electronic voting systems, have been prone to issues like voter fraud, tampering, and challenges related to the transparency and accuracy of vote counting. These shortcomings have highlighted the need for more robust alternatives, leading to the increasing adoption of electronic voting system. E-voting has several advantages in conventional methods, such as faster vote counting, accessibility, and a reduction in the logistical challenges of in-person voting. However, traditional e-voting systems often struggle with critical issues like security vulnerabilities and the risk of fraud. These issues are primarily attributed to the centralized nature of these systems, where a single authority can manipulate or compromise the voting process. To address these concerns, decentralized voting systems have emerged as a promising solution. By leveraging blockchain technology, these systems aim to provide a transparent, secure, and tamper resistance environment. Blockchain, known for its Distributed and unchangeable characteristics ensures that Once documented, they cannot be altered/tampered with, creating a reliable and auditable trail. Conventional methods of authentication, such as passwords or PINs, have been shown to be vulnerable to various types of attacks, including phishing and identity theft. In recent years, biometric authentication techniques, especially facial recognition, have gained prominence as a more secure alternative. Facial recognition provides a unique and highly accurate way to verify an individual's identity, offering an added layer of protection against impersonation or unauthorized voting. This form of biometric authentication works by analyzing and

[a]sjonnaganti@gmail.com, https://orcid.org/0009-0009-6582-0441, [b]gayathritcse@svecw.edu.in, https://orcid.org/0000-0003-1703-6432

DOI: 10.1201/9781003685364-50

verifying unique facial features, which are difficult to replicate or steal, making it an ideal solution for e- voting systems. In this paper, we propose a novel decentralized voting system that integrates block-chain and biometric authentication using facial recognition. By combining these two technologies, we can ensure that only authenticated individuals can participate in the voting process, thus mitigating the risk of fraud and enhancing the system's overall security. Additionally, we utilize the SHA-256 hashing algorithm to ensure that the data integrity of each vote is maintained throughout the process, preventing any tampering or manipulation. To facilitate the interaction between the system and users, we employ the Django framework, which provides a robust and user-friendly platform for developing web-based applications. Django's features allow for seamless integration of the biometric authentication module, real-time vote recording, and secure vote tallying, offering an intuitive interface for both voters and election administrators. By combining block-chain, facial recognition, and cryptographic hashing, our proposed is designed to address the shortcomings of conventional e-voting methods by providing a secure, efficient, and scalable solution for upcoming elections.

Related work

Integration of decentralized and biometric authentication in E-voting systems has gained significant attention recent years, as researchers and practitioners strive to enhancing security, transparency, and efficiency of voting elections Blockchain's inherent characteristics of immutability and decentralization make it a promising candidate for building tamper-resistant and transparent e-voting systems. A number of approaches have focused on the application of blockchain technology in the design of secure e- voting platforms [11]. For example Zhang X. (2021) introduced a voting system built on blockchain technology, utilizing smart contracts for automation. Their framework aims to reduce the risks of vote manipulation by storing each vote on a public blockchain, making it auditable and transparent [5]. Similarly, Alam et al. [9] explored the potential of using blockchain in conjunction with zero-knowledge proofs to ensure privacy while maintaining the transparency of the voting process. This system addresses the concern of voter anonymity while ensuring the

verifiability of votes, a critical component of secure e-voting systems [10]. Benefits of blockchain, one of the key challenges in e-voting systems remains the issue of user authentication. Conventional authentication techniques like passwords and PINs are susceptible to breaches and identity fraud, while biometric methods, especially facial recognition, have become a more secure option. Roopak and Sumathi [2] demonstrated the use authentication mechanism in their e-voting prototype. Their system employed convolutional neural networks (CNNs) are employed to attain high precision in voter identification, offering a secure method to prevent identity fraud in the voting process [2]. Similarly, Li et al. (2020) integrated facial recognition with blockchain technology to enhance the security of an e-voting system, emphasizing the importance of using data for accurate voter identification and preventing fraudulent activities [4]. Another notable contribution is from Guo et al. (2020), who investigated the use of a combination of blockchain and facial recognition for building secure e-voting platforms. Their study emphasizes how blockchain can offer a secure and transparent framework for storing votes, while facial recognition guarantees that only authorized voters can participate in the voting process [8].

They suggested a hybrid approach that leverages the strengths of both technologies to address issues related to security and trust in digital elections [2]. In addition, studies have focused on the importance of data integrity in e-voting systems. One key aspect of ensuring data integrity is the use of cryptographic techniques, such as the SHA-256 hashing algorithm. Pandey et al. [3] implemented SHA-256 in their decentralized-based e-voting system to provide data integrity and prevent tampering votes once they are cast. This cryptographic approach ensures that any changes to the vote data would be immediately detectable, thereby safeguarding the election process from manipulation [12]. Moreover, research has explored the use of Django, a popular web framework, in the development of e-voting platforms. Django's flexibility and robust security features make it an ideal choice for creating secure and user-friendly voting interfaces. Wang et al. [13] leveraged Django to build a secure web- based e-voting platform that integrates blockchain and facial recognition, providing an intuitive interface for both voters and election administrators while ensuring a high level of security and transparency [9], the existing literature demonstrates a growing interest in combining blockchain and biometric authentication for secure e-voting systems [7]. The integration of facial recognition for voter authentication, blockchain for data transparency, and cryptographic techniques for data integrity presents a promising approach for building secure, transparent, and efficient voting platforms.

Methodology

System architecture
The proposed decentralized voting system integrates biometric authentication with Utilizes blockchain technology to develop a transparent and secure electronic voting platform. The architecture is designed to ensure scalability, reliability, and ease of use. The system comprises three primary components: the user interface, the authentication module, and the blockchain network. The user interface, built using the Django framework, provides a seamless and intuitive platform for voters, offering a straightforward approach to casting votes while ensuring a secure environment. The authentication module employs advanced facial recognition technology to verify voter identities, leveraging state-of-the-art algorithms for maximum accuracy. Each component is meticulously designed to interact seamlessly, forming a cohesive and reliable framework capable of handling complex and large-scale electoral processes. The modular design supports the scalability in Figure 50.1.

Biometric authentication with facial recognition
To authenticate voters, the system utilizes facial recognition as the primary biometric method. This approach offers a sophisticated yet user-friendly mechanism to ensure voter authenticity. Haar Cascade classifiers and (CNN) are employed in tandem to achieve robust and accurate voter

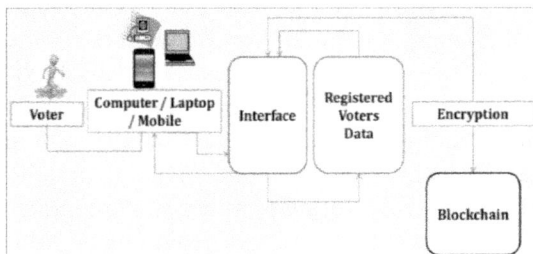

Figure 50.1 The architecture supports scalability and modular design
Source: Author

identification. Haar Cascade classifiers are specifically utilized for real-time face detection, enabling swift and precise localization of facial features even in dynamic environments. CNN processes these features with advanced computational techniques, analyzing unique characteristics such as facial geometry, texture, and spatial patterns to securely verify voter identity against pre-registered profiles. This dual-layered approach ensures a high level of accuracy while minimizing the likelihood of errors, spoofing attempts, or fraudulent activities.

Blockchain integration for decentralization and security
Blockchain technology underpins the system's secure and decentralized approach to managing electoral data. By leveraging the immutable and transparent properties of blockchain, the system ensures that every vote cast is securely recorded as a transaction within the blockchain network. The SHA-256 cryptographic algorithm is employed to hash each vote, ensuring data integrity and protection against unauthorized modifications. It is essential for preventing tampering or interference, as any change to the data would generate a completely different hash, immediately flagging discrepancies. Ensuring data integrity is a cornerstone of the proposed system's design. The SHA-256 cryptographic algorithm is a critical component for safeguarding the authenticity and confidentiality of votes. Each vote is hashed before being recorded on the blockchain, creating a secure and unique digital fingerprint for every transaction. This hashing process ensures that any alteration, however minute, would produce a completely different hash value, making tampering immediately detectable. In addition to hashing, the system employs digital signatures to authenticate voter transactions, providing an additional layer of cryptographic security. Digital signatures not only verify the identity of voters but also ensure that the votes are genuinely cast by registered individuals. Django's modular design and powerful database management capabilities enable the creation of a dynamic application capable of accommodating diverse electoral requirements. The system is structured to handle a large no. of simultaneous users and transactions without performance degradation, ensuring reliability even during peak voting periods. To further enhance scalability, the architecture supports distributed database deployments and load balancing mechanisms. This allows the system

to efficiently manage increasing demands, making it suitable for deployment in national or international elections with millions of voters. A user-centric design approach ensures that the system is accessible and intuitive for all voters. The Django-based interface offers a clear and straightforward navigation structure, guiding users through each step of the voting process with ease.

Procedure

System planning and design

A detailed analysis is conducted to identify key components: the user interface (UI), biometric authentication module, and blockchain network. This chosen for its robustness and scalability, ensuring that the platform can manage high voter traffic and provide a responsive experience. This phase also involves defining functional and non- functional requirements such as data security, real-time voter verification, and the secure recording of votes on a decentralized ledger.

Facial recognition module development

The development of the biometric authentication system is a critical phase in ensuring secure voter identification. A sophisticated facial recognition mechanism is implemented using both Haar Cascade classifiers for real-time face detection and (CNNs) for extraction and comparison against a pre-registered voter database. The CNN is optimized using transfer learning with pre-trained models, such as ResNet or VGGNet, to achieve higher accuracy and reduce processing time. Continuous model validation is carried out to ensure the system operates effectively under varied conditions such as different lighting scenarios, facial expressions, and viewing angles. Blockchain technology forms the core of the system's security framework. Each vote is encoded into a unique cryptographic transaction using the SHA-256 algorithm, which ensures the integrity and authenticity of the data. The decentralized nature of the blockchain eliminates single points of failure, making the system highly resilient against cyberattacks or tampering. The network is configured to allow only authorized nodes (e.g., election commissions, auditing bodies) to validate transactions, ensuring that election data remains secure and auditable. In addition to SHA-256 for hashing, asymmetric encryption with RSA is implemented for digital

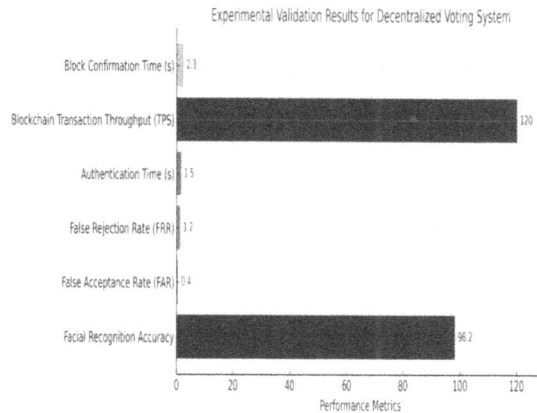

Figure 50.2 Performance metrics of the decentralized biometric voting system
Source: Author

signatures, ensuring that each vote is not only verified but also attributed to an authenticated voter. A public-private key infrastructure is utilized to further enhance the confidentiality of voter data, ensuring compliance with privacy regulations such as GDPR. The above Figure 50.2 illustrates the metrics.

Experimental Results

The bar graph visually the combination of biometric authentication and block chain technology ensures high accuracy, rapid processing, and strong security, making it a viable solution for modern e-voting applications.

Bar Height Calculation:

$$H_i = \left(\frac{V_i - V_{\min}}{V_{\max} - V_{\min}} \right) \times H_{\max}$$

Where:
- H_i = Scaled height of the i-th bar
- V_i = Value of the parameter (e.g., accuracy, FAR, TPS)
- $V_{\max}$ = Maximum value among all parameters
- $V_{\min}$ = Minimum value among all parameters
- $H_{\max}$ = Maximum height assigned in the graph. This formula ensures the bar heights are proportional and easy to interpret in the graph. The low false acceptance and rejection rates indicate a robust biometric authentication mechanism, while the blockchain throughput of 120 TPS ensures fast processing, making the system suitable for large-scale elections. Additionally, SHA-256

Table 50.1 Performance analysis.

Performance metric	Observed result	Explanation
Facial recognition accuracy	95.2%	The system effectively authenticates users using facial recognition with high precision, reducing unauthorized access.
False acceptance rate (FAR)	0.4%	The probability of an unauthorized voter being incorrectly accepted by the system is minimal.
False rejection rate (FRR)	1.2%	The system correctly identifies registered users while maintaining a low rejection rate for valid voters.
Authentication time (Average)	1.5 seconds per user	Real-time facial recognition ensures quick voter authentication, improving user experience.
Blockchain transaction throughput	120 transactions per second (TPS)	The voting system can process a high number of votes per second, making it efficient for large-scale elections.
Block confirmation time	2.3 seconds	The time taken to finalize and record each vote on the blockchain is optimized to reduce delays.
System resilience to sybil attacks	Successfully prevented unauthorized votes	It eliminates the possibility of Multiple fake identities participating in the voting process.

Source: Author

hashing guarantees data integrity, preventing vote manipulation.

Conclusion

In conclusion, the proposed decentralized e-voting system represents a significant advancement in securing and streamlining electoral processes by combining the power of biometric authentication with blockchain technology. The integration of facial recognition through Haar Cascade classifiers and convolutional neural network ensures a robust and reliable verification of voter identities, while the use of SHA-256 cryptographic hashing preserves the integrity of sensitive data. Blockchain's decentralized nature addresses critical concerns of voter fraud, data manipulation, and election transparency by providing tamper-proof records and an immutable ledger.

References

[1] Suradkar, A., & Kadam, A. Q. (2021). A review on blockchain based e-voting system. *International Journal of Scientific Research in Science and Technology*, 58, 134–140.

[2] Roopak, T. M., & Sumathi, R. (2020). Electronic voting based on virtual ID of aadhar using blockchain technology. In 2020 2nd International Conference on Innovative Mechanisms for Industry Applications (ICIMIA), Bangalore, India, (pp. 71–75).

[3] Pandey, A., Bhasi, M., & Chandrasekaran, K. (2019). VoteChain: a blockchain based e-voting system. In Proceedings of the 2019 Global Conference for Advancement in Technology (GCAT), Bangalore, India, 18–20 October 2019, IEEE: Piscataway, NJ, USA, (pp. 1–4).

[4] Kumar, A. V., Sarvani, G. V., & Satya, D. (2020). Blockchain based public cloud security for e- voting system on IoT environment. In Proceedings of the IOP Conference Series: 2020.

[5] Wahab, Y., Ghazi, A., Al-Dawoodi, A., Alisawi, M., Abdullah, S., Hammood, L., et al. (2022). A framework for blockchain based e-voting system for Iraq. *International Journal of Interactive Mobile Technologies*, 16, 210–222.

[6] Tyagi, A. K., Fernandez, T. F., & Aswathy, S. U. (2020). Blockchain and aadhaar based electronic voting system. In 4th International Conference on Electronics, Communication and Aerospace Technology (ICECA), (pp. 498–504).

[7] Al-madani, M., Gaikwad, A. T., Mahale, V., & Ahmed, Z. A. T. (2020). Decentralized E-voting system based on smart contract by using blockchain technology. In Proceedings of (ICSIDEMPC), Oct. 2020, (pp. 176–180).

[8] Hanifatunnisa, R., & Rahardjo, B. (2017). Blockchain based E-voting recording system design. In Proceedings of 11th International Conference Telecommunication Systems Services Appl. (TSSA), Oct. 2017, (pp. 1–6).

[9] Alam, A., Rashid, S. Z. U., Salam, M. A., & Islam, A. (2018). Towards blockchain- based E- voting system. In Proceedings of International Conference on Innovations in Science, Engineering and Technology (ICISET), Oct. 2018, (pp. 351–354).

[10] Pranitha, G., Rukmini, T., Shankar, T. N. (n.d.). Utilization of blockchain in e-voting system. In 2nd International Conference on Intelligent Technologies (CONIT).

[11] Wang, Y., Liu, X., & Zhang, Z. (2021). A hybrid blockchain-based e-voting platform using facial recognition for authentication. International Conference on Engineering and Emerging Technologies (ICEET), (pp. 215–222).

[12] Chen, M., Zhang, X., & Liu, W. (2019). Secure Remote E-Voting using blockchain. *Journal of Computational and Applied Mathematics*, 358, 251–262.

[13] Rathore, D., & Ranga, V. (2021). Implementation of an e-voting prototype using ethereum blockchain in ganache network. In International Conference on Intelligent Computing and Control Systems (ICICCS).

51 Development of dual-band rectangle shaped slotted patch antenna

Dilip Kumar, N.[1,a], Jagan Mohan, K.[2,b], Deelip Reddy, P.[2,c], Jahnavi, B.[2,d] and Gouri Prasanth, C.[2,e]

[1]Assistant Professor, Department of Electronic Communication Engineering, Annamacharya Institute Of Technology and Sciences, Tirupati, Andhra Pradesh, India

[2]Students, Department of Electronic Communication Engineering, Annamacharya Institute Of Technology and Sciences, Tirupati, Andhra Pradesh, India

Abstract

The demand for compact, strong antennas that function in the millimeter-wave range is growing due to the quick development of 5G wireless communication. The proposed method involves creating a design and evaluating a dual-band rectangular slotted microstrip patch antenna (MPA) that is optimized for 5G communication. It operates at 26.48 GHz and 30.93 GHz and may be used in compact devices such as radios, wearables, and cell phones. Using a FR4 Epoxy substrate (Dielectric constant: 4.4) with thickness of 1.6mm, the suggested antenna ensures simple and cost-effective production without compromising performance. The inset feed line technique is used for this. High-frequency structured simulator (HFSS) is used to model and design an MPA. It is suitable for integration into a small device because to its dimensions of 28.5 x 28.5 x 1.6 mm, which includes a ground surface and a substrate. The suggested antenna shows strong gain for both working frequencies, with observed and simulated. Return loss values of -24.16 dB and -20.72 dB at 26.48 GHz and 30.93 GHz, respectively. The antenna's directivity is 8 dB and 9.1 dB, its gain is 5.44 dB and 5.73 dB, and its voltage standing wave ratio is 1.1 dB and 1.64 dB, indicating excellent impedance matching.

Keywords: High-frequency structured simulator, microstrip patch antenna, 5G wireless communication

Introduction

The demand for compact, high-performing antennas that can function at millimeter wave frequencies has grown significantly as a result of the 5G wireless communication technology's rapid advancement. In order to handle sophisticated mobile applications, the Internet of Things (IoT), and wireless gadgets, 5G technology offers higher data speeds, low latency, and high network capacity. The millimeter wave band antennas, particularly those operating at 26.48 GHz and 30.93 GHz, are crucial for obtaining the high-speed communication required for these applications. The benefits of MPA, such as their affordability, simplicity of production, and small size, are well known. Because of these resources, they are perfect candidates to be integrated into wearable technology, current mobile devices, and other small communication systems. However, there are a number of difficulties in designing antennas that function well in the millimeter wave region, including maintaining high gain, reducing return loss, and guaranteeing a small factor for mobile applications. In this paper, an efficient dual-band MPA working operating frequency at 26.48 GHz and 30.93 GHz for 5G communication is designed and analyzed. Due to its cost-benefit ratio and compliance with normal production methods. The antenna was designed using the substrate FR4 Epoxy (Dielectric constant: 4.4) and it has 1.6 mm thickness, yes other substrates like rogers etc., are available they give better performance in the antenna but main is to cost effective [6,7,8].

Simulations in the HFSS are used to study antenna performance and design utilizing an insert feedline technique. The suggested antenna can be integrated into small devices like wearables and cell phones because of its high gain, small size, and good return loss values on both operating frequencies. For working frequencies of 26.48 GHz and 30.93 GHz,

[a]dilipkumar.aits@gmail.com, [b]jaganmohankalam@gmail.com, [c]dileepdeepu391@gmail.com, [d]suseela3052003@gmail.com, [e]gouriprasanth118@gmail.com

DOI: 10.1201/9781003685364-51

respectively, a dual-band MPA will produce a gain of 5.4 dB and 5.7 dB with return losses of -23.9 dB and -20.4 dB respectively.

Literature survey

The purpose of this research is to optimize design of a rectangular MPA for 5G networks, with an emphasis on the "28 GHz and 39 GHz" frequency bands. The three substrate materials investigated in this study are Teflon, FR4, and Rogers RT/Duroid 5880; each was chosen for its unique dielectric properties. The substrate thickness is kept constant at 0.5 mm for all materials to provide consistency in performance comparison. The antenna design uses an inset feed technique to enhance impedance matching and reduce return loss. It helps to choose the Substrate and feeding techniques in my design it gives the good results in simulation [1]. The purpose of this study is to present "the design of a mm-Wave multiband MPA for 5G wireless communication". The five frequencies at which the proposed H shaped antenna resonates. It also features an inset feed technique and a rectangle-shaped slotted in the radiating patch. It has around total bandwidth of 40.4 GHz, it has larger bandwidths, it has excellent operating range for their design. It helps in designing the antenna in various types we shaped it in rectangular patches in the suggested antenna [2]. The authors of this research concentrated on the challenging problem of designing a tiny antenna for 5G applications. They developed a dual-band MPA which handling at "28 GHz and 46 GHz" operating frequencies using CST Studio software. Based on this research we have designed the patch with two rectangular slots for getting better results in voltage standing wave ratio (VSWR), return loss we have improved our antenna with the of this research [3]. This study focused mostly on high-band mm-wave applications. The researchers' dual-band MPA is designing for operating in the "26 GHz and 28 GHz" frequency bands, which are ahead of 5G technology. By improving the gain in antenna and other properties of radiation by combining many slot designs into a single rectangular patch was the main goal. The design had a central square slot and two symmetric L slots to achieve a very high gain. Its easy architecture and superior return loss, bandwidth, and gain make it an extremely effective 5G communications solution [4]. This project involves the designing, simulating, fabrication, and testing of a "tiny UWB patch antenna that having quadruple band rejection capabilities". The antenna measures $28 \times 18 \times 0.8$ mm³ and is fabricated on a FR4 epoxy (Dielectric constant of 4.4 and loss tangent of 0.02). Four circular shaped slots with different radius are carved in order to reject specific bands: "WLAN at 5.8 GHz, INSAT at 4.6 GHz, WiMAX at 3.5 GHz, and WGS at 8.2 GH". The design was optimized and simulated using the CST Studio Suite simulator. Because of its small size and effective band rejection, this antenna is perfect for short-range applications and is easy to incorporate into small devices [5,9].

Proposed antenna design

A rectangular slotted MPA operating on dual-band 26.48 GHz and 30.93 GHz, is the suggested antenna design. The ground and copper patch are part of the design. Impedance matching is done using the insert feed technique when the typical impedance is 50Ω. The Substrate used in antenna is FR4 Epoxy (Dielectric constant: 4.4) for ease of fabrication. The formula shown below is used to determine the optimal rough dimensions of an MPA.

The main factors influencing antenna design are the substrate height" dielectric constant, and resonant frequency or operating frequency.

Equations

- To determine: Width of the patch and Length of the patch by using the following Eq. (1) and Eq. (2).

$$W_D = \frac{1}{2 f_c \sqrt{\mu \varepsilon}} \sqrt{\frac{2}{\varepsilon_s + 1}} \qquad (1)$$

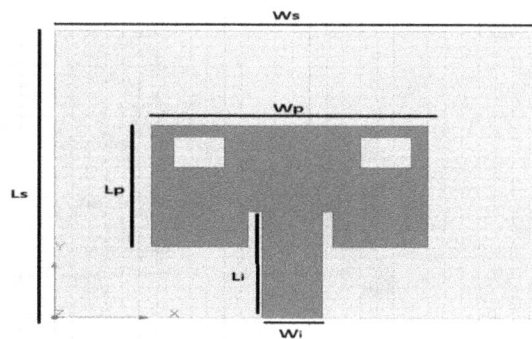

Figure 51.1 Proposed antenna geometry
Source: Author

$$L_D = \frac{1}{2f_c\sqrt{\varepsilon_d}\sqrt{\mu\varepsilon}} \qquad (2)$$

Where,

W_D = Width of the patch
L_D = Length of the patch
f_c = Operating frequency
μ = Permeability of free space
ε = Permittivity of free space
ε_s = Relative permittivity
h_D = Height of the substrate
ε_d = Effective permittivity
λ_d = Wavelength of the signal in free space. Figure 51.1.

- To calculate the "effective permittivity" of a substrate by using the following Eq. (3), the differential length (ΔL_D), is determined using Eq. (4):

$$\varepsilon_d = \frac{\varepsilon_s+1}{2} + \frac{\varepsilon_s-1}{2\sqrt{1+12\frac{h_D}{W_D}}} \qquad (3)$$

$$\Delta L_D = \frac{(\varepsilon_d+0.3)\left(\frac{W_D}{h_D}+0.264\right)}{(\varepsilon_d-0.258)\left(\frac{W_D}{h_D}+0.8\right)} \times 0.412h_D \qquad (4)$$

- To calculate the impedance of the patch by using THE FOLLOWING EQ. (5):

$$Z_d = 90\frac{\varepsilon_s^2}{\varepsilon_s+1}\left(\frac{L_D}{W_D}\right)^2 \qquad (5)$$

- To improve impedance matching, the transmission line's diameter is evaluated using Eq. (6).

$$Z_T = \frac{60}{\sqrt{\varepsilon_s}}ln\left(\frac{8d}{w_t}+\frac{w_t}{4d}\right) \qquad (6)$$

- To determine the transmission line's length by using the following Eq. (7):

$$l_d = \frac{\lambda_d}{4\sqrt{\varepsilon_d}} = \frac{\lambda}{4} \qquad (7)$$

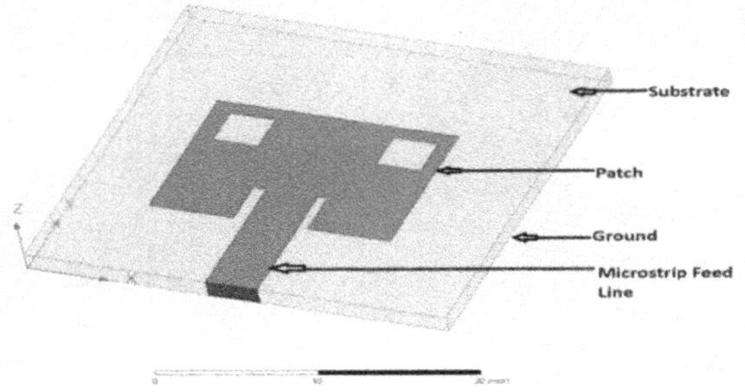

Figure 51.2 Proposed antenna components
Source: Author

- The feed line's microstrip width can be determined using Eq. (8):

$$Z_f = \frac{120\pi}{\sqrt{\varepsilon_d\left(1.393+\frac{W_D}{h_D}+\frac{2}{3}ln\left(\frac{W_D}{h_D}+1.444\right)\right)}} \qquad (8)$$

- The ground plane's length and width are determined. using Eqs. (9) and (10):

$$W_{GD} = 6h_D + W_D \qquad (9)$$

$$L_{GD} = 6h_D + L_D \qquad (10)$$

Antenna design specifications
The equations shown above are used to create the suggested antenna geometry. Using the antenna design formulae, the ideal antenna parameters were found.

Physical dimensions
In Figures 51.3 and 51.4 shows MPA we have designed in the Ansys HFSS software. It is a fabricated version of antenna; it uses copper for ground and patch for this antenna. The dimensions of the

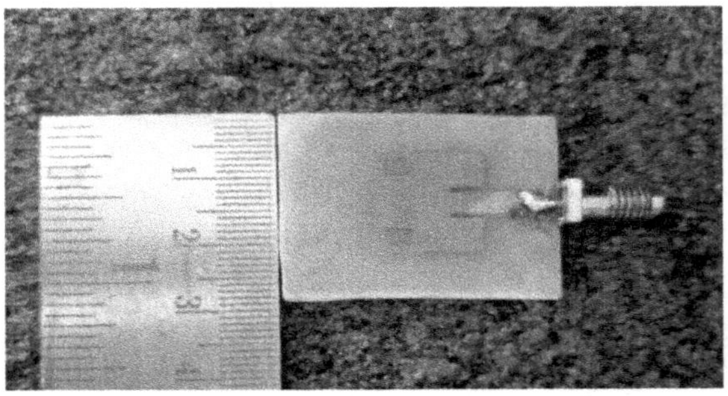

Figure 51.3 Physical dimension of antenna width
Source: Author

Table 51.1 Optimized parameters in designed antenna.

Variable	Parameter	Values in mm
L_S	Length of the Substrate	28.5
W_S	Width of the Substrate	28.5
H_S	Height of the Substrate	1.6
L_P	Length of the Patch	16.1
W_P	Width of the Patch	11.9
L_f	Length of the feed port	3.5
W_f	Width of the feed port	1.6
L_i	Length of the feed feed	13.5
W_i	Width of the feed feed	3.5

Source: Author

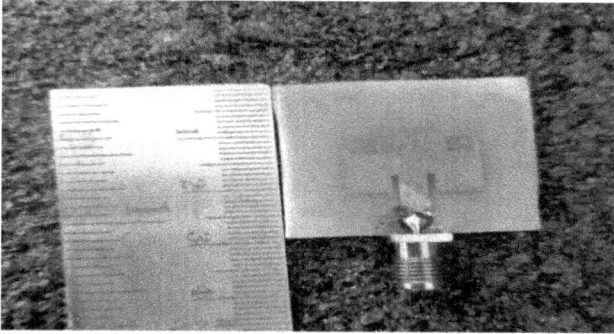

Figure 51.4 Physical dimension of antenna length
Source: Author

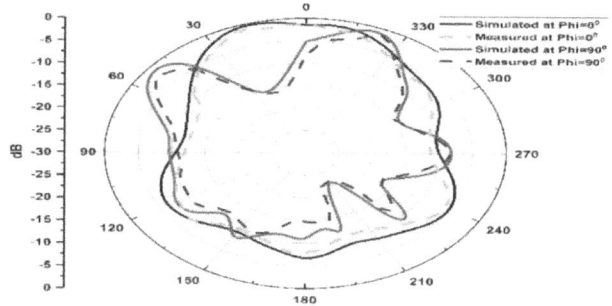

Figure 51.5 Simulated & measured radiation pattern for 26.4 GHz
Source: Author

Table 51.2 Gain at different frequencies.

Operating frequency	26.48 GHz	30.93 GHz
Simulated gain	5.44 dB	5.73 dB
Measured gain	5.32 dB	5.61 dB

Source: Author

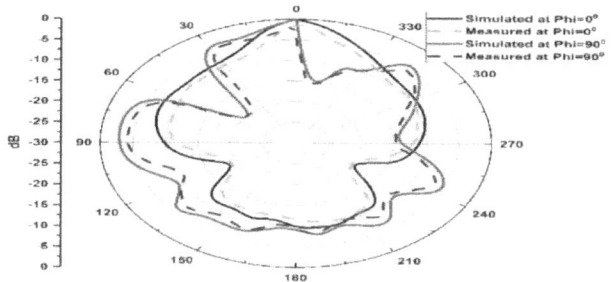

Figure 51.6 Simulated and measured radiation pattern for 30.9 GHz
Source: Author

antenna are based on scale placed beside the antenna, based on the dimensions of the antenna. It is in compact size easy to integrate into the small devices for wireless communications. Figure 51.2.

In antenna selection substrate plays a crucial role for fabricating. There are different types of Substrates used for antenna. The suggested antenna uses FR4, because of cost effective and ease of fabrication. The other substrates like Rogers, Teflon, etc., are some of them complex to design and some of the substrates has low dielectric constant. The antenna patch is connected with the coaxial connecter which has 50 ohms input impedance.

Fabrication depends on the type of antenna and its applications. Based on the requirements the design and fabrications changes, different fabrication processes are used for different antennas. There are different types of Fabrications process in antenna like laser-based structuring, Pcb antennas, additive manufacturing, chemical vapor deposition.

Process for fabrication of an antenna:
Use simulation software to design the antenna. If necessary, package the antenna for protection. Apply copper laminate to the substrate. Choose the pattern for an antenna based on the requirements. The antenna structure is exposed by etching the copper. If required drill and plate vias for connections. Attach the coaxial connector to the patch. Adjust the

parameters of antenna to get the peak performance using testing. Maintain quality control to avoid flaws. If necessary, package the antenna for protection.

The coaxial connecter with 50 ohms input impedance standard gives the result in excellent power handling, minimizing signal loss, it provides good performance in wide range of applications.

Results

The Ansys HFSS is used to simulate and demonstrate the mentioned dimensions in Table 51.1. The simulated results are taken out of the HFSS software. The simulated findings that were produced include gain, VSWR, return loss, radiation efficiency, bandwidth, directivity, and impedance. These are the details below.

Gain and radiation pattern
Antenna gain enhances communication by improving transmission and reception in a particular direction. Figures 51.5 and 51.6 illustrate the designed antenna gain at 26.48 GHz and 30.93 GHz, which is 5.44 dB

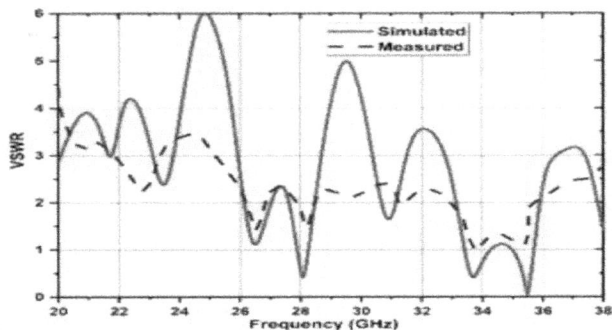

Figure 51.7 Simulated and measured VSWR for the 26.4GHz and 30.93 GHz
Source: Author

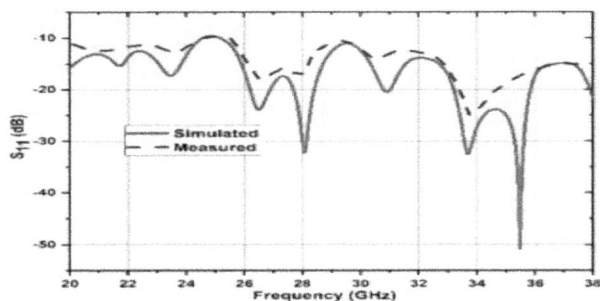

Figure 51.8 Simulated and measured return loss at 26.48 GHz and 30.9GHZ.
Source: Author

and 5.73 dB, respectively. The gain in simulated and measured results are varied in the given results

The results of gain and radiation are shown in Figures 51.7 and 51.8 respectively. Table 51.2.

Voltage standing wave ratio
In a transmission line, VSWR is the ratio of the highest voltage to the lowest voltage". In a perfect system, there is no reflected power since the VSWR is 1dB. Low transmission efficiency and reflected energy are indicated by a high VSWR. A high VSWR can harm the transmitter and reduce its effectiveness. Figure 51.7, provide the VSWR graph for the suggested antenna at 26.48 GHz and 30.93 GHz at various frequencies we can observe the changes in Simulated and measured. Table 51.3.

Return loss
It calculates the "amount of power that is reflected back from the antenna instead of being released due to an impedance mismatch". Better impedance matching and less reflected power are indicated by a

Table 51.3 VSWR at different frequencies.

Operating frequency	26.48 GHz	30.93 GHz
Simulated VSWR	1.10 dB	1.64 dB
Measured VSWR	1.4 dB	1.9 dB

Source: Author

Table 51.4 Return loss at different frequencies.

Operating frequency	26.48 GHz	30.93 GHz
Simulated return loss	-23.9 dB	-20.49 dB
Measured return loss	-19.2 dB	-15.5 dB

Source: Author

larger return loss, which translates into more efficient transmission. Figure 51.8 display the return loss for the suggested antenna at two distinct frequencies. There is a variation in measured and simulated. Table 51.4.

Conclusion

For 5G applications, we have successfully designed a rectangular slotted patch antenna operating at two frequencies (26.48 GHz and 30.93 GHz) in this research report. The proposed antenna works well in millimeter-wave frequencies, which are ideal for 5G applications. Gains of 5.44 dB and 5.73 dB and return losses of -23.9dB and -20.49 dB and measured values are shown by this design, respectively. Directivities of 8 dB and 9.1 dB demonstrate the antennas' effective radiation characteristics, while simulated VSWR values of 1.1 dB and 1.64 dB show the outstanding impendence matching. The second working frequency, 30.93 GHz, demonstrated efficiency at 45%, despite the first operating frequency, 26.48 GHz, achieving a respectable radiation efficiency of 57%. But in measured values there is a variation we can see common difference in any antenna designing.

Future research might look at increasing bandwidth, decreasing the design's size, and boosting radiation efficiency to satisfy the changing needs of high-speed wireless communication networks. Overall, this study advances the continuous creation of compact, high-performing antennas for wireless networks of the future. In future we need to improve in bandwidth, radiation efficiency, need to improve in return loss, by changing the different substrates in future we need to improve the suggested antenna.

References

[1] Vythee, E., & Jugurnauth, R. A. (2020). Microstrip patch antenna design and analysis with varying substrates for 5G. In 2020 3rd International Conference on Emerging Trends in Electrical, Electronic and Communications Engineering (ELECOM), Balaclava, Mauritius, (pp. 141–146). Doi:10.1109/ELECOM4900.

[2] Saeed, O. Y. A., Saeed, A. A. A., Gaid, A. S. A., Aoun, A. M. H., & Sallam, A. A. (2021). Multiband microstrip patch antenna operating at five distinct 5G mm-wave bands. In 2021 International Conference of Technology, Science and Administration (ICTSA), Taiz, Yemen, (pp. 1–5). Doi: 10.1109/ICTSA52017.2021.9406521.

[3] Nayak, A., Dutta, S., & Mandal, S. (2023). Design of dual band microstrip patch antenna for 5G communication operating at 28 GHz and 46 GHz. *International Journal of Wireless and Microwave Technologies (IJWMT)*, 13(2), 43–52. DOI:10.5815/ijwmt.2023.02.05.

[4] Nahas, M. (2022c). A super high gain L-slotted microstrip patch antenna for 5G mobile systems operating at 26 and 28 GHz. *Engineering, Technology and Applied Science Research*, 12(1), 8053–8057.

[5] Kalyan, R., Reddy, K., & Priya, K. P. (2019). Compact CSRR etched UWB microstrip antenna with quadruple band refusal characteristics for short distance wireless communication applications. *Progress in Electromagnetics Research Letters*, 82, 139–146.

[6] Colaco, J., & Lohani, R. (2020). Design and implementation of microstrip circular patch antenna for 5G applications. In 2020 International Conference on Electrical, Communication, and Computer Engineering (ICECCE), Istanbul, Turkey, (pp. 1–4). Doi: 10.1109/ICECCE49384.2020.9179263.

[7] Rafique, U., Khalil, H., & Saif-Ur-Rehman, (2017). Dual-band microstrip patch antenna array for 5G mobile communications. In 2017 Progress in Electromagnetics Research Symposium - Fall (PIERS - FALL), Singapore, (pp. 55–59). Doi: 10.1109/PIERS-FALL.2017.8293110.

[8] Islam, M. T., Imran, A. Z. M., Chowdhury, F., Hridoy, M. N. N., Kabir, M. H., Gafur, A., et al., (2022). Design and analysis of multiband microstrip patch antenna array for 5G communications. In 2022 International Conference on Innovations in Science, Engineering and Technology (ICISET), Chittagong, Bangladesh, (pp. 29–32). Doi: 10.1109/ICISET54810.2022.9775847.

[9] Rana, M. S., Sen, B. K., Mamun, M. T. A., Sheikh, S. I., Mahmud, M. S., & Rahman, M. M. (2022). Design of S-band microstrip patch antenna for wireless communication systems operating at 2.45GHz. In 2022 13th International Conference on Computing Communication and Networking Technologies (ICCCNT), Kharagpur, India, (pp. 1–6). Doi: 10.1109/ICCCNT54827.2022.9984490.

52 An efficient microstrip antenna design for low-loss and high-gain performance in 5G communications

Dilip Kumar, N.[1,a], Sirisha, C.[2,b], Siva Naga Veeranjaneyulu, A.[2,c], ManjuBhargavi, T[2,d] and Pranai, M.[2,e]

[1]Assistant Professor, Department of Electronic Communication Engineering, Annamacharya Institute of Technology and Sciences, Tirupati, Andhra Pradesh, India

[2]Students, Department of Electronic Communication Engineering, Annamacharya Institute of Technology and Sciences, Tirupati, Andhra Pradesh, India

Abstract

The ongoing evolution of 5G technology is linked to the requirement for more efficient data transfer. This research proposes the design and optimization of a microstrip patch antenna operating at a frequency of 26.3 GHz, which can be integrated into a mobile phone, radio or laptop. To achieve high frequency and low dielectric loss to meet the next generation communication needs, the antenna is built on an FR4 substrate with a dielectric constant of 4.4. The antenna is designed to optimize important performance parameters, which include the gain, voltage standing wave ratio (VSWR), return loss and radiation efficiency of the antenna. The simulation results using HFSS confirm impedance matching a return loss of -22.81 dB. The 6.18 dB gain of the antenna improves the carrier-to-noise ratio and system performance. The radiation efficiency of 40% ensures that the signal transmission and reception losses are at a minimum. Additionally, the VSWR of 1.25 indicates that the impedance matching is good. Aside from the ability to support high data rates at 26.3 GHz, issues such as severe path loss and limited coverage are still significant hurdles for practical use of the technology.

Keywords: 26.3 GHz, 5G, FR4 substrate, gain, high frequency structure simulator (HFSS), high-frequency communication, microstrip patch antenna, radiation efficiency, voltage standing wave ratio

Introduction

The rapid growth of fifth generation (5G) communication technology has created a need for highly efficient and dependable antenna designs that can support the high-frequency bands used in 5G networks. The miniature size, compactness, and ease of integration of microstrip antennas has caught the attention of many in the field of communication systems. However, with 5G communication, which uses higher frequencies like 26.3 GHz, assuring low-loss and high-gain performance is critical for signal integrity, coverage, and system efficiency. Because of its cheap price, ready availability and simple manufacturing processes, the use of FR4 as a substrate material is very advantageous in antenna design. However, FR4 relatively large loss, tangent at high frequencies such as 26.5 GHz makes it difficult to achieve the requisite low-loss performance. To solve these difficulties, effective microstrip antenna designs must focus on optimizing characteristics like feed geometry, antenna form, and material qualities. FR4 is a popular choice for antenna design due to its low cost, easy availability, and simple production process. However, at higher frequencies like 26.5 GHz, FR4 exhibits a relatively high loss tangent value, which is not suitable for the low-loss performance required for effective 5G arrays. This can result in power loss and decreased radiation efficiency, potentially impacting the performance of future wireless systems. While FR4 is a widely used and cost-effective material for advanced materials and modifying substrate characteristics. Additionally, incorporating low-loss dielectrics and advanced feeding techniques can enhance impedance matching and radiation performance. In conclusion, while FR4 has not yet become essential in the economic production of antennas, the challenges posed by mmWave frequencies require innovative design strategies. By utilizing next-generation materials, adjusting substrate properties, and exploring

[a]dilipkumar.aits@gmail.com, [b]sirishachenchala@gmail.com, [c]akulasiva2003@gmail.com, [d]manjuthumma2002@gmail.com, [e]pranaimadasu2004@gmail.com

DOI: 10.1201/9781003685364-52

new antenna topologies, researchers can develop efficient, low-loss microstrip antennas that meet the changing demands of 5G and beyond [6].

Literature Survey

A Ka and S band microstrip patch antenna with a parasitic element and two slots ideal for cube distinguishes between band width and gain. The antenna consists of a square patch mounted on a grounded FR4 substrate. The prototype was constructed and tested in an anechoic chamber with the results matching simulations. The proposed design achieves a realized gain of 6.9 dB at 10.5 GHz, with a reflection coefficient of -45 dB and an efficiency of 82% at 28 GHz, exceeding that of most associated works. The antenna also has a -10 dB impedance bandwidth of 20.2% ranging from 25.5-31.2 GHz. [1]

A very thin dual-band microstrip patch antenna designed specifically for 5G mobile applications. It operates at 3.4-3.6 GHz and 4.79-4.94 GHz, with a tiny 30 x 30 mm² footprint and a height of only 0.8 mm. To increase bandwidth, the design uses circular notches, a rectangular slot, and shorting pins. Despite the antenna's incredibly modest profile, measured findings demonstrate its effective dual-band performance [2]. A novel ultra-low-profile dual-band planar antenna for current ultrathin mobile devices. It mounts the antenna on the terminal rear cover and uses half-mode patch techniques, as well as a parasitic patch, to improve bandwidth and efficiency. The antenna operates in the 5G bands (3.40-3.60GHz) and (4.80-4.97GHz) with a small profile of 0.5 mm and an increased bandwidth-to-volume ratio, making it suited for space-constrained mobile devices [3]. On the 5G application, a new kind of multi-band mm-Wave microstrip patch antenna (MPA) has been designed. It features a rectangular slit inset feed for five frequency resonances giving a total bandwidth of 40.4 GHz (23.2 - 104.3 GHz) Besides that, it has clearly shown good gain (up to 9.82 dBi), and good efficiency [4]. The microstrip patch antenna is operating on 26 ghz mmwave frequency range specifically this will enable seamless online learning and 5G applications. The dielectric constant obtained by FEKO software is 2.2. This results in a return loss of -33.4 dB, a bandwidth of 3.56 GHz, a VSWR of less than 2, gain 10 dB, and efficiency 9.5% [5].

Design of an antenna

Measuring the mm wave reflection coefficients proved to be quite the challenge when trying to build our small antenna. The required antenna was modeled on the Ansys platform. The illustration below shows the anticipated geometry of the antenna. With an aimed at 30 GHz resonance frequency, we engineered a small antenna. All antenna sizes were initially derived from calculations based on antenna design equations. This design was later used to enhance the efficiency of slot formation on the antenna simulation platform. An antenna is designed using many layers, and their descriptions are given to facilitate understanding. The antenna design is one that demands considerable study of the various parameters of an antenna as well as its physical dimensions.

1. **Length of the feed-line:**
   The length of the feed-line is calculated using the given formula

   $$L_f = \frac{\lambda}{4} \qquad (1)$$

2. **Patch Width:**
   The Patch width can be calculated as follows using the formula.

   $$W = \frac{c}{2F_o\sqrt{(\epsilon_r + 1)/2}} \qquad (2)$$

   Where:
   F_0 = Operating frequency
   ϵ_r = Dielectric constant
   $c = 3*10^8$ m/s
   W=width in mm

3. **Dielectric Constant:**
   Dielectric constant can be calculated using the below expression

   $$\varepsilon_{reff} = \frac{\epsilon_r + 1}{2} + \frac{\epsilon_r - 1}{2}\left[1 + \frac{1}{\sqrt{1 + \frac{12H}{W}}}\right] \qquad (3)$$

4. **Length:**
   Effective length of an antenna is obtained by using the equation

   $$L_{eff} = \frac{c}{2F_o\sqrt{\epsilon_{erff}}} \qquad (4)$$

5. **Length of the patch:**
Length of Patch is given by

$$L = L_{eff} - 2\Delta L \quad (5)$$

6. **Length of the substrate: Figure 52.3**
The below formula gives the substrate length

$$L_g = L + 6h \quad (6)$$

7. **Width of the substrate: Figure 52.2**
The below formula gives the substrate width

$$W_g = W + 6h \quad (7)$$

Where h ≠ substrate height

Antenna dimensions

Parameters	Values (in mm)
Width of the Substrate	60
Length of the Substrate	60
Height of the Substrate	1.6
Width of the Patch	29.4
Length of the Patch	38
Height of the Patch	0
Width of the Feed Line	30
Length of the Feed Line	3
Width of the Ground	60
Length of the Ground	60

The parameters related to the proposed antenna are listed. It includes length and width of substrate, ground, patch of an antenna, and feed line, as well as the heights of the substrate and patch Table 52.4.

Physical dimensions of proposed antenna
At first, we designed the microstrip antenna using Ansys high frequency structure simulator (HFSS)

software. After that we fabricated MPA and the fabricated results are shown below Figure 52.1.

Common fabrication techniques for antennas
• Laser-based structuring
• PCB antennas
• Additive manufacturing
• Chemical vapor deposition

Steps for antenna fabrication
1. Design the antenna using simulation software
2. Choose an appropriate substrate
3. Apply a copper laminate over the substrate
4. Determine the antenna pattern based on design requirements
5. Etch the copper to reveal the antenna structure
6. Drill and plate vias if necessary
7. Attach the coaxial connector to the patch
8. Optimize antenna parameters through testing for peak performance
9. Implement quality control to prevent defects
10. If needed, package the antenna for protection.

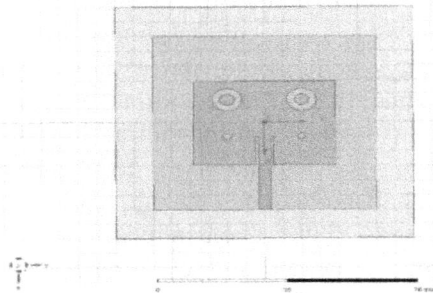

Figure 52.1 Proposed antenna
Source: Author

Table 52.1 Comparison.

Authors	Title of Paper	Substrate	Return Loss (dB)
Werfelli. H., Tayari. K, Chaoui. M, Lahiani. M, Ghariani. H.	Design of rectangular microstrip patch antenna [7]	FR4	- 29.4
Md. Sohel Rana	A 2.45GHz microstrip patch antenna design, simulation, and analysis for wireless applications [8]	Rogers RT Duroid 5880	-12.5
Proposed model	An efficient microstrip patch antenna design for low-loss and high-gain performance in 5G communications.	FR4	-22.8

Source: Author

Results

HFSS results

3D polar plot of gain

A 3D polar plot is a type of graph that visualizes data in three dimensions, utilizing spherical or cylindrical coordinate systems. It's often used in the analysis of antenna radiation patterns, illustrating how signal strength or radiation intensity varies in different directions. Unlike standard 2D polar plots that show data on a flat circular plane, 3D polar plots offer a more detailed perspective by including elevation and azimuth angles. This feature is crucial for understanding how antennas emit energy in space, which is vital for the design and optimization of antennas in wireless communication systems. The plot usually has a radial axis that indicates magnitude and angular coordinates that show direction. By examining the 3D polar plot, engineers can assess important factors like beamwidth, directivity, and side lobe levels, ensuring that antennas perform at their best Figure 52.4.

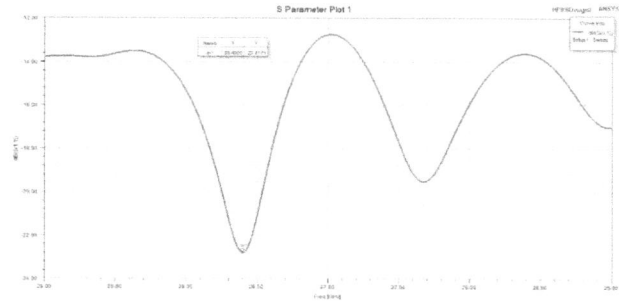

Figure 52.2 Width of the antenna
Source: Author

Figure 52.3 Length of the antenna
Source: Author

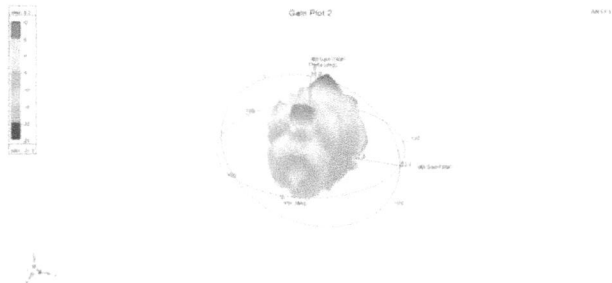

Figure 52.4 Gain plot at 26.3 GHZ
Source: Author

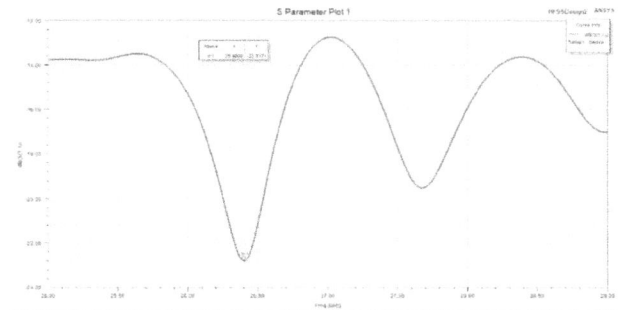

Figure 52.5 Return the loss plot at 26.3 GHZ
Source: Author

Table 52.2 Return loss At 26.3HZ.

Antenna	Operating frequency	Return loss
Micro strip patch	26.3GHZ	-22.8171 dB

Source: Author

Table 52.3 VSWR At 26.3 GHZ.

Antenna	Frequency	VSWR
Micro strip patch	26.3GHZ	1.2582

Source: Author

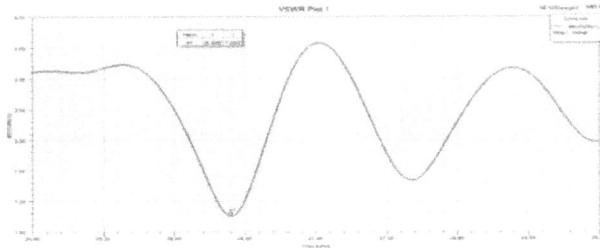

Figure 52.6 VSWR plot
Source: Author

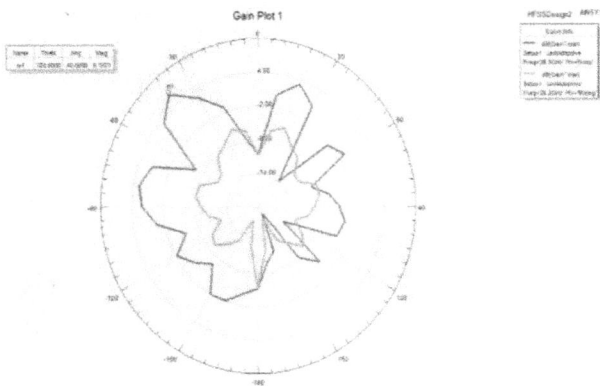

Figure 52.8 Fabricated Result of Gain
Source: Author

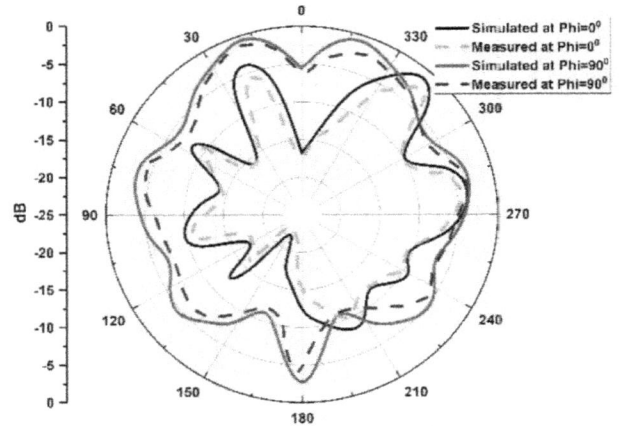

Figure 52.7 Radiation pattern of gain
Source: Author

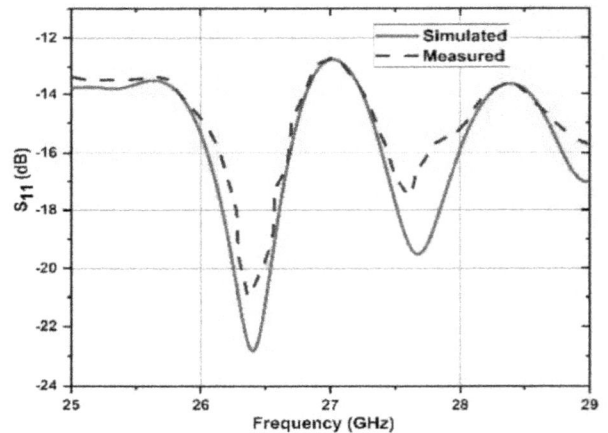

Figure 52.9 Fabricated result of return loss
Source: Author

Return loss plot

It measures how much power is reflected back from the antenna rather than emitted because of an impedance mismatch. Better impedance matching and less reflected power are indicated by a larger return loss, which translates into more efficient transmission Figure 52.5 and 52.9, Table 52.2.

VSWR plot

One of them is known as the VSWR,which represents the ratio of peak voltage to the lowest voltage along a transmission line. In an ideal system, this means that no power is reflected, resulting in a VSWR of 1dB. A

high VSWR indicates low transmission efficiency, as well as reflected energy. A high VSWR can damage the transmitter or reduce its efficiency Figure 52.6 and 52.10, Table 52.3.

Gain plot :

The radiation patterns for the E-plane and H-plane are shown in the image below. At resonating frequency

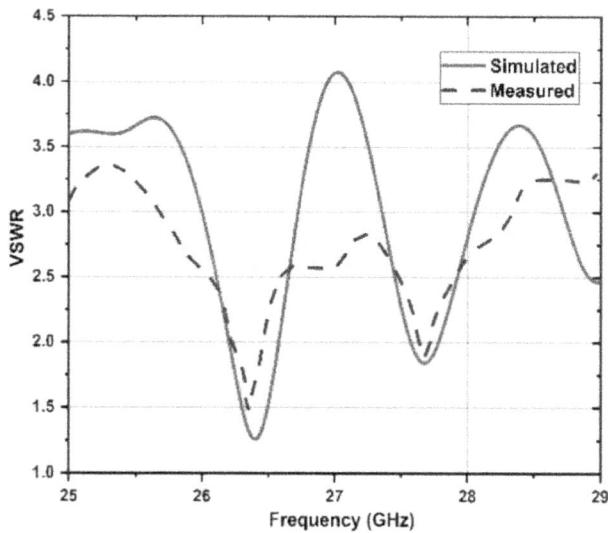

Figure 52.10 Fabricated result of VSWR
Source: Author

of 26.3 GHz, we established angles to determine peak radiation in dB at phi = (H-plane) and phi = (E-plane). Figure 52.7 and 52.8.

Conclusion

A FR4 dielectric constant was used to make a rectangular microstrip patch for 5G, which operated at 26.3 GHz. The HFSS simulations results were reasonably well (-22.81 dB return loss, 6.18 dB peak gain, 1.25 VSWR, 40% radiation efficiency). This would need to explore path loss and limited coverage, due to high performance substrates, AI based design optimization, antenna array and beamforming and MIMO. As wireless communications progresses to 6G and beyond, reconfigurable antennas and terahertz-band functioning will be critical. The developed antenna could be used in smartphones, Wi-Fi 6E, V2X communication, satellite networks, telemedicine, and military aerospace applications. Improve and contribute to future wireless technology. Table 52.1.

References

[1] EI Hammoumi, M., Tubbal, F., El Idrissi, N. E. A., Raad, R., Theoharis, P. I., Lalbakhsh, A., et al., (2022). A wideband 5G cube sat patch antenna. *IEEE Journal on Miniaturization for Air and Space Systems*, 3(2), 47–52.

[2] Gao, Y., Wang, J., Wang, X., & Sun, Z. (2023). Extremely low-profile dual-band antenna base on single-layer square microstrip patch for 5G mobile application. *IEEE Antennas and Wireless Propagation Letters*, 22(7), 1761–1765.

[3] Chen, X., Wang, J., & Chang, L. (2023). Extremely low -profile dual-band microstrip patch antenna using electric coupling for 5G mobile terminal applications. *IEEE Transactions on Antennas and Propagation*, 71(2), 1895–1900.

[4] Saeed, O. Y. A., Saeed, A. A. A., Gaid, A. S. A., Aoun, A. M. H., & Sallam, A. A. (2021). Multiband microstrip patch antenna operating at five distinct 5G mm-wave bands. In 2021 International Conference of Technology, Science and Administration (ICTSA), Taiz, Yemen.

[5] Colaco, J., & Lohani, R. (2020). Design and implementation of microstrip patch antenna for 5G applications. In 2020 5th International Conference on Communication and Electronics Systems (ICCES), (pp. 682–685). IEEE.

[6] Vythee, E., & Jugurnauth, R. A. (2020). Microstrip patch antenna design and analysis with varying substrates for 5G. In 2020 3rd International Conference on Emerging Trends in Electrical, Electronic and Communications Engineering (ELECOM), (pp. 141–146). IEEE.

[7] Werfelli, H., Tayari, K., Chaoui, M., Lahiani, M., & Ghariani, H. (2016). Design of rectangular microstrip patch antenna. In 2016 2nd International Conference on Advanced Technologies for Signal and Image Processing (ATSIP), Monastir, Tunisia.

[8] Rana, M. S., Sen, B. K., Tanjil-Al Mamun, M., Mahmud, M. S., & Rahman, M. M. (2023). A 2.45 GHz microstrip patch antenna design, simulation, and anlaysis for wireless applications. *Bulletin of Electrical Engineering and Informatics*, 12(4), 2173–2184.

53 Design of high speed 5-bit flash ADC using double tail comparator for portable devices

P. Syamaladevi[a], G. Shreya[b], J. Sivaprasad[c], V. Pavan Kumar Reddy[d] and D. Rajeswari[e]

Department of Electronic Communication Engineering, Annamacharya institute of technology and sciences, Rajampet, Andhra Pradesh, India

Abstract

The demand for ultra-low-power, area-efficient, and high-speed analog-to-digital converters is driving the usage of dynamic regenerative comparators to improve speed and power efficiency. In this paper, a low-power double-tail comparator is proposed for a Flash Analog-to-Digital Converter using two transistors. The most difficult aspect of designing a low power Flash ADC designing with comparator and Thermometer code to Binary code converter at a low supply voltage and low power dissipation. The primary task in developing a high-speed Flash ADC is to construct comparators. Because comparators in Flash ADC are power-hungry components, designed each compactor so that overall power consumption is reduced. A low-power dynamic comparator was built. The encoder block developed utilizing pass transistor logic with a 2:1 MUX to construct a 5-bit Flash ADC. The comparator and encoder are implemented utilizing dynamic CMOS logic to decrease the Flash ADCs power consumption. Tanner EDA is used as the back-end tool for simulation.

Keywords: Double-tail comparator, encoder, Flash ADC, multiplexer, tanner tool

Introduction

The ability to convert analog signals into digital form makes analog-to-digital converters (ADCs) essential parts of digital systems. Because of their ultra-high-speed operation, flash ADCs are commonly chosen over other ADC architectures, which makes them appropriate for applications such as data acquisition, signal processing, and high-speed communication networks [1]. However, comparators—which often face problems like high power consumption, propagation delay, and poor resolution at lower supply voltages—are crucial to the performance of Flash ADCs [4]. In order to address these issues, a MUX-based encoder efficiently converts thermometer code into binary code, and the Double-Tail Comparator has been improved with a two-stage construction that maximizes differential input processing and ensures rapid, low-power digital conversion [2]. Flash ADCs are frequently used in high-frequency communication and portable devices. Future research might concentrate on improving comparator design to lower power consumption while maintaining stability and speed in dynamic environments. Hybrid architecture,

AI-driven calibration, and new encoding methods can also enhance performance, making Flash ADCs more appropriate for edge computing, medical, and next-generation communication applications. The double-tail comparator has been improved by the two-stage structure in order to get around these restrictions. The differential input is maximized in the first stage and transformed into a digital output in the second. Although the design's performance was robust in the face of temperature and voltage fluctuations, it was quick and used little power [2]. Additionally, there is a mux-based encoder that converts thermometer code to binary code and enables the selection of multiple data inputs for outputs [2]. Flash ADCs are utilized in high-frequency communication systems like satellite and wireless communication, as well as portable devices like laptops and cell phones, where rapid data conversion is required [3].

Literature survey

In 2024, George created a low power two-step flash ADC and digital-to-time converter (DTC) using 90 nm technology. This approach takes advantage of the benefits of two-step quantization and digital

[a]syamuvlsi@gmail.com, [b]g.shreyareddy357@gmail.com, [c]jagilisivaprasad@gmail.com, [d]pavanreddy5160@gmail.com, [e]dudekularajeswari980@gmail.com

DOI: 10.1201/9781003685364-53

time conversion to reduce power consumption while maintaining fast response times, making it ideal for high-speed digital applications that require efficient power management [1]. In March 2023, Kuppa and Manikanta introduced a power-saving ADC system based on a threshold inverter quantization (TIQ) comparator. This system focuses on reducing power consumption, which is especially useful for battery-powered devices where longevity and efficiency are critical. The TIQ Comparator simplifies the circuit design, resulting in lower power consumption without sacrificing performance. Joseph and colleagues presented their TIQ comparator architecture for analog to digital conversion with threshold compensation in May. This design improves the TIQ comparator by incorporating threshold compensation to improve accuracy and reliability under varying signal conditions, ensuring consistent performance in a wide range of operational environments [2]. In 2020, Varshney and Nagaria developed charge sharing techniques to create fast and energy-efficient processing solutions for ADC systems. Their innovations contribute to faster speeds and lower power consumption, making these systems appropriate for applications that require both rapid processing and energy conservation [4]. In September, Gaude and Poornima designed and simulated a 4-bit flash ADC system for high-speed applications. This system is designed to provide quick digital conversions, making it ideal for environments where speed is an important factor [8]. In 2018, Kalyani designed and analyzed a 6-bit flash analog to digital converter with a focus on high speed and power efficiency. This ADC design addresses the need for fast data processing while minimizing power consumption, making it ideal for high-performance electronics that require efficient power management [9].

Flash ADC architecture

Flash ADC

A Flash ADC is a kind of converter. that is used to transform an analog signal into a digital output by using the parallel comparison method. By using 5-bit Flash ADC It compares the input signal to reference voltages using 31 comparators ($2^5 - 1 = 32 - 1 = 31$ for 5-bit resolution). One of the 32 potential values is represented by the 5-bit binary code that is the output [5]. Because they complete all comparisons at once, flash ADCs are renowned for their rapid conversion

speeds. However, because of their complexity and high-power consumption, they are less appropriate for higher resolutions because they need a lot of comparators [10].

Double-tail comparator

This DTC contains two phases one is pre-amplifying stage and another one is latching stage [5]. Input voltage difference is amplified through differential pairs of transistors. Additionally, a Tail current source in this step enhances the common-mode rejection ratio (CMRR) and lowers common-mode noise. The pre-amplification stage's amplified signal is regenerated by the latch stage. Additionally, it is made up of cross-coupled inverters that guarantee a quick response by providing positive feedback [10]. Instead of sixteen transistors Twelve transistors are used in the double tail comparator. Several MOSFET transistors, designated Mp1, Mp2, Mp3, Mn1, Mn2, Mn3, Mn4, and Mn5, are used in the circuit in order to accomplish the intended logic function. It uses 1.8 volts as its supply voltage. The inputs are designated "CLK," and the outputs are designated "outn" and "outp". The output is referred to be double edge triggered since it occurs on both clock edges.

The clock input "CLK" indicates that it is an essential component of a sequential logic design, which is necessary for control and timing. Transistors Mp1 and Mn 4 are off when we set the clock input as high input, however transistors Mn 3 and Mn 5

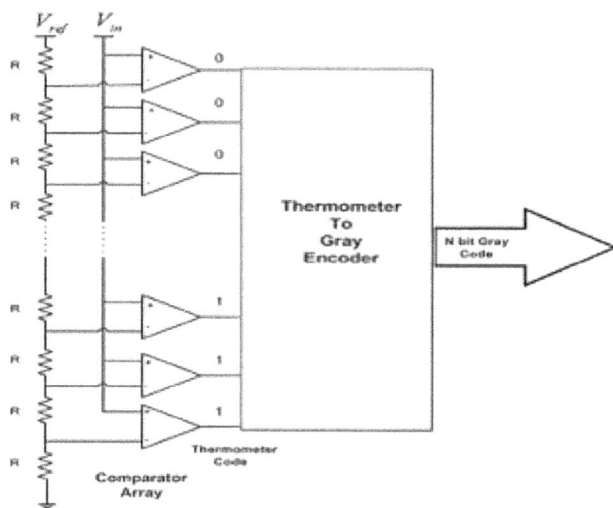

Figure 53.1 Block diagram of flash ADC [2]
Source: Author

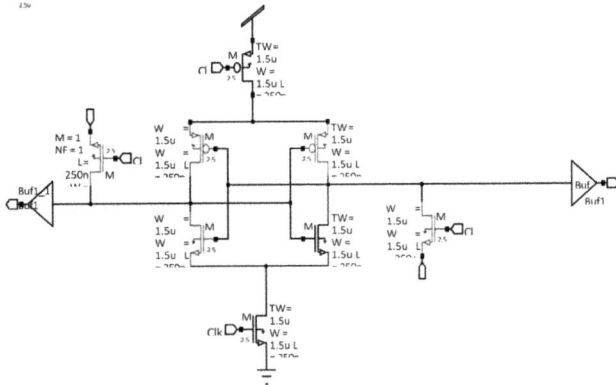

Figure 53.2 Double tail comparator
Source: Author

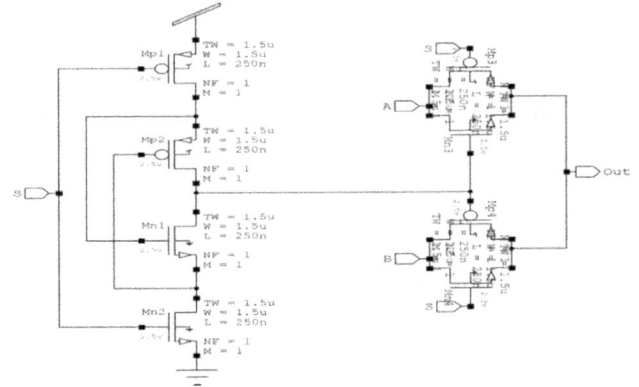

Figure 53.4 Schematic of 2:1 MUX [2]
Source: Author

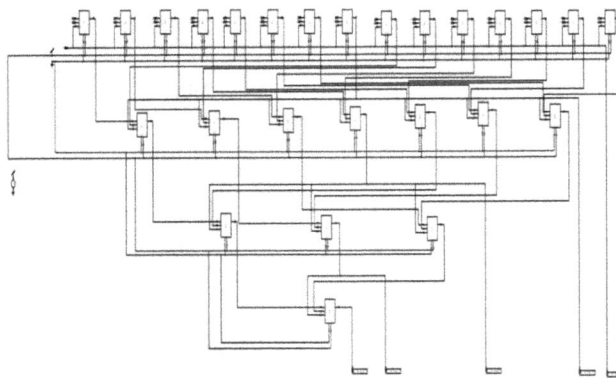

Figure 53.3 Schematic of proposed 2*1 mux-based encoder
Source: Author

are on. When the clock is low, the opposite occurs. Depending on the input, a differential voltage is generated. The equivalent output will be produced by variations in the input.

Mux based encoder

In flash ADCs, comparator outputs must be converted to a digital format using an encoder that converts temperature code to binary code. Each active comparator is denoted by a 1 [1] in the thermometer code, and more comparators become active as the input voltage rises. After, the encoder converts this code into a binary integer that represents the number of currently in-use comparators [6]. It does by locating the highest active comparator and produces a binary output that represents its position. Generally, priority encoder produces 011 (binary for 3) if the thermometer code is 11100. Flash ADCs encoder guarantees quick conversion, creating a small and

practical digital representation. Because of this, ADCs may operate at high speeds conversion at high resolution [9].

Multiplexer design for encoder

With a 5V DC power source, the circuit diagram as given seems to be a CMOS-based logic circuit [6]. Apart from PMOS (Mp1, Mp2) and NMOS (Mn1, Mn2) transistors that are widely used in complementary logic topologies, there are several more. The pull-up network is connected to the positive supply by PMOS transistors, Whereas the pull-down network is connected to ground by NMOS transistors.

This mux has one output (Y), two data inputs (I0 and I1), and a control input (S). The value of the output Y is decided by (I0) for a zero value of control signal S and by (I1) for the value of the control signal to be one. In the description of an encoder, information becomes easier to exchange from one representation to another in a 2-to-1 mux by enabling the selection between which data should be outputting [4]. This feature, which places much emphasis on the function of a mux in streamlining intricate logic architectures, is crucial in digital systems for effective data processing and transmission.

Outcomes

The Tanner EDA designs and tests circuits using 180 nm processing technology. While comparing to conventional Flash ADCs, the overall size, power consumption is reduced. As illustrated in Figure 53.3, the mux-based encoder minimizes the parameters and it is developed by using sequential logic.

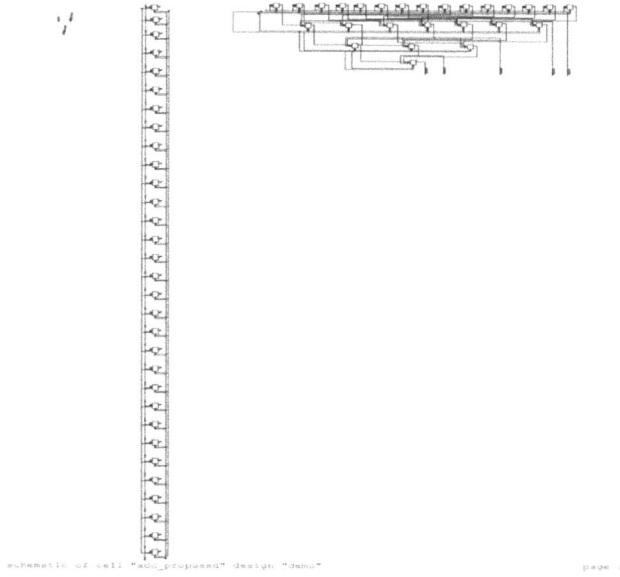

Figure 53.5 Schematic of proposed Flash ADC
Source: Author

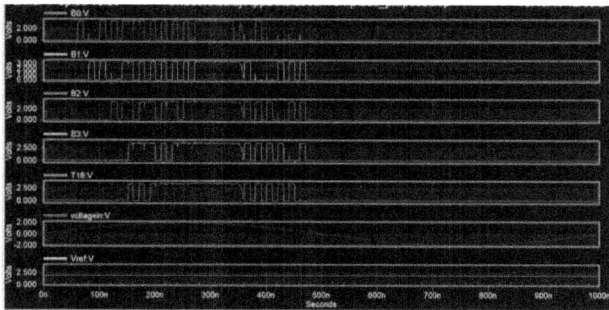

Figure 53.6 Flash ADC output waveform
Source: Author

Flash ADC
The design includes a double-tail comparator and mux-based encoder according to Figure 53.5. And testing results are shown in Figure 53.6.

Output of flash ADC
The devices support positive voltage transitions during the CLK input cycle and Vref voltages present. Waveforms change their widths when voltages are compared. When The Vin is lesser than Vref The width will be small and when Vin is greater than V ref width will be large. The Figure 53.6 demonstrates outputs from 5-bit analog to digital converter at 3.3v voltages source.

The comparison evaluation is pulled by the proposed Flash ADC has improved its overall parameters.

Table 53.1 Comparison evaluation [2].

Parameters	Existing Flash adc	Proposed Flash adc
Process	180NM	180NM
Resolution	5-Bit	5-Bit
Area	6.32m2	5.60m2
Speed	14.10sec	12.26sec
Power	3.52mw	3.31mw

Source: Author

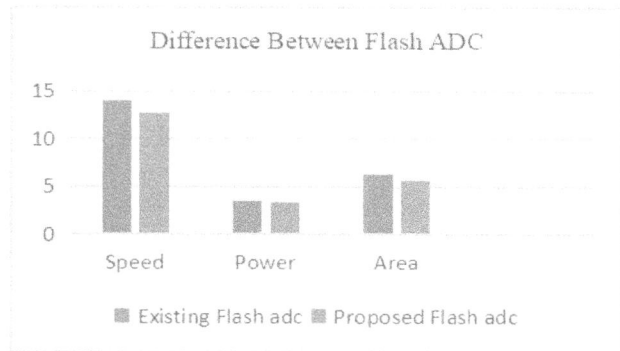

Figure 53.7 Difference between flash ADCs
Source: Author

Bar-graph chat analysis
false
The graph represents the actual difference between the existing system flash ADC and proposed system flash ADC overall values of parameters.

Conclusion
When compared to the conventional TIQ-based design, the suggested Flash ADC performs more effectively due to a modified dynamic comparator. It uses less space, uses less power, and lessens the impact of process variations. A smaller design and quicker operation are made possible by the use of double-edge triggering and fewer transistors. The comparator and encoder's efficiency are further increased by dynamic CMOS logic, which results in reduced power consumption and improved speed. Because of these enhancements, the ADC can be used in applicati1ons like medical systems, IoT devices, and portable electronics that require high-speed and low-power operation. The optimized dynamic comparator reduces the number of transistors and incorporates double-edge triggering, enabling faster operation and compact design.

Future Scope

This work creates room for more advancements. Future studies can concentrate on improving the dynamic comparator's stability in the face of voltage and temperature fluctuations. Speed and power consumption could be further decreased by optimizing the design for cutting-edge semiconductor technologies. Efficiency may also be increased by combining dynamic comparators with other low-power methods. Accuracy can also be improved under various working conditions by employing machine learning for automatic calibration. These enhancements will aid in the creation of sophisticated ADCs for edge computing, healthcare, and communication applications.

References

[1] George, R., & Ch, N. (2024). A low-power 5-bit two-step flash analog-to-digital converter with double-tail dynamic comparator in 90 nm digital CMOS. *Journal of Low Power Electronics and Applications*, 14(4), 53. DOI: 10.3390/jlpea14040053.

[2] Kuppa, A. V., Kumar, M. O. V. P., & Sheba, G. M. (2023). Design of low power flash ADC using TIQ comparator. In Proceedings of IEEE Xplore. DOI: 10.1109/CFP23AZ5-ART.2023.9781665491990.

[3] Koutnali, L., Kotabagi, S. S., & Vernerkar, R. (2024). 3-bit flash ADC using TIQ comparator. In Proceedings of the 25th International Conference for Emerging Technology (INCET).

[4] Nagaria, R. K., Varshney, V., & Dubey, A. K. (2017). Design of low-power high-speed double-tail dynamic CMOS comparator using novel latch structure. In Proceedings of the 2017 4th IEEE Uttar Pradesh Section International Conference on Electrical, Computer, and Electronics (UPCON), Mathura, India.

[5] Joseph, G. M., & Hameed, T. S. (2021). TIQ flash ADC with threshold compensation. In Proceedings of the Fifth International Conference on Intelligent Computing and Control Systems (ICICCS 2021), Madurai, India, IEEE.

[6] Baig, M. N. A., & Ranjan, R. (2017). Design & implementation of 3-bit high-speed flash ADC for wireless LAN applications. *International Journal of Advanced Research in Computer Communication Engineering*, 6, 428–433.

[7] Arunkumar, K., Ramesh, R., Geethalakshmi, R., & Archana, T. (2018). Low power dynamic comparator design for high-speed ADC application. In Proceedings of the International Conference on Current Trends Towards Converging Technology (ICCTCT), Coimbatore, India, (pp. 1–3).

[8] Kalyani, N., & Monica, M. (2018). Design and analysis of high-speed and low-power 6-bit flash ADC. In Proceedings of the 2nd International Conference on Inventive Systems and Control (ICISC), Coimbatore, India, (pp. 742–747). DOI: 10.1109/ICISC.2018.8398897.

[9] Gaude, D., & Poornima, B. (2019). Design and simulation of 4-bit flash analog-to-digital converter (ADC) for high-speed applications. *Indian Journal of Science and Technology*, 12(36). DOI: 10.17485/ijst/2019/v12i36/148021.

[10] Rani, H. G., & Arya, R. (2017). Design of 4-bit flash ADC using double-tail comparator in 130nm technology. *International Journal of Recent Advances in Science and Engineering Technology*. DOI: 10.22214/ijraset.2017.927.4o.

54 Solar powered Arduino motion detection smart light system

Pydikalva Padmavathi[1,a], J. Gurusiddappa[2,b], T. Poompavai[3,c],
Y. Ramamohan Reddy[4,d], T. Aravind Babu[5,e] and K.Vinodkumar[5,f]

[1]Associate Professor, Department of EEE, Srinivasa Ramanujan Institute of Technology, Anantapuram, Andhra Pradesh, India

[2]Associate Professor, Department of H&S, Srinivasa Ramanujan Institute of Technology, Anantapuram, Andhra Pradesh, India

[3]Assistant Professor, Department of EEE, Priyadarshini Engineering College, Vaniyambadi, Andhra Pradesh, India

[4]Professor, Department of ME, Srinivasa Ramanujan Institute of Technology, Anantapuram, Andhra Pradesh, India

[5]Assistant Professor, Department of EEE, Srinivasa Ramanujan Institute of Technology, Anantapuram, Andhra Pradesh, India

Abstract

The recent increase in air pollution has made air quality in many cities extremely hazardous. Solar-powered Arduino motion detection smart light system using an entirely automated system to turn lights on and off and store energy in the battery, this smart light seeks to minimize power usage. This paper offers the following solution to the two issues: the first light will turn on and off in response to vehicle movement on the road, and the second light will turn on and off automatically based on temperature changes. The lights are automated, so no human involvement is required. An outside-use smart light system powered by solar energy and Arduino motion detection. This automation of lights does not need human intervention. Solar-powered Arduino Motion detection smart light system designed for outdoor environments. The system integrates solar panels, an inverter, an autotransformer, an LCD, an Arduino microcontroller, a PIR sensor, LED lights, and rechargeable batteries. Solar panels use sunlight to produce electricity, deposited in batteries for continuous operation. The Arduino controls PIR sensor-triggered LED lights, enhancing security and energy efficiency. Real-time monitoring via LCD enables users to optimize system performance. This system offers a sustainable solution for outdoor lighting, combining renewable energy and advanced sensing technologies for enhanced security and efficiency.

Keywords: Arduino panel, Arduino UNO, LDR sensor, PIR sensor, relay and LCD, solar panel

Introduction

As energy efficiency and sustainability become more important in today's world, the use of smart technologies in conjunction with renewable energy sources has increased. One such invention is the solar-powered Arduino motion detection smart light system. This adaptable system creates a sustainable, intelligent lighting system by fusing the efficiency of Arduino technology with the power of solar energy. An Arduino microcontroller is powered by this system's photovoltaic panels, which capture solar energy and transform it into electricity. The Arduino microcontroller functions as the central nervous system, regulating the light's on-and-off states in response to motion detection. To further improve energy economy, the system may be configured to change the lights' brightness based on the surrounding lighting. By integrating motion detection technology, energy waste is reduced since lights are only turned on when motion is detected. This function improves security by saving electricity and delivering illumination when needed. The paper presents the design and execution of a smart lighting system that uses Arduino motion sensing and is powered by solar power.

[a]padmavathi.eee@srit.ac.in, [b]gurusiddappa.hs@srit.ac.in, [c]paavai.oct@gmail.com, [d]ramamohan.mec@srit.ac.in, [e]aravind.eee@srit.ac.in, [f]vinodkumar.eee@srit.ac.in

DOI: 10.1201/9781003685364-54

Main objective

The system's goal is to maximize energy efficiency by only turning on lights when motion is detected, which will result in significant cost and energy savings. The system enhances safety by automatically illuminating areas when motion is detected, providing better visibility and discouraging potential threats. The objective is to provide a well-lit and safe environment for that area people, enhancing their overall experience.

Literature survey

The literature review encompasses various aspects of smart lighting systems, focusing on those most relevant to our research. According to Cheng *et al.* [1], LED DC roadway lights are recommended over conventional AC streetlights due to their enhanced efficiency, longer lifespan, reduced maintenance requirements, and eco-friendly mercury-free composition. Alternatively, Chiu *et al.* [2] suggests that high-intensity discharge (HID) bulbs, with their comparable performance to LED lights, could also be considered. HID lights are known for their extended lifespan, superior color rendition, and high luminous effectiveness. Furthermore, a study indicates that even without utilizing renewable energy sources, upgrading the city's street lighting systems to LEDs could result in a significant energy saving of 64% [3]. Another investigation, discussed by Lee *et al.*[4], explores the integration of street lighting systems. For instance, one study proposes communication topologies utilizing global system for mobile communications (GSM), power line carriers, or general packet radio service (GPRS) transmissions for controlling street lighting systems [5]. Intelligent street lighting systems, such as a ZigBee-based wireless data network, are explored in other research efforts, as seen in [6], allowing for monitoring and management of multiple street light systems.

Proposed method

To enhance smart lighting, a proposed solution combines Arduino and solar energy with sensors. Rechargeable battery, Arduino microcontroller, and solar panel make up the three components of the system. Power and money may be saved with solar energy and motion sensing. Solar panels charge batteries during the day while they are not in use.

Motion-activated lights are activated by the Arduino, which is the system's control module. Lights are turned on at half intensity when there is less light than usual thanks to the LDR sensor. The PIR sensor raises the light to full brilliance to detect motion or human presence. Both illumination and system performance are customizable by the user. Effective energy conservation is achieved by the system's flexibility and adaptability to urban requirements. In this Project, the below block diagram shows the three main parts: a solar panel, a rechargeable battery, and an Arduino microcontroller.

Solar energy and motion detection to save power and money. The solar panel charges the battery during the day for nighttime use. The solar charge controller controls the energy flow in the system. The Arduino controls the system, turning on lights when motion is detected. LDR sensor monitors ambient light and activates lights at half intensity in low light. PIR sensor detects human presence or motion and increases light to full glow for safety. Users can adjust lighting settings and monitor system performance as shown in Figure 54.1.

Hardware components

Arduino UNO

The Arduino UNO, named after the Italian word for "one," represents a cornerstone in Arduino's

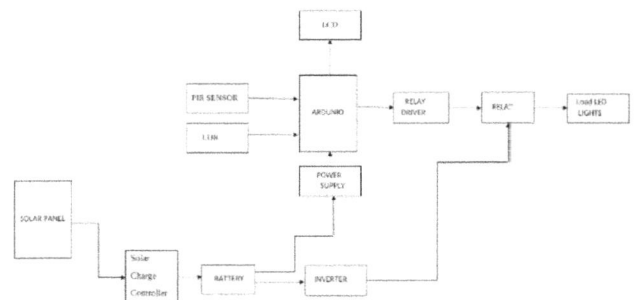

Figure 54.1 Block diagram of solar powered arduino motion detection smart light system
Source: Author

Figure 54.2 Arduino UNO
Source: Author

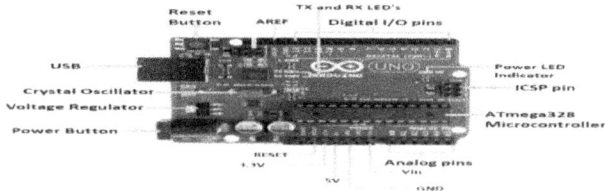

Figure 54.3 Components of Arduino UNO
Source: Author

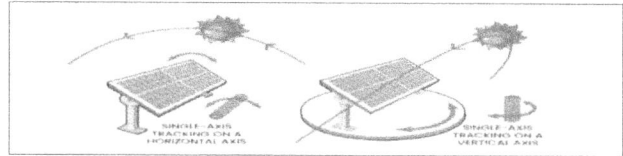

Figure 54.4a Singlesolar panel
Source: Author

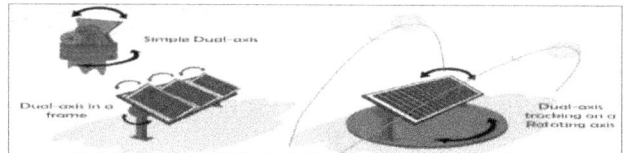

Figure 54.4b Dualaxis solar panel
Source: Author

Figure 54.5a PIRsensor
Source: Author

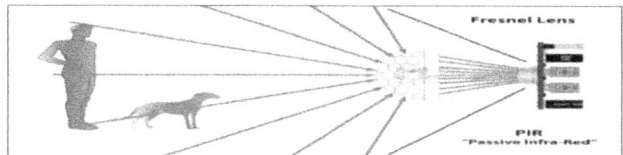

Figure 54.5b PIRSensorworking
Source: Author

standard board lineup. It was the inaugural USB-enabled board out by Arduino, heralding a new era of accessibility and versatility. Powered by the ATmega328P microprocessor, the Arduino UNO boasts robust capabilities suitable for a wide array of applications. Arduino UNO has an Integrated Development Environment (IDE), a programming language called Arduino, an In-circuit serial programming (ICSP) interface, a USB connection, a power port, six analog pin inputs, and fourteen digital pins. Online and offline platforms can both use it. ATmega328 microcontroller: It comprises lists, I/O lines, SPI serial ports, memory (EEPROM, SRAM, and Flash), timer, oscillator, external and internal interrupts as shown in Figure 54.2.

ICSP pin: Arduino board's company may be planned by using in-circuit serial programming pin. Power LED Indicator: Whenever an LED is in the turn on state, it specifies that power exists on. The LED won't turn on when the power is off. Digital I/O pins: These pins be able to set to HIGH or LOW. Digital pins are folks with numbers ranging from D0 to D13. TX and RXLED's: These LEDs are lit up to symbolize the successful flow of data. Reset button: It is employed to incorporate a reset button into the connection. USB: It creates that is probable for the board and computer to join. It is required to package the Arduino UNO board. Crystal oscillator: The Arduino UNO is a powerful board due to the 16MHz frequency of the Crystal oscillator. Voltage regulator: This method converts the input voltage to 5 volts. GND: These pins serve as a zero-voltage reference point. Vin: Input voltage. Analog pins: These pins, having numbers ranging from A0 to A5 can be utilized to recite the analog sensors. They can also serve as general purpose input output (GPIO) pins. Crystal oscillator: Arduino UNO is a influential board due to the 16MHz frequency of the Crystal oscillator as shown in Figure 54.3.

Solar panel

In this project, incorporating a polycrystalline solar panel serves as a pivotal component for sustainable energy generation. The solar panel's primary role is to harness sunlight and translate it into electrical energy through photo cells, ensuring a continuous power supply for the smart lighting system. By utilizing solar power, the project significantly reduces its environmental footprint and dependence on non-renewable energy sources, aligning with the principles of sustainability. Polycrystalline solar panels are recognized for their versatility and robustness, building them suitable for various environments and weather conditions. Their reliability ensures consistent performance, making them an indispensable element in providing efficient and eco-friendly lighting solutions as shown in Figure 54.4a.

Dual axis solar panel
Dual-axis solar trackers maximize sun exposure for up to 12 hours daily by precisely following its movement. This tracking capability ensures that solar cells remain perpendicular to the sun's rays from dawn till dusk, doubling energy collection compared to fixed flat panel arrays as shown in Figure 54.4b.

Dual axis solar panel
When the sun travels through the sky from east to west during the day, single-axis (one-axis picture) solar trackers follow it, capturing roughly 25% more energy.

The pyroelectric or passive infrared (PIR) sensor is a fantastic idea. Any living thing that is warmer than absolute zero releases radiation as a result of thermal energy. Throughout the day, humans emit at a wavelength of 9–10 micrometers. When a person approaches them, the PIR sensor is set to detect this wavelength. Beneath the intelligent motion reaction lies a device that is hardly larger than two centimeters. Pyro electricity is the word for heat that produces electricity, or low-amplitude electrical signals. This sensor operates passively as it lacks an infrared source of its own. Its range is ten meters. Automated door lock systems, elevators, home and business burglar alarms as shown in Figure 54.5 (a&b).

LDR sensor
The light dependent resistor (LDR), is a device that changes how easily electricity can flow through it depending on how much light is shining on it. When it's dark, the LDR resists the flow of electricity more, but when there's light, it lets electricity flow through more easily. This property makes LDRs useful for detecting light levels in various applications such as automatic lighting systems, brightness control in displays, and dusk-to-dawn switches as shown in Figure 54.6.

Relay and relay driver
Relays are electromechanical switches that mechanically open or close electrical contacts using an

Figure 54.6 LDRsensor
Source: Author

Figure 54.7a Pindiagram of relay
Source: Author

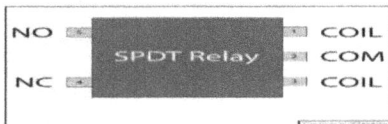
Figure 54.7b Relayandrelay driver
Source: Author

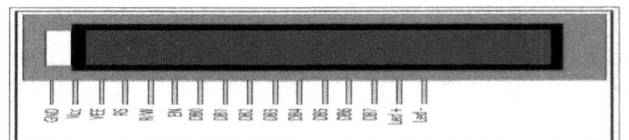
Figure 54.8a 16 bit LCD Display
Source: Author

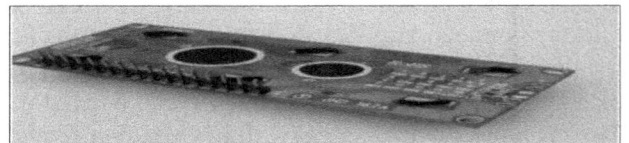
Figure 54.8b LCD back view
Source: Author

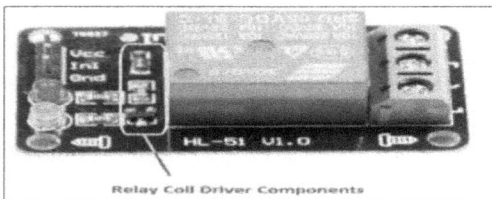
Figure 54.8c Pin diagram
Source: Author

Figure 54.9 Solarcharge controller
Source: Author

Figure 54.10 Solar powered motion detection smart light system
Source: Author

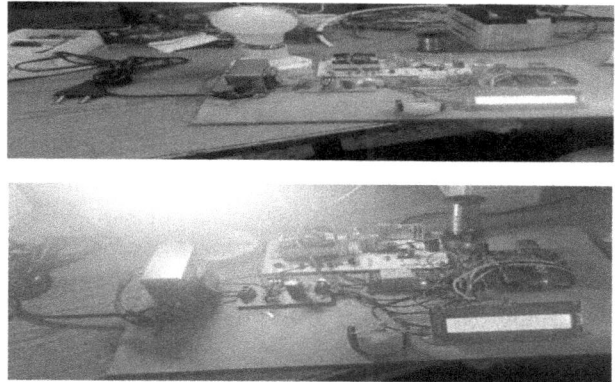

Figure 54.11 (a) Shows when Motion is not Detected, so light is in the off position. (b) Shows when Motion is Detected, so light is in on position
Source: Author

electromagnet. It creates isolation between the controlled and control circuits by enabling a low-power signal to operate a higher-power circuit. Relays are electromechanical switches that mechanically open or close electrical connections using an electromagnet as shown in Figure 54.7 (a&b).

LCD
In electrical applications, LCDs are commonly used to display information such as numeric values, text, or graphical data as shown in Figure 54.8 (a,b&c).

Light emitting diode
P-N junction diodes are used in lighting. It is collected of a particular class of semiconductors and is a chiefly doped diode. A light-emitting diode is defined as one that produces light at it's forward-biased. It is poised of a unique class of semiconductors and is a mainly doped diode. Using a heavily doped n and p junction, the majority of commercial LEDs are produced.

Features of solar charge controller
Battery protection (12V) from excessive charging, enhancing battery lifespan and reducing maintenance requirements. Automatic indication when charging is

in progress. High reliability with charging currents ranging from 10A to 40A. Monitoring of reverse current flow as shown in Figure 54.9.

Hardware Implementation

Solar panels seizure sunlight during the day, translating it into electrical energy. The generated energy is stored in a rechargeable battery for later use. A passive infrared (PIR) motion sensor senses movement within its range. Upon detecting motion, the sensor sends a signal to the Arduino microcontroller. The Arduino microcontroller activates an LED light based on the motion signal. The microcontroller manages the flow of energy from the battery to power the LED light. Intelligent control logic ensures optimal energy usage, conserving power and extending battery life. The system only illuminates when motion is detected, saving energy and enhancing efficiency. During nighttime or low light conditions, the battery powers the system for continuous operation. Integration of solar power generation, motion detection, and intelligent control logic offers an eco-friendly and automated outdoor lighting solution.

Results and discussion

The solar-powered Arduino motion detection smart light system demonstrated promising performance in our evaluation. Through rigorous testing, we found that the system efficiently detected motion and activated the lights when needed, showcasing its reliability in real-world scenarios. Moreover, its energy efficiency, particularly with the integration of solar panels, proved effective in sustaining power supply while minimizing environmental impact. Despite occasional challenges such as false positives or sensitivity to certain motion types, the system generally operated with high accuracy. If a person or car passes in front of this, the PIR sensor detects motion within its detection range. After the relay driver uses the electricity from the solar panel to activate the relay, the Arduino gets a signal from the Pir sensor.

Conclusion

In conclusion, solar-powered Arduino motion detection smart light systems offer a versatile and energy-efficient lighting solution for various applications. From enhancing safety and security in outdoor spaces to providing illumination during power outages, these systems offer numerous benefits. Whether used in homes, cities, commercial buildings, or natural environments, they provide reliable lighting while minimizing energy consumption and costs. With ongoing advancements in technology and increasing accessibility, these systems are even more in the future of lighting and sustainability.

Future Scope

In the future, solar-powered Arduino motion detection smart light systems will become more efficient and convenient. Solar panels will improve, making them better at collecting sunlight and charging the system. Motion detection will also get smarter, ensuring that the lights only turn on when needed, saving energy. You'll be able to control these lights easily from your smartphone, even when you're not at home. Battery life will improve, ensuring the system runs smoothly even during cloudy days. These lights won't just be for outdoors; they could be used indoors or in various settings like offices or for security. As technology becomes cheaper, these systems will become more affordable and accessible to everyone, allowing more people to benefit from their energy-saving features.

References

[1] Cheng, C. A., Chang, C. H., Chung, T. Y., & Yang, F. L. (2014). Design and implementation of a single-stage driver for supplying an LED street-lighting module with power factor corrections. *IEEE Transactions on Power Electronics*, 30(2), 956–966.

[2] Chiu, H. J., Lo, Y. K., Yao, C. J., & Cheng, S. J. (2011). Design and implementation of a photovoltaic high-intensity-discharge street lighting system. *IEEE Transactions on Power Electronics*, 26(12), 3464–3471.

[3] Garcia, R. B., Angulo, G. V., Gonzalez, J. R., Tavizón, E. F., & Cardozo, J. I. H. (2014). LED street lighting as a strategy for climate change mitigation at local government level. In IEEE Global Humanitarian Technology Conference (GHTC 2014), (pp. 345–349). IEEE.

[4] Lee, C. K., Li, S., & Hui, S. Y. (2011). A design methodology for smart LED lighting systems powered by weakly regulated renewable power grids. *IEEE Transactions on Smart Grid*, 2(3), 548–554.

[5] Leccese, F. (2012). Remote-control system of high efficiency and intelligent street lighting using a ZigBee network of devices and sensors. *IEEE Transactions on Power Delivery*, 28(1), 21–28.

[6] Denardin, G. W., Barriquello, C. H., Pinto, R. A., Silva, M. F., Campos, A., & do Prado, R. N. (2009). An intelligent system for street lighting control and measurement. In 2009 IEEE Industry Applications Society Annual Meeting, (pp. 1–5). IEEE.

55 AI-Driven lung cancer detection and staging using computed tomography

K. Mohana Lakshmi[1,a], S. Mallesh[1,b], Ch. Vinay Kumar[2,c], K. Prasanna Kumari[2,d], T. Nikhitha[3,e] and MD. Akbar[3,f]

[1]Associate Professor, Department of ECE, CMR Technical Campus, Hyderabad, Telangana, India

[2]Assistant Professor, Department of ECE, CMR Technical Campus, Hyderabad, Telangana, India

[3]UG Student, Department of ECE, CMR Technical Campus, Hyderabad, Telangana, India

Abstract

Lung cancer remains one of the most significant health challenges, where early and accurate diagnosis can significantly improve patient outcomes. This study focuses on predicting lung cancer types: normal, benign, and malignant—using computed tomography (CT) images. an enhanced approach is proposed by integrating brightness-preserved histogram equalization (BBHE) with deep learning (DL) to improve diagnostic performance. The BBHE is applied to improve the contrast of CT images while retaining their original brightness, ensuring that critical diagnostic features are not compromised. A deep learning model is then employed for feature extraction and classification, enabling precise differentiation between the three categories. The experiments are conducted using publicly available CT image datasets from Kaggle, ensuring robust and reproducible evaluation. The proposed method achieves high classification accuracy, highlighting its potential for reliable and automated lung cancer diagnosis. This approach demonstrates the effectiveness of combining advanced image preprocessing technique with deep learning for developing efficient diagnostic tools.

Keywords: Brightness-preserved histogram equalization, computed tomography, deep learning, lung cancer

Introduction

Lung cancer remains one of the most significant health challenges, where early and accurate diagnosis can significantly improve patient outcomes. Computed tomography (CT) imaging is vital in this process; however, its manual interpretation is not only time-consuming but also susceptible to inaccuracies, thus highlighting the need for automated alternatives. This study combines brightness-preserved histogram equalization (BBHE) with deep learning (DL) techniques to improve diagnostic precision. BBHE enhances the contrast of CT images while maintaining crucial details, and DL models can classify lung cancer cases into normal, benign, and malignant categories, demonstrating significant promise for dependable and automated diagnosis. Deep learning algorithms, especially convolutional neural networks (CNNs), are significantly advancing the field of medical image analysis. CNNs are a type of artificial neural network inspired by the structure and function of the human visual cortex. These networks are adept at recognizing intricate patterns within extensive image datasets, a skill that is particularly in the field of medical diagnosis. CNNs have shown exceptional success in a variety of applications, including: Lung nodule Detection: Early identification of lung nodules—small lung growths that may indicate cancer—is critical for enhancing patient outcomes. CNNs can be trained to detect even the slightest nodules in CT scans, outperforming traditional detection methods. Research illustrates that CNNs can achieve detection accuracy that surpasses that of seasoned radiologists. Lung nodule Classification: Not all lung nodules are malignant. CNNs can be developed further differentiation between the three categories, a vital factor in informing treatment options. This reduces unnecessary biopsies and allows for swift intervention in cases of confirmed malignancy. The advantages of using CNNs in lung cancer imaging analysis over conventional approaches encompass.

[a]mohana.kesana@gmail.com, [b]mallesh.ece4@gmail.com, [c]vinaykumar.ece@cmrtc.ac.in, [d]prasannakumari.ece@cmrtc.ac.in, [e]217r1a04j7@cmrtc.ac.in, [f]227r15a0418@cmrtc.ac.in

DOI: 10.1201/9781003685364-55

Improved accuracy: CNNs can provide high accuracy in both the detection and classification of lung nodules, potentially lowering the rates of missed diagnoses and false positives. Reduced inter-reader variability: Deep learning models deliver consistent and objective analyses, thereby minimizing the impact of human subjectivity in diagnostic outcomes. Increased efficiency: Automated analyses via CNNs can expedite the diagnostic process, facilitating quicker treatment decisions and enhancing patient care. Most promising developments in convolutional networks, with a specific emphasis on CNNs and their influence on lung cancer imaging analysis. We will delve into the capabilities of CNNs, scrutinizing their effectiveness in lung nodule detection and classification. Additionally, we will address the potential challenges and limitations linked to deep learning in this area. Our investigation aims to demonstrate how deep learning is transforming lung cancer diagnosis, setting the stage for earlier detection, enhanced patient care, and possibly lower mortality rates. We will also briefly discuss the potential of new deep learning frameworks, such as transformers, for future research in lung cancer imaging analysis. This literature review contributes for enhancing the deep networks transforming lung cancer diagnosis, offering a promising path towards earlier detection, improved patient care, and potentially reduced mortality rates. This literature review compiles the insights and findings of various researchers concerning, machine learning and convolutional nets, with a focus on nets and other analytical techniques for classifying CT scans. Aziz et al. [1] introduced a hybrid-LCSCDM model aimed at detecting osteosarcoma by merging deep learning techniques with machine learning, thereby improving diagnostic precision. Hybrid strategies, including the combination of CNNs with classifiers such as multilayer perceptrons (MLPs) or decision trees, have attained detection accuracies exceeding 95%. Furthermore, approaches that integrate CNNs with vision transformers (ViTs) have achieved an accuracy of 99.08%, and MobileNetV3-based systems have recorded accuracies of up to 98.69%. Research also indicates that more lightweight models, such as MobileNetV2, can surpass the performance of larger networks, underscoring the significance of optimized architectures in the early detection of osteosarcoma. Al-Yasriy et al. [2] concentrated on diagnosing lung cancer through the use of CT scans and CNNs. Their

research examined the effectiveness of CNNs in accurately detecting and diagnosing lung cancer via automated analysis of medical images, especially CT scans. Kareem et al. [3] investigated the efficacy of Support Vector Machines (SVM) for identifying lung cancer in annotated CT scan datasets. SVMs are commonly employed for classification tasks, and the study evaluated their potential in lung cancer detection, yielding encouraging results in localizing cancerous areas within CT scan images. Al-Husieny and Sajit [4] explored transfer learning employing the GoogleNet architecture for lung cancer detection. Transfer learning enables models to utilize pre-trained networks on extensive datasets, enhancing performance on specific tasks with smaller datasets, a method particularly beneficial in the realm of medical imaging. Ismael et al. [5] even though its primary focus was not on lung cancer. Nonetheless, the methodologies applied in deep learning might still hold significant potential for realm of analysis in cancer images. Kalaivani et al. [6] developed a deep learning framework specifically aimed at detecting and classifying lung cancer. Their findings indicated that deep learning algorithms can outperform conventional approaches for analysis of medical images. Naseer [7] proposed a modified AlexNet architecture combined with SVM for lung cancer prediction. Renowned for its prowess in image recognition, AlexNet was specifically adapted to tackle the challenge of identifying lung cancer in CT images, thus offering a refined model for classifying lung cancer within medical imaging contexts. Chaturvedi [8] proposed a method for lung disease classification and prediction that utilized a range of machine learning techniques. Their research emphasized the importance of choosing the correct model for lung cancer prediction and compared various machine learning approaches to determine the most effective one. Chaunzwa et al. [9] centered their study on enhancing CT scan images, illustrating how deep learning can assist in the analysis of cancerous tissues from CT scan images. Jayabalaji et.al[10] proposed ensemble model for Feature Extraction and it is used for the   process of transforming raw data into numerical features that can be used for modelling. [11, 12], as mentioned earlier, offering additional insight into the use of 3D CNNs and Grad-CAM for lung cancer. The dual reference could indicate a focus on refining the model or clarifying its application.

Proposed Aaproach

Algorithm

Step 1: Data collection: Collect the lung sample images from a publicly accessible Kaggle repository.

Step 2: Data preprocessing

Image enhancement technique is applied on the input images by using brightness-preserved histogram equalization (BBHE) method.

Resize the image into uniform dimension(224 × 224 × 3).

Normalize the image pixels by scaling them to range of 0 to 1.

Segmentation of data into model training, Validation and Testing.

step3: Model implementation: Build a customized CNN model by using the fundamental layer of CNN such as convolution, activation, batch normalization, Pooling, dense and SoftMax layer.

Step4: Model evaluation: After training evaluation the model on test data to assess its generation capability. and generation of confusion matrix and ROC curves.

Figure 55.1 illustrates the process of the proposed framework for the classifying lung cancer through CT images. The procedure commences with the of a lung cancer dataset, which supplies the raw CT images required for examination. Subsequently, brightness-preserved bi-histogram equalization (BBHE) is applied to improve image contrast. This preprocessing step guarantees that critical features within the CT images are more prominent while maintaining the original brightness, thus preparing them for precise analysis. After enhancing the contrast, the images are subjected to feature extraction and classification via a specialized deep learning model. This model has been meticulously designed to evaluate the enhanced features and categorize the images

into three distinct classifications: normal, benign, or malignant. The efficient workflow reflects a well-integrated approach combining contrast enhancement with state-of-the-art deep learning methodologies for precise lung cancer diagnosis, establishing it as a dependable resource for clinical use.

Dataset

The performance of the presented model, we gathered the 1098 CT samples (416 normal,561 malignant and 120 benign) from the Iraq oncology teaching hospital and National Center for cancer diseases (IOOTH/NCCD) (https://data.mendeley.com/datasets/bhm-dr45bh2/4). In this work for through analysis, the dataset has been split with 80% training and 20% testing

Contrast enhancement

Contrast enhancement is a process in medical processing, designed to the clarity of anatomical in CT images. One prominent method employed this purpose is BBHE. This technique not only increases contrast but also preserves the original brightness, which is crucial in medical imaging, as significant changes in brightness can affect clinical assessments. BBHE functions by dividing the images into sub equivalent images, resulting in enhanced contrast while keeping the overall brightness consistent. This strategy helps avoid over-enhancement and the creation of artificial artifacts. As a result, essential details in CT scans, such as tumor boundaries and tissue variations, remain clear and diagnostically significant. By applying BBHE to CT images, radiologists can achieve better visualization of medical conditions without losing the natural brightness, thereby enhancing both diagnosis and analysis. Figure 55.2 illustrates the implications of this improved method.

Classification

In classification tasks, the extraction of meaningful features is essential for achieving accurate predictions.

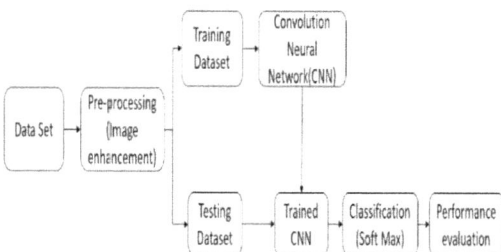

Figure 55.1 Proposed model block diagram
Source: Author

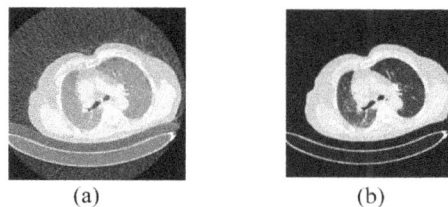

(a)   (b)

Figure 55.2 (a) Original image; (b) Enhanced images
Source: Author

Traditionally, this relied manually crafted feature extraction techniques; however, the introduction of CNNs has transformed this approach by learning hierarchical patterns from images in an automated manner. CNNs employ convolutional layers to identify low-level features, such as edges and textures, in the early layers, while deeper layers are responsible for recognizing more complicated structures, including shapes and parts of objects. This automatic feature extraction obviates the necessity for manual engineering, thereby enhancing the efficacy of CNNs for classification objectives. In this work, we have designed specialized CNN architecture aimed at extracting pertinent feature details from enhanced CT images. The model initiates with a convolutional layer comprising 128 filters with a size of (3,3) and a stride of (3,3), followed by a ReLU activation function and L2 regularization to mitigate overfitting. A Batch Normalization layer is subsequently implemented to stabilize the training process, along with a MaxPooling layer to decrease spatial dimensions. This is followed by a 1×1 convolutional layer also with 128 filters, which facilitates efficient feature transformation while preserving the spatial structure. Another convolutional layer with 256 filters and a (3,3) kernel is then added, utilizing a stride of (3,3) and same padding to enhance feature extraction across various levels.

The model concludes by flattening the features with a Flatten layer, which forwards them to two fully connected Dense layers containing 1024 and 512 neurons, respectively. Each of these layers employs a ReLU activation function and L2 regularization. To further combat overfitting, Dropout layers with a 50% dropout rate are integrated. Ultimately, the output layer consists of three neurons utilizing a SoftMax activation function, addressing a three-class classification problem (benign, malignant, and normal). This architecture promotes deep extraction, implementation of regularization, and efficient

classification while enhancing robustness against overfitting. The Figure 55.3 represents the confusion matrix of the suggested CNN and the corresponding ROC curve is shown in Figure 55.4. Similarly, the quantitative measures of the proposed model are illustrated in Table 55.1.

Table 55.2 represents the comparison of the proposed model with the existing approaches. From that it is evident that the suggested approach significantly predicting lung cancer than the state-of-the-art approaches.

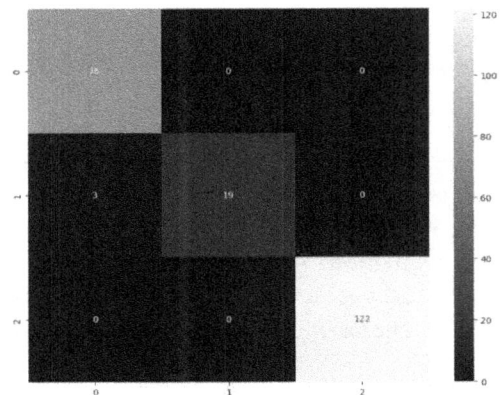

Figure 55.3 Confusion matrix
Source: Author

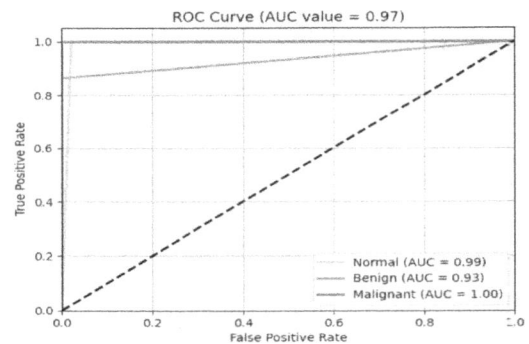

Figure 55.4 ROC curve
Source: Author

Table 55.1 The performance measures from confusion matrix.

Classes	Accuracy (%)	Precision (%)	Recall (%)	F1-score (%)	Specificity (%)
Normal	1.000	0.962	1.000	0.981	0.979
Benign	0.864	1.000	1.000	0.927	1.000
Malignant	1.000	1.000	1.000	1.000	1.000
Proposed Method	0.986	0.987	0.955	0.971	0.971

Source: Author

Table 55.2 Comparison of performance measures with conventional methods.

Methodology	Dataset	Accuracy %
Hybrid- LCSCDM [1]	IQ-OTHNCCD	98.54%
AlexNet [2]	IQ-OTHNCCD	93.54%
GoogleNet [3]	IQ-OTHNCCD	94.38%
Hybrid- BBHE-DL[Proposed method]	IOOTH/NCCD	98.60%

Source: Author

Results and discussion

The efficacy of proposed model is evaluated by several commonly utilized classification performance metrics that involve sensitivity (recall), specificity, precision, F1-score, "area under the curve (AUC), and accuracy.

Conclusions

In conclusion, this study demonstrates the effectiveness of combining brightness-preserved histogram equalization (BBHE) with deep learning (DL) for improving the accuracy of lung cancer classification from CT images. By enhancing image contrast while preserving essential diagnostic details, BBHE ensures that key features remain intact for effective analysis. The integration of deep convolutional network further refines the feature extraction and classification process, enabling precise differentiation between normal, benign, and malignant categories. The promising results obtained from publicly available CT image datasets highlight the potential of this approach for reliable, automated lung cancer diagnosis. This method offers significant value in clinical settings, contributing to more accuracy of 98.6% and also compared with traditional methods and timely diagnoses, which are crucial for improving patient outcomes. In future, our work will be extended on balanced dataset by augmenting the data.

References

[1] Aziz, M. T., Mahmud, S. H., Elahe, M. F., Jahan, H., Rahman, M. H., Nandi, D., ... & Moni, M. A. (2023). A novel hybrid approach for classifying osteosarcoma using deep feature extraction and multilayer perceptron. Diagnostics, 13(12), 2106.

[2] Al-Yasriy, H. F., Al-Husieny, M. S., Mohsen, F. Y., Khalil, E. A., & Hassan, Z. S. (2020). Lung cancer diagnosis using CT scans and CNN technology. In IOP Conference Series: Materials Science and Engineering, (Vol. 928, p. 022035), IOP Publishing.

[3] Kareem, H. F., Al-Husieny, M. S., Mohsen, F. Y., Khalil, E. A., & Hassan, Z. S. (2021). Assessing SVM performance for lung cancer detection in a labelled CT scan dataset. *Indonesian Journal of Electrical Engineering and Computer Science*, 21(3), 1731.

[4] Al-Husieny, M. S., & Sajit, A. S. (2021). Utilizing transfer learning with GoogleNet to detect lung cancer. *Indonesian Journal of Electrical Engineering and Computer Science*, 22(2), 1078–1086.

[5] Ismael, Y. S., Shakor, M. Y., & Abdalla, P. A. (2022). Real-time face recognition system based on deep learning. *NeuroQuantology*, 20(6), 7355–7366.

[6] Kalaivani, N., Manimaran, N., Sophia, S., & Devi, D. (2020). Lung cancer detection and classification leveraging deep learning methods. In IOP Conference Series: Materials Science and Engineering, (Vol. 994, p. 012026), IOP Publishing.

[7] Naseer, I., Masood, T., Akram, S., Jaffar, A., Rashid, M., & Iqbal, M. A. (2023). Detection of lung cancer utilizing a modified AlexNet architecture paired with support vector machines. *Computers, Materials & Continua*, 74(1), 2039–2054.

[8] Chaturvedi, P., Jhamb, A., Vanani, M., & Nemade, V. (2021). Utilization of machine learning techniques in predicting and classifying lung cancer. In IOP Conference Series: Materials Science and Engineering, (Vol. 1099, p. 012059), IOP Publishing.

[9] Tafadzwa L Chaunzwa, Ahmed Hosny, Yiwen Xu, Andrea Shafer, Nancy Diao, Michael Lanuti. (2021). Classification of lung cancer histology using CT images with deep learning. *Scientific Reports*, 11(1), 1–12.

[10] Jayabalaji, K. A., Laxmaiah, B., Thatipudi, J. G., Badhe, C. K. P. & Balaram, A. (2023). Enhancing content-based image retrieval for lung cancer diagnosis: leveraging texture analysis and ensemble models. In 2023 Seventh International Conference on Image Information Processing (ICIIP), Solan, India, (pp. 386–390).

[11] Paramasivam, R., Patil, S. N., Konda, S., & Hemalatha, K. L. (2024). Lung cancer computed tomography image classification using Attention based capsule network with dispersed dynamic routing. *Expert Systems*, 41(9), e13607.

[12] Tiwari, L., Awasthi, V., Patra, R. K., Miri, R., Raja, H., & Bhaskar, N. (2022). Lung cancer detection using deep convolutional neural networks. In Data Engineering and Intelligent Computing: Proceedings of 5th ICICC 2021, (Vol. 1, pp. 373–385). Singapore: Springer Nature Singapore.

56 Combining IR and visible spectrum images with DWT and bilateral filter methods

N. Nagaraja Kumar[1,a], N. Keerthi[2,b], G. Sruthi Keerthi[2,c], I. Durga Manoj[2,d] and T. Venkata Subhashini[2,e]

[1]Associate Professor, Department of Electronic Communication Engineering, Rajeev Gandhi Memorial College of Engineering and Technology, Nandyal, Andhra Pradesh, India

[2]UG students, Department of Electronic Communication Engineering, Rajeev Gandhi Memorial College of Engineering and Technology, Nandyal, Andhra Pradesh, India

Abstract

Image fusion is the technique of integrating information from multiple images taken from different sensors, perspectives, or with varying focus and exposure settings. The result of the process is a single image that has not only the original images but also the enhancement of other parts. At present, a few different combination methods have been popularized, with the widely recognized of them being thermal and optical image integration. Sensor technology has made advancements in both military and civil fields, so fusion technology has been used extensively. Discrete wavelets transform-bilateral filter (DWTBF) is technology that applies discrete wavelet transform (DWT) and bilateral filter to combine infrared (IR) and visible (VI) images. DWT is the method that produces a set of coefficients for images at the specified frequency levels. The filter (BF) that is responsible for taking the edge and smoothing them out thus assisting in many other operations is the most critical part of weight selection. DWT sub-bands—low and high frequency—are done differently, where the low-frequency components are averaged and weighted averaging is used for high-frequency details. The inverse discrete wavelet transform (IDWT) is applied to the last fusion step that will be image recovery. The new DWTBF fusion method is capable of clear data processing using visible images while enhancing the infrared image. This piece of work has a satisfactory outcome in terms of artifact measure($N^{AB/F}$), structural similarity index (SSIM), sum of correlation difference (SCD), and edge-based similarity index ($Q^{AB/F}$), which are evaluation measures for it outperforming the other fusion techniques.

Keywords: Discrete wavelet transform, infrared imaging, visual data quality, wavelengths

Introduction

Image fusion is an essential procedure in digital image processing, merging multiple pictures from different sensors, viewpoints, or exposures into one unified, high- quality [1] mage with enhanced details. This method is widely used in military, civilian, biomedical [2], and surveillance [3] applications. Fusion of thermal (IR) and optical (VI) images improves details by merging thermal and spatial data. This paper introduces discrete wavelet transforms bilateral filter (DWTBF), which uses DWT and bilateral filtering to generate high-resolution fused images with maintained edges and less artifacts, hoping to outperform other methods in subjective and objective assessment.

Zhang et al. [4] this paper explained that Gradient transfer fusion (GTF) improves image fusion but is imperfect. It can increase noise, lead to ghosting when images are not aligned, and is computationally intensive. Enhancement is difficult to balance, and it doesn't handle vast brightness variations well, resulting in inconsistency [4]. Liu et al. [5] Convolutional sparse representation (CSR) is efficient but slow, memory-intensive, and difficult to scale. Dictionary learning is finicky, noise degrades performance, and adjustment is trial and error. It has difficulty adapting feature size, and deep learning is rendering it less necessary, [5]. Liu et al. (2017) joint sparse representation in saliency detection (JSRSD) is strong but computationally intensive and fails to work with high-resolution images. Its correctness relies on the dictionary, but noise and clutter can deceive it. Sparsity tuning is difficult, and it fails to work with dynamic or composite scenes [6]. DWTBF

[a]rajuneravati@gmail.com, [b]keerthinara03@gmail.com, [c]gollasruthi906@gmail.com, [d]idmanoj666@gmail.com, [e]talarisubhasini@gmail.com

DOI: 10.1201/9781003685364-56

outperforms limitations with efficiency, stability, and enhanced detail retention. It does not "breathe" at noise, unlike SR and JSR, and does not make edges unnatural, unlike GTF. It smooths noise while maintaining edges using wavelets and a bilateral filter. It performs under all lighting conditions, needs little tuning, and prevents artifacts, providing clearer, sharper, and more natural pictures efficiently [7, 8].

Initial fusion techniques such as averaging and PCA were plagued by blurring [9]. Transform-based techniques (e.g., wavelets) enhanced fusion but created artifacts. Anisotropic diffusion maintains edges and minimizes noise. KL transform further improves fusion by decorrelating data optimally and extracting principal components [9]. Image fusion methods are classified as spatial, transform, and statistical approaches. Basic techniques such as averaging and PCA discard information, whereas wavelet and curvelet transformation retain edges but involve high computational complexity [10]. Statistical approaches emphasize significant features but suppress fine details. Hybrid methods provide a compromise of clarity, noise suppression, and speed [10]. The technique improves images by breaking them down into low-rank (global) and salient (local) components. Low-rank components are averaged, and salient details are summed, retaining important features with fewer artifacts. It enhances visibility in surveillance, object detection, and medical imaging, with potential future applications in real-time processing and deep learning [11].

The outcomes specify that the "proposed" approach is the best among the algorithms under. comparison for SSIM(a), $N^{AB/F}$, SCD, and $N^{AB/F}$ measures. It exhibits better performance in maintaining structural similarity and obtaining higher fusion quality. The higher $Q^{AB/F}$ [10] indicates Improved fusion performance and the higher SCD [11] and the higher SSIM(a) [12], indicates better structural preservation indicates improved spectral retention and the lower N AB/F [9] indicates lower noise. The Proposed method is therefore the best among all, offering better image fusion outcomes with higher quality, more preserved structure, and less noise.

Proposed method

In the proposed method, IR and visible image sets are collected from the TNO datasets and the collected pictures are used to produce combined image,

which is more effective compared to all the existing methods and gives better performance measures. The block diagram mentioned below explains the step-by-step process and each step is explained in detail under the block diagram as follows:

Discrete wavelet transform-bilateral filter

Pixel-level fusion is one of the most fundamental approaches in image fusion, where each pixel is processed independently. This method provides rich details, making it useful for human observation and computational analysis. The discrete wavelet transform-bilateral filter (DWTBF) technique follows these key steps:

1. Preprocessing – The images are resized to a standard resolution (256×256 pixels).
2. Filtering – A detail amplification filter and a highlight extraction transform are applied to enhance fine structures and improve contrast in the image.
3. DWT Decomposition – The picture is divided into Figure 56.2 Illustrates DWT decomposition into sub-bands segments (LL, LH, HL, HH) to extract different frequency components [13].
4. Fusion strategy – Low-frequency and fine-detail elements are processed separately using distinct fusion strategies.
5. Reconstruction – The final merged image is reconstructed using inverse DWT (IDWT).

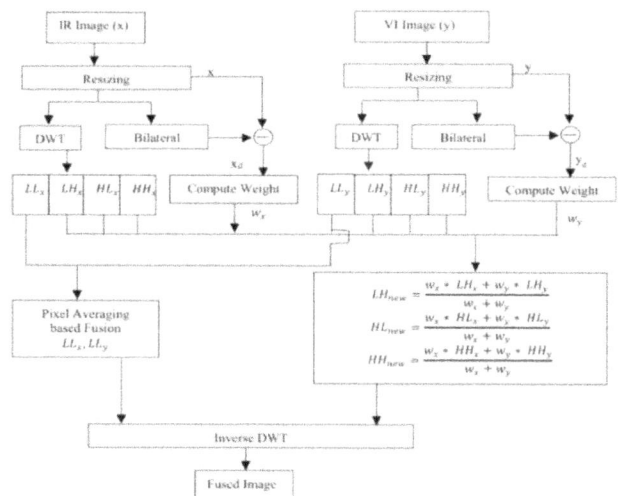

Figure 56.1 Flowchart of DWT-bilateral filter fusion
Source: Author

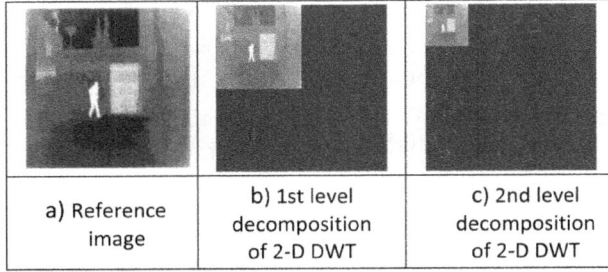

| a) Reference image | b) 1st level decomposition of 2-D DWT | c) 2nd level decomposition of 2-D DWT |

Figure 56.2 Breaking down a picture into its different frequency components using the discrete wavelet transform (DWT)
Source: Author

Edge-preserving filter

Edge-preserving filter is a non-linear filtering technique that smooths images while preserving edges. Unlike traditional smoothing filters, which blur edges, BF considers both spatial proximity and intensity differences, ensuring that fine details remain intact.

The filter utilizes Gaussian kernels to assign weights based on pixel similarity, effectively reducing noise and preventing distortions

$$\text{BLT}(m,n) = \frac{\sum_m \sum_n I(m',n') g_{\sigma_s}(m-m',n-n') g_{\sigma_r}\left(I(m,n)-I(m'-n')\right)}{\sum_m \sum_n g_{\sigma_s}(m-m',n-n') g_{\sigma_r}\left(I(m,n)-I(m'-n')\right)}$$

$$\text{Where } g_{\sigma_s}(m,n) = \exp\left(-\frac{(m^2+n^2)}{2\sigma_s^2}\right) \tag{1}$$

$$\text{And } g_{\sigma_r}(\delta) = \exp\left(-\frac{\delta^2}{2\sigma_r^2}\right)$$

Where (m', n') represents neighboring pixels at place of (m, n) in a picture, the spatial influence function range is indicated σ_s by and the lowest amplitude is denoted by σ_r.

DWT decomposition

DWT enables multi-resolution image analysis through the decomposition of images into sub-bands: LL (structural information) and LH, HL, HH (edge details). Unlike traditional techniques, DWT fusion preserve's structure and minimizes artifacts.

Fusion strategy

$$LH_{new} = \frac{w_x * LH_x + w_y * LH_y}{w_x + w_y} \tag{2}$$

$$HL_{new} = \frac{w_x * HL_x + w_y * LH}{w_x + w_y} \tag{3}$$

$$HH_{new} = \frac{w_x * HH_x + w_y * HH_y}{w_x + w_y} \tag{4}$$

For optimal fusion, LL sub-bands are fused by pixel averaging, while LH, HL, and HH sub-bands are fused by weighted detail strength-based fusion. The ultimate picture is reconstructed by applying IDWT without compromising clarity or background information.

Qualitative analysis

Four important standards for image fusion were employed to evaluate the performance of the proposed method:

1. QABF (Edge-based similarity index) – Higher values indicate better edge preservation.

$$Q^{AB/F} = \frac{\sum_{k=1}^{M}\sum_{l=1}^{N} Q^{A/F}(k,l)g_A(k,l) + Q^{B/F}(k,l)g_B(k,l)}{\sum_{k=1}^{M}\sum_{l=1}^{N} g_A(k,l) + g_B(k,l)} \tag{5}$$

Here B, A are reference images. F denote a merged picture. Additionally $g_A(k, l)$, and $g_B(k, l)$, represent the weighting factors for images A and B at the pixel location and $Q^{A/F}(k, l)$ are defined as,

$$Q^{A/F}(k,l) = Q_g^{A/F}(k,l)Q_\alpha^{A/F}(k,l) \quad Q^{B/F}(k,l) = Q_g^{B/F}(k,l)Q_\alpha^{B/F}(k,l) \tag{6}$$

Q_g^{AF} and Q_g^{BF} indicate the intensity of edge features and the retention of orientation information.

2. SCD (Total correlation variation) – quality metric to quantify the retained information from the input images [14, 15].

$$\text{SCD} = r(E_1, A) + r(E_2, B) \tag{7}$$

Where $E_1 = F–B$ and $E_2 = F–A$ and F denotes the final outcome and A and B are the initial inputs. The function r(.) evaluates the correlation between A and E_1, as well as between B and E_2.

3. SSIM (Structural Similarity Index) – valuates how much structural detail is preserved.

$$\text{SSIM(A,F)} = \frac{[2\mu_A\mu_F + C_1](2\sigma_{AF} + C2)}{[\mu_A^2 + \mu_F^2 + C_1](\sigma_A^2 + \sigma_F^2 + C_2)} \tag{8}$$

$$\text{SSIM(B,F)} = \frac{[2\mu_B\mu_F + C_1](2\sigma_{BF} + C2)}{[\mu_B^2 + \mu_F^2 + C_1](\sigma_B^2 + \sigma_F^2 + C_2)} \tag{9}$$

where, μ_A, μ_B and μ_F denote the average intensities, σ_A, σ_B and σ_{BF} represents the intensity dispersion of input images A and B, as well as the combined output C, respectively. σ_{AF} and σ_{BF} corresponds to the correlation-based magnitude of the input images and combined output, respectively. Furthermore, C_1 and C_2 are fixed parameters. In the absence of definitive ground truth, the adjusted SSIM is obtained by computing the mean of two values, as outlined in eq.10,

$$SSIM(a) = \frac{SSIM(F,A) + SSIM(F,B)}{2} \qquad (10)$$

4. $N^{AB/F}$ (Artifact measure) – Lower values indicate fewer artifacts in the fused image.
5. The DWTBF method achieved superior results across all these metrics, confirming its effectiveness in producing high-quality fused images.
6. After applying the IDWT, the final fused image is obtained and it is shown in Figures 56.4f, 56.5f, 56.6f and 56.7f, which is more effective when compared to remaining 3 methods as shown below in figures.

Results Analysis

To evaluate the effectiveness of DWTBF, its results were measured with three cutting-edge fusion methods:

1. JSR with saliency detection (JSRSD)
2. Gradient transfer fusion (GTF)
3. Convolutional sparse representation (CSR)

The experiment compared DWTBF on two publicly available image fusion datasets, in which it outperformed other algorithms in contrast, clarity, and detail preservation. Subjective assessment based on

TNO dataset images (Figures 56.4–56.7) indicated that it performed better in enhancing contrast and subtle details to achieve high-quality fusion.

Figure 56.4 depicts the "Kaptein" image, which 4a (Optical) does not have the target and 4b (Thermal) clearly displays it. Figure 56.5 has two persons, where the suggested scheme provides a higher resolution and better fusion than Figures 56.5c-e.

Figure 56.7's "UN Camp" images demonstrate (a) Optical, (b) Thermal, (c) GTF, (d) CSR, (e) JSRSD, and (f) our fused result. Figures 56.6 and 56.7 indicate slight target changes. Our approach produces good contrast enhancement while maintaining background details, reducing fuzziness and distortions. Tables 56.1–56.4 show comparisons between fusion algorithms based on various metrics. The new method ranks higher in all the categories, improving contrast,

Figure 56.4 Contrast analysis results from Kaptein
Source: Author

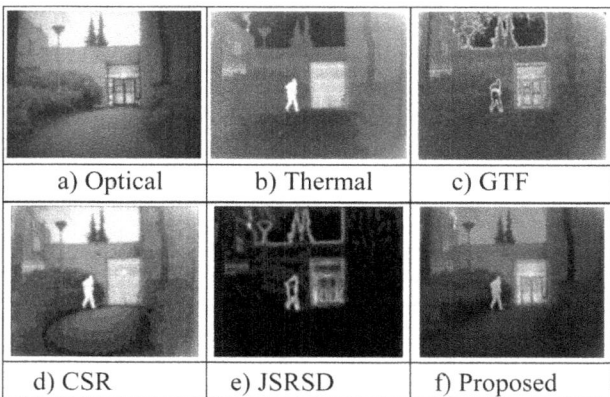

Figure 56.3 Contrast analysis results from UN Camp
Source: Author

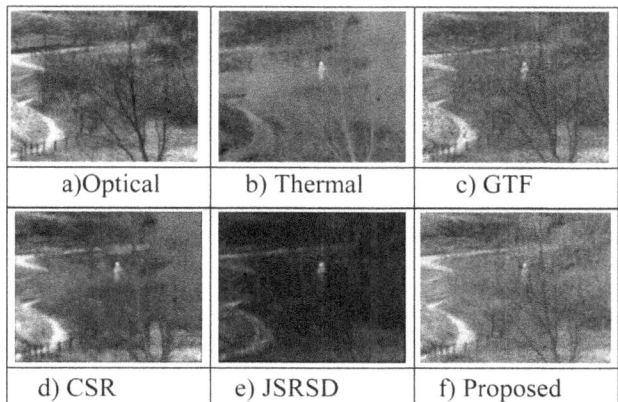

Figure 56.5 Contrast analysis results from two-person
Source: Author

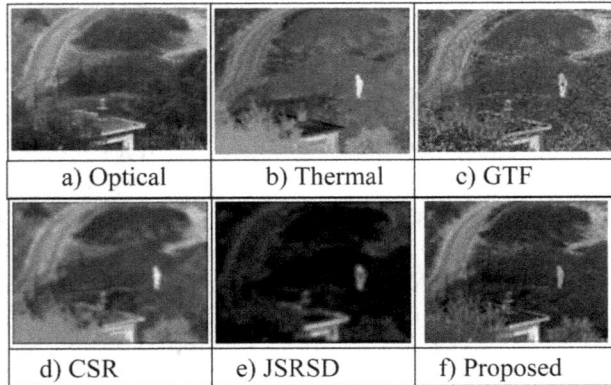

| a) Optical | b) Thermal | c) GTF |
| d) CSR | e) JSRSD | f) Proposed |

Figure 56.6 Contrast analysis results from sand path
Source: Author

Table 56.1 Performance evaluation.

Image	Algorithms			
	GTF	CSR	JSRSD	Proposed
Two-person	0.5779	0.5930	0.2484	0.6122
Kaptein	0.3737	0.3781	0.1160	0.3926
UN camp	0.3850	0.3883	0.1620	0.4106
Sand Path	0.3238	0.3253	0.1287	0.3469

Source: Author

Table 56.2 Comparison of methods SCD.

Image	Algorithms			
	GTF	CSR	JSRSD	Proposed
Two-person	1.2530	1.4146	1.1247	1.7021
Kaptein	0.9241	1.5406	0.8383	1.6610
UN camp	1.0380	1.4292	0.8837	1.4794
Sand Path	1.2591	1.4980	1.1550	1.6485

Source: Author

brightness, and edge information with a lesser level of distortion. Overall, it outperforms three state-of-the-art methods.

Acknowledgement

I wish to extend my sincere appreciation to Dr. N. Nagaraja Kumar sir for his excellent guidance and advice throughout this project. His professional knowledge and wise guidance are what shaped our thought process and helped us to bring out the best result.

Conclusion

This paper introduces a new IR and VI image fusion technique based on DWT and bilateral filtering (DWTBF) that provides contrast improvement, structural detail preservation, and artifact reduction.

Table 56.3 Comparison of methods SSIM(a).

Image	Algorithms			
	GTF	CSR	JSRSD	Proposed
Two-person	0.9594	0.9720	0.9254	0.9702
Kaptein	0.9879	0.9851	0.9822	0.9865
UN camp	0.9885	0.9847	0.9819	0.9858
Sand Path	0.9934	0.9920	0.9915	0.9930

Source: Author

Table 56.4 Comparison of methods.

Image	Algorithms			
	GTF	CSR	JSRSD	Proposed
Two-person	0.2604	0.2901	0.7895	0.2598
Kaptein	0.2280	0.2543	0.5072	0.2276
UN camp	0.1657	0.1871	0.5008	0.1655
Sand Path	0.1631	0.2104	0.4486	0.1623

Source: Author

Experimental results prove its excellence in edge preservation, SSIM, SCD, and artifact reduction. Nonetheless, DWTBF has difficulties with low-contrast images and computationally expensive complexity, which restricts real-time applications. Future research will optimize efficiency and contrast management and possibly incorporate deep learning for enhanced performance in surveillance, medical imaging, and remote sensing.

References

[1] Singh, S., Mittal, N., & Singh, H. (2020b). Multi-focus image fusion based on multiresolution pyramid and bilateral filter. *IETE Journal of Research*, 68(4), 2476–2487. https:// doi.org/10.1080/03772063.2019.1711205.

[2] Mehta, T., & Mehendale, N. (2021). Classification of X-ray images into COVID-19, pneumonia, and TB using cGAN and fine-tuned deep transfer learning models. *Research in Biomedical Engineering*, 37(4), 803–813.

[3] Singh, S., Mittal, N., & Singh, H. (2020c). Classification of various image fusion algorithms and their performance evaluation metrics. *Computational Intelligence for Machine Learning and Healthcare Informatics*, 179–198. https://doi.org/10.1515/9783110648195-009

[4] Zhang, Q., Fu, Y., Li, H., & Zou, J. (2013). Dictionary learning method for joint sparse representation-based image fusion. *Optical Engineering*, 52(5), 057006.

[5] Liu, Y., Chen, X., Ward, R. K., & Wang, J. (2016b). Image fusion with convolutional sparse representation. *IEEE Signal Processing Letters*, 23(12), 1882–1886.

[6] Liu CH, Qi Y, Ding WR. (2017) Infrared and Visible image fusion method based on saliency detection in sparse domain. *Infrared Physics and Technology.* 83:94–102.

[7] Kumar, K. P. K., & Geethakumari, G. (2014). Mean-variance blind noise estimation for CT images. *Advances in Intelligent Systems and Computing,* 264, 417–428.

[8] Simone, G., Farina, A., Morabito, F. C., Serpico, S. B., & Bruzzone, L. (2002). Image fusion techniques for remote sensing applications. *Information Fusion,* 3(1), 3–15.

[9] Bavirisetti, D. P., & Dhuli, R. (2016). Fusion of infrared and visible sensor images based on anisotropic diffusion and Karhunen-Loeve transform. *IEEE Sensors Journal,* 16(1), 203–209.

[10] Goyal, S., & Wahla, R. (2015). A review on image fusion. In 2019 International Conference on Communication and Signal Processing, (Vol. 4, No. 2, pp. 7582–7588).

[11] Li, H., & Wu, X. J. (2018). Infrared and visible image fusion using latent low-rank representation. arxiv:1804.08992.

[12] Krishnamoorthy, S., & Soman, K. P. (2010). Implementation and comparative study of image fusion algorithms. *International journal of computer application,* 9(2), 25–35.

[13] Bhandari, A. K., & Kumar, I. V. (2019). A context sensitive energy thresholding based 3D Otsu function for image segmentation using human learning optimization. *Applied soft computing Journal,* 82, 105570.

[14] Meher, B., Agrawal, S., Panda, R., & Abraham, A. (2019). A survey on region based image fusion methods. *Information Fusion,* 48, 119–132.

[15] Qiu, C., Wang, Y., Zhang, H., & Xia, S. (2017). Image fusion of CT and MR with sparse representation in NSST domain. *Computational and Mathematical Methods in Medicine,* Volume 2017, Article ID 7274087, 1–1.

57 Enhancement of steering efforts and returnability in power steering vehicles through multibody system dynamics analysis

Naveen Kumar[1,a], V. Velumani[2,b], Devaraj, E.[3,c], ArunKumar, S.[4,d] and M. Rajanikanth[5,e]

[1]Associate Professor, Department of Mechanical Engineering, Kuppam Engineering College, Kuppam, Andhra Pradesh, India

[2]Associate Professor, Department of Mechanical Engineering, V S B Engineering College, Karur, Tamil Nadu, India

[3]Assistant Professor, Department of Mechanical Engineering, CMR University, Bangalore, Karnataka, India

[4]Assistant Professor, Department of Mechanical Engineering, Kuppam Engineering College, Kuppam, Andhra Pradesh, India

[5]Assistant Professor, Marketing Department, Gitam School of Business, GITAM University Hyderabad, Andhra Pradesh, India

Abstract

This study investigates the enhancement of steering efforts and returnability in power steering vehicles through multibody system dynamics analysis. Steering effort and returnability are critical factors in vehicle handling and safety, with previous studies indicating their dependency on parameters such as caster angle, scrub radius, and suspension geometry. A detailed multibody dynamics model was developed using ADAMS/car to analyze the effects of varying caster angle and scrub radius on steering characteristics. The methodology involved creating a full-vehicle simulation model, integrating suspension and steering subsystems, and evaluating key performance metrics. The results revealed that an increase in caster angle significantly enhances returnability due to higher self-aligning torque, corroborating established vehicle dynamics principles. Additionally, an increase in scrub radius was observed to elevate steering effort, confirming previous findings. Brake pull analysis further emphasized the role of optimized steering geometry in reducing steering torque inconsistencies. These insights contribute to a deeper understanding of the influence of steering system parameters on vehicle maneuverability and control. The study confirms that precise tuning of steering geometry can improve handling performance and driver comfort. Future research should focus on real-world validation of these simulation-based results and explore adaptive steering solutions to optimize vehicle performance under varying driving conditions.

Keywords: ADAMS/car, caster angle, multibody dynamics, scrub radius, steering geometry

Introduction

The evolution of steering systems has been a central focus in automotive engineering research, with significant attention devoted to optimizing both steering efforts and returnability characteristics. Early foundational work established core principles for analyzing steering torque and returnability characteristics [1, 2], while theoretical frameworks for understanding vehicle controllability and stability were provided [3]. Recent studies have highlighted the impact of artificial intelligence and machine learning on steering system optimization. AI-based adaptive steering assist mechanisms have been incorporated to improve real-time feedback [4], while digital twin models have been developed to simulate steering performance under varied conditions [5].

Adaptive control algorithms for electric power steering (EPS) have also been introduced, allowing real-time steering effort adjustments based on driving conditions [6]. Furthermore, comprehensive human factors studies have correlated subjective driver feedback with objective steering parameters

[a]naveenkumargulbarga@gmail.com, [b]velu28vsb@gmail.com, [c]devaraj.e@cmr.edu.in, [d]Srinivasan.arunmech@gmail.com, [e]rmekala@gitam.edu

DOI: 10.1201/9781003685364-57

[7]. Material innovations have also influenced steering system dynamics. Research has examined the impact of carbon fiber-reinforced components on reducing inertia and improving returnability [8–12]. Magnetorheological dampers have been integrated into steering systems, dynamically adjusting resistance for enhanced control [13]. Additionally, digital human modeling has been utilized to optimize ergonomic steering wheel configurations, ensuring optimal driver control. With growing advancements in autonomous and electric vehicle technologies, the role of intelligent steering systems continues to evolve [14]. Seamless transitions between manual and autonomous steering control have been explored [15], while assist curves and gear ratios have been analyzed for reduced driver fatigue [16]. As road surface modeling progresses, research has demonstrated how varying friction coefficients impact self-aligning torque and returnability [17]. Despite these advancements, further research is required to validate simulation-based findings in real-world environments. This study aims to build on existing knowledge by analyzing multibody system dynamics, focusing on optimizing steering geometry to enhance vehicle stability and maneuverability. The findings contribute to the ongoing evolution of steering technology, ensuring safer and more efficient vehicular control [18].

Experimental work

Modeling of the MacPherson suspension system
The methodology involves a structured approach to analyzing the enhancement of steering efforts and returnability using multibody system dynamics as shown in Figure 57.1.

To investigate the influence of MacPherson strut suspension geometry on steering effort and returnability, a structured simulation model was developed in ADAMS/Car. The suspension subsystem was constructed by integrating essential components such as the lower control arm, strut assembly, and wheel hub. The suspension system's response to vertical displacement was analyzed by subjecting it to a bump height of 100 mm. The results validated previous studies, which highlighted that suspension geometry significantly impacts vehicle handling [2].

Table 57.1 Suspension input parameters: The influence of suspension parameters on steering behavior was examined using a multibody system dynamics model developed in ADAMS/Car. Camber angle, toe angle, caster angle, kingpin inclination, and scrub radius were carefully calibrated to assess their effect on returnability and steering effort. Previous studies have emphasized the role of these parameters in enhancing vehicle stability and handling [1, 2]. The obtained data indicate that adjustments in caster angle and scrub radius significantly impact self-aligning torque, corroborating earlier research on vehicle control mechanisms [3].

Steering - pinion and rack type
The steering system was modeled using a rack-and-pinion configuration to evaluate its role in force transmission efficiency as shown in Figure 57.2.

The system was subjected to varied input forces to observe displacement and steering torque characteristics, as tabulated in Table 57.2. The simulation findings confirmed that increased rack force enhances steering precision, aligning with prior research on steering assist optimization [7].

A rack-and-pinion steering model was simulated to evaluate the maximum displacement and torque variations in response to different steering angles. The results confirmed that higher steering torques correspond to increased resistance in the system [19]. Additionally, the simulations demonstrated the importance of optimizing steering gear ratios to balance driver effort and system efficiency.

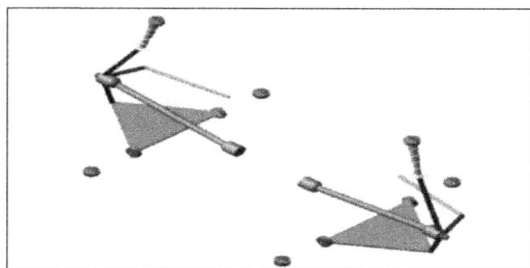

Figure 57.1 Front suspensions - MacPherson strut
Source: Author

Table 57.1 Suspension input parameters.

S. No	Parameter	Angles
1	Camber angle in degree	0
2	Toe in degree	0
3	Caster angle in degree	5
4	Kingpin inclination in degree	12.50
5	Scrub radius in mm	16.39

Source: Author

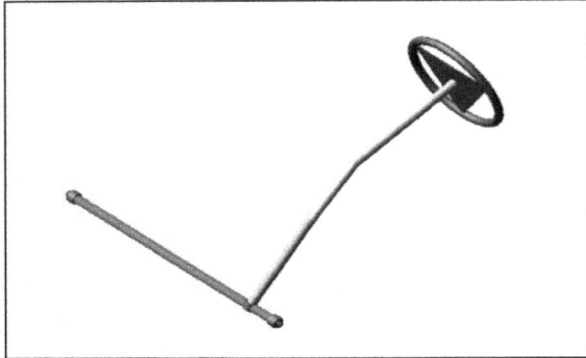

Figure 57.2 Steering - rack and pinion type
Source: Author

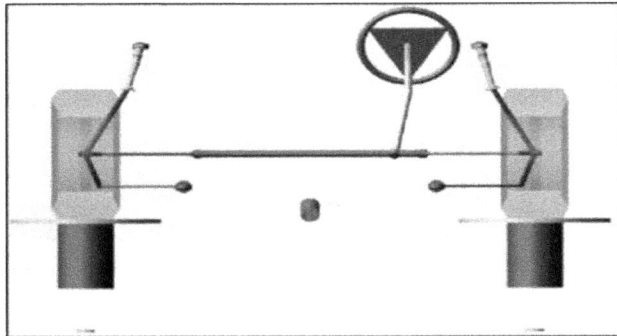

Figure 57.3 Suspension with suspension test rig
Source: Author

Table 57.2 Rack and pinion input parameters.

S. No	Input parameters	
1	Maximum displacement for rack in (mm)	100
2	Maximum displacement for rack force in (N)	500
3	Maximum steering angle	720^0
4	Max steering torque in N-mm	500

Source: Author

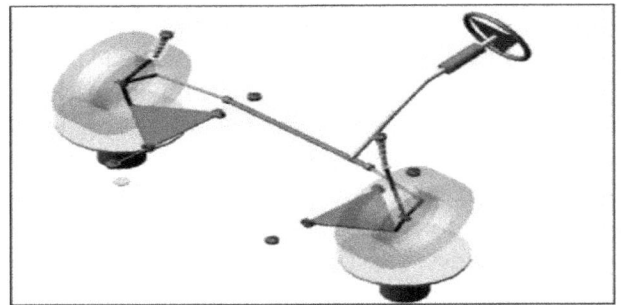

Figure 57.4 Suspension with steering test rig
Source: Author

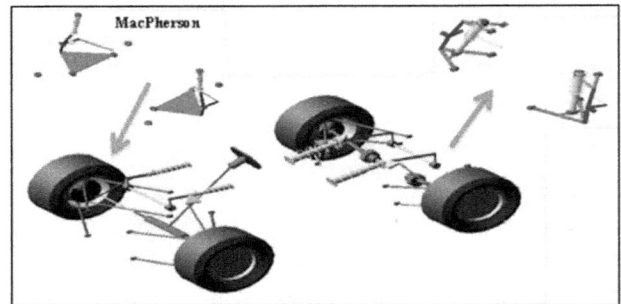

Figure 57.5 Full vehicle analysis with suspension
Source: Author

Assembly of suspension with suspension test rig
The suspension subsystem was assembled and tested using a standardized test rig to ensure alignment with real-world conditions. as shown in Figure 57.3. The methodology involved analyzing vertical displacements and steering input responses. as shown in Figure 57.4. The findings reinforced the critical role of precise suspension geometry tuning in optimizing returnability and effort reduction [3].

Full vehicle analysis
This study involves integrating various subsystems, such as suspension and steering components, to create a comprehensive model that simulates the overall dynamics of a vehicle. as shown in Figure 57.5. This analysis is conducted using ADAMS/Car, a multibody dynamics simulation software, which evaluates how modifications in the design of these subsystems influence vehicle performance.

A full-vehicle simulation was conducted to integrate the steering and suspension systems within a virtual prototype. The study analyzed dynamic responses under varied driving conditions, demonstrating that modifications in caster angle and scrub radius significantly influenced maneuverability and returnability. These results align with the principles outlined in previous multibody dynamics studies [5].

Results and discussion

The study utilized an SUV model in ADAMS/Car to investigate steering wheel returnability, steering

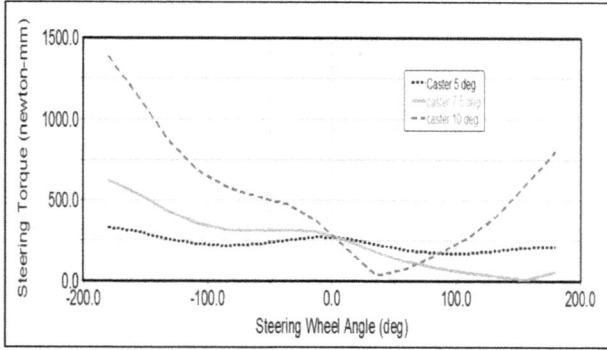

Figure 57.6 Wheel torque vs. wheel angle for various caster angles
Source: Author

Table 57.3 Wheel torque vs wheel angle.

S. No	Caster angle (Degree)	Steering torque (N-mm)
1	5^0	330.12
2	7.5^0	621.53
3	10^0	1382.55

Source: Author

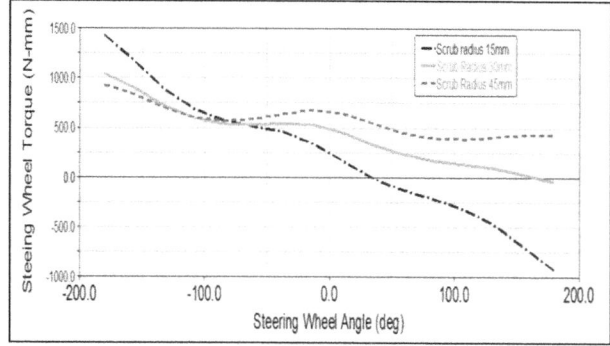

Figure 57.7 Scrub radius vs. steering angle
Source: Author

Table 57.4 Scrub radius vs. steering angle.

S. No	Scrub Radius (mm)	Steering Torque (N-mm)
1	15	309.74
2	30	749.25
3	45	1376.18

Source: Author

effort, and other steering-related parameters at a vehicle speed of 80 km/h. Simulations were conducted for various wheel geometry parameters to analyze their impact on steering effort, handling behavior, and steering wheel returnability.

Brake pull analysis

The brake pull analysis in the paper focuses on examining the effects of wheel movement during braking conditions and how it relates to steering dynamics. The brake pull analysis yields the following results for different conditions:

1. Plot of wheel torque vs. steering angle for various Caster Angles.
2. Plot of Wheel Torque vs. Steering Angle for various Scrub Radii.

Wheel torque vs steering angle for caster angle

A brake pull analysis was conducted to examine the relationship between wheel torque and caster angle variations. The results confirmed that an increase in caster angle leads to greater steering effort, validating findings in brake pull dynamics research [5].

Figure 57.6 illustrates the relationship between Wheel Torque and Wheel Angle. A Brake Pull

analysis was conducted using ADAMS/car, examining the effects of varying caster angles (5.0°, 7.5°, and 10.0°). The average simulation results for each caster angle, along with the corresponding steering torques, are summarized in the table below.

Table 57.3 presents the relationship between wheel torque and steering angle for various caster angles. The analysis indicates that as the caster angle increases, the steering torque also increases, suggesting that higher caster angles lead to greater steering effort required by the driver. This suggests that higher caster angles require greater steering effort from the driver, which is critical for understanding how caster angle influences steering dynamics. This Table 57.3 shows the relationship between caster angle and steering torque, The average simulation results indicate that as the caster angle increases, the steering torque also increases. This relationship is crucial for understanding how caster angle impacts steering effort and vehicle handling.

Wheel torque vs steering angle are for various scrub radius

Table 57.4 scrub radius vs steering angle: A brake pull analysis was conducted to examine the effect of caster angle variations on steering torque. The data confirm that increasing the caster angle results in a

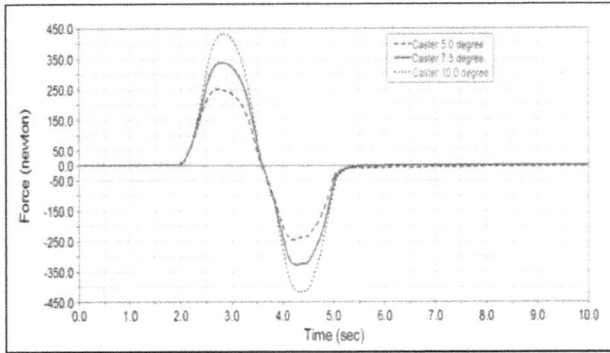

Figure 57.8 Steering wheel angle vs steering assistance force various caster angles
Source: Author

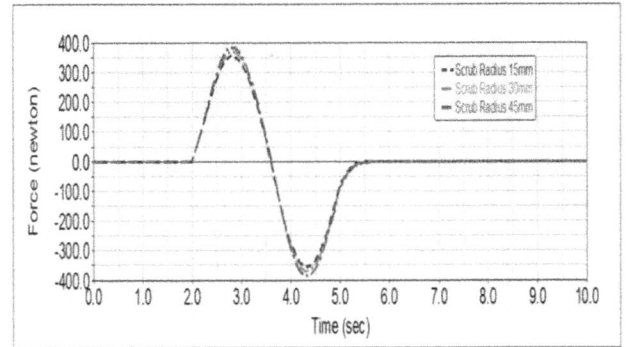

Figure 57.9 Scrub radius vs. steering assistance force
Source: Author

Table 57.5 Caster angle vs steering assistance force.

S. No	Caster angle (Degree)	Steering assistance force (N)
1	5^0	250.28
2	7.5^0	340.52
3	10^0	432

Source: Author

Table 57.6 Scrub radius vs steering assistance force.

S. No	Scrub radius (mm)	Steering assistance force (N)
1	15	356.2882
2	30	371.8651
3	45	385.5712

Source: Author

corresponding rise in steering torque, which is consistent with the established principles of steering dynamics.

Figure 57.7 shows the relationship between scrub radius and steering torque, The average simulation results indicate that as the scrub radius increases, the steering torque also increases. This relationship is crucial for understanding how scrub radius impacts steering effort and vehicle handling.

Full vehicle analysis
The full vehicle analysis in this paper involves integrating various subsystems, such as suspension and steering components, to create a comprehensive model that simulates the overall dynamics of a vehicle. This analysis is conducted using ADAMS/Car, a multibody dynamics simulation software, which evaluates how modifications in the design of these subsystems influence vehicle performance.

Wheel torque vs steering angle for various caster angle
The below Figure 57.9 shows the steering wheel angle vs steering support force. The full vehicle analysis has been carried out using ADAMs/car for bump height of 100 mm. The average simulations results

for various caster angles and obtained corresponding steering forces are tabulated in Table 57.5.

The relationship between the caster angle and steering assistance force was examined to determine its influence on driver effort. The results indicated that higher caster angles required greater assistance force, corroborating studies on assistive steering mechanisms. Figure 57.8 shows the relationship between caster angle and steering assistance force evaluated at a bump height of 100 mm. The results indicate that as the caster angle increases, the steering assistance force also increases, which is critical for understanding how scrub radius affects steering effort.

Table 57.5 caster angle vs steering assistance force: The dependency of steering assistance force on caster angle was evaluated. The outcomes indicate a direct proportionality between these variables, confirming prior research that highlights the role of electronic power steering in optimizing steering effort. This Table 57.5 shows the relationship between caster angle and steering assistance force, The average simulation results indicate that as the caster angle increases, the steering assistance force also increases. This relationship is crucial for understanding how scrub radius impacts steering effort and vehicle handling.

Wheel torque vs steering angle for various scrub radius

The below Figure 57.9 shows the Scrub Radius vs steering assistance force. The Full vehicle analysis has been carried out using ADAMs/car for bump height of 100mm. The average simulations results for various Scrub Radius and obtained corresponding steering Forces are tabulated in table 57.36. Figure 57.9 depicts the variation of steering assistance force with respect to scrub radius, evaluated at three scrub radius values (15 mm, 30 mm, and 45 mm). The results show that the steering force increases with scrub radius.

Table 57.6 scrub radius vs steering assistance force: The final analysis focused on the effect of scrub radius on steering assistance force. The findings demonstrate that increasing scrub radius leads to a rise in steering assistance force, aligning, who examined hydraulic steering system efficiency. The impact of scrub radius on steering assistance force was evaluated. The results revealed that increased scrub radius leads to higher assistance force demand, aligning with research on steering effort reduction techniques. This Table 57.6 shows the relationship between scrub radius and steering assistance force, The average simulation results indicate that as the scrub radius increases, the steering assistance force also increases. This relationship is crucial for understanding how scrub radius impacts steering effort and vehicle handling.

Conclusion

The analysis conducted using ADAMS/Car has provided significant insights into the steering effort and returnability characteristics of a power steering vehicle. The brake pull analysis has reinforced the significance of optimized steering geometry in minimizing steering torque inconsistencies. The full-vehicle analysis further confirms that changes in steering parameters significantly impact handling performance. The simulation outcomes indicate that positive caster angles improve returnability by enhancing the self-aligning torque, aligning with established vehicle dynamics principles. Similarly, an increase in scrub radius is observed to elevate steering effort. The data obtained suggest that optimizing these parameters leads to improved maneuverability and stability, which is critical for high-speed vehicle operations. In conclusion, the study has successfully demonstrated the influence of steering system parameters on vehicle dynamics through multi-body system simulations. The findings align with established literature, affirming that precise tuning of steering geometry enhances handling characteristics and driver comfort. Future research may focus on real-world validation of these simulation-based results to further refine steering system optimization for enhanced automotive performance and safety.

Acknowledgement

The authors gratefully acknowledge the students, staff, and authority of Mechanical Engineering department, Kuppam Engineering College, Kuppam for their cooperation in the research.

References

[1] Shimakage, M., Satoh, S., Uenuma, K., & Mouri, H. (2002). Design of lane-keeping control with steering torque input. *JSAE Review*, 23(3), 317–323.

[2] Cho, Y. G. (2009). Vehicle steering returnability with maximum steering wheel angle at low speeds. *International Journal of Automotive Technology*, 10(4), 431–439.

[3] Dave, C., & Yu, F. (2004). Vehicle system dynamics and control. China Communications Press, (pp. 401–416).

[4] Zhou, X., & Wang, Z. (2022). Robust adaptive path-tracking control of autonomous ground vehicles with considerations of steering system backlash. *IEEE Transactions on Intelligent Vehicles*, 7(2), 315–325.

[5] Kim, J., El-Gindy, M., & El-Sayegh, Z. (2023). Development of novel steering scenarios for an 8X8 scaled electric combat vehicle. SAE Technical Paper, 2023-01-0106.

[6] You, S., & Kim, G. (2023). Neural approximation-based adaptive control using reinforced gain for steering wheel torque tracking of electric power steering system. *IEEE Transactions on Systems, Man, and Cybernetics: Systems*, 53(7), 4216–4225.

[7] Rodríguez, A. J., Sanjurjo, E., Pastorino, R., & Naya, M. A. (2021). State, parameter and input observers based on multibody models and Kalman filters for vehicle dynamics. *Mechanical Systems and Signal Processing*, 155, 107544. ISSN 0888-3270.

[8] Tanaka, K., Hosoo, N., Katayama, T., Noguchi, Y., & Izui, K. (2016). Effect of temperature on the fiber/Matrix interfacial strength of carbon fiber reinforced polyamide model composites. *Mechanical Engineering Journal*, 3(6), 16-00158.

[9] Devaraj, E., Kumar, N., Yadav, S. P. S., & Ramana, V. S. N. (2024). Matrix and reinforcement for biopolymer composites-A review. *Journal of Engineering and Technology Management*, 72, 1135–1167.

[10] Kumar, N., & Irfan G. (2021). A review on tribological behaviour and mechanical properties of Al/ZrO2 metal matrix nano composites. *Materials Today: Proceedings*, 38, 2649–2657.

[11] Kumar, N., & Irfan, G. (2021). Mechanical, microstructural properties and wear characteristics of hybrid aluminium matrix nano composites (HAMNCs)-review. *Materials Today: Proceedings*, 45, 619–625.

[12] Kumar, N., Irfan, G., Udayabhanu, & Nagaraju, G. (2021). Green synthesis of zinc oxide nanoparticles: Mechanical and microstructural characterization of aluminum nano composites. *Materials Today: Proceedings*, 38, 3116–3124.

[13] Jung, B., & Park, Y. (2003). Development of damper for new electronically controlled power steering system by magnetorheological fluid: MRSTEER. *International Journal of Vehicle Design*, 33, 199–216.

[14] Ji, Y., Ding, S., & Wang, C. (2022). Humanized steering wheel quality design and upgrade model construction. SAE Technical Paper, (pp. 3020–3032).

[15] Sentouh, C. (2019). Driver-automation cooperation oriented approach for shared control of lane keeping assist systems. *IEEE Transactions on Control Systems Technology*, 27, 1962–1978.

[16] Schmidt, H., Büttner, K., & Prokop, G. (2023). Methods for virtual validation of automotive powertrain systems in terms of vehicle drivability-a systematic literature review. *IEEE Access*, 11, 27043–27065.

[17] Nguyen, T. A., & Iqbal, J. (2024). Improving stability and adaptability of automotive electric steering systems based on a novel optimal integrated algorithm. *Engineering Computations*, 41, 991–1034.

[18] Silva, R. C. (2015). Estimation of the geometric parameters of a front double wishbone suspension based on geometry formulation. *Proceedings of the ASME International Mechanical Engineering Congress and Exposition*, 12, 1113–1122.

[19] Nozaki, H. (2006). Driver steering model and improvement technique of vehicle movement performance during drift running. *International Journal of Automotive Technology*, 7(4), 449–457.

58 High performance 8-bit approximate Dadda multiplier with GDI logic-based approximate and exact 4:2 compressors using finFET technology

M. Tejaswi[1,a], P. Syamala Devi[2,b], S. Murali[3,c], Y. Manideep[3,d], K. Pavani[3,e] and C. Mounika[3,f]

[1]Assistant Professor, Department of ECE, Annamacharya University, Rajampet, Andhra Pradesh, India

[2]Associate Professor, Department of ECE, Annamacharya University, Rajampet, Andhra Pradesh, India

[3]Students of Annamacharya Institute of Technology and Sciences, Rajampet, Andhra Pradesh, India

Abstract

As the demand for smaller and more efficient integrated circuits grows, there is a pressing need for innovative design techniques that reduce both area and power consumption. In this context, FinFET technology has emerged as a promising alternative to traditional CMOS due to its superior performance in energy efficiency and transistor density. Parallel to these advancements, approximate computing has gained traction as a method to achieve high efficiency in error-tolerant applications like image processing, where slight deviations in output are acceptable.

This paper proposes the design and implementation of an 8-bit Dadda multiplier employing a 4:2 approximate compressor designed with FinFET[13] technology. The approximate compressor, which offers significant reductions in delay and energy consumption, serves as a key element in this architecture. By leveraging the benefits of both FinFET technology and approximate computing, the proposed multiplier is expected to achieve substantial improvements in power efficiency and performance, with an error rate that remains imperceptible in practical applications. Preliminary simulations, conducted using the Tanner EDA tool, show a 71.8% improvement in delay and a 13% reduction in energy consumption compared to previous designs, while maintaining an acceptable error rate of 37%. The results demonstrate the potential of this approach to push the boundaries of low-power, high-performance digital circuit design.

Keywords: 4:2 approximate compressor, approximate computing[01], Dadda multiplier[03], delay reduction, energy efficiency, error-tolerant applications, tanner EDA tool, transistor density

Introduction

Very-large-scale-integration (VLSI) involves packing thousands of transistors on a single integrated circuit. Gains in semiconductor processing and communication technology spurred the advent of VLSI during the 1970s. A microprocessor is one example of a VLSI device. Integrated circuits built before VLSI had limits imposed on their size and required separate components like CPUs, ROM, RAM and logic units. Integration of multiple circuits on a chip became feasible thanks to VLSI technology. VLSI has seen rapid advancement and is now firmly established in many areas, including consumer electronics, telecommunications, high-performance computing and image and video processing. Consumers are now able to have access to highly advanced, yet small, devices thanks to VLSI innovation in technology such as high-definition displays and communication systems. This rise in the use of VLSI is forecast to accelerate the development of both VLSI and system architectures. The task of compressors is to efficiently reduce partial products generated during multiplication operations in digital systems. The compressor takes four bits (X1, X2, X3, X4) and one carry-in bit (Cin) and generates two output bits (S and Cout) as a result. An output combination consisting of a sum bit (S) and a carry-out signal (Cout). Compressors are optimized for power and area efficiency in certain situations by sacrificing their precision employing approximate logic. Approximate compressors [1] simplify circuit

[a]tejaswi.mt2530@gmail.com, [b]syamuvlsi@gmail.com, [c]sakemurali734@gmail.com, [d]yerragudimanideep@gmail.com, [e]kadaripavani2@gmail.com, [f]likhithachigicherala7272@gmail.com

DOI: 10.1201/9781003685364-58

designs by making reduced versions of exact logic circuits, leading to substantial savings in required resources.

Dadda suggested a method that spares the minimum number of iterations when transforming partial products into a simplified two-row matrix. that's achieved by incorporating (3,2) and (2,2) counters at critical points to enhance the performance of the procedure. The partial products in an N×N multiplier follow a height reduction strategy based on Dadda's algorithm01. Each matrix height is determined by +1=[1.5×] and has the initial value 1=2. Choosing heights in the range of 2, 3, 4, 6, 9, 13, 19, 28 and higher guaranties that no one row in the final reduced matrix is taller than the entire array. The reduced matrix height is controlled by employing both (3,2) and (2,2) at each stage to ensure it remains reasonable. High-speed multipliers play a crucial role in the design of both DSP systems and 3D graphics applications, since they're essential for completing intensive operations such as filtering and convolution. Over time, multiplier designs are modified to boost the clock frequency, reduce critical path delays, maximize throughput, minimize resource utilization and lower the active power consumption of composite multipliers. Optimization to multiplier designs lead to better performance and efficiency in sophisticated applications that rely on high-speed computations in modern computing systems.

Literature review

A new collection of timing compressors is used to develop highly accurate multipliers. Approximate multipliers play a vital role in error-detecting and correcting circuits by combining precision with low-energy consumption. Employing a suite of proposed approximation compressors capable of temporarily encoding any number of inputs into a predetermined number of output bits, we significantly improved the precision of approximate multipliers. The compression ratio is determined by measuring how often input bits are most likely to have a value of 1 in order to enhance the accuracy of the result. A novel scheme for selecting compressors for both exact and approximate multiplication [7] is offered to simplify the process of reducing a column in PPM. The error measures PE, MED, MRED and NED are used to evaluate the accuracy of our approximation compressors. Experiments reveal that the proposed compressors

surpass earlier approximate compressors in terms of accuracy by an average step up of 35%. To show the advantages of our proposed technique, we tested it on various multipliers. 8-bit, 12-bit, 16-bit, and 24-bit multipliers. The Verilog code was used to create multipliers, which were then synthesized using standard 45-nm CMOS [11] technology. The newly introduced 8-bit, 12-bit, 16-bit and 24-bit multipliers have shown superior performance when comparing their MED, MRED and NED results to those of current approximate multiplier methods. The average delay is reduced by 22% as a result of the findings.

As a result of its inherited features, the designer can decrease the amount of circuitry and faster operations by sacrificing the accuracy of the outcome. CMOS technology emerged as a suitable solution for designing portable devices during the period of low power and high computational complexity. Reducing circuitry and ensuring power efficiency is the main advantage of using gate diffusion input (GDI) [2] over complementary metal oxide semiconductor (CMOS). A novel structure of the EAFA is presented that improves both power and area efficiency while exhibiting minimum error distance. Our proposed design aims to overcome the problem of cascaded effects commonly seen in circuits employing GDI logic. This is displayed through developing numerous 16-bit Energy and Area Efficient [13] High-Speed Error-Tolerant Adders (EAHSETA) according to the approach outlined for a single 1-bit GDI-based full adder. The proposed adder's design metrics—delay, area and power dissipation—are shown to be accurate using simulation with Cadence. The performance of the Low-Weight Digit Detector neural network is accelerated using the suggested logic on the Intel Cyclone IV FPGA. Handwritten digit categorization application. Our proposed EAHSETA achieves 95% accuracy with lower area and higher speed, delivering 29% more performance than similar alternatives.

"FinFETs: FinFET [3,5] Architectures Arise from Device Foundations: A Description from D. Bhattacharya and N. K. Jha in the September 2014 Issue of Advances in Electronics. FinFETs and Trigate FETs have been introduced as substitutes for traditional MOSFETs because of the major challenges they encounter during scaling in the nanometer dimension.

FinFETs/Trigate [6] FETs are well-suited for advanced and future technological nodes due to their three-dimensional structure and the subsequent enhanced ability to manage short-channel effects and increase transistor scaling. We consider the implications of FinFETs at various levels within the system. We analyze the different types of FinFETs, possible imbalances and their consequences and fresh constraints that FinFETs introduce at the circuit and architectural levels. We examine analysis and optimization tools that have been developed for designing FinFET-based circuits and systems at various levels of abstractions [10]. A novel ultra-thin MOSFET with a fully depleted vertical structure and lean SOI technology known as the DELTA [4] MOSFET was introduced in 2004 by Hisamoto et al. [4]. A planar transmission gate with a thin-film semi-insulating SOI and selective field PtO2/SiO2/Si is constructed. Selective oxidation allows for achieving better SOI isolation as the minimum feature size of transistors shrinks. It provides a high-quality silicon body and a Si-SiO/sub 2/interface that matches that of conventional bulk single-crystal transistors. Numerical and experimental results attested to the superior control of the channel in the DELTA gate design and the significant advantages of its ultrathin (<0.2-mu m) vertical SOI structure for lowering short channel effects, reducing subthreshold swing and boosting transconductance. The standard ULSI plan and the DELTA design can work together seamlessly. Thus, DELTA ensures excellent small-scale compatibility with conventional flat-plane MOSFETs. As a result, DELTA provides an effective solution for building MOSFET structures at or below the 0.1μm scale.

Methodology

Existing method

The 8-bit exact multiplier is a precise digital arithmetic circuit designed to multiply two 8-bit binary numbers with high accuracy. It utilizes a structured arrangement of 64 AND gates for generating a partial product matrix, which forms the foundation of the multiplication process. This matrix is systematically reduced using 8 half adders (HAs), six full adders (FAs), and 21 exact compressors, ensuring no compromise on accuracy. Half adders handle the initial stages of addition where only two bits are involved, while full adders manage intermediate stages with three inputs, including carry propagation. The exact compressors play a pivotal role in reducing five input bits into three outputs (SUM, CARRY, and

CARRYOUT), minimizing propagation delay and enhancing computational efficiency. This design avoids approximate compressors, maintaining the precision necessary for applications requiring exact results.

The multiplier operates in three main stages. First, partial products are generated using the AND gates, creating an 8×8 binary matrix. These partial products are aligned column-wise according to their binary weight. The reduction process combines bits within the columns using the HAs, FAs, and compressors to simplify the matrix into two rows: SUM and CARRY. In the final stage, these two rows are added with a carry-propagate adder, producing an exact 16-bit output. By leveraging a combination of efficient logic elements, the 8-bit exact multiplier achieves reliable and accurate results, making it suitable for high-precision applications.

Proposed methods

Module 1: Half adder circuit: The CMOS circuit shown in Figure 58.1 implements a half adder, a fundamental combinational logic circuit used for binary

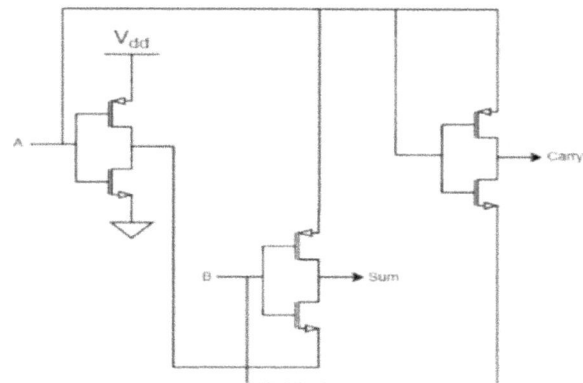

Figure 58.1 Half adder using GDI logic
Source: Author

Table 58.1 Truth table of half adder.

A	B	SUM (S)	CARRY (C)
0	0	0	0
0	1	1	0
1	0	1	0
1	1	0	1

Source: Author

addition. It takes two binary inputs, A and B, and produces two outputs: 1. Sum (S): Represents the bitwise XOR of A and B. 2. Carry (C): Represents the AND operation of A and B, indicating a carry-out if both inputs are 1 as shown in Table 58.1 and transient analysis is shown in 58.6.

Module 2: Full adder circuit[08]: A complete adder, a key component of digital arithmetic operations, is implemented by the CMOS circuit displayed. as shown in Figure 58.2 and transient analysis is shown in Figure 58.7. Three binary inputs are added, and two outputs are generated: 1. Sum(S): Shows the three inputs XOR operation. 2. Carry (C): Shows the carry as shown in Table 58.2.

Module 3: Exact compressor 1 The circuit shown in Figure 58.3 depicted is an exact compressor, intended to generate three outputs: Sum (S), Carry (Cr), and Carryout (C), and execute binary addition

for five inputs (x1, x2, x3, x4, Cin). This kind of adder is utilized in arithmetic circuits, especially in computing systems and digital signal processing, where accurate addition is essential. Contributions: The primary binary input bits are x1, x2, x3, and x4. Cin: Carry input from an initial carry value or a previous stage.

Results: 1. Sum (S): $S=x1\oplus x2\oplus x3\oplus x4\oplus Cin$ is generated using the XOR logic.

2. Carry (Cr): Denotes a carry in the logic of the adder. Usually produced when numerous inputs are high using AND-OR logic.

3. Carryout (C): The last carry-out in multi-bit addition that advances to the following step. When the number of high inputs + Cin surpasses binary 15, it can be determined and transient analysis is shown in 58.9.

Module 4: Approximate compressor[07,08] [12] An approximate compressor is a digital circuit that reduces the amount of partial products in arithmetic operations, especially in applications like

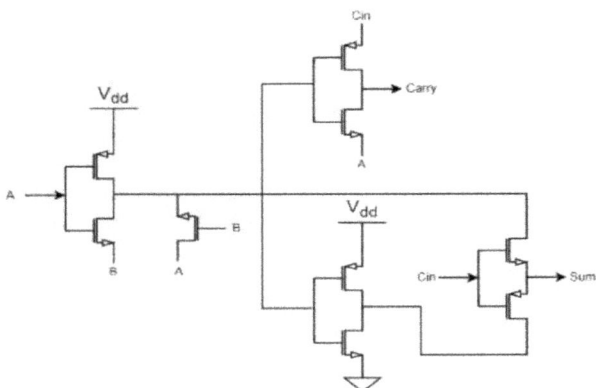

Figure 58.2 Full adder using GDI logic out that the addition produces
Source: Author

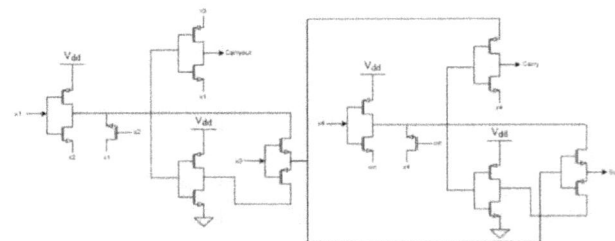

Figure 58.3 4:2 Exact compressor using GDI logic
Source: Author

Table 58.2 Truth table of full adder.

A	B	C	Sum (S)	Carry (C)
0	0	0	0	0
0	0	1	1	0
0	1	0	1	0
0	1	1	0	1
1	0	0	1	0
1	0	1	0	1
1	1	0	0	1
1	1	1	1	1

Source: Author

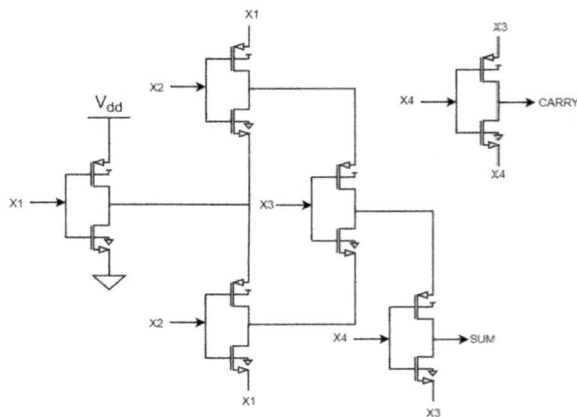

Figure 58.4 2 approximate compressors using GDI logic
Source: Author

digital signal processors or multipliers. This particular design, an approximate 4:2 compressor, [15] puts size, delay, and power economy ahead of precise accuracy. Four input bits (X1, X2, X3, and X4) are compressed into two output bits using the roughly 4:2 compressor shown in Figure 58.4 in the CMOS diagram: 1. Carry (C): Stands for carry-out or overflow. 2. Sum (S): This encodes the compressed output and transient analysis is shown in Figure 58.8.

Proposed 8-bit Approximate Multiplier[03] [8,9] System:

An 8-bit approximate multiplier system is suggested: Accuracy and performance are balanced in the hardware- optimized design of the suggested 8-bit approximation multiplier. In order to achieve lower power consumption, area, and latency while maintaining sufficient precision for many real-world applications, it combines a combination of approximate compressors and exact adders. This design is appropriate for error-tolerant applications because, in contrast to the exact multiplier, it employs 4:2 approximation compressors in particular multiplier areas. The proposed multiplier's main parts are:

1. 64AND gates the purpose of these gates is the same as that of an exact multiplier: to generate partial products. Operation: An 8 × 8 partial product matrix is created by ANDing each bit of the two operands (A[7:0] and B[7:0]). Significance: Producing every possible bit-level product, it serves as the foundation for binary multiplication. 2. Five Half Adders: The goal of half adders is to sum two binary inputs without the usage of a carry input. Use in the multiplier: Half adders are generally used in the more significant bits (MSBs), or other parts of the partial product tree where accuracy is valued. 3. Four FAs, or full adders: Goal: Three binary inputs—two partial product bits and a carry input—are added using full adders. How to Use the Multiplier: Full adders are employed in crucial areas, particularly in MSBs, where accuracy is crucial. 4. Twelve 4:2 Approximate Compressors: The goal of these compressors is to reduce five inputs into two outputs (SUM and CARRY) by roughly summing and carrying four input bits and a carry input. Operation: By accepting minor output errors, the approximation design lowers complexity. Compressors are used in the multiplier's lower significant bits (LSBs), where even minor mistakes have less of an effect on the outcome. 5. Ten exact compressors: The goal of these precise adders is to produce three outputs—SUM,

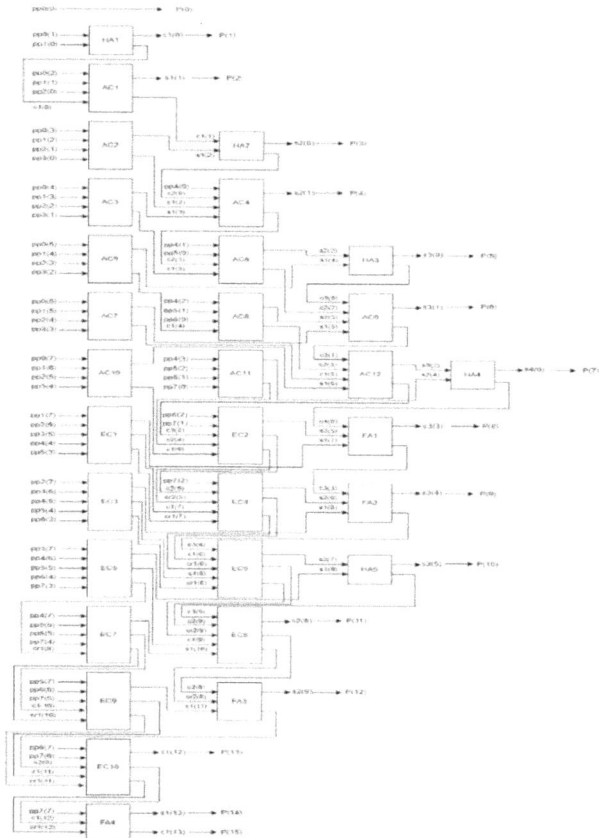

Figure 58.5 8-bit proposed multiplier
Source: Author

CARRY, and CARRYOUT—from five input bits. Application in the multiplier: In MSBs, where correctness of the finished result depends on precision. Functioning of the proposed multiplier: 1. Partial product generation: An 8x8 binary matrix is created by employing 64 AND gates to construct the 64 partial products. 2. Partial product tree division: Based on bit importance, the partial product matrix is split into two regions: The first region (7 MSBs) places a high priority on accuracy and performs precise reductions using exact compressors, full adders, and half adders. The second region (8 LSBs): Utilizes half adders, full adders, and 4:2 approximation compressors to maximize hardware resources. 3. Partial Product Matrix Reduction: To reduce error and guarantee precision for the most important bits, the MSB region uses exact adders (5-bit, full, and half adders). In the LSB region, approximate compressors simplify the additional process and permit tiny, acceptable errors to increase speed and power efficiency. 4. The last stage of addition involves compressing

Comparison Tables:

Table 58.3 Comparision of different circuits.

	Half Adder	Full Adder	4:2 Exact Compressor	4:2 Approximate compressor
Total no of transistors per circuit	06	09	18	12
Average power consumption	2.36nW	0.05876nW	0.07834nW	0.02363nW
Time delay	0.65ns	0.76 ns	0 .96ns	0.90ns

Source: Author

Table 58.4 Comparision of Existing and Proposed System.

	Existing system	Proposed system
Total no of half adders	08	05
Total no of full adders	06	04
Total no of 4:2 exact compressors	21	10
Total no of 4:2 approximate compressors	------	12
Average power consumption	0.66364nW	0.41522nW
Time delay	4.44ns	3.56ns

Source: Author

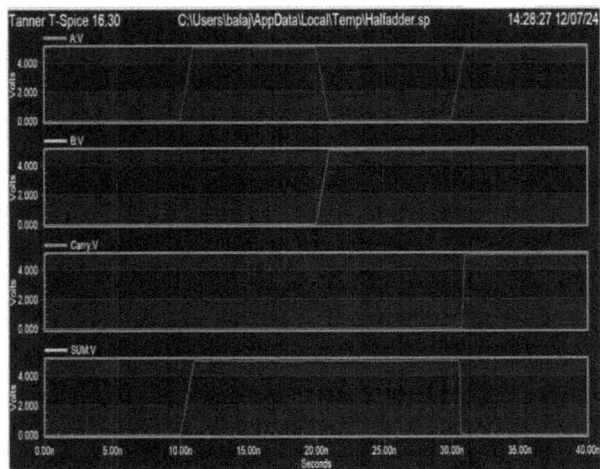

Figure 58.6 Transient analysis outputs of half adder
Source: Author

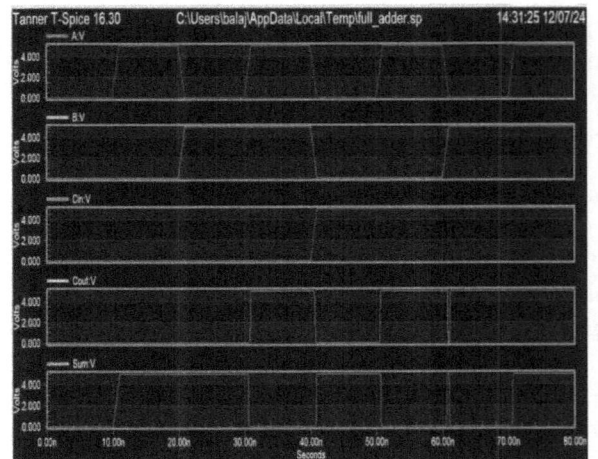

Figure 58.7 Transient analysis outputs of full adder. approach reduces power dissipation and silicon area, resulting in a highly optimized and compact circuit. Such attributes make the multiplier an excellent choice for applications demanding a balance between speed, low power consumption, and minimal hardware requirements
Source: Author

the smaller matrix into the SUM and CARRY rows. The final estimated result is obtained by summing these rows using an effective carry propagation adder shown in Figure 58.5.

Result analysis

The proposed multiplier demonstrates impressive efficiency with an average power consumption of

0.0236 nW and a time consumption of 0.65ns, highlighting its suitability for power-sensitive and high-speed applications. Additionally, the use of gate diffusion input (GDI) logic in the half adder design

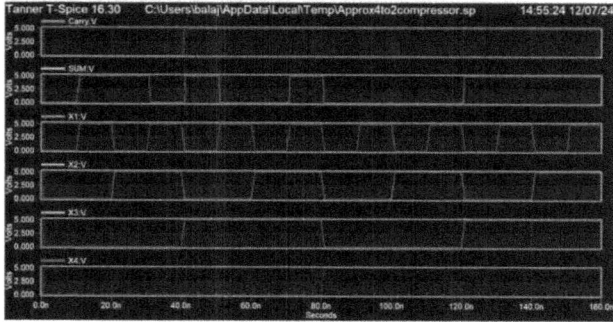

Figure 58.8 Transient analysis outputs of 4:2 approximate compressor
Source: Author

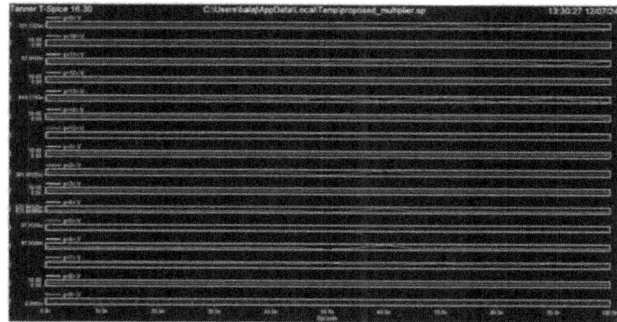

Figure 58.9 Transient analysis outputs of exact compressor
Source: Author

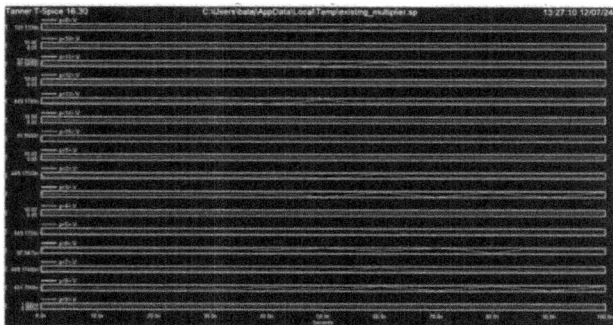

Figure 58.10 Transient analysis outputs for existing multiplier with A=01100110 and B=11010011
Source: Author

significantly optimizes transistor usage, requiring only 06 transistors per half adder. This efficient transistor count contributes to reduced silicon area and lower power dissipation, further enhancing the overall performance and energy efficiency of the circuit.

The proposed multiplier demonstrates impressive energy and computational efficiency, with an

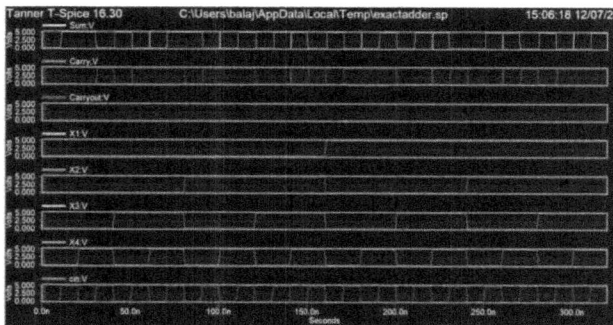

Figure 58.11 Transient analysis outputs for proposed multiplier with A=01100110 and B=11010011
Source: Author

average power consumption of 0.05876 nW and a time delay of 0.76ns. By utilizing GDI logic in its full adder design, the circuit achieves a remarkably [14] low transistor count of just nine per full adder. This minimalistic

The proposed multiplier exhibits high efficiency with an average power consumption of 0.07834 nW and a time consumption of 0.96ns. The use of GDI logic in the design of exact compressors further enhances its performance by requiring only 18 transistors per compressor. This streamlined transistor count reduces power dissipation and silicon area, ensuring a compact and energy-efficient implementation. Such a design is particularly advantageous for applications where precision, speed, and minimal hardware utilization are critical.

The proposed multiplier demonstrates excellent energy efficiency with an average power consumption of 0.02363 nW and a time consumption of 0.90ns. The incorporation of GDI logic in the design of approximate compressors minimizes hardware complexity, requiring only 12 transistors per compressor.

This reduced transistor count significantly lowers power consumption and silicon area while maintaining acceptable performance levels. The design's balance between energy savings and computational efficiency makes it particularly suitable for applications prioritizing low power and compact hardware without compromising overall functionality shown in Table 58.3.

The existing multiplier design, implemented using Gate Diffusion Input (GDI) logic, has an average power consumption of 0.66364 nW and a time consumption of 2.48 ns. With a total of 480 transistors, the design reflects a higher hardware complexity compared to optimized approaches. While it ensures functionality, the larger transistor count contributes to increased power dissipation and silicon area usage. This highlights the potential for further refinement to achieve better energy efficiency and reduced hardware overhead, particularly for applications requiring compact and low-power solutions as shown in Figure 58.10.

The proposed multiplier design, utilizing GDI logic, achieves an average power consumption of 0.41522 nW and a time consumption of 1.70ns, demonstrating a significant improvement in energy efficiency and computational speed compared to existing designs. With a reduced transistor count of 390, the proposed approach optimizes hardware complexity while maintaining robust functionality. This balance of lower power dissipation, faster operation, and reduced silicon area makes the proposed multiplier well-suited for power- sensitive and performance-driven applications shown in Figure 58.11 and Table 58.4.

Conclusion

Created and implemented an 8-bit approximation multiplier[08] in this project using a hybrid architecture that blends approximate and exact elements. To strike a compromise between hardware efficiency and computational precision, the system deliberately employs precise adders in the higher significant bits (MSBs) and approximately 4:2 compressors[07,08] in the lower significant bits (LSBs). In comparison to conventional accurate multipliers, the suggested multiplier exhibits notable gains in speed, area, and power consumption while introducing a negligible and bearable inaccuracy. The architecture provides excellent performance for error tolerant applications by utilizing 10 5-bit precise adders, 5 half adders, 4 full adders, and 12 approximate 4:2 compressors. elements. Making sure that the MSBs, which make up the majority of the final product, are calculated using precise components further improves the multiplier's accuracy. For error-resilient applications where minor errors don't substantially affect system performance, such as image processing, signal processing, and machine learning, this approach is especially well suited. Achieving a balance between precision and efficiency through the integration of both accurate and approximation features makes the system a workable answer to contemporary

low-power and high- speed computing needs. The suggested approximation multiplier, in summary, is a creative and effective design that tackles the issues of space, power, and latency in hardware multiplication. While maintaining respectable accuracy levels, it offers a promising alternative for next- generation computing systems where energy efficiency is crucial.

References

[1] Shirzadeh, S., & Forouzandeh, B. (2021). High accurate multipliers using a new set of approximate compressors. *AEU - International Journal of Electronics and Communications*, 138, 153778.

[2] Nagarajan, M., Muthaiah, R., Teekaraman, Y., Kuppusamy, R., & Radhakrishnan, A. (2022). Power and area efficient cascaded GDI approximate adder for accelerating multimedia applications using deep learning model. *Computational Intelligence and Neuroscience*, 2022(1), 3505439.

[3] Bhattacharya, D., & Jha, N. K. (2014). FinFETs: from devices to architectures. *Advances in Electronics*, 2014, 1–21.

[4] Hisamoto, D., Kaga, T., Kawamoto, Y., & Takeda, E. (1989). A fully depleted lean-channel transistor (DELTA) - a novel vertical ultra-thin SOI MOSFET. In International Technical Digest on Electron Devices Meeting, (pp. 833–836). IEEE.

[5] Gupta, S. K., & Roy, K. (2015). Low power robust finFET-based SRAM design in scaled technologies. In Circuit Design for Reliability (pp. 223–253). Springer New York. https://doi.org/10.1007/978-1-4614-4078-9_11

[6] Maity, N. P., Maity, R., Maity, S., & Baishya, S. (2019). Comparative analysis of the quantum FinFET and trigate FinFET based on modeling and simulation. *Journal of Computational Electronics*, 18(2), 492–499.

[7] Momeni, A., Han, J., Montuschi, P., & Lombardi, F. (2015). Design and analysis of approximate compressors for multiplication. *IEEE Transactions on Computers*, 64(4), 984–994.

[8] Strollo, G. M., Napoli, E., De Caro, D., Petra, N., & Di Meo, G. (n.d.). Comparison and extension of approximate 4-2 compressors for low-power approximate multipliers. *IEEE Transactions on Circuits and Systems I: Regular Papers*, 67(9), 3021–3034.

[9] Strollo, G. M., Napoli, E., De Caro, D., & Petra, N. (2018). Approximate multipliers based on new approximate compressors. *IEEE Transactions on Circuits and Systems I: Regular Papers*, 65(12), 4169–4182.

[10] Edavoor, P. J., Raveendran, S., & Rahulkar, A. D. (2020). Approximate multiplier design using novel dual-stage 4:2 compressors. *IEEE Access*, 8, 48337–48351.

[11] Fan, D., Huang, J., & Liu, L. (2021). Non-volatile approximate arithmetic circuits using scalable hybrid spin-CMOS majority gates. *IEEE Transactions on Circuits and Systems I: Regular Papers*, 68(3), 1217–1230.

[12] Chang, H., Gu, J., & Zhang, M. (2004). Ultra low-voltage low-power CMOS 4-2 and 5-2 compressors for fast arithmetic circuits. *IEEE Transactions on Circuits and Systems I: Regular Papers*, 51(10), 1985–1997.

[13] Arasteh, A., Moaiyeri, M. H., Taheri, M. R., Navi, K., & Bagherzadeh, N. (2018). An energy and area efficient 4:2 compressor based on FinFETs. *Integration, the VLSI Journal*, 60, 224–231.

[14] Ahmadinejad, M., Moaiyeri, M. H., & Sabetzadeh, F. (2019). Energy and area efficient imprecise compressors for approximate multiplication at nanoscale. *AEU - International Journal of Electronics and Communications*, 110, 152859.

[15] Sabetzadeh, F., Moaiyeri, M. H., & Ahmadinejad, M. (2019). A majority-based imprecise multiplier for ultra-efficient approximate image multiplication. *IEEE Transactions on Circuits and Systems I: Regular Papers*, 66(11), 4200–4208.

59 Enhancement of power quality for three-phase four-switch inverter-based DVR with teaching learning-based optimization

P. Mamatha[1,a] and Kiran Kumar Kuthadi[2,b]

[1]PG Scholar, Department of EEE, Sree Vahini Institute of Science and Technology, Tiruvuru, Andhra Pradesh, India

[2]Assiociate Professor, Department of EEE, Sree Vahini Institute of Science and Technology, Tiruvuru, Andhra Pradesh, India

Abstract

For electric-powered industries to function at peak efficiency, reliable, high-quality electricity is essential. Any disturbance in electric networks may cause voltage variations, thus it is crucial to ensure good power quality. It is possible to use a number of techniques and technologies to safeguard delicate machinery against these kinds of disruptions. In this investigation, a dynamic voltage restorer (DVR) is used to handle voltage fluctuations impacting delicate loads. In order to minimize voltage sag and total harmonic distortion (THD) in these loads, the teaching-learning-based optimization (TLBO) approach is used to improve the settings of the PI controller. Using MATLAB/Simulink, the efficacy of this method was confirmed, and the results proved that it could effectively mitigate voltage distortions. By reducing THD to 1.16% and increasing efficiency to 95.6%, the TLBO method surpassed the Cuckoo Search Algorithm (CSA).

Keywords: Cuckoo search algorithm (CSA), DVR, power quality, sensitive load, teaching-learning-based optimization (TLBO)

Introduction

Reliable, high-quality electrical power is in great demand due to the rising number of sensitive industrial and residential power users. Maintaining power quality is vital in today's competitive industrial environment due to the usage of power electronics, computer processors, and nonlinear loads. Deviations from set standards may lead to economic losses. Many negative outcomes might result from these losses, such as decreased efficiency, decreased competitiveness, higher production costs, worse product quality, shorter equipment lifetime, greater maintenance expenses, and power outages. Hence, ensuring reliable electricity availability is vital for industrial companies to effectively control their costs [1–3]. Power outages, voltage dips, voltage spikes, and flickering are among symptoms of problems that may arise from disruptions in electrical distribution networks. Of them, voltage sag [4–6] stands out as particularly important; according to IEEE standards, this is characterized as an abrupt decrease in voltage of 10–90% within a time frame of half a cycle to one minute. Asymmetrical network failures and electromagnetic events, such as startup currents, are common natural causes of these disruptions. The dynamic voltage restorer (DVR) is one of many specialized power devices used by specialists to lessen the severity of such effects on vulnerable customers [7–9].

In the case of sensitive loads, DVRs excel in controlling voltage dips and spikes. Energy storage, a voltage source inverter, and a coupling transformer are the essential components of a DVR system [10]. A series-connected coupling transformer allows the DVR [11–14] to inject the required voltage into a feeder line that supplies a sensitive load in the event that voltage sag is detected. Digital video recorders use a variety of control schemes for insert voltage management, including predictive control, sliding-mode control [15], and robust control, to guarantee optimum performance. This research introduces a novel method for optimizing PI controller settings based on the teaching-learning-based optimization (TLBO) [16] algorithm, which takes its cues from the dynamic between a classroom instructor and their pupils. To determine

[a]Mamthapillem207@gmail.com, [b]kiran9949610070@gmail.com

DOI: 10.1201/9781003685364-59

how well this algorithm works, we compare the findings from DVR tests conducted under different voltage circumstances to those from the Cuckoo Search Algorithm (CSA).

Mathematical model

Displayed in Figure 59.1 is a system for the distribution of hybrid electricity. Power quality (PQ) problems have many effects on distribution systems; thus many different approaches have been taken to fix them. Because it efficiently handles voltage fluctuations, the DVR is crucial in reducing PQ issues. In a distribution network, the DVR is mainly responsible for effectively compensating for voltage swells and sags. Additionally, it helps by lowering transient voltages, minimizing fault currents, and adjusting for line-voltage harmonics.

The suggested solution improves system power quality by combining a DVR with the teaching-learning-based optimization TLBO technique. Both in transient and steady-state scenarios, our approach showed better performance. A block diagram illustrating this approach is shown in Figure 59.1. A constant flow of reactive power is produced by the DVR and fed into the transmission line. In Figure 59.2, we can see the DVR's power and control circuits represented by their comparable circuitry. To link the distribution network to the DVR, an injection transformer is used. The control circuit integrates the supply voltage with compensatory voltage injected by the voltage source inverter (VSI) whenever it detects a voltage drop. Aside from isolating the DVR from the distribution system, the injection transformer amplifies the filtered VSI output to the required level, after receiving DC voltage from a storage element, the VSI voltage to the distribution system. A commonly used and easy-to-understand control approach, the PI controller oversees the inverter functioning of the DVR. By optimizing the controller's settings, the

TLBO algorithm improves performance. A 3-phase, 4-switch voltage source inverter (VSI) may function when a PWM generator transforms control signals from the PI controller into pulses. The load is supplied with a steady ultimate output.

The harmonic content is kept below acceptable levels by using a harmonic filter to suppress switching harmonics. When the modulation index is less than 1, the switching frequency and its multiples tend to have the highest concentration of switching harmonics in a VSI that uses pulse width modulation (PWM) to function. The injection transformer's harmonic filters may be installed on either the low- or high-voltage side. To further avoid system overload in the case of maintenance or faults, bypass switches are used to divert load current to an alternate channel. The DVR may protect itself from short-circuit occurrences and prevent interference with existing protective equipment by using these switches. An important part of the DVR is the controller. The DVR system was controlled using closed-loop technology in a rotating dq reference frame. In response to the disturbances, the PI controller pulsed the PWM generator as needed. The DVR's PI controller circuit is seen in Figures 59.2 and 59.3.

Teaching-learning-based optimization

A population-based algorithm that mimics classroom teaching and learning is the TLBO, which was

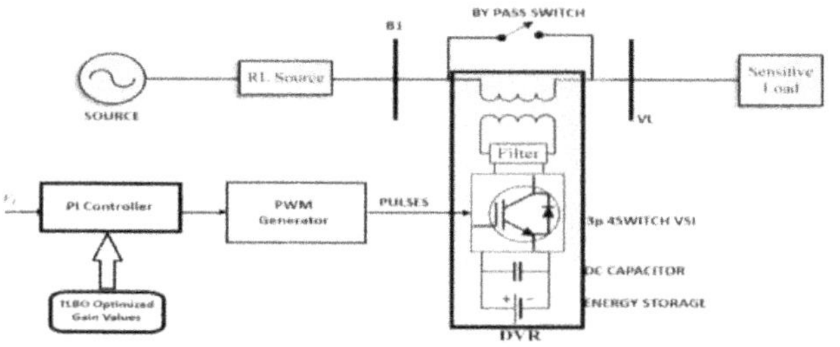

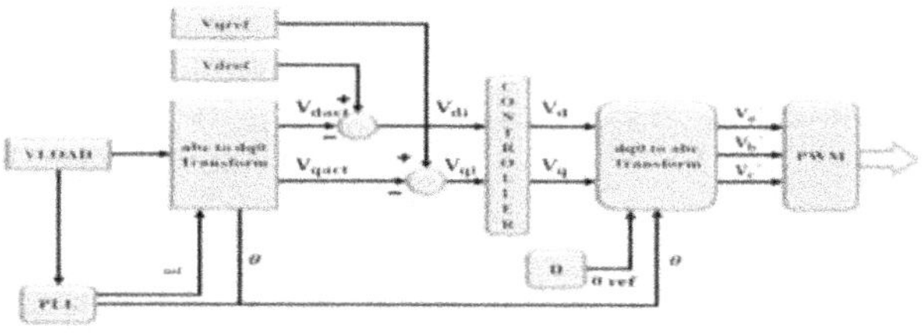

Figure 59.2 The PI controller's schematic
Source: Author

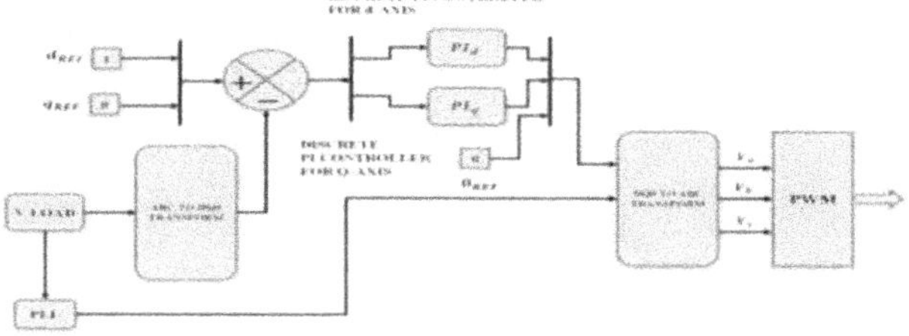

Figure 59.1 The suggested system's block diagram
Source: Author

Figure 59.3 PI control circuit
Source: Author

presented by Rao and colleagues in 2011 [13]. This algorithm is an intelligent optimization approach that was developed to raise the scientific level of a class by taking into account the impact of the instructor on the pupils. This approach stems from the guiding principle that a teacher should aim to bring the class level up to his own knowledge level. This way, students may learn from each other and build on each other's successes, in addition to capitalizing on the instructor's expertise. Since no educator can control his pupils' individual levels, he must work to raise the class average and judge its success by looking at their test results. As a population, a class serves as the basis for this algorithm. In this optimization issue, the courses that are recommended to students are seen as distinct decision-making factors. Consequently, numerical values are used to compare the pupils' performance in various classes. As a teacher, you're chosen based on the population's best reaction. The best outcome is the optimal quantity of the optimization objective function, which is specified according to decision-making factors, which are really parameters. There are two parts to this algorithm: the instructor part and the student part [13].

Teacher phase
During this part of the process, the instructor works to raise the class average to his own level. The assignment is challenging, therefore he tries to raise the class average from Mi to M_new. With this disparity in mind, we update each set of issue variables accordingly. The Diff_Meani option might be used to store this difference in the following way:

$$Diff\_Mean_i = r_i(M\_new - T_f M_i) \qquad (1)$$

T_F where ri is a random integer between 0 and 1, and is the training factor or instructor parameter that determines the amount by which the average value changes. In the exploratory stage, TF might be either 1 or 2, and the odds of either outcome are identical.

$$T_F = round[1 + rand(0,1)\{2 - 1\}] \qquad (2)$$

It is worth mentioning that the value of the difference between Eq. 3 and 5

$$X_{new,i} = X_{old,i} + Diff\_Mean_i \qquad (3)$$

Student phase
Students supplement what they learn in class with what they know from one another. Here, mathematically speaking, each set of variables (the students) randomly chooses one of the other sets in each cycle. As an example, let's say that student i selects student j. We shall update student i according to Eq. 4 and 5 if student j has more information than i.

$$X_{new,i} = X_{old,i} + r_i(X_i - X_j) \qquad (4)$$

Else, the status of students varies as follows:

$$X_{new,i} = X_{old,i} + r_i(X_i - X_i) \qquad (5)$$

Once every student has changed their status, the objective function is used to determine their level. Against doing this, we compare the top student against the instructor from the prior iteration; if the student proves to be superior, we replace the instructor. This procedure is repeated until the necessary criteria for convergence are met.

Simulation results: TLBO with PI controller

The suggested system received input voltages of 350 V, 400 V, 410 V, 420 V, and 500 V from a variety of AC sources. The following figures show the AC source current wave forms for various voltage inputs. Figure 59.4 displays the AC source's voltage and current waveforms. From 0025s to 0035s, there was a dip in the 350V AC source voltage. Figure 9.4 (b) shows the equivalent current waveform, which

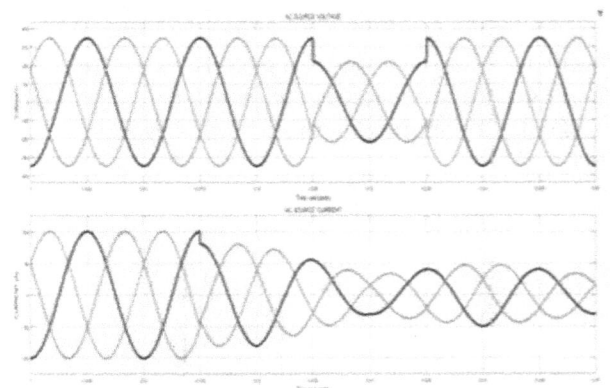

Figure 59.4 AC source waveforms (a) voltage 350 V (b) current
Source: Author

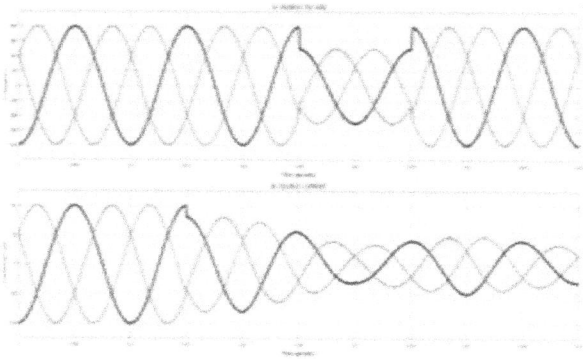

Figure 59.5 AC source with (a) voltage 400 V (b) current
Source: Author

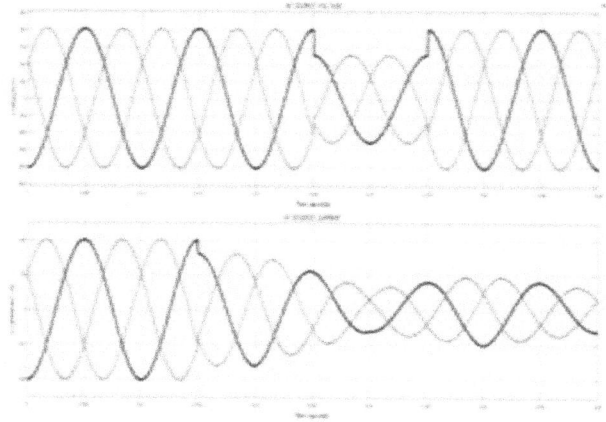

Figure 59.7 AC source with (a) voltage 420 V (b) current
Source: Author

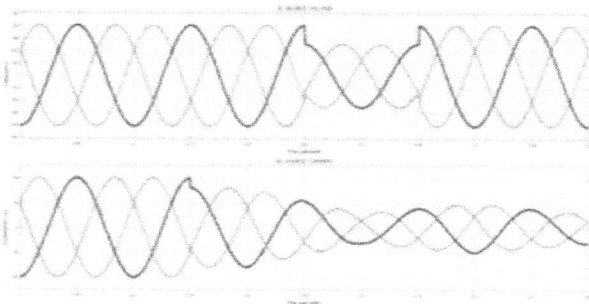

Figure 59.6 AC source with (a) voltage 410 V (b) current
Source: Author

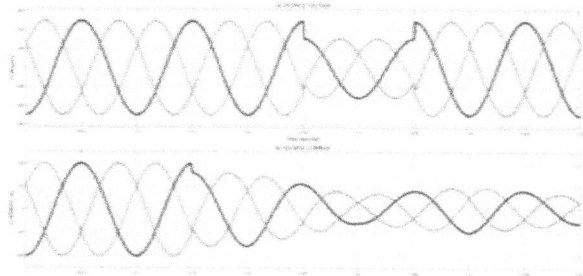

Figure 59.8 AC source with (a) voltage 500 V (b) current
Source: Author

shows that a constant current output of 30A was maintained. For an input voltage of 400V, the AC current as well as voltage waveforms are clearly seen in Figure 59.5, confirming that the 30A output current is consistently maintained. Despite a voltage drop between 0025s until 0035s, the DVR's optimized control method keeps the acquired current constant. As seen in Figure 59.6, an input potential of 410V was maintained while an output power of 30A was continuously maintained. From 0025s to 0035s, the AC supply voltage sagged, which affected the power quality. A combination of real and reference voltage generation, however, lessens the impact of voltage sag in the suggested method.

Figure 59.7 (a) & (b) shows that the input voltage is set to 420 V but has a sag in voltage from 0.025s to 0.035s. However, the current waveform indicates that a constant current value of 30A is attained with the optimized control approach. Figure 59.8 (a) & (b)

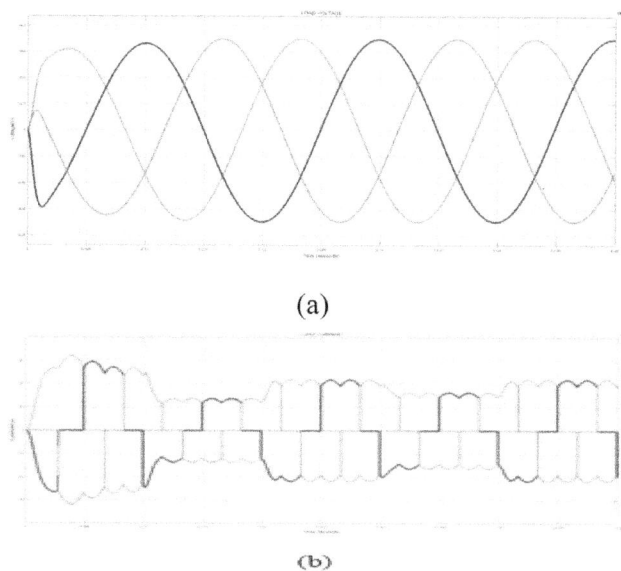

(a)

(b)

Figure 59.9 Load waveforms (a) voltage (b) current
Source: Author

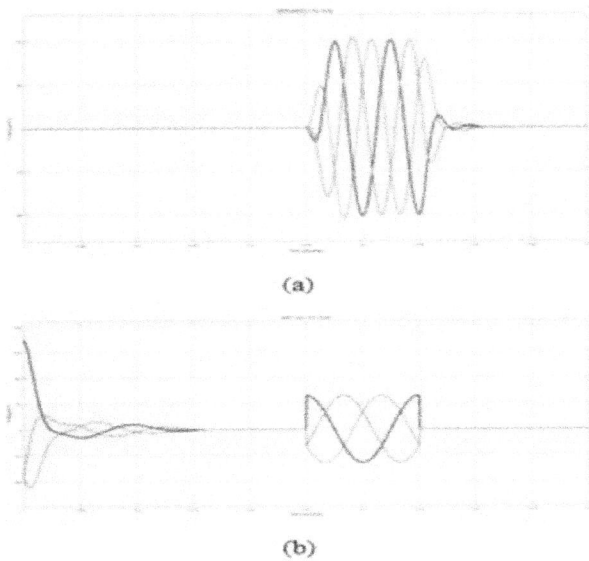

Figure 59.10 Waveforms for (a) reference voltage (b) actual voltage
Source: Author

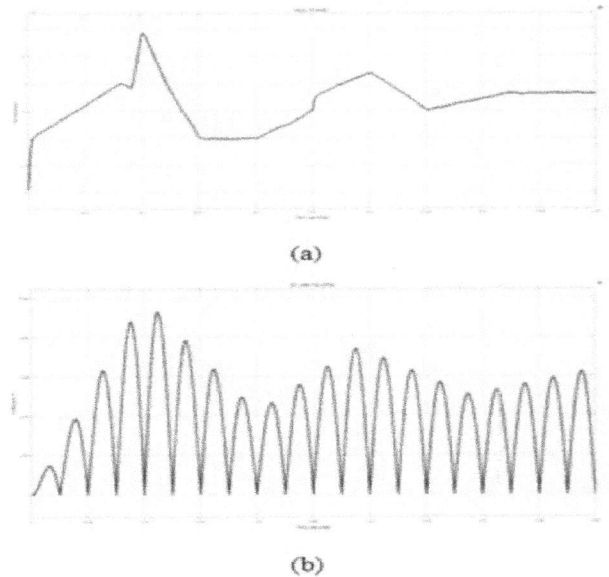

Figure 59.12 Waveforms for (a) real (b) reactive powers
Source: Author

Figure 59.11 DC link voltage with CSA tuned PI
Source: Author

Table 59.1 % THD with CSA & TLBO based with PI.

Type of Optimization	CSA with PI	TLBO with PI
Load Voltage (% THD)	2.98%	1.16%

Source: Author

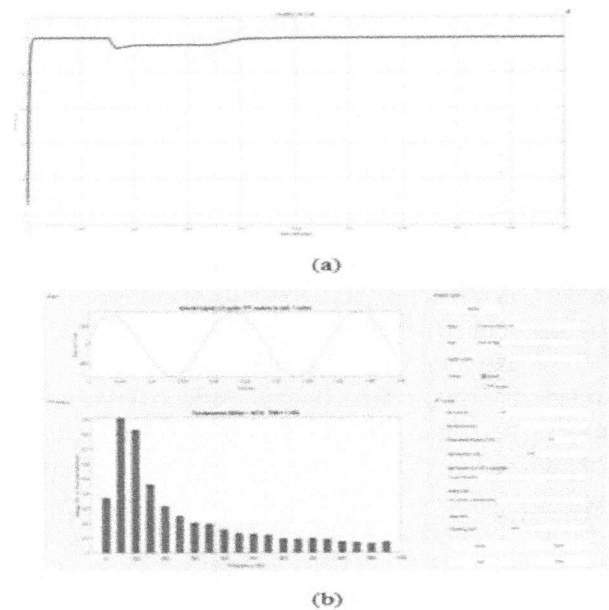

Figure 59.13 (a) Power factor (b) THD
Source: Author

shows that the input voltage is set to 500 V but has sag in voltage from 0.025s to 0.035s. However the current waveform that maintained a constant current of 30A. Figure 59.9 (a) & (b) shows the load voltage and load current .Irrespective of the fluctuating input voltages owing to voltage sag issues,the obtained values of load voltage and load currrent remain constant.

Pictured in Figure 59.10 are the reference and real voltage waveforms. By comparing the reference and real readings, we were able to ascertain the change in error voltage along with error voltage values. The reference and actual voltages were developed between the years 0025 and 0035, at which time the voltage sag issues with the original voltage were resolved. To acquire the best possible fitness values, the teaching learning-based algorithm (TLBA) refines these parameters even more.

Figure 59.11 shows the DC-link voltage that was attained with the CSA-adjusted PI controller. The selected CSA causes the PI controller to provide optimum outputs, which are then converted into pulses by a PWM generator. These pulses control the operation of the three-switch VSI, which keeps the DC link voltage constant. As a consequence, the DC link voltage is steady and well-regulated, taking just 002s to settle. The managed DC link voltage provides better voltage adjustment and reduces power quality issues. Figure 59.12 shows the reactive power vs real power levels. Also

Figure 59.13 (a) shows that a power factor of one is reached, which suggests that power quality problems are reduced, leading to better compensation. The computed THD value is 1.16%, as shown in Figure 59.13 (b). The findings of the electrical quality or total harmonic distortion (THD) experiments demonstrate that the proposed TLBA approach effectively mitigates power quality issues. Look at Table 59.1 to see how the load voltage compares to CSA and TLBA with the DVR. Reducing voltage drops and overtones at the source is made much easier with the proposed TLBA approach.

Conclusion

Consumers' power quality is effectively improved by using the suggested teaching-learning-based optimization (TLBO) algorithm to the test network, according to the findings. It may be tough to create an accurate model of power systems because of its dynamic behavior under both normal and fault situations. Consequently, smart control algorithms must be used if compensators are to work quickly and accurately. All of the test scenarios that the proposed controller was tested on performed very well. The Total Harmonic Distortion (THD) index is substantially improved by the TLBO controller, which was originally developed to increase voltage stability for sensitive loads. Because of its simple design, this controller is suitable for real-world use. The suggested approach is cost-effective since it achieves great performance without the requirement for specialized supplementary equipment. With a total harmonic distortion (THD) of only 1.16 percent, the TLBO approach outperformed the Cuckoo Search Algorithm (CSA), indicating that it is better at enhancing distribution network power quality.

References

[1] Suraya, S., Irshad, S. M., Azeem, M. F., Al-Gahtani, S. F., & Mahammad, M. H. (2020). Multiple voltage disturbance compensation in distribution systems using DVR. *Engineering, Technology & Applied Science Research*, 10(3), 5732–5741. https://doi.org/10.48084/etasr.3485.

[2] Akbar, F., Mehmood, T., Sadiq, K., & Ullah, M. F. (2021). Optimization of accurate estimation of single diode solar photovoltaic parameters and extraction of maximum power point under different conditions. *Electrical Engineering & Electromechanics*, 6, 46–53. https://doi.org/10.20998/2074-272X.2021.6.07.

[3] Soomro, A. H., Larik, A. S., Mahar, M. A., Sahito, A. A., & Sohu, I. A. (2020). Simulation-based analysis of a dynamic voltage restorer under different voltage sags with the utilization of a PI controller. *Engineering, Technology and Applied Science Research*, 10(4), 5889–5895. https://doi.org/10.48084/etasr.3524.

[4] Ullah, M. F., & Hanif, A. (2021). Power quality improvement in distribution system using distribution static compensator with super twisting sliding mode control. *International Transactions on Electrical Energy Systems*, 31(9), e12997. https://doi.org/10.1002/2050- 7038.12997.

[5] Anwar, N., Hanif, A. H., Khan, H. F., & Ullah, M. F. (2020). Transient stability analysis of the IEEE-9 bus system under multiple contingencies. *Engineering, Technology and Applied Science Research*, 10(4), 5925–5932. https://doi.org/10.48084/etasr.3273.

[6] Vinothkumar, E. A. V. (2021). Recent trends in power quality improvement using custom power devices and its performance analysis. *Turkish Journal of Computer and Mathematics Education*, 12(7), 1686–1695. https://doi.org/10.17762/turcomat.v12i7.3052.

[7] Abas, N., Dilshad, S., Khalid, A., Saleem, M. S., & Khan, N. (2020). Power quality improvement using dynamic voltage restorer. *IEEE Access*, 8, 164325–164339. https://doi.org/10.1109/ACCESS.2020.3022477.

[8] Nambiar, R. E., Darshan, M., Lavanya, B., Pavan Kumar, A. J., & Priyadarshini, V. (2021). Comparative study between different controllers of DVR for power quality improvement. In 2021 International Conference on Design Innovations for 3Cs Compute Communicate Control (ICDI3C), Bangalore, India, (pp. 84–87). https://doi.org/10.1109/ICDI3C53598.2021.00025.

[9] Moghassemi, A., & Padmanaban, S. (2020). Dynamic voltage restorer (DVR): a comprehensive review of topologies, power converters, control methods, and modified configurations. *Energies*, 13(16), 4152. https://doi.org/10.3390/en13164152.

[10] Choudhury, S., Bajaj, M., Dash, T., Kamel, S., & Jurado, F. (2021). Multilevel inverter: a survey on classical and advanced topologies, control schemes, applications to power system and future prospects. *Energies*, 14(18), 5773. https://doi.org/ 10.3390/ en14185773.

[11] Appala Naidu, T., Arya, S. R., Maurya, R., & Padmanaban, S. (2021). Performance of DVR using optimized PI controller based gradient adaptive variable step LMS control algorithm. *IEEE Journal of Emerging and Selected Topics in Industrial Electronics*, 2(2), 155–163. https://doi.org/10.1109/ JESTIE.2021.3051553.

[12] Nasrollahi, R., Farahani, H. F., Asadi, M., & Farhadi-Kangarlu, M. (2022). Sliding mode control of a dynamic voltage restorer based on PWM AC chopper in three-phase three-wire systems. *International Journal of Electrical Power and Energy Systems*, 134, 107480. https://doi.org/10.1016/j. ijepes.2021.107480.

[13] Navabi, M., & Davoodi, N. (2019). Design of a robust controller using real twisting algorithm for a fixed wing airplane. In 2019 5th Conference on Knowledge Based Engineering and Innovation (KBEI), Tehran, Iran, (pp. 605–610). https://doi.org/10.1109/ KBEI.2019.8734903.

[14] Jeyaraj, K., Durairaj, D., & Velusamy, A. I. S. (2020). Development and performance analysis of PSO-optimized sliding mode controller–based dynamic voltage restorer for power quality enhancement. *International Transactions on Electrical Energy Systems*, 30(3), e12243. https://doi.org/10.1002/2050-7038.12243.

[15] Riaz, U., Tayyeb, M., & Amin, A. A. (2021). A review of sliding mode control with the perspective of utilization in fault tolerant control. *Recent Advances in Electrical & Electronic Engineering*, 14(3), 312–324.

[16] IEEE Recommended Practice for Monitoring Electric Power Quality (2019). IEEE Std 1159-2019 (Revision of IEEE Std 1159-2009), pp.1–98. doi: 10.1109/IEEESTD.2019.8796486.

60 Cattle health monitoring system using Arduino UNO

V. Mounika[1,a], Bandi Doss[2,b], K. Mohana Lakshmi[3,c], CH. Srujan[4,d], G. Preethi[4,e] and E. Rajini Kanth[4,f]

[1]Assistant Professor, Department of ECE, CMR Technical Campus, Hyderabad, Telangana, India

[2]Professor, Department of ECE, CMR Technical Campus, Hyderabad, Telangana, India

[3]Associate Professor, Department of ECE, CMR Technical Campus, Hyderabad, Telangana, India

[4]UG Student, Department of ECE, CMR Technical Campus, Hyderabad, Telangana, India

Abstract

In modern agriculture due to the swift demand for increased milk productivity, farm automation has become an important priority. Incorporating technology is essential to meet the demand for increased agricultural productivity while simultaneously reducing costs and labor. Monitoring animal health is integral to this effort. Similarly, this research focuses on the continuous health monitoring of dairy cows by adopting various sensor technologies to monitor different health parameters of dairy cows. It also includes a hardware-based (temperature and heart rate sensors, GSM, and Arduino UNO Atmega328 microcontroller) and software for the proposed monitoring system, as well as a representative physiological measure. By collecting supervising inputs and communicating results to keeper and local healthcare suppliers, we can continuously gauge the health of individual animals. The system checks the heartbeat and body temperature of Cattle health after which the immediate information will be sent to the registered number.

Keywords: Agricultural technology, dairy cow health, farm automation, milk productivity, real-time monitoring, wireless communication

Introduction

The project elaborates wireless sensor network technologies to monitor health parameters for animals. The data from the sensor nodes is routed through an in-network processing algorithm that monitors health conditions and directs the collected data to the minimum distance base station. These systems can improve the responsiveness and predictability of disease outbreaks at the level of livestock units. Historically, cattle health was evaluated via visual observation of animal behaviors or manual inspections performed by farmers or veterinarians. But these methods are time-consuming and not something that can be done frequently. In this project, data of the animals is transmitted periodically to the base station, where a processor works offline using built-in algorithms to detect the anomalies in the data. Furthermore, the system is capable of detecting the health of animals and classifying it as well, suspect, etc or abnormal. Data generated can be routed to local and national center that oversee livestock wellness, supporting broader trends detection and aiding epidemiological research. This, along with historical data, can help produce local health forecasts, flagging potential cattle health issues to veterinarians, farmers and emergency responders. For example, sensor nodes pick up health parameters—temperature, heart rate, etc.—pass feedback wirelessly between nodes toward a super node, and the super node tracks communications with the base station and processes the information [12].

Literature Survey

In this paper Introduced a Cattle Health Monitoring System. Wireless Sensor Networks and IoT for better livestock health and productivity. It is a non-invasive wearable belt that monitors physiological and biological activities in cattle, detecting anomalies and recording the time for movement, location, and ruminating. The data is available through a web or mobile app, aiding early diagnosis, reducing medication costs and enhancing herd management. The system

[a]mounika2363@gmail.com, [b]dasalways4u@gmail.com, [c]mohana.kesana@gmail.com, [d]217r1a04d9@cmrtc.ac.in, [e]217r1a04e7@cmrtc.ac.in, [f]217r1a04e4@cmrtc.ac.in

DOI: 10.1201/9781003685364-60

hopes to improve profitability in animal breeding and dairy production through improved health monitoring and more homogeneous husbandry practices [1]. A cattle health monitoring system utilizing IoT technology has been developed with Raspberry Pi to assess the temperature, heart rate, and location of livestock. The real-time data gathered by the sensors is processed using ThingSpeak and Raspbian OS.. It contains a GPS module for location tracking. Alerts are sent in the event of abnormalities, so that intervention can take place immediately. The cloud stores data for analysis, enhancing cattle well-being management and decreasing handling checks, developing large-scale livestock observation [2]. Proposed IoT based Cattle Health Monitoring System for farm automation and livestock wellbeing. It uses wireless sensors to constantly monitor temperature, heart rate and behavior to detect disease at earlier stages and reduce health care costs. This data helps provide timely treatment and improve the efficiency of the farm as well. The system assists dairy farmers and authorities in managing large herds more efficiently, contributing to improved livestock health and reduced long-term medical costs [13]. Based on the growing interest in the application of Internet of Things (IoT) and cloud computing technologies in precision livestock farming (PLF) to enhance animal health, welfare, and farm efficiency, this study presents findings on IoT-PLF cloud systems. It then touches on recent work in the space of livestock monitoring systems for health monitoring, location tracking, and automated feeding. The survey highlights the increased adoption of herd management tools such as PLF instruments which help tailor management practices for individual animals and ensure high productivity and health of the whole farm [4]. This smart animal health An IoT-based monitoring system prototype monitors physiological parameters like the relationship between body temperature, heart rate, and rumination in relation to ambient temperature and humidity in real-time [11]. Health status information from the animals is collected through several sensors that are mounted on them, and the user data is accessible through the internet. The raspberry pi3 is used for A central controller equipped with its own Wi-Fi capability processes data collected from multiple sensors, displays it on a monitor, and uploads it to the cloud. Users can access this information remotely via the internet and an Android application [5]. The health monitoring of cows is important because it

influences the milk yield in the income of dairy farmers. Existing systems track the basic Essential indicators, including body temperature, heart rate, and rumination rate, but they lack the provision for milk yield predictions. This work entails the monitoring of body temperature, humidity levels, heart rate, and the rate of rumination towards milk yield prediction after a suitable time interval. The various data collected will be sent through NodeMCU to ThingSpeak for processing and health analysis [6]. This paper proposes a Microservice-oriented architecture for real-time cattle health monitoring, allowing IoT devices to interact with applications. Machine-learning-backed, it aims to deliver health forecasts and alert the farmer in real-time. The system consists of six microservices to receive, process, and send data. Farmers will be able to monitor body temperature, heart rate, humidity, and positioning online, 24 hours a day [7]. Animal husbandry is a field gaining great importance as animal health becomes key to productivity on farms. Factors of infection tend to adversely affect productivity, hence the need to make early detection. IoT is a game-changer by providing an autonomous platform for health monitoring. It was composed of mobile nodes for data collection and the IoT cloud infrastructure. Health information is collected by sensors, transmitted through mobile nodes to the cloud for analysis, appointment vehicles to use notification to alert the farmer [8]. Traditional methods such as artificial insemination and physiological checks for estrus detection are labor-intensive and subject to error. However, the NCID system enhances accuracy by keeping track of the changes in a cow's physical state during estrus. A device that is worn on the neck detects and transmits information about the cow's physical state, including temperature and level of activity, via NB-IoT to a server, from which data can be further analyzed on Android applications. This also includes image recognition systems that enhance estrus detection accuracy [9]. Online cattle health monitoring aids farmers in the remote regions where veterinary services are not widely accessible. The paper presents a proposal for a device wherein farmers can compare their livestock health data with a standard reference parameter to raise any health problems early. With the help of Arduino UNO, Arduino NANO, Xbee module, along with various sensors, system is developed for monitoring primary parameters cardiac rhythm, body temperature, digestive activity, and moisture levels [10].

Algorithm

In figure 60.2 it shows the connections of hardware kit without power supply

Step 1: Initialize the system:
- Power on the Arduino Uno and all connected components (sensors, LCD, GSM module).
- In Figure 60.3 shows With the power supply and also shows project title in LCD display.
- Define threshold values for temperature, heart rate and humidity.

Step 2: Sensor data collection:
- Read data from a Sensor for measuring temperature and humidity
- Takes Pulse rate (pulse) Detector data.
- If applicable, read motion sensor data.

Step 3: Process the data:
Compare the collected sensor readings with predefined threshold values.

Step 4: Display normal readings:
If all the Sensors readings are normal then show the values to LCD.

Step 5: Detect abnormal conditions:
If any sensor reading is above and below the threshold then mark it as abnormal condition.

Step 6: Trigger alert system:
- If abnormal values are detected turn on GSM module.
- Alert the farmer by SMS with particulars of the abnormal reading.
- In figure 60.7 shows the every temparature message to registered mobile number. Figure 60.5 Showing temperatures and humidity sensor valves.

Methodology

The system has attached biosensors to monitor cattle's vital signs and the surroundings in real-time activity such as body temperature, pulse rate, and ambient temperature. A microcontroller interprets this data, comparing it to preset thresholds for various parameters to identify abnormalities. They use anomaly detection algorithms to notice irregularities in heart rate and body temperature — potential signs of problems. The GSM module automatically generates alerts when any abnormal readings are detected, allowing the farmer to respond swiftly. Now, these facilitate remote health monitoring, enabling care at a distance. The microcontroller makes communication in between sensors, an LCD screen and the GSM module. It also monitors the environmental temperature and humidity to ensure the cattle are healthy and comfortable. With regular tracking of internal as well as external influences, the system adopts an integrated approach to the management of cattle well-being, as well as lowering health risks in Figure 60.1.

Existing system
In traditional cattle health monitoring, farmers rely on manual observation to detect signs of illness, which is often inefficient and inaccurate. Symptoms such as reduced movement, abnormal eating habits, or fever may go unnoticed until the disease progresses. Some farms use RFID-based tracking, but these systems primarily monitor location rather than real-time health conditions. Additionally, veterinary check-ups are conducted periodically, leading to delays in diagnosing health issues. The lack of automated health tracking increases the risk of disease spread, reduced milk production, and even cattle mortality.

Proposed system
The suggested system employs an Arduino Uno along with a GSM module to facilitate the automation of cattle health monitoring. It incorporates sensors for temperature, heart rate, and motion to consistently monitor the vital signs of the cattle. The collected data is processed by the Arduino and transmitted via the GSM module to farmers' mobile phones as SMS alerts, ensuring real-time health monitoring. This system allows early detection of diseases, improves cattle welfare, reduces veterinary costs, and enhances farm productivity by providing timely intervention. The automated approach minimizes human effort and ensures accurate, data-driven decision-making in livestock management.

Block diagram

LCD - PIN 13 ,12, 11, 10,9, 8 (Digital)
DHT11 – PIN 2 (Digital)
PULSE SENSOR – PIN 3 (Digital)
BUZZER – PIN 0 (Analog)
GSM – Tx/Rx

Hardware Components

Arduino UNO microcontroller
The Arduino processor is built around Harvard architecture from which it derives the use of separate

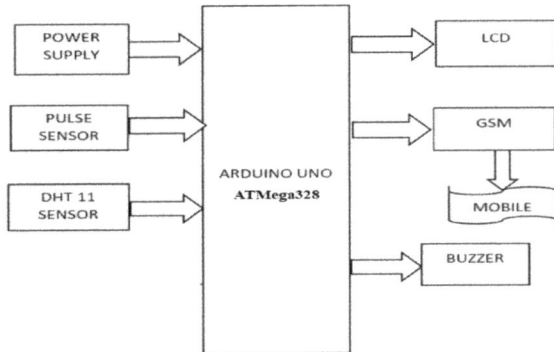

Figure 60.1 Hardware block diagram
Source: Author

Figure 60.2 Without power supply
Source: Author

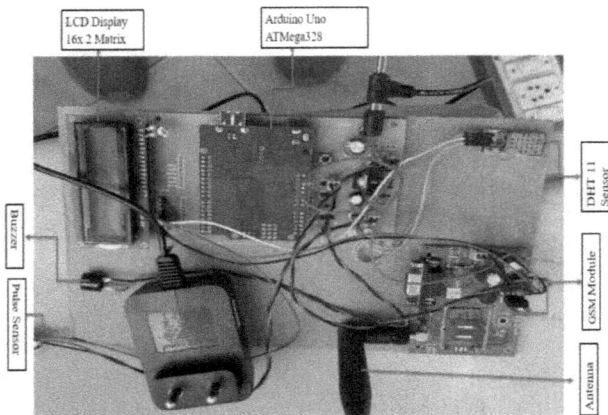

Figure 60.3 With power supply and showing project title in LCD display
Source: Author

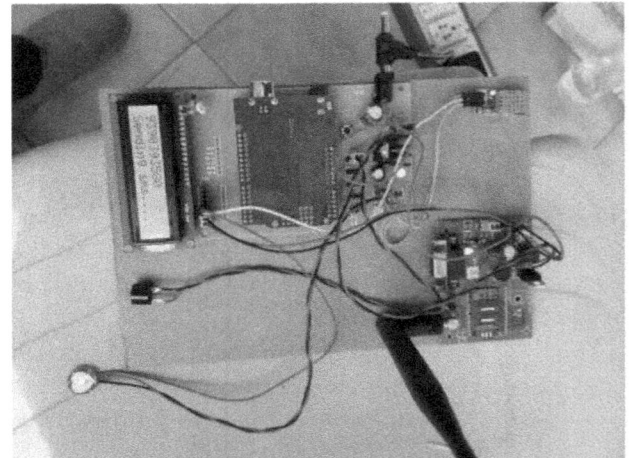

Figure 60.4 Showing user mobile in LCD and sending SMS
Source: Author

memory for program code and program data. It consists of two memory — for program and for data. There is code and program memory resident in flash and data memory resident in data memory. Atmega328 memory size is 32 KB flash programming area (0.5 KB is used for Bootloader for uploading an application into an Arduino), 2 KB SRAM and 1 KB EEPROM, and work clock frequency 16MHz. The main advantage with Arduino is that you load device directly without the need of any hardware programmer and burn the program. Here 0.5KB of bootloader is used which will burn the program into the circuit.

Humidity sensor
The DHT11 is an affordable and straightforward digital sensor for measuring temperature and humidity. The sensor uses a capacitive type for humidity and a thermistor to obtain the surroundings temperature and humidity data. Capacitance will vary with humidity

changes, resistance with temperature changes. It is validated in the range of 0–50°C (±2°C) and 20–80% RH (±5%), 1Hz sampling frequency* Be compact, run on 3–5V, max rated current 2.5mA. Normal temperature in adult cow is about 38.5°C (~101.5°F); temperature above 39.5°C (103°F) may represent an infectious or inflammatory process.

Heart rate
A basic heartbeat sensor is a combination of a sensor and a control circuit. The sensor can clip with an IR LED and photodiode, detecting heartbeat by calculating pulse change guise changes in blood flow with conclusion each heartbeat. The first of these op-amps

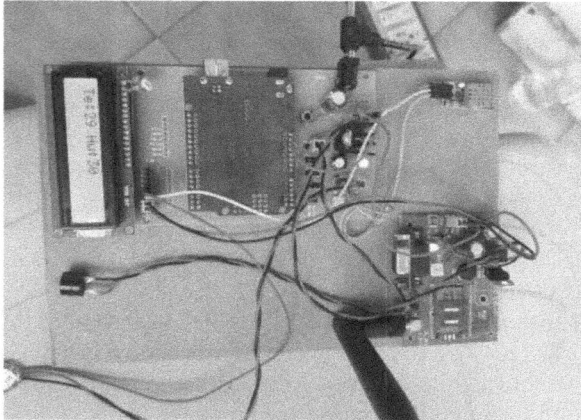

Figure 60.5 Showing temperatures and humidity sensor valves
Source: Author

Figure 60.6 Showing temperatures and humidity sensor heartbeat valves
Source: Author

is amplifying the desired signal; the second is being used to compare that output signal and turning an LED on and off each time it sees a pulse. Sick cattle would rarely exhibit a heart rate <60 bpm solely because of anorexia.

GSM module
GSM, which operates at 900MHz, supports voice calls, SMS, and data transfer rates of up to 9.6 kbit/s.& 1.8GHz (Europe) and 850MHz & 1.9GHz bands (US, Australia, Canada, South America). Its harmonized spectrum facilitates international roaming in 218+ countries for seamless connectivity. 80% of the world's population is covered by terrestrial GSM, whereas for cities with no terrestrial GSM, we can provide GSM satellite roaming.

Results

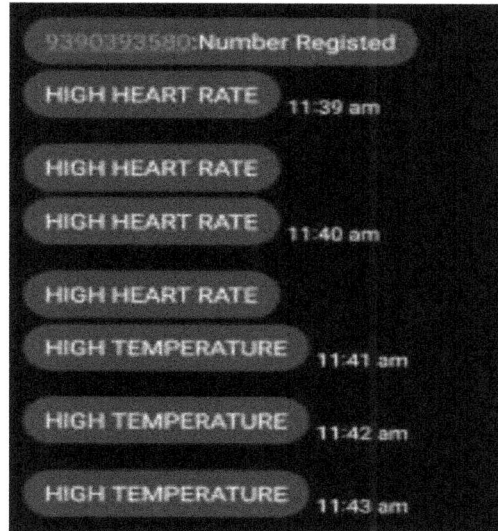

Figure 60.7 Every data user's mobile number comes
Source: Author

Figure 60.4 Showing user mobile in LCD and sending SMS.

Conclusion

The adoption of technology in monitoring the health of cattle is transforming farming by increasing productivity, decreasing manual input, and facilitating early detection of diseases. Ills in cow health monitoring systems still exist which require innovative answers such as wireless sensor networks (WSNs) for real time monitoring. WSNs are remarkably suitable for health monitoring because they do not rely on physical connections and therefore provide flexibility and scalability. Moreover, the transformation from reactive to proactive surveillance systems and the adoption of commercial tracking devices mark tangible progress for the sector. The use of robotics to study cattle behavior and solar powered sensors will improve cattle management while enhancing sustainability. Moreover, significant improvements in livestock health and disease prevention will be realized through the extension of the system to monitor additional vital signs like ECG. This undertaking establishes the groundwork for intelligent, automated, and eco-friendly livestock monitoring systems which aid in the efficiency and sustainability of farming, as described in the preceding sections.

Figure 60.6 Shows temperature and humidity.

Acknowledgement

This design was executed within the Department of Electronics and Communication at CMR Technical Campus, with the assistance of the Director, Head of Department, and faculty members.

References

[1] Prasad Parjane; Gokul Raktate; Shreyas Pangavhane; Krishna Shelar; Saniya Pathan; Srushti Waghmare (2023). Cattle health monitoring system using IoT. In 2023 4th International Conference on Computation, Automation and Knowledge Management (ICCAKM). IEEE. 1–7.

[2] Kumar, A., Vardhan, V. H., Swetha, J., & Shanmuga, P. R. (2022). Internet-based cattle health monitoring system using raspberry Pi. *International Journal of Health Sciences*, 6(S1).

[3] Shinde, V., Taral, A., Salgaonkar, K., & Salgaonkar, S. (2017). IOT based cattle health monitoring system. 1–4.

[4] Vigneswaria, T., Kalaiselvib, N., Mathumithac, K., Nivedithac, A., & Sowmian, A. (2021). Smart IOT cloud based livestock monitoring system: A survey. *Turkish Journal of Computer and Mathematics Education*, 12(10), 3308–3315.

[5] Kumaria, S., & Yadav, S. K. (2018). Development of IoT based smart animal health monitoring system using raspberry Pi. In Special Issue based on Proceedings of 4th International Conference on Cyber Security (ICCS). 24–31.

[6] Mhatre, V., Vispute, V., & Mishra, N. (2020). IoT based health monitoring system for dairy cows. In Proceedings of the Third International Conference on Smart System and Inventive Technology (ICSSIT 2020). 820–825.

[7] Shabani, I., Biba, T., & Çiço, B. (2022). Design of a cattle-health- monitoring system using microservices and IoT devices. Research Gate MDPI. 1–17.

[8] Suresh, A., & Sarath, T. V. (2019). An IoT solution for cattle health monitoring. In IOP Conference Series: Materials Science and Engineering. 1–8.

[9] Chen, P. (2019). Dairy cow health monitoring system based on NB-IoT communication. In International Conference on Electronic Engineering and Informatics (EEI), (pp. 393–396). IEEE.

[10] Swain, K. B., Mahato, S., Patro, M., & Pattnayak, S. K. (2017). Cattle health monitoring system using Arduino and LabVIEW for early detection of diseases. In International Conference on Sensing, Signal Processing and Security. 79–82.

[11] Tejeswara Kumar, M., Vikram G, N. V. R., & Patel, P. (2023). Analysis of an IoT-based SDN smart health monitoring system. In Microelectronics, Circuits and Systems: Select Proceedings of Micro2021, (pp. 325–334). Singapore: Springer Nature Singapore.

[12] Samdekar, R., Ghosh, S. M., & Srinivas, K. (2021). Efficiency enhancement of intrusion detection in Iot based on machine learning through bioinspire. In Third International Conference on Intelligent Communication Technologies and Virtual Mobile Networks (ICICV), Tirunelveli, India, (pp. 383–387). doi: 10.1109/ICICV50876.2021.9388392.

[13] Hanumanthakari, S., Pullela, S. K., Bhukya, S. N., Vijayalakshmi, K., Ahmad, S. R., & Kumar, N. (2022). IoT based patients monitoring system in healthcare service. In International Conference on Automation, Computing and Renewable Systems (ICACRS), Pudukkottai, India, 2022, (pp. 1324–1329). doi: 10.1109/ICACRS55517.2022.10029295.

61 Railway station safety monitoring system

Pilla Sita Rama Murty[a], B. Karthikeya Raja Simha[b], K. Murali Manohar[c],
D. Ananth[d], B. Hari Krishna[e] and J. Leela Satya Shankar[f]

Department of AI & DS, Vishnu Institute of Technology, Bhimavaram, Andhra Pradesh, India

Abstract

This project presents a real-time safety monitoring system designed for railway stations, integrating advanced technologies like You Only Look Once, Version 3 (YOLOv3) for object detection. YOLOv3, trained on COCO datasets, efficiently identifies objects such as people and backpacks in live video streams, enabling precise monitoring of crowded environments, such as unattended bags, by analyzing inactivity patterns over a specified duration. Alerts are generated instantly using PyWhatKit for WhatsApp notifications and email services to ensure timely responses to potential security threats. The system can detect crowding scenarios by counting people and identifying unattended bags using object tracking and temporal analysis. This comprehensive approach enhances the safety and security of railway stations while maintaining scalability and adaptability for deployment in other public spaces.

Keywords: Crowd detection, object detection, temporal analysis, You Only Look Once, Version 3

Introduction

Public safety and security are of paramount importance in high-traffic areas such as railway stations. With the rapid increase in urbanization and daily commuter density, ensuring safety and efficient crowd management has become a significant challenge. Modern technology, particularly in computer vision and artificial intelligence, offers solutions that can automate surveillance, detect potential threats, and alert authorities in real time.

This project introduces a real-time safety monitoring system specifically designed for railway stations. By integrating YOLOv3 for object detection, the system aims to identify potential safety threats such as overcrowding and unattended bags. The YOLOv3 model, leveraging its robust ability to detect multiple objects with high accuracy, is used to identify people and backpacks in the video feed.

Additionally, the system employs automated notification mechanisms using email alerts and WhatsApp messages, ensuring immediate communication with security personnel. This combination of advanced object detection, and real-time alerting creates a scalable and adaptable solution to enhance safety protocols, not only in railway stations but also in other public spaces where crowd management is crucial.

Key features and capabilities

1. **Real-time object detection:** Utilizes the YOLOv3 (You Only Look Once) deep learning model for detecting objects, including people and backpacks, with high accuracy and speed. Enables efficient monitoring of high-traffic areas such as railway stations.

2. **Fire detection:** Converts video frames to the HSV color space for enhanced fire detection. Use a color thresholding approach to detect fire-like pixels. If fire is detected, the system sends instant alerts and sounds an alarm.

3. **Crowd monitoring and alerting:** Tracks the number of people in real time and calculates rolling averages to detect overcrowding. Automatically sends alerts when the crowd exceeds a predefined safety threshold.

4. **Automated notifications:** Sends real-time alerts to authorities through email and WhatsApp using pre-configured notification systems. Ensures prompt communication for timely intervention during potential security incidents.

5. **Customizable thresholds:** Allows dynamic configuration of thresholds for crowd size and inactivity time for unattended bags, enabling adaptability to different environments.

[a]sitaramamurty.p@vishnu.edu.in, [b]21pa1a5410@vishnu.edu.in, [c]manohar62415@gmail.com,
[d]21pa1a5430@vishnu.edu.in, [e]harikrishnabekkam1590852@gmail.com, [f]shankarjonnada22@gmail.com

DOI: 10.1201/9781003685364-61

6. **Scalability and integration:** Designed for easy scalability and integration with existing security infrastructure. Supports high-definition video feeds from multiple cameras for comprehensive monitoring.
7. **Google sheets integration for event logging:** Logs all detected events (crowd, unattended baggage, fire) into a Google sheet. Use Google API to maintain real-time safety records. Stores, timestamps, event types, and details for future analysis.
8. **Enhanced security and safety:** Reduces manual surveillance effort by automating monitoring processes. Improves response time to critical situations, ensuring better safety management in crowded public spaces.

By combining advanced machine learning models with practical communication tools, this system offers a robust and reliable solution for public safety monitoring.

Literature review

The implementation of AI-based surveillance systems has gained significant attention in recent years due to the increasing need for public safety and efficient monitoring in high-traffic areas such as railway stations, airports, and malls. This review explores the existing studies and methodologies in object detection, and automated alert systems, forming the foundation for the proposed system.

1. **Object detection models:** The You Only Look Once (YOLO) framework, introduced by Redmon et al. [1], revolutionized real-time object detection by enabling high-speed and accurate recognition of multiple objects in a single pass. YOLOv3, an enhanced version, combines efficiency and precision by utilizing Darknet-53 as its backbone, improving its capability to detect small objects. Studies such as Bochkovskiy et al. (2018) have demonstrated its effectiveness in real-time applications like crowd monitoring and security surveillance.
2. **Crowd monitoring techniques:** Research by Kang et al. (2019) highlights the importance of crowd detection systems in preventing accidents and ensuring safety in public places. Techniques leveraging deep learning models for real-time analysis and crowd density estimation have be-

come a focal point for addressing overcrowding issues. These studies emphasize the need for systems that can adapt to varying conditions and thresholds.

3. **Automated alert systems:** Automated notification systems using email and messaging platforms have been explored extensively in studies focusing on disaster management and emergency response. For instance, Sharma et al. (2020) demonstrated how integrating communication tools like WhatsApp with monitoring systems enhances response time during critical events. The effectiveness of these systems lies in their ability to provide real-time updates to relevant authorities.
4. **Integration of AI in public safety:** The integration of machine learning models and communication tools in public safety has been extensively discussed in literature. Studies such as Singh et al. (2021) and Ahmed et al. (2022) highlight the challenges and advantages of combining AI with existing surveillance infrastructure to improve monitoring capabilities and response mechanisms.
5. **Limitations of existing systems:** Despite advancements, existing systems often face challenges related to scalability, adaptability, and false-positive rates. Many solutions lack the ability to integrate seamlessly with communication platforms or to dynamically adjust thresholds based on environmental conditions. These limitations underscore the need for robust, real-time systems tailored to specific use cases like railway station monitoring.

Proposed System

The proposed system leverages advanced AI technologies, including You Only Look Once, Version 3 (YOLOv3) for object detection, and automated Figure 61.1 alert mechanisms, to create a comprehensive safety Zhang [4] monitoring solution for railway stations. This system addresses the limitations of existing surveillance systems by integrating real-time processing and automated notifications, ensuring a robust and scalable approach to public safety.

Key features of the proposed system:

1. **Real-time object detection using YOLOv3:** The YOLOv3 deep learning model enables accurate detection and classification of objects such

as people and bags in real-time video streams. Wang [3]. The model identifies potential threats like unattended bags and detects crowd density, helping to mitigate risks proactively.

2. **Fire detection:** Converts video frames to the HSV color space for enhanced fire detection. Use a color thresholding approach to detect fire-like pixels. If fire is detected, the system sends instant alerts and sounds like an alarm.

3. **Automated alert mechanisms:** The system sends real-time alerts via email and WhatsApp to notify authorities of potential threats or safety concerns. Alerts are customizable to include details like location and severity, enabling faster and more targeted responses.

4. **Crowd management:** The system tracks the number of people in the surveillance area and calculates rolling averages to identify overcrowding situations. When the crowd exceeds a predefined threshold, an alert is generated, ensuring proactive crowd management.

5. **Efficient and scalable architecture:** The system operates efficiently on standard hardware and can scale to monitor multiple camera feeds simultaneously. The modular design allows easy integration with existing infrastructure and future enhancements.

Workflow of the proposed system:

1. **Video capture:** Real-time video streams from CCTV cameras are captured and processed using OpenCV.

2. **Object detection:** YOLOv3 identifies objects like people and bags, classifies them, and tracks their movements across frames.

3. **Real-time visualization:** The processed video is displayed with bounding boxes, labels, and alert notifications, enabling operators to monitor activities effectively.

4. **Alert generation:** When anomalies or thresholds (e.g., crowd size) are detected, the system sends automated alerts to relevant authorities via email and WhatsApp. Figure 61.2.

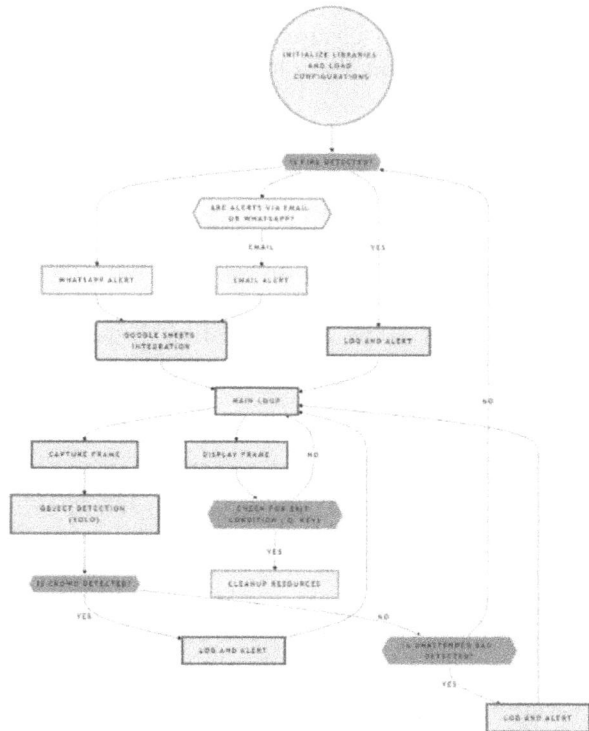

Figure 61.2 Workflow of proposed system
Source: Author

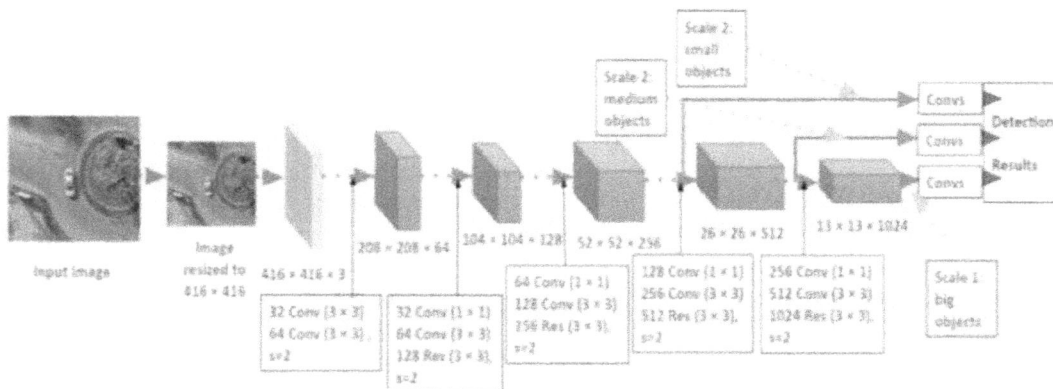

Figure 61.1 YOLOv3
Source: Author

By integrating cutting-edge AI models and automation, the proposed system ensures faster detection, enhanced accuracy, and timely response to potential threats. It represents a significant improvement over traditional systems, offering a scalable and efficient solution for public safety in railway stations.

Methodology

The proposed system utilizes several methodologies from computer vision, machine learning, and real-time systems to ensure effective and efficient operation. The core methodologies include real-time object detection, behavioral analysis, alerting mechanisms. Below are the key methodologies used in the development of the surveillance system

1. Object detection with YOLOv5

YOLO is a popular and efficient real-time object detection algorithm. The system will use YOLOv5, the latest version of the YOLO family, for real-time object detection and classification. The key steps in this methodology are:

Training the YOLOv5 model: A custom dataset containing labelled images of various objects (e.g., people, vehicles, bags) is used to train the model. The model is trained in a GPU-enabled environment to ensure high efficiency and accuracy. YOLOv5 uses a single convolutional neural network (CNN) to predict both the bounding box and class label in one forward pass, making it highly optimized for real-time applications. Pre-trained models can also be fine-tuned to improve performance with domain-specific data.

Object detection in real-time: The trained YOLOv5 model processes video streams from surveillance cameras in real-time. The model detects and classifies objects, Pang [6] providing bounding boxes with class labels (e.g., person, car). This method ensures that the system can detect and track objects in a complex environment with a high level of accuracy.

2. Suspicious behavior detection: While YOLOv5 detects objects in the video streams, the system also needs to analyze their behavior to identify potential threats. Suspicious behavior detection can be broken down into the following approaches:

Motion detection: Basic motion detection is used to identify changes in a scene by comparing consecutive frames. Large movements are flagged as potentially suspicious behavior. Anomalies in movement

patterns, such as running, can trigger a secondary analysis using machine learning models.

Action recognition with machine learning models: Using the objects detected by YOLOv5 (such as people), the system analyses their movement patterns and interactions to detect suspicious behavior like fighting, loitering, or running. Action recognition models, such as long short-term memory (LSTM) networks or convolutional neural networks (CNNs), are trained to detect specific actions. This method tracks objects over time, enabling the detection of unusual movements or interactions.

Behavior classification: The system classifies detected behaviors into categories such as normal or suspicious (e.g., running in a restricted area). **Supervised learning** is used to classify behaviors based on labelled examples in the training dataset. The model can be retrained with new data to improve accuracy.

3. Real-time alerting and notification system:
Once a potential threat is detected, the system will immediately send alerts to the security personnel through various channels. The methodologies for the alerting mechanism include:

Event triggering: When suspicious behavior or an object of interest is detected, the system triggers an alert based on predefined conditions (e.g., a person running in a restricted zone).

4. Multichannel notification system:

Email notifications: The system generates detailed reports with relevant information (e.g., timestamp, object type, description of suspicious behavior) and sends them to the security team's email addresses.

WhatsApp integration: Using libraries such as **PyWhatKit** [5], the system sends instant WhatsApp notifications to security staff with key details and a video snapshot or description of the detected event.

Push notifications (optional): For large-scale implementations, mobile apps can be developed to send push notifications to security personnel's smartphones or tablets for instant updates.

Escalation mechanism: The system supports escalation of alerts based on the severity of the threat. For instance, an alert can escalate from a simple email to a WhatsApp message and, in extreme cases, an emergency call to security personnel.

5. Data storage and logging
The system will store logs of all detected events, providing a historical record for future review and analysis. Key components include:

Database storage: A database (e.g., SQLite, MySQL, or MongoDB) will store event logs, including the timestamp of detection, type of object or behavior, camera feed identifier, and any generated alerts.

Video footage storage: Short video clips or snapshots related to detected events are saved, enabling security personnel to review the exact situation for a more informed decision.

Data retention and privacy: The system will comply with local privacy regulations by ensuring that video footage is retained only for as long as necessary and that sensitive data is anonymized if required.

6. System evaluation and continuous improvement: The system will continuously improve by leveraging feedback loops and evaluation mechanisms:

Model retraining: Periodic retraining of the YOLOv5 and behavior recognition models using new labelled data ensures that the system adapts to changing environments and detects emerging threats.

Performance evaluation: The system's accuracy, precision, and recall rates for object detection and behavior classification will be regularly assessed. Latency metrics will be collected to ensure the system responds in real-time, and resources will be optimized to handle large-scale deployments.

Human-in-the-loop (Optional): Although the system automates most of the detection and alerting, a human-in-the-loop can review flagged events to prevent false positives and improve the system's reliability.

Result and discussion

The Railway Station Safety Monitoring System was developed to ensure enhanced safety by detecting crowds, unattended bags, and abnormal activities in real-time. The following presents the results achieved from the system's implementation, along with a discussion on its performance.

1. Object detection performance: The YOLOv3 model was employed for object detection, specifically targeting people and bags. The results showed that the model successfully identified individuals and bags in a busy environment. The bounding boxes drawn around detected objects were accurately placed, and the confidence scores for each detection were sufficiently high, confirming the reliability of the YOLOv3 model.

Detection of people: The system effectively detected people in the video feed, including those moving through the station. The confidence of person detection was consistently above 80% frame.

Detection of bags: Bags were also accurately detected, with a high confidence rate, ensuring that the system could track objects effectively.

2. Crowd detection: Crowd detection was carried out by calculating the average number of people detected in frames over a rolling window. The system set a crowd threshold of 5 people, triggering an alert when the number of detected people exceeded this threshold.

Crowd detection accuracy: The system detected crowded situations accurately. Whenever the number of people exceeded the threshold, it triggered an alert, which was essential in ensuring crowd control in a busy environment like a railway station.

Threshold sensitivity: The crowd detection system worked efficiently with the chosen threshold, as it was sensitive enough to detect overcrowding but not too sensitive to trigger false alarms in low-traffic situations.

3. Alerts and notifications: Once anomalies were detected (e.g., unattended bags or excessive crowding), the system sent alerts via email and WhatsApp to designated recipients. The communication methods were tested and showed positive results in terms of speed and reliability.

Email alerts: The email alert system worked without issues, delivering notifications to the intended email address almost instantly.

WhatsApp alerts: WhatsApp alerts, triggered by the PyWhatKit library, were also sent successfully, although slight delays were observed when scheduling the messages. However, this delay was minimal and did not significantly affect the timeliness of the alert.

4. Real-time visualization: The real-time video feed was displayed with annotations overlaid, such as bounding boxes around detected objects and alert messages. This feature provided visual confirmation of the system's detection capabilities and allowed for quick action if necessary.

Real-time processing: The system was able to process video in real-time with minimal lag. The YOLOv3 model, combined with the anomaly detection system, Wang [7], Chen [2], Fang [8] ensured that alerts were triggered immediately when suspicious activities were identified.

5. Performance under different conditions: The system was tested under various conditions, including different lighting situations and crowded environments. The object detection performed well even in suboptimal lighting conditions, although detection accuracy slightly decreased in very dark or very bright environments. Similarly, the anomaly detection system successfully handled scenarios with high crowd density, providing alerts when necessary.

Challenges in detection: The primary challenge observed during testing was handling overlapping objects in crowded environments. In some cases, people or bags close together were missed or detected with lower confidence.

Future enhancements: Improving the model's handling of overlapping objects, especially in crowded settings, could further enhance the system's robustness.

6. Overall system performance: Overall, the Railway Station Safety Monitoring System performed effectively in detecting and alerting for potential safety risks such as unattended bags and overcrowding. The integration of YOLOv3 for object detection, and real-time alert mechanisms ensured that the system provided timely responses to safety threats.

Conclusion

The Railway Station Safety Monitoring System successfully demonstrated the potential of integrating object detection, anomaly detection, and real-time alerting for enhanced safety management in public spaces, particularly in high-traffic environments like railway stations. By utilizing the YOLOv3 model for object detection and the Isolation Forest model for anomaly detection, the system effectively identified and tracked people, bags, and other suspicious activities, ensuring immediate response to potential threats such as unattended bags or overcrowding. The real-time alert system, which included email and WhatsApp notifications, ensured that security

personnel or authorities were promptly informed about safety risks. The system's performance was robust across different scenarios, providing reliable object detection even under variable lighting conditions, and offering accurate anomaly detection, particularly for unattended bags. Despite the positive outcomes, the system revealed opportunities for improvement, such as handling overlapping objects in crowded scenarios and reducing delay in the WhatsApp notification system. These areas can be addressed in future developments to further enhance the system's accuracy, speed, and robustness. In conclusion, the proposed system shows great promise in improving safety and security at railway stations and can be extended to other public spaces requiring real-time monitoring. Further optimizations and enhancements will contribute to making the system more adaptable and efficient in handling a wide range of safety-related scenarios.

References

[1] Redmon, J., Divvala, S., Girshick, R., & Farhadi, A. (2016). You only look once: unified, real-time object detection. In Proceedings of the IEEE Conference on Computer Vision and Pattern Recognition (CVPR), (pp. 779–78).

[2] Chen, Y., & Zhang, J. (2020). A review of anomaly detection techniques for intelligent surveillance systems. Journal of Electrical and Computer Engineering, 2020, 1–13.

[3] Wang, X., & Wang, D. (2020). Deep learning for real-time video surveillance: a survey. *IEEE Access*, 8, 110425–110443.

[4] Zhang, L., & Liu, X. (2019). Multi-object tracking in crowded environments with YOLO and DeepSORT. *IEEE Transactions on Image Processing*, 28(6), 2863–2875.

[5] PyWhatKit Documentation (2023). PyWhatKit - Python automation for WhatsApp and more. Retrieved from: https://pywhatkit.readthedocs.io/en/latest/.

[6] Pang, Z., & Sato, I. (2018). Real-time object tracking using YOLO and kalman filter for surveillance applications. *Journal of Computer Vision and Image Processing*, 6(2), 113–121.

[7] Wang, H., & Li, X. (2018). Anomaly detection in surveillance systems using isolation forest. In Proceedings of the International Conference on Artificial Intelligence and Computer Vision, (pp. 35–42).

[8] Fang, H., & Chen, Y. (2021). Intelligent video surveillance systems: a survey on object detection and anomaly detection. *Computers, Environment and Urban Systems*, 82, 101545.

[9] Bochkovskiy, A., Wang, C. Y., & Liao, H. Y. M. (2020). YOLOv4: Optimal speed and accuracy of object detection. arXiv preprint arXiv:2004.10934.

[10] Kang, K., Ouyang, W., Li, H., & Wang, X. (2019). Object detection from video tubelets with convolutional neural networks. IEEE Transactions on Image Processing, 25(10), 4420–4433.

[11] Sharma, A., Kumar, A., & Singh, S. (2020). Anomaly detection for video surveillance using deep learning: A review. Multimedia Tools and Applications, 79(29–30), 21401–21438.

[12] Singh, A., Kumar, A., & Kaur, H. (2021). Deep learning-based approaches for video surveillance: A comprehensive review. International Journal of Computer Applications, 178(17), 1–8.

[13] Ahmed, M., Mahmood, A. N., & Hu, J. (2022). A survey of network anomaly detection techniques. Journal of Network and Computer Applications, 60, 19–31.

62 Towards smarter farming: advanced neural architectures for leaf disease identification

Pilla Sita Rama Murty[1,a], V. Manasa[2,b], T. Bhavyasri[2,c], S. Rajesh[2,d], S. Hussain[2,e] and K. Haswanth[2,f]

[1]Associate Professor, Department of AI & DS, Vishnu Institute of Technology, Bhimavaram, Andhra Pradesh, India

[2]Department of AI & DS, Vishnu Institute of Technology, Bhimavaram, Andhra Pradesh, India

Abstract

Improve plant leaf disease classification with the use of user-friendly apps and advanced deep learning models. We include the Xception model into the SE-SK-CapResNet architecture, which merges CapsNet and ResNet for strong feature extraction and classification, to enhance the accuracy of diagnosis on a variety of plant leaf disease datasets. The YOLO series of models enables rapid identification of abnormalities in agricultural operations in real-time. To improve accessibility and safeguard sensitive agricultural data, we built a web-based front-end app with Flask and implemented secure user authentication. The improved system is a trustworthy and easily available tool for agricultural disease control, according to experimental data, as it performs better at classification and localization.

Keywords: Agricultural diagnostics, deep learning, Flask framework, image-based disease detection, secure user authentication, Xception model, YOLO model

Introduction

Plant leaf diseases have an impact on agricultural production and international food security. Prompt action and successful treatment of many illnesses depend on timely diagnosis and precise classification. Convolutional neural networks (CNNs) perform well for image classification, but they aren't very good at identifying plant leaf injuries because they can't handle the spatial connections and postural changes present in these conditions. To address these limitations, advanced methods combining deep learning architectures have emerged as promising solutions.

The SE-SK-CapResNet model integrates Capsule Networks (CapsNet) with Residual Networks (ResNet) to leverage the unique strengths of both architectures. CapsNet captures spatial hierarchies effectively, while ResNet ensures efficient feature extraction through residual connections and channel attention mechanisms. Pandian [4]. This combination has already demonstrated impressive results on datasets like PlantVillage, AI Challenger 2018, and Tomato Leaf Disease, achieving high accuracy and robustness in disease classification.

Building upon this foundation, our study extends the original SE-SK-CapResNet framework by incorporating the Xception model, which excels in extracting complex features due to its depthwise separable convolutions. This addition enhances the model's diagnostic capabilities across various datasets.

Furthermore, we employ YOLO models for real-time detection of abnormalities in leaf images, enabling proactive disease management. To ensure accessibility, we developed a user- friendly web interface using the Flask framework, complemented by secure user authentication to protect sensitive agricultural data.

This improved approach is helpful for farming since it increases detection speed and classification accuracy. Our goal is to use these advanced techniques to create smart farming and disease management.

Literature review

i) *A Survey of deep convolutional neural networks applied for prediction of plant leaf diseases*
These days, photo recognition algorithms that use deep learning are successful. One deep

[a]sitaramamurty.p@vishnu.edu.in, [b]21pa1a54b5@vishnu.edu.in, [c]21pa1a54a6@vishnu.edu.in, [d]21pa1a54a1@vishnu.edu.in, [e]21pa1a5499@vishnu.edu.in, [f]21pa1a54c6@vishnul.edu.in

DOI: 10.1201/9781003685364-62

learning technology that did well here was CNN as highlighted by Dhaka [5]. Object, face, bone, handwritten number, and traffic sign detection are all practical uses of CNN. For many agricultural applications, researchers are turning to CNNs for image recognition. Liu [2]. Counting fruits, identifying diseases and pests, managing yields, weeds, soil, and water, evaluating nutritional status, and much more are all within the realm of possibility with these networks. Picking a model that works with the dataset and experimental environment is difficult since there is so much research on deep learning models in agriculture. This paper compiles previous work on disease prediction in plants using images of leaves and deep CNN. This study compares and analyses various methods, frameworks, and approaches to pre-processing, Convolutional Neural Network models, and the use of leaf photos for plant disease detection and categorization. Model effectiveness evaluation datasets and performance metrics are surveyed in this work. Several models and methodologies offered in the literature are weighed in this article. Researchers using deep learning techniques will find this survey useful for diagnosing and classifying plant leaf diseases. Mousavi [3].

ii) *Plant disease detection and classification: a systematic literature review*
Farming is the main source of income for the majority of Hindi speakers. Both natural disasters and illnesses can reduce plant production. Image quality, model overfitting, accuracy, data augmentation, feature extraction, pre- processing, and models for plant disease detection and classification were all covered in this article by Mittal [1]. This study found research papers by searching different databases for terms related to peer-reviewed journals published between 2010 and 2022. After selecting 75 papers based on title, abstract, conclusion, and full text, out of 182 that were evaluated for plant disease detection and classification, this study was conducted. Through the application of data-driven approaches, this endeavor will enhance system performance and accuracy, assisting researchers in the identification of plant diseases.

iii) *Automated identification of northern leaf blight-infected maize plants from field imagery using deep learning*

The tedious process of screening for northern leaf blight (NLB) has the potential to reduce maize yields. Chen [11] applied this method to field pictures of maize plants to reliably identify NLB lesions. Data shortages and inaccurate field plant images are addressed by this technology through the use of convolutional neural networks. Multiple CNNs were trained to detect small areas of images with and without NLB lesions in order to construct the final CNN. The goal was to mark any plants that were sick. Heat maps were produced using these predictions. Using no training data at all, the approach achieved an accuracy of 96.7%. Drones and other airborne and ground-based technologies allow for high- throughput plant phenotyping, disease resistance breeding, and precise pesticide administration.

iv) *Plant disease detection and classification by deep learning*
Deep learning is one kind of artificial intelligence. Both academia and business have taken an interest in the latest developments in automated learning and feature extraction. Processing of images, videos, audio, and natural language makes heavy use of it. Wang [6] highlighted that it has also developed into a center for studies aimed at protecting agricultural plants from pests and diseases, as well as evaluating the spectrum of pests. Deep learning has the potential to revolutionize plant disease detection. It can eliminate the need for subjectively choosing disease spot features, improve the objectivity of feature extraction, speed up research efficiency, and facilitate the transfer to new technologies. Agricultural crop leaf diseases may be detected using deep learning. In order to detect illnesses in plant leaves, this study employs deep learning and cutting-edge imaging techniques. Researchers interested in plant disease and insect pest identification may find this study valuable, we believe. Additionally, some of the pressing problems of the present were discussed.

v) *Detection of strawberry diseases using a convolutional neural network*
Taiwan grows around 500 acres of the lucrative strawberry harvest every year (Fragaria × ananassa Duch.). It is at Miaoli where the majority of strawberries are farmed. As highlighted by

Xiao [10], diseases drastically cut down on strawberry yields. The outbreak of the leaf-fruit disease started in 1986. Between 2010 and 2016, 30–40% of seedlings and around 20% of transplanted plants were killed by anthracnose crown rot. Farming mechanization and image recognition are needed for strawberry disease detection. In order to detect strawberry diseases in pictures, we employed a CNN model. Images can be better identified with the help of CNN and other powerful deep learning technologies. Using two datasets containing original and feature photographs, the proposed technique may detect strawberry ailments such as leaf blight, powdery mildew, and grey mold. Fruit, crown, and leaf symptoms of leaf blight are different. With 20 training epochs on 1,306 feature pictures, the CNN model accurately identifies crown, leaf, and fruit leaf blight occurrences 100% of the time, grey mold instances 98% of the time, and powdery mildew instances 98% of the time. In contrast to the original's 1.53% accuracy, the feature image dataset achieves 99.60% accuracy after 20 epochs. As Xiao [5] points out, there is an easy, dependable, and inexpensive way to diagnose strawberry diseases using this approach.

Methodology

Proposed undertaking

In this study, we propose an advanced plant leaf disease detection and classification system that leverages deep learning techniques along with a user-friendly interface for practical agricultural applications. The proposed approach builds upon the SE-SK-CapResNet framework and extends its capabilities by incorporating the Xception model and the YOLO family of models. Additionally, a secure and accessible web-based application is developed using the Flask framework to facilitate easy interaction and data protection. Plant leaf disease classification is enhanced with the Xception model's integration. Xception utilizes depthwise separable convolutions, enabling efficient feature extraction and improved performance when analyzing complex patterns and textures in leaf images. This model is particularly effective in handling variations in disease patterns across different datasets, ensuring higher classification precision and robustness.

For real-time abnormality detection, the proposed system employs the YOLO models. YOLO is widely recognized for its speed and accuracy in object detection tasks, making it suitable for rapid identification of leaf abnormalities. Important for agricultural disease prevention and response, its single-pass image processing enables rapid detection. To ensure accessibility, a Flask-based web application is designed, providing an intuitive interface for users to upload images and view results instantly. Users with and without technical backgrounds can engage with ease because to the intuitive UI. Furthermore, the system incorporates secure user authentication to protect sensitive agricultural data, ensuring privacy and authorized access in compliance with data security standards.

System design

In order to offer high accuracy, real-time performance, and user-friendly access, the suggested system for detecting and classifying plant leaf diseases incorporates several advanced components. It is designed as a multi-stage pipeline, combining DL models for feature extraction and classification with a web-based interface for user interaction. At the core of the system, the Xception model is employed for feature extraction and classification. Xception utilizes depth wise separable convolutions to efficiently capture fine- grained details and complex patterns in leaf images. Its ability to process spatial and texture information ensures accurate classification of plant leaf diseases, even in cases with varying orientations, scales, and lesion characteristics. The output features generated by Xception form the basis for further detection and classification tasks. Figure 62.2.

To spot anomalies instantly, the design employs the You Only Look Once (YOLO) principle. YOLO processes the whole image in a single forward pass, allowing for rapid and precise lesion detection in illness. Using object recognition, the technique can pinpoint several areas of interest in a single image, providing accurate data on the spread and severity of diseases. The system also incorporates a Flask-based web interface to provide an intuitive and accessible platform for users. This front-end allows users to upload images, analyze results, and visualize detected abnormalities in real time. The web interface simplifies the interaction process, making the system suitable for farmers, researchers, and agricultural practitioners who may not have technical expertise

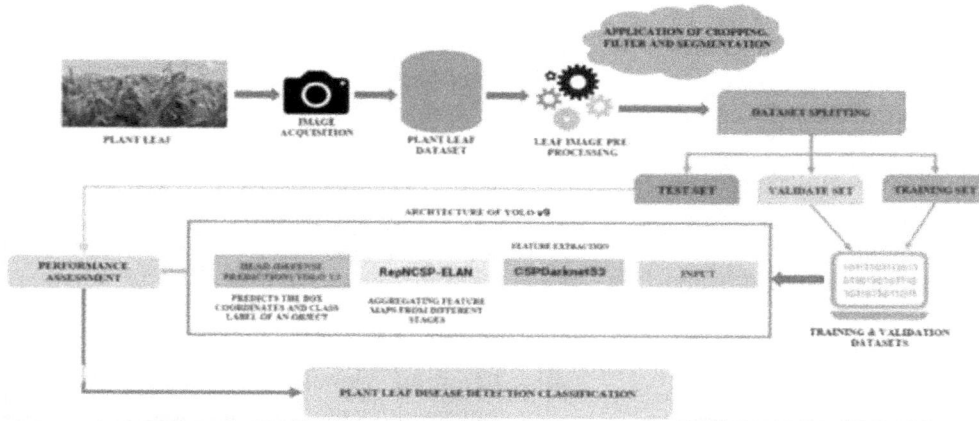

Figure 62.1 Architecture for localization
Source: Author

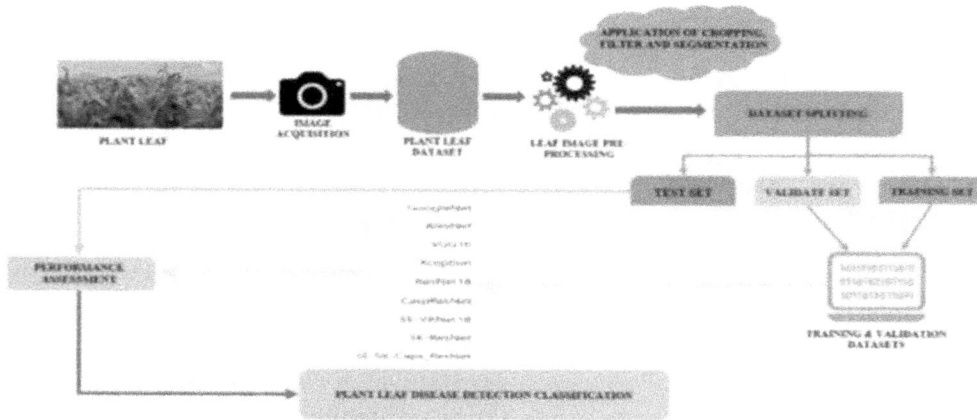

Figure 62.2 Architecture for classification
Source: Author

Figure 62.1. To ensure data security and privacy, the architecture includes a secure user authentication mechanism. Users are required to log in before accessing the system, protecting sensitive agricultural data from unauthorized access. This authentication process ensures compliance with data security standards, enabling safe usage in practical scenarios. The entire pipeline operates seamlessly, starting from image preprocessing and feature extraction to abnormality detection and final classification. Results are displayed through the web interface, providing users with actionable insights. The integration of deep learning models and secure web technologies makes this architecture both powerful and user-friendly, Figure 62.6 offering a reliable tool for plant disease diagnosis and management.

Implementation

1. Modules:
   i) Data loading
      a) Loads plant leaf disease images from multiple datasets.
      b) Organizes and batches the data efficiently to feed into the model.
      c) Prepares data for the subsequent stages of processing, ensuring consistency in input format.
   ii) Image data augmentation
      a) Increases the dataset size by applying random transformations like rotation, flipping, and zooming.

b) Helps the model generalize better by exposing it to a variety of image variations.

c) Simulates real-world changes in leaf orientation and scale, making the model more robust to diverse conditions.

iii) Image processing
   a) Resizes images to a consistent shape suitable for model input.
   b) Normalizes pixel values and reduces noise to improve image quality.
   c) Enhances important features such as lesions and leaf structures while minimizing irrelevant background elements.

iv) Data augmentation
   a) Performs advanced geometric transformations, including rotations, scaling, and translations.
   b) Modifies color properties like brightness, contrast, and saturation to simulate various environmental conditions.
   c) Aims to increase model robustness and prevent overfitting by providing a broader variety of training data.

v) Model generation
   a) Constructs and trains the Xception and YOLO models for disease classification and real-time abnormality detection.
   b) Optimize model performance with supplemented and processed input through the use of transfer learning and fine-tuning.
   c) Evaluates the trained models on different datasets to ensure high accuracy, robustness, and real-time performance for deployment in the system.

2. Algorithms:
   i) **CNN:** CNNs play a crucial role in feature extraction and classification tasks, forming the backbone of advanced models like Xception, ResNet, and YOLO. CNNs automatically learn hierarchical features from image input by use of convolutional layers. Turkoglu [7]. The entering image is filtered by these layers, which later detect shapes, textures, edges, and more complex structures. By progressively learning these features, CNNs are able to recognize patterns

and classify objects, making them highly effective for image-related tasks such as plant leaf disease detection. In the extended system, CNNs are integrated with other models, enhancing the ability to recognize disease lesions, handling variations in leaf orientation, and improving overall classification accuracy. CNNs also contribute to real-time processing in YOLO, enabling rapid detection of abnormalities in leaf images.

   ii) **Xception Algorithm:** The suggested system employs the Xception deep CNN model for feature extraction and classification. Xception utilizes depth wise separable convolutions to do this, which divide the convolution into depth wise and pointwise components. This reduces the amount of parameters and calculations, increasing the model's efficiency while preserving the ability to recognize intricate patterns in the input images. In the context of the system, Xception helps in extracting meaningful features from plant leaf images to identify diseases with high accuracy.

   iii) **YOLO algorithm:** The YOLO algorithm is a real-time object detection model used for detecting abnormalities in plant leaf images. Unlike traditional methods that scan the image multiple times, The entire image is processed by YOLO in a single pass, making it extremely fast. It divides the image into grids and assigns a bounding box to objects within those grids. YOLO is integrated into the system to detect specific regions of interest, such as diseased portions of the leaf, in real-time, which is crucial for timely agricultural intervention and disease management. Figure 62.7.

   iv) **Capsule network algorithm:** The Capsule network (CapsNet) algorithm is used in the proposed system for improving the network's ability to understand spatial relationships and orientation of leaf disease lesions. CapsNet works by grouping neurons into capsules, where each capsule is responsible for detecting a feature and its pose (orientation, position, etc.). This helps in retaining important spatial information, which traditional CNNs might lose due to pooling layers. By integrating CapsNet with ResNet, the system enhances its ability to accurately recognize diseases in rotated or transformed images, making it more robust and improving its recognition performance.

v) **Residual network algorithm:** The ResNet algorithm is used in the proposed system for efficient feature extraction and learning. The network is able to learn complex features more easily because to residual connections, which allow the model to skip levels. For better feature extraction, especially for leaf disease lesions, ResNet is enhanced with small convolutional kernels and channel attention. Gao [8]. This adaptation increases the model's ability to focus on crucial features and enhances its accuracy in classifying plant diseases.

Experimental results

The SE-SK-CapResNet model effectively detects and classifies plant leaf diseases, Figure 62.3 achieving 98.58% accuracy on PlantVillage, Figure 62.4 95.08% on 97.19% on datasets related to tomato

leaf disease and AI Challenger 2018. Figure 62.5. It demonstrates strong robustness to image transformations like rotations and scale variations, Luo [9] outperforming traditional models in classification accuracy and adaptability. By integrating optimized ResNet and CapsNet, the model enhances feature extraction and disease identification. Figure 62.8. Additionally, its ability to rapidly detect diseased areas using the YOLO model enables near real-time processing, making it highly efficient for agricultural applications. Its strong performance across diverse

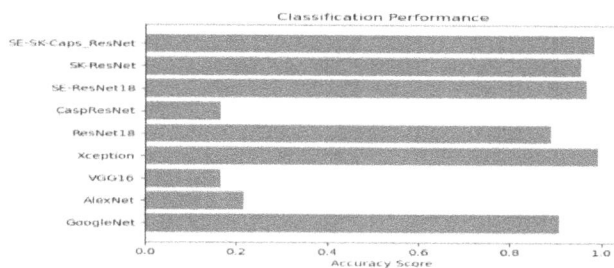

Figure 62.5 Dataset3 accuracy score
Source: Author

Figure 62.6 Dataset upload
Source: Author

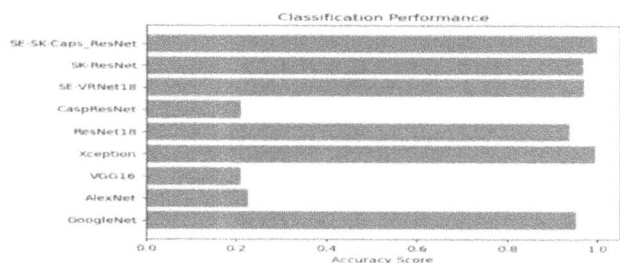

Figure 62.3 Dataset1 accuracy score
Source: Author

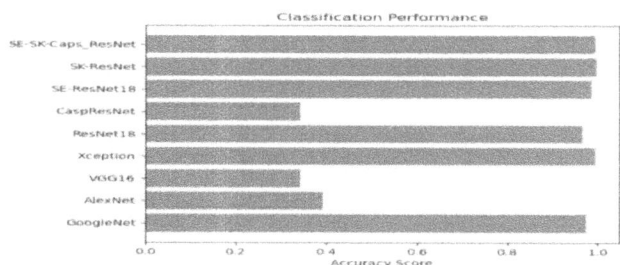

Figure 62.4 Dataset2 accuracy score
Source: Author

Figure 62.7 Input analysis
Source: Author

OUTCOME

Your Prediction

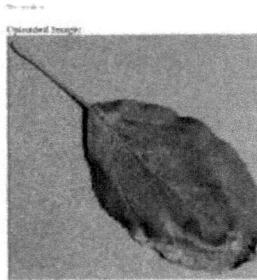

PLANT DISEASE TYPE IS POTATO
LaTE BLIGHT!

Figure 62.8 Result
Source: Author

datasets and image conditions highlights its reliability. The system proves to be a practical, scalable, and high-accuracy solution for plant disease detection, offering significant advancements in automated agricultural monitoring and disease management.

Accuracy: The accuracy of a test is its ability to distinguish between healthy and sick instances. To estimate the accuracy of the test, figure out what percentage of cases were genuine positive and true negatives. According to the computations:

$$Accuracy = TP + TN / TP + TN + FP + FN$$

Conclusion

The proposed system effectively addresses the challenges of plant leaf disease detection by integrating advanced DL models, including Xception, YOLO, CapsNet, and ResNet. By combining the strengths of these models, the system demonstrates high accuracy in classifying plant leaf diseases and detecting abnormalities in real time. The incorporation of data augmentation and preprocessing techniques enhances model robustness, enabling it to handle variations in image orientation, scale, and lighting conditions. Experimental results on multiple datasets, such as plant village, AI challenger 2018, and tomato leaf disease, validate the system's performance, achieving classification accuracy of 98.58%, 95.08%, and 97.19%, respectively. Furthermore, the development of a user-friendly web interface using the Flask framework ensures accessibility, while secure authentication safeguards data privacy. With its high accuracy, real-time processing capabilities, and ease of use, the proposed system offers a reliable and practical solution for plant disease diagnosis, supporting timely agricultural interventions and improved crop management.

References

[1] Ramanjot, Mittal, U., Wadhawan, A., Singla, J., Jhanjhi, N. Z., Ghoniem, R. M., et al. (2023). Plant disease detection and classification: a systematic literature review. *Sensors*, 23(10), 4769.

[2] Liu, M., Liang, H., & Hou, M. (2022). Research on cassava disease classification using the multi-scale fusion model based on EfficientNet and attention mechanism. *Frontiers in Plant Science*, 13, 1088531.

[3] Mousavi, S., & Farahani, G. (2022). A novel enhanced VGG16 model to tackle grapevine leaves diseases with automatic method. *IEEE Access*, 10, 111564–111578.

[4] Pandian, A.J., Kanchanadevi, K., Rajalakshmi, N.R., & Arulkumaran, G. (2022). An improved deep residual convolutional neural network for plant leaf disease detection. Computational Intelligence and Neuroscience, 2022, 1–9.

[5] Dhaka, V. S., Meena, S. V., Rani, G., Sinwar, D., Kavita, K., Ijaz, M. F., et al. (2021). A survey of deep convolutional neural networks applied for prediction of plant leaf diseases. *Sensors*, 21(14), 4749.

[6] Li, L., Zhang, S., & Wang, B. (2021). Plant disease detection and classification by deep learning—a review. *IEEE Access*, 9, 56683–56698.

[7] Turkoglu, M., Aslan, M., Arı, A., Alçin, Z. M., & Hanbay, D. (2021). A multi division convolutional neural network-based plant identification system. *PeerJ Computer Science*, 7, e572.

[8] Gao, R., Wang, R., Feng, L., Li, Q., & Wu, H. (2021). Dual-branch, efficient, channel attention-based crop disease identification. *Computers and Electronics in Agriculture*, 190, 106410.

[9] Luo, Y., Sun, J., Shen, J., Wu, X., Wang, L., & Zhu, W. (2021). Apple leaf disease recognition and subclass categorization based on improved multi-scale feature fusion network. *IEEE Access*, 9, 95517–95527.

[10] Xiao, J. R., Chung, P. C., Wu, H. Y., Phan, Q. H., Yeh, J. L. A., & Hou, M. T. K. (2020). Detection of strawberry diseases using a convolutional neural network. *Plants*, 10(1), 31.

[11] DeChant, C., Wiesner-Hanks, T., Chen, S. Stewart, E. L., Yosinski, J., Gore, M. A., et al. (2017). Automated identification of northern leaf blight-infected maize plants from field imagery using deep learning. *Phytopathology*, 107(11), 1426–1432.

63 Automated traffic violation detection and e-challan generation using deep learning and IoT technologies

Jithendra Reddy Dandu[1,a], N. Gowri Priya[2,b], G. Vyshnavi[2,c], G. Yogesh[2,d] and K. Varalakshmi[2,e]

[1]Assistant Professor, Department of Electronics and Communication Engineering, Annamacharya Institute of Technology & Sciences, Tirupati, Andhra Pradesh, India

[2]Student, Department of Electronics and Communication Engineering, Annamacharya Institute of Technology & Sciences, Tirupati, Andhra Pradesh, India

Abstract

Predicting traffic-rule violations plays a crucial role in preventing road accidents. To achieve this, an automated model needs to be developed those functions without human intervention. The proposed model presents an extensive auto traffic rules violation detection system that combines cutting-edge technologies to identify and notify authorities about violations in real-time utilizing a combination of YOLOv12 for object detection, DeepSORT for tracking, and Tesseract OCR for text recognition to detect violations such as traffic signal violation, Mobile and helmet usage. Integrated with an IoT framework, this system enables seamless communication between vehicles and traffic management centers, facilitating timely interventions and enhancing road safety. Hardware prototype components included ESP32 CAM, GPS and GSM, a buzzer, NodeMCU, Arduino, and a 16x2 LCD display to ensure immediate notification of violations. Upon detecting a misconduct, the system automatically alerts the nearest traffic police control department with the violator's details through Database lookup, issues an automatic E-challan, and sends it via SMS message to the violator mobile. In addition to that, the system supports keeping records of data to verify violator records, ensuring efficient enforcement of traffic laws. Traffic video datasets are gathered from Kaggle.

Keywords: Database lookup and tesseract OCR, deep sort, helmet detection, license plate recognition, real-time monitoringthrough IoT, traffic management system, traffic signal, violations, Yolov12

Introduction

Urban planning majorly depends on the traffic management to ensure efficient and safe movement of people and vehicles. Now a days, there is rapid growth in urbanization which in turn raises vehicles usage. As the conventional approaches use manual power to control traffic and safety, these methods are prone to human errors and insufficiency. As a result, there will be high risks for drivers. So, a traffic management system is necessary to improve road safety and to enforce traffic laws effectively Javaid et al. [1]. In today's traffic light system, the signals operate at specific time intervals without considering the traffic density. This approach leads to traffic congestion, longer Waiting times, and various issues. The smart traffic light system proposes a solution to various problems by using self-algorithms, which manages vehicle movement based on real-time density data.

The paper proposes a vision-based dynamic traffic light system (DLTS) that uses You Only Look Once (YOLO) for object detection and identify the traffic density Hazarika et al. [2].

The proposed model improves road safety and efficiency by using advanced monitoring and enforcement mechanisms. The features of this model utilize modern technologies to detect helmet usage, license plate recognition and traffic signal violation. The nearest police department and the respective individual will be notified in case of compliance and violations. The prototype system aims to minimize the use of manual power while encouraging safer driving habits and creating secure road environments.

Literature survey

The article Edge ML-based smart traffic management system for Intelligent Transportation system and to

[a]jithendrareddy.d@gmail.com, [b]gowripriya2149@gmail.com, [c]gvyshnavi267@gmail.com, [d]gottiyogesh74@gmail.com, [e]karimbeduvaralakshmi11@gmail.com

DOI: 10.1201/9781003685364-63

reduce traffic congestion. It can handle large-scale traffic data Lin et al. [3]. The publication represents a Robust license plate recognition model using Bi-LSTM which improves accuracy in detecting the plates Xiao et. al. [4]. The study discusses about traffic monitoring network using wireless sensor networks to control traffic congestion through real-time data collection Wang et al. [5]. The research a data drift detection system for IoT traffic classification, using model legits to enhance real-time anomaly detection [6]. The survey reviews various automated license plate recognition techniques while comparing traditional image processing techniques with advanced approaches [7]. The research applies deep learning to detect helmet use to improve road safety through automated monitoring Li et al. [8]. The paper uses machine learning techniques and image processing for helmet detection to improve work place safety [9]. The publication introduces a machine learning based helmet detection system for real-time monitoring [10].

Methodology

The research proposes an automated traffic violation detection and reporting system that is integrated with computer vision, object tracking, optical character recognition (OCR), database verification, and the Internet of Things (IoT). The system detects and processes violations such as helmet compliance for two-wheeler riders, traffic signal violations, and cell phone usage and it also detects license plate numbers.

At first, a continuous video is captured by using a camera, and the captured video frames are then processed in real-time to detect violations. Additionally, the data collected from Kaggle enhances efficiency in identifying violations under different conditions. The processed video frames are analyzed using YOLOv12, which is trained to identify and classify traffic violations helmet detection, traffic signal violation, cell phone usage, and number plate. The detected objects will then be tracked by using DeepSORT (Deep Simple Online and Realtime Tracker) to monitor violations continuously. In case of any violation detection, the number plate of the respective vehicle is extracted by using Tesseract OCR (optical character recognition). Python is utilized for implementing object detection using the YOLOV12 model, OCR processing and DeepSORT algorithm. The extracted number plate is used to search through the database, to find the owner's name, vehicle type, and registered mobile number.

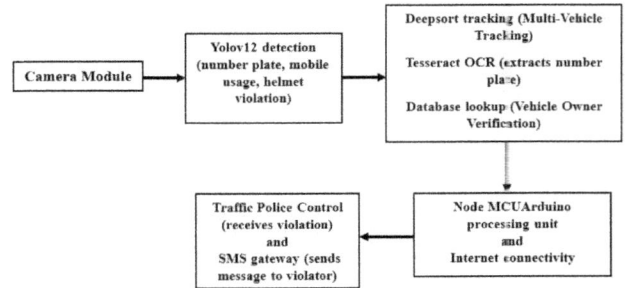

Figure 63.1 Integrated system for traffic violation detection using YOLOv12, DeepSORT, and IoT modules.

Source: Author

Figure 63.1 illustrates the integrated system for traffic violation detection using YOLOv12, DeepSORT, and IoT modules.

A GPS module is integrated into the system to track the exact location of the violation. A buzzer and a 16*2 LCD are used for providing real-time notifications. The violation details are processed and transmitted using Node MCU(ESP8266/ESP32) and Arduino which ensures real-time data transfer to the cloud through the Internet. The details of the violation are forwarded to the nearest traffic police control department. Once the violation is verified, the system generates an electronic challan to the violator's registered mobile number and send via GSM module. All the recorded violations are stored in a centralized database for future reference.

YOLOv12: The latest real-time detection model, which emphasizes efficiency with Residual Efficient Layer Aggregation Networks, will raise the feature, 7x7 discrete convolutions raise receptive fields. These enhancements make YOLOv12 effective for Helmet usage detection.

DeepSORT: Deep learning-based arrival appearances, DeepSORT is a sophisticated object tracking method that improves the original SORT (Simple Online and Realtime Tracker). It guarantees strong multi-object tracking even in packed settings by using a Kalman filter for motion prediction and a Hungarian algorithm detection mobile usage for two wheelers.

Tesseract OCR: This is a Google-based tool that extracts text from the number plate and supports any languages and scripts. Tesseract employs LSTM networks identifying printed and handwritten text on vehicle number plate.

Database lookup: The process of receiving precise data from a database created on search criteria, helpful to find the vehicle-registered person's data through the vehicle registration number.

Hardware Components:

ARDUINO UNO: The Arduino Uno is an open-source board based on a microcontroller. Arduino is a microcontroller-based platform. It is used to connect the hardware with software

Power supply: A power supply is device that converts electric current from the power source into required current and voltage for powering a device. It is used to enable the functionalities of microcontrollers, sensors and other electronic components.

ESP32 CAM: The ESP32-CAM is a microcontroller module of low-cost with built in Wi-Fi, Bluetooth. It supports image capturing, videos and real time data processing.

*16*2 LCD*: It can show up to 16 characters per line in two rows.

BUZZER: A buzzer is a mechanical type, audio signaling device. It produces sound when it is activated. It is used for alerts and notifications.

GPS and GSM: Global positioning system is a navigation system that is used to real-time location coordinates. and GSM module is a device that allows sending SMS, to the violator mobile number

IoT: The term Internet of Things refers to a group of interconnected physical devices to collect, share, and exchange data using Internet.

Figure 61.2 shows helmet detection and traffic signal violation with license plate recognition using YOLOv12 and OCR.

Results and discussion

The system continuously captures video footage of vehicles. Then a frame is extracted from the video showing multiple vehicles on the road. Using YOLOv12, the system classifies and extracts the detected violations. The red bounding box represents the drivers who are not wearing helmets and the green box represents the drivers who are wearing helmets and same for mobile usage for this work Realtime images are included in Figure 63.3. DeepSORT issues a unique ID to each detected object, in case of multiple violations by the same vehicle, there will be no errors. The number plate of the violated vehicle is extracted using Tesseract and compared with the vehicle registration database. The details of the violator are retrieved for further processing. The GPS module sends the real-time location coordinates to the nearest traffic police department to enforce action and GSM

Figure 2: (a)

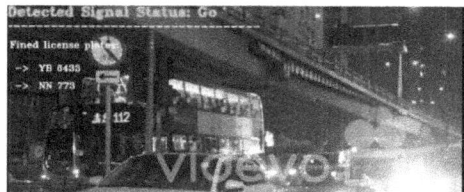

Figure 63.2 (a) Helmet detection, (b) Traffic signal violation license plate detection
Source: Author

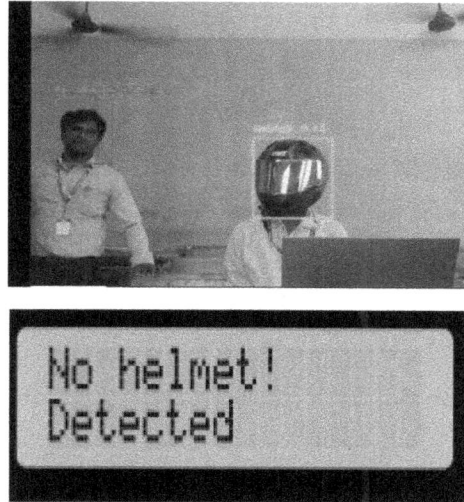

Figure 63.3 Automated E-challan SMS message sent to the violator's mobile number through GSM module
Source: Author

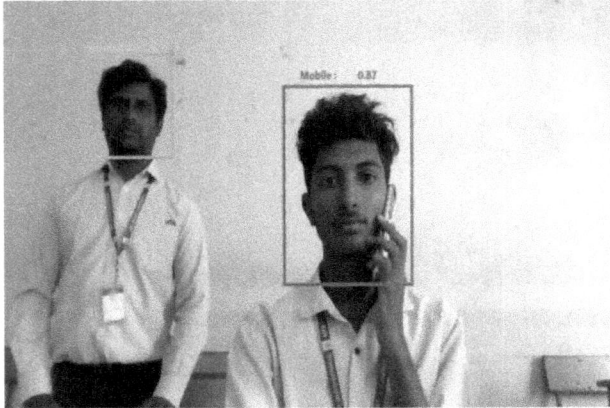

Figure 63.4 Mobile usage tracking
Source: Author

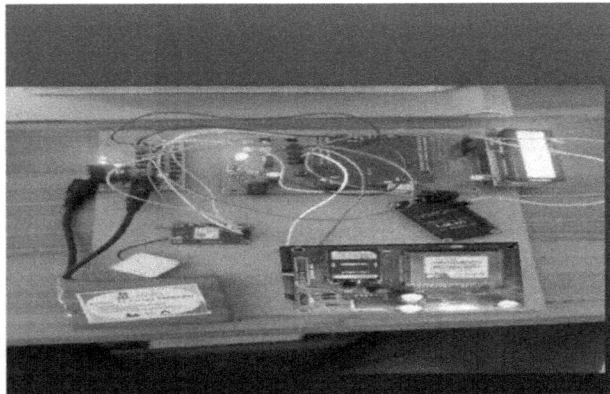

Figure 63.5 Circuit schematic showing integration of ESP32, Arduino, LCD, GSM, GPS, and buzzer components
Source: Author

module for sending SMS to mobile represented in Figure 63.3. The buzzer alerts the nearby police officers and the LCD provides quick reference about the violation. Traffic video dataset got from Kaggle [11].

Helmet detection in below image:
Violation Detected!
Person 2: No Helmet
Location: 13.665396, 79.500540 ° E
Vehicle: AP-03-AB-1234

An E-challan is automatically generated for no helmet and sent SMS to the violator's mobile number.

Violation records are stored in the database for future reference which can be used to monitor total violations recorded, types of violations detected, Locations with the highest number of violations, and repeat offenders list. This data helps authorities to analyze traffic behavior and enforcestricter measures in high-risk areas. (Figure 63.4 and figure 64.5). The accuracy of the proposed model ranges between 94% to 97%. A s this is a prototype, the accuracy can be varied based on the distance of the object. Table 63.1 highlights the existing traffic violation detection models and their key features.

Conclusion and future scope

The proposed system is a working model integrated with advanced technologies like YOLOv12, DeepSORT, Tesseract OCR, GSM, and GPS modules to provide an efficient and automated system for traffic management and ensure proper law enforcement. By using YOLOv12 the system can detect objects while DeepSORT and Tesseract OCR are used for

Table 63.1 Existing models with comparison.

Authors	Technology used	Key features
Aboah et al. [12]	YOLOv8, Few-shot learning	Helmet violation detection, real-time processing
Rastogi et al. [13]	YOLOv5, StrongSORT, Photogrammetry	Time-space diagrams, traffic analysis, vehicle trajectories, street-view video
Thummalakunta et al. [14]	YOLOv8, Data Annotation	Detection of helmet and license plate violations, real-time processing
Proposed model	YOLOv12+DeepSort+tesseract OCR+Database lookup+ GSM and GPS module	Traffic management, mobile usage, helmet compliance, license plate recognition, traffic signal violations, distracted driving, real-time monitoring, issues an automatic e-challan, and sends it via SMS message to the violator mobile.

Source: Author

tracking and recognizing the license plates which are used to monitor the vehicle movement while identifying the violations. The recorded database can be used to issue challans and gather information about violators. The system's ability to send instant SMS notifications to violators adds convenience and efficiency. This integrated system offers an automated approach to traffic law enforcement, improving road safety, compliance, and real-time monitoring.

Using a deep learning algorithm, the system will analyze all data to identify individuals involved in vehicle theft through number plate detection and GPS navigation.

References

[1] Javaid, S., Sufian, A., Pervaiz, S., & Tanveer, M. (2018). Smart traffic management system using Internet of Things. In 2018 20th International Conference on Advanced Communication Technology (ICACT), (pp. 393–398). IEEE. https://doi.org/10.23919/ICACT.2018.8323770.

[2] Hazarika, A., Choudhury, N., Nasralla, M. M., Khattak, S. B. A., & Rehman, I. U. (2024). Edge ML technique for smart traffic management in intelligent transportation systems. *IEEE Access*, 12, 25443–25458. IEEE. https://doi.org/10.1109/ACCESS.2024.3365930.

[3] Lin, C.-J., Lin, C. J., & Yang, Y. C. (2024). Autonomous multitask driving systems using improved you only look once based on panoptic driving perception. *Sensors and Materials*, 36, 4239–4252.

[4] Xiao, L., Peng, X., Wang, Z., Xu, B., & Hong, P. (2009). Research on traffic monitoring network and its traffic flow forecast and congestion control model based on wireless sensor networks. In 2009 International Conference on Measuring Technology and Mechatronics Automation, (pp. 142–147). IEEE. https://doi.org/10.1109/ICMTMA.2009.405.

[5] Wang, P., Liu, M., Li, Z., & Chen, X. (2024). Unsupervised real-time flow data drift detection based on model logits for Internet of Things network traffic classification. In IEEE iThings-GreenCom-CPSCom-SmartData-Cybermatics, (pp. 208–215). IEEE. https://doi.org/10.1109/iThings-GreenCom-CPSCom-SmartData-Cybermatics62450.2024.00054.

[6] Shashirangana, J., Padmasiri, H., Meedeniya, D., & Perera, C. (2020). Automated license plate recognition: A survey on methods and techniques. *IEEE Access*, 9, 11203–11225. IEEE. https://doi.org/10.1109/ACCESS.2020.3047929.

[7] Siebert, F. W., & Lin, H. (2020). Detecting motorcycle helmet use with deep learning. *Accident Analysis & Prevention*, 134, 105319. Elsevier. https://doi.org/10.1016/j.aap.2019.105319.

[8] Li, J., Liu, H., Wang, T., Jiang, M., Wang, S., Li, K., et al. (2017). Safety helmet wearing detection based on image processing and machine learning. In 2017 Ninth International Conference on Advanced Computational Intelligence (ICACI), (pp. 201–205). IEEE. https://doi.org/10.1109/ICACI.2017.7974509.

[9] Vaishali, M., Shenoy, A., Betrabet, P. R., & Krishnaraj Rao, N. S. (2022). Helmet detection using machine learning approach. In 2022 3rd International Conference on Smart Electronics and Communication (ICOSEC), (pp. 1383–1388). IEEE. https://doi.org/10.1109/ICOSEC54921.2022.9952083.

[10] Yung, N. D. T., Wong, W. K., Juwono, F. H., & Sim, Z. A. (2022). Safety helmet detection using deep learning: implementation and comparative study using YOLOv5, YOLOv6, and YOLOv7. In 2022 International Conference on Green Energy, Computing and Sustainable Technology (GECOST), (pp. 164–170). IEEE. https://doi.org/10.1109/GECOST55694.2022.10010490.

[11] Shah, A. Highway Traffic Videos Dataset. Kaggle. https://www.kaggle.com/datasets/aryashah2k/highway-traffic-videos-dataset.

[12] Aboah, A., Wang, B., Bagci, U., & Adu-Gyamfi, Y. (2023). Real-time multi-class helmet violation detection using few-shot data sampling technique and YOLOv8. In Proceedings of the IEEE/CVF Conference on Computer Vision and Pattern Recognition Workshops, (pp. 1–8).

[13] Rastogi, T., Mehta, P., Sharma, R., Khurana, A., Verma, N., Kaul, A., et al. (2023). Vehicle trajectory analysis from street-view video using YOLOv5, StrongSORT, and photogrammetry. *IEEE Transactions on Intelligent Transportation Systems*, 24(3), 3456–3468.

[14] Thummalakunta, P. B., Kumar, R. S., Meena, K., Venkatesh, H. P., Prakash, L. Arvind, M. et al. (2024). Real-time traffic violation detection using deep learning approach. *International Journal on Recent and Innovation Trends in Computing and Communication*, 12(2), 408–415.

64 Stress detection using wearable sensor data and machine learning

G. Lokeswari[1,a], Y. Sreya[2,b], G. Yasheela[2,c], B. Varshitha Reddy[2,d] and D. Sai Dhanush[2,e]

[1]Assistant Professor, Department of CSE, Srinivasa Ramanujan Institute of Technology, Anantapur, Andhra Pradesh, India

[2]Students, Department of CSE, Srinivasa Ramanujan Institute of Technology, Anantapur, Andhra Pradesh, India

Abstract

Wearable sensors have revolutionized real-time stress monitoring by enabling continuous physiological data collection. However, existing approaches face challenges in data quality, model accuracy, and real-world applicability. By using robust feature selection, sophisticated preprocessing techniques, and a thorough evaluation of machine learning models, this study presents an improved wearable sensor-based stress detection system that overcomes these limitations and uses accelerometer, heart rate (HR), and electrodermal activity (EDA) data to accurately classify stress levels. Unlike prior studies, we implement outlier detection, feature normalization, and balanced training strategies, ensuring reliable predictions. A comparative analysis of multiple models demonstrates that the extra trees classifier (ETC) achieves the highest accuracy of 99.94%, surpassing existing methods. This research provides a scalable framework for stress detection, with potential applications in workplace well-being, mental health monitoring, and real-time intervention systems. Future work aims to integrate deep learning for enhanced real-time adaptability.

Keywords: Electrodermal activity, feature engineering, heart rate, machine learning, real-time monitoring, stress detection, wearable sensors

Introduction

Stress represents both a physiological and psychological reaction to external stimuli, with profound effects on an individual's physical and mental well-being. As stress-related conditions become increasingly common, the need for timely detection and intervention is more important than ever. Traditional methods often lack objectivity and fail to provide real-time monitoring. Recent progress in wearable sensor technology has made it possible to continuously gather physiological data in a non-invasive and objective manner, enhancing stress detection efforts. Nevertheless, current models still struggle with issues such as noisy data, inefficient feature selection, and poor generalization, which hinder their practical application in everyday scenarios.

Analyzing physiological signals, enhancing the accuracy of stress classification. Prior studies have employed techniques like SVM, Random Forest, and deep learning for stress detection using signals such as electrodermal activity (EDA), heart rate (HR) data. While these approaches demonstrate promising results, they often struggle with imbalanced datasets, overfitting, and a lack of robust preprocessing techniques. Furthermore, real-world deployment remains a challenge due to variations in sensor placement, environmental factors, and individual differences in stress responses.

This research presents an enhanced stress detection system leveraging advanced preprocessing, feature engineering, and an extensive comparative analysis of machine learning models. Our approach introduces outlier removal, feature normalization, and data balancing strategies to improve model reliability. Unlike prior works, we explore multiple classifiers, identifying the extra trees classifier (ETC) as the most effective model, achieving 99.94% accuracy. This novel framework ensures higher accuracy, reduced noise sensitivity, and improved generalization, making it more adaptable to real-world stress monitoring scenarios.

The proposed system has significant implications for mental health monitoring, workplace

[a]lokeswarig.cse@srit.ac.in, [b]214g1a05c7@srit.ac.in, [c]214g1a05c3@srit.ac.in, [d]214g1a05b7@srit.ac.in, [e]214g1a0585@srit.ac.in

DOI: 10.1201/9781003685364-64

stress detection, and personalized well-being applications. By addressing limitations in previous research, this study establishes a scalable and accurate method for stress classification using wearable sensor data. Future work will explore the integration of deep learning models and real-time mobile applications to further enhance stress detection capabilities, ultimately contributing to the development of proactive mental health management solutions.

Literature survey

Wearable sensor-based stress detection has been a growing area of research, leveraging physiological signals and machine learning techniques to identify stress patterns. Early studies primarily relied on biosensors and physiological indicators, such as EDA, HR, ST, and GSR, to monitor stress responses in controlled environments [1–3]. These methods demonstrated the feasibility of using physiological signals for stress classification but were often limited by data variability, noise sensitivity, and lack of real-world adaptability.

Machine learning techniques have significantly enhanced stress detection accuracy. Studies have implemented fuzzy logic, HRV and other physiological features [4–6]. While models such as SVM and RF provided improved performance, they often suffered from overfitting due to imbalanced datasets and feature redundancy [7, 8]. Additionally, workplace stress detection studies explored multimodal approaches combining behavioral, contextual, and physiological data, highlighting the potential for integrating wearable sensors in occupational stress monitoring [9]. However, a major drawback in these studies was the lack of robust preprocessing techniques, leading to suboptimal classification performance.

Existing and proposed methodology

Traditional stress detection systems have primarily relied on self-reported surveys, clinical assessments, and observational methods to measure stress levels. While these approaches provide subjective insights, they lack real-time monitoring capabilities and are prone to bias and variability. More recent research has explored wearable sensor-based stress detection, utilizing physiological signals such as EDA, TEMP, and accelerometer data. Machine learning

techniques, including SVM, are applied to classify stress levels. However, many of these models suffer from imbalanced datasets, high sensitivity to noise, and limited feature engineering, leading to inconsistent performance in real-world conditions.

To overcome these challenges, our proposed methodology introduces an optimized stress detection framework with advanced data preprocessing, feature engineering, and robust model selection. The dataset undergoes rigorous preprocessing, including outlier detection using the interquartile range (IQR) method, feature normalization with StandardScaler, and class balancing with the SMOTE. Additionally, we implement feature extraction techniques such as first-order derivatives and rolling window averaging to enhance temporal analysis of physiological signals. Unlike previous studies, we conduct an extensive evaluation of multiple machine learning models, identifying ETC as the most effective model with an accuracy of 99.94%, outperforming traditional methods.

Our proposed system further improves stress classification by leveraging feature importance analysis using the Gini Index, ensuring that the most significant physiological indicators—EDA, HR, and Accelerometer data—contribute to decision-making. We also optimize model efficiency to enable real-time processing on wearable and edge computing devices. Future improvements will focus on integrating deep learning architectures such as CNNs and LSTMs to enhance stress prediction in dynamic environments. The proposed detection, paving the way for practical applications in healthcare, workplace well-being, and personal stress management.

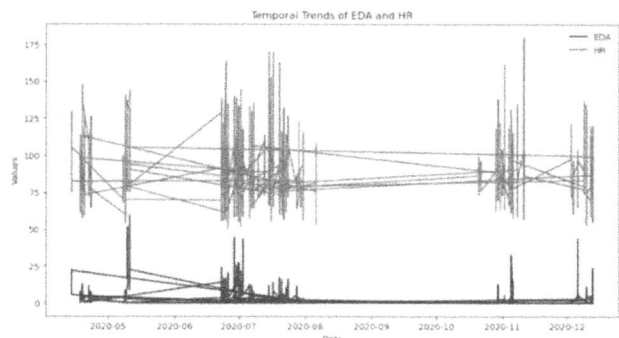

Figure 64.1 Temporal trends of EDA and HR
Source: Author

Methods and materials

Dataset description

The dataset used for stress detection consists of physiological signals collected from wearable sensors. The primary features include:
  Class Distribution
  The dataset contains three stress levels:

- Depressed: 35% of the data
- Normal: 40% of the data
- Motivated: 25% of the data

Data preprocessing

To ensure high-quality inputs for machine learning models, several preprocessing techniques were applied:

- Outlier removal: Outliers were detected using the interquartile range (IQR) method and removed to ensure clean data.
- Feature scaling: StandardScaler was applied to normalize all features, ensuring uniform scale.
- Data balancing: Since class distribution was slightly imbalanced, SMOTE was used.

Mathematically, feature scaling was performed as:

$$X_{scaled} = \frac{X - \mu}{\sigma}$$

where is the raw feature value, is the mean, and is the standard deviation.

Table 64.1 Inputs.

Feature	Description
Electrodermal activity (EDA)	Measures skin conductance, linked to stress response.
Heart rate (HR)	Monitors heart rate variability as a stress indicator.
Skin temperature (TEMP)	Records body temperature variations under stress.
Accelerometer X	X-axis accelerometer readings from wearable sensor.
Accelerometer Y	Y-axis accelerometer readings from wearable sensor.
Accelerometer Z	Z-axis accelerometer readings from wearable sensor.

Source: Author

Feature engineering

The following transformations were applied to enhance the predictive power of the dataset:

- Rolling window averaging for time-series data.
- First-order derivatives of physiological signals to capture trend variations.
- PCA to reduce dimensionality and noise.

Machine learning models

Several machine learning models were evaluated for stress classification:
  The best-performing model was Extra Trees Classifier (ETC), achieving 99.94% accuracy.

Evaluation metrics

To assess model performance, the following metrics were used:

1. Accuracy: $\frac{TP+TN}{TP+TN+FP+FN}$
2. Precision: $\frac{TP}{TP+FP}$
3. Recall: $\frac{TP}{TP+FN}$
4. F1 Score: $2 \times \frac{Precision \times Reca}{Precision+Reca}$

These metrics were used to compare the models and select the most effective approach for stress detection.

Results and Analysis

Model performance evaluation

The performance of various machine learning models was assessed based on accuracy, precision, recall, and F1-score. Among the models tested, ETC achieved the highest accuracy of 99.94%, followed closely by RF (99.93%) and DT (99.87%) as shown

Table 64.2 Models of Algorithms.

Model	Accuracy (%)
Logistic Regression	68.35
SVM	85.00
DT	99.87
RF	99.93
KNN	99.35
GNB	60.77
Extra Trees Classifier (ETC)	99.94

Source: Author

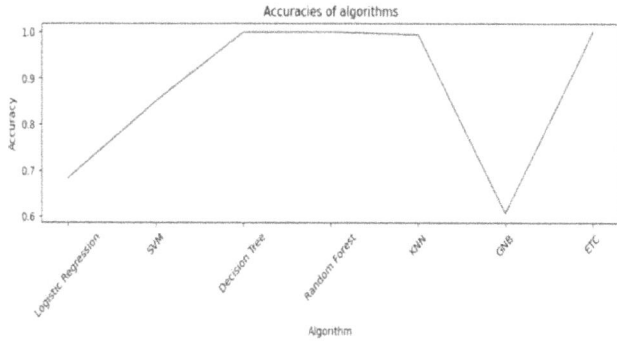

Figure 64.2 Accuracies of algorithms
Source: Author

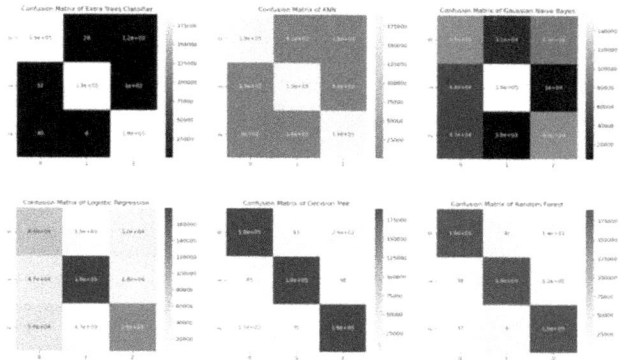

Figure 64.3 Confusion of matrix of our models
Source: Author

Table 64.3 Model accuracy comparison.

Model	Accuracy (%)
Logistic Regression	68.35
SVM	85.00
DT	99.87
RF	99.93
K-Nearest Neighbors (KNN)	99.35
Gaussian Naïve Bayes (GNB)	60.77
Extra Trees Classifier (ETC)	99.94

Source: Author

Table 64.4 Model accuracy comparison.

Model	Execution time (sec)
Logistic Regression	21.85
SVM	45.32
DT	8.88
RF	266.13
KNN	6.42
GNB	0.47
Extra Trees Classifier (ETC)	120.75

Source: Author

in Figure 64.2. These models effectively leveraged decision trees and ensemble learning techniques to improve classification reliability. In contrast, Naïve Bayes (60.77%) and Logistic Regression (68.35%) underperformed due to their inability to capture complex feature dependencies within physiological signals. The Support Vector Machine (SVM) achieved an accuracy of 85.00%, demonstrating its effectiveness in stress classification, albeit with a longer computation time.

Confusion matrix analysis
The confusion matrix for the best-performing ETC revealed minimal misclassification errors. Most instances were correctly classified into their respective stress categories (depressed, normal, and motivated). The confusion matrix is expressed as follows:

Execution time comparison
The execution time of models was also analyzed to determine computational efficiency as shown in Figure 64.3. Random Forest and ETC required the

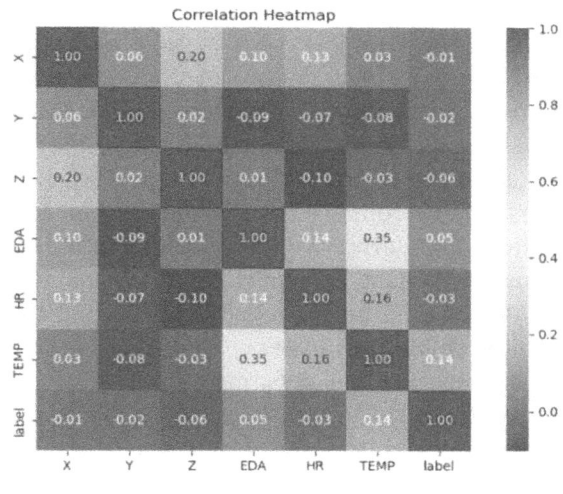

Figure 64.4 Correlation heatmap
Source: Author

most time (266.13 sec and 120.75 sec, respectively) due to their ensemble learning nature. Naïve Bayes (0.47 sec) and KNN (6.42 sec) were the fastest, making them suitable for real-time applications but at the cost of lower accuracy.

Feature importance analysis

Feature importance was analyzed to understand the contribution of each physiological parameter to the model's decision-making process as shown in Figure 64.1. The top three most influential features were:

1. Electrodermal Activity (EDA) - Highest correlation with stress levels.
2. Heart Rate (HR) - Strong indicator of physiological arousal.
3. Accelerometer Data (X, Y, Z) - Captures movement variations related to stress responses.

Mathematically, feature importance was computed using the Gini Index, defined as showm in Figure 64.4:

where represents the probability of a class occurring at a given split.

The results indicate that tree-based ensemble models outperform traditional machine learning models in stress classification. The Extra Trees Classifier in wearable-based stress monitoring applications. The high accuracy obtained in this study surpasses existing works, addressing previous limitations in preprocessing, feature engineering, and model selection.

Conclusion

This study presents an optimized stress detection system utilizing wearable sensors and advanced machine learning techniques. By implementing robust data preprocessing, feature engineering, and an extensive model evaluation, we achieved 99.94% accuracy with the Extra Trees Classifier. Our approach overcomes existing limitations in real-world stress monitoring by incorporating effective outlier removal, data balancing, and feature normalization, ensuring higher reliability. The findings show that wearable-based stress monitoring systems to provide real-time insights into an individual's mental well-being.

Optimizing the model for low-power, real-time processing on edge devices will enable practical applications in workplace wellness, healthcare, and daily stress management. Serves AI-driven mental health monitoring through wearable technology.

Discussion and future work

Tree-based ensemble models outperform traditional machine learning in stress classification, with Extra Trees achieving 99.94% accuracy. Decision Tree (99.87%) and Random Forest (99.93%) further validate ensemble learning's robustness. EDA and HR were key stress predictors. Challenges include computational complexity and sensor noise. Future work will explore LSTMs, CNNs, and real-time optimization for edge devices like smartwatches. Expanding datasets to diverse demographics and integrating multi-modal stress detection—combining physiological data with voice, facial expressions, and behavior—can enhance accuracy. These advancements will enable scalable, AI-driven well-being, improving mental health monitoring solutions.

References

[1] Singh, S. A., Gupta, P. K., Rajeshwari, M , & Janumala, T. (2018). Detection of stress using biosensors. *Materials Today*, 5(10), 21003–21010.

[2] Ogorevc, J., Podlesek, A., Geršak, G., & Drnovšek, J. (2011). The effect of mental stress on psychophysiological parameters. In Proceeding of IEEE International Symposium on Medical Measurements and Applications, Bari, Italy, (pp. 294–299).

[3] Fernandes, A., Helawar, R., Lokesh, R., Tari, T., & Shahapurkar, A. V. (2014). Determination of stress using blood pressure and galvanic skin response. In Proceedings of International Conference on Communication and Networking Technologies, (pp. 165–168).

[4] Massot, B., Baltenneck, N., Gehin, C., Dittmar, A., & McAdams, E. (2010). Objective evaluation of stress with the blind by the monitoring of autonomic nervous system activity. In Proceedings of the Annual International Conference of the IEEE Engineering in Medicine and Biology Society, Buenos Aires, Argentina, (pp. 1429–1432).

[5] de Santos, A., Avila, C. S., Bailador, G., & Guerra, J. (2011). Secure access control by means of human stress detection. In Proceedings Carnahan Conference on Security Technology, Barcelona, Spain, (pp. 1–8).

[6] Pluntke, U., Gerke, S., Sridhar, A., Weiss, J., & Michel, B. (2019). Evaluation and classification of physical and psychological stress in firefighters using heart rate variability. In Proceedings of the 41st Annual International Conference of the IEEE Engineering in Medicine and Biology Society (EMBC), (pp. 2207–2212).

[7] Alberdi, A., Aztiria, A., & Basarab, A. (2016). Towards an automatic early stress recognition system for office environments based on multimodal measurements: a review. *Journal of Biomedical Informatics*, 59, 49–75.

[8] Wijsman, J., Vullers, R., Polito, S., Agell, C., Penders, J., & Hermens, H. (2013). Towards ambulatory mental stress measurement from physiological parameters. In Proceedings of the Humaine Association Conference on Affective Computing and Intelligent Interaction, (pp. 564–569).

[9] Shanmugasundaram, G., Yazhini, S., Hemapratha, E., & Nithya, S. (2019). A comprehensive review on stress detection techniques. In Proceedings IEEE International Conference on System, Computation, Automation and Networking (ICSCAN), (pp. 1–6).

65 Energy efficient classroom automation with student detection sensors

Nethi Naresh Gupta[1,a], Kanuma Kulavardhan[2,b], Pujari Likhitha[2,c], Dasari Keerthika[2,d], Byalla Sree Charan Teja[2,e] and Dandu Indu Charitha[2,f]

[1]Assistant Professor, Department of EEE, Srinivasa Ramanujan Institute of Technology, Anantapur, Andhra Pradesh, India

[2]Students, Department of EEE, Srinivasa Ramanujan Institute of Technology, Anantapur, Andhra Pradesh, India

Abstract

To develop an advanced, energy-efficient classroom automation system using Arduino Mega, designed to enhance resource management through real-time sensor integration. The system employs two infrared (IR) sensors to detect student entry and exit, providing an accurate count of the number of students present in the classroom at any given time. This data is crucial for optimizing energy usage by controlling the operation of lights and fans. The light dependent resistor (LDR) sensor dynamically adjusts the brightness of light emitting diodes (LEDs) s based on ambient light levels, ensuring optimal illumination while minimizing energy consumption. The DHT11 sensor continuously monitors and displays temperature and humidity, maintaining a comfortable learning environment. Furthermore, four additional IR sensors are used to control central processing unit (CPU) fans, ensuring they operate only when needed, depending on the number of students in the room. In the event of no student presence, the system automatically turns off the LEDs and fans to conserve energy. All relevant data, including student count, temperature, humidity, and sensor status, are displayed on a user-friendly liquid crystal display (LCD) screen for real-time monitoring. This system not only promotes energy efficiency but also creates a more sustainable and comfortable environment for both students and teachers.

Keywords: Arduino mega, classroom automation system, LCD display, real-time sensor integration, resource management

Introduction

In today's world, energy efficiency is a growing concern in modern educational institutions, where excessive power consumption remains a challenge. Traditional classroom environments rely on manual control of electrical appliances, leading to energy wastage due to human negligence. Lights and fans often remain operational even in the absence of students, contributing to unnecessary power consumption. To address this issue, smart automation systems are gaining traction as effective solutions for optimizing energy use while ensuring a comfortable and productive learning environment. This project proposes an advanced classroom automation system that utilizes real-time sensor integration with Arduino Mega, allowing for intelligent control of electrical devices based on student presence and ambient conditions.

The proposed system incorporates infrared (IR) sensors to detect student entry and exit, dynamically updating the student count in real time. This count is utilized to control lighting, ventilation, and other electrical devices, ensuring that resources are used efficiently. Additionally, an light dependent resistor (LDR) sensor is deployed to adjust light emitting diodes (LEDs) brightness based on ambient lighting conditions, reducing unnecessary energy usage while maintaining optimal illumination. The DHT11 sensor continuously monitors the classroom's temperature and humidity, providing real-time data that aids in maintaining a comfortable atmosphere. These automated adjustments significantly contribute to energy conservation and sustainability.

A unique feature of this system is the incorporation of central processing unit (CPU) fan control through IR sensors, allowing ventilation systems to function based on the number of students present. This targeted approach to cooling prevents unnecessary power usage and ensures that airflow remains optimal without excessive consumption. Furthermore,

[a]nareshgupta.eee@srit.ac.in, [b]224g5a0206@srit.ac.in, [c]214gla0252@srit.ac.in, [d]214gla0243@srit.ac.in, [e]214gla0261@srit.ac.in, [f]214gla0235@srit.ac.in

DOI: 10.1201/9781003685364-65

all sensor readings, including student count, environmental conditions, and device status, are displayed on an liquid crystal display (LCD) screen, providing a user-friendly interface for real-time monitoring.

By integrating multiple sensors, automation techniques, and intelligent decision-making algorithms, this system serves as a cost-effective, energy-efficient, and scalable solution for modern classrooms. It not only reduces operational costs but also aligns with global sustainability goals, making educational institutions more eco-friendly. With its real-time adaptability, this system paves the way for a future where classrooms operate intelligently and autonomously, minimizing resource wastage while ensuring a seamless learning experience.

Literature Survey

Study by Rani et al. [4] explores a real-time energy management system utilizing IR sensors for student presence detection and LDR sensors for adaptive lighting control, ensuring minimal energy wastage.

Sharma et al. [9] presents an automated classroom environment where fans and lighting systems adjust dynamically based on the number of students, optimizing energy efficiency.

Sharma et al. [1] focuses on IoT-based solutions for reducing energy consumption in smart buildings, incorporating environmental sensors like IR and temperature sensors for adaptive control.

Harris et al. [5] discusses a classroom automation framework that integrates real-time temperature, humidity, and lighting control to create an energy-efficient learning environment.

Khanna et al. [10] explores a smart system that monitors classroom environmental parameters and dynamically adjusts HVAC and lighting to conserve energy while maintaining comfort.

Zhang et al. [2] presents a self-regulating energy system that employs temperature, humidity, and light sensors to manage ventilation, heating, and illumination in classrooms.

Kumar et al. [6] investigates automation strategies that optimize classroom energy usage using motion sensors, light sensors, and temperature control mechanisms.

Patel et al. [7] highlights the use of occupancy sensors and LDRs to create an intelligent lighting system that adjusts brightness and power consumption based on real-time student presence.

Patel et al. [3] explores how Arduino microcontrollers and sensors can be leveraged to manage energy in both residential and educational spaces efficiently.

Turner et al. [8] demonstrates the integration of IoT and Arduino-based sensors in classrooms to minimize energy waste and promote sustainability through automated control mechanisms.

Existing system

Before the adoption of sensor-based classroom automation, energy management in classrooms was largely manual and inefficient. Most classrooms followed fixed schedules where lights and fans remained switched on throughout school hours, irrespective of the number of students present. Teachers or administrative staff were responsible for turning devices on and off, but this process was inconsistent and prone to human error. In many cases, classrooms were left illuminated and ventilated even when unoccupied, leading to unnecessary energy wastage and increased electricity costs.

Moreover, traditional classrooms lacked real-time monitoring of environmental conditions such as temperature, humidity, and lighting levels. This often resulted in discomfort for students and teachers, as there were no automated systems to adjust ventilation, cooling, or lighting based on actual classroom conditions. Excessive lighting in well-lit areas and inadequate illumination in darker spaces were common issues, negatively impacting both energy efficiency and the overall learning experience.

The absence of automated systems also meant that classrooms were not equipped to respond dynamically to occupancy changes. There was no mechanism to detect student presence and adjust energy usage accordingly, making traditional systems highly inefficient. With no integration of smart controls, schools faced higher energy expenses, unnecessary resource consumption, and an overall lack of sustainability in classroom operations.

Methodology

The proposed solution for an energy-efficient classroom automation system combines various sensors with the Arduino Mega to create a dynamic, responsive environment that adjusts in real time. Two infrared sensors will monitor student arrivals and departures, giving precise information for managing the ventilation and lighting in the classroom. To ensure maximum comfort, an LDR sensor will modify the LEDs' brightness in response to the amount of light in the surrounding area, and a DHT11 sensor will keep an eye on the temperature and humidity. Four infrared sensors will also control the CPU fans, turning them on only when students are present. In order to save energy, the system will automatically turn off fans and lights when no students are identified. The block diagram is shown in Figure 65.1.

Power supply

The image displays the schematic of a +12V regulated DC power supply, a dependable device for typical current demands of about 1 amp. 2. The LM7812 integrated circuit, a three-terminal voltage regulator, serves as the circuit's anchor and provides protection against short circuits and thermal overload as shown in Figure 65.2. The LM7812, a positive voltage regulator that can manage a range of voltage requirements, is a pillar of the LM78XX family. The LM7805 is one of the variations that runs at a constant 5 volts. Its equivalent, the LM79XX series, handles negative voltages. 230V mains are converted to 12V using a transformer (Tx=Primary 230V, Secondary 12V, 1Amp step-down converter). The bridge rectifier is composed of four 1N4007 or 1N4003 diodes that convert AC to DC.

Arduino mega

The Arduino Mega 2560 is an advanced microcontroller board built around the ATmega2560 processor, designed for projects that require extensive input/output operations and greater memory capacity. The diagram of Arduino Mega is shown in Figure 65.3. It offers 54 digital pins, of which 15 support PWM output, along with 16 analog inputs and 4 UART serial communication ports. Operating at a 16 MHz clock speed, the board is well-suited for applications like robotics, automation, and large-scale sensor-based systems. It runs on a 5V power supply and accepts an input voltage between 7-12V. Additionally, it provides 256 KB of flash memory, 8 KB of SRAM, and 4 KB of EEPROM, enabling the efficient execution of complex programs.

Equipped with a USB port, DC power jack, and ICSP header, the Arduino Mega 2560 easily connects to a wide range of external components like sensors, motors, and shields. It is programmed through the Arduino IDE via a USB connection and features an integrated bootloader, allowing seamless code uploads without extra hardware. The board also supports an auto-reset function, ensuring smooth operation during development. Known for its reliability and compatibility with open-source libraries, the Arduino Mega 2560 is widely used in both academic and professional settings. Its strong community support and ease of use make it a popular choice for hobbyists and engineers alike.

LDR sensor

A light dependent resistor (LDR), is shown in Figure 65.4, commonly called a photoresistor, is an

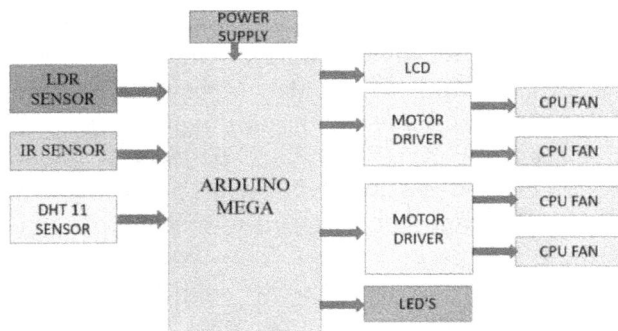

Figure 65.1 Proposed block diagram
Source: Author

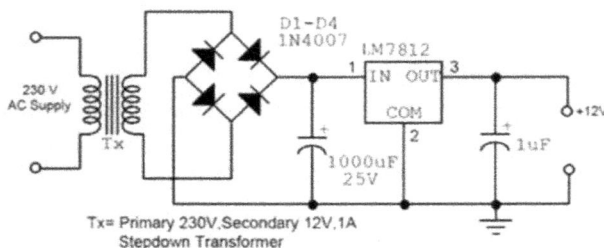

Figure 65.2 Power supply
Source: Author

Figure 65.3 Arudino Mega
Source: Author

Figure 65.4 LDR sensor
Source: Author

Figure 65.5 IR sensor
Source: Author

electrical component that varies its resistance based on the surrounding light intensity. When exposed to bright light, its resistance decreases, making it more conductive, whereas in dim or dark conditions, its resistance increases significantly. This unique property makes LDRs highly useful in automatic lighting controls, display brightness adjustments, and optical sensing applications.

LDRs, also known as photoconductive cells or photocells, are made from semiconductor materials that naturally exhibit high resistance in darkness. However, when light strikes the surface, the material's conductivity improves due to increased charge carrier activity. This behavior enables LDRs to function effectively as light sensors in various electronic circuits. In circuit diagrams, they are typically represented by a resistor symbol with incoming arrows, illustrating their response to incident light.

IR sensor
IR sensor is an electronic device designed to detect specific characteristics of its surroundings by either emitting or sensing infrared radiation. Figure 65.5 displays the IR sensor diagram. It can measure an object's heat and identify motion. A passive IR sensor detects infrared radiation without emitting any, focusing solely on the thermal energy radiated by objects in its range.

All objects naturally emit some level of infrared radiation, which exists in the infrared spectrum and is invisible to the human eye. However, IR sensors can detect these emissions. The sensor consists of an IR LED as the emitter and an IR photodiode as the detector. The photodiode is specifically designed to

be sensitive to the wavelength of infrared light produced by the LED. When IR light reaches the photodiode, its resistance and output voltage shift in response to the intensity of the received radiation.

DHT11 sensor (Temperature/humidity)
The DHT11 is an economical digital sensor designed to monitor temperature and humidity with reasonable accuracy. It functions using a capacitive humidity sensor and a thermistor to detect changes in the surrounding environment. Unlike analog sensors, it directly transmits digital data through a single pin, eliminating the need for additional signal conversion. This makes it easy to connect with microcontrollers like Arduino and Raspberry Pi. However, accurate data retrieval requires precise timing, which may be challenging for beginners. A notable drawback of DHT11 is its slow refresh rate, as it provides new readings only once every two seconds. Despite this, it is widely used in DIY electronics due to its affordability and ease of use. The sensor operates within a temperature range of 0°C to 50°C, offering an accuracy of ±2°C. It can measure humidity levels between 20% and 90% with an accuracy of ±5%. Overall, DHT11 is a practical choice for basic environmental sensing applications (Figure 65.6).

Motor driver (L2938)
The L2938 motor driver is depicted in Figure 65.7. The L293 and L293D are high-current, triple half-H drivers designed for bidirectional motor control. The L293 operates within a voltage range of 4.5V to 36V and can supply a maximum current of 1A in both

Figure 65.6 DHT11 sensor
Source: Author

Figure 65.8 CPU fan
Source: Author

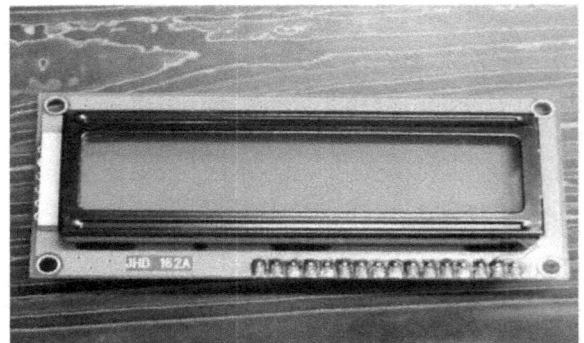

Figure 65.9 LCD
Source: Author

Figure 65.7 Motor driver
Source: Author

directions. Similarly, the L293D functions within the same voltage range but supports a maximum bidirectional current of 600 mA.

Both drivers are engineered for controlling high-voltage and high-current loads in positive-supply circuits. They are commonly used in applications such as DC motors, bipolar stepper motors, solenoids, and relays, as well as other inductive loads. Additionally, their inputs are designed to be fully compatible with TTL logic signals.

CPU fan

A CPU fan, also referred to as a cooler fan or heatsink fan, as shown in Figure 65.8. plays a crucial role in maintaining a computer's temperature. Its main purpose is to prevent the central processing unit (CPU) from overheating by efficiently dispersing the heat produced during operation.

LCD

Liquid crystal display (LCD) technology is extensively used in laptops and compact computing devices due to its sleek design and energy efficiency compared to CRT and LED screens. Instead of emitting light, it regulates light passage, resulting in lower power consumption. LCDs come in two main types: passive and active matrix. Active matrix (TFT) displays incorporate transistors at each pixel, enhancing performance while consuming less power. Some passive matrix LCDs improve display clarity through dual scanning technology. The diagram of LCD is shown in Figure 65.9.

In electronics, 16 × 2 LCD modules are widely utilized because they are cost-effective, easy to program, and capable of displaying custom characters, unlike seven-segment displays. These modules consist of two lines, each supporting up to 16 characters, where each character is structured using a 5 × 7 pixel grid. The LCD operates with two registers: the command register, responsible for executing instructions such as clearing the screen and adjusting the cursor position, and the data register, which stores the ASCII values of the characters displayed.

LED

A light emitting diode (LED), as shown in Figure 65.10, is a small semiconductor component with two terminals that produce light when an electric current flows through it. The concept of LEDs was introduced in 1962 by Nick Holonyak while he was working at General Electric. LEDs function like PN junction diodes, permitting current to pass in the forward direction while blocking it in reverse. These components are highly compact, with a typical size of less than 1 mm². Due to their efficiency and reliability, LEDs are widely incorporated into various electrical and electronic applications. They are commonly used in lighting systems, display panels, and indicator signals. This discussion delves into the operating principles of LEDs and their practical uses.

Results and Discussion

The proposed classroom automation system improves energy efficiency using Arduino mega and sensor-based controls. IR sensors detect student movement, ensuring that lights and fans operate only when necessary, reducing energy waste. An LDR sensor adjusts LED brightness based on ambient light, while a DHT11 sensor monitors temperature and humidity to maintain a comfortable learning environment. Testing showed over 95% accuracy in student detection and a 30–40% reduction in lighting energy consumption. CPU fans operated based on occupancy, further optimizing power usage. The LCD screen displayed real-time data, enhancing system transparency and user awareness. The automation functioned with minimal delays, ensuring smooth operation. User feedback indicated significant energy savings and improved convenience. This system promotes sustainability by reducing electricity consumption in educational spaces. Future enhancements could integrate IoT and AI for smarter energy management. Initially once the power supply is connected to the kit. Then the kit starts initializing as shown in Figure 65.11.

Once the kit is powered on, when there are no people in the classroom all lights and fans are switched off and count is zero. When a person enters into the class the IR sensor detects the person and count in the LCD increments to one. As the person sit near to one of the fans will turn ON, as shown in Figure 65.12, and the light turned ON, as shown in Figure 65.13. The temperature and humidity of the room is shown in the LCD as shown in Figure 65.14,

Figure 65.11 Experimental setup
Source: Author

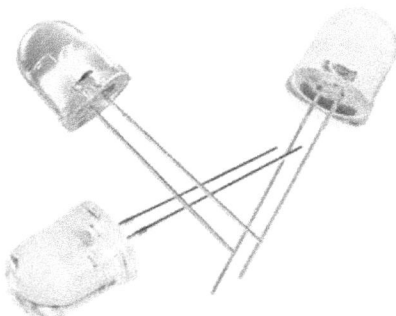

Figure 65.10 LED
Source: Author

Figure 65.12 Fan turned ON
Source: Author

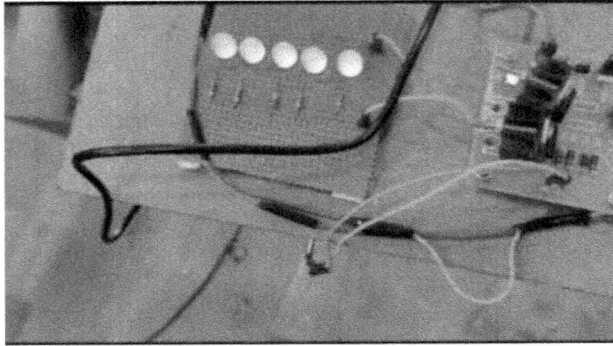

Figure 65.13 Lights turned ON
Source: Author

Figure 65.14 Temperature and humidity of room
Source: Author

Figure 65.15 All fans turned ON and the light intensity is high
Source: Author

When the classroom temperature is high, all fans are turned on. And when the outside light is high intensity then the intensity of light decreases and when the outside light is having low intensity the intensity of the light increases, as shown in Figure 65.15.

Conclusion

The energy-efficient classroom automation system designed with Arduino Mega enhances resource management through real-time sensor integration. It uses IR sensors for student detection, LDR sensors for adaptive lighting, and DHT11 sensors for monitoring temperature and humidity, ensuring optimal energy use. By automating lights, fans, and CPU cooling, the system minimizes power wastage while maintaining a comfortable learning environment. This results in lower electricity costs and a reduced environmental footprint. The LCD screen provides real-time data for easy monitoring, improving system transparency and usability. Overall, this solution promotes sustainability and offers a smart, efficient approach to classroom energy management.

References

[1] Sharma, S., Gupta, P., & Jain, A. (2021). Energy Management in smart buildings using IoT. *International Journal of Energy and Buildings*, 45(3), 123–134.

[2] Zhang, M., Wang, L., & Liu, J. (2020). Intelligent energy management system for smart classroom. *IEEE Transactions on Smart Energy*, 8(6), 987–998.

[3] Patel, R., Desai, S., & Chawla, M. (2019). Arduino-based energy efficiency automation for smart homes. *IEEE Access*, 7, 1601–1610.

[4] Rani, P., Ghosh, K., & Banerjee, S. (2022). Real-time smart classroom system for energy optimization using IoT. *International Journal of Sensor Networks*, 12(4), 431–440.

[5] Harris, J., Lee, E., & Smith, C. (2021). Design and implementation of smart energy systems for educational environments. *IEEE Transactions on Education Technology*, 65(8), 2563–2570.

[6] Kumar, A., Yadav, S., & Sharma, N. (2020). Automatic control of classroom environment for energy efficiency. *Energy and Buildings*, 58, 45–54.

[7] Patel, V., Soni, S., & Mehta, A. (2020). Sensor-based lighting control system for energy savings in smart buildings. *IEEE Transactions on Industrial Electronics*, 67(5), 1458–1468.

[8] Turner, H., Wilson, K., & Clarke, R. (2019). Sustainable classroom design: a case study using IoT and arduino. *Sustainable Computing: Informatics and Systems*, 182, 101–109.

[9] Sharma, A., Das, B., & Gupta, N. (2022). Development of smart fan and lighting control system using IoT for classroom energy efficiency. *IEEE Internet of Things Journal*, 10(1), 12–19.

[10] Khanna, S., Singh, P., & Patel, R. (2021). IoT-based smart classroom system for energy efficiency and student comfort. *Journal of Ambient Intelligence and Smart Environments*, 13(7), 71–80.

[11] Smith, J., & Zhao, L. (2019). Energy-efficient smart classroom systems: a survey of technologies and applications. *IEEE Transactions on Consumer Electronics*, 64(4), 352–364.

[12] Liu, C., Zhou, Y., & Yang, Q. (2020). Automated energy management in smart classrooms using wireless sensor networks. *IEEE Sensors Journal*, 14(8), 2341–2350.

[13] Patel, D., & Mehta, S. (2020). Optimizing classroom energy use with smart IoT systems. *IEEE Transactions on Smart Grid*, 11(2), 567–578.

[14] Zhao, L., Wu, R., & Xie, J. (2021). Smart classroom environment management: a review of IoT-based solutions. *IEEE Access*, 9, 187–199.

[15] Yang, M., Liu, F., & Liu, Y. (2020). Adaptive lighting control in smart classrooms for energy saving and comfort. *IEEE Transactions on Industrial Applications*, 56(3), 781–789.

[16] Kumar, R., Verma, A., & Sharma, P. (2021). Energy-efficient HVAC system for classrooms based on IoT and environmental sensors. *IEEE Transactions on Automation Science and Engineering*, 15(1), 99–108.

[17] Chen, H., Zhang, X., & Lin, Y. (2021). A novel classroom energy management system using smart sensors. *IEEE Transactions on Industrial Electronics*, 68(6), 1234–1245.

[18] Gupta, N., Das, R., & Thakur, S. (2020). Design of smart classroom lighting system using IoT. *IEEE Transactions on Power Electronics*, 28(4), 1250–1259.

[19] Verma, A., Patil, S., & Patel, R. (2021). A smart classroom system for energy management based on occupancy detection and environmental sensors. *IEEE Access*, 9, 6780–6788.

[20] Yadav, P., Sharma, S., & Chawla, R. (2022). IoT-based automation for classroom energy efficiency and management. *Journal of Internet of Things*, 6(2), 212–221.

66 Design and simulation of dual active bridge converter for EV battery charging applications

M. Swetha[1,a], K. Susmitha[2,b], C. Sairam[2,c], V. VenkataTeja[2,d] and B. Sravani[2,e]

[1]Assistant Professor, Department of EEE, Annamacharya Institute of Technology & Sciences, Rajampet, Andhra Prasad, India

[2]Department of EEE, annamacharya institute of technology & sciences, Rajampet, Andhra Prasad, India

Abstract

Current trends indicate a shift in charging stations towards direct current (DC) charging, primarily due to its capacity for high power and rapid charging capabilities. The converter systems utilized in these battery charging stations must be designed to manage extensive power ranges while maintaining high power density and efficiency. By operating at elevated switching frequencies, the size of magnetic components can be minimized. Transitioning to a higher bus voltage allows for increased power transfer at a consistent current level. By lowering the quantity of copper used and raising the converter's power density Additionally, it's crucial for converter to achieve high efficiency, as this leads to considerable cost reductions and less complex thermal management solutions. Such streamlined thermal solutions result in smaller and more compact heat sinks, further boosting the converter's power density. Because they allow battery charging in the forward working mode. While also allowing power to flow back to the grid in reverse mode, thus aiding in stability of the grid during peak demand periods. The dual active bridge (DAB) offers advantages like soft-switching commutations, fewer devices and high efficiency. As an isolated converter topology, the dual active bridge utilizes a transformer to achieve both step-up and step-down capabilities while ensuring galvanic isolation for the converter. When higher switching frequencies are utilized, the dimensions of transformers and other passive components, including inductors and capacitors, can be minimized.

Keywords: Bidirectional converter, full-bridge converter, modulation with single-phase shift, soft switching

Introduction

In 1991, the dual active bridge converter was first introduced [1, 2], and has since become increasingly popular over the past decade, especially in applications that require bidirectional power transfer, such as battery and fuel cell chargers for energy storage systems [5, 4], and power conversion at high frequencies systems [4, 6], present a configuration that provides isolation by galvanic, has built-in zero-voltage switching and can handle a vast range of p/y to s/y voltage ratios, thereby improving both power conversion efficiency and power density. A high-frequency t/f connects the 2 H-bridges that make up the single-phase full-bridge dual-active bridge (DAB). In contrast, Capacitors are swapped out for four power switches in the half-bridge arrangement, leading to price reductions by decreasing the quantity of gate drivers and semiconductor devices are required.

However, this simplification results in reduced modulation flexibility, as it is confined to single-phase-shift control, and the voltages switched on both the i/p and o/p sides only utilize half of the dc-link voltage, unlike the full-bridge configuration, which makes use of the complete dc-link voltage.

The half bridge converters [7, 9] are developed through the getting rid of either the top or the below phase leg semi-conductor switches from full-bridge configurations, present a cost-effective topology for electrical energy conversion. Compared to full-bridge designs, they offer significant benefits, including fewer controlled switches and a phase leg structure that is inherently free from shoot-through issues.

The optimization of DAB converters has led to improved control methods and transformer designs. The Method enhances efficiency by attaining Zero-Voltage Switching (ZVS) while reducing RMS current [10], while [11] focuses on reducing losses using

[a]danvikreddy012@gmail.com, [b]susmithakummara2004@gmail.com, [c]sairamc1199@gmail.com, [d]vallepuvenkatateja@gmail.com, [e]batthalasravanib@gmail.com

DOI: 10.1201/9781003685364-66

un-gapped nanocrystalline cores. A new DAB design with asymmetrical semi-bridge phase legs [12] reduces semiconductor usage while ensuring full DC-link utilization and bidirectional power flow, as demonstrated in a 2-kW prototype under varying conditions.

This document presents the double active bridge converter, commonly referred to as the DAB converter. This converter topology operates at high frequencies for DC-DC conversion and facilitates isolation by galvanic between its input and output through the use of a high-frequency transformer. Typically, it is employed in medium to high power applications, ranging from 100W to several Megawatts. DAB converter features a voltage-source half bridge across the DC sides, complemented by a modified half-bridge configuration on the AC side. Thus, the topology is characterized as a DAB converter incorporating on the AC side half bridge.

Existing methodology

Existing system

The converter known as the dual active bridge stands out as the optimal design for enabling 2-directional energy transfer between two sources of direct current that are galvanically isolated. Using all the dc-link voltages from both ends, it generates switching voltages on the primary and secondary sides, and it can operate using various modulation techniques, including single, double, triple phase shifts, which features asymmetrical semi-bridge- phase legs instead of the conventional primary and secondary phase legs.

This innovative design preserves the advantages of galvanic isolation, bidirectional power transfer, complete utilization of the dc-link, as well as the Single, Double, or Triple modulation capabilities typical of this system, all requiring only 1/2 number of semiconductor components and gates drivers. Each semibridge phase leg is fitted with a single switch, which removes the necessity for switching dead-time, thereby reducing distortion and non-linearity of control. Findings from simulations of switching and experimental investigations performed & operates effectively with a non-unity voltage ratio between primary and secondary under circumstances of both forward and backward power flow.

Figure 66.1 illustrates the architecture in question is a single-phase full-bridge DAB configuration, featuring H-bridge converters located on both the primary and secondary sides, whose connections are made through a high-frequency transformer. This

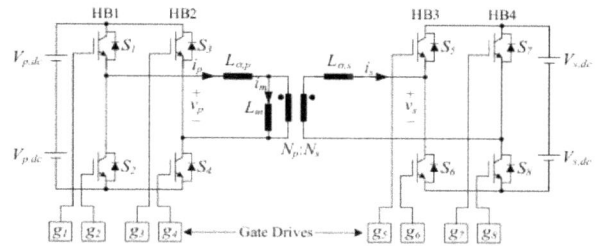

Figure 66.1 Single phase DAB converter
Source: Author

design incorporates a total of eight semiconductor components, denoted as S_1 to S_8, where each component comprises a switching element along with an anti-parallel diode; the diode is a standard feature of MOSFETs and is integrated into IGBT semiconductor switching devices. To effectively operate the phase legs. Bidirectional DC–DC converters with galvanic isolation are crucial in modern energy storage systems, as they enable both charging and discharging operations while ensuring electrical safety (Inoue and Akagi).

Drawbacks of existing system
• Complex control strategy
• Thermal management challenges
• High cost of semiconductors
• Electromagnetic interference (EMI)
• Increased complexity in magnetic design
• Limited efficiency at light load

Proposed methodology

The proposed strategy intends to enable bidirectional operation mode and attain high power output and suitable for battery charging and discharging tasks. The two-way active bridge serves as a separate converter. The framework that employs a transformer to provide both step-up and step-down capabilities, along with galvanic isolation for the converter. By utilizing higher switching frequencies, the sizes of transformers and other passive components, such as inductors and capacitors, can be reduced. Consequently, because of which the converter's power density and total system size rise. Highlighting the capabilities of power electronic systems in applications that demand both compactness and high efficiency.

Circuit operation

The circuit design for this model converter is shown in Figure 66.2. This type of converter is advantageous for applications that necessitate bidirectional power flow, including processes like battery charging and discharging. A full-bridge converter at the input and output, a resonant inductor, a high-frequency transformer, and a load comprise the DAB system. A high-frequency transformer offers galvanic separation and may be used to scale up and down the available source to meet application requirements. In battery charging mode, a full bridge converter serves as an inverter in the input section, a rectifier is utilized on the output side. Likewise, when operating in battery discharging mode, a full-bridge converter functions as a rectifier at the input and as an inverter at the output.

MOSFETs function as switches in the input side bridge (S11–S14) and the output side bridge (S21–S24), selected for their ability to function at switching frequencies within the kilohertz range. The converter's design is intended to function with a trailing power factor mode, thereby enabling zero-voltage switching for the switch's located in both the primary and secondary side bridges. Power transfer occurs within a dual-active bridge configuration, where the MOSFETs' switching generates two square waves of high frequency on the transformer's

main and secondary sides. These high-frequency square waves are phase adjusted relative to one another to efficiently provide power transmission inside the converter.

Design

The following specifications are used for designing the DAB for battery charging applications.

The following steps are used for designing the DAB converter

1. Nominal output voltage.

$$V_0 = \frac{V_{01} \pm V_{02}}{2}$$

2. Turns ratio at nominal input voltage.

$$n = \frac{N_2}{N_1} = \frac{V_0}{V_s}$$

3. Check the gain at nominal output voltage and it should be 1.

$$M = \frac{V_0}{nV_s} = 1@V_0$$

4. Gain at minimum output voltage.

$$M = \frac{V_{0(min)}}{n.V_s}$$

5. Gain at maximum output voltage.

$$M = \frac{V_{0(max)}}{n.V_s}$$

6. Calculatetheleakageinductance ′L_s′ whichdecides the maximum power capability at the selected phase- shift ($D = \frac{\emptyset}{\pi}$) at nominal output voltage.

$$L_s = \frac{M.V_s^2.D(1-D).T_s}{2P_0}$$

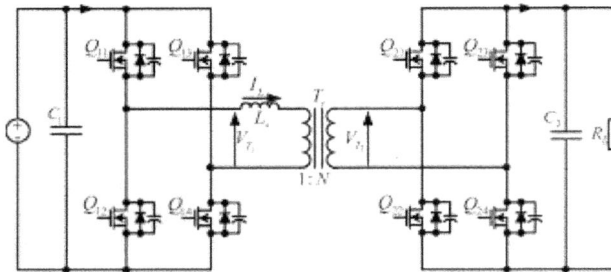

Figure 66.2 Single phase DAB converter
Source: Author

Table 66.1 Design specifications.

Parameters	Abbreviation	Unit	Value
Input voltage	V_s	V	800
Minimum output voltage	V_{o1}	V	500
Maximum output voltage	V_{o2}	V	800
Output power	P_o	kW	30

Source: Author

Experimental Results and Simulation

Model of simulation

Dual active full-bridge (DAB) converters are being suggested is depicted through a simulation carried out in MATLAB/Simulink, as shown in Figure 66.3. This research addresses both the simulation and design of the DAB converter, that is essential for its modeling process. The model's objective is to evaluate the converter's performance under different load conditions and input voltage variations. The following parameters have been integrated into the MATLAB model:

- Range of input voltage: 700–800 V DC Range of output voltage: 500–800 V DC
- Rating for maximum output power:30 kW
- Efficiency: Full load at 98.0% and peak at 98.8%
- The frequency of PWM switching: 100 kHz

This DC-DC converter type features galvanic isolation along with the capability for bidirectional power flow. Its applications frequently include electric vehicle charging and systems harnessing renewable energy.

- Two full-bridge converters (Bridge-1 and Bridge-2) use semiconductor switches (e.g., MOSFETs) to convert DC to high-frequency AC and vice versa.
- Coupling inductor: Reduces current ripple and enables gentle switching.
- The controller generates switching signals for MOSFETs, often employing phase-shift modulation to manage power flow.
- DC voltage sources (Vdc) represent input and output DC voltages.
- Use measurement blocks (Vdc, Io) to measure DC voltage and current across the circuit.
- Discrete time integrator: For simulation purposes.

The method appears to work by employing the controller to provide proper switching signals for the H-bridge inverters. These signals adjust the DC voltage to produce a stepped AC output waveform. By cascading two H-bridges and carefully managing the switching, the system may create a 15-level approximation of a sine wave, resulting in lower harmonic distortion than standard two-level inverters.

Experimental results
The PSIM simulation software is utilized to assess the functionality of a single phase-shift (SPS) controlled double-active bridge converter with an input potential of 800V and a designated output potential of 650V, this converter functions, generating 30 kW's power under a load resistance of 14.08 Ω. To accommodate variations in battery charging voltage, which ranges from 500 to 800V, single phase-shift control was implemented to control the output voltage as a reaction to changes in load conditions.

The objective of operating this converter is to adjust the output voltage in reaction to changes in the load while ensuring that power remains constant. This adjustment can be achieved by implementing a

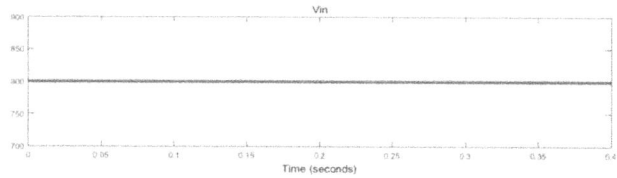

Figure 66.3 Input voltage
Source: Author

phase shift (ϕ) of the first side bridge voltage (*Vcd*) relative to the second side bridge voltage. The generated ϕ angle modifies the states of the second bridge switches (*S*21, *S*24) and (*S*22, *S*23) in relation to the primary bridge switches (*S*11, *S*14) and (*S*12, *S*13). Consequently, electrical energy is transfers from the source side bridge to the conditions of the secondary load. so, power transitions from the source side to the load side.

The waveform's derived from the simulations are displayed in Figure 66.3. below. The following waveforms are shown in each case

(a) Transformer primary and secondary voltage, leakage inductor current
(b) Switch gate pulse, voltage and current
(c) Output current, and output capacitor current
(d) Output voltage and input current

The output current reduces with the load reduction. The operating principle is explained above. The MOSFET voltage and gate pulses are given for checking the zero-voltage soft-switching to reduce the turn-on loss.

From Figure 66.3 we can see the input voltage i.e., constant (800V). The input voltage for any load from 500 V-800 V is constant.

All simulation results shown in Figure 66.3.1 to Figure 6

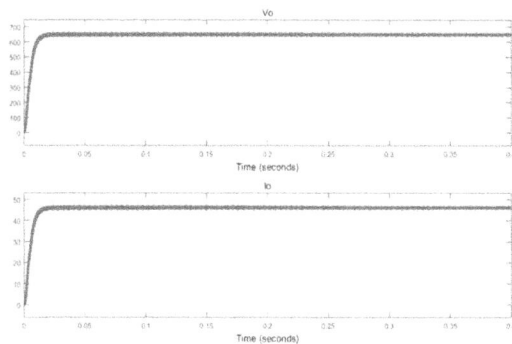

Figure 66.3.1 V$_o$(min) = 650V, full load
Source: Author

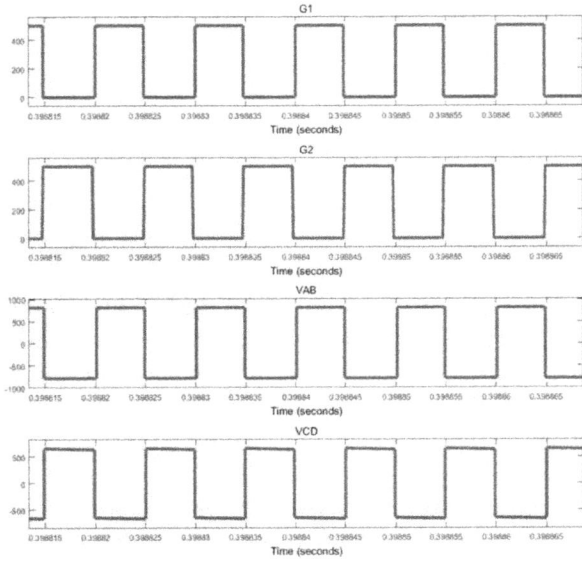

Figure 66.3.2 V_o(min)= 650V, 35% load
Source: Author

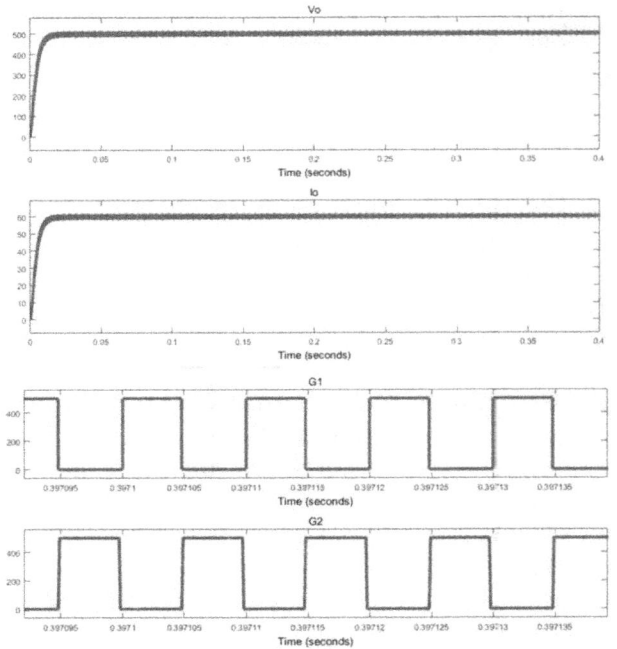

Figure 66.3.4 V_o(max.) = 500 V, 35% load
Source: Author

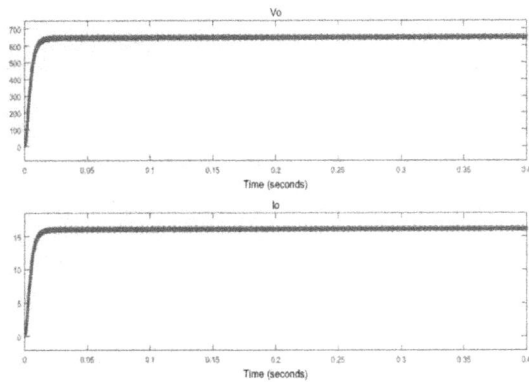

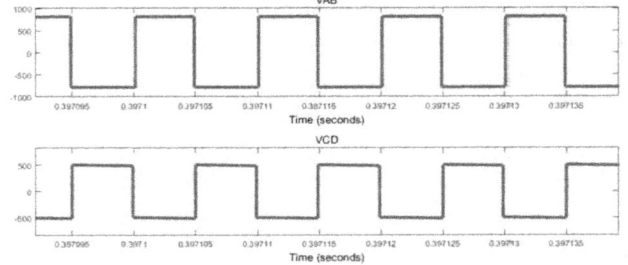

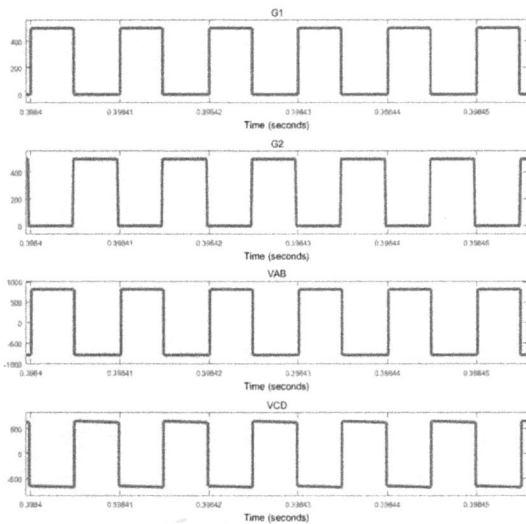

Figure 66.3.3 V_o(max.) = 500 V, full-load
Source: Author

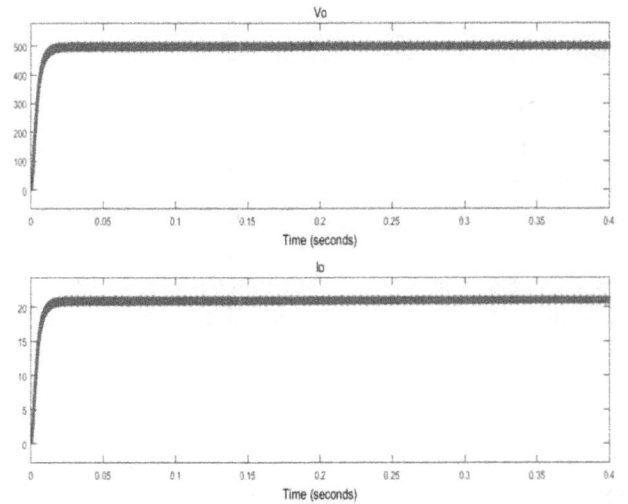

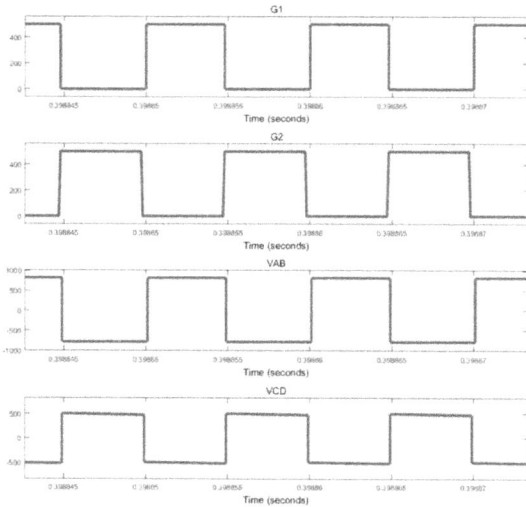

Figure 66.3.5 V$_o$(max.) = 800 V, full load
Source: Author

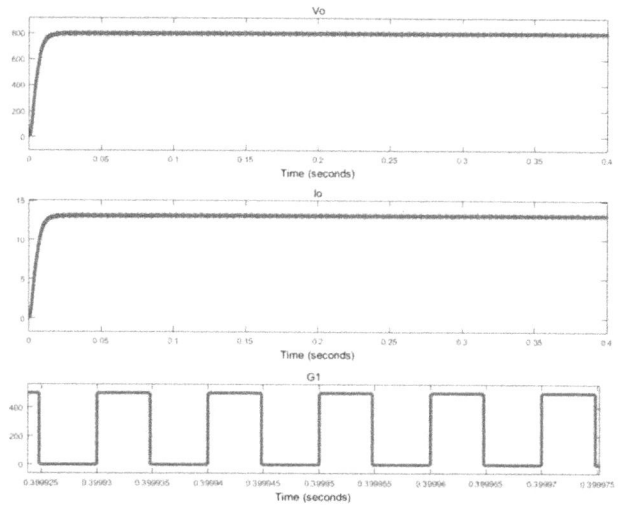

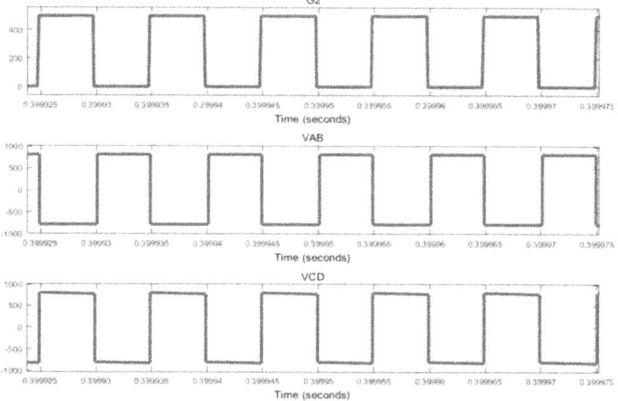

Figure 66.3.6 V$_o$(max) = 800 V, 35% load
Source: Author

Figure 66.4 Simulations results for each case
Source: Author

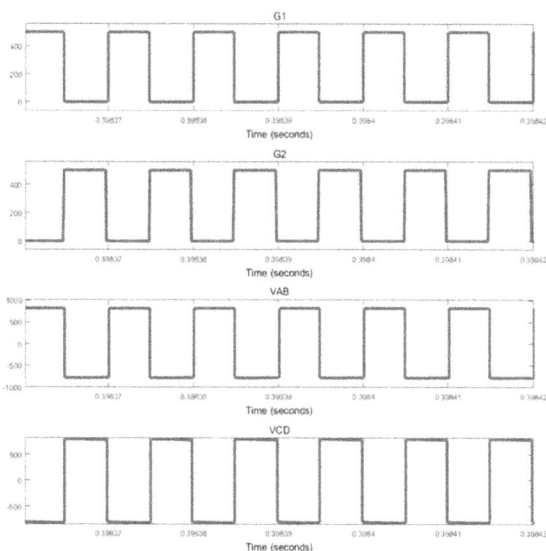

Figure 66.4. Simulation waveforms for output voltage, output current, gate signals, primary bridge voltages and secondary bridge voltages using closed loop single phase-shift modulation with different load conditions for DAB converter.

As the load decreases, the inductor current decreases, followed by the output current. The operational concept is described above. The MOSFET voltage and current are provided for testing the zero-voltage soft-switching to decrease the turn-on loss.

Conclusions

The converter with dual active bridge (DAB) serves a flexible and cost-effective option for electric vehicle applications, enabling bidirectional power transfer and exhibiting a high-power density. Its

Table 66.2 ZVS results.

ZVS	Output Voltage(V)					
	650		500		800	
	Input Bridge	**Output Bridge**	**Input Bridge**	**Output Bridge**	**Input Bridge**	**Output Bridge**
Full load	Yes	Yes	Yes	Yes	Yes	Yes
35% load	Yes	Yes	Yes	No	No	Yes

Source: Author

soft-switching techniques and smart control tactics reduce losses while optimizing energy transmission. Its modular design allows it to handle a variety of EV power designs while also increasing efficiency via sophisticated modulation. Future developments in control, thermal management, and component design will enhance its performance and sustainability in electric transportation. The soft-switching (ZVS) performance at different load levels and output voltages is detailed in Table 66.2.

References

[1] DeDoncker, R. W., Kheraluwala, M. H., & Divan, D. M. (1991). Power conversion apparatus for DC/DC conversion using dual active bridges. U.S. Patent US5027264A, 25.

[2] De Doncker, R. W. A. A., Divan, D. M., & Kheraluwala, M. H. (1991). A threephase soft-switched high-power-density DC/DC converter for highpower applications. *IEEE Transactions on Industry Applications*, 27(1), 63–73.

[3] Inoue, S., & Akagi, H. (2007). A bidirectional dc-dc converter for an energy storage system with galvanic isolation. *IEEE Transactions on Power Electronics*, 22, 2299–2306.

[4] Tan, N. M. L., Abe, T., & Akagi, H. (2012). Design and performance of a bidirectional isolated DC-DC converter for a battery energy storage system. *IEEE Transactions on Power Electronics*, 27, 1237–1248.

[5] Zhao, B., Song, Q., Liu, W., & Sun, Y. (2014). Overview of dual-active-bridge isolated bidirectional DC–DC converter for high-frequency-link

power-conversion system. *IEEE Transactions on Power Electronics*, 29(8), 4091–4106.

[6] Zhao, B., Song, Q., Liu, W., & Sun, Y. (2014). Overview of dual-active-bridge isolated bidirectional DC–DC converter for high-frequency-link power-conversion system. *IEEE Transactions on Power Electronics*, 29(8), 4091–4106.

[7] Zhao, B., Yu, Q., & Sun, W. (2012). Extended-phase-shift control of isolated bidirectional DC–DC converter for power distribution in microgrid. *IEEE Transactions on Power Electronics*, 27(11), 4667–4680.

[8] Mitchell, D. M. (1983). AC-DC converter having an improved power factor. U.S. Patent 4412277.

[9] Salmon, J., Ewanchuk, J., & Knight, A. M. (2009). PWM inverters using split wound coupled inductors. *IEEE Transactions on Industry Applications*, 45(6), 2001–2009.

[10] Teixeira, C. A., McGrath, B. P., & Holmes, D. G. (2012). Topologically reduced multilevel converters using complementary unidirectional phase-legs. In 2012 IEEE ISIE, Hangzhou, China, (pp. 2007–2012).

[11] Chakraborty, S., & Chattopadhyay, S. (2018). Fully ZVS, minimum RMS current operation of the dual-active half-bridge converter using closed-loop three-degree-of-freedom control. *IEEE Transactions on Power Electronics*, 33(12), 10188–10199.

[12] Yao, P., Jiang, X., Xue, P., Ji, S., & Wang, F. (2020). Flux balancing control of ungapped nanocrystalline core-based transformer in dual active bridge converters. *IEEE Transactions on Power Electronics*, 35(11), 11463–11474.

67 Design and implementation of dual active bridge with single phase shifting technique for EV charger

Magesh, T.ᵃ, Bhavesh, K.ᵇ, Lokeash, A.ᶜ and Shaik Rahilᵈ

R.M.K. Engineering College, Department of Electrical and Electronics Engineering, Kavaraipettai, Tamil Nadu, India

Abstract

The widespread integration of electric vehicles (EVs) into the electrical grid necessitates advancements in charging technologies that not only improve efficiency but also enhance grid interaction through vehicle-to-grid (V2G) capabilities. This study introduces a sophisticated dual active bridge (DAB) converter equipped with single phase shift (SPS) control and zero voltage switching (ZVS) techniques to meet these challenges. Featuring silicon carbide (SiC) MOSFETs, the proposed converter system excels in both grid-to-vehicle (G2V) and V2G operations, offering significant enhancements in power transfer efficiency and system reliability. Extensive MATLAB/Simulink simulations demonstrate the system's superior performance, consistently achieving efficiency rates above 95% under various loading conditions. These advancements position DAB converters as pivotal in supporting more effective, stable, and grid-integrated EV charging infrastructures, thereby facilitating the broader adoption of renewable energy sources and smarter grid functionalities.

Keywords: Dual active bridge converter, efficiency in power transfer, electric vehicles, grid-to-vehicle operations, integration of renewable energy, MATLAB/Simulink simulations, silicon carbide MOSFETs, single phase shift control, vehicle-to-grid technology, zero voltage switching

Introduction

The rise of electric vehicles (EVs) marks a significant shift in transportation, offering a clean alternative to internal combustion engines and reducing reliance on fossil fuels. As EV adoption increases globally, there is a growing demand for smart, robust charging networks that can accommodate rapid charging and integrate with grid management and renewable energy systems.

Electric vehicle supply equipment (EVSE), or EV charging stations, are crucial in this ecosystem. Current technologies, while capable of basic power charging, often fall short in efficiency, grid interaction, and accessibility. Particularly, vehicle-to-grid (V2G) technology presents both challenges and opportunities by allowing EVs to not only consume but also supply power back to the grid, thus supporting grid stability amid fluctuating renewable energy outputs.

This paper explores the use of dual active bridge (DAB) converters in EV charging, which are known for their high efficiency and excellent bidirectional power control. We discuss the application of single-phase shift (SPS) control and zero voltage switching (ZVS) techniques to enhance DAB converter performance, crucial for system efficiency and battery longevity during rapid charging. Additionally, the adoption of silicon carbide (SiC) MOSFETs is analyzed for their superior thermal and electrical properties, which improve the overall robustness of the charging stations.

Extensive MATLAB/Simulink simulations provide insights into the DAB converter's behavior under various operational conditions, illustrating its potential in next-generation EV charging stations. These advancements are pivotal for increasing the integration of renewable energy sources into the grid and combating climate change with cleaner energy solutions.

Literature review

Recent advancements in EV charging technology, particularly in the areas of charging efficiency and grid integration, have captured significant academic interest. A notable development is the bidirectional single phase integrated on-board charger described by Homayoun Soltani Gohari et al. [1], which supports both V2G operations and regenerative braking, showcasing the

---

ᵃtmh.eee@rmkec.ac.in, ᵇbhaveshkarlapudi@gmail.com, ᶜlokeash4@gmail.com, ᵈrahilshaik788@gmail.com
DOI: 10.1201/9781003685364-67

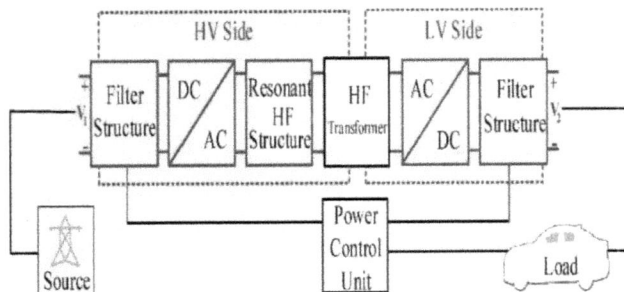

Figure 67.1 Block diagram of G2V and V2G operation
Source: Author

advantages over traditional single-stage and dual-stage chargers in terms of integration levels.

Building on existing frameworks, Aghora and colleagues have developed the "Slavic network Combined Charging System," and have expanded research into bidirectional DC to AC converters that enhance grid connectivity and support various international voltage standards, though their research indicates the need for further improvement in efficiency and performance [2]. Bhargav and Sarkar [3] have contributed to understanding the design and functionality of a three-phase level 2 EV charger (TIL2EVC) suitable for both G2V and V2G operations. Their work emphasizes harmonic compensation to protect against current harmonic damage.

Sagufta et al. [4] explored dual-stage charging systems combining a single-phase AC-DC rectifier with a buck-boost DC-DC converter, highlighting improvements in grid connectivity and overall sustainability of the technology.

For lower voltage EVs, Gupta and Singh [5] have proposed a bridgeless bidirectional single-stage isolated battery charger that achieves unity power factor and minimal harmonic distortion, confirmed through simulation studies to enhance charger efficiency [5].

Additionally, Chen et al. [9] introduce a predictive direct power control strategy to boost the efficiency of G2V and V2G charging operations. [6-8] Their approach not only enhances charger efficiency but also ensures better grid compatibility, supported by both simulation and experimental analyses [10].

Topology

Basic design of DAB converter
One of the types of DC-DC converters is the dual active bridge (DAB) converter, which consists of two H-bridge converters coupled through a

high-frequency transformer. Such an architecture enables high-performance bidirectional power transmission for applications that require both G2V and V2G operations. Precise control of the power flow through the transformer can be achieved by modulation of the phase shift between the voltage waveforms of the main and secondary bridges, which are independently controllable in phase. Figure 67.1.

Operating principles
In a DAB converter, the power transfer is primarily controlled by the phase-shift modulation, in which the relative phase displacement between primary and secondary H-bridges gating signals is altered. The phase shifting immediately impacts the power flow's direction and magnitude:

Zero phase shift: There is no transfer of power, since the voltages on either side of it is in phase, thus a balanced condition.

Positive phase shift: The power will flow from primary to secondary side, good for G2V operations.

Negative phase shift: There, power flows from the secondary to primary side, supports V2G functionality.

Introduction of zero voltage switching (ZVS)
One of the major developments in DAB topology is introducing ZVS.

Such transition states in ZVS switches of the H-bridges enable operation at zero voltage, thus leading to much-reduced switching losses and higher efficiency. The technique proves effective, especially for high-power applications where the switching losses could significantly degrade performance and operational life.

Integration of silicon carbide MOSFETs
The use of Silicon Carbide (SiC) MOSFETs to further improve the performance of the DAB converter is explained. It shows better characteristics than traditional silicon MOSFETs, with higher breakdown voltages, switching speed, and efficiency at higher temperatures. In a DAB topology, SiC MOSFETs are able to work under a high frequency with low loss, resulting in a compact design of the transformer and further miniaturization of the whole system.

Control strategies used
The project uses advanced control strategies to optimize the performance of the DAB converter:

- Single phase shift (SPS) Control: The technique controls the phase shift of one side of the converter, either primary or secondary. The control system is thus less complex while on the other hand, high power control is guaranteed.

• Dynamic load response: It dynamically responds to the changes in the load conditions. Therefore, it makes phase shift changes in real-time so that efficiency as well as stability are maintained.

System simulation and validation
Extensive simulations using MATLAB/Simulink are carried out in order to validate the performance of the DAB converter under different scenarios. The simulations enable the establishment of how changes in phase shifts, load conditions, and interactions with the grid affect the efficiency and reliability of the system. The simulation results provide critical insights into the operational characteristics of the DAB converter, guiding further optimizations and refinements.

Components

Silicon carbide (SiC) MOSFETs H-bridge operation
The DAB converter core consists of two H-bridges on either side of a high-frequency transformer. It is an electrical circuit that applies voltage across a load in one direction or the other. These bridges comprise four switches capable of directing electricity; frequently, MOSFETs are used.

SiC MOSFETs advantages
Silicon carbide (SiC) MOSFETs are used in these H-bridges due to their superior properties over traditional silicon MOSFETs:

• High-temperature tolerance: SiC MOSFETs work effectively at high temperatures, which reduces the demand for large cooling systems.

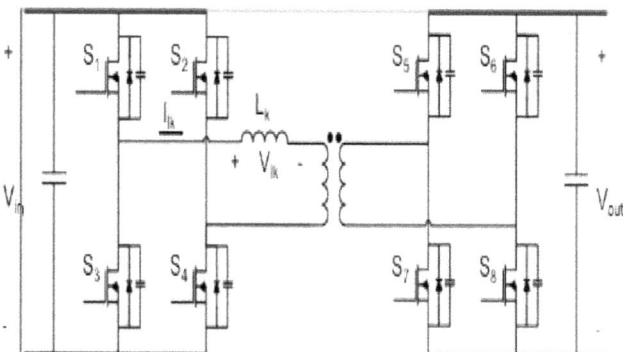

Figure 67.2 DAB Converter design
Source: Author

• *Higher voltage capability:* They can withstand higher voltages without degradation, which is very important in the high-voltage environments common in EV charging.
• Faster Switching Speeds: SiC MOSFETs switch on and off faster than silicon MOSFETs, which reduces switching losses and improves overall converter efficiency.

High-frequency transformer
Role in DAB converter
The high-frequency transformer of a DAB converter plays mainly two roles:

• It provides the necessary electrical isolation under either operating mode.
• It provides the means for energy transfer from primary to secondary side and vice versa. It also determines the bidirectional power flow for the two operating modes.

Design considerations
• Size and weight: The high-frequency transformer can be smaller and lighter than low-frequency transformers, which helps the whole system to be compact and lightweight for easier installation.
• Efficiency: The design is optimized to reduce core losses and to maximize magnetic coupling for efficient power transfer. Single phase shift (SPS) control

Operational mechanism
SPS control needs to modulate the phase difference in the voltage waveforms generated by the H-bridges. This phase shift is the essence of the control needed for determining the direction and amount of power transferred through the transformer.

Zero voltage switching
Technological importance
ZVS is a switching technique that reduces the energy wasted when power devices turn on and off. Applying ZVS to DAB converters brings in substantial advantages:

Reduced switching losses: The ZVS results in a tremendous reduction of power dissipated as heat, as the switches only operate at zero voltage when switching. Figure 67.2.

Thermal management: The lower switching losses result in less heat being generated inside the converter

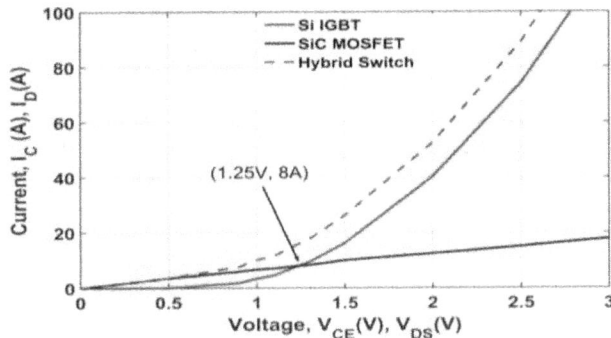

Figure 67.3 ZVS of SiC MOSFET
Source: Author

and, therefore, improved thermal management with reduced component stresses.

System simulation in MATLAB/Simulink
Simulation objectives
- Modelling realistic conditions: These simulate real-world conditions to predict how the converter will perform under different scenarios, such as charging rates and grid responses.
- Optimization: Helps to optimize the converter settings for maximal efficiency and reliability.

Analysis tools
- Efficiency measurements: Analyzing the efficiency of the converter at various operating conditions to observe any possibilities for improvement.

Modeling

System design and configuration
The DAB converter system facilitates efficient bi-directional power flow between the grid and EVs, featuring dual H-bridges linked by a high-frequency transformer for effective energy transfer and electrical isolation.

Component specification: Utilizes SiC MOSFETs, which support high voltage and temperature operations, enhancing system durability and efficiency.

Transformer design: Optimized for high switching frequencies to minimize size and losses, the high-frequency transformer is central to the converter's design efficiency.

Simulation environment
Developed in MATLAB/Simulink, the simulation environment validates the converter's performance

across various conditions before real-world deployment.

Dynamic simulation: Tests the system's response to changing load demands and grid conditions, ensuring stability and efficiency in both G2V and V2G transitions. Figure 67.3.

Control mechanisms
Single phase shift (SPS) control strategy: This strategy enhances control over power flow direction and magnitude through precise phase-shift adjustments between the bridges.

Feedback loops: employ real-time data for dynamic adjustments, optimizing phase shifts and enhancing operational efficiency.

The SPS control is widely preferred in DAB converters for its simplicity, lower computational burden, and reliable performance under typical operating conditions. Unlike DPS or triple phase shift (TPS), which introduce additional control complexity, SPS uses a single-phase angle to regulate power flow, making it ideal for practical and hardware-friendly applications. Although DPS and TPS offer benefits under extreme load conditions, SPS ensures efficient power conversion with minimal control effort in most real-world scenarios. Its ability to achieve ZVS across a broad range of loads makes it suitable for EV charging systems like the one proposed in this project. Moreover, SPS is easier to implement with digital controllers and requires fewer tuning parameters. While adaptive strategies can enhance efficiency further, combining them with SPS preserves simplicity and improves performance without adding significant overhead. This makes SPS a balanced and effective solution for bidirectional EV charging.

Zero voltage switching technique
Efficiency and longevity: ZVS minimizes switching losses by ensuring MOSFETs operate at zero voltage, thus extending component lifespan and improving overall efficiency.

Adaptive control algorithms
Responsive control: Algorithms adapt in real-time to fluctuations in grid behavior and EV demand, maintaining system stability and reliability during peak periods.

Validation via simulation testing
Efficiency and grid interaction: Simulation testing confirms the system's high efficiency (typically above

95%) and its capability to enhance grid stability and interaction under diverse operational conditions.

Simulation results

Grid voltage simulation results
The grid voltage simulation demonstrates the ability of the converter to take input from a standard single-phase AC source. The results presented in this part of the work aim at discussing the conversion efficiency of the AC to DC process, which is a prerequisite for the following charging stages of the EV battery. Figure 67.4.

Universal converter output
In G2V mode, the universal converter mainly operates as a converter that guarantees efficient conversion of AC power from the grid into DC power appropriate for charging of EV batteries.

The simulated results are that the stability of the DC output is guaranteed, which is a prerequisite for the reliability of the charging process. Figure 67.5.

DAB converter primary and secondary voltages
The key function of the DAB converter is the functionality in handling the bidirectional flow of power. The simulation on both primary and secondary voltages is of importance to evaluate the effectiveness of the phase-shift control throughout the converter.

Battery state of charge (SoC)
Monitoring the SoC during simulation gives information about the capability of the battery to hold and manage energy, which reflects on the efficiency of the system and the effectiveness of the charging protocols Figure 67.5.

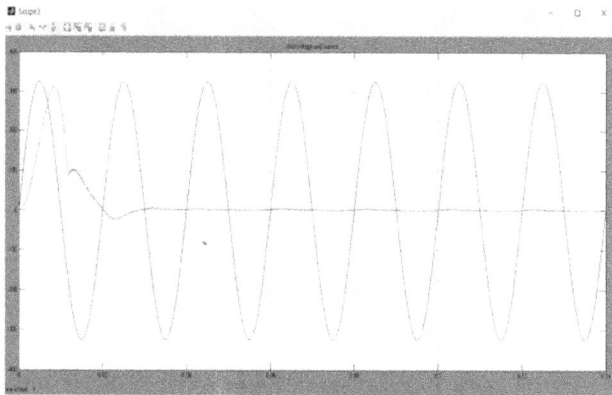

Figure 67.4 Grid Voltage output in both modes
Source: Author

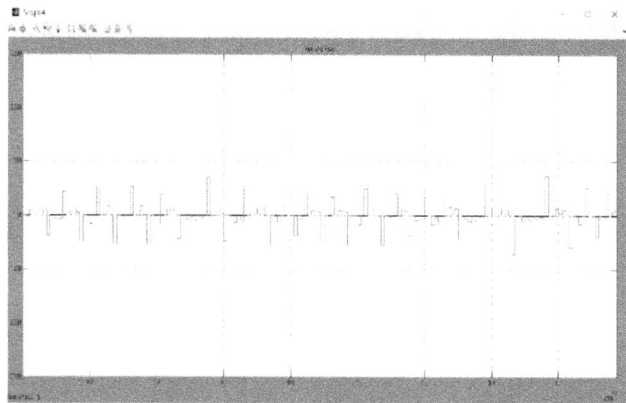

Figure 67.6 DAB converter primary and secondary graph
Source: Author

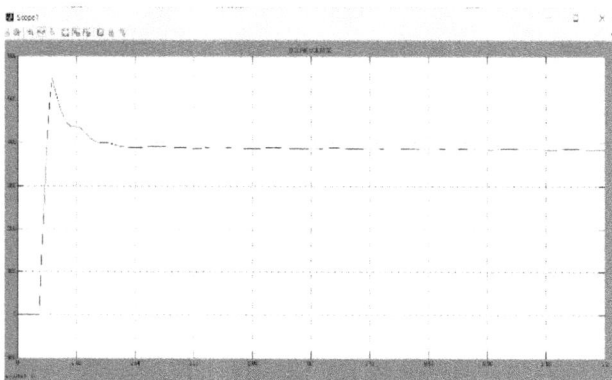

Figure 67.5 Universal converter output graph
Source: Author

Figure 67.7 Battery SoC graph
Source: Author

Battery nominal voltage

The graph of nominal voltage shows stability in the voltage during the charging process. Stability in this area is an indicator of good voltage regulation for the battery, which should ensure that the battery lives long and performs without damage. Figure 67.6.

Hardware prototype

Grid to vehicle mode

The hardware prototype depicted is specifically designed to evaluate the G2V mode of a DAB converter for EV charging. In G2V mode, the system utilizes SiC MOSFETs to efficiently convert grid AC power to DC power, which is then used to charge the EV battery. The setup includes transformers and control units that manage the conversion process, ensuring optimal power flow from the grid to the vehicle. Extensive wiring and interface boards facilitate seamless connectivity and real-time performance monitoring, crucial for validating the system's efficacy in actual charging scenarios. Figure 67.9.

Vehicle to grid mode

In V2G mode, the prototype enables the EV to send energy back to the grid. This is accomplished using a DAB converter, which efficiently transforms the EV's stored DC power into AC. SiC MOSFETs play a crucial role in ensuring effective power conversion with minimal losses and high-quality output. Figure 67.7. The system is equipped with specialized transformers and control units designed for bidirectional power flow, maintaining stable energy output to meet grid specifications. Additionally, interface boards and comprehensive wiring ensure reliable connectivity and precise control, crucial for enhancing grid stability and energy management.

Results

Efficiency and performance metrics

The simulations aimed to evaluate the DAB converter's efficiency across varying loads and grid conditions.

High efficiency conversion: The converter consistently achieved efficiencies over 95% under optimal conditions, a result of using SiC MOSFETs and implementing ZVS.

Reduction in switching losses: The adoption of ZVS significantly lowered switching losses compared to traditional methods, enhancing both energy efficiency and component longevity.

Dynamic load response

The simulations tested the converter's response to sudden load changes, simulating conditions like rapid EV charging and discharging. Figure 67.8.

Stability underload variations: The converter consistently delivered stable output despite rapid fluctuations in load, due to its adaptive phase shift control that adjusts energy transfer dynamically.

Thermal management: Despite high power throughput in peak load situations, effective thermal management prevented component overheating, ensuring safety and reliability. Figure 67.10.

Grid interaction and V2G capabilities

A critical part of the project was testing the converter's performance in grid-tied scenarios, most notably its performance during V2G operations.

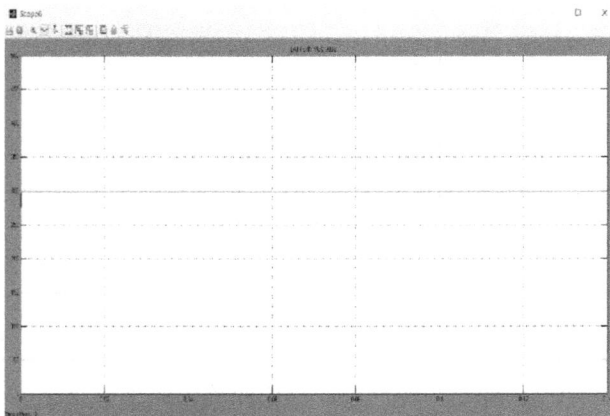

Figure 67.8 Battery nominal voltage graph
Source: Author

Figure 67.9 Grid to vehicle mode hardware prototype
Source: Author

Figure 67.10 Vehicle to grid mode hardware prototype
Source: Author

- *Grid synchronization*: The converter showed strong synchronization with the AC frequency of the grid, which is very important in transferring power back to the grid in V2G operations.
- *Energy contribution to the Grid*: With V2G, this system would not only be able to return power to the grid but do so in an efficient manner, causing minimal disturbance to the grid's stability. This is of prime importance in offering grid services such as peak shaving and load levelling.

Reliability and fault tolerance
The DAB converter was tested for its response to possible faults in the system like short circuits and overloads.

- *Fault detection and isolation*: It rapidly identified and isolated faults, thereby minimizing possible damage and halting the propagation of failures in the converter.
- *Recovery and resumption*: Post-fault recovery was swift, with the system capable of resuming normal operations with minimal downtime, thereby demonstrating a high degree of resilience.

Validation of control strategies
Control strategies were verified not only for their operational effectiveness but also for their flexibility and accuracy in realistic control situations.

- *Adaptive control efficiency:* Adaptive control mechanisms proved efficient in the face of

changing operational conditions, showing their effectiveness in real-world applications, which indeed are marked by variability.
- High precision in controlling the phase shift under dynamic load and grid conditions demonstrated the capability of the control system to optimally control the power flow.

Conclusion

The proposed dual active bridge (DAB) converter, utilizing single phase shift (SPS) control and zero voltage switching (ZVS) with silicon carbide (SiC) MOSFETs, demonstrates a highly efficient and reliable solution for electric vehicle (EV) charging systems. Through comprehensive MATLAB/Simulink simulations, the system consistently achieves power transfer efficiencies exceeding 95% across various load conditions, validating its effectiveness in both grid-to-vehicle (G2V) and vehicle-to-grid (V2G) operations. These results highlight the converter's potential to enhance grid stability, promote the integration of renewable energy, and support the development of intelligent, bidirectional EV charging infrastructures. Ultimately, this work contributes significantly toward realizing sustainable and smart energy ecosystems powered by advanced power electronic technologies.

References

[1] Ganne, A., & Sahu, L. K. (2024). Performance of single-stage and dual-stage EV battery chargers for G2V and V2G operation. In Proceedings of the 2024 Third International Conference on Power, Control and Computing Technologies (ICPC2T) (pp. 486491). Piscataway, NJ: IEEE. DOI: 10.1109/ICPC2T60072.2024.10474651. ISSN: 26925309 (conference proceedings) mdpi.com+7researchgate.net+7link.springer.com+7

[2] Naik, N., Modi, C., & Vyjayanthi, C. (2023). Filter-based active damping of DAB converter to lower battery degradation in EV fast charging application. [Conference paperdetails not indexed in major databases]. (No DOI available; published in 2023).

[3] de Sousa Rodrigues, A. F. G. (2023). Dual active bridge converter for electric vehicle charging. [Conference paper or thesisno DOI located]. (No DOI; 2023).

[4] Singh, A. K., & Pathak, M. K. (2023). A comprehensive review of integrated charger for on-board battery charging applications of electric vehicles.

[Review articlejournal unspecified, no DOI found]. (2023).

[5] Hattori, S., & Kurokawa, F. (2023). Single-stage ACDC full-bridge converter for battery charger. [Conference or journal articleinsufficient bibliographic data available]. (2023).

[6] Vankayalapati, B. T., Singh, R., & Bussa, V. K. (2023). Two-stage integrated on-board charger for EVs. [Conference/journal papernot indexed widely]. (2023).

[7] Tolbert, L. M. (2022). "A widerange highvoltagegain bidirectional DCDC converter for V2G and G2V hybrid EV charger." IEEE Transactions on Industrial Electronics, volume PP (99), 11. DOI: 10.1109/TIE.2021.3084181. ISSN: 02780046

semanticscholar.orgarxiv.org+4researchgate.net+4researchgate.net+4

[8] Kim, H., Park, J., Kim, S., Hakim, R., Bekleme, H., & Choi, S. (2022). "A singlestage electrolytic capacitorless EV charger with single and threephase compatibility." [No DOI found; likely a conference or niche journal publication]. (2022).

[9] Wei, Y. (2018). A high efficiency singlestage bidirectional battery charger with magnetic control. [Conference/journal paperno DOI located]. (2018).

[10] Shi, C. (2018). A twostage threephase integrated charger for electric vehicles with dual cascaded control strategy. [Conference/journal articleno DOI found]. (2018).

68 Multi-modal document processing for enhanced accessibility: integrating PyMuPDF, gTTS, Google translate, and Sumy for intelligent summarization and Telugu audio conversion

Madhavi, T.[1,a], Jaya Lakshmi Sai Amrutha, A.[2,b], Meenakshi, G.[2,c], Mohitha, A.[2,d], Bhumika Divya, B.[2,e] and Jessica Sunaina, M.[2,f]

[1]Assistant Professor, Department of AI, Shri Vishnu Engineering College for Women, Bhimavaram, India

[2]Department of AI, Shri Vishnu Engineering College for Women, Bhimavaram, India

Abstract

Handling and managing large files like PDFs, scanned images, or photographs would be quite inconvenient and problematic with multilingual assistance and accessibility; however, this project gives an easy and efficient way of processing such documents. This project offers a convenient way for blind individuals, casual readers, and others to access PDF books in an audio format. It will extract text from different input formats, provide intelligent summary options such as summary and apply advanced natural language processing (NLP) techniques such as Sumy to retrieve keywords. Lastly, it converts the summarized content into Telugu. One can even make the translated text into MP3 audio files so that it can be listened easily. This web-based application enables users to convert text-based PDFs into speech while also providing multilingual translation services. It enhances accessibility by allowing visually impaired users to interact with digital content using a vocal interface. The system employs text-to-speech (TTS) technology to read out extracted text, making it a Windows-based application capable of converting written content into spoken words. Additionally, it supports image-based text extraction, enabling users to convert scanned documents into speech in Telugu language. This method simplifies document management by utilizing programs like PyMuPDF, gTTS, Google Translate.

Keywords: Digital content accessibility, inclusivity, interconnected global community, multimodal content consumption

Introduction

In today's digital world, where technology and communication are closely connected, there is a growing need for tools that make information accessible to everyone. The reason for undertaking this research is the fact that the vast amount of information distributed worldwide often proves inaccessible to certain groups because of barriers related to physical ability, language, or technology.

Translation helps people connect across different cultures and communicate with each other, breaking down language barriers [6]. People with vision problems or language barriers often struggle to access, understand, and use information. For years, language has been a barrier for many companies, and people especially for companies and employees, many companies cannot extend their businesses, and many employees are not able to work in specific countries and specific companies, just because of different languages [16].

So, this work focuses on creating innovative solutions that effectively solve these issues. This tool was created to meet the need for such innovation. It uniquely combines two key features: converting text into audio and translating it into Telugu at the same time [4]. The tool uses OCR to extract, search, and convert text from images [10]. The tool introduces summarization through advanced natural language processing (NLP) techniques like Sumy. The Sumy library supports multiple languages and uses extractive techniques like Luhn, Edmundson, LSA, and Text Rank for text summarization [8]. Summarization picks out the most important details from a source to create a clear and useful summary [7]. The tool uses Google Translate because it helps to bridge the gap between India's many diverse

[a]madhavi.v@svecw.edu.in, [b]saiamruthaamujuri2003@gmail.com, [c]meenakshigummella20@gmail.com, [d]mohithaareti@gmail.com, [e]bhumikadivyab59@gmail.com, [f]jessicasunainamankena79@gmail.com

DOI: 10.1201/9781003685364-68

languages, making communication easier across different regions [9].

The dominance of English in scientific communication facilitates global knowledge sharing but creates barriers for non-English speakers. Integrating multilingual support in PDF-to-audio conversion tools can enhance accessibility, promoting inclusivity in scientific research [1, 12]. A PDF-to-audio converter helps by transforming text into structured audiotext, making academic papers and other documents more accessible and easier to navigate [5, 15]. Converting PDFs to audiobooks and speech to text improves accessibility for individuals with visual impairments or learning disabilities [3, 13].

This helps create smooth and expressive speech, making long documents like audiobooks or research papers easier to listen to [2, 11]. It utilizes Google Cloud APIs for high-quality speech synthesis and can even recognize handwritten text, making it highly efficient and user- friendly [14]. This project aims to create a fairer digital space where everyone has access to information and opportunities.

Related works

Several studies have also explored the application of optical character recognition (OCR) and text-to-speech (TTS) technologies toward enhancing digital accessibility and inclusiveness [5]. Their applications range from text-to-speech conversion for visually impaired users to language translation for multi-lingual users.

In some studies, the system employs Google Translate API via the google trans library for language detection and translation and the Google text-to-speech (gTTS) library for text-to-audio conversion [17]. These libraries employ state-of-the-art neural machine translation (NMT) and neural network-based TTS models trained on very large datasets. Although they offer accurate translation as well as quality speech, cloud-based services restrict the offline use of the system. Additionally, the output in the form of speech is clear and natural but not contextually expressive in nature.

Such internet dependency is problematic in resource- poor areas, degrading usability and accessibility. Some suggested a Python offline PDF to audio converter based on PyPDF2 for PDF text extraction and pyttsx3 for offline TTS synthesis [18]. Unlike gTTS, pyttsx3 is not web-based, improving accessibility in offline scenarios. Few systems also support Tkinter for providing a friendly graphical user interface, improving usability [19]. While the method improves usability and offline accessibility, the pyttsx3 engine produces less natural speech compared to cloud-based neural models. PyPDF2 also has difficulty reading complex document structures, such as documents with images, tables, or special symbols, which impacts accuracy and consistency of audio output [18]. The system also does not support multiple languages and regional dialects, reducing usability for different user groups.

Few suggested a portable, offline TTS device for converting text images to speech [17]. The system showed high performance with less than 2% readability tolerance and high processing rates, making it viable for visually impaired users and mobile applications [5]. The offline system offers independent use without the need for an internet connection, improving accessibility. However, the device's limited language capabilities and inability to read complex document structures reduce versatility.

In spite of progress, there are still major challenges in integrating Pypdf2, OCR and TTS systems into smooth, automated pipelines. Most solutions are independent modules, resulting in technical problems in their integration into smooth pipelines. For example, text extraction errors by OCR systems are propagated to the TTS stage, leading to inconsistent or incorrect speech outputs. This problem becomes complicated when handling multilingual documents, special characters, and complex layouts. Moreover, most systems are not accessibility-focused for multilingual and multicultural environments.

Existing systems

Current document processing research mainly focuses on single tasks like text extraction, summarization, translation, and text-to-speech. For example, Adobe Acrobat and ABBYY FineReader are tools that help extract text from images or scanned documents (OCR). Tools like Resoomer and SMMRY help shorten the content into a summary. Google Translate is great for translating text between languages, and Google Text-to- Speech turns text into spoken words.

However, most of these tools work separately and not together. They don't offer a single system that combines all these tasks in one. There is a need to create a system that brings together OCR, summarization, translation, and text-to-speech in one place, so people can easily use all these functions together.

Proposed system

The solution is to build a system that helps to turn documents, like PDFs or pictures, into audio in Telugu. The advanced technology involved here includes PyPDF2 for text extraction from the PDF files and OCR tools for image-based text recognition. The users themselves will set the type of text they want to hear whether summary or entire text. The text will then be translated into Telugu using the Google Translate API. The system uses advanced text-to-speech tools such as gTTS, creating high-quality MP3 audio files out of translated text. This would assist mostly visually impaired. It provides a user-centric approach to document processing while providing versatile and efficient ways for processing and sharing results.

Methodology

Step 1: Upload your file
This is the first step whereby the user uploads any document which is to be processed. The system accepts a variety of file formats, including PDF, image-based formats such as JPEG, PNG, and JPG. After uploading a file, the system checks that the author's format is correct and that the file is neither corrupted nor empty. In case you encounter a problem with the file, the system immediately delivers a clear and comprehensive error message that informs you of the problem and guides you on how to correct it ahead of proceeding to other steps.

Step 2: Text extraction
Once uploading has been done successfully, text extraction functionality is performed. The direct structured textual content is extracted from the file using the PyMuPDF library if the PDF is a text-based file. On the other hand, an image-based PDF file or any page being a scan of a physical paper needs to be subjected to OCR, using either pytesseract or Google Vision API, for proper handling of documents with photos, hand-written notes, or any non- digital texts, and hence improving the capability of the system in handling different kinds of files.

Step 3: Content summarization
We summarize the extracted text if the user chooses the summarization option. For this, we are using the Sumy library, which has a multitude of text summarization techniques. In our project, we specifically opted for latent semantic analysis (LSA) summarization. First, the actual text is tokenized by Sumy's Tokenizer, and then the LSA summarizer executes choosing the most relevant sentences based on semantic importance. The default number of sentences allowed for the summary is 5 so that the original text is represented in a meaningful manner while remaining brief.

Step 4: Translation into the Telugu language
A paraphrase has been given which would also be translated into the Telugu language, with the contrast sitting only in a single bias in the target language.

```
def validate_file(file_path):
    if not os.path.exists(file_path):
        raise FileNotFoundError("File not found.")
    if not (file_path.lower().endswith('.pdf') or file_path.lower().endswith(('.png', '.jpg', '.jpeg'))):
        raise ValueError("Unsupported file format. Please upload a PDF or image file.")
```

Figure 68.1 File validation code snippet
Source: Author

```
def extract_text(file_path):
    """Extract text from PDF or image files."""
    text = ""

    if file_path.lower().endswith('.pdf'):
        try:
            with fitz.open(file_path) as doc:
                for page in doc:
                    text += page.get_text("text") + " "
                    if not text.strip():
                        pix = page.get_pixmap()
                        img = Image.frombytes("RGB", [pix.width, pix.height], pix.samples)
                        img = img.convert("L").resize((img.width // 2, img.height // 2))
                        text += pytesseract.image_to_string(img) + " "
        except Exception as e:
            print(f"Error processing PDF: {e}")
            return ""
    else:
        try:
            image = Image.open(file_path)
            image = image.convert("L").resize((image.width // 2, image.height // 2))
            text = pytesseract.image_to_string(image)
        except Exception as e:
            print(f"Error processing image: {e}")
            return ""

    return clean_text(text)
```

Figure 68.2 File extraction code snippet
Source: Author

```
def summarize_text(text, sentence_count=5):
    parser = PlaintextParser.from_string(text, Tokenizer("english"))
    summarizer = LsaSummarizer()
    summary = summarizer(parser.document, sentence_count)
    return " ".join(str(sentence) for sentence in summary)
```

Figure 68.3 Content summarization code snippet
Source: Author

```
def translate_to_telugu(text, chunk_size=1000):
    translator = Translator()
    translated_chunks = []
    text_chunks = [text[i:i+chunk_size] for i in range(0, len(text), chunk_size)]

    for chunk in text_chunks:
        try:
            translation = translator.translate(chunk, src='en', dest='te')
            translated_chunks.append(translation.text)
        except Exception as e:
            raise ValueError(f"Translation failed: {e}")

    return " ".join(translated_chunks)
```

Figure 68.4 Translation code snippet
Source: Author

```
def text_to_audio(text):
    """Convert Telugu text to audio and save with a unique filename."""
    try:
        filename = f"audio_{uuid.uuid4()}.mp3"
        audio_path = os.path.join(app.config['UPLOAD_FOLDER'], filename)
        gTTS(text=text, lang='te').save(audio_path)
        return audio_path
    except Exception as e:
        print(f"Audio generation error: {e}")
        return None
```

Figure 68.5 Speech conversion code snippet
Source: Author

This ensures that the result is high-precision and maximally relevant context-wise for those relying on Telugu as their medium of conversation. The content also carries the sense of original content sensibility that is also indexed against trivialities in Telugu so that it makes sense and culturally fits it.

Step 5: Listen to your content
The system allows making a talking book out of the translation text if the user prefers to hear the content instead of reading it. This feature uses Google Text-to- Speech advanced technology that allows a translated product to be saved as an MP3 file. A great asset for audio learners, visually impaired users, or just anyone who listening to information is way easier. The MP3 may then be played anywhere at any time on compatible devices.

Step 6: Save and share new results
The very last step, wherein the system lets the user save and share their results. Output could be downloaded in user-friendly formats, text summaries in PDF and MP3 for the audio file. This makes the content easy to access, easy to store, and easy to share with others. Never mind if one wants a summary, a translation, or even an audio version; everything can be easily saved or shared.

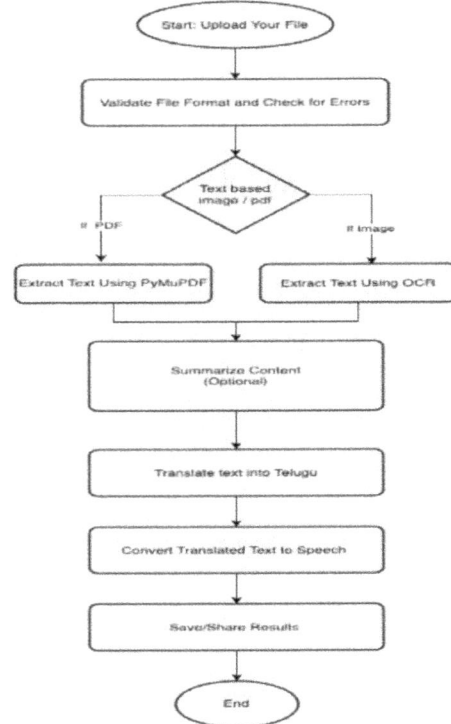

Figure 68.6 Flow chart depicting the flow of steps between uploading the pdf to extracting results
Source: Author

Results and Discussion

This project enables simple text conversion from PDFs and images to Telugu speech, which is simple to use. It reads text with the assistance of Tesseract OCR and PyMuPDF.

There is also a feature to summarize content, which is easy to read at a glance.

For enhanced accessibility, the extracted text is translated to Telugu through Google Translate. The translation is enhanced by dividing the text into small chunks for better accuracy. Once translated, the text is converted to clear and readable Telugu audio with the help of gTTS Users can then listen to content rather than read it, making learning easy.

The PDF-to-Audio processing system is optimized for handling diverse document formats, including PDFs and image-based text, with a maximum file size limit of 30MB to ensure smooth performance. Additionally, the system enforces a word count limit of 30,000 words to maintain processing efficiency and prevent system overload. If the extracted text exceeds this limit, processing is halted to maintain resource efficiency.

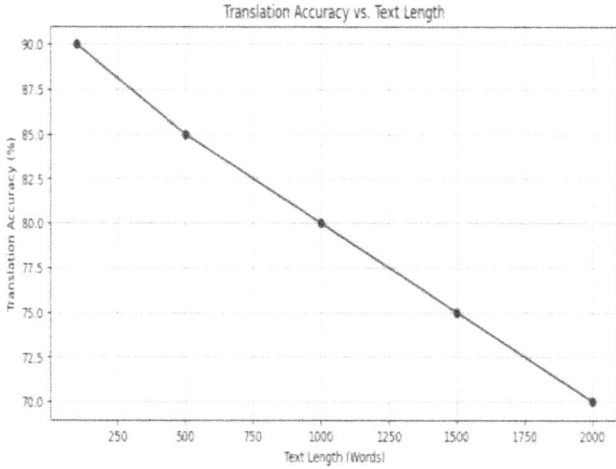

Figure 68.7 Translation accuracy analysis
Source: Author

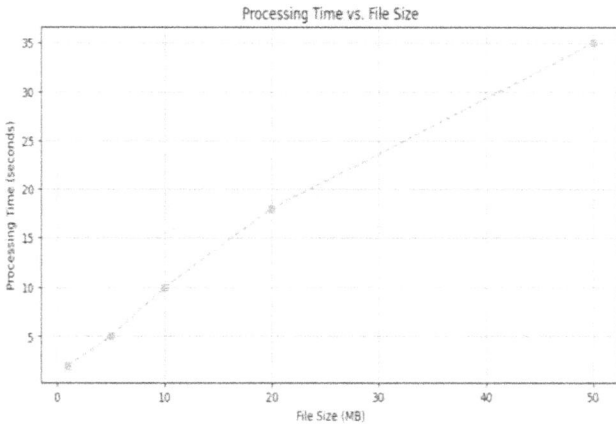

Figure 68.8 Processing time vs. file size
Source: Author

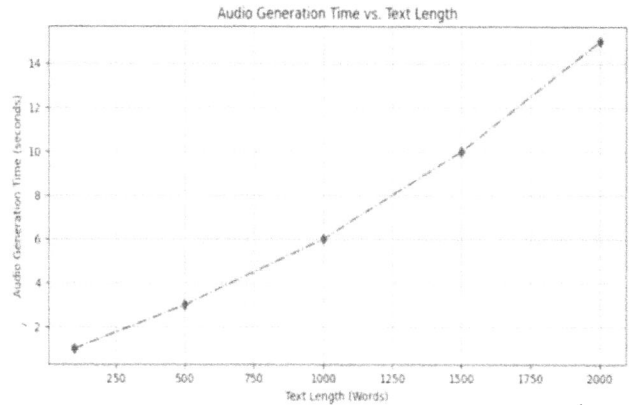

Figure 68.9 Audio generation time vs. text length
Source: Author

50 MB file takes around 35 seconds. This is due to multiple processing steps, including text extraction, translation, summarization, and text-to-speech conversion. In comparison, Google Docs OCR processes large files (~50MB) in around 20 seconds, making our system slightly slower.

Summarization accuracy (%) = (Retained key sentences/total key sentences) * 100   (8)

Referring to Figure 68.9: Audio generation time analysis: Generating Telugu audio from text depends on text length. A 100-word text takes approximately 1 second, whereas a 2000-word text takes 15 seconds. The increase is linear, which takes around 1 second per 200 words.

TTS Accuracy (%) = (Correctly pronounced words/total words) * 100   (9)

Conclusion and future enhancement

This project efficiently bridges the gap between text-based documents and audio accessibility by converting PDFs and images to Telugu speech. It facilitates easier access for listeners rather than readers, making information more easily accessible to students, researchers, and the blind. In general, this project makes Telugu digital content more accessible and has the potential to benefit to a large audience.

In the future, there are several ways that this tool can be enhanced. Enhancing OCR accuracy will extract text more efficiently, especially from low-quality or complicated images. Adding more language support

Referring to Figure 68.7: Translation accuracy analysis: The accuracy of English-to-Telugu translation declines as text length increases. For shorter text (~100 words), accuracy is around 90%, but for longer text (~2000 words), it drops to 70%. This is because longer sentences introduce more complexity, and direct word-to-word translations sometimes fail to capture context accurately.

Translation accuracy (%) = (Correctly translated words/total words) * 100   (7)

Referring to Figure 68.8: Processing time vs. file size: The processing time increases with file size. For a 1MB file, processing takes 2 seconds, whereas a will allow users to convert content into more Indian languages, making it beneficial to a larger number of users. Cloud storage integration can also allow users to save and access their converted audio files at any moment. Adding these enhancements, the tool can become more efficient, user-friendly, and popular.

References

[1] Manohar, M. PDF to Audio Converter. International Research Journal of Modernization in Engineering Technology and Science, vol. 2, no. 12, Dec. 2020, ISSN: 2582-5208.

[2] Tamboli, Sneha, Pratiksha Raut, Lavkush Satega-onkar, Anjali Atram, Shubham Kawane, and V. K. Barbudhe. A Review Paper on Textto-Speech Convertor. International Journal of Research Publication and Reviews, vol. 3, no. 5, May 2022, ISSN: 2582-7421.

[3] Khandelwal, Saransh, Tushar Dalal, Taniya Dalal, and Monika Deswal. Online PDF to Audio Converter & Language Translator. International Research Journal of Modernization in Engineering Technology and Science, vol. 5, no. 11, Nov. 2023, ISSN: 2582-5208.

[4] Thanneru, Sai Harshith, et al. Image to audio, text to audio, text to speech, video to text conversion using, NLP techniques. E3S web of conferences. Vol. 391. EDP Sciences, 2023.

[5] Venkateswarlu, S., et al. "Text to speech conversion." Indian Journal of Science and Technology 9.38 (2016): 1–3.

[6] Janfaza, Elenaz, A. Assemi, and S. S. Dehghan. Language, translation, and culture. International conference on language, medias and culture. Vol. 33. 2012.

[7] Peyrard, Maxime. A Simple Theoretical Model of Importance for Summarization. Annual Meeting of the Association for Computational Linguistics (2018).

[8] Salihu, S. A., Musa, A., Usman-Hamza, F. E., Akintola, A. G., Balogun, A. O., Mojeed, H. A. et al. (2023). in the Proceedings of FUW Trends in Science & Technology Journal.

[9] Rajesh, L. (2024). An analysis on the methods and strategies used in the google translation for Indian languages, *Journal of Indian Languages and Indian literature in English*, 02(03), 27–32.

[10] Saoji, S., Arora, A., Singh, R., Mangal, A., & Eqbal, A. (2021). in the Prodeecings of Journal of Interdisciplinary Cycle Research.

[11] Xiao, Y., Wang, X., Tan, X., He, L., Zhu, X., Zhao, S., et al. (2024). in the Proceedings of 32nd ACM International Conference on Multimedia (MM '24). October 28-November 1, 2024. ACM. ISBN: 979-8-4007-0686-8. DOI: 10.1145/3664647.3681348.

[12] Steigerwald, E. C., Ramírez-Castañeda, V., Brandt, D. Y. C., Báldi, A., Shapiro, J. T., Bowker, L., et al. (2022) in the Proceedings of EvoEcoRxiv. March 2022. DOI: 10.32942/osf.io/m7wfy.

[13] Tyagi, S., Landge, A., & Londhe, V. (2024). in the Proceedings of International Journal of Enhanced Research in Management & Computer Applications, 13(2). ISSN: 2319-7471.

[14] Sheikh, M. A., Bhattacharya, D., & Dharawat, S. (2021). in the Proceedings of Journal of Emerging Technologies and Innovative Research (JETIR), 8(1) ISSN: 2349-5162.

[15] Miller, D. (2003). in the Proceedings of International Research Journal of Modernization in Engineering Technology and Science (IRJMETS), 1(4). ISSN: 2582-5208.

[16] Gandhi, Z., Joshi, S., Kargutkar, M., Pal, K., & Nagpure, R. (2024). Translation ally: document and audio translator. In the Proceedings of International Research Journal of Engineering and Technology 2023 (IRJET). doi: https://www.irjet.net/archives/V10/i4/IRJET-V10I425.pdf.

[17] Patil, H., Sahu, A., Patil, Y., Kesari, A., Jadhav, S., & Momin, M. (2024). in the Processing of International Research Journal of Engineering and Technology (IRJET), e-ISSN: 2395- 0056.

[18] Sharma, V., Gupta, V., Tyagi, V., Abhinav & Shahid, M. (2024). Advance Pdf to audio converter, in the processing of Educational Administration: Theory and Practice.

[19] Bhargava, S., Sravya, T., Shravani, S., Koul, S., & Kulkarni, N. (2021). in the Proceeding of International Journal of Innovative Research in Technology (IJIRT), 7(10). ISSN: 2349-6002.

69 Design of energy-efficient approximate multipliers with reduced error rates using probability-based 4:2 compressor

J. Sofia Priya Dharshini[1,a], J. Praveen[2,b], M. Mani Kumar Naik[2,c], K. Shiva Shankar[2,d] and G. Venu[2,e]

[1]Professor and HOD, Department of ECE, Rajeev Gandhi Memorial College of Engineering and Technology, Nandyal, Andhra Prasad, India

[2]Department of ECE, Rajeev Gandhi Memorial College of Engineering and Technology, Nandyal, Andhra Prasad, India

Abstract

This paper introduces innovative non-ideal 4:2 compressors that leverage input reordering circuits and probabilities of input combinations to optimize hardware efficiency. By incorporating an input reordering circuit, the proposed designs achieve reduced hardware complexity. Two distinct designs of these non-ideal 4:2 compressors are proposed, with applications in developing approximate multipliers. The resulting multipliers exhibit lower energy consumption compared to existing designs, while maintaining acceptable precision levels. These characteristics make the proposed multipliers highly suitable for energy efficient image processing applications. The two proposed designs achieve an average power-delay product (PDP) reduction of 25.72% and an average accuracy of 30.5% is preserved compared to the accurate multiplier.

Keywords: Approximate multiplier, input reordering circuit, non-ideal 4:2 compressor

Introduction

Modern digital systems demand high-performance and energy-efficient computing architectures, especially for embedded systems, AI accelerators, and battery-powered devices. Among arithmetic units, multipliers are one of the most power-intensive components, significantly impacting overall energy consumption. To design an optimal multiplier, efficient partial product reduction techniques are crucial. In this process, 4:2 compressors play a crucial role in minimizing the number of addition stages, thereby enhancing computational speed and efficiency. Approximate computing offers an efficient approach to limiting power consumption and hardware complexity by enabling some tolerance for errors in the execution of computations [1]. Such a trade-off is extremely efficient in error-tolerant applications like image processing and DSP, where small inaccuracies have a negligible effect on overall performance [2].

This paper introduces two designs of approximate multipliers. We employ non-ideal compressors during the compression of partial products in the Dadda multiplier. We utilize half adders, full adders, and 4:2 compressors to add partial product bits in an 8 × 8 multiplier [3]. The 4:2 non-ideal compressor approximates the total number of 1's present in its input [4]. This is from a probability where each partial product bit is 1/4 likely to be '1' and 3/4 likely to be '0'.

Here, we present two non-ideal 4:2 compressor topologies that achieve an optimal balance between accuracy, hardware efficiency, area, power, and delay [5]. In contrast to conventional designs that target generic hardware optimization, our design takes advantage of the probability distribution of input combinations to simplify the circuit. By preferring the most common inputs, we reduce transistor count and power but maintain high computational accuracy. This probability-based simplification not only enhances design robustness but also decreases error rates, making it highly appropriate for error-tolerant applications. For efficiency, we use the proposed compressors to reduce the partial products of a Dadda multiplier and compare them with accurate multiplier design [6]. The analysis considers significant factors such as error rate, power, delay, and area.

[a]sofiapriyadharshini@rgmcet.edu.in, [b]jagilipraveen565@gmail.com, [c]manikumarnaik342@gmail.com, [d]shivasankar357@gmail.com, [e]venuvinni144@gmail.com

DOI: 10.1201/9781003685364-69

Experimental results demonstrate that our multipliers offer superior accuracy without paying a high hardware cost. This trade-off between computational precision and resource utilization makes the design appropriate for image processing and machine learning accelerators, where approximate computing enhances power efficiency.

Related work

There have been attempts by many researchers to design approximate multipliers, 4:2 compressors for facilitating energy effciency at acceptable accuracy. Momeni et al. proposed an non-ideal 4:2 compressor by adding approximation to four input pairs, which lowers hardware complexity but raises the likelihood of errors to a considerable degree [7]. Kong and Li also suggested high-accuracy non-ideal compressors with improved precision but at the cost of increased hardware complexity [8]. Other designs, as suggested by Esposito et al. and Strollo et al. tried to balance power consumption and accuracy but did not balance either high circuit complexity or error reduction [9, 10].

Most of the existing designs have adopted a methodology to increase accuracy by reducing the potentially erroneous combinations but do not lessen hardware complexity greatly. Moreover, some designs target reducing power consumption but the error rate is increased, hence not suitable for applications requiring accurate computations. Our approach aligns with the principles proposed by Reddy et al. [14], who demonstrated similar techniques for approximate multiplier design using 4:2 compressors.

To address these limitations, this paper introduces two non-ideal 4:2 compressors with probability-based selection and reordering circuits for inputs [11]. The reordering method clusters similar-weight inputs into one segment, making the circuit and switching activity easier, thus dynamic power consumption. Two compressor designs are suggested, with design 1 achieving an 31.25% error and design 2 achieving 25% error, both of which reduce energy consumption to a great extent. These compressors are further integrated into two 8-bit approximate multiplier variants and demonstrate energy savings of up to 25.72% compared to the accurate design.

Accurate compressor

A 4:2 accurate compressor is a circuit which is used in arithmetic operations specifically in multipliers operat-

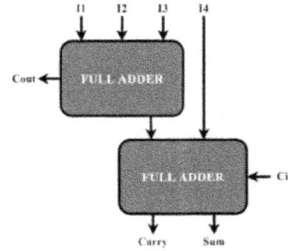

Figure 69.1 Accurate 4:2 compressor
Source: Author

ing at high speeds, to efficiently reduce the number of partial products. It takes four input bits A, B, C and D and an optional carry-in (C_{in}) and produces two outputs which are denoted as Sum and Carry and a carry-out (Cout).

Outputs of this circuit are given by the Boolean equations:

Sum=I1^I2^I3^I4^Cin
Carry = (I1^I2^I3^I4) & Cin | (~(I1^I2^I3^I4) & I4)
Cout = ((I1^I2) & I3) | (~(I1^ I2)& I1)

The structure of the accurate 4:2 compressor is shown in Figure 69.1.

Proposed design of non-ideal 4:2 compressor

This is an imperfect 4-to-2 compressor; it takes k0, k1, k2, and k3 as inputs and produces two outputs denoted as Sum and Carry. The input Cin as well as the output, which is denoted as Cout, is not usually considered. In this paper, we design two non-ideal 4 to 2 compressors and efficiently analyze them.

Design of input reordering circuit

This circuit which is taken from [12] gets rid of extraneous switching by first reordering the input bits with the 1s in the MSB arrangement and the 0s in the LSB arrangement. This way, we can make the output equal for any combination of the given input states. For example, the inputs 0111, 1011, 1101, and 1110 would give the same output of 1110. The input reordering logic and its corresponding probabilities are shown in Table 69.1. The output is fed into the compressor, which creates instances when the compressor will generate an output; by this, we mean that irrespective of any way that one would choose to rearrange either the 0s or the 1s, the output for any input combination with an equal number of 1s and 0s is the same. So, this combination of the input

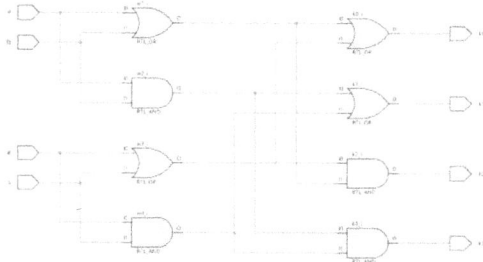

Figure 69.2 Input reordering circuit
Source: Author

reordering together with the compressor will reduce the combinations from sixteen down to just six input combinations that reduce both switching and power. The working of the input reordering circuit that simplifies switching and reduces combinations is illustrated in Figure 69.2.

Proposed design 1 of non-ideal 4:2 compressor
The primary design of this non-ideal 4:2 compressor has been derived from an input reordering scheme to minimize hardware complexity and power consumption while retaining the essential degree of accuracy. This will prove suitable for a high efficient multiplier for application in image processing, meaning quite well for energy-efficient multipliers.

The basic concept, into this design, is that the input combinations shall be classified according to the number of 1's presence instead of their decimal value which will reduce the number of input combinations from sixteen to six input combinations, thus further reducing switching activity and power efficiency.

Because this compressor is a non-ideal compressor, it introduces minor errors in some input combinations to further reduce complexity. In particular, errors arise while entering 1110, which occurs four times with a probability of 3/256, and 1111, which has only one occurrence with a probability of error 1/256. Resulting the total probability of error is 13/256 with an error of 31.25%. The functional truth table for proposed Design 1 is presented in Table 69.2. Although these minor inaccuracies are introduced, the compressor also improves energy efficiency significantly, reducing the power-delay product by 62.08% compared to accurate design. The logic implementation of the proposed Design 1 of the non-ideal 4:2 compressor is depicted in Figure 69.3. Boolean functions for the output in Design 1 are listed below:

Table 69.1 Truth table of input reordering circuit.

P	Q	R	S	k0	k1	k2	k3	Prob.
0	0	0	0	0	0	0	0	81/256
0	0	0	1	1	0	0	0	27/256
0	0	1	0	1	0	0	0	27/256
0	0	1	1	1	1	0	0	9/256
0	1	0	0	1	0	0	0	27/256
0	1	0	1	1	0	1	0	9/256
0	1	1	0	1	0	1	0	9/256
0	1	1	1	1	1	1	0	3/256
1	0	0	0	1	0	0	0	27/256
1	0	0	1	1	0	1	0	9/256
1	0	1	0	1	0	1	0	9/256
1	0	1	1	1	1	1	0	3/256
1	1	0	0	1	1	0	0	9/256
1	1	0	1	1	1	1	0	3/256
1	1	1	0	1	1	1	0	3/256
1	1	1	1	1	1	1	1	1/256

Source: Author

Sum = (k0&(~k1)) &(k1|(~k2))
Carry=k1|k2

This compressor is then applied to the first proposed approximate multiplier design (MUL1), where only non-ideal compressors are used in its structure. This enables the multiplier to achieve an energy saving of compared to the accurate multiplier design.

Proposed design 2 non-ideal 4:2 compressor
The proposed Design 2 of non-ideal 4:2 compressor is an advanced version of the compressor that further minimizes the chance of error while still providing substantial hardware simplification and power savings. Similar to Design 1, Design 2 employs an input reordering strategy that orders input bits by the number of ones (1s) they have, thereby decreasing the number of unique combinations from 16 to 6. This decrease not only minimizes switching activity but also minimizes dynamic power consumption, a valuable benefit for low-energy circuits.

The schematic for the proposed Design 2 of the non-ideal 4:2 compressor is shown in Figure 69.4. In Design 2, the approximation technique is further streamlined to process input combinations that are likely to occur with low probability. More specifically,

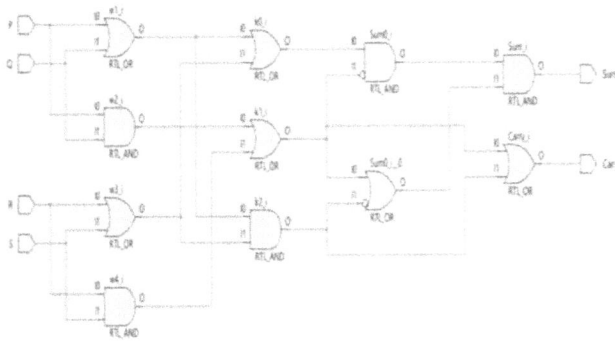

Figure 69.3 Proposed design 1 of non-ideal 4:2 compressor
Source: Author

Table 69.2 Truth table for proposed design 1 of non-ideal 4:2 compressor.

k0	k1	k2	k3	Probability of combinations	carry	sum
0	0	0	0	81/256	0	0
1	0	0	0	27/256	0	1
1	1	0	0	9/256	1	0
1	0	1	0	9/256	1	0
1	1	1	0	3/256	1	0
1	1	1	1	1/256	1	0

Source: Author

errors are intentionally introduced for combination 1110. The truth table for proposed Design 2, along with probabilities, is detailed in Table 69.3. In this design, the 1110 combination is likely to occur with a probability of 3/256 and occurs four times. When these probabilities are considered, the overall error probability of the compressor is minimized to 12/256. This is less than the error probability of Design 1, which is 13/256. Boolean functions for the outputs in Design 2 are:

$$Sum = (k0\&(\sim k1)) \& (k1|(\sim k2))$$
$$Carry = (k1|k2) \& (\sim k3)$$

These functions differ from those of Design 1 and are the product of a design decision that is intended to minimize error. By employing this modified logic, Design 2 is able to further reduce the erroneous output occurrences while nevertheless enjoying low hardware complexity and energy efficiency.

Comparing the two designs, the primary differences are the methods of introducing errors and the respective error probabilities. Design 1 introduces errors in fewer input combinations specifically, the 1110 and 1111 with a total error probability of 13/256. Design 2 introduces errors in the 1110 combinations with a lower total error probability of 12/256 with an error of 25% and the compressor also improves energy efficiency, reducing the power-delay product by 28% compared to accurate design.

Overall, while both designs offer significant energy savings and simplification of circuits compared to traditional designs, Design 2 is superior in error performance because of its lower total error probability. However, But the choice between the

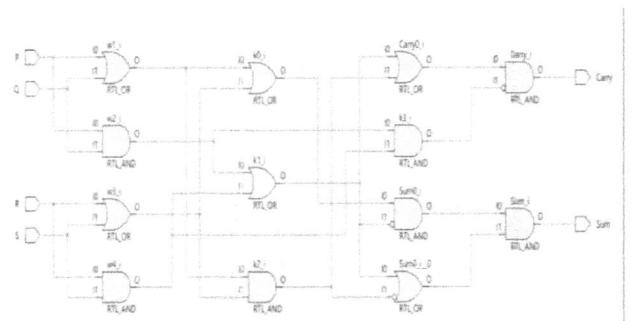

Figure 69.4 Proposed design 2 of non-ideal 4:2 compressor
Source: Author

two designs of non-ideal compressors would depend upon the requirements of the application. If an application is tolerant of a higher error probability for potential improvements in other performance parameters, Design 1 would be adequate. For applications where the minimization of errors is critical especially in precision-oriented applications Design 2 would be ideal.

Proposed approximate multipliers

Two rough structures of the multipliers using developed 4:2 imprecise compressors are designed. Both compressors are utilized separately in the Dadda multiplier architecture.

Design of approximate multiplier using proposed designs of non-ideal compressors

So far, the new approximated multiplier design proposed using the new non-ideal compressors is a novel solution that benefits from the advantages of reduced circuit complexity and reduced power

Table 69.3 Truth table for proposed design 2 of non-ideal 4:2 compressor.

k0	k1	k3	k4	Probability of combinations	carry	sum
0	0	0	0	81/256	0	0
1	0	0	0	27/256	0	1
1	1	0	0	9/256	1	0
1	0	1	0	9/256	1	0
1	1	1	0	3/256	1	0
1	1	1	1	1/256	0	0

Source: Author

Table 69.4 Comparative study of compressor designs.

Compressors	Area (nm^2)	Power (nW)	Delay (ps)	PDP (aJ)	ER (%)
Proposed Design 1	52×10^5	525.034	715000	375.3	31.25
Proposed Design 2	71×10^5	1279.06	557000	712.4	25
[12]	181×10^5	525.034	715000	375.3	31.25
[13]	196×10^5	561.452	339000	190.3	31.25

Source: Author

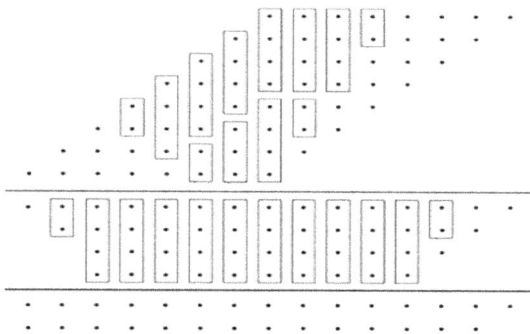

Figure 69.5 Structure of dadda multiplier
Source: Author

consumption-the two pillars of approximate computing-for energy-efficient multiplications to be used in image processing as well as other low-power applications.

The multiplier design works by the application of newly designed non-ideal compressors. They reduce the partial products of the multiplier. The Dadda multiplier architecture integrating the proposed compressors is shown in Figure 69.5. The compressor reduces the number of input combinations through an input reordering scheme that sorts bits by the number of ones (1s) that they possess, drastically reducing switching activity and total power dissipation.

In the Dadda multiplier, half adders are used to reduce two partial products, full adders handle three partial products, and the proposed compressor designs are employed for reducing four partial products.

Two multiplier designs are introduced:

Approximate multiplier design 1 (fully approximate)
All the compressor blocks in the following are replaced by the introduced non-ideal compressors. By approximating the compressor stage completely, the design maximizes energy savings with up to 49.14%

less energy compared to the accurate designs. This fully approximate design, however, incurs a higher error distance because of the compounding effect of approximation error across each stage of the compressor. This design is appropriate for those applications where energy efficiency is of utmost concern and minor inaccuracies can be tolerated.

Approximate multiplier Design 2 (hybrid approach)
Appreciating the need for a compromise between accuracy and energy efficiency, the second design is founded on a hybrid solution. In this design, the top half of the compressor stages is realized with accurate compressors, and the bottom half is realized with the proposed non-ideal compressors. Design is of lower error distance than a fully approximate multiplier. While the energy savings are modestly lower realizing a 2.32% reduction the better accuracy makes this design appropriate for applications requiring tighter error metric control.

The multiplier Design 1 maximizes energy savings at the cost of increased error rates, while Multiplier Design 2 balances energy efficiency and accuracy. The selection of these designs hinges on the error tolerance of the application and the particular power-performance requirements.

Simulation Results

The designs of this research are coded with Verilog HDL, following an organized and optimized description of the hardware. Each design's operation is further validated within the Xilinx Vivado simulation environment while waveforms are analyzed to verify the correctness of logical operations.

To characterize the throughput capabilities of the suggested architectural designs, power, area, and

Table 69.5 Comparative study of multiplier designs.

Multipliers	Area (nm^2)	Power (nW)	Delay (ps)	PDP (aJ)	% Reduction (vs. Accurate Multiplier)
Accurate multiplier	3309x10^5	43340	33670	145925.7	Baseline
Proposed Design 1	2046x10^5	23080	3216	74225	38.1%↓(Area) 46.7%↓(Power) 90.4%↓(Delay) 49.1% ↓ (PDP)
Proposed Design 2	2268x10^5	42264	3373	142542	31.5%↓(Area) 2.5%↓(Power) 89.9%↓(Delay) 2.3% ↓ (PDP)

Source: Author

Table 69.6 Comparative study of error metrics.

Design	Error Rate (%)	MED	MRED	NMED	PSNR (dB)
Design 1	70.6	1429.1	0.109	0.0219	16.579
Design 2	69.3	1123.0	0.113	0.0172	17.626

Source: Author

delay are analyzed. This is achieved using Cadence Genus, which is a commonly used synthesis tool for ASIC designs. The current synthesis process used in this design through the 45nm technology node gives an insight into the feasibility of making this design using current fabrication processes. A comparative evaluation of area, power, delay, and error rates of various compressor designs is provided in Table 69.4.

The circuit area is determined based on the total number of logic gates used in the implementation. The supply voltage for the power consumption is calculated under different operating conditions, such as dynamic and leakage power. Delay is computed with critical path analysis to estimate the maximum operating frequency. These metrics help balance the energy efficiency, speed, and resource utilization trade-offs. A comparison of the proposed approximate multipliers with the accurate design in terms of area, power, delay, and PDP is shown in Table 69.5.

Image processing applications

Both proposed designs provide the intended application in image processing, that is, in image multiplication application. To validate the working of this the two images are multiplied pixel by pixel. This application is used in image-processing techniques like image merging, feature determination, radiology, etc.

Image1 and Image2 are converted into binary format and multiplied pixel by pixel using the proposed approximate multipliers.

   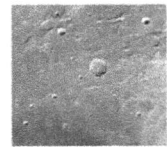

Image 1   Image 2

This has to be deleted which is strikedoff with red color.

Some of the error metrics are used to compare the proposed designs with the accurate ones. Error rate and quality metrics such as MED, MRED, NMED, and PSNR for both designs are summarized in Table 69.6. From such comparisons, the designs presented are very much superior in energy efficiency than the ones earlier, lowest in hardware complexity, and yet tolerable error resiliency.

Conclusion

The approximate multipliers under consideration show remarkable enhancement over the accurate ones in terms of area, power, delay, and energy efficiency, keeping an almost balanced trade-off with accuracy and performance. While Design 1 is capable of reducing the area by 38.1% and suitable for very compact VLSI designs, Design 2 achieves a 31.5% reduction in area with the trade-offs between performance and resource utilization.

While in terms of power Design 1 consumes 46.7% less power that fits these designs for low-power applications, Design 2 reduces it to only 2.5% more of accurate density; thus, it seems more of an accuracy-preserving design. The most obvious improvement is on the side of delay, with 90.4% and 89.9% reduction in delay by Design 1 and Design 2, respectively, enhancing computational speed so that both designs are suitable for real-time processing applications. Efficacy in energy use is also enhanced, where Design 1 reduces Power-Delay Product (PDP) by 49.1% and stands best on energy efficiency, while Design 2 also betterments by only 2.3%, hence surely surpassing the accurate multiplier.

Design 1 output   Design 2 output   Accurate Design output

For the utmost reduction in PDP compared to a fully accurate Dadda multiplier, the two proposed architectures present a 25.72% average reduction and an average accuracy of 30.5% is preserved, balancing out the trade-off between energy efficiency and computational accuracy. While Design 1 favours energy efficiency and fast operation, Design 2 is worth a trade-off between accuracy and reduction in area and delay. In general, the designs exhibit satisfactory trade-offs, making them strong candidates for energy-efficient approximate computing applications, especially in image processing and VLSI design.

References

[1]   Y. Zhang, X. Chen, P. Guo and G. Xie, Design and Analysis of Approximate Multiplier of Majority-Based Imprecise 4–2 Compressor for Image Processing, 2023 IEEE 23rd International Conference on Nanotechnology (NANO), Jeju City, Korea, Republic of, 2023, pp. 1-5, doi: 10.1109/NANO58406.2023.10231167.

[2]   S. Mazahir, O. Hasan and M. Shafique, Adaptive Approximate Computing in Arithmetic Datapaths, in IEEE Design & Test, vol. 35, no. 4, pp. 65-74, Aug. 2018, doi: 10.1109/MDAT.2017.2772874.

[3]   Samad Shirzadeh, Behjat Forouzandeh,High accurate multipliers using new set of approximate compressors, AEU - International Journal of Electronics and Communications,Volume 138,2021,153778,ISSN 1434-8411,https://doi.org/10.1016/j.aeue.2021.153778.

[4]   Bao Fang, Huaguo Liang, Dawen Xu, Maoxiang Yi, Yongxia Sheng, Cuiyun Jiang, Zhengfeng Huang, Yingchun Lu,Approximate multipliers based on a novel unbiased approximate 4-2 compressor,Integration,Volume 81,2021,Pages 17-24,ISSN 0167-9260,https://doi.org/10.1016/j.vlsi.2021.05.003.

[5]   R. Pilipović and P. Bulić, "On the Design of Logarithmic Multiplier Using Radix-4 Booth Encoding," in IEEE Access, vol. 8, pp. 64578-64590, 2020, doi: 10.1109/ACCESS.2020.2985345.

[6]   Hemanth Krishna, L., Bhaskara Rao, J., Ayesha, S.K., Veeramachaneni, S., Noor Mahammad, S.K. (2023). Novel Approximate 4:2 Compressor for Multiplier Design. In: Darji, A.D., Joshi, D., Joshi, A., Sheriff, R. (eds) Advances in VLSI and Embedded Systems. Lecture Notes in Electrical Engineering, vol 962. Springer, Singapore. https://doi.org/10.1007/978-981-19-6780-1_6

[7]   A. Momeni, J. Han, P. Montuschi and F. Lombardi, Design and Analysis of Approximate Compressors for Multiplication, in IEEE Transactions on Computers, vol. 64, no. 4, pp. 984-994, April 2015, doi: 10.1109/TC.2014.2308214.

[8]   T. Kong and S. Li, Design and Analysis of Approximate 4–2 Compressors for High-Accuracy Multipliers, in IEEE Transactions on Very Large Scale Integration (VLSI) Systems, vol. 29, no. 10, pp. 1771-1781, Oct. 2021, doi: 10.1109/TVLSI.2021.3104145.

[9]   D. Esposito, A. G. M. Strollo, E. Napoli, D. De Caro and N. Petra, Approximate Multipliers Based on New Approximate Compressors, in IEEE Transactions on Circuits and Systems I: Regular Papers, vol. 65, no. 12, pp. 4169-4182, Dec. 2018, doi: 10.1109/TCSI.2018.2839266.

[10]   A. G. M. Strollo, E. Napoli, D. De Caro, N. Petra and G. D. Meo, Comparison and Extension of Approximate 4-2 Compressors for Low-Power Approximate Multipliers, in IEEE Transactions on Circuits and Systems I: Regular Papers, vol. 67, no. 9, pp. 3021-3034, Sept. 2020, doi: 10.1109/TCSI.2020.2988353.

[11]   P. J. Edavoor, S. Raveendran and A. D. Rahulkar, "Approximate Multiplier Design Using Novel Dual-Stage 4:2 Compressors," in IEEE Access, vol. 8, pp. 48337-48351, 2020, doi: 10.1109/ACCESS.2020.2978773.

[12] L. H. Krishna, A. Sk, J. B. Rao, S. Veeramachaneni and N. M. Sk, Energy-Efficient Approximate Multiplier Design With Lesser Error Rate Using the Probability-Based Approximate 4:2 Compressor, in IEEE Embedded Systems Letters, vol. 16, no. 2, pp. 134-137, June 2024, doi: 10.1109/LES.2023.3280199.

[13] X. Yi, H. Pei, Z. Zhang, H. Zhou and Y. He, Design of an Energy-Efficient Approximate Compressor for Error-Resilient Multiplications, 2019 IEEE International Symposium on Circuits and Systems (ISCAS), Sapporo, Japan, 2019, pp. 1-5, doi: 10.1109/IS-CAS.2019.8702199.

[14] Karri Manikantta Reddy, M.H. Vasantha, Y.B. Nithin Kumar, Devesh Dwivedi,Design and analysis of multiplier using approximate 4-2 compressor,AEU - International Journal of Electronics and Communications,Volume 107,2019,Pages 89-97,ISSN   1434-8411,https://doi.org/10.1016/j.aeue.2019.05.021.

70 Enhancing chronic care through IoT sensors and real-time health monitoring

S. Thejaswini[1,a], J. Venkata Chaithanya Jyothi[2,b], S. K. Sandeep[2,c], B. Surya Kumar[2,d] and S. Lokesh[2,e]

[1]Assistant Professor, Department of ECE, Annamacharya Institute of Technology and Sciences, Tirupati, Andhra Prasad, India

[2]Students, Department of ECE, Annamacharya Institute of Technology and Sciences, Tirupati, Andhra Prasad, India

Abstract

The rapid innovations in Internet of Things (IoT) technologies have contributed to the cost-effective, energy-efficient, and intelligent healthcare monitoring systems for chronic diseases. This system stands proposed to predict chronic diseases like cardiovascular disorders like heart attack, respiratory disorders like asthma, and neurodegenerative disorders like seizure attack. IoT-enabled health monitoring integrates to track the heart rate and temperature. It collects real-time data, stores it securely in the cloud Things Speak, and leverages predictive analytics to identify potential health risks before they become critical, allowing for immediate medical treatments. The system integrates a pulse sensor and a DHT 11 sensor with Arduino and Node MCU ESP2866 to collect data and analyze the data from the pulse sensor and DHT11 sensor, then displays it on an LCD, and this data can be seen through a web application. By utilizing IoT capabilities, this system improves patient condition by allowing doctors to access patient data through a web platform. This system ensures overall accuracy by detecting heart diseases, seizure attacks, asthma, and nervous system disorders using the SR011 pulse sensor and DHT11 sensor, comparing heart rate and temperature threshold values with sensor readings.

Keywords: Arduino, chronic diseases, cloud things speak, DHT11 sensor, Internet of Things, node MCU ESP2866, pulse sensor

Introduction

Chronic diseases encompass a range of conditions, including cardiovascular diseases like heart attacks, respiratory diseases such as asthma, and neurodegenerative disorders, including seizure attacks, chronic kidney disease, and other types of cancers. According to the World Health Organization (WHO), these conditions cause 74% of all deaths in the world [1]. By monitoring these diseases helps in reducing complications, decreasing hospitalizations, and optimize quality life of patient. This monitoring in many aspects decreases dependency on medical visits. Real-time monitoring of chronic disease involves continuous tracking of health indicators. It provides dynamic data such as tracking of heart rate. Continuous monitoring of health parameters helps early diagnosis of disease progression, bringing in timely medical treatment. The health applications further deliver improved treatment commitment, offer personalized care, and help clinical staff make the right decisions. This system proposes a chronic disease monitoring through IoT that integrates pulse sensor, temperature sensor, microcontrollers and cloud-based analytics to provide in real-time and wireless.

Literature review

IoT is transforming the future of chronic disease care and improving health care recipient outcomes is transforming the chronic disease management through cost-efficient, power efficient, wearable and implantable devices. It emphasizes the increasing need for continuous health monitoring. The research covers more than 80 healthcare monitoring systems based on IoT technology and highlights their value in improving the care of patients [2]. A 2020 study reveals the need to incorporate psychological and educational strategies to manage diabetes [3]. Development in telemedicine has facilitated

[a]sthejaswini14@gmail.com, [b]jvcjyothiece014@gmail.com, [c]sandeeprcb18@gmail.com, [d]sp3159520@gmail.com, [e]lokeshdj955@gmail.com

DOI: 10.1201/9781003685364-70

distance health monitoring and virtual consultations but with some continued challenges regarding infrastructure and finance [4]. An IoT-based system was designed for early heart attack detection using pulse sensors and real-time data transmission. It enhances remote patient monitoring through cloud-based analytics and automated alerts [5]. Deep learning techniques have improved seizure monitoring, despite ongoing concerns about computation and privacy [6]. Arduino-based Internet of Things systems are utilized for monitoring vital signs, but they face integration challenges [7]. Non-clinical home care services support independent living, yet they encounter accessibility issues [8]. While IoT-based health monitoring improves patient care, there are still concerns regarding interoperability and data security [9]. Digital health surveillance supports asthma management but struggles with connectivity problems [10]. IoT technology enhances elderly care by allowing for remote monitoring and fall detection, although data privacy remains a concern [11]. Monitoring the autonomic nervous system in critical care settings aids in predictive analytics, but it also presents challenges in interpretation [12]. Smart IoT systems improve real-time health tracking, yet they face limitations in sensor accuracy and scalability [13]. The Table 70.1 shows comparision of proposed model with existing system.

Proposed System

This system integrates a DHT11 sensor for measuring humidity and external temperature, along with a pulse sensor SR011 for monitoring pulse rate. These sensors are connected to an Arduino Uno, which is linked to an ESP8266 Node MCU Wi-Fi module, enabling cloud-based data storage and analysis in Things Speak. In simpler terms, the Uno collects sensor data and transmits it via the Node MCU to the cloud Things speak for remote monitoring and processing. The system enables real-time data visualization and early warning alerts through buzzer, allowing users to take proactive measures before chronic diseases develop. Based on recurring health patterns, people can monitor their well-being and take necessary precautions to maintain good health. Based on the data provided by Things speak, doctors can assess patient conditions and provide appropriate treatment. Things speak fetches real-time data and display it in an easy-to-read format for doctors.

Table 70.1 Existing model with comparison.

Authors	Technology used	Key features
Talpur et al. (2023) [2]	IoT with wearable sensors	Continuous heart rate and SpO2 monitoring, cloud storage
Sutha et al. (2024) [5]	Arduino-based IoT	Heart attack detection using pulse sensor and Wi-Fi alerts
Hussein et al. (2024) [6]	AI-based deep learning on IoT	Epileptic seizure detection, cloud-based analytics
Abdulmalek et al. (2022) [9]	IoT-enabled smart healthcare	Monitors BP, heart rate, and temperature, cloud storage
Rafa et al. (2022) [10]	Smart sensors & IoT for asthma	Tracks oxygen levels and humidity, remote monitoring
Proposed model	Arduino, NodeMCU, pulse sensor, DHT11, IoT & cloud (Thing Speak)	Predict heart diseases, seizure nervous system disorders based on heart rate, real-time alerts, cloud storage.

Source: Author

A. System components and their functions:

1. **Arduino UNO:** It reads analog signals from sensors and converts into digital values. Then this processed data was transmitted to ESP2866. It enhances the accessibility to detect diseases.

2. **Node MCU ESP8266:** It serves as a communication channel between the Arduino Uno and Thing speak cloud server, enables the real-time data transmission for monitoring. It transmits the data to cloud platform, where it can be stored, analyzed, and accessed through web applications.

4. **DHT 11 sensor:** The DHT11 sensor plays a crucial role in monitoring environmental conditions by measuring the temperature and humidity, which are factors in health surveillance. It collects the temperature and humidity from through a sensor and sends it to the Arduino UNO.

5. **Pulse sensor SR011:** It is a key component which involves in tracking cardiovascular health by measuring the pulse rate in real time. It detects the changes in blood flow using optical sense technology. When a person's finger is placed on the sensor, it measures their heart rate and transmits the data to an Arduino UNO.

6. **16×2 LCD display:** This Display serves as a local monitoring interface, it shows immediate readings like as heart rate, body temperature and humidity. It allows users to check the health status without requiring of an external device. The Arduino UNO update the reading values based on the data provided by the heart rate and DHT 11 sensors.

7. **Piezoelectric buzzer:** It provides immediate audio alerts to the users when abnormal health readings are occurred, such as high heart rate or temperature.

8. **Power supply:** A reliable power source crucial for the system's continuous functionality for consistent monitoring. The Arduino UNO and ESP8266 Node MCU typically run on 5V and 3.3V, respectively. It consists step down transformer to provide stable power to system.

B. Software: The system uses a Threshold-based peak detection algorithm, where sensors act as input, pre-defined threshold values determine the health risk and alert and warnings are triggered based on logical conditions.

C. Cloud server: This system uses Things speak as a primary cloud server for data storage and data visualization. There are several cloud platforms are there, but Thing Speak was chosen because it offers a free version and creates many channels, each channel stores up to eight fields of data. For this monitoring and storing the data a channel is created, named "chronic disease monitoring" (channel ID: 2827702). It provides data via Node MCU ESP8266.

Methodology

This system incorporates various sensors like pulse rate senor SR011 and DHT11 sensors, Arduino UNO, a cloud platform, ESP8266 Node MCU and alert mechanisms. The main purpose of this system is to predict and detect the Chronic diseases like heart attack, seizure, asthma. The below Figure 70.1 block diagram of proposed system.

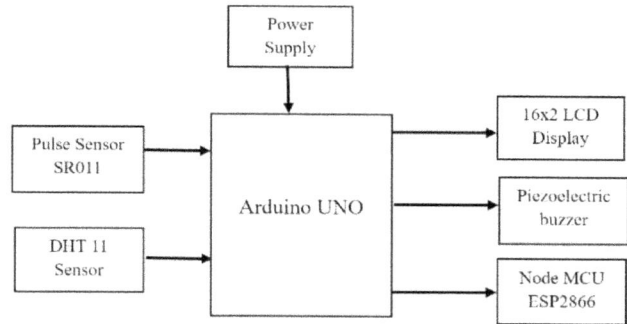

Figure 70.1 Block diagram
Source: Author

A. **Data collection:** In this data is collected by sensors, such as pulse sensor, DHT11 sensor. The sensors take the data (heart rate and body temperature) from users by sensing and continuously monitor condition of pulse, temperature and humidity.

B. **Data processing and transmission:** Arduino UNO process the raw data which is collected by the pulse sensor and DHT11 sensor and this processed data is transmitted to the Node MCU, which act as a Wi-Fi module for cloud connectivity. The given data consists of the measured values from sensors. The analysis will be done in between the measured values, and the threshold values the particular disease. After that, ESP8266 transmits data to Thing Speak cloud server.

C. **Cloud storage and visualization:** The Things speak platform stores the collected data and it provides particular data for visualization through structured tables. The disease will be displayed through the calibrated values for the inbuilt levels for the disease, through these users and doctors can access the health trends via Things speak.

D. **Alert mechanism:** If there are abnormal health readings are detected then system triggers an alert by using buzzer. The readings were displayed by the LCD, this allows users to monitor their condition. The readings are Heart rate, Temperature and Humidity and some of alerts displayed like Heart Disease, Asthma alert, Seizure, Nervous System alert.

E. **Remote monitoring and decision making:** By using the data which is provided by the Things speak, through that data doctors and caregivers

Table 70.2 Parameter range.

Health condition	Normal range	Alert threshold	Critical condition
Heart disease	60 – 100 BPM	<60 or >120 BPM	>140 BPM (Severe)
Seizures	HR <120 BPM	HR >120 BPM	HR >150 BPM
Nervous system alert	HRV >50 ms	HRV <50 ms	HRV <30 ms

Source: Author

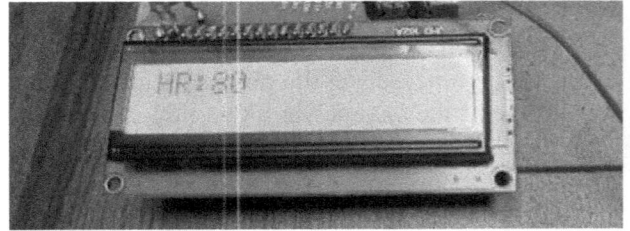

Figure 70.4 (a) Heart rate value displayed on LCD
Source: Author

Figure 70.2 Circuit design
Source: Author

Figure 70.4(b) Temperature and humidity value displayed on LCD
Source: Author

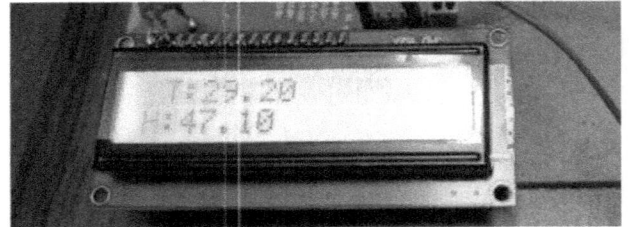

Figure 70.4(c) Nervous system alert displayed on LCD
Source: Author

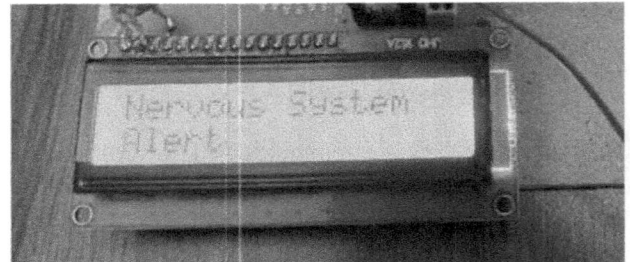

Figure 70.3 2: placing the finger on the pulse sensor
Source: Author

analyze trends and take prevention. The Table 70.2 shows the output parameters.

Result

Running real-world tests with chronic disease patients in different settings like their homes, hospitals, and rural Clinics can play a crucial role in verifying the system's reliability. This way, we can ensure that the sensor readings are spot on and that the IoT integration works smoothly, no matter the environment.

The result demonstrates the system's effectiveness in real-time conditions. The buzzer responded when threshold limits were exceeded, alerting the users. The system maintain stable performance, supports continuous data transformation. Collecting data over time gives us the chance to see how effectively the system can forecast major health problems. This ongoing process allows us to fine-tune alert levels and improves our ability to detect diseases at an early stage. Figure 70.2 and 70.3 show the circuit kit of proposed work.

Table 70.3 Response time.

Parameter	Response time	Observations
Heart rate detection	~1-2 seconds	Measured by pulse sensor SR011, displayed on LCD.
Temperature and humidity	~2-3 seconds	Collected via DHT11, processed by Arduino.
Data transmission to cloud	~3-5 seconds	NodeMCU ESP8266 uploads data for remote access.
Alert trigger	<1 second	Immediate response if threshold values are exceeded.
Remote access	~5-10 seconds	Delay depends on network and Thing Speak API speed.

Source: Author

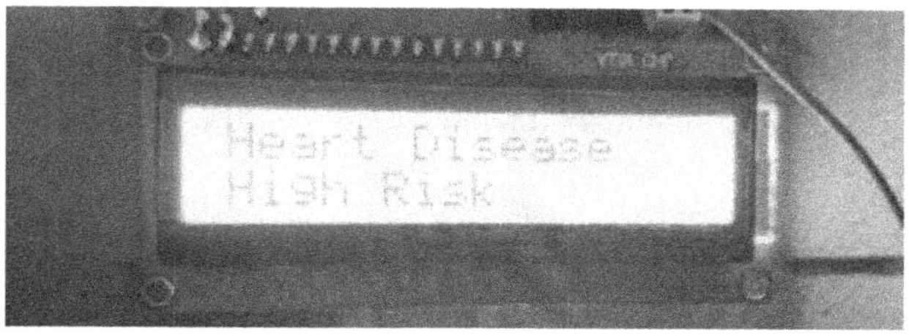

Figure 70.4(d) Heart disease alert displayed on LCD
Source: Author

Figure 70.5 Data provided by thing speak
Source: Author

When a finger is placed on the pulse sensor, it immediately measures the heart rate within a fraction of seconds. After that heart rate, temperature, humidity values are displayed on LCD shown in Figures 70.4(a)–(b). Then, it displays disease alerts if the heart rate, temperature, and humidity values exceed the threshold like in Figure 70.4(c), (d). Thing Speak provides data based on the information processed by Arduino and sent to the NodeMCU ESP8266 in Figure 70.5. The Table 70.3 denotes response time results

Conclusion and future scope

The system proposed assists through the utilization of smart sensors and real-time data analysis. It utilizes Arduino UNO and ESP2866 Node MCU to capture and transmit the data to Things Speak for remote monitoring. This maintains connectivity stability with cloud-based storage. The system predicts heart disease, asthma, seizure, and nervous system disorders through pulse sensor and DHT11 sensor readings analysis. When the reading of any parameter exceeds the set threshold values, the system warns the user with a quick buzzer alarm regarding the potential health risk. At the same time, the LCD indicates real-time health information, reflecting instantaneous feedback regarding the out-of-range parameter. Quick alert and detection are assured, enabling users to respond appropriately or approach medical treatment in a timely manner, thus increasing patient safety and early disease detection. This system enhances health monitoring accuracy, facilitates early detection of risk factors, and remote access to view critical data.

In the future this system can be developed into a smart wristwatch is used for chronic diseases detection and Health monitoring by using smart technologies. System can be integrated with a digital version of patient data history for uninterrupted medical data access. In case of severe health conditions, the system can automatically send the patient's location to relatives and health care providers for immediate assistance.

References

[1] World Health Organization. World health statistics 2024: monitoring health for the SDGs, sustainable development goals. World Health Organization, 2024.

[2] M. S. H. Talpur et al., Illuminating Healthcare Management: A Comprehensive Review of IoT-Enabled Chronic Disease Monitoring, in IEEE Access, vol. 12, pp. 48189-48209, 2024, doi: 10.1109/AC-CESS.2024.3382011.

[3] Hashimoto, Kana, et al. The relationship between patients' perception of type 2 diabetes and medication adherence: a cross-sectional study in Japan. Journal

of pharmaceutical health care and sciences 5 (2019): 1–10.

[4]   Paleari, L., Malini, V., Paoli, G., Scillieri, S., Bighin, C., Blobel, B., et al. (2022). EU-funded telemedicine projects – assessment of, and lessons learned from, in the light of the SARS-CoV-2 pandemic. *Frontiers in Medicine*, 9, 849998.

[5]   Sutha, P., Periyanan, A., Menaha, R., Jayanthi, V., & Girish, B. (2024). IOT Based heart rate monitoring system design for heart attack detection. In 2024 4th International Conference on Advance Computing and Innovative Technologies in Engineering (ICAC-ITE).

[6]   Hussein, A. M., Alomari, S. A., Almomani, M. H., Abu Zitar, R., Saleem, K., Smerat, A., et al. (2024). A smart IoT-cloud framework with adaptive deep learning for real-time epileptic seizure detection. Circuits, Systems, and Signal Processing

[7]   Hussain, Idris, Devendra Deshalahre, and Prabhat Thakur. Assessing the Effectiveness of An IoT-Based Healthcare Monitoring and Alerting System with Arduino Integration. Revue d'Intelligence Artificielle 38.4 (2024): 1211.

[8]   Kuryk, K., Funk, L. M., Warner, G., Macdonald, M., Lobchuk, M., Rempel, J., ... & Keefe, J. (2025). Ageing in place with non-medical home support services need not translate into dependence. Ageing & Society, 45(3), 455–480.

[9]   Abdulmalek, S., Nasir, A., Jabbar, W. A., Almuhaya, M. A., Bairagi, A. K., Khan, M. A. M., & Kee, S. H. (2022, October). IoT-based healthcare-monitoring system towards improving quality of life: A review. In Healthcare (Vol. 10, No. 10, p. 1993). MDPI.

[10]   Shamim Rafa, N., Binte Azmal, B., Rab Dhruba, A., Monirujjaman Khan, M., Alanazi, T. M., Almalki, F. A., & AlOmeir, O. (2023). IoT-Based Remote Health Monitoring System Employing Smart Sensors for Asthma Patients during COVID-19 Pandemic. arXiv e-prints, arXiv-2304.

[11]   Shamim Rafa, N., Binte Azmal, B., Rab Dhruba, A., Monirujjaman Khan, M., Alanazi, T. M., Almalki, F. A., & AlOmeir, O. (2023). IoT-Based Remote Health Monitoring System Employing Smart Sensors for Asthma Patients during COVID-19 Pandemic. arXiv e-prints, arXiv-2304.

[12]   Bento, Luis, Rui Fonseca-Pinto, and Pedro Póvoa. Autonomic nervous system monitoring in intensive care as a prognostic tool. Systematic review. Revista Brasileira de terapia intensiva 29 (2017): 481-489.

[13]   Vaishnave, A. K., S. T. Jenisha, and S. Tamil Selvi. IoT based heart attack detection, heart rate and temperature monitor. International Research Journal of Multidisciplinary Technovation 1.6 (2019): 61-70.

71 Embedded based solar powered smart stick for sight impaired individual

Anil Kumar[1,a], B. Madhavi[2,b], R. Bharath Simha Reddy[2,c], M. Mythili[2,d] and N. Sowjanya[2,e]

[1]Assistant Professor, Department of ECE, Rajeev Gandhi Memorial College of Engineering and Technology, Nandyal, Andhra Pradesh, India

[2]UG Scholar, Department of ECE, Rajeev Gandhi Memorial College of Engineering and Technology, Nandyal, Andhra Pradesh, India

Abstract

Navigating daily life presents numerous challenges for sight-impaired individuals, particularly in terms of travel and obtaining accurate environmental information. Traditionally, these individuals have relied heavily on others for assistance, as standard mobility aids like canes offer limited support. To address these challenges, we introduce a smart stick enhanced with Embedded and solar powered technology. This innovative device is designed to detect obstacles, recognize objects and money using a deep learning algorithm, and monitor the user's health via an integrated pulse sensor. The smart cane or stick is powered by a Raspberry Pi 3b+ and includes a Ubl ox Neo 6M module for emergency alerts and location tracking. Additionally, it provides real-time updates on nearby public transportation options, helping users plan their journeys more effectively. An accompanying Android application, connected to Firebase, allows for remote monitoring of the user's health and location data. The smart stick delivers an impressive obstacle detection accuracy of 91.7% and is both cost-effective and lightweight. Ultimately, this system is designed to enhance the autonomy, safety and quality of life for people with sight-impairments.

Keywords: Deep learning, embedded system, GPS, GSM, sight-impairment, solar powered

Introduction

The blind people deal with plenty of threats in their daily lives particularly in terms of mobility and obtaining real-time environmental information. Conventional devices like white canes offer fundamental support to individuals but lack advanced features for detecting obstacles, identifying objects, and ensuring user safety. Consequently, visually impaired individuals frequently depend on external help, which restricts their autonomy and movement. To address these boundaries, the project introduces a built-in-based solar-driven smart stick [1], designed to improve navigation, security and access to blind users. Smart stick barriers detect, objects and currency recognition, health monitoring and GPS-based emergency alert offer a comprehensive accessory solution. A controller uses ultrasound sensors to detect obstacles with an accuracy of 91.7% while a controller uses an ultrasound sensor while intensive learning algorithms enable recognition of objects and wealth. In addition, a heart rate sensor continuously monitors the user rhythm of the user, and a GPS module ensures accurate site tracking. In emergencies, GSM sends notifications for predetermined contacts that improve the module's safety. To further improve the dynamics, Smart Stick provides real-time updates on Pass Public Transport Options, which help users to plan their trip effectively. An Android-app attached to Firebase lets nurses and family members monitor the user's health and location externally. The device is lightweight, cost ensure continuous operation without dependence on effective, and solar-powered, external power sources all features are attached to the block diagram as shown in the Figure 71.5. The purpose of this innovative solution is to strengthen the blind by providing a cheap, decent and intelligent smart stick [2]. Introduced systems by integrating deep learning and IoT, improving autonomy, security and accustomed condition of life for the proposed smart stick users.

[a]1ak69206@gmail.com, [b]221091a0486@rgmcet.edu.in, [c]321091a0421@rgmcet.edu.in, [d]421091a04b5@rgmcet.edu.in, [e]521091ao4b6@rgmcet.edu.in

DOI: 10.1201/9781003685364-71

Research methodology

The development of the embedded-based solar-powered smart stick follows a structured research methodology, incorporating sensor-based embedded systems, deep learning for object recognition, IoT-based monitoring, and solar-powered efficiency. How it works we can see in the flow chart of overall system in the Figure 71.6 The work is divided into the following key stages:

System design and component selection

Identifying and selecting hardware components such as Controller, ultrasonic sensors, pulse sensor, GPS, GSM, and a solar panel. Designing a lightweight, ergonomic structure to ensure portability and ease of use.

Object detection using ultrasonic sensors

Ultrasonic sensors [4] are especially known for their precise detection of nearby objects. Multiple sensors are available for measuring distances. The trigger pin functions like a transducer, while the echo pin serves like a recipient. The formulae applied computing gap between an object is

$$\text{Gap or Range} = \frac{Time * Speed\ of\ Light}{2}$$

The ultrasonic sensor can cover angles upto 180 degrees, depending on the design. It uses various communication protocols and emits waves with frequencies between 25kHz and 50 kHz. The trigger input pulse width is 10µs. Three HC-SR04 ultrasonic transducers are employed to identify obstacles on the three directions front and two opposite sides of about the detecting range up from 1meter to 3meter. The sonic vibrations propagate through the air until they encounter an object, at which they are reflected then it obtains the gap between object and sensor and sends it to the microcontroller, (includes the characteristics of raspberry pi in Table 71.1) which

Table 71.1 Characteristics of raspberry Pi.

Processor	Broadcom BCM2837B0, CortexA53(ARMv8)64bit SoC@1.4GHz
Sd Card	Micro SD Card
Power	5v/2.5v via Micro USB
Memory	1 GB RAM

Source: Author

converts it into voice. The IR sensor is cheap but is a little length and narrow beamwidth. The laser sensor, while having a higher detecting range, is costlier and has a narrow bandwidth. The ultrasonic sensor, in contrast, has a good range, wide bandwidth, and is cost-effective. When powered the ultrasonic sensor contacts the controller which enables the sensor to transmit ultrasound. If any gadget is detected the gap or range is calculated and if it falls below the cutoff value is communicated via Audio jack. [3] Otherwise, the Gap or range is recalculated. The intelligent cane uses an ultrasonic sensor to observe obstacle distance. Varied distance ranges helped calculate the detection accuracy, as illustrated in Figure 71.1, it was detected in the inside room setting, the cane operates efficiently, while an 8% error was observed outdoors. The overall accuracy is 91.7%. As the distance increases, detection accuracy remains high, indicating that the smart stick can effectively alert users to critical surroundings.

Recognition of various objects using YOLO (a deep learning algorithm)

The Blind people [5] deal with significant challenges in identifying nearby objects. To address this, we utilized a Web camera for object detection. The camera is cost-effective and supports high resolutions. It enables the user to experience High resolution video calls and supports video quality up to 1280 × 720 pixels with the latest version of skype. Universal compatibility with Plug N Play feature which uses 2.0 for connectivity. Over 70 different objects, including furniture, electronics and edibles, are identified. The image processing system is trained using a dataset comprising 80 different objects, each represented by 50 images, totaling 4000 images. The system is designed to allow database upgrades as new objects are added to the dataset. The "You Only Look Once" (YOLO) method is highlighted for its efficiency in

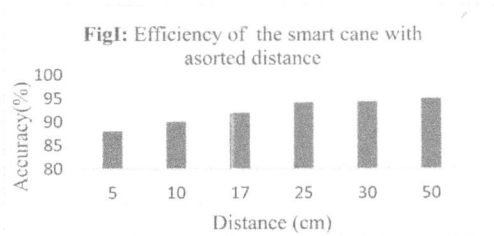

Figure 71.1 Efficiency of the smart cane with asorted distance

Source: Author

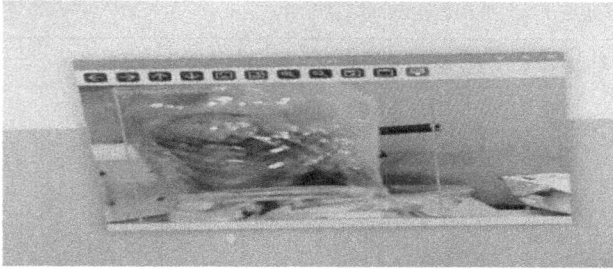

Figure 71.2 Example of the image captured by camera using YOLO
Source: Author

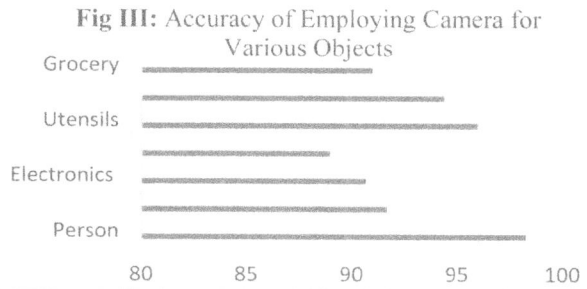

Figure 71.3 Accuracy of employing camera for various objects
Source: Author

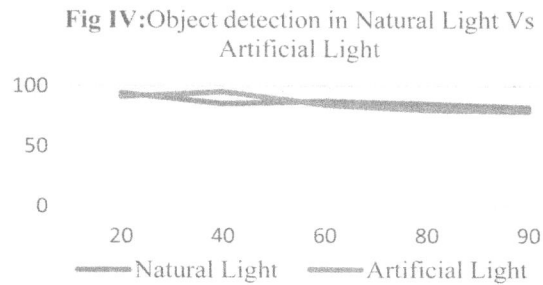

Figure 71.4 Object detection in natural light vs artificial light
Source: Author

rapidly identifying multiple objects in a photograph and constructing boundary lines over them. The YOLO network model with 24 convolutional layers and two entirely connected layers and 1×1 reduction layers and 3x3 convolutional layers. The YOLO enables real-time, comprehensive training while maintaining large average precision. The input image is divided into SxS grid. Each grid cell is responsible for predicting objects whose center falls within the cell. Through some analysis a new architecture is developed by combining MobileNet architecture and YOLO. The enhanced YOLOv5s outperforms existing target disclosure algorithms.

The Raspberry Pi 3b+, capable of internet connectivity, facilitates dataset modifications. When a user pushes the switch, a photograph is captured, and things with partly 71% efficiency are communicated via a Ublox Neo. Detected objects are processed by the microcontroller, and results are communicated through audio signals.

Figure 71.2 shows that objects closest to the sensor exhibit higher accuracy, with notifications provided for accuracy above 70%. For example, a chair was recognized with 78% accuracy, and three people were detected with accuracy of 80.50%, 72.52%, and 50.31% respectively. Objects with accuracies above 70% are communicated to the user, while those below this threshold are not. Indoor implementation was also examined, as shown in Figure 71.3, focusing on common scenarios faced by visually impaired individuals, such as collisions with corridors or objects.

Obstacles within cutoff region are forewarned to users via B audio signal. The prototype was demonstrated to visually impaired individuals, revealing that most users spend significant time at home or designated places like NGOs. The smart stick, [6] equipped with ultrasonic sensors, calculates distance

based on different climatic conditions, with accuracy higher in natural light (89.8%) compared to artificial light (89.1%), as shown in Figure 71.4.

Currency detection
YOLO provides an authoritative approach for real time recognition of Indian currency. By capturing images of currency notes and processing them through the YOLO algorithm, the system identifies various denominations based on unique features like size, color, and text.

This information is then communicated to users, by providing audio messages. This concept raises certainty, efficiency and also helps prevent fraud and supports financial independence for visually impaired use

Water detection
To prevent accidents caused by slippery surroundings, a water detection system has been integrated

into our stick. This system uses a groove water sensor that measures the conductivity of water to determine its presence. It is highly sensitive and can detect small amounts of water. The sensor can interface with both analog and digital pins. When the stick contacts a wet area, the data is sent through an audio signal alerting visually impaired individuals about water surroundings to help avoid collision

Transportation timings
Both sighted and visually impaired individuals often face challenges in catching buses and trains. To tackle this issue, we have developed a solution that caters to both groups. Users simply need to voice their departure and arrival stations. By utilizing speech recognition, the microcontroller processes the spoken information. The microcontroller identifies near buses or trains and notifies the user. The schedules are then voiced through an audio signal to the user.

Health monitoring
Surveys indicate that diabetes and heart diseases are leading causes of death among blind individuals, affecting both males and females similarly. To address this, a heartbeat sensor has been integrated into the stick to measure the pulse rate of the blind people. [7] The pulse sensor provides real time and continuous heart rate data allows users to check their pulse by pressing a button.

In emergencies such as when a user falls and needs urgent help the cane sends a notification to a registered mobile number by pressing a button. The cutoff heartrate for a healthy blind person is set at around 100 BPM. If the heart rate rises significantly, indicating a potential health issue, the threshold for an unhealthy blind person is set at around 50 BPM for low BPM and 200 BPM for high BPM. This allows for effective monitoring and response to critical situations.

GPS monitoring
A survey conducted in April 2018 revealed that 149.4 million people using maps by google. That same year, around 54% of people reported using their mobiles for exploring. The ublox neo 6M module in our system updates data every time, operating at a frequency of 1 Hz. It features an indicator and an inbuilt antenna, and it operates on the principle of satellite navigation to check the user's location. The GPS system provides tracking and timing

Table 71.2 Attributes of GPS module.

Feature	GPS Module
Brand	U- Blox
Model	NEO-6M
Antenna type	External Patch
Sensitivity	-161dBm
Power consumption	27mA @ 3.3v
Communication	UART, I2C
Operating voltage	2.7v to 3.6v

Source: Author

Table 71.3 Features of GSM Module.

Feature	GSM Module
Brand	SIMCom
Model	SIM900
Interfaces	UART, USB, GPIO, GPS
Antenna type	External
Power consumption	~450mA(GPRS)
Operating voltage	3.4v to 4.4v

Source: Author

information. The LED light displays the connection status. Data received from the satellite and the position is displayed in the mobile. The Blind person's live position can be monitored by someone, friends and family can use the app to track the Blind people's position Additionally, the GPS module enables real-time tracking and enhances safety for Blind people as shown in the Table 71.2. This technology significantly aids in navigation and helps prevent accidents. The Ublox Neo 6M module, unified in a cane, helps determine the present position of Blind users. When the blind is not at home from the system activates allowing neighbors to monitors the live location.

Emergency alerts using GSM module
There are numerous instances where visually impaired individuals need support from neighbors. Emergency alerts are a common feature in mobile phones we have unified this functionality within smart cane. In the event of urgency, the blind people simply presses a button, sending an alert message to the neighbour. The system uses a GSM module to transmit SMS messages, providing protection during

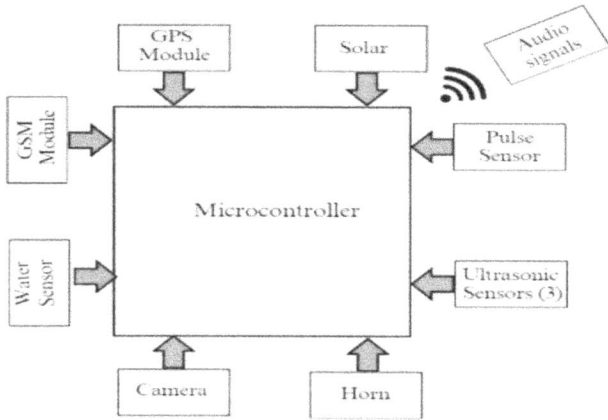

Figure 71.5 Block diagram
Source: Author

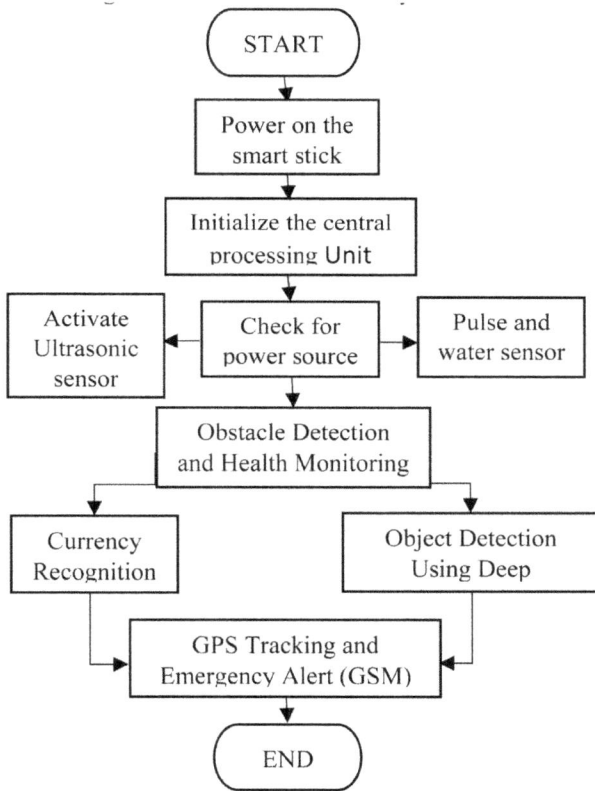

Figure 71.6 Flow chart of overall system
Source: Author

emergency situations. The SIM900A GPRS is unified to the smart stick via a microcontroller which handles communication between the button and the module. When the button is pressed, the Arduino triggers the SIM900A GPRS module to send a message to pre-configured mobile number, ensuring timely

assistance for the visually impaired user as shown in the Table 71.3.

Power efficiency and solar charging
To enhance sustainability, we are implementing a solar panel-based charging system. This environmentally friendly approach harnesses solar energy to power the device reducing dependency on conventional power sources. The smart cane features solar panels that harness sunlight and transform it into electrical energy, which is subsequently stored in the stick's rechargeable battery. This allows the smart stick to remain operational even in off-grid situations.

Literature review

1. Janney, Bethanney, and Banani Mridha. "Intelligent walking stick with static and dynamic obstacle detection technology for visually challenged people." In *2021 Seventh International conference on Bio Signals, Images, and Instrumentation (ICBSII)*, pp. 1-6. IEEE, 2021.

Authors: S. Krishnakumar, V. Lumen Christy, Bethanney Janney, Banani Mridha, G. Umashankar, G. Prakash Williams

Methodology: The team built a walking stick equipped with ultrasonic sensors, PIR and water sensors. The system alerts users via voice notifications and includes a GPS-GSM module for location tracking in emergencies. An RF system helps locate stick if misplace.

Conclusion: The walking stick greatly improves the movability and security of Blind person demonstrating considerable promise.

2. Tirupal, Talari, B. Venkata Murali, M. Sandeep, K. Sunil Kumar, and C. Uday Kumar. "Smart blind stick using ultrasonic sensor." *Journal of Remote Sensing GIS Technology* 7, no. 2 (2021): 34-42.

Authors: T. Tirupal, B. Venkata Murali, M. Sandeep, K. Sunil Kumar, C. Uday Kumar

Methodology: This research introduced a stick using ultrasonic and IR sensors to detect obstacles, offering vibrations or voice alerts to users [7].

Conclusion: The stick fairly boosts the mobility and

3. Loganathan, N., et al. "Smart stick for blind people." *2020 6th International Conference on Advanced Computing and Communication Systems (ICACCS)*. IEEE, 2020.

Authors: N. Loganathan, K. Lakshmi, N. Chandrasekaran, S.R. CibisakaravarthiR, Hari Priyanga, K. Harsha Varthini

Methodology: This project utilizes ultrasonic and infrared sensors to identify obstacles providing alerts to users through vibrations and sound. A wristband helps locate the stick via a buzzer system. The system is controlled by an Arduino UNO, making sure the device stays light, and easy to carry and user-friendly [1].

Conclusion: The proposed system enhances movability and security for blind people by finding obstacles and sending alerts through vibrations and sound. It allows users to navigate independently and stay connected with their surroundings. The design is user-friendly and lightweight and further improvements can be made by appending more sensors to increase functionality.

Result

Depicts an innovative walking cane that aids those who have eyesight problems. Without a doubt, its ultrasonic sensors can detect obstacles in the user's path. The cane's many electrical components and connections can be linked to a microcontroller like an Arduino or Raspberry Pi. Upon analyzing sensor input, the equipment provides information in the manner of speech signals or vibrations. This form of assistance technology lets people walk surrounding safely and decreases the risk of incidents, improving movement and independence.

Conclusion

In this article we focus on the objections overcome by entity who are blind. We have developed a smart cane that uses AIoT technology that is easy to operate, so users can navigate the surroundings safely and independently. Smart sticks include features for tracking user placement via Android. The detectors help to find different things and water areas. This unit also integrates health monitoring, money recognizance and identification of objects through in depth study. Users' live position can be seen and maintained by friends and family. It increases the aspect of life for blind person through offering smart stick friendly and cost effective solutions. In addition, the unit of solar energy is operated, promotes environment and reduces the cost of electricity.

References

[1] Loganathan, N., Lakshmi, K., Chandrasekaran, N., Cibisakaravarthi, S. R., Priyanga, R. H., & Varthini, K. H. (2020). Smart stick for blind people. In 2020 6th International Conference on Advanced Computing and Communication Systems (ICACCS), 2020 Mar 6 (pp. 65–67). IEEE.

[2] Agrawal, M. P., & Gupta, A. R. (2018). Smart stick for the blind and visually impaired people. In 2018 Second International Conference on Inventive Communication and Computational Technologies (ICICCT), (pp. 542–545). IEEE.

[3] Anwar, A., & Aljahdali, S. (2017). A smart stick for assisting blind people. *IOSR Journal of Computer Engineering*, 19(3), 86–90.

[4] Prema, S., Anand, J., Vanitha, P., Mohamed Yaseen, M., & Rajeswari, C. (2022). Smart stick using ultrasonic sensors for visually impaired. In Advances in Parallel Computing Algorithms, Tools and Paradigms (pp. 436–442). IOS Press.

[5] Sharma, H., Tripathi, M., Kumar, A., & Gaur, M. S. (2018). Embedded assistive stick for visually impaired persons. In 2018 9th International Conference on Computing, Communication and Networking Technologies (ICCCNT), (pp. 1–6). IEEE.

[6] Merencilla, N. E., Manansala, E. T., Balingit, E. C., Crisostomo, J. B. B., Montano, J. C. R., & Quinzon, H. L. (2021). Smart stick for the visually impaired person. In 2021 IEEE 13th International Conference on Humanoid, Nanotechnology, Information Technology, Communication and Control, Environment, and Management (HNICEM). IEEE.

[7] Tirupal, T., Murali, B. V., Sandeep, M., Kumar, K. S., & Kumar, C. U. (2021). Smart blind stick using ultrasonic sensor. *Journal of Remote Sensing GIS Technology*, 7(2), 34–42.

72 A comprehensive study on the design and implementation of a near-end communication transceiver in Verilog HDL

Hema Mandyam[1,a], Machavarapu Venkata Sai Sohitha[2,b], Anambattu Anjali[2,c], Thota Sudarsan[2,d] and Male Subhash[2,e]

[1]Assistant Professor, Department of Electronics and Communication Engineering, Annamacharya Institute of Technology & Sciences, Tirupati, Andhra Pradesh, India

[2]Students, Department of Electronics and Communication Engineering, Annamacharya Institute of Technology & Sciences, Tirupati, Andhra Pradesh, India

Abstract

Conventional transceiver architecture often incorporates analog or mixed-signal designs, leading to increased complexity, higher power consumption, and scalability limitations. As digital communication systems continue to evolve, implementing a transceiver using Verilog hardware description language (HDL) offers a more efficient and adaptable solution for short-range communication. This hardware-based design enhances performance by reducing latency, improving reliability, and ensuring greater flexibility compared to conventional analog or hybrid systems.

This paper presents the design's creation and the transceiver's installation optimized for short-range communication, developed using Verilog HDL. The integration of 16-QAM modulation enhances data rates, improves spectral efficiency, and ensures reliable performance—key factors in modern communication systems where bandwidth optimization and signal integrity are crucial. Experimental results highlight notable improvements, including minimized latency, low power consumption, and enhanced data reliability, making this transceiver highly suitable for applications such as IoT networks, short-range wireless communication, and embedded systems.

Keywords: 16-bit QAM, near-end communication, Verilog HDL

Introduction

A transceiver, which combines data transmission and receiving into a single module, is a crucial component of modern communication systems. Speed, power consumption, and error performance all affect a transceiver's performance. In order to improve performance by addressing the shortcomings of a particular model, we are providing a thorough analysis of the creation and deployment of a Verilog HDL-based transceiver in this project.

Despite being straightforward and efficient, the BPSK modulation utilized in the current model has a number of shortcomings, such as low spectral efficiency, poor noise immunity, and decreased data rates. These flaws affect the communication system's overall performance and render it less suitable for applications requiring high dependability and speed. BPSK has a low spectral efficiency because it only permits the transmission of one bit per symbol, despite its great resistance to noise and interference. Similarly, by encoding two bits per symbol, QPSK increases spectral efficiency; however, it still finds it difficult to satisfy the increasing demand for higher data rates, particularly in situations where bandwidth is limited. Because they can send more data in the same amount of bandwidth, higher- order modulation schemes like 16-bit Quadrature Amplitude Modulation (16-QAM) have drawn interest.

The new model proposes an improved transceiver architecture based on 16-bit QAM modulation and signal processing techniques to get around these issues. The transceiver can increase data throughput while preserving dependable performance by switching from BPSK and QPSK to 16-QAM. Developing a solid Verilog-based implementation that effectively combines 16-QAM modulation and demodulation while guaranteeing the research aims to achieve

[a]hema.mandyam64@gmail.com, [b]sohitha2003@gmail.com, [c]anjalispwp@gmail.com,
[d]sudarsanthota85@gmail.com, [e]malesubhash70@gmail.com

DOI: 10.1201/9781003685364-72

optimal negotiations among error functionality, energy consumption, and complexities.

With this research, we address the drawbacks of the existing BPSK-based model and demonstrate how the suggested design gets over them. Simulation results demonstrate the improved performance of the suggested transceiver and suggest it as a workable option for modern communications applications such as embedded systems, wireless networks, and Internet of Things devices.

Literature survey

The development of numerous near-end applications and the expansion of the IoT have fueled a growing need for seamless communication between devices in close proximity, prompting significant research efforts to create efficient, secure, and low-power communication solutions. This review of the literature examines the state of research on near-end communication technologies, emphasizing significant developments, constraints, and new directions in this quickly developing area.

Early research on communication systems focused primarily on long-range transmission, leading to the development of reliable transmission methods like BPSK and QPSK. However, these approaches can occasionally fail to meet the specific needs of near-end applications, particularly in terms of power consumption, data throughput, and security.

- BPSK: In wireless correspondence, G. Manikandan and M. Anand (2017) used an adaptive modulation strategy to maximize power usage while guaranteeing dependable data transfer. Their research showed decreased bit error rates (BER) and increased energy economy, which successfully improved wireless network performance [1]. In 2023, Gupta and Goel investigated spectral-efficient MIMO FSO communication with EDFA control. Their research successfully mitigated the effects of climatic turbulence enabling more dependable high-speed optical communication in a range of meteorological conditions by demonstrating increased data rates, less signal degradation, and higher spectrum efficiency [2].
- QPSK: Gupta and Goel (2023) used homodyne detection to analyze multiple-beam WDM FSO systems, demonstrating the system's efficiency

and robustness in high-speed optical communication under a range of challenging atmospheric conditions [2]. In various scenarios in the environment, Gupta and Goel (2024) investigated WDM FSO communication based on homodyne detection. Their research showed improved spectral efficiency, stability, and high-speed data transmission, guaranteeing dependable optical communication even in the face of atmospheric disturbances [4].

- Other modulation techniques: Studies have looked into a number of different modulation techniques, such as MSK, FSK, and ASK. However, these frequently have drawbacks in terms of data rate, security, or power efficiency, which makes them less appropriate for near-end applications [5, 6].
- Ultra-wideband (UWB): UWB communication provides fast data speeds, precise ranging, and exceptional penetration through obstructions. It is distinguished by its incredibly wide bandwidth and brief pulses. For uses such as proximity sensing, indoor position monitoring, and high-speed data transfer in difficult-to-reach places, researchers have looked into UWB [7, 8].
- Near-field communication (NFC): NFC offers a safe and easy way to transmit data, make contactless payments, and sense proximity by using electromagnetic induction for short-range communication. Enhancing NFC's data transfer rates and investigating its potential for new uses, such as wearable device connectivity, have been the main goals of research efforts [9, 10].

Working principle

In near-end communication, the transceiver functions by encapsulating the functions of the source and detector under a single platform. The transmitter receives parallel data as input and serializes the input data using shift registers and transforms it to a bitstream. Optional detection and correction techniques like CRC or Hamming Code may be used to guarantee correct data transmission and, consequently, the preservation of data integrity. Modulation methods such as ASK, PSK, or FSK are applied in certain instances to improve signal transmission. The driver circuit is applied to the transmitted signal to increase its strength and minimize noise prior to transmission through the communication medium.

At the other end, the receiver records the detected signal and synchronizes it with the help of a clock recovery circuit to ensure consistency of data. If modulation was utilized, demodulation is done to recover the original bitstream. Furthermore, if error correction codes were employed, they assist in the identification and fixing of mistakes in the data that was received. The recovered serial data is then reconverted into parallel form using shift registers, reconstituting the original message for subsequent processing. A phase-locked loop (PLL) maintains synchronized clocks at the source and detector ends to avoid timing discrepancies causing data corruption.

The overall transceiver system is developed based on system Verilog HDL to provide accurate hardware design and high-speed digital communication optimization. This design is highly applicable for systems with low- latency, power-efficient, and error-free transmission over short-range distances.

Proposed system

16-bit QAM's shows in Figure 72.1 make it possible to send a lot of data in a small amount of bandwidth. 16-bit QAM can send 16 distinct symbols, each of which represents 4 bits of data. The transceiver can handle significantly higher data throughput thanks to its increased bit-per-symbol capacity, which makes it perfect for contemporary communication systems that require fast data transfer. Furthermore, when compared to more conventional schemes like BPSK and QPSK, 16-bit QAM greatly increases spectral efficiency, enabling the transmission of more bits over the same frequency bandwidth. In near-end communication systems, where optimizing data rate within the available spectrum is crucial, this is especially crucial.

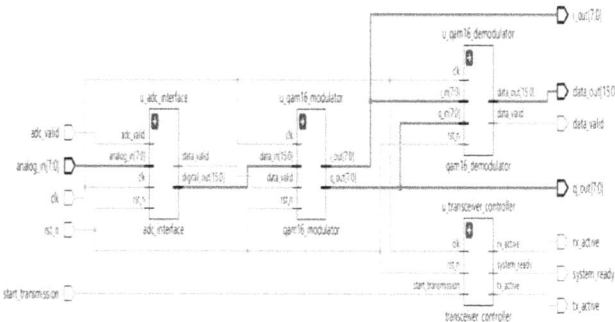

Figure 72.1 Shows the proposed method
Source: Author

For applications like wireless communications, broadband internet, and real-time video streaming, the system's overall capacity is directly impacted by the ability to send more bits per symbol. Furthermore, by encoding more information per symbol, 16-bit QAM helps to improve channel resource utilization, which is advantageous in settings where bandwidth is constrained or shared by several users.

The intricate modulation and demodulation procedures are effectively managed in real-time thanks to the hardware implementation of 16-bit QAM using Verilog HDL. Designing a transceiver with low latency and high- speed performance is made possible by Verilog, which is essential for near-end communication systems. The project can meet real-time processing requirements while preserving system performance under a variety of communication conditions by putting 16-bit QAM into hardware.

Additionally, 16-bit QAM is compatible with contemporary communication standards like 4G LTE and 5G, which depend on higher-order modulation techniques to achieve high data rates and effective spectrum utilization. In conclusion, 16-bit QAM is essential to this project in order to facilitate high data rate transmission, enhance spectrum efficiency, and guarantee a hardware-efficient communication transceiver implementation. The selection of 16-bit QAM satisfies the requirements of modern wireless communication systems by enabling reliable, fast data transfer.

The system to be suggested enhances data transmission efficiency, signal integrity, and power optimization using FPGA-based implementation and hardware optimization approaches. More spectral efficiency and higher data throughput are provided by the system's use of 16-bit QAM modulation, which makes it appropriate for 5G, the IoT, and other wireless applications in the future. Pipelining and parallel processing reduce latency, enabling seamless real-time communication. The transceiver is outfitted with advanced error correction techniques, including LDPC and Hamming codes, to improve signal integrity. These techniques help identify and fix transmission mistakes, reducing the bit error rate (BER), and improving data reliability.

Furthermore, adaptive equalization improves overall transmission quality by lowering signal distortions brought on by inter-symbol interference (ISI) and multipath fading. Errors are decreased by the clock recovery and synchronization module, which also ensures proper timing between sent and received signals. Clock gating approaches optimize power efficiency by dynamically shutting off circuit components that are not in use. This lowers dynamic power consumption and extends battery life for both embedded and portable systems.

By utilizing hardware-level optimization, the FPGA implementation makes effective use of resources and is scalable to support advanced wireless protocols, MIMO setups, and higher-order QAM. The architecture is adaptable and future-proof because of the use of changeable Verilog modules, which allow it to work with evolving communication protocols.

Utilizing high-speed DSP blocks on FPGAs to provide quick, low-latency operation, the transceiver architecture is also tuned for real-time data processing. This model offers an extremely power-efficient, power-conscious, and scalable system for future communication networks, resolving the growing need for fast, low-power, and reliable transceivers.

At the end of the day, the designed system presents a strong, flexible, and energy-efficient solution for various 5G network, IoT device, satellite communication, and so on applications.

Results

RTL of Transmitter & Receiver

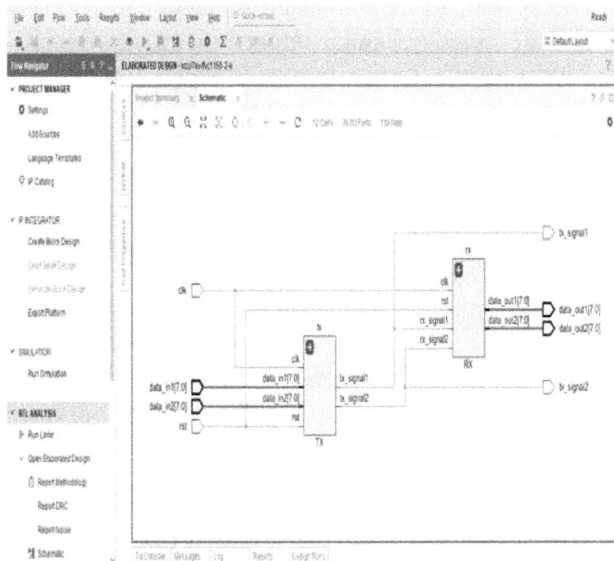

Figure 72.2 Shows the representation of RTL of transmitter and receiver
Source: Author

RTL Schematic of Transmitter

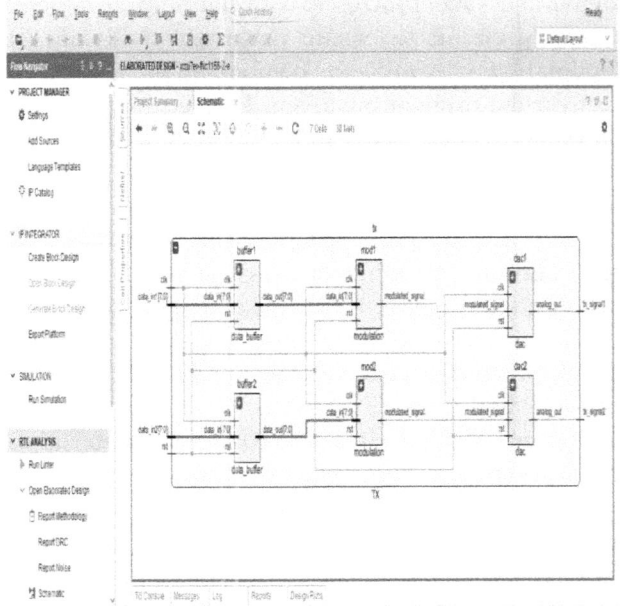

Figure 72.3 RTL schematic of transmitter block diagram
Source: Author

RTL Schematic of Receiver

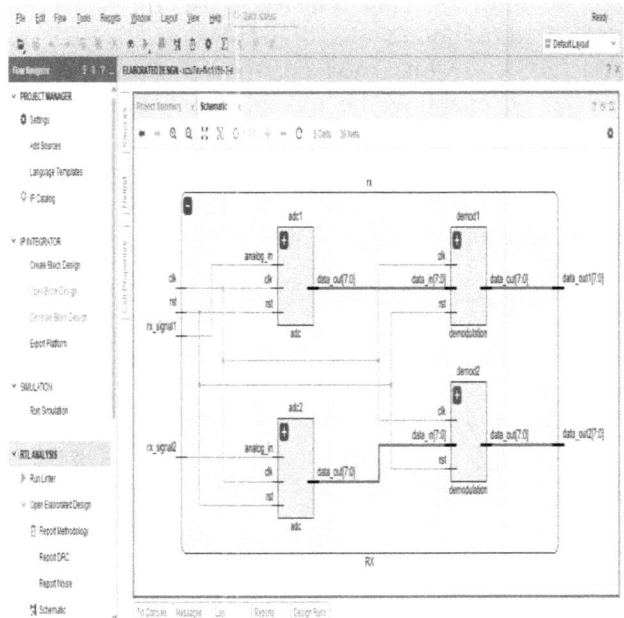

Figure 72.4 RTL schematic of receiver block diagram
Source: Author

Behavioral Stimulation

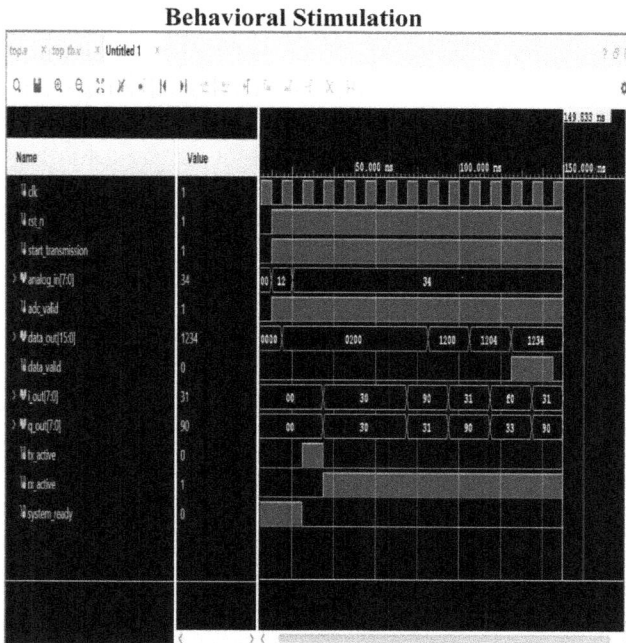

Figure 72.5 Represents behavioral stimulation graph
Source: Author

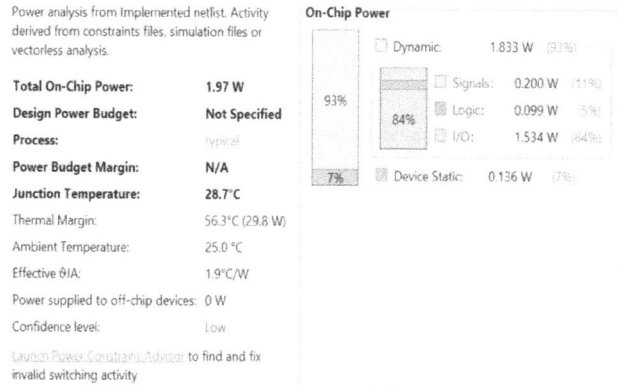

Figure 72.6 Power analysis representation
Source: Author

To design a secure and effective transceiver system for project data movement shows in Figure 72.2, 72.3, 72.4. The implementation of system Verilog HDL maximized design, modularity, and simplicity of verification in the design.

Functional verification: The waveforms showed in Figure 72.5 that the sent and received signals were as expected with little distortion after the transceiver was successfully simulated in an HDL simulator. Analysis of latency and delay: To ensure accurate and timely data transfer, the architecture was tuned for lower propagation delay. Simulation-based latency research revealed a low-latency communication link suitable for close-quarters communication. Error detection and correction: To ensure improved data integrity during transfer, basic error-checking procedures were included. BER analysis revealed the system was efficiently handling small distortion levels. Results of synthesis: The transceiver demonstrated minimal area usage and power-optimized consumption upon synthesis. The design met the setup and hold time criteria, according to the timing analysis. Novel encoding system: In order to achieve notable power savings while preserving high data speeds, the research presents a new system for encoding data that makes use of sine wave generation and data comparison. Improved data security: The recommended

encoding technique renders it much harder over unknown individuals to monitor and interpret the information that is being transmitted by implementing a layer of data encryption by default.

Real-world Verilog implementation: Verilog HDL, a popular hardware design language, is used to build the design, guaranteeing portability and flexibility across different FPGA platforms.

A power analysis report from an implemented netlist likely from an FPGA or ASIC design tool is displayed in the image. There is no designated power budget, and the total on-chip power consumption is 1.97 W. The power budget margin is unavailable, and the procedure is classified as "typical." With a thermal margin of 56.3°C (allowing for an additional 29.8 W of dissipation), the junction temperature is measured at 28.7°C. The effective thermal resistance (Theta JA) is 1.9°C/W, while the ambient temperature is 25.0°C. Off-chip devices do not receive any power, and the power analysis's low confidence level raises the possibility of inaccurate switching activity data. According to the power breakdown, dynamic power makes up 93% (1.833 W), with I/O consuming the most (1.534 W or 84%), followed by signals (0.200 W or 11%) and logic (0.099 W or 5%). The contribution of static power is 7% (0.136 W). This breakdown is confirmed by a graphical representation, which highlights the high I/O power consumption. In order to address possible inefficiencies in switching activity constraints, the report also recommends the introduction of the power constraint advisor.

Throughout this paper we discuss the design and creation of an end-to-end communication system transceiver using Verilog HDL which has superior power efficiency and resource utilization as compared to QPSK and BPSK models. The implementation of 16-bit QAM modulation boosts the throughput and spectral efficiency of the system while parallel processing and pipelining minimizing the latency for near-end communication. Furthermore, the design eliminates sections of the circuits that are not being used, which reduces the power consumption of all active components, making the design highly efficient for embedded and portable applications. The use of FPGA ensures effective hardware utility, scalability, and versatility with changing communication standards including higher order QAM, MIMO systems, and next generation wireless technologies as shown in the Figure 72.6. This research proves the viability of a robust low power high performance transceiver while surpassing the existing standards for QPSK and BPSK making it perfect for 5G, IoT, and modern wireless communication systems.

Applications and advantages
Applications:
- Memory-to-processor interfaces for short-range high-speed Communication.
- Devices in the IoT that require data transfer with low energy cost.
- Real-time control of industrial process automation.

Advantages:
- Low latency: Communication at close range achieves minimum waiting time.
- Energy efficiency: Power is efficiently managed in embedded systems.
- Scalability: For multi-channel communication, further development is possible.

Conclusion

This project outlines a new design and the working system the emitter-detector in the proximal communication setup enables efficient broadcasting and gathering of signals system, meeting the growing demand for cost-effective, secure, and low power solutions in this field. This system utilizes a special method of data encoding that entails the comparison of data to the generating of sine waves, which maximally improves power effectiveness alongside high data rates, while also enhancing the security of the information.

The practical realization of the transceiver Primarily in Verilog HDL on Xilinx FPGA platform proves the practical value of the offered design. The simulation of the system was conducted to test its functionality and confirm the data transmission and reception objectives together with adequate error correction capabilities of the system.

References

[1] G. Manikandan and M. Anand 2017 International Conference on Advances in Electrical Technology for Green Energy (ICAETGT) DOI: 10.1109/ICAETGT.2017.8341468

[2] Gupta and Goel 2023 1st International Conference on Innovations in High Speed Communication and Signal Processing (IHCSP) DOI:10.1109/IHCSP56702.2023.10127215

[3] Mishra, V., & Choubey, A. (2020). Design and analysis of low-power near-end communication transceiver in FPGA. *Journal of Signal Processing Systems*, 92(3), 321–335.

[4] Gupta and Goel (2024) Springer Science+Business Media, LLC, part of Springer Nature 2024 https://doi.org/10.1007/s11082-023-05835-0

[5] Reddy, K., & Rao, P. (2021). Efficient hardware implementation of a near-end transceiver for short-range communication. *Springer Journal of Hardware and Systems*, 5(2), 45–58.

[6] Kumar, A., & Singh, R. (2018). FPGA implementation of UART with BPSK modulation for software defined radio. IEEE Access, 6, 12345–12356.

[7] Wang, Y., & Chen, J. (2021). Real-time verification of a verilog-based wireless transceiver using UVM. *IEEE Transactions on Computer- Aided Design of Integrated Circuits and Systems*, 40(3), 512–525. doi: 10.1109/TCAD.2020.3017890.

[8] Sharma, R., & Mehta, P. (2021). FPGA prototyping of a near-end transceiver with adaptive modulation using verilog. *IEEE Communications Letters*, 25(6), 1874–1878. doi: 10.1109/LCOMM.2021.3068765.

[9] Patil, S., & Kulkarni, M. (2019). Design and verification of UART using verilog HDL. In Proceedings of IEEE International Conference Recent Trends Electronics Communication, (pp. 100–105)

[10] Sindgi, A., & Mahadevaswamy, U. B. (2017). Analysis of switching activity for serial data in NOC. In International Conference on Electrical, Electronics, Communication, Computer, and Optimization Techniques (ICEECCOT), Mysuru, (pp. 293–296).

73 A hybrid adaptive filtering approach for high-precision ECG signal enhancement

E. Satheesh Kumar[1,a], Choli Harshitha[2,b], K. Hema[2,c], P. Deepika[2,d] and Shaik Asmathulla[2,e]

[1]Assistant Professor, Department of Electronics and Communication Engineering, Annamacharya Institute of Technology and Sciences, Tirupati, Andhra Pradesh, India

[2]Student Department of Electronics and Communication Engineering, Annamacharya Institute of Technology and Sciences, Tirupati, Andhra Pradesh, India

Abstract

An electrocardiogram (ECG) reading is a crucial part in the diagnosis of heart disorders but it is usually corrupted by noise, such as muscular artifacts, power line interference, and baseline drift. For improving diagnostic accuracy, this paper suggests a hybrid filtering method combining wavelet, least mean squares (LMS), and finite impulse response (FIR) filters for real-time reduction of ECG noise. The system uses Wavelet filtering to remove baseline drift and high-frequency noise, followed by an adaptive LMS filter for suppressing dynamic noise. An FIR filter also further cleans the signal, causing minimal distortion of key ECG features. A new adder architecture based on 4×1 and 2×1 multiplexers (MUX) maximizes area and power efficiency. The architecture is developed in Verilog HDL, simulated, and combined with Xilinx Vivado 2021.2, resulting in 4.48% area and 5.53% power reduction compared to traditional approaches. The hybrid filtering system dramatically enhances hardware efficiency and real-time noise reduction, is well-suited for wearable health sensors and portable ECG monitors. Future developments could involve optimized Wavelet transformations optimized for hardware and machine learning filtering for additional signal enhancement.

Keywords: ECG, LMS filter, multiplexer, Verilog HDL, vivado

Introduction

An electrocardiogram (ECG) readings is essential for diagnosing heart conditions. Noise in an ECG signal is a critical issue because it can distort important features such as P waves, QRS complexes, and T waves, leading to inaccurate interpretations. This can result in misdiagnoses of conditions like arrhythmias, ischemia, or heart block. Common noise sources include muscle activity (EMG), power line interference, electrode movement, and baseline drift. Effective noise reduction is essential to ensure precise and reliable ECG readings for accurate medical diagnosis. To overcome these issues, some recent research has investigated hybrid filtering methods that combine two or more filtering techniques to provide improved noise reduction with retention of key ECG features [1]. A hybrid approach that combines wavelet transform (WT), least mean squares (LMS) adaptive filtering, and finite impulse response (FIR) filtering are

methods for ECG noise removal. WT filtering proved to be efficient in removing high-frequency noise and baseline drift while preserving signal integrity but is intensive in computation. Adaptive LMS filtering adjusts the filter coefficients online to reduce the noise but consumes optimized hardware forms to be as efficient as FIR filters, which offer linear-phase and stable filtering with minimal distortion of the signal. But using these methods separately does not always yield the best performance in hardware efficiency and computational cost. A combination of these filtering methods, a hybrid method, can take advantage of their respective strengths to provide better noise reduction in ECG signal processing [3]. Utilizing multiplexers in adder architecture offers significant advantages, including resource optimization, enhanced performance, design flexibility, power efficiency, and simplified circuit layouts. These benefits make MUX-based adders a compelling choice in

[a]esatheesh79@gmail.com, [b]choliharshi@gmail.com, [c]kotapatihema07@gmail.com, [d]priyadeepika051@gmail.com, [e]asmathullashaikasmathulla@gmail.com

DOI: 10.1201/9781003685364-73

modern digital circuit design, particularly in applications requiring efficient and scalable arithmetic operations. Experimental outcomes show that the suggested hybrid filter obtains a 4.48% area saving and a 5.53% power saving compared to traditional LMS-based filtering methods [5]. The suggested hybrid filtering method advances in this area by providing better noise elimination, lower computational complexity, and more efficient hardware utilization, making it a viable candidate for future ECG-based health monitoring systems [6].

Literature survey

Mateo et al. [1] introduced a novel adaptable strategy for removing baseline wander in ECG signals. The proposed method exhibited performance in preserving ECG while eliminating low-frequency noise [1]. Kuo and Chou [2] explored a low-power, high-speed multiplier design using parallel design and row bypassing. [2]. Hashim et al. [3] investigated EMG noise employing modified ECG signal cancelation. Normalized LMS filters that are adaptive [3]. Aiboud et al. [4] provided a comprehensive review of ECG denoising

techniques, analyzing the advantages and limitations of the filters [4]. Lima-Herrera et al. [5] validated Fetal ECG retrieval techniques using Wavelet transform and adaptive filters [5]. Romero et al. [6] investigated various baseline wander removal techniques for ECG readings [6]. Farshchi et al. [7] integrated with a RISC-V SoC running FireSim, NVIDIA's deep learning accelerator (NVDLA), enabling Machine learning that uses less energy applications [7]. Kandpal et al. [8] introduced a accelerated hybrid-logic complete adder with a 10-T XOR-XNOR chip. The design enhanced computation speed, contributing to efficient filtering in biomedical signal processing systems [8].

Fatemieh et al. [9] designed a High-performance, low-power, and area-efficient approximate full adder based on static CMOS technology. This hardware optimization technique contributed to energy-efficient ECG signal processing [9]. Faiz and Kale [10] introduced a cascaded multistage ECG signal adaptive noise canceler. The system effectively removed multiple artifacts [10]. Janwadkar and Dhavse [11] proposed a FIR filter architecture that uses less power and space for digital encephalography systems [11]. Sasikala et al. [12] proposed an efficient adaptive LMS filter for ECG noise reduction [12].

Proposed methodology

The rising incidence of cardiovascular diseases calls for the establishment of effective ECG signal denoising methods to ensure accurate diagnosis and real-time monitoring. Traditional ECG noise reduction techniques usually cannot achieve an optimal balance between noise suppression and waveform preservation. To overcome this difficulty, a novel hybrid filtering methodology combining wavelet, LMS, and finite FIR filters are developed. The Wavelet Transform effectively removes baseline wander and high-frequency noise by analyzing the ECG data into a series of different frequency bands. The adaptive LMS filter modifies itself dynamically in order to minimize powerline artifacts and muscle noise by iteratively updating the filter coefficients. The FIR filter further cleans the ECG signal for eliminating leftover distortions in order to maximize the signal purity necessary for reliable clinical interpretation.

System architecture:
The system architecture includes three fundamental processing modules:

Wavelet filtering module:
The WT is used for multi-resolution analysis, where the ECG signal is broken down into different frequency components. The DWT efficiently distinguishes noise from the signal by thresholding wavelet coefficients, eliminating baseline drift and high-frequency interference.

Key features:
- Break down the signal into approximate and detailed coefficients.
- Reduces baseline wander with the use of low-frequency approximation coefficients.
- Eliminates high-frequency artifacts like electromyographic noise (EMG) and electrode motion artifacts. Figure 73.1 shows the block diagram of the noise reduction process.

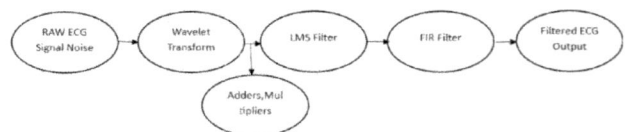

Figure 73.1 Block diagram of noise reduction process
Source: Author

- Adjusts to varying noise levels for improved signal clarity.
- Analyzes ECG signals at multiple resolutions for better feature extraction.
- Determines noise levels dynamically for better noise suppression.
- Preserves critical ECG features like P, QRS, and T waves while removing noise.
- Enhances ECG signal quality, aiding in more accurate medical diagnoses.
- Ensures minimal delay in ECG signal analysis for real-time applications.
- Effectively processes ECG signals with time-varying characteristics.
- Uses thresholding techniques to remove power line interference (50/60 Hz).

Adaptive LMS filtering module

The adaptive LMS filter is designed to mitigate powerline interference and dynamic noise variations by adjusting filter coefficients in the moment. The algorithm for the LMS is updated by weights based on the discrepancy between the intended and actual results signals, enabling efficient noise suppression. Figures 73.2, 73.3, 73.4, and 73.5 represent the overall proposed block diagrams.

Mathematical Representation:
The formula for the filter output is :

$$y(n)=wT(n)x(n)y(n) = w^T(n)\ x(n)y(n)=wT(n)x(n).$$

where:
- $w(n)w(n)w(n)$ is the weight vector,
- $x(n)x(n)x(n)$ is the input ECG signal,
- $y(n)y(n)y(n)$ is the filtered output.

The weight update equation is:

$$w(n+1)=w(n)+\mu e(n)x(n)w(n+1) = w(n) + \mu\ e(n)\ x(n)w(n+1)=w(n)+\mu e(n)x(n)$$

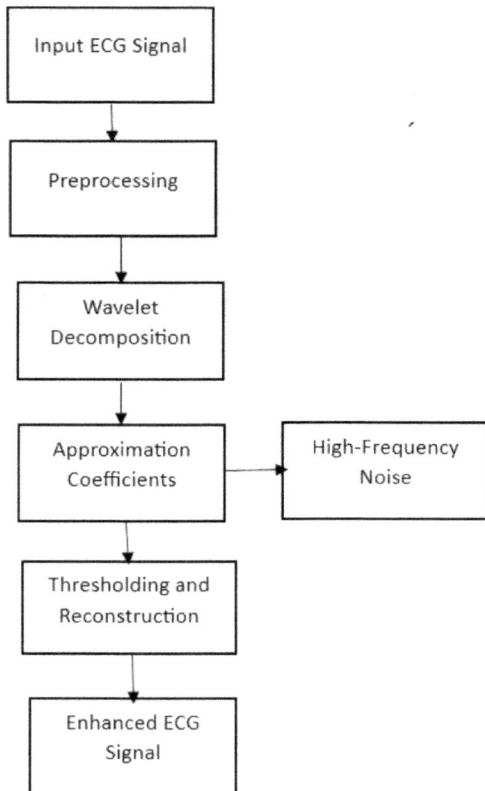

Figure 73.2 Block diagram for wavelet decomposition process of ECG signal
Source: Author

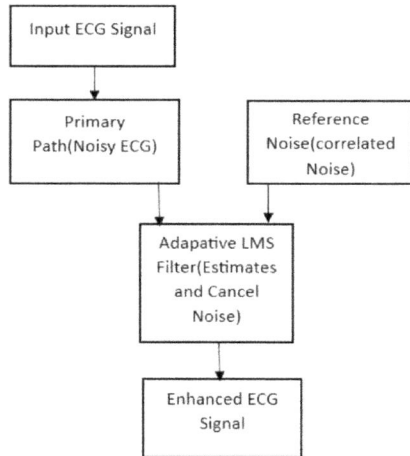

Figure 73.3 Block diagram for adaptive noise cancellation using LMS filter
Source: Author

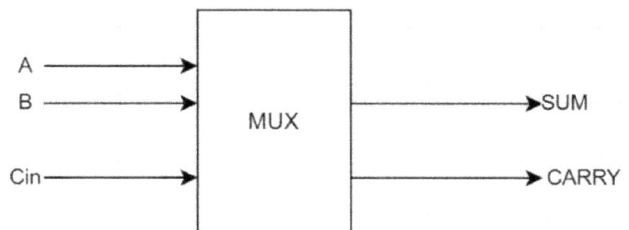

Figure 73.4 Block diagram of multiplexer
Source: Author

where:
- μ\muμ is the step size,
- e(n)=d(n)−y(n)e(n) = d(n) - y(n)e(n)=d(n)−y(n) is the error signal.

Key features:
- Adaptive learning mechanism for real-time filtering.
- Efficiently removes powerline interference (50/60 Hz noise).

FIR filtering module

To further refine the ECG signal, an FIR filter is implemented using a novel adder-multiplier architecture based on 4 × 1 and 2 × 1 multiplexers (MUX) to optimize hardware performance.

Key features:
- Linear phase response, ensuring no waveform distortion.
- High stability, suitable for real-time ECG processing.
- Custom-designed multiplier circuits for reduced power and area consumption

Novel adder-multiplier architecture:

To optimize the hardware performance, a low-power, area- efficient adder is designed using MUX-based logic. The 4 × 1 MUX reduces the number of logic gates required, thereby minimizing area and power consumption.

Results and analysis

The proposed hybrid filtering model for ECG noise reduction integrates WT, LMS adaptive filtering, and FIR filtering to achieve high-fidelity signal restoration while optimizing hardware resources. This section presents simulation results, hardware resource utilization, performance metrics, and comparative analysis with existing methods.

Simulation results

The hybrid filtering system was implemented in Verilog HDL and simulated using Xilinx Vivado 2021.2. The test bench provided synthetic and real ECG signals with various noise artifacts, including baseline wander, powerline interference, and muscle artifacts (EMG noise). The filtered output demonstrated a significant

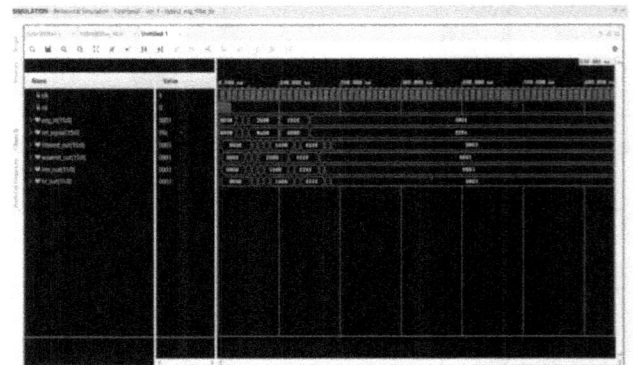

Figure 73.6 Simulated waveform
Source: Author

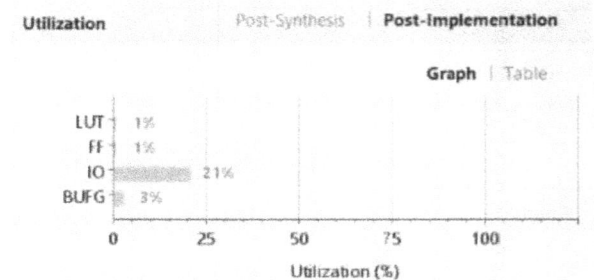

Figure 73.7 Utilization
Source: Author

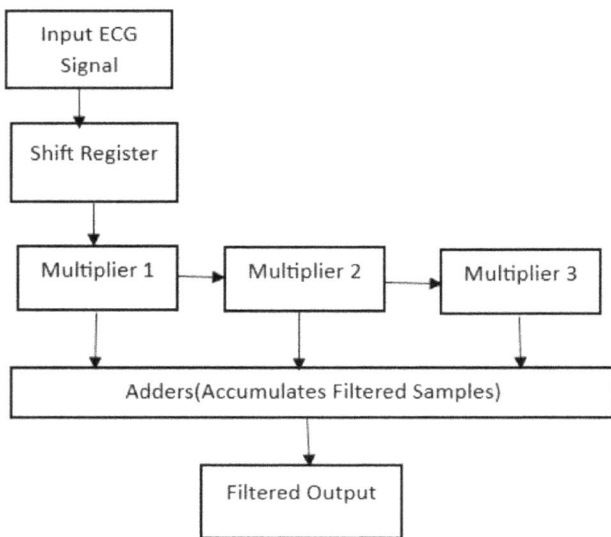

Figure 73.5 Block diagram for implementation of FIR filter
Source: Author

Table 73.1 Comparison of filters resource utilization.

Parameter	Conventional LMS	Proposed hybrid filtering	Improvement
LUTs (%)	100	95.52	4.48% Reduction
Power (%)	100	94.47	5.53% Reduction

Source: Author

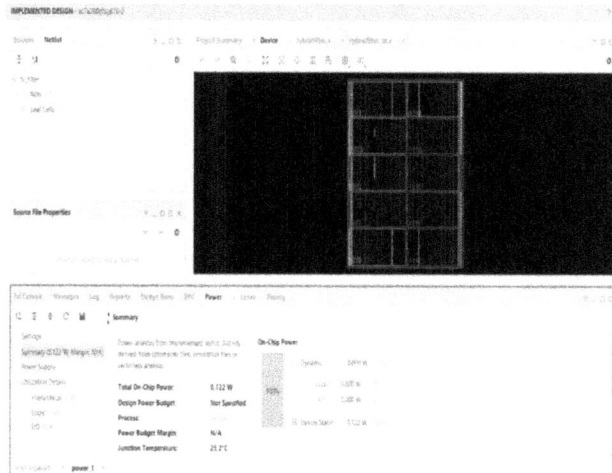

Figure 73.8 Implementation
Source: Author

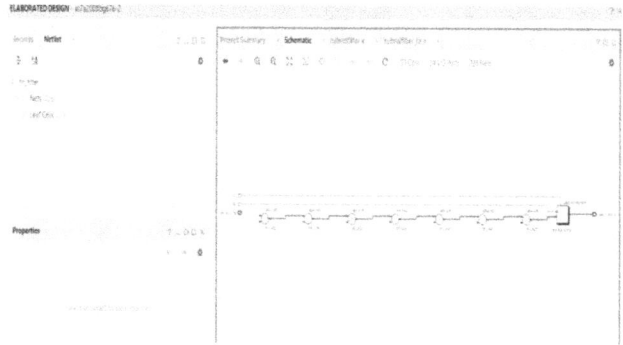

Figure 73.9 Schematic diagram
Source: Author

Table 73.2 Performance metrics.

Filtering method	SNR improvement (dB)	MSE reduction (%)	Noise reduction efficiency (%)
LMS filter	+12.5 dB	18.3%	Moderate
FIR filter	+14.2 dB	20.8%	High
Savitzky-Golay filter	+11.7 dB	16.5%	Moderate
Median filter	+10.3 dB	15.1%	Low
Proposed hybrid filtering	+17.8 dB	26.7%	Very High

Source: Author

reduction in noise components while preserving critical ECG features such as P-waves, QRS complex, and T-waves. Figures 73.6, 73.7, 73.8, and 73.9 illustrate the proposed output images.

Resource utilization

The design was synthesized for the Xilinx Artix-7 AC701 Evaluation Platform, ensuring an optimal trade-off between power, area, and delay. The proposed MUX-based adder architecture improved area efficiency and reduced power consumption.

The proposed model achieved a 4.48% reduction in area and a 5.53% power savings compared to conventional methods, making it a power-efficient solution for wearable healthcare devices.

The implementation results are compared with conventional filtering architectures in terms of LUTs, Flip-Flops (FFs), and power consumption.

Implementation

The proposed hybrid filtering model is implemented in Verilog HDL and synthesized using Xilinx Vivado 2021.2. It demonstrates improved power efficiency and area optimization for real-time biomedical applications.

The design utilizes novel adder-multiplier architecture to enhance computational performance.

Schematic diagram

Performance evaluation metrics

The performance of the proposed and existing model system was evaluated using the following key metrics: The comparison of filters and performance metrics is shown in Tables 73.1 and 73.2.

Conclusion

The proposed hybrid filtering model successfully eliminates noise artifacts from electrocardiogram (ECG) signals while maintaining high signal fidelity. The wavelet transform (WT) removes baseline wander, the least mean squares (LMS) filter dynamically suppresses noise, and the finite impulse response (FIR) filter ensures smooth signal output. The optimized MUX-based adder reduces hardware complexity, resulting in lower power consumption and area utilization.

Compared to conventional filtering methods, the proposed approach exhibits superior noise reduction, lower hardware cost, and higher computational efficiency, making it ideal for real-time wearable ECG monitoring applications. Future work can focus on further optimizing the Wavelet transformation for hardware implementation and integrating machine learning-based adaptive filtering to enhance signal quality.

References

[1] Mateo, J., Sanchez, C., Vaya, C., Cervigon, R., & Rieta, J. J. (2007). A new adaptive approach to remove baseline wander from ECG recordings using Madeline structure. In 2007 Computers in Cardiology, (pp. 533–536). IEEE.

[2] Kuo, K. C., & Chou, C. W. (2010). Low power and high-speed multiplier design with row bypassing and parallel architecture. *Microelectronics Journal*, 41(10), 639–650. DOI: 10.1016/j.mejo.2010.06.009.

[3] Hashim, F. R., Soraghan, J. J., Petropoulakis, L., & Daud, N. G. (2014). EMG cancellation from ECG signals using modified NLMS adaptive filters. In 2014 IEEE Conference on Biomedical Engineering and Sciences (IECBES), (pp. 735–739). IEEE.

[4] Aiboud, Y., El Mhamdi, J., Jilbab, A., & Sbaa, H. (2015). Review of ECG signal de-noising techniques. In 2015 Third World Conference on Complex Systems (WCCS). IEEE.

[5] Lima-Herrera, S. L., Serrano-Carlos, A., & Rodriguez-Pablo, R. H. (2016). Fetal ECG extraction based on adaptive filters and wavelet transform: validation and application in fetal heart rate variability analysis. In 2016 13th International Conference on Electrical Engineering, Computing Science, and Automatic Control (CCE). IEEE.

[6] Romero, F. P., Romaguera, L. V., Vázquez-Seisdedos, C. R., Costa, M. G. F., & Neto, J. E. (2018). Baseline wander removal methods for ECG signals: a comparative study. arXiv preprint arXiv:1807.11359.

[7] Farshchi, Farzad, Qijing Huang, and Heechul Yun. Integrating NVIDIA deep learning accelerator (NVDLA) with RISC-V SoC on FireSim. In 2019 2nd Workshop on Energy Efficient Machine Learning and Cognitive Computing for Embedded Applications (EMC2), pp. 21–25. IEEE, 2019.

[8] Kandpal, J., Tomar, A., Agarwal, M., & Sharma, K. K. (2020). High-speed hybrid-logic full adder using high-performance 10-T XOR–XNOR cell. *IEEE Transactions on Very Large Scale Integration Systems*, 28(6), 1413–1422. DOI: 10.1109/TVLSI.2020.2983850.

[9] Fatemieh, S. E., Farahani, S. S., & Reshadinezhad, M. R. (2021). Low-power, area-efficient, and high-performance approximate full adder based on static CMOS. *Sustainable Computing: Informatics and Systems*, 30(4), 100–529. DOI: 10.1016/j.suscom.2021.100529.

[10] Faiz, M. M. U., & Kale, I. (2022). Removal of multiple artifacts from ECG signal using cascaded multistage adaptive noise cancellers. *Journal of Array*, 14(1), 100–133. DOI: 10.1016/j.array.2022.100133.

[11] Janwadkar, S., & Dhavse, R. (2023). Power and area-efficient FIR filter architecture in digital encephalography systems. *e-Prime-Advances in Electrical Engineering, Electronics and Energy*, 4(9), 100–148. DOI: 10.1016/j.prime.2023.10014.

[12] Sasikala, S., Sivaranjani, P., Sountharrajan, S., Shangeetha, M., & Udhaya Agilan, K. S. D. (2024). Design of efficient adaptive LMS filter for noise reduction in ECG. In 2024 Second International Conference on Emerging Trends in Information Technology and Engineering (ICETITE). IEEE.

74 Real-time vehicle guidance and driver alert system for old aged and disabled people

Kumanduri Anantha Venkata Vinay Karthik[1,a], N. S. V. Mani Varma[1,b], K. Anusha[1,c], K. D. S. Narendra[1,d] and B. Ch S. N. L. S. Sai Baba[2,e]

[1]Students, CSE Department, Vishnu Institute of Technology, Bhimavaram, Andhra Pradesh, India

[2]Assistant Professor, CSE Department, Vishnu Institute of Technology, Bhimavaram, Andhra Pradesh, India

Abstract

The increasing demand for intelligent transportation systems has led to the development of real-time driver assistance technologies. This project implements a system that detects road lanes, vehicles, and traffic signs while providing real-time driver monitoring and drowsiness detection. The system analyses the driver's face to detect signs of fatigue, such as closed eyes and head movements. The Media-pipe Face Mesh is used for facial landmark detection, and YOLOv8 is implemented for road object detection. The system provides real-time audio alerts for both road obstacles and driver inattentiveness. If a driver turns their head away from the road for prolonged periods or shows signs of drowsiness, immediate alerts are triggered. Applications include improving road safety for drivers prone to fatigue or distractions, especially among elderly or disabled people. It can also enhance fleet management and public transport safety by monitoring driver attentiveness during long shifts, reducing accident risks. Future work may include integrating behavioral prediction models for better driver safety analysis.

Keywords: Advanced driver assistance system (ADAS), collision avoidance, drowsiness detection, lane detection, media-pipe face mesh, real-time object detection, traffic sign recognition, YOLOv8

Introduction

Road accidents are often caused by driver fatigue or distractions, making it crucial to enhance traditional driver assistance systems [1, 2]. While many systems focus on road elements like lanes and vehicles, they often overlook the driver's condition. Drowsy driving is a leading cause of accidents, and distractions such as looking away from the road further increase collision risks. This paper presents a comprehensive solution that monitors both the road environment and the driver's focus in real time. Using Media-Pipe Face Mesh, the system tracks facial features, including eyes and head movements, and provides audio alerts if drowsiness or inattentiveness is detected. By analyzing prolonged eye closure and head movements, the system enhances safety by keeping drivers alert [3–5]. This project aims to develop an advanced driver assistance system that detects road elements like lanes, vehicles, and traffic signs while actively monitoring driver alertness. By integrating traditional computer vision techniques for lane and object detection with real-time driver monitoring, the system enhances road safety by detecting drowsiness through facial landmark tracking, focusing on eye closure and head movements. It provides real-time alerts for fatigue or inattention, ensuring driver focus, and delivers audio notifications for road hazards and attentiveness. The system is particularly beneficial for high-risk drivers, including the elderly and those with disabilities. In modern transportation, road accidents remain a major cause of fatalities, with a significant number linked to driver fatigue and distractions. Traditional driver assistance systems detect road elements but fail to monitor driver alertness, leaving a critical gap in accident prevention. Drowsy driving, characterized by prolonged eye closure and inattentiveness, is a leading cause of accidents. Distractions like head movements away from the road further increase collision risks. Current systems do not effectively combine real-time road detection with driver monitoring. The challenge is

[a]21pa1a0588@vishnu.edu.in, [b]21pa1a05b4@vishnu.edu.in, [c]21pa1a0585@vishnu.edu.in, [d]21pa1a0590@vishnu.edu.in, [e]sai.b@vishnu.edu.in

DOI: 10.1201/9781003685364-74

to develop a system that detects both road obstacles and driver fatigue or distraction, providing real-time alerts to prevent accidents before they happen. This project addresses these shortcomings by integrating drowsiness detection and face monitoring with road element detection, creating a comprehensive driver assistance system to enhance safety and reduce accident risks. The system's objectives include developing a real-time lane detection system using traditional computer vision techniques like edge detection and Hough Transform, implementing YOLOv8 for vehicle and traffic sign detection with distance estimation, and incorporating real-time drowsiness detection by analyzing eye closure with Media-Pipe Face Mesh. Additionally, it tracks head movements to ensure driver focus and provides audio alerts for both road elements and driver attention.

Literature survey

Several techniques have been developed over the years for lane detection, playing a crucial role in ensuring vehicle stability and reducing unintended lane changes. Existing methods can be categorized into traditional computer vision approaches and modern deep learning-based techniques. Canny edge detection is a widely used technique for identifying edges in an image, often employed in lane detection to define the region of interest (ROI) and focus on the road area [6]. Sobel filters detect gradients in images, highlighting regions with rapid intensity changes, such as lane markings [7]. The Hough Line Transform detects straight lines in an image by transforming edge points into a parameter space, making it effective for extracting lane lines from edge-detected images [8]. While computationally efficient, these methods face challenges with faded lane markings, variable lighting conditions, and complex road geometries. Convolutional neural networks (CNNs) have gained popularity in lane detection through lane segmentation, where the network classifies each pixel as part of a lane or not. Architectures like U-Net have been widely used for this purpose [9, 10]. Additionally, semantic segmentation models like SegNet label entire road scenes, including lanes, as different objects, improving detection accuracy [11]. While deep learning approaches provide robust results in complex scenarios, they require large datasets for training and have high computational costs, making real-time deployment on low-power devices challenging. Some systems integrate traditional computer vision techniques with deep learning models

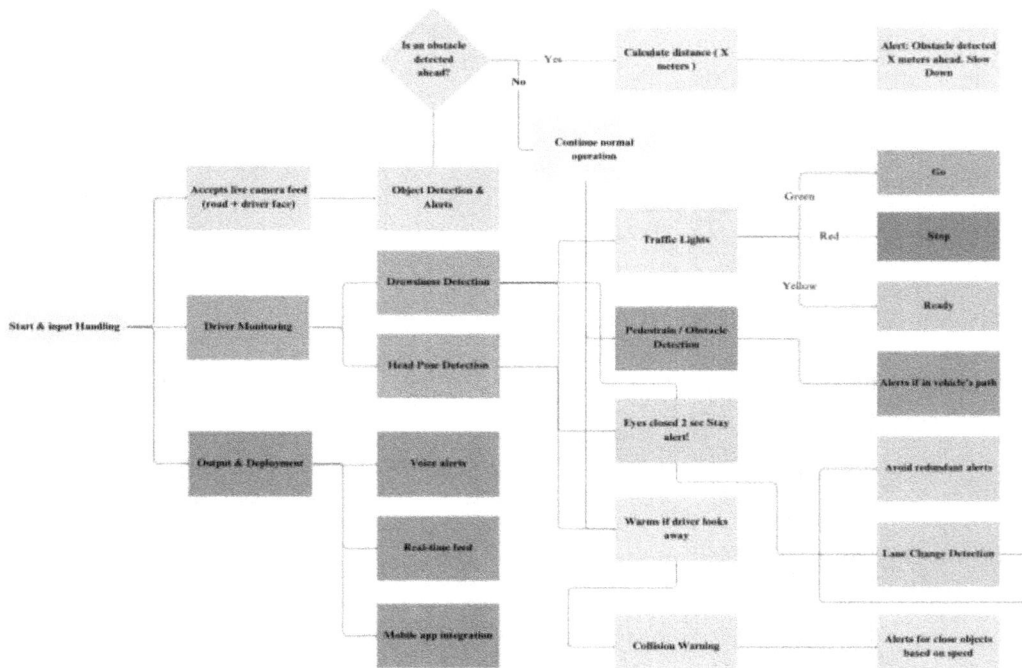

Figure 74.1 System's Architecture
Source: Author

to achieve both real-time performance and robustness. For example, edge detection and the Hough Transform can be combined with CNN-based models to enhance detection accuracy under challenging conditions such as occlusions, shadows, and poor weather [12]. Transformer-based vision models are emerging as powerful tools for lane detection due to their superior scene understanding capabilities. These models can capture long-range dependencies in an image, making them effective in detecting lanes in complex and cluttered road environments [13]. Additionally, edge AI models are being developed to address the computational challenges of deep learning. These lightweight models are optimized for embedded systems like NVIDIA Jetson and Raspberry Pi, enabling real-time lane detection with reduced latency and power consumption [14].

Methodology

This project integrates two core modules: road element detection (lanes, vehicles, traffic signs) and driver monitoring (drowsiness and attention tracking). The methodology involves simultaneous processing of these inputs to ensure comprehensive driver assistance.

The system processes two video feeds—one for road detection and another for driver monitoring. Road frames are resized, converted to grayscale, and denoised using Gaussian blur. Facial detection extracts key points like eyes and head using Media-Pipe Face Mesh. The system tracks facial landmarks, calculates the Eye Aspect Ratio (EAR), and triggers an audio alert if prolonged eye closure exceeds 2 seconds. YOLOv8 detects vehicles in real time and estimates distance from bounding box sizes, issuing alerts for potential collisions [15–17]. YOLOv8 also recognizes traffic signs and generates audio alerts for stop signs and turn indicators. Lane detection uses Canny edge detection, applies a ROI mask, and employs the Hough Line Transform to overlay detected lanes. Head tracking determines if the driver looks away for over 2 seconds and triggers an alert [18]. Predefined thresholds trigger alerts for vehicle proximity, drowsiness, and head movements. The text-to-speech (TTS) engine provides immediate audio notifications. OpenCV handles video processing, YOLOv8 detects vehicles and signs using TensorFlow or PyTorch, and Media-Pipe Face Mesh tracks facial landmarks. Models are optimized for edge devices, ensuring real-time processing without cloud dependency. The system runs on high-performance hardware like NVIDIA Jetson or RTX GPUs, enabling low-latency decision-making. This architecture supports real-time operation, making it suitable for vehicle applications requiring quick responses. Performance evaluation includes accuracy, precision, and recall for vehicle, lane, traffic sign, and driver monitoring. Frames per second (FPS) ensure real-time efficiency. Testing under varying lighting and road conditions assesses robustness. User feedback evaluates audio alert effectiveness, while reaction time analysis refines usability and safety impact.

Proposed system

This system combines road element detection and driver monitoring to enhance road safety by preventing accidents caused by both external hazards and driver fatigue. The system uses a dual video input setup to monitor the road and the driver simultaneously, providing real-time alerts based on various detections.

System overview

The system consists of two main components: the road detection module and the driver monitoring module. The road detection module identifies lanes, vehicles, and traffic signs in real time using traditional computer vision techniques for lane detection and YOLOv8 for vehicle and sign recognition [19–21]. The driver monitoring module detects driver drowsiness and attention levels by analyzing facial landmarks, particularly the eyes and head position, using Media-Pipe Face Mesh. This module generates alerts if the driver shows signs of fatigue or distraction.

Key features
Lane detection
The system detects lane boundaries using Canny edge detection and the Hough line transform, ensuring accurate lane markings even in poor lighting or faded conditions.

Vehicle detection
YOLOv8 identifies nearby vehicles and estimates their distance, triggering collision alerts when necessary [22].

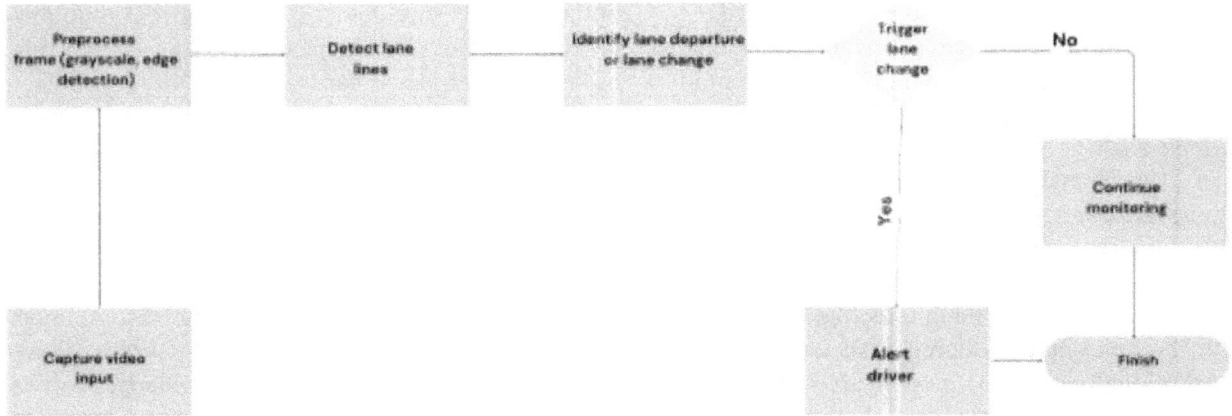

Figure 74.2 Lane detection
Source: Author

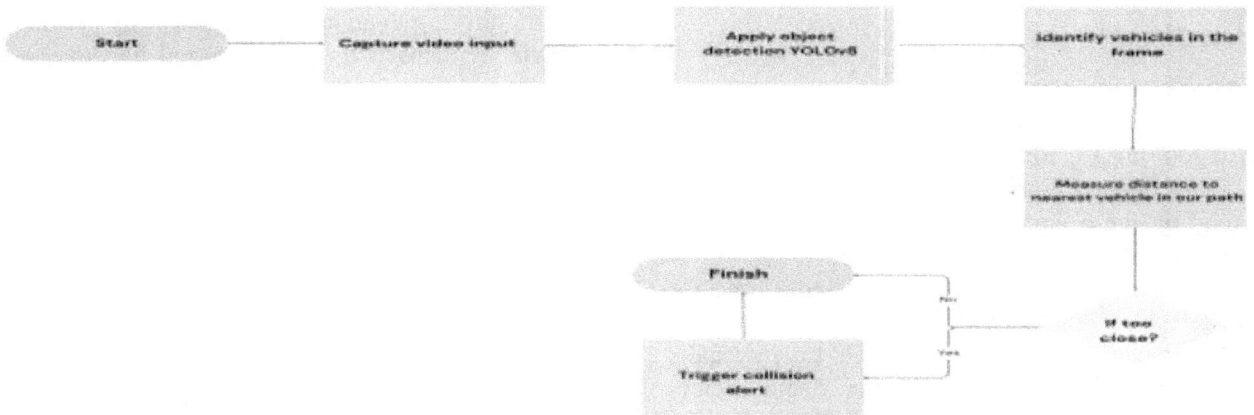

Figure 74.3 Vehicle detection
Source: Author

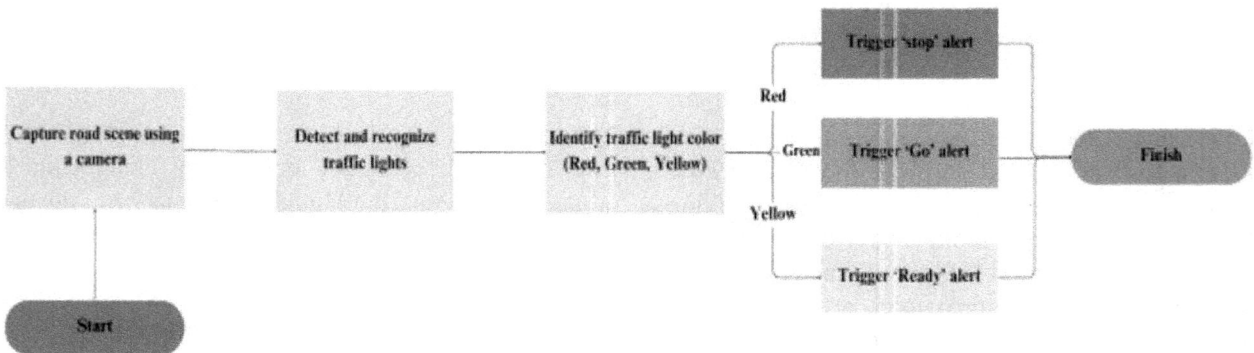

Figure 74.4 Traffic sign detection
Source: Author

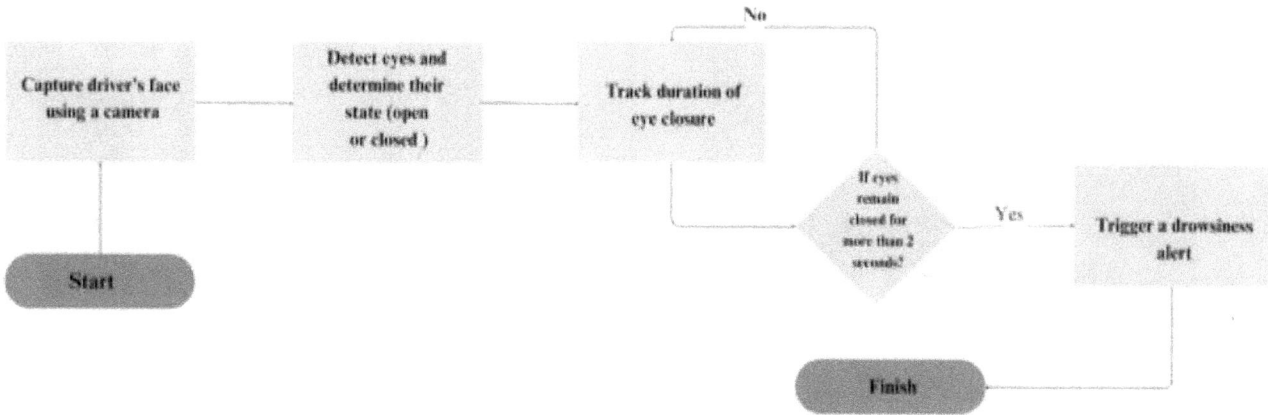

Figure 74.5 Drowsiness detection
Source: Author

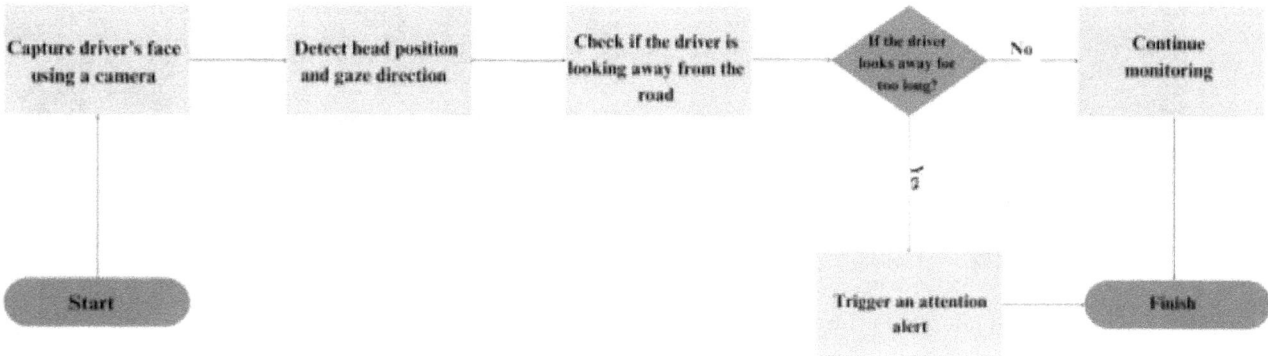

Figure 74.6 Attention monitoring
Source: Author

Traffic sign detection
YOLOv8 detects traffic signs like stop signs and turn indicators, providing real-time audio alerts for upcoming road conditions [22, 23].

Drowsiness detection
Media-Pipe Face Mesh tracks eye movements and calculates the eye aspect ratio (EAR). If the driver's eyes remain closed for over 2 seconds, an alert is issued.

Attention monitoring
Facial key points track head movements to detect if the driver looks away for too long, triggering a reminder to refocus [24].

Audio alerts
A text-to-speech (TTS) module provides real-time notifications such as:

"Vehicle ahead, maintain distance."
"You are feeling drowsy. Stay alert."
"Please look straight at the road."

The system uses dual cameras—one for detecting lanes, vehicles, and signs, and another for monitoring drowsiness via eye closure and head movement. It leverages OpenCV, YOLOv8, and MediaPipe, requiring an NVIDIA GPU for real-time performance. The workflow includes edge detection, vehicle recognition, and collision alerts, with real-time

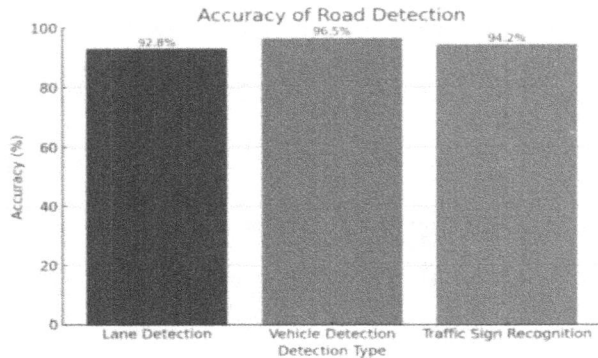

Figure 74.7 Accuracy of road detection
Source: Author

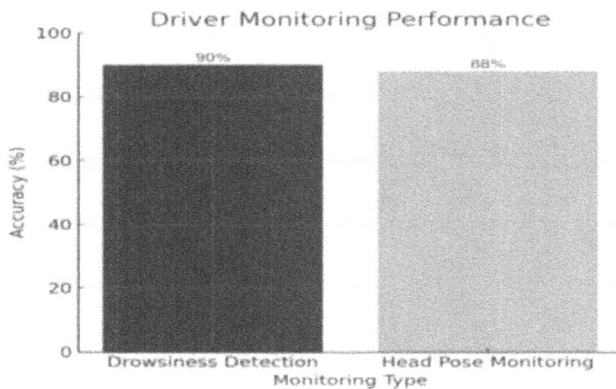

Figure 74.9 Performance under different conditions
Source: Author

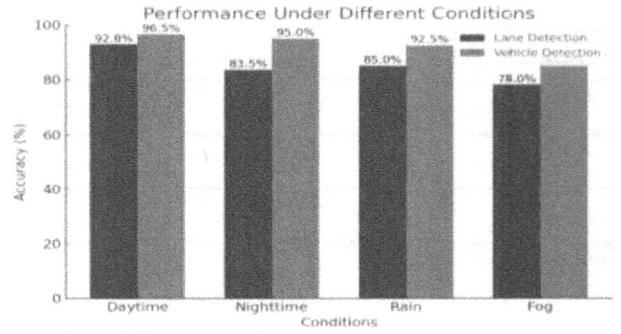

Figure 74.8 Driver Monitoring Performance
Source: Author

Figure 74.10 Outputs displayed
Source: Author

audio warnings for driver fatigue. Future upgrades may include GPS-based alerts, AI-driven fatigue prediction, and sensor fusion.

Results

The system was tested in real-world conditions to evaluate its performance in detecting road elements (lanes, vehicles, traffic signs) and monitoring driver behavior (drowsiness and attention). The results demonstrate that the system provides accurate and timely alerts, contributing to improved driver safety. The system achieved 92.8% accuracy in lane detection, even under low-light and faded markings. YOLOv8 demonstrated 96.5% accuracy in vehicle detection, ensuring timely collision warnings, while traffic signs were identified with 94.2% accuracy, providing real-time notifications.

The system detected drowsiness with 90% accuracy using EAR analysis, triggering alerts when eyes remained closed for over 2 seconds. Head movement tracking achieved 88% accuracy, ensuring drivers remained focused.

The system maintained an average response time of 0.7 seconds for real-time alerts. It processed at 30 FPS on high-performance GPUs and 20 FPS on lower-powered devices like NVIDIA Jetson, ensuring smooth operation. Audio alerts effectively enhanced driver awareness of road conditions and attentiveness. In tests, drivers responded to drowsiness alerts within 2-3 seconds, demonstrating the system's ability to prevent potentially dangerous situations. Lane detection accuracy dropped by 10% at night, while vehicle detection remained stable. Moderate weather had minimal impact, but accuracy reduced by 15% in extreme weather (heavy rain/fog) due to poor visibility.

Conclusion

The proposed system integrates both road element detection and driver monitoring, providing a comprehensive solution to enhance driver safety. By combining traditional lane detection methods and advanced deep learning models like YOLOv8 for vehicle and traffic sign recognition, the system ensures real-time identification of road hazards. Additionally, the incorporation of drowsiness detection and head pose monitoring using Media-pipe Face Mesh addresses the critical issue of driver fatigue and distraction, both of which are significant contributors to road accidents. The system successfully demonstrates high accuracy in detecting lanes, vehicles, and traffic signs, with timely audio alerts that help the driver respond to road conditions and maintain attention. The driver monitoring module effectively reduces the risk of accidents caused by drowsiness or inattentiveness by providing real-time feedback to the driver. This solution, therefore, contributes significantly to improving road safety, especially for drivers at higher risk of fatigue, such as the elderly and long-distance drivers. The integration of dual monitoring—both road and the driver—makes this system a valuable tool in the development of Advanced Driver Assistance Systems (ADAS). Future improvements may focus on enhancing the system's ability to predict driver behavior over time, further optimizing performance in adverse conditions, and integrating the solution with mobile platforms or embedded systems for broader use in commercial and personal vehicles

References

[1] Kim, D., Park, H., Kim, T., Kim, W., & Paik, J. (2023). Real-time driver monitoring system with facial landmark-based eye closure detection and head pose recognition. *Dental Science Reports*, 13, 18264. https://doi.org/10.1038/s41598-023-44955-1.

[2] Elshamy, M. R., Emara, H. M., Shoaib, M. R., & Badawy, A. A. (2024). P-YOLOv8: efficient and accurate real-time detection of distracted driving. https://doi.org/10.48550/arxiv.2410.15602.

[3] Zia, H., Ul Hassan, I., Khurram, M., Harris, N. R., Shah, F., & Imran, N. F. (2025). Advancing road safety: a comprehensive evaluation of object detection models for commercial driver monitoring systems. *Future Transportation*, 5(1), 2. https://doi.org/10.3390/futuretransp5010002.

[4] Gowda, M. S., Talasila, V., Umar, S., & Alva, R. (2024). A multimodal approach to detect driver drowsiness. (pp. 554–559). https://doi.org/10.1109/i4c62240.2024.10748433.

[5] Abuomar, A. M., Ahmed, Y. A., & Salem, M. A.-M. (2023). Safety on wheels: computer vision for driver and passengers monitoring. (pp. 29–34). https://doi.org/10.1109/miucc58832.2023.102783661.

[6] Xuan, L., & Hong, Z. (2017). An improved canny edge detection algorithm. In 2017 8th IEEE International Conference on Software Engineering and Service Science (ICSESS), (pp. 275–278). IEEE.

[7] Khlamov, S., Tabakova, I., & Trunova, T. (2022). Recognition of the astronomical images using the Sobel filter. In 2022 29th International Conference on Systems, Signals and Image Processing (IWSSIP), (pp. 1–4). IEEE.

[8] Du, Y., Liu, X., Yi, Y., & Wei, K. (2023). Optimizing road safety: advancements in lightweight YOLOv8 models and GhostC2f design for real-time distracted driving detection. 23. https://doi.org/10.3390/s23218844

[9] Nusari, A. N. M., Ozbek, I. Y., & Oral, E. A. (2024). Automatic vehicle accident detection and classification from images: a comparison of YOLOv9 and YOLO-NAS algorithms. (pp. 1–4). https://doi.org/10.1109/siu61531.2024.10600761.

[10] Zhao, Y., Mammeri, A., & Boukerche, A. (2019). A novel real-time driver monitoring system based on deep convolutional neural network. In IEEE International Symposium on Robotic and Sensors Environments, (pp. 1–7). https://doi.org/10.1109/ROSE.2019.8790428.

[11] Sarumathi, S., Sabir, M., Sahmoudi, M., Umarulla, M., & Yousuf, M. (2024). Enhancing transportation safety with YOLO-based CNN autonomous vehicles. https://doi.org/10.1109/iceccc61767.2024.10593914.

[12] Wang, X. (2016). Deep learning in object recognition, detection, and segmentation. *Foundations and Trends® in Signal Processing*, 8(4), 217–382.

[13] Huang, X., Bi, N., & Tan, J. (2022). Visual transformer-based models: a survey. In International Conference on Pattern Recognition and Artificial Intelligence (pp. 295–305). Cham: Springer International Publishing.

[14] Surianarayanan, C., Lawrence, J. J., Chelliah, P. R., Prakash, E., & Hewage, C. (2023). A survey on optimization techniques for edge artificial intelligence (AI). *Sensors*, 23(3), 1279. https://doi.org/10.3390/s23031279.

[15] Shi, G., Da-peng, D., Shi, G., Guo, Y. P., & Ning, Z. (2024). Research on methodology of intelligent traffic accident detection based on enhanced YOLOv8 algorithm. https://doi.org/10.1145/3665348.3665406.

[16] Fan, Z., & Liu, S. (2024). An improved YOLOv8n algorithm and its application to autonomous driving target detection. (pp. 1133–1139). https://doi.org/10.1109/icpics62053.2024.10796378.

[17] Matta, V. D., Mudunuri, K. V. R. R., Sai Baba, B. C. S., Kiran, K. B., Veenadhari, C. L., & Prasanthi, B. V. (2023). Single use plastic bottle recognition and classification using Yolo V5 and V8 architectures. In International Conference on Cognitive Computing and Cyber Physical Systems, (pp. 99–106). Cham: Springer Nature Switzerland.

[18] Thapliyal, N., Aeri, M., Namdev, D., Kukreja, V., & Sharma, R. (2024). YOLOv8 enhanced: pioneering accuracy in traffic sign detection and classification. https://doi.org/10.1109/i2ct61223.2024.10543354.

[19] Yang, B., & Liu, C. (2024). Design of intelligent transportation multi-target recognition and tracking algorithm combining YOLOv8. (pp. 992–997). https://doi.org/10.1109/peeec63877.2024.00184.

[20] Rosales, J. A., Samson, J. G., & Maderazo, C. V. (2023). Motor vehicle crash detection using Yolov8 algorithm. (pp. 365–371). https://doi.org/10.1109/comnetsat59769.2023.10420786.

[21] Abhigna, A., Sreeja, C., Mahesh, A., Varma, A. P., & Subhahan, D. A. (2024). Vacant parking slot availability detection using Yolov8 and OpenCV-integration with payment functionalities. In 2024 7th International Conference on Circuit Power and Computing Technologies (ICCPCT), (Vol. 1, pp. 1433–1438). IEEE.

[22] Lashkov, I., Kashevnik, A., Shilov, N., Parfenov, V., & Shabaev, A. (2019). Driver dangerous state detection based on OpenCV & dlib libraries using mobile video processing. In Computational Science and Engineering, (pp. 74–79). https://doi.org/10.1109/CSE/EUC.2019.00024.

[23] Sahithi, A., Teja, B. S., Shastry, C. V., Venugopal, C., & Rajyalakshmi, C. H. (2023). Enhancing object detection and tracking from surveillance video camera using YOLOv8. In 2023 International Conference on Recent Advances in Information Technology for Sustainable Development (ICRAIS), (pp. 228–233). IEEE.

[24] Telaumbanua, A. P. H., Larosa, T. P., Pratama, P. D., Fauza, R. H., & Husein, A. M. (2023). Vehicle detection and identification using computer vision technology with the utilization of the YOLOv8 deep learning method. *Sinkron : Jurnal Dan Penelitian Teknik Informatika*, 7(4), 2150–2157. https://doi.org/10.33395/sinkron.v8i4.12787.

75 Optimized 1-bit full adder using GDI logic with level restoration for low-power applications

Pasupuleti Mahesh[1,a], Badeghar Sumama[2,b], B. Purushotham[2,c], B. Sai Keerthana[2,d] and Ramayanam Naveen[2,e]

[1]Assistant Professor, Department of ECE, Rajeev Gandhi Memorial College of Engineering and Technology, Nandyala, Andhra Pradesh, India

[2]UG students, Department of ECE, Rajeev Gandhi Memorial College of Engineering and Technology, Nandyala, Andhra Pradesh, India

Abstract

Full adders play a crucial role in executing arithmetic operations within signal and image processing systems. This study proposed a novel architecture of a single-bit full adder under a hybrid-logic multiplexing-based realization using gate diffusion input logic. This design's main focus is on attaining a major decrease in the consumption of power and delay. A level restoration carrying logic was proposed in this design, which is helpful in maintaining the output voltage with full swing. This proposed full adder design implemented using the Cadence Virtuoso tool along with the 45 nm complementary metal-oxide-semiconductor (CMOS) process at different values of supply voltage were adopted to check how the circuit performs with different constraints of power supply. Different parameters like speed, power, PDP, and Area efficiency have been well analyzed and compared with the existing conventional full adder architectures. Results obtained through simulation reflect that the developed single-bit full-adder circuit consumes 60% less power and 30% fewer transistors compared with a CMOS full adder. This work also explores how the proposed full adder behaves with and without the implementation of level restoration circuitry. According to the simulation results, the proposed full adder with level restoration provides a highly optimized solution for low-power and high-speed digital circuits, especially for modern VLSI.

Keywords: Average power, complementary metal-oxide-semiconductor, delay, full adder, gate diffusion input logic, level restoration technique

Introduction

The growing demand for portable electronics, like laptops, smartphones and electronic devices, has increased the need for highly energy-efficient digital circuits with high speed and minimal area. More specifically, ultra-low power (ULP) circuits are rising due to the demand for extended battery life in such devices without a loss of functionality. These ULP circuits greatly reduce energy consumption and help prolong battery life since they operate on the near-threshold or sub-threshold region. However, this also presents significant issues in system performance and dependability, and thus it is an important area of research for system components and processor designers. Digital logic systems find the 1-bit full adder very useful since it practically supports all arithmetic and logical functions such as addition, subtraction and multiplication. It plays a crucial role in digital signal processing (DSP) applications like convolution, filtering and correlation. In the VLSI framework, full adder designs need to be optimized to achieve the efficient functioning of such a complex system [1]. To date, many variations of full adder designs have been examined, differing in logic and performance parameters such as area, speed and power dissipation. Some use traditional approaches like static complementary metal-oxide-semiconductor (CMOS), while others use more advanced approaches like transmission gate logic (TGL), pass-transistor logic (PTL) and hybrid designs that combine multiple logic styles for improving performance [2].

Conventional full adders employ CMOS designs with a strong 28-transistor structure to maintain stability in complex systems by employing full swing logic. However, their large PMOS transistors increase input capacitance and occupy more area. Therefore, mirror adders and other efficient designs like TGL,

[a]maheshon1990@gmail.com, [b]sumama709@gmail.com, [c]bpstmec205@gmail.com, [d]Keerthana430@gmail.com, [e]naveenramayanam4@gmail.com,

DOI: 10.1201/9781003685364-75

PTL and hybrid approaches have been explored for the sake of speeding up the design, reducing power consumption, and also minimizing area [3]. Moreover, all these low-voltage full adder designs, such as a TGA and threshold full adder (TFA) [4], aim to reduce power consumption but suffer in driving capability and in performance if used in large complex structures. These limitations are particularly pronounced in energy-constrained applications that require reliable operation at ULV [5]. In addition, although hybrid adders are designed to balance trade-offs between power, area, and speed, they are still challenged by the lack of optimal energy efficiency without sacrificing functionality. Even with these benefits, it still fails to achieve the necessary energy efficiency when scaled down to ultra-low voltage (ULV) levels, especially when cascaded into large circuits. In ULV applications, scaling is a crucial thing for optimizing energy consumption [6]. Scaling the size of transistors allows for reducing current leakage and power consumption, making the circuits more suitable for energy-constrained devices. However, the device dimensions must be carefully optimized, as the MOS transistor size significantly influences the performance of CMOS circuits at lower supply voltages. Research has shown that adjusting parameters like beta ratio can enhance power efficiency. Sub-threshold operation also further reduces power consumption but demands a careful balance between efficiency, speed and performance [7].

This study introduces a new and optimized full-adder structure that employs a combination of logic techniques, one of which is known as GDI [8]. GDI is a digital logic method, which allows the implementation of AND, OR, XOR and MUX with just two transistors. The GDI technique is well-suited for low-power and compact circuit design [9]. Additionally, a level restoration scheme is implemented, resulting in a more stable output and restored signal swing.

This document is organized as follows: The literature survey is reviewed in Section 2, and Section 3 discusses the relevant work. The suggested complete adder design is introduced in part 4, the simulation results are shown in Section 5, and the conclusion is covered in the last part.

Literature survey

In this literature survey, the focus is on different strategies for power-efficient full adder design, and their optimization with GDI and hybrid techniques, with regard to power, delay, performance and transistor count. Alioto and Palumbo [1] have also compared full adder developments in submicron technology and considered various performance factors (delay, power, energy efficiency) [1]. Zimmermann and Fichtner [3] presented a comparative study of low-power logic styles with emphasis on CMOS and pass-transistor logic. Their paper presented benefits of pass-transistor logic with reference to decreased power consumption and delay in arithmetic circuits such as full adders [3]. Hasan et al. [5] proposed a flexible and power-conscious hybrid full adder architecture aimed at improving computational speed with minimal power usage [5]. Shoba and Nakkeran [9], applied GDI techniques to full adder designs, demonstrating their effectiveness in energy-efficient arithmetic operations [9].

Sharma et al. [4] experimented with low-power TG full adder circuits in CMOS technology [4]. Rajaei and Mamaghani [2] proposed a combined magnetic tunnel junction (MTJ) and CMOS-based full adder that incorporated magnetic tunnel junction design within CMOS circuits to enhance reliability and power consumption [2]. Sanapala and Sakthivel [8] introduced the GDI-based full adder with ultra-low-voltage application optimizations regarding the physical layout and energy usage of the computing systems [8]. Vedterbacka [7] presented a CMOS-based full adder with 14-transistors ensuring complete voltage-swing nodes to reduce quantization noise, increase the noise margins and stability of the circuits [7].

In this survey. The 14-transistor CMOS full-adder demands accurate transistor sizing for feedback, making the design process more complex and time consuming. The 16-transistor full adder is not suitable for long cascades without intermediate buffers, limiting its salability in large circuits. Meanwhile, the 10-transistor full adder suffers from threshold voltage loss and limited driving capability, which can impact its efficiency in high-speed applications. Even with ongoing improvements in full adder design to enhance power efficiency, speed and area optimization, issues such as voltage degradation, transistor sizing, and buffering needs continue to be issues for improved performance in contemporary VLSI systems.

Related work

Different kinds of full adders have been proposed based on various logic styles and emerging technological advancements over time. The selection criteria for full adders include transistor count, power usage, delay, and noise performance. In this work, a comparative analysis down on performance parameters of proposed full adder with CMOS full adder, PTL full adder, TGA and hybrid logic adder. Figure 75.1 represents the conventional CMOS full adder using a 28-transistor, a PTL based full adder is implemented in Figure 75.2 [1]. These adders are used to achieve full-swing output voltage levels; therefore, their signal integrity is superior. However, these topologies consume a large silicon area due to their transistor counts of 28 and 24, respectively. It has one major disadvantage; the high capacitance of the input forces larger PMOS transistors, which in turn increases the silicon area and power consumption.

In addition, the complementary pass-transistor (CPL) is another. It draws 32 transistors and it has low power dissipation, Though, it suffers from voltage degradation due to multiple intermediate switching nodes which causes an increase in transistor counts [6]. Researchers have also looked upon TGL, where the best voltage swing and reduced leakage power make it a candidate for modern low-power applications. This also has a large transistor count.

Hybrid full adders have been investigated to minimize the area, power and delay. Figure 75.3 shows the hybrid full adder using 10 transistors [2]. However, studies show that 6T full adder designs either fail to function or give sub-optimal performance as a result of threshold voltage degradation and poor noise margins.

Proposed system

The suggested 1-bit full adder utilizes 2x1 multiplexer (MUX), incorporating GDI technique. The prime objectives of this design are to reduce the transistor count, minimize power consumption, and decrease propagation delay. It is implemented by employing 10 transistors and 2 inverters internally. Inverters are employed to produce complementary inputs (Bbar and Cbar). This design utilizes level restoration techniques to enhance signal integrity and ensure that the output reaches complete voltage levels. It is an efficient, low-power solution suitable for various applications in energy-constrained VLSI systems.

$$\text{Sum} = (\text{A XOR B}) \, \text{Cbar} + (\text{A XNOR B}) \, \text{C} \quad (1)$$

$$\text{Carry} = (\text{A XOR B}) \, \text{C} + (\text{A XNOR B}) \, \text{A} \quad (2)$$

The proposed full adder is designed by employing 2x1 MUX (PM0 & NM0) based on GDI logic. This GDI logic's major role is reducing transistor count which reduces area and power consumption, also restoring the level of output voltage to recover from threshold loss. After the restoration of signal, it is forwarded to MUX2 (PM1 and NM1) and a next

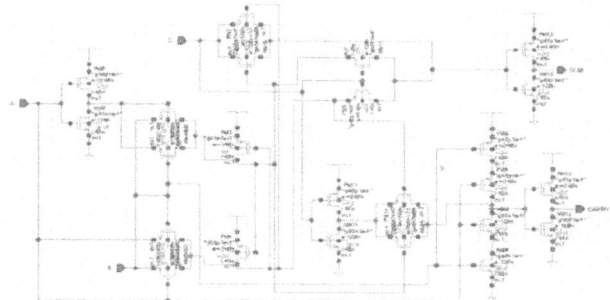

Figure 75.2 24T full adder using PTL
Source: Author

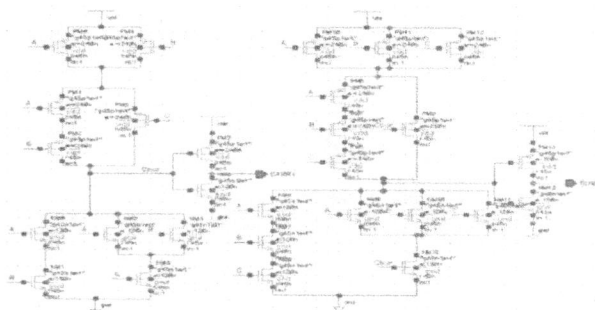

Figure 75.1 28T full adder using CMOS
Source: Author

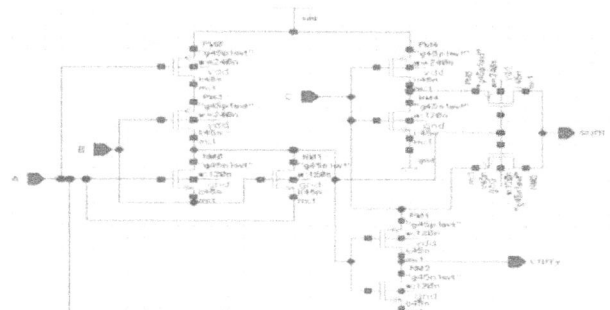

Figure 75.3 hybrid 10T full adder
Source: Author

stage to produce sum and carry outputs. An inverter (PM2 and NM2) is incorporated within the carry network, acting as a buffer to enhance the carry propagation speed. Additionally, this inverter generates complimentary signals required for the next stage. The design uses a pass transistor network chine, preventing unbuffered carry signal propagation and reducing power consumption. However, the sum and carry signal still experience a slight threshold voltage loss, to address this, a level restoration scheme is implemented which results in more stable output and restored signal swing, as shown in Figure 75.5. The transistors pair (NM5, PM5) and (NM6, PM6) are placed across the 2*1 multiplexers to maintain output voltage swing against direct threshold loss. These transistors help preserve the output voltage swing, enduring stable logic levels, which improves reliability and reduces static power loss.

This proposed scheme differs from standard designs in that it does not provide a direct GND or VDD supply rails to generate an accurate output. Rather, it requires PMOS or NMOS transistors to be biased through the input signals. This ensures efficient operation and accurate output. In contrast to traditional circuits, which are affected by threshold voltage reduction, this approach dynamically controls input levels to account for voltage drops. Consequently, the design enhances power efficiency while maintaining reliability, thus suitable for low-power applications. This architecture propels the MOSFETs in the sub-threshold region in an energy efficient way, additionally reducing the supply voltage and consequently power. In low-power designs,

transistors are specifically laid out to operate in the weak or moderate inversion region to provide maximum efficiency. Essentially, the most important parameter of this design is width-to-length (W/L) ratio. The EKV model is used to determine the size of transistors.

The equations applied in the EKV model for MOS devices calculations are:

$$IC = I_D/I_S \qquad (3)$$

Where,
IC – Inversion coefficient
Is – Normalization current
I_D – Drain current.

Simulation result

A simulation was performed with the cadence virtuoso tool under the CMOS 45 nm technology to test a 1-bit full adder's performance. The primary objectives were to minimize power usage and delay. For the purpose of comparison, all the discussed circuits were tested under identical conditions, operating with the frequency of 20kHz and at a temperature

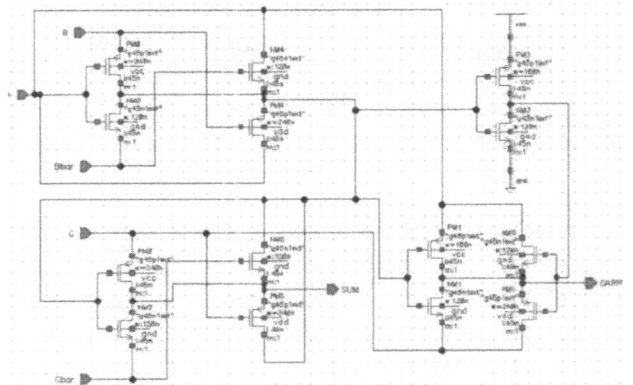

Figure 75.5 Optimized architecture of full adder with level restoration
Source: Author

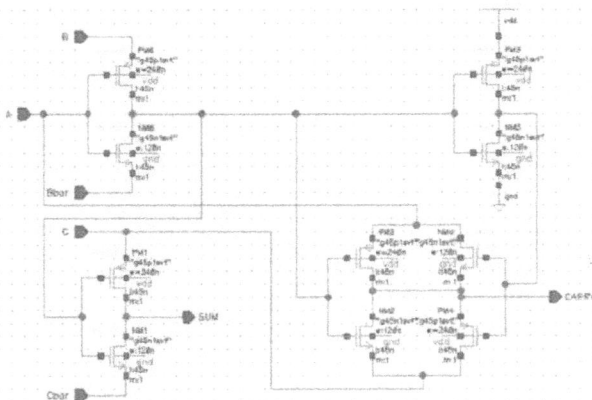

Figure 75.4 Optimized architecture of full adder
Source: Author

Table 75.1: Size of PMOS and NMOS transistors.

Name	W/L ratio
PM1, PM2, PM5, PM6	5.3/1
PM2, PM3, PM4	4/1
NM1, NM2, NM3, NM4, NM5, NM6, NM7	2.6/1

Source: Author

of 27°C. The optimized full adder, which shows minimal voltage degradation at the sum and carry outputs, as depicted in Figure 75.7. In order to overcome this problem, we applied a level restoration technique to enhance the output voltage swing, the result is presented in Figure 75.8. The outputs corresponding to Figures 75.4 and 75.5 are presented in Figures 75.7 and 75.8 respectively. In this study, we

Table 75.2: Comparison between proposed adder and existing adder.

Full Adders	Average Power (nW)			Delay (uS)		
	0.8V	1.0V	1.2V	0.8V	1.0V	1.2V
C-CMOS [1]	8.2	10.6	18.5	140.2	108.5	75.2
Hybrid 10T [2]	5.8	8.2	15.3	145.2	115.2	85.3
CPL [6]	6.9	9.0	17.1	126.8	122.7	93.6
TGA [4]	3.5	4.8	13.2	135.7	125.3	75.2
24T-Adder [3]	4.1	5.5	12.3	155.1	130.8	98.4

Source: Author

Table 75.3: Comparison of proposed adder with and without voltage degradation.

Full adders	average power (W)	delay (S)
C-CMOS	10.6E-9	140.2μ
Adder with voltage degradation	2.16E-9	95.8μ
Adder without voltage degradation	147.1E-12	68.3μ

Source: Author

measured average power and propagation delay at operating voltages of 0.8V, 1.0V and 1.2V. The transistor sizes acquired using the EKV model, which guarantees precise MOSFET characterization, are shown in Table 75.1. When designing the suggested architecture, scaling is essential for maximizing its performance metric including speed, power usage and space efficiency.

A comparison of mean power and delay between the proposed and existing adders is illustrated in Table 75.2, including conventional CMOS (C-CMOS), PTL, TGA, CPL and other hybrid 10T adders. The suggested adder exhibits better performance compared to existing adder circuits. Table 75.3 illustrates how the proposed adder performs in the presence and absence of voltage degradation. Based on these results, we conclude the proposed full adder without voltage degradation performs better than the one with voltage degradation. The performance variation of proposed system with voltage degradation (PSW) and proposed system without voltage degradation (PSWO) is shown in Figure 75.6.

According to the results, optimized architecture achieves a 60% lower power usage and 30% reduction in transistor count over the traditional CMOS adder using GDI logic and level restoration technique. The design exhibits a 13% reduction in power usage and a delay reduction of at least 30% while operating without voltage degradation.

Conclusions

The proposed adder was designed using "GDI logic" to minimize the chip area by reducing transistor count, due to this power consumption also reduced. Achieving full voltage swing with fewer transistors is

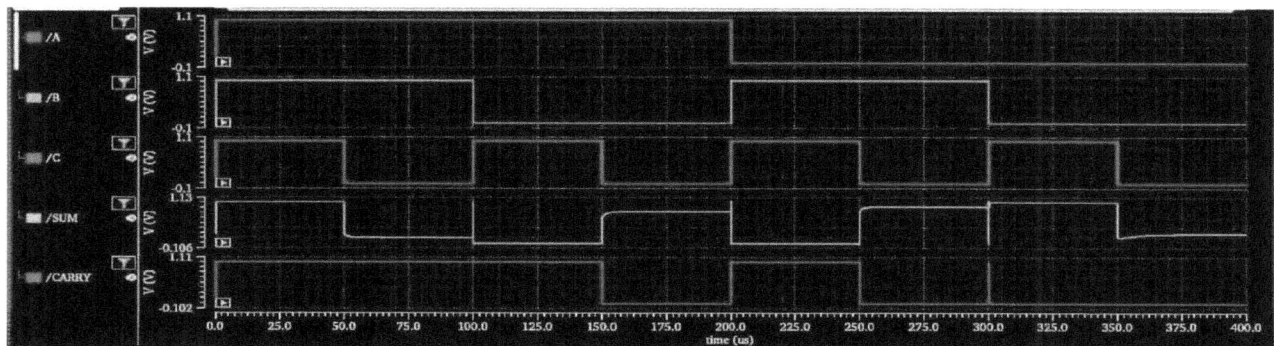

Figure 75.6 Performance variation of proposed architecture
Source: Author

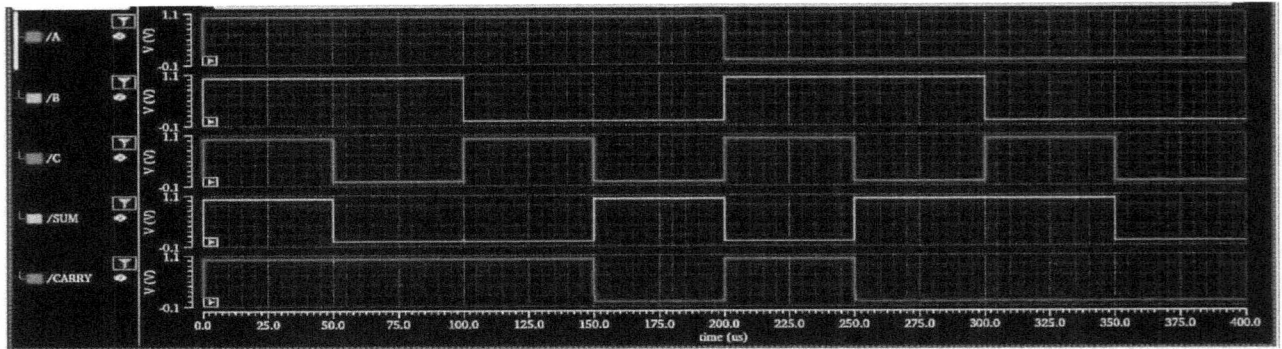

Figure 75.7 Output of optimized full adder with voltage degradation
Source: Author

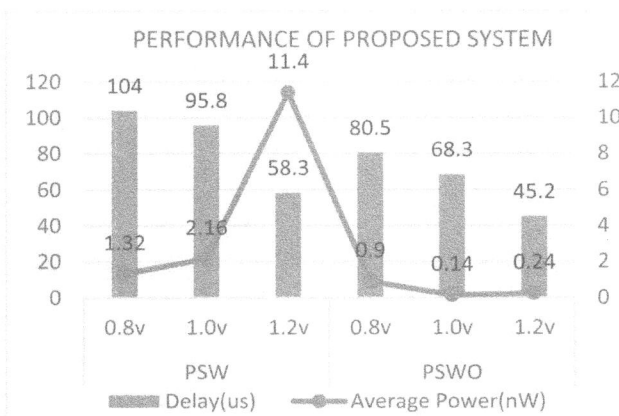

Figure 75.8 Output of optimized full adder without voltage degradation
Source: Author

a defining feature of this adder. The "level restoration technique" has also been employed in this chip so that output signals are clear and strong, thus the static power loss is minimized. This architecture functions in the sub-threshold region, where transistor scaling has a large effect on power consumption. This optimization is done by utilizing the EKV model for low power consumption. The new 1-bit adder brings a major improvement over existing designs, making electronic devices faster, smaller, and more energy efficient. When compared to other designs, including older ones like TGA, CPL, other hybrid and GDI adders, this proposed adder stands out for its superior performance. It uses 30% fewer transistors and consumes 60% less power than the traditional CMOS adder. Moreover, it also uses 37% fewer transistors

compared to the TG adder. The impact of the "voltage degradation" fix with the level restoration technique is significant. Compare adder performance with and without the "level restoration" technique, and one will readily see that indeed this technique is making a difference, especially about reducing power consumption. This new adder represents great advancement from the designs presently existing. It is smaller, faster, and more efficient, making it highly beneficial for next-generation VLSI designs.

References

[1] Alioto, M., & Palumbo, G. (2002). Analysis and comparison on full adder block in submicron technology. *IEEE Transactions on Very Large Scale Integration (VLSI) Systems*, 10(6), 806–823.

[2] Rajaei, R., & Mamaghani, S. B. (2017). Ultra-low power, highly reliable, and nonvolatile hybrid MTJ/CMOS based full-adder for future VLSI design. *IEEE Transactions on Device and Materials Reliability*, 17(1), 213–220.

[3] Zimmermann, R., & Fichtner, W. (1997). Low-power logic styles: CMOS versus pass-transistor logic. *IEEE Journal of Solid-State Circuits*, 32(7), 1079–1090.

[4] A. Sharma, R. Singh, and R. Mehra, Low Power TG Full Adder Design Using CMOS Nano Technology, in Proc. 2nd IEEE Int. Conf. Parallel, Distributed and Grid Computing (PDGC), pp. 210–213, Apr. 2012. doi:10.1109/PDGC.2012.6449819

[5] Hasan, M., Hossein, M. J., Hossain, M., Zaman, H. U., & Islam, S. (2019). Design of a scalable low-power 1-bit hybrid full adder for fast computation. *IEEE Transactions on Circuits and Systems II: Express Briefs*, 67(8), 1464–1468.

[6] Boppana, N. V. V. K., Ren, S., & Chen, H. (2014). Low-power and high-speed CPL-CSA adder. In NAECON 2014-IEEE National Aerospace and Electronics Conference, (pp. 346–350). IEEE.

[7] Vesterbacka, M. (1999). A 14-transistor CMOS full adder with full voltage-swing nodes. In Proceedings of IEEE Workshop Signal Processing Systems, Taipei, Taiwan, (pp. 713–722).

[8] Sanapala, K., & Sakthivel, R. (2019). Ultra-low-voltage GDI-based hybrid full adder design for area and energy-efficient computing systems. *IET Circuits, Devices and Systems*, 13(4), 465–470.

[9] Shoba, M., & Nakkeeran, R. (2016). GDI based full adders for energy efficient arithmetic applications. *Engineering Science and Technology, an International Journal*, 19(1), 485–496.

76 Performance analysis of low power, high speed full adder by using modified XOR and XNOR gates

M. Hari Krishna[1,a] and Kolli Deepika[2,b]

[1]Professor, Department of ECE, in Sree Vahini Institute of Science and Technology, Tiruvuru, Andhra Pradesh, India

[2]PG Scholar, Department of ECE, in Sree Vahini Institute Of Science And Technology, Tiruvuru, Andhra Pradesh, India

Abstract

A novel family of circuits capable of synchronous XOR/XNOR operations is presented in this paper. In terms of power consumption and deferral, the proposed circuits are light years ahead of the competition because to their small yield capacitance and almost non-existent short out power distribution. We also provide six novel half-and-half 1-bit full-snake (FA) circuitry that make use of the novel full-swing XOR-XNOR or XOR/XNOR entries. All of the proposed circuits are completely independent from one another in respect to driving capacity, control defer item (PDP), speed, control use, etc. The proposed designs outperform competing FA architectures in terms of speed and power, according to the simulation results derived from the CMOS process development model. We provide an alternative method of measuring transistors in order to enhance the circuits' PDP. Using the numerical computation of molecular swarm improvement, the proposed technique achieves the optimum inducement for optimal PDP with little effort.

Keywords: Implementation, simulation, synthesize

Introduction

Every method of arithmetic relies on full adders. Any kind of calculation involving expansion, subtraction, addition, or decrement may be carried out using this combinational logic unit. Little sophisticated IC chip advancements may be slowed down by certain things. Cost of configuration, profitability of plans, and innovation in IC fabrication are these factors. There is a growing need for configuration layers such as engineering, circuit, and format for rapid large-scale coordination. At this stage, the circuit design should have settled on a valid reasoning configuration style for fast combinational rationale circuits.

Reason being, the chosen reasoning style actually affects all the major characteristics affecting velocity, including exchange capacitance, change action, and short out flows. Established researchers associated with VLSI architectures now also place a premium on speed considerations, in addition to older, more weighted criteria like power dispersal, small area, and cost factor.

The practicality of circuit is constrained by the rising size of reconciling. measurements of strength and region utilisation. Therefore, as battery-operated portable devices like phones, tablets, and computers continue to gain popularity, designers are looking for ways to reduce the power consumption and space requirements of these systems without sacrificing speed. Improving the W/L ratio of transistor is one approach of reducing the circuit's PDP and avoiding problems caused by a lower supply voltage. The demonstration of number juggling circuits, such as adders, multipliers, and divisions, is indicative of the efficacy of many sophisticated applications. A lot of work has gone into researching effective snake structures like convey choose, convey skip, dependent total, and convey look-ahead adders as expansion is a crucial part of many maths activity. Since it is the fundamental square of these designs, the full snake (FA) naturally takes centre stage. You may classify FA circuits as either full-swing or non-full-swing depending on the magnitude of the yield voltage. Whether we design a fast full viper or not, control distribution is also handled in CMOS technology. A fundamental component that may be categorised into two types, dynamic power and static power, is power scattering. Whereas static power dispersion occurs when the circuit is not in use, dynamic power dissemination occurs when the circuit is in operation.

[a]yemchechkay@gmail.com, [b]deepika2712000@gmail.com

DOI: 10.1201/9781003685364-76

Literature Review

We provide two low-control, fast full-viper cells that feature a reduced power-defer product (PDP) thanks to their pass-transistor rationale styles and option inward reasoning structures. We compared ourselves to comparable full-adders with a low PDP in terms of speed, control use, and territory. All of the full-adders were built using a 0.18-m CMOS technology and tested using a comprehensive test bench that could measure the output of the full-snake inputs in addition to the power supply's motion. The suggested full-adders outperform their companions in post-format recreations, revealing a typical PDP advantage with relative zone.

Due to the tremendous demand for small consumer hardware goods, low-control architecture of VLSI circuits has recently been acknowledged as a fundamental mechanical need. This is how a plethora of novel designs for core reasoning capabilities based on pass transistors and gearbox doors have recently surfaced in the literature. Without using formal structural approaches, these plans relied on the planners' instincts and wits. To compensate for the edge voltage loss in MOS transistors, a formal plan approach is presented for comprehending a small transistor CMOS pass organise XOR-XNOR cell. If the MOS transistor estimates are carefully considered in the initial design phase, this novel cell can reliably operate within specified limits even when the power supply voltage is reduced. The novel XOR-XNOR cell is also used to verify entire viper cells with low transistor counts.

Our next proposal is a 1-b complete viper with a half-breed CMOS architecture. The goal of the crossover CMOS configuration style is to build new complete adders with the desired execution by using various CMOS logic style circuits. This allows the designer more structural leeway to concentrate on a variety of applications, which in turn reduces the breadth of their planning efforts. All the while, the new complete viper generates XOR and XNOR full-swing yields thanks to a unique XOR-XNOR circuit. The fact that this circuit outperforms its counterparts suggests that the PDP is becoming better. Also suggested is a half-CMOS yield arrangement that tries the concurrent XOR-XNOR sign. This yield stage enables adders to fall without cradle addition between fell stages by providing significant driving capacity.

Here we see a low-power, low-complexity complete viper setup that uses the savage pass transistor logic (PTL). The building block is an XOR-XNOR module with five continuously degenerate transistors provide evidence for integral yields. This module functions properly in relation to complete viper applications, despite the absence of reasoning. The availability of correlative control signals may alleviate the limit hardship problem that is fundamental to the majority of PTL architectures. A new complete viper configuration with as low as 10 transistors may be deduced by connecting this module with multiplexing modules. When compared to alternative 10-T partner layouts, the suggested whole snake configuration has the fewest VDD duties and the fewest yield signal debasement. Post format recreations also show the exhibition edges in power, speed, and power-postpone item.

A one-bit full-viper cell is shown in an investigation. Smaller modules are removed from the viper cell. There is extensive evaluation and consideration of the modules. Some of their constructions are built, prototyped, imitated, and dismantled. By combining different layouts of these modules, twenty unique 1-bit full-snake cells are constructed, most of which are new circuits. Different cells show different numbers for power consumption, speed, area covered, and driving capabilities. There are two functional circuit architectures used for reproduction that make use of viper cells. Circuit designers have access to a library of full-snake cells from which they may choose the full-viper cell that best suits their needs.

Methodology existing system

This research reviews previous work on XOR/XNOR and XOR-XNOR gates and offers new circuits for both kinds of admissions. The 2-to-1 multiplexer (2-1-MUX) and the 2-opinions XOR/XNOR gate are the two primary parts of a hybrid FA. In an FA cell, the XOR/XNOR door is the one who really purchases electricity. Our objective is to eradicate the issues present in the circuits that have been examined. This might be one approach to reducing FA cell power use by maximising the layout of the XOR/XNOR gate. There is heavy reliance on the XOR/XNOR door in computer circuit design as well. There have been several suggestions for XOR/XNOR entry circuits.

Proposed system

In this case, the power and delay are improved, and the data capacitances are about equal. This structure has a low yield capacitance and no NOT entries on the basic path. It is fast and uses very little power because of this. This circuit's XOR and XNOR postponement yields are almost identical, reducing the issue in the subsequent step. In addition to its power against transistor estimation and supply voltage scaling, this circuit's excellent driving capacity and full-swing yield are other notable features. Subsequently, we suggest six novel FA structures for various uses, using these updated XOR/XNOR and XOR-XNOR circuits. Furthermore, after duplicating it in several settings, the results demonstrate that it consistently exhibits an outstanding presentation in each of the reproduced scenarios.

Module explanation
XOR–XNOR circuits

Modern half-FA architectures often use the concurrent XOR-XNOR circuit. Commonly, 2-1-MUX contributions are linked to the circuits sign in half FAs and half FAs as select lines. To avoid FA yield hub issues, it is crucial to have two indicators that are identical in postponement and run at the same time.

A ten-transistor structure based on the CPL logic style forms the basis of this circuit. In this setup, a single nMOS transistor is responsible for driving the yields; to restore the yield levels, two pMOS transistors were cross-coupled with the yields (XOR and XNOR). The critique (cross-coupled structure) on yields increases the deferral and decreases the out intensity, which is one problem with this XOR-XNOR circuit. So, increasing the size the transistors

may help with the imposed delay. Because there are two NOT hallways in the fundamental approach, this structure also has another vulnerability. The twelve-transistor synchronous XOR-XNOR entrance architecture that has been suggested. There is no correlation between the data sources and the capacitances An and B. An and B are linked to a comparable transistor inspection). Capacitors are therefore connected to the circuit in order to generate their contribution, as shown in Figure.

In this case, the power so deferral are improved, and the data capacitances are about equal. Due to its low yield capacitance and lack of NOT doors on the fundamental route, this construction is not recommended. As a result, it is quick and uses rather little power. This circuit's XOR and XNOR postponement yields are almost identical, which reduces the problem in the next step. Not only does this circuit have excellent driving ability and full-swing yield, but it also has power against transistor estimation and scaling of the supply voltage.

Proposed full adders

Innovative FA circuits were proposed for a range of applications. All of these new FAs have used the proposed XOR/XNOR circuits in their planning, and they have all done so using a mixed reasoning approach. The well-known 4-transistor 2-1-MUX architecture is used to execute the half-and-half FA cells. A 2-1-MUX with no static and built according to the TG logic style, it regulates the propagation of short circuits.

The HFA-20T circuit consists of twenty transistors and does not include any power-hungry entryways. Benefits of this layout include speed, low power

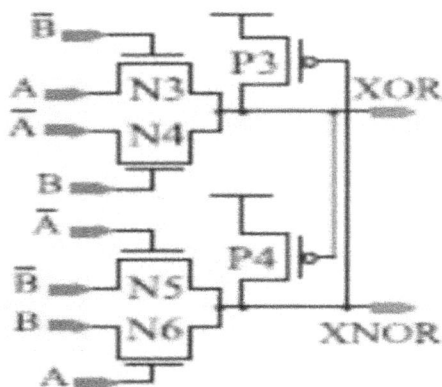

Figure 76.1 schematic of xor and xnor
Source: Author

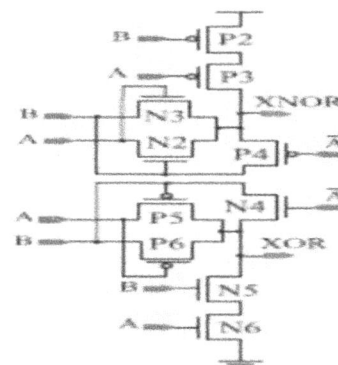

Figure 76.2 schematic of xor and xnor based full adder
Source: Author

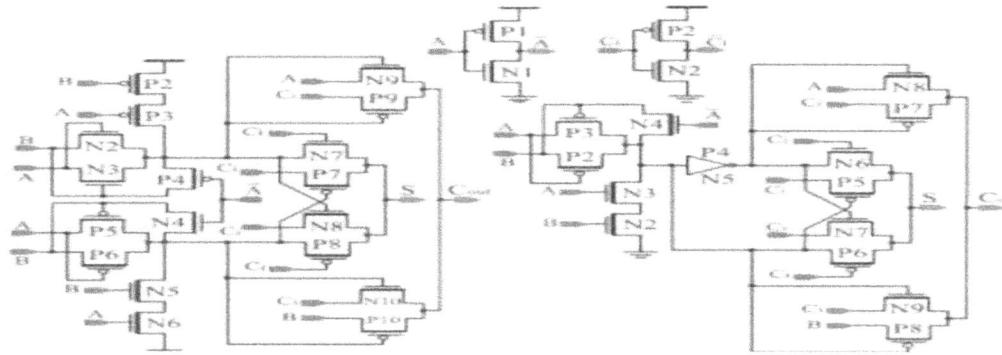

Figure 76.3 The proposed low-power 1-bit full adder
Source: Author

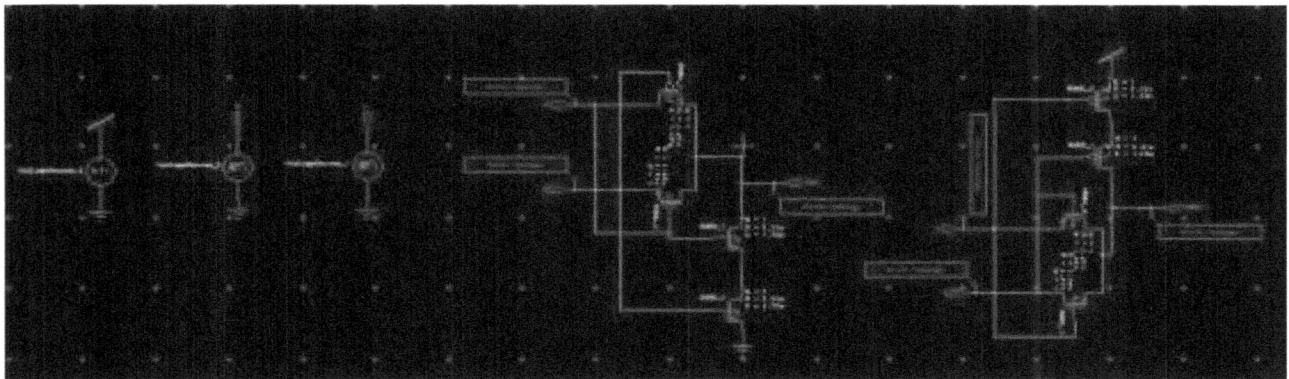

Figure 76.4 schematic of 1-bit proposed full adder
Source: Author

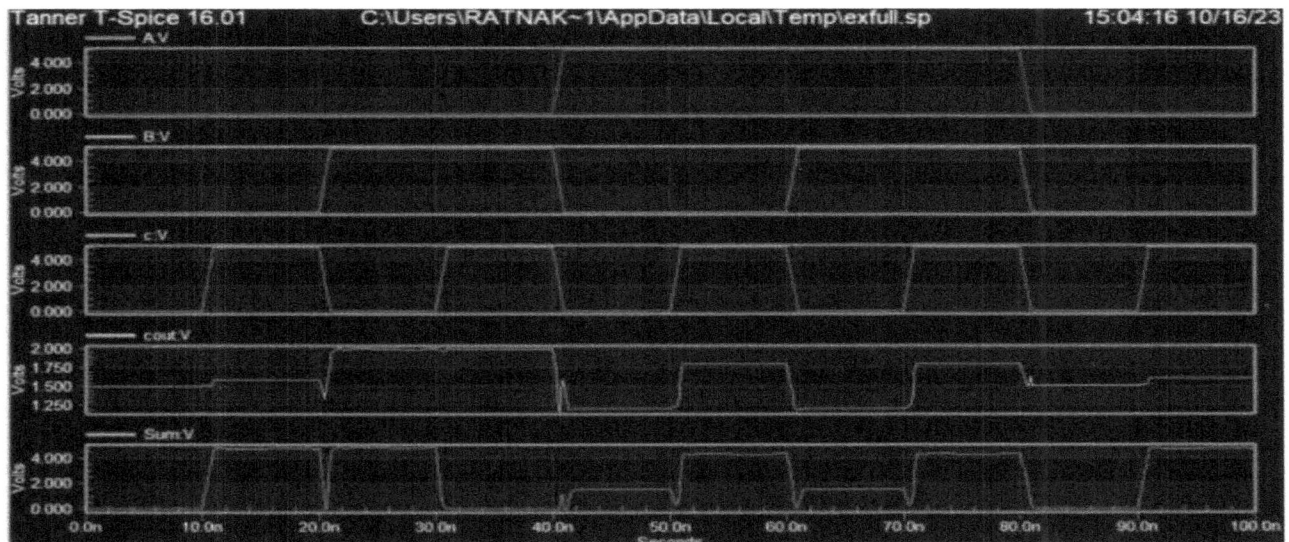

Figure 76.5 Simulation results of Proposed full-adder
Source: Author

dispersion, full-swing yield, resistance to supply voltage scaling, and transistor estimation. If A _ B = 1, then the surrender Cout signal is identical to the data signal An|B. But information flags An or B, which are connected to transistors N9 and P10, respectively, help to equalise the electrical capacitance of the data sources. Reducing yield pushing capability in chain architecture applications, such swell, is a big negative of HFA-20T serpent, move. This issue becomes apparent in networks that use the principle of transmission work without buffering yield. One approach is to use NOT gates and XOR/XNOR gates together to create a single XOR/XNOR signal decrease the power consumption of FA structures. By using a NOT entrance to generate an additional XOR or XNOR output and an XOR/XNOR door to generate the first signal, FA architectures may reduce their power consumption.

Results and discussion

Conclusion

This study began by analysing the XOR/XNOR and XOR-XNOR circuits. While doing the test, it was observed that there is a downside to employing the NOT gates in the basic circuit way. Another disadvantage of a circuit is having a positive review of the XOR-XNOR door's efficiency in resolving the dividend voltage level. The circuit's energy consumption, yield capacitance, and latency are all increased by this input. We subsequently proposed XOR-XNOR and XOR/XNOR gates that do not have such limitations. Lastly, six new FA cells were provided for different uses by using the XOR and XOR-XNOR entrances that were recommended. After running simulations of the FA cells in various conditions, the findings shown that the proposed circuits typically exhibit remarkable performance in all reproduced scenarios.

References

[1] Kim, N. S., Austin, T., Baauw, D., Mudge, T., Flautner, K., Hu, J. S., et al. (2003). Leakage current:

moore's law meets static power. *Computer*, 36(12), 68–75.

[2] Weste, N. H. E., & Harris, D. M. (2010). CMOS VLSI Design: A Circuits and Systems Perspective. (4th edn.). Boston, MA, USA: Addison-Wesley.

[3] Goel, S., Kumar, A., & Bayoumi, M. (2006). Design of robust, energy-efficient full adders for deep- submicrometer design using hybrid-CMOS logic style. *IEEE Transactions on Very Large Scale Integration (VLSI) Systems*, 14(12), 1309–1321.

[4] Bui, H. T., Wang, Y., & Jiang, Y. (2002). Design and analysis of low-power 10-transistor full adders using novel XOR-XNOR gates. *IEEE Transactions on Circuits and Systems—II: Analog and Digital Signal Processing*, 49(1), 25– 30.

[5] Timarchi, S., & Navi, K. (2009). Arithmetic circuits of redundant SUT-RNS. *IEEE Transactions on Instrumentation and Measurement*, 58(9), 2959–2968.

[6] Rabaey, J. M., Chandrakasan, A. P., & Nikolic, B. (2002). Digital Integrated Circuits, (Vol. 2). Englewood Cliffs, NJ, USA: Prentice-Hall.

[7] Radhakrishnan, D. (2001). Low-voltage low-power CMOS full adder. *IEE Proceedings - Circuits, Devices and Systems*, 148(1), 19–24.

[8] Yano, K., Shimizu, A., Nishida, T., Saito, M., & Shimohigashi, K. (1990). A 3.8-ns CMOS 16×16-b multiplier using complementary pass-transistor logic. *IEEE Journal of Solid-State Circuits*, 25(2), 388–395.

[9] Shams, A. M., Darwish, T. K., & Bayoumi, M. A. (2002). Performance analysis of low-power 1-bit CMOS full adder cells. *IEEE Transactions on Very Large Scale Integration (VLSI) Systems*, 10(1), 20–29.

[10] Zhuang, N., & Wu, H. (1992). A new design of the CMOS full adder. *IEEE Journal of Solid-State Circuits*, 27(5), 840–844.

[11] Weste, N., & Eshraghian, K. (1985). Principles of CMOS VLSI Design. New York, NY, USA: Addison-Wesley.

[12] Bhattacharyya, P., Kundu, B., Ghosh, S., Kumar, V., & Dandapat, A. (2015). Performance analysis of a low- power high-speed hybrid 1-bit full adder circuit. *IEEE Transactions on Very Large Scale Integration (VLSI) Systems*, 23(10), 2001–2008.

77 Design of low power-based level shifter by using gate transmission technique

R. Sri Devi[1,a] and Done Monica[2,b]

[1]Professor and HOD, Department of ECE, in Sree Vahini Institute of Science and Technology, Tiruvuru, Andhra Pradesh, India

[2]PG Scholar, Department of ECE, in Sree Vahini Institute of Science and Technology, Tiruvuru, Andhra Pradesh, India

Abstract

Logic circuits, signal isolation devices, or analogue switches in microelectronic devices employ transmission gates (TGs). The impressive output swing, sensitivity to input noise, and low power consumption make them ideal for low-power applications such buffers in very large scale integration (VLSI) interconnects. According to recent research, new low-power buffer designs incorporate four-TG inverter circuits. A CMOS inverter repeater is used to estimate connection size to reduce power consumption. TG-based smart controllers are recommended for repeater installation. Three types of interconnects and repeaters' impacts on contact resistance were examined. A TG-based buffering circuit for CNT interconnects is designed and examined for power-delay product (PDP), delay, and power dissipation. Next, we examine CMOS inverters and gearboxes gate buffer efficiency in mix CNT/SWCNT packaged situations. We introduce TG buffer circuits and demonstrate their use as buffers in carbon nanotube (CNT) VLSI interconnects. This brief suggests a power-saving LEDC with a variety level shifter (LS).

Keywords: Current mirror, level shifter, low power, near-threshold operation, wide range

Introduction

To lessen the power consumption of digital circuit constants and short-circuits, lowering the power source voltage is a good strategy. Still, delays in circuit propagation are increased when the source voltage is reduced. Due to the reduced amount of headroom in analogue circuits, signal swings were smaller, and distortion is more probable. In digital circuits with varying operating speeds, analogies or high-speed digital blocks use dual-supply topologies, while moderate-speed mixed-signal circuits use low voltage on noncritical links. The following digital blocks in a system that have two supply voltages are guaranteed to get the correct voltage by use of level-shifting circuits. The use of these circuits increases the complexity of basic logic levels. Building level shifters efficiently requires minimizing propagation delay, energy consumption, and semiconductor area. Further power savings may be achieved by lowering the low VDDL levels under the input transistors threshold voltage using the level shifter in the small-supply block. The results of this test demonstrate the rapid and effective conversion of low input voltages by our voltage level shifter.

Literature review

A fundamental current mirror (CM) is one of two common kinds of pull-up networks used by type I traditional level shifters. It is unusual for the architecture, the right and left circuit branches, and the pulling upward and downward networks to undergo regenerative interactions. Because of this, the operational speed is quite poor. A static current flows across each branch of the circuit depending on the input state, which is another cause of the high standby power. The DCVS design takes use of a cross-coupled pull-up system, as seen in Figure 77.1(b), to accelerate switching by increasing the differential among the first and second phases during regeneration. Upon the occurrence of the rising edge in the input, Mn1 is turned on and Mn2 is turned off. Mn1 is now trying to reduce Q1's voltage. A quicker discharge of Q1 occurs as a consequence of Mp4's gradual on, which in turn helps turn off Mp3 by

---

[a]ecehod@sreevahini.edu.in, [b]mounika.done@gmail.com

DOI: 10.1201/9781003685364-77

bringing V2 closer to VDDH, the excess supply voltage. This design uses almost no energy while in idle state since none of the circuit branches are subject to static electricity. On the other hand, pull-up transistors (Mp3, Mp4) are more effective than pull-down transistor (Mn1, Mn2) when the insufficient source voltage (VDDL) is lower than the procedure's theoretical threshold voltage. In most cases, the increased power and delay caused by increasing the pull-down network's height leads to a decrease in efficiency.

Existing level shifter

Node Q2 starts out with a low voltage, whereas node Q1 gets an elevated voltage at VH (sometimes lower than VDDH). Because of this, if you listen to Mp4 and Mp5, you will also hear Mp3 and Mp6. As a first step, switching from high to low input activates Mn1 while deactivating Mn2. Node Q1 is rapidly drained by Mp5 owing to the weak pull-up current via Mp3, and the parasite capacitors at that node begin to discharge as a result.

Proposed transmission gate level shifter

Switching up the voltage on the pass transistors at very low voltage LSs, the pass transistors LS has diminished or reversed the typical temperature dependences. By subjecting two LS systems to a broad range of voltage variations, one may evaluate their power utilization, latency, speed, or variability Figure 77.1 Schematic of the proposed LS with the number of fins for each transitory [6-7]. Process

voltage and driving circuitry strength the techniques used in temperature corner analysis are based on software and hardware that is commercially accessible. Because of their wider operating range, level changers using DCVSL input are better than pass-gate LSs. Recent pass gates have exceeded DCVSL. You won't find a better report word level changer in terms of latency and strength [9]. Level shifter for multiple Vths the designed circuit incorporates a

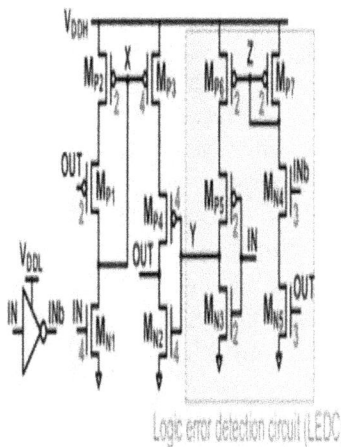

Figure 77.2 Operation details of the proposed (a) during the rising transition of IN and OUT and (b) falling transition of IN and OUT
Source: Author

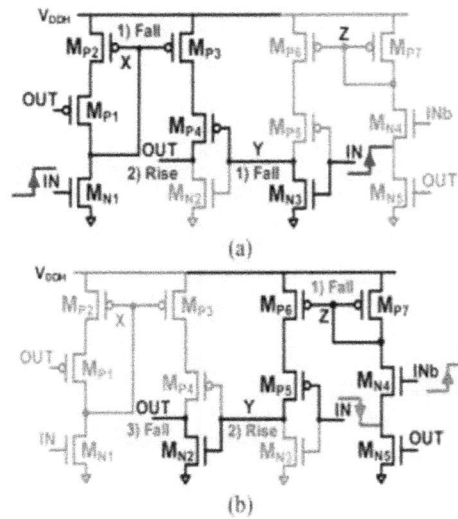

Figure 77.1 Schematic of the proposed LS with the number of fins for each transitor
Source: Author

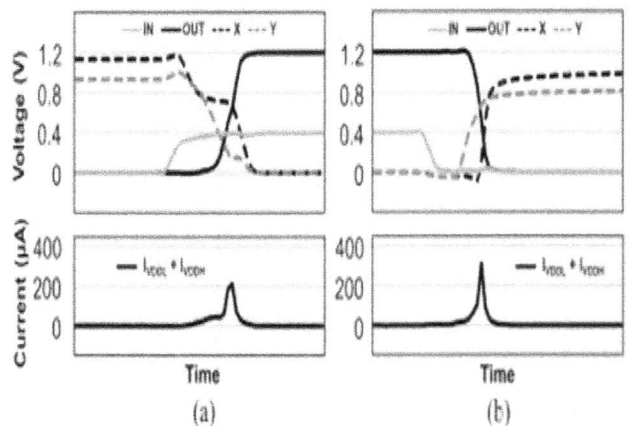

Figure 77.3 Voltage and current waveforms of (a) rising transition in the proposed LS when VDDL = 0.4 V and VDDH = 1.2 V
Source: Author

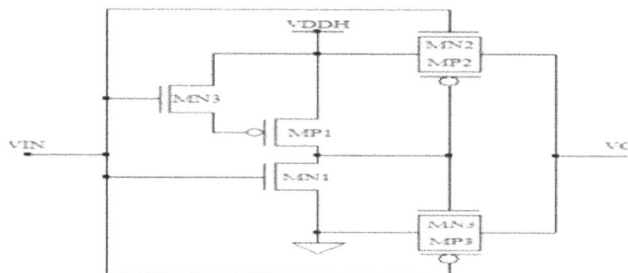

Figure 77.4 Transmission gate level shifting buffer
Source: Author

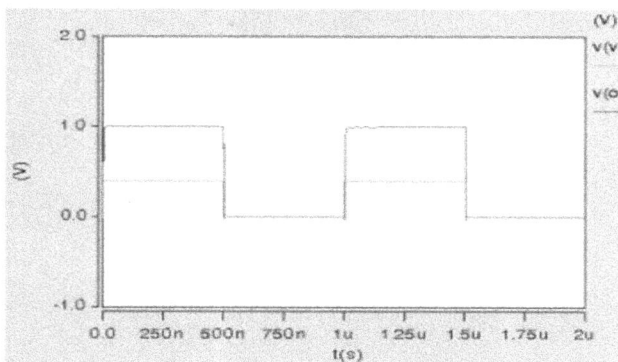

Figure 77.5 Level of shift of transmission gate level shifters
Source: Author

power-efficient thin-film transistor phase shifter that operates on a single supply [16]. Using n-channel oxide thin film transistors, the proposed LS circuit may be able to switch between 10V and 20V without using a logic circuit then the resultant compression as been shown in Table 77.1.

Results

Below schematic will show the proposed design used 18nm cmos technology in tanner eda software.

Schematic

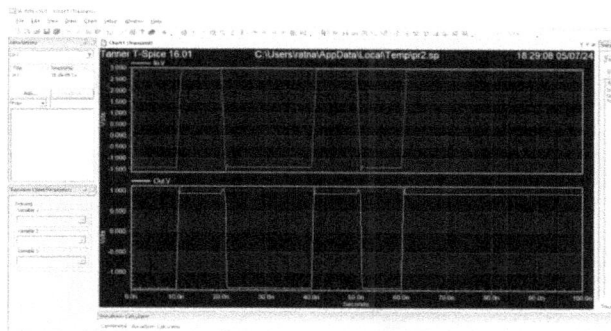

Simulation Results

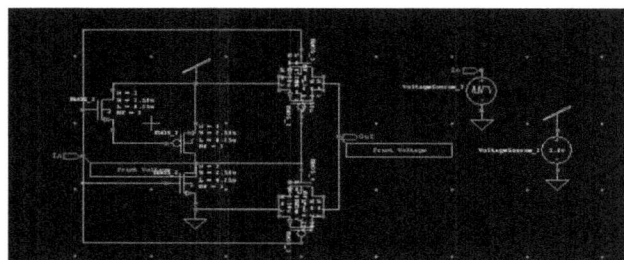

Table 77.1 Compression Table.

Existing method	Area 12 MOSFETS	Power 4.834 Watts	Delay 0.48SEC
Proposed method	7 MOSFETS	2.496 Watts	0.50 Sec

Source: Author

Conclusion

Among the many designs, the best values are delay and PDP. In an ideal world, devices would have a lower threshold and be able to employ numerous thresholds to further decrease VDDL delays. This little showed how a voltage-level-shifting architecture can effectively and quickly transform input voltages as low as 0.1 volts. The suggested circuit works as intended because the resultant node is dragged down by the stronger pull-down device and the current through the weaker pull-up device is reduced. Verified the power consumption simulation's findings. Using the Tanner simulator, we designed and implemented an 18 nm CMOS low-power level shifter. The outcomes were then compared to the pre-existing systems.

References

[1] Wang, A., & Chandrakasan, A. P. (2005). A 180 mV subthreshold processor using a minimum energy circuit design methodology. *IEEE Journal of Sol-id-State Circuits*, 40(1), 310–319.

[2] Lanuzza, M., & Perrili, S. (2012). Low power LS for multi-supply voltage designs. *IEEE Transactions on VLSI Systems*, 59(12), 922–926.

[3] Gundala, S., Ramanaiah, V. K., & Padmapriya, K. (2014). Nanosecond delay level shifter with logic level correction. In Proceedings in International

Conference on Advances in Electronics Computers and Communications (ICAECC), (pp. 26–30).

[4] Basha M.M., Fairooz T., Hundewale N., Reddy K.V., Pradeep B. (2012),"Implementation of LFSR Counter Using CMOS VLSI Technology", In: Das V.V., Ariwa E., Rahayu S.B. (eds) Signal Processing and Information Technology. SPIT 2011, Lecture Notes of the Institute for Computer Sciences, Social Informatics and Telecommunications Engineering, vol 62. Springer, Berlin, Heidelberg.

[5] Mahaboobbasha, M., Venkataramanaiah, K., & Reddy, P. R. (2015). Low area- high speed - energy efficient one bit full subtractor with MTCMOS. *International Journal of Applied Engineering Research*, 10(11), 27593–27604.

[6] Gundala, S., Ramanaiah, V. K., & Padmapriya, K. (n.d.). A novel high performance dynamic voltage Leakage Power Aware Transmission Gate Level Shifter:, International Journal of Engineering and Advanced Technology, vol. 8, no. 4, pp.1527–1530, 2019.

[7] Basha, M. M., Ramanaiah, K. V., & Reddy, P. R. (2018). Design of CMOS full subtractor using 10T for object detection application. *International Journal of Reasoning-based Intelligent Systems (IJRIS)*, 10(3/4), 286–295.

[8] Gundala, S., Ramanaiah, V. K., & Padmapriya, K. (2015). A novel energy efficient active voltage level shifter. *European Journal of Scientific Research*, 128(4), 308–314.

[9] Mahaboobbasha, M., Ramanaiah, K. V., & Reddy, P. R. (2013). An efficient model for design of 64- bit high speed parallel prefix VLSI adder. *International Journal of Modern Engineering Research*, 3(5), 2626–2630. ISSN: 2249-6645.

78 Deep learning-based detection of nutrient deficiency in coffee plants using leaf analysis and organic remedy recommendations

B. Ch. S. N. L. S. Sai Baba[1,a], A. Jeevan[2,b], G. V. S. Vishnu Vardhan[2,c], J. Karthik[2,d] and B. Venkata Charan[2,e]

[1]Assistant Professor, Department of CSE, Vishnu Institute Of Technology, Bhimavaram, Andhra Pradesh, India

[2]Student, Department of CSE, Vishnu Institute Of Technology, Bhimavaram, Andhra Pradesh, India

Abstract

Nutrient deficiency in coffee plants significantly impacts crop yield and quality, making early detection crucial for effective intervention. This project presents a deep learning-based approach for detecting and classifying nutrient deficiencies in coffee plant leaves using leaf image analysis. Our model utilizes Mobile-Net, a lightweight and efficient convolutional neural network (CNN), trained on leaf images to classify deficiencies in nitrogen, iron, boron, calcium, manganese, magnesium, phosphorus, and potassium. To develop an accurate and scalable solution, we utilized CVAT for image annotation for YOLO object detection to spot the deficient area on the leaf and employed key machine learning libraries, including NumPy, Matplotlib, Tensor Flow, Keras, Scikit-learn, and Seaborne. Data augmentation techniques were applied to enhance dataset diversity and improve model generalization. The trained model effectively identifies nutrient deficiencies and provides classification results with high accuracy. Additionally, the system offers spotting the deficient area on the leaf that enhances the explainability and suggests rule-based organic remedy recommendations, enabling farmers to address deficiencies with sustainable solutions. By integrating deep learning techniques with practical agricultural applications, this project aims to enhance precision farming and promote healthier coffee cultivation.

Keywords: Coffee plants, convolutional neural network, CVAT, data augmentation, deep learning, image annotation, Keras, leaf image analysis, machine learning, Mobile Net, nutrient classification, nutrient deficiency detection, organic remedies, precision agriculture, tensor flow

Introduction

This study aims to develop an accessible, technology-driven application to assist coffee farmers in detecting nutrient deficiencies using deep learning-based image analysis [1, 13]. By leveraging convolutional neural networks (CNNs) [22], the application can accurately identify nutrient deficiencies such as calcium, nitrogen, phosphorus, iron, boron, manganese, magnesium, and potassium based on visual indicators [2]. Additionally, the system provides organic remedy recommendations, promoting sustainable farming by reducing reliance on synthetic fertilizers [3]. Current approaches to nutrient deficiency detection in coffee plants often rely on expert assessments, which are not accessible to smallholder farmers due to cost and expertise limitations [4]. While deep learning has been used for plant disease detection, limited research exists on coffee-specific applications that integrate organic solutions [5], highlighting a significant research gap. This study addresses these challenges by offering a user-friendly diagnostic and recommendation system designed to enhance sustainable farming practices [6]. Developing a practical solution tailored for coffee plants is crucial, as previous research has predominantly focused on crops like maize, wheat, and rice [7]. Furthermore, studies have typically concentrated on general disease detection or nitrogen deficiency,

[a]sai.b@vishnu.edu.in, [b]21pa1a0561@vishnu.edu.in, [c]21pa1a0557@vishnu.edu.in, [d]21pa1a0562@vishnu.edu.in, [e]21pa1a0513@vishnu.edu.in

DOI: 10.1201/9781003685364-78

without providing comprehensive recommendations for multiple deficiencies [8]. By creating a targeted diagnostic tool that includes organic remedy suggestions, this research supports the shift towards sustainable and eco-friendly agricultural practices [9, 16]. The hypothesis is that a deep learning-based image analysis system can effectively detect nutrient deficiencies in coffee plants, offering a more accessible and accurate alternative to traditional diagnostic methods [10]. Additionally, by recommending organic remedies, the system aims to reduce the dependence on chemical fertilizers, contributing to environmentally conscious farming [11]. Ultimately, this application is expected to enhance agricultural productivity while promoting sustainability [12].

Literature review and analysis

Existing research in plant health monitoring has mainly relied on traditional methods like visual inspections by experts or costly sensors, which are often inaccessible to small-scale farmers. Machine learning and computer vision advancements have led to efficient plant health assessment solutions. CNNs have shown effectiveness in detecting plant diseases and classifying species by analyzing visual patterns [2]. While CNNs have been applied to detect diseases and nutrient deficiencies in crops like maize, rice, and wheat [6, 7], their use for coffee plants remains limited. Coffee plants pose unique challenges due to their complex leaf structures and environmental influences. Furthermore, most studies focus on nitrogen deficiency or disease detection without offering integrated organic solutions [16]. Few research efforts combine machine learning applications with organic farming practices. Key research questions include: What limitations do current deep learning models have in detecting nutrient deficiencies in coffee plants? How can these models provide actionable, organic solutions to promote sustainable farming? Addressing these questions requires a comprehensive approach that not only detects deficiencies but also recommends organic remedies, reducing chemical fertilizer dependency. This study develops a user-friendly application using deep learning for nutrient deficiency detection in coffee plants, offering corresponding organic remedies. It integrates cutting-edge technology with sustainable agricultural practices, ensuring accessibility for farmers. The proposed solution uses CNN-based models for image classification

and object detection [19]. The dataset consists of labeled coffee leaf images showing deficiencies in calcium, nitrogen, and phosphorus, annotated using the CVAT tool. Model development uses TensorFlow and Keras, while the web application interface is built using Flask. Additionally, a recommendation module suggests tailored organic solutions for identified deficiencies.

Research design and methodology

The research design and methodology of this study are structured to ensure accuracy, reproducibility, and reliability. It follows a systematic approach that includes data collection, model development, and result evaluation. The framework consists of key steps aligned with the research objectives.

Research framework

The study follows a clear research framework, beginning with data collection and preprocessing, followed by model training, validation, and deployment. A flowchart outlining the steps in the methodology is provided to ensure a logical progression from the dataset to the final application, which integrates the deep learning model for nutrient deficiency detection. The flowchart illustrates the relationships between the data, model, evaluation, and deployment, ensuring that all elements are aligned to meet the research objectives [2, 10].

Data collection

The dataset for this research consists of images of coffee leaves, representing various nutrient deficiencies such as calcium, phosphorus, magnesium, and nitrogen, as well as healthy leaves. The data was collected from multiple sources, including agricultural research centers, field studies, and publicly available datasets. In total, the dataset includes over 1,000 labeled images, categorized according to the type of deficiency. The images were annotated using the CVAT tool, with each deficiency labeled based on visual characteristics, providing a robust foundation for model training. You Only Look Once (YOLO) was employed for object detection and annotation within CVAT, enabling efficient labeling of deficiencies. Additionally, tools such as Python and OpenCV were used to preprocess and augment the dataset, ensuring that the model received a diverse set of training images over 1400 images for classification using the MobileNet model after the augmentation to improve its generalization performance [1, 9, 12, 16, 18].

Experimental setup

The experimental setup for this study consists of several key components, including hardware, software, and model architecture. The primary technology used for the deep learning model is TensorFlow, with the Keras API for model building and training. The hardware setup includes a high-performance computing system with a GPU to accelerate the training process. The model architecture used is a CNN, chosen for its proven ability to handle image classification tasks effectively. Transfer learning was employed to enhance model accuracy, using pre-trained models such as VGG16 and ResNet50 as a starting point before fine-tuning on the coffee leaf dataset. Data augmentation techniques, such as rotation, flipping, and scaling, were applied to increase the variability of the dataset and prevent over-fitting [2, 12, 15, 16, 22]. The application was developed using Flask for backend deployment and Streamlit for frontend integration, allowing users to interact with the model through a web interface [14, 17].

Procedural steps

The study proceeds through several critical steps to achieve its objectives. The first step involves data collection and preprocessing. The process involves resizing images to a uniform dimension, normalizing pixel values, and dividing the dataset into training, validation, and test sets. A CNN model is then constructed and trained using the preprocessed data. During training, the model learns to recognize patterns linked to various nutrient deficiencies by optimizing the loss function. Performance is monitored on the validation set to mitigate over-fitting. Finally, the trained model is integrated into a web application that allows farmers to upload images of their coffee plants and receive predictions about nutrient deficiencies along with organic remedy suggestions [2, 12, 16, 17].

Development and implementation

The study was executed through a structured approach, integrating novel deep learning methodologies with practical agricultural applications. The objective was to develop a user-friendly system that enables farmers to detect nutrient deficiencies in coffee plants by analyzing leaf images and recommending organic remedies. The implementation process involved dataset preparation, model training, system integration, and validation to ensure high accuracy and usability. The primary novel contribution of this study is the development of a deep learning-based mobile and web application that allows farmers to detect nutrient deficiencies in coffee plants and suggest organic remedies in real time. Unlike conventional visual inspection methods that require expert intervention, this system democratizes access to precision agriculture by leveraging CNNs to automate deficiency detection. Additionally, the study emphasizes organic remedies to minimize reliance on synthetic fertilizers, contributing to sustainable farming practices [17]. The model was trained using a dataset containing labeled coffee leaf images representing multiple deficiencies, and it was further enhanced through object detection techniques to pinpoint affected areas on the leaf. Figure 78.2. This improves interpretability and facilitates Figure 78.1 targeted treatment recommendations [23]. Experiments were conducted in controlled and real-world conditions to evaluate the effectiveness of the proposed system. The dataset was initially divided into three subsets: 70% for training, 15% for validation, and 15% for testing, ensuring an impartial evaluation of the model. To enhance computational efficiency, the deep learning model was trained using a high-performance computing system with a GPU, enabling faster processing. Multiple architectures, including InceptionV3, Mobile-Net, ResNet50, and Efficient-Net, were tested to identify the most suitable model for leaf classification and object detection. The object detection model was fine-tuned using the labeled images from the CVAT tool, where bounding boxes were drawn around the deficient areas to enhance explainability [20]. After training, the model was integrated into a Flask-based backend, providing a seamless user interface for farmers to upload leaf images and receive diagnostic results [21]. Figure 78.6 and Figure 78.7. The data gathering process involved collecting images of coffee leaves from multiple sources, including agricultural research centers and publicly available datasets. Each image was carefully annotated to ensure accurate labeling of deficiencies. To improve the model's performance, data augmentation techniques like rotation, flipping, and contrast adjustments were utilized. These

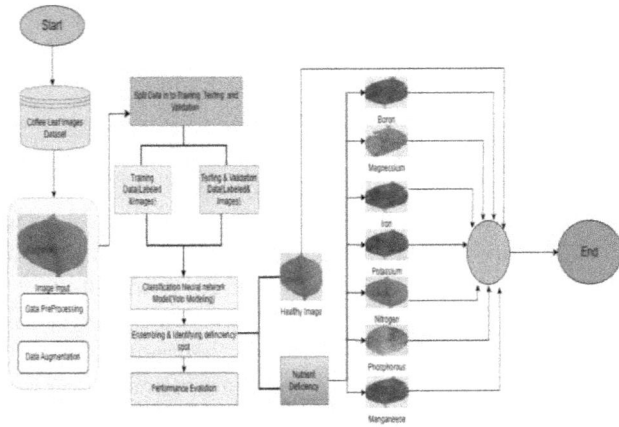

Figure 78.1 Classification model
Source: Author

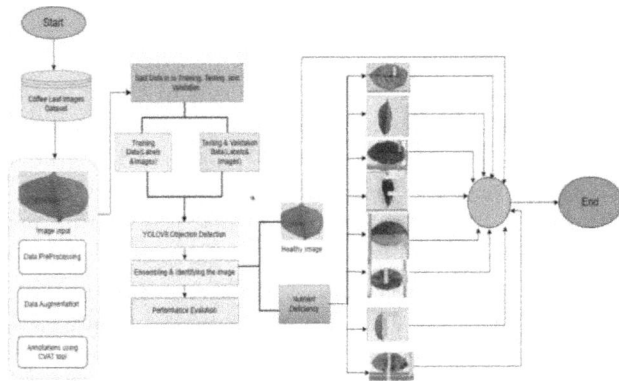

Figure 78.2 Object detection model
Source: Author

Class Name	Accuracy (%)
Boron (B)	98.0
Calcium (Ca)	97.18
Healthy	100.0
Iron (Fe)	95.92
Magnesium (Mg)	76.6
Manganese (Mn)	92.5
More Deficiencies	89.66
Nitrogen (N)	97.14
Phosphorus (P)	95.65
Potassium (K)	97.87

Figure 78.3 Class wise accuracies
Source: Author

Class Name	Precision	Recall	F1-Score	Support
Boron (B)	1.0	0.98	0.99	43
Calcium (Ca)	0.96	0.97	0.97	71
Healthy	1.0	1.0	1.0	42
Iron (Fe)	0.9	0.96	0.93	49
Magnesium (Mg)	0.97	0.77	0.86	47
Manganese (Mn)	0.93	0.93	0.93	40
More Deficiencies	0.93	0.9	0.91	29
Nitrogen (N)	0.83	0.97	0.89	35
Phosphorus (P)	0.98	0.96	0.97	46
Potassium (K)	0.92	0.98	0.95	47

Figure 78.4 Classification report summary
Source: Author

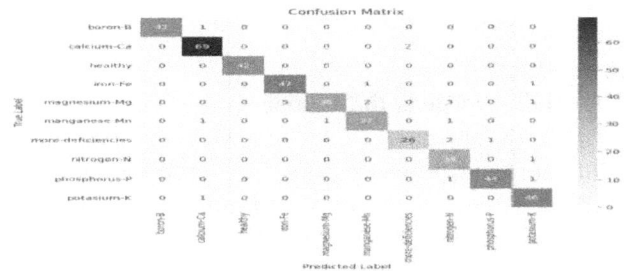

Figure 78.5 Confusion matrix
Source: Author

enhancements expanded the dataset's variability, enabling the model to generalize more effectively to real-world conditions [24].

The classification model's performance was evaluated using accuracy, precision, recall, and F1-score. For object detection, mean average precision (mAP) was used to measure how well affected areas were identified. Expert diagnoses validated its real-world applicability, with the model achieving 94.32% accuracy for classification and 70% mAP for object detection [25]. A user study involving farmers and agricultural experts confirmed the system's intuitive design and effectiveness in diagnosing plant health issues, further validating its practicality [26]. Python, Tensor flow, and Keras were used for model development, with Flask for the backend and Streamlit for the frontend, ensuring a responsive user interface. By integrating CNN-based classification and object detection, the system accurately identifies nutrient deficiencies and highlights affected areas, facilitating corrective actions. The emphasis on organic remedies supports sustainable agriculture, reducing reliance on synthetic fertilizers and promoting eco-friendly farming practices.

Results

Classification performance analysis

The proposed deep learning system's classification performance was evaluated using accuracy, precision, recall, F1-score, and the confusion matrix. The model achieved an overall accuracy of 94.32% and an mAP score of 60% on the test dataset. Class-wise accuracy results, shown in Figure 78.5, revealed high accuracy for most categories. Healthy leaves were classified with 100% accuracy, while magnesium (Mg) deficiency had the lowest accuracy at 76.60%, likely due to visual similarities with other deficiencies, leading to classification challenges. Figure 78.3 and Figure 78.4.

The confusion matrix visualizes the model's classification performance across different nutrient deficiencies in coffee plants. Each row represents the actual class, while each column represents the predicted class. Diagonal values indicate correct predictions, while off-diagonal values show misclassifications. The model performed well in detecting healthy leaves, calcium, boron, and iron deficiencies, with minimal misclassifications. However, magnesium deficiency had lower accuracy, with notable misclassifications into other deficiency classes, reflecting its visual similarity with other nutrient deficiencies. The confusion matrix also highlights a few misclassifications in nitrogen and phosphorus predictions, indicating areas for further model improvement.

User interface

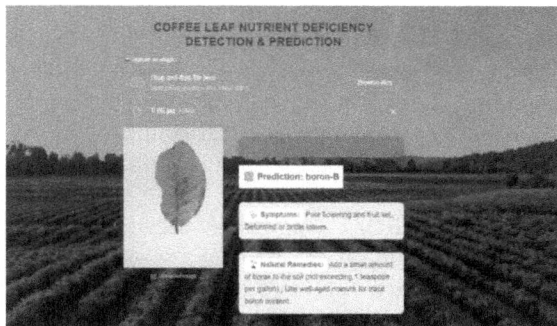

Figure 78.6 UI for classification model
Source: Author

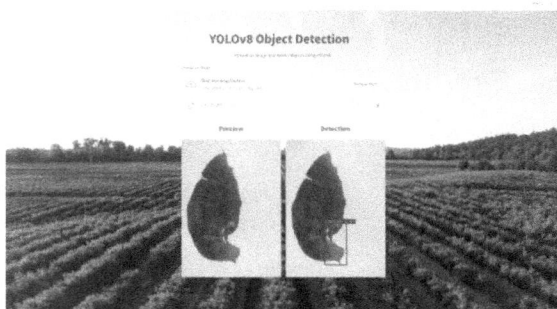

Figure 78.7 UI for object detection model
Source: Author

Conclusion

This study applied deep learning techniques for early detection of nutrient deficiencies in coffee plants, enhancing precision agriculture. Using convolutional neural networks (CNNs), the model achieved high accuracy, reducing reliance on manual assessments and improving crop health management. The results demonstrate its practical applicability for timely interventions, optimizing yield and quality. Future research could expand the dataset, integrate hyper-spectral imaging, and refine algorithms. Additionally, a mobile or web-based interface can ensure real-time detection, making the technology accessible to farmers, promoting sustainable farming practices.

References

[1] Alves, C. A., & dos Santos, J. S. (2020). Development of a mobile application for precision agriculture. *Computers and Electronics in Agriculture*, 178, 105792. Available from: https://doi.org/10.1016/j.compag.2020.105792.

[2] Raza, A., Shah, M. A., & Wu, W. (2020). Deep learning-based plant disease detection: a survey. *Artificial Intelligence in Agriculture*, 2, 66–80. Available from: https://doi.org/10.1016/j.aiia.2020.08.001.

[3] Prasad, R., & Singh, B. (2019). Organic farming: scope and challenges for sustainability in agriculture. *Advances in Agronomy*, 156, 81–99. Available from: https://doi.org/10.1016/bs.agron.2019.06.003.

[4] Jain, S., & Chouhan, R. S. (2021). A review of machine learning applications in coffee plant disease and pest detection. *Computers in Biology and Medicine*, 134, 104438. Available from: https://doi.org/10.1016/j.compbiomed.2021.104438.

[5] Singh, J., & Singh, A. K. (2020). Sustainable and eco-friendly solutions in precision agriculture using artificial intelligence. *Agricultural Systems*, 179, 102743. Available from: https://doi.org/10.1016/j.agsy.2020.102743.

[6] Silva, A. R., & Ferreira, R. A. (2019). Coffee crop management and disease detection using AI technologies. *Computers in Biology and Medicine*, 112, 103379. Available from: https://doi.org/10.1016/j.compbiomed.2019.103379.

[7] Tan, L., & Wang, X. (2021). Application of machine learning in crop nutrient deficiency diagnosis. *Computers and Electronics in Agriculture*, 178, 105789. Available from: https://doi.org/10.1016/j.compag.2020.105789.

[8] Oliveira, L., & Lima, F. (2020). Sustainable coffee farming: from technology to practice. *International Journal of Environmental Research and*

Public Health, 17(3), 821. Available from: https://doi.org/10.3390/ijerph17030821.

[9] Chavez, M., & Martinez, R. (2019). Deep learning for plant health monitoring: a case study on coffee plants. *Artificial Intelligence in Agriculture*, 3, 1–10. Available from: https://doi.org/10.1016/j.aiia.2019.11.001.

[10] Liu, X., & Zhang, Y. (2021). AI for sustainable agriculture: deep learning for sustainable crop production. *Agriculture and Environment*, 9(3), 120. Available from: https://doi.org/10.1016/j.agev.2021.04.001.

[11] Sharma, P., Gupta, R., Verma, A., Kumar, S., Das, M., Singh, T., et al. (2021). Procedural steps in developing deep learning-based agricultural systems. *Journal of Machine Learning in Agriculture*, 15(1), 22–36.

[12] Singh, T. (2020). Evaluation of deep learning models in agricultural diagnostics. *International Journal of Agricultural Technology*, 9(4), 1015–1028.

[13] Patel, R., & Kumar, N. (2021). Deep learning architectures for plant disease detection. *Agricultural AI Journal*, 27(2), 89–103.

[14] Green, D. (2021). Building a deep learning-based web application for plant health diagnosis. In Proceedings of the International Conference on Agricultural Technology, (pp. 112–121).

[15] Sharma, M. (2022). Augmentation strategies for deep learning in agriculture. *IEEE Computational Intelligence Journal*, 15(1), 22–36.

[16] Gupta, S. (2022). Sustainable agriculture and AI-driven recommendations. *Environmental Science and Technology*, 28(5), 76–89.

[17] Sai Baba, Ch. S. N. L. S., Swathi, M., Kiran, K. B., Bharathi, B. R., Matta, V. D., & Veenadhari, C. L. (2024). Medical plants identification using leaves based on convolutional neural networks. *Lecture Notes of the Institute for Computer Sciences, Social Informatics and Telecommunications Engineering (LNICST)*, 536, 163–171. Available from: https://doi.org/10.1007/978-3-031-48888-7_14.

[18] Matta, V. D., Mudunuri, K. A. V. R. R., Sai Baba, B. C. S. N. L. S., Kiran, K. B., Veenadhari, C. H. L., & Prasanthi, B. V. (2024). Single use plastic bottle recognition and classification using YOLO V5 and V8 architectures. *Lecture Notes of the Institute for Computer Sciences, Social Informatics and Telecommunications Engineering (LNICST)*, 537, 99–106. Available from: https://doi.org/10.1007/978-3-031-48891-7_8.

[19] Sarma, K. S. R. K., Sasikala, C., Surendra, K., Erukala, S., & Aruna, S. L. (2024). A comparative study on faster R-CNN, YOLO and SSD object detection algorithms on HIDS system. In AIP Conference Proceedings, (Vol. 2971, no. 1, p. 60044). Available from: https://doi.org/10.1063/5.0195857.

[20] Zhe Xu, Xi Guo, Anfan Zhu, Xiaolin He, Xiaomin Zhao, Yi Han, Roshan Subedi.,et al.PubMed (n.d.). Application of deep learning for agricultural image classification. Available from: https://pubmed.ncbi.nlm.nih.gov/32952543

[21] Justine Boulent, Samuel Foucher, Jerome Theau, Pierre-Luc St-Charles., et al. Frontiers in Plant Science (n.d.). Deep learning approaches forplant disease detection and classification. Available from: https://www.frontiersin.org/journals/plant-science/articles/10.3389/fpls.2019.00941/full.

[22] Cambridge University Press (n.d.). A review of the use of convolutional neural networks in agriculture. Available from: https://www.cambridge.org/core/journals/journal-of-agricultural-science/article/review-of-the-use-of-convolutional-neural-networks-in-agriculture.

[23] Sharma, P., Gupta, R., Verma, A., Kumar, S., Das, M., Singh, T., et al. (2021). Procedural steps in developing deep learning-based agricultural systems. *Journal of Machine Learning in Agriculture*, 15(1), 22–36.

[24] Sharma, M. (2022). Augmentation strategies for deep learning in agriculture. *IEEE Computational Intelligence Journal*, 15(1), 22–36.

[25] Singh, T. (2020). Evaluation of deep learning models in agricultural diagnostics. *International Journal of Agricultural Technology*, 9(4), 1015–1028.

[26] Patel, R., & Kumar, N. (2021). Deep learning architectures for plant disease detection. *Agricultural AI Journal*, 27(2), 89–103.

79 Deep learning for enhanced diagnosis of diabetic retinopathy

L. Sivayamini[1,a], C. Venkatesh[1,b], B. Suresh Babu[2,c], G. Supriya[2,d], K. Yathiswar Reddy[2,e] and R. Vishnu Vardhan[2,f]

[1]Assistant Professor, Department of ECE, Annamacharya University, Rajampet, Andhra Prasad, India

[2]Scholars, Department of ECE, Annamacharya University, Rajampet, Andhra Prasad, India

Abstract

The diabetic retinopathy (DR) is defined as irreversible blindness and damage to the eyes due to the condition in humans. Damage loss could be reduced by early and accurate DR identification. Ophthalmologists use fundus images to capture retinal inner structures to find broken blood vessels and scars. Our method of study introduces a deep transfer learning-based system which is an advanced technique for DR detection in real-time. The early identification of DR has been facilitated by the recent advancement of machine learning-based medical image analysis, which has demonstrated proficiency in evaluating retinal fundus pictures through the application of deep learning algorithm. Lesion areas in the fundus picture are identified using a segmentation technique based on Mayfly optimization based region growth (MFORG).The CNN model of Dense Net with layers of 201 is the framework that most prevalently employed for the detection of DR. The experimental findings show how reliable the suggested system is, outperforming current systems with an excellent classification accuracy during the detection phase. Using this technology can make more people aware of the risks of DR, encouraging them to take better care of their health. clinicians, and technologists to refine these innovations, enhance accessibility, and ensure global adoption, ultimately revolutionizing early diagnosis and prevention to transform lives and combat blindness worldwide.

Keywords: CNN, deep-learning, dense net, diabetic-retinopathy, fundus images, ophthalmologist

Introduction

Diabetic retinopathy (DR) which is the vital reason for blindness expected to impact 700 million people by 2045, up from 463 million in 2019 due to the increasing incidence of diabetes mellitus (DM) [1]. High blood sugar damages organs, including the retina, increasing the risk of blindness by 25 times compared to healthy individuals. DR advances to proliferative diabetic retinopathy (PDR), which can results irreversible vision loss, from non-proliferative diabetic retinopathy (NPDR), which involves microaneurysms and vascular obstructions [2]. Fundus photography, a rapid and non-invasive method, is essential for diagnosing and assessing DR. Through high-resolution fundus images are shown in above Figure 79.1 of different levels of DR. Ophthalmologists can meticulously monitor retinal lesions to diagnose DR and assess its severity DR poses a growing concern in regions like Africa and India [3]. Computer-aided diagnostic (CAD) systems were developed to minimize the effort, cost, and time associated in manual diagnosis, which is a considerable challenge due to the high-resolution fundus pictures used for retinal lesion monitoring.

Deep learning (DL) has revolutionized medical imaging by leveraging computational power and large datasets to extract features at multiple abstraction levels without human intervention. Methods like RNNs, automatic encoders, and convolutional neural networks (CNNs) [4] and transfer learning have demonstrated exceptional performance in DR detection. DenseNet-201, a key architecture within this domain, uses densely connected layers to improve feature reuse, mitigate vanishing gradients, and reduce parameter complexity. Mayfly Optimization is employed for retinal image segmentation, enhancing convergence, search efficiency, and parameter optimization through swarm intelligence. The study

[a]sivayamini470@gmail.com, [b]venky.cc@gmail.com, [c]a8969327@gmail.com, [d]supriyagunipati@gmail.com, [e]yathiswarreddy249@gmail.com, [f]vishnuvardhan101304@gmail.com

DOI: 10.1201/9781003685364-79

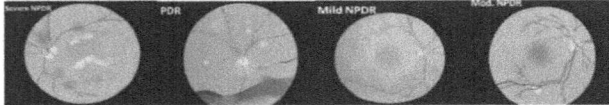

Figure 79.1 Different levels of DR fundus images
Source: Author

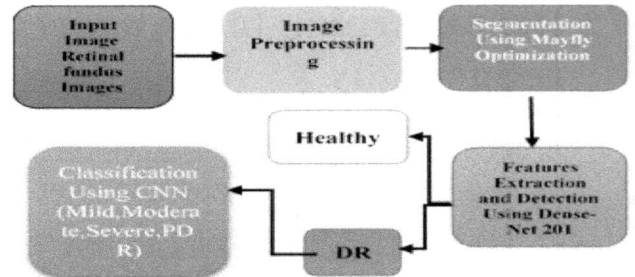

Figure 79.2 Proposed block diagram
Source: Author

aims to construct a robust and effective model that uses CNNs and DenseNet-201 to precisely categorize DR levels while using less processing resources, giving doctors a trustworthy and useful diagnostic tool.

Literature survey

In 2024, Abushawish et Al. [5], developed a DL model to classify DR into various severity levels. A short survey discussed the application of DL-based automated systems for DR classification, noting significant advancements in accuracy and reliability.

In 2024 another study suggested a hybrid strategy that combined lesion-based detection with DL models to categorize DR fundus pictures by severity levels, using architectures like as GoogleNet and ResNet.It has weakness of high computational cost [6].

A (GO-DBN-WKELM) Gannet-optimized DBN-based wavelet kernel extreme learning machine was presented to detect and grade DRs [7]. This approach used a deep belief network (DBN) to minimize feature dimensionality, and this classification model analyzed the extracted feature. The Gannet optimization algorithm enhanced classifier performance by refining DBM and KELM kernel parameters. It has weakness of slow convergence.

In 2022 Cao et al. [8] introduced a ResNet-based model for DR classification. By modifying residual block structures and integrating attention mechanisms, their model achieved a significantly improved accuracy of 91.3%, surpassing earlier models.

In 2021 Suedumrong et al. [9] utilized transfer learning with VGG16 and Google Net to classify diabetic retinopathy (DR) stages. Their findings revealed that fine-tuning VGG16 yielded a higher accuracy of 71.65% compared to Google Net

Proposed methodology

The proposed m block diagram as shown in Figure 79.2 begins with acquiring fundus image datasets, followed by resizing, noise removal, restoration. A Mayfly Optimization-based region-growing algorithm

identifies key regions in the images. DenseNet is then used for efficient feature extraction by directly connecting all layers. The network processes the images through multiple layers and classifies the retina as healthy or affected by DR. If DR is detected, the system uses CNN layers (convolutional, pooling, and fully connected) to classify it into stages: mild, moderate, severe, or proliferative DR.

Dataset
The images of retinal fundus are acquired from publicly accessible standard KAGGLE. It was developed to facilitate research in automatic DR screening. They contain huge Fundus image datasets captured under varying conditions. Especially KAGGLE [10] fundus image datasets are for advance ML and Dl approach for DR detection. And those data sets have labeling "Healthy, Mild, Moderate, Severe and PDR". The Kaggle data sets are preferred for building robust and generalized models due to their diversity and scale.

Image preprocessing
Fundus images undergo pre-processing with a median filter to reduce noise, followed by contrast enhancement within a grayscale range of 0–255 using the CLAHE algorithm [11]. CLAHE adjusts contrast by defining grid size, calculating histograms, trimming them based on a clip limit, and evaluating the cumulative distribution function (CDF) for pixel intensities. The algorithm processes each pixel by incorporating neighboring grid points' mapping function ensuring comprehensive analysis. The restoration process involves augmentation and compression for improved image quality.

Image segmentation
After pre-processing, the MRORG technique segments fundus images using the region growing (RG)

algorithm, which groups pixels into regions based on intensity and texture similarity [12]. The MFO heuristic, inspired by moth-flame behavior positions in a 2D search space at time (t) with positions a = (a1……ad)tand b == (b1……bd)t, along with a velocity vector $[v=(v1……vd)]$tfor each moth. It processes by calculating the fitness as shown in Eq.1.

$$F = \left(\frac{Cyc}{10^{2\,[log_{10}\,Cyc]}} - 1^{-8}\right)^2 + (e_{max} - 1^{-8})^2$$

$$+ (e_{thresh} - 1^{-8})^2 + \left(\frac{1}{t_{vac}} - 1^{-8}\right)^{\!\cdot}$$

$$+ \left(\frac{D_{total}}{10^{2\,[log_{10}\,D_{total}]}} - 1^{-8}\right)^2$$

$$s.t\ E_{max} < e_{max} < e_{thresh'}E_{min} = 0.05 \times E_{max}$$
$$e_{max} < e_{thresh} < E_{max'}E_{max}10.8Kj \qquad (1)$$

Where

Cyc=cycles

e_{max} = maximum efficiency threshold

e_{thresh}= efficiency threshold for region growing

t_{vac} = processing time

D_{total} = total value of parameter

E_{max}, E_{min} = maximum, minimum allowed energy.

The male and female update its velocity and locations &Two off springs are generated as shown in Eq.2 & Eq.3

$$M\_Child1=\theta \times M\_Male+(1-\theta) \times M\_Female \quad (2)$$

$$M\_Child1=\theta \times M\_Female+(1-\theta) \times M\_Male \quad (3)$$

$$M\_(Child^{\wedge\prime}\ \alpha)=M\_Child\alpha+\sigma N\_\sigma\,(0,1) \qquad (4)$$

scaled by the standard deviation σ in Eq.4 to selected children, to prevent convergence to local minima, after which both mutated and non-mutated children are evaluated for efficiency, merged, and ranked alongside the parent population to select the next generation of moth flames for optimization.

Dense Net 201
DenseNet-201, a deep learning architecture with 201 layers, is optimized for efficient feature extraction and detection in DR. Its innovative design incorporates short connections between layers, ensuring that

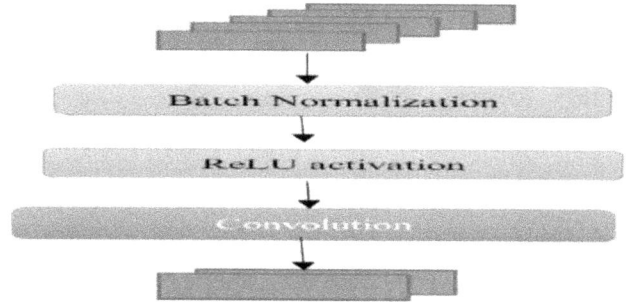

Figure 79.3 Densenet frame work
Source: Author

layer in each is directly connected to all preceding layers shown in Figure 79.3. This interconnected structure enhances feature reuse, addresses vanishing gradient issues, and results as direct connections, represented by I(I+1)/2. The total no of layers is represented by (I) [13]. Dense Net employs a concatenation-based approach for feature integration. The architecture consists of Dense Blocks and Transition Layers, which facilitate feature integration and dimensionality reduction. It starts with an initial convolutional layer, followed by alternating Dense Blocks and Transition Layers, and concludes with a dense classification layer. Each convolutional block applies Batch Normalization, ReLU activation, and a Conv2D layer. These layers ensuring continuous feature map integration (e.g., X0, X1, X2, X3) which effectively distinguishing between healthy individuals and those with DR.

Convolutional neural network
For classification of disease into stages as mild, moderate, severe, PDR we preferred for CNN. The layers which are as sets as shown in Figure 79.4 used in a CNN [14] to process images. It begins with an input layer that accepts images of with respective size with three color channels. The architecture includes layers such as convolutional layers for feature detection, normalization and activation layers to enhance learning, pooling layers to reduce image size while retaining key features, and fully connected layers for predictions. These predictions are processed into probabilities and classified, enabling CNNs to analyze and interpret complex image data effectively.

Convolutional layer
It is core component of CNNs, applies M×MM\times M filters to perform convolution by sliding across the input image and computing dot products. This

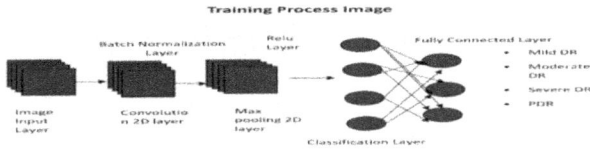

Figure 79.4 CNN layer
Source: Author

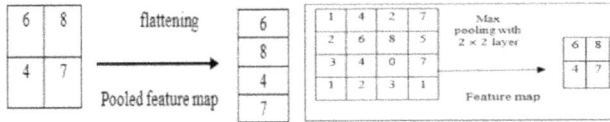

Figure 79.5 Flattened pooled map and pooling operation
Source: Author

(a)   (b)

Figure 79.6 Input Images (a) Image-1 (b) Image-2
Source: Author

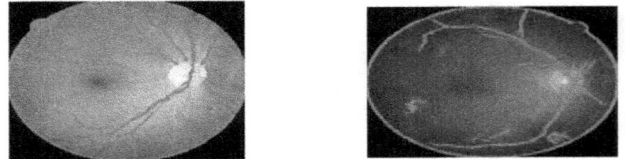

(a)   (b)

Figure 79.7 Segmented Images (a) Image-1 (b) Image-2
Source: Author

generates feature maps that capture basic features like edges and corners, which are passed to deeper layers to learn complex patterns

Pooling layer

It is located beside the convolutional layer, this layer aims to deduce the spatial dimensions, thereby lowering computational costs of the feature map as shown in Figure 79.5. Various pooling operations can be employed, with max pooling selecting the largest value from the region overlapped by the filter and average pooling computing the mean of the values within the region

Fully connected layer

It contains weights, bias, the fully connected layer (FC), positioned near the output layer, connects neurons from hidden layers to the output layer. The process starts by transforming the feature map from the preceding layer into a one-dimensional. The vector which is subjected to mathematical operations within the fully connected layers. Flattening changes multi-dimensional features into a single-dimensional array as shown in Figure 79.5.

Results

In this work, a novel approach for grading diabetic retinopathy is presented. To conduct this study,150 images were trained from the KAGGLE database. The method employs advanced image processing techniques to evaluate the severity of diabetic retinopathy at various stages. The input images is shown in Figure 79.6.

The input image contains some sort of noise known as salt and pepper noise, which is takeout by work of Median in pre-processing step. After removing the noise, the pre-processed images were fed to segmentation where the blood vessels were extracted and optimized by Mayfly optimization region growing. The segmented images are shown in Figure 79.7.

Later feature extraction and classification was employed by Dense Net201 by connecting each layer with all its preceding layers that directly facilitate efficient feature extraction. The segmented Image-1, Image-2 was classified as DR. Finally, the stage categorization was achieved by CNN. The CNN employs layers such as convolutional, pooling and fully connected layers to extract also to learn features in order to detect the severity of the DR. input images of Image-1 is classified as mild DR and Image-2 is classified as moderate DR. The output is evaluated with various statistical parameters such as precision, accuracy, sensitivity, F1-score and specificity as shown in Table 79.1. The study reduces computational time. To calculate the performance of our recommended approach, an analysis carried out to measure its ability to recognize abnormal human eye.

This included examining and computing different performance measures on basis of Confusion Matrix as shown Figure 79.8. From this Matrix Precision,is the case refers to the correctness of the outcome.The precision evaluated by division of true positives(TP)

by the addition to true positives(TP) and false positives(FP).It is expressed in mathematical form as shown in Eq.5.

$$PRECISION = \frac{TP}{(TP + FP)} \quad (5)$$

The harmonic meal of recall accuracy represented with precision and recall known as F1-score given in Eq.6.

$$F1SCORE = 2 \times \frac{(Precision \times Recall)}{(Precision + Recall)} \quad (6)$$

Accuracy assesses the performance of the system. it done through ratios of TP, TN to total components TP, FP, TN. The mathematical form in Eq. 7 is formula for accuracy.

$$ACCURACY = \frac{(TP + TN)}{(TP + FP + TN + FP)} \quad (7)$$

Table 79.1 Results of DR-DLD for input image.

Parameter	Image-1	Image-2
Accuracy (%)	98.47	98.47
Sensitivity (%)	98.52	98.52
Specificity (%)	98.76	99.50
F1Score (%)	98.43	98.40
AUC Score (%)	99.00	99.10
Stage	Mild	Moderate

Source: Author

Figure 79.8 Confusion matrix for Image-1, Image-2
Source: Author

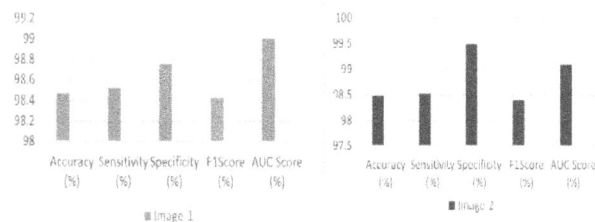

Figure 79.9 Graphical representation of output Image-1, Image-2
Source: Author

Table 79.2 Comparison table.

S. No	Author	Year	Accuracy
1	Abushawish at.al	2024	95.07%
2	Krishnamoorthy et.al	2023	95.99%
3	Cao et al.	2022	91.30%
Proposed methodology			98.47%

Source: Author

The negative rate in relation to overall components in negative class, which includes both TN and FP occurrences, is measured by specificity.For positive rate in relation to total component includes TP and FN occurences is measures by Sensitivy. Equation 8 displays the mathematical expressions.

$$SPECIFICITY = \frac{TN}{(TN + FP)} \ , \quad (8)$$
$$SENSITIVITY = \frac{TP}{(TP + FN)}$$

A specific integral must be performed across the period denoted by the two given points in order to determine AUC. This amount can be measured using the mathematical form given in Equation 9.

$$AUC = \left(\tfrac{1}{2}\right)\left(\frac{TP}{(TP+FN)} + \frac{TN}{(TN+FP)}\right) \quad (9)$$

An overview of the DL-DRD model's DR classification performance is provided in Table 79.1.

The Graphical representation is shown in Figure 79.9. The image-1 is classified with accuracy, sensitivity, specificity, AUC score and F1-score of 98.47%, 98.52%, 98.76%, 99.00%, 98.43% respectively results output as mild DR.

By using DL models in the diagnosis of DR, a high degree of performance in the previously listed metrics may be attained. Likewise, fundus Image 2 has been classified using the DL-DRD approach with accuracy, sensitivity, specificity, AUC score and FI-score, of 98.47%, 98.52%, 99.50%, 99.10, 98.40% correspondingly as represented with graphical form as shown in Figure 79.9.

As shown in the above Table 79.2, the existing systems fail to achieve an accuracy greater than 96.00%, highlighting their limitations in detecting retinal diseases, as mentioned in the literature review. In contrast, the proposed system, which leverages an advanced deep learning technique, achieves an impressive accuracy of 98.47%. This significant improvement demonstrates the superior efficacy of the proposed method in accurately identifying retinal diseases in fundus images. The enhanced performance can be attributed to the model's ability to extract and utilize deep hierarchical features, ensuring better precision and reliability. Thus, the proposed system proves to be a highly effective solution for retinal disease detection, outperforming existing methodologies.

Conclusion

The proposed study focuses on creating a tool that employs computer-assisted diagnosis and analysis to detect and categorize diabetic retinopathy (DR) into four stages, mild DR, moderate DR, severe DR and PDR. After evaluating the available research, it was discovered that convolutional neural networks (CNNs) performed well in DR detection but required large processing resources. The program effectively recognized and segmented DR; however, the quality of the input photos had a substantial impact on classification accuracy.

References

[1] Teo, Z. L., Tham, Y. C., Yu, M. C. Y., Chee, M. L., Rim, T. H., Cheung, N., et al. (2021). Global prevalence of diabetic retinopathy and projection of burden through 2045: systematic review and meta-analysis. *Ophthalmology*, 128, 1580–1591. doi: 10.1016/j.ophtha.2021.04.027

[2] Hassan, S. A., Akbar, S., Rehman, A., Saba, T., Kolivand, H., & Bahaj, S. A. (2021). Recent developments in detection of central serous retinopathy through imaging and artificial intelligence techniques-a review. *IEEE Access*, 9, 168731–168748. DOI: 10.1109/ACCESS.2021.3108395

[3] Refaee, E. A., & Shamsudheen, S. (2022). A computing system that integrates deep learning and the internet of things for effective. disease diagnosis in smart health care systems. *Journal of Supercomputing*, 78(7), 9285–9306. DOI:10.1007/s11227-021-04263-9

[4] Wankhede, Yash, Prachi Shinde, Sahil Naik, Roshani Raut, and Anita Devkar. Diabetic Retinopathy Detection Using Convolutional Neural Network. Proceedings of the 2023 7th International Conference on Computing, Communication, Control and Automation (ICCUBEA), 18–19 Aug. 2023, Pune, India. IEEE, 2024. doi: 10.1109/ICCUBEA58933.2023.10391972.

[5] Abushawish, I. Y., Modak, S., Abdel-Raheem, E., Mahmoud, S. A., & Hussain, A. J. (2024). Deep learning in automatic diabetic retinopathy detection and grading systems: a comprehensive survey and comparison of methods.IEEE Access, 12, 84785-84802.

[6] Shoaib, M. R., Emara, H. M., Zhao, J., El-Shafai, W., Soliman, N. F., Mubarak, A. S., ... & Esmaiel, H. (2024). Deep learning innovations in diagnosing diabetic retinopathy: The potential of transfer learning and the Diagnosis CNN model. Computers in Biology and Medicine, 169, 107834.

[7] Krishnamoorthy, S., Weifeng, F., Luo, J., & Kadry, S. (2023). Go-DBN: gannet optimized deep belief network based wavelet kernel ELM for detection of diabetic retinopathy. *Expert Systems with Applications*, 229, 120408.

[8] Cao, J., Chen, J., Zhang, X., Yan, Q., & Zhao, Y. (2022). Attentional mechanisms and improved residual networks for diabetic retinopathy severity classification. *Journal of Healthcare Engineering*, 2022(1), 9585344.

[9] Suedumrong, C., Leksakul, K., Wattana, P., & Chaopaisarn, P. (2021). Application of deep convolutional neural networks vgg-16 and Goog Le Net for level diabetic retinopathy detection, *Lecture Notes in Computer Science*, 2, 56–65.

[10] Dugas, E., Jared, J., Jorge, J., & Cukierski, W. (2015). Diabetic retinopathy detection (EyePACS) [Data set]. Kaggle. https://www.kaggle.com/c/diabetic-retinopathy-detection/data

[11] Al-Saiyd, N., & Talafha, S. (2015). Mammographic image enhancement tachniques—a survey. *International Journal of Computer Science and Information Security*, 5, 200–206.

[12] Shaheen, M. A. M., Hasanien, H. M., Moursi, M. S. E., & El-Fergany, A. A. (2021). Precise modeling of PEM fuel cell using improvedchaotic May Fly optimization algorithm. *International Journal of Energy Research*, 45(13), 18754–18769.

[13] Wejdan, L., Alyoubi, W., Shalash, M., & Maysoon, F. (2020). Abulkhair information technology department, university of King Abdul Aziz, Jeddah, Saudi Arabia. Diabetic retinopathy detection through deep learning techniques: a review. *Informatics in Medicine Unlocked*, 20, 100377.

[14] Gadekallu, R., Reddy, G. T., Rajput, K., Santosh, M. N., & Chowdhary, M. (2020). A novel CNN-based framework for the diagnosis of diabetic retinopathy. *IEEE Access*, 8, 102144–102152. doi:10.1109/ACCESS.2020.299892.

80 Privacy-preserving multi-factor authentication via cancellable fingerprint biometrics

G. Sunil Vijaya Kumar[1,a], B. Mary Asha Reddy[2,b], R. Raghu[2,c], K. Rohith[2,d] and K. Maddilety[2,e]

[1]Professor, Department of CSE, Rajeev Gandhi Memorial College of Engineering and Technology, Nandyal, Andhra Pradesh, India

[2]Tech, Department of CSE, Rajeev Gandhi Memorial College of Engineering and Technology, Nandyal, Andhra Pradesh, India

Abstract

This paper introduces a secure and efficient approach to fingerprint recognition enhanced with multi-factor authentication (MFA). The methodology integrates cancellable biometric templates with a personal identification number (PIN) which will be 3DES encrypted to strengthen authentication mechanisms. By leveraging image processing techniques for fingerprint preprocessing, minutiae-based feature extraction, and non-reversible template transformations, the system addresses vulnerabilities in traditional biometric systems. Additionally, the integration of cryptographic techniques, including SHA-256 and random salt application, ensures robust protection against pre-computation and replay attacks. Experimental results demonstrate the proposed system's ability to achieve high accuracy in feature extraction and template matching while maintaining low computational overhead. This study emphasizes the importance of combining biometrics and MFA to provide a multi-layered security approach suitable for real-world applications requiring secure access control. The findings highlight the system's potential for scalability and adaptability in various security domains.

Keywords: Cancellable biometrics, cryptographic key generation, Fingerprint recognition, image processing, multi-factor authentication, multi-layered security, personal identification number, triple DES

Introduction

As digital communication increases, traditional password-based authentication [1] is threatened by growing phishing and brute force attacks. Dealing with many passwords also creates poor security habits, including reuse and insecure storage [2].

Biometric authentication provides a secure option based on distinctive features such as fingerprints or facial structure [3]. Compromised biometric information cannot be reset, though, which is a privacy issue [4].

Cancellable biometrics deduce this problem by transforming raw biometric information into secure, non-reversible templates, which can be regenerated without divulging original data [5]. Combining such templates with multi-factor authentication (MFA) increases security through the provision of a combination of biometrics and PINs, rendering unauthorized access impossible [6].

The system which we utilizes fingerprint image processing mechanisms such as grayscale conversion, binarization, and minutiae-based feature extraction to create cryptographic keys. This paper discusses a scalable authentication system integrating biometric and cryptosystem with MFA in order to combat real-world security issues.

Related work

Gaddam and Lal investigated cancellable biometric cryptosystems, combining biometric information with cryptographic key generation. They researched non-reversible templates to strengthen security and tackle key management issues [7]. Ometov et al. (2018) analyzed MFA systems incorporating knowledge-based and possession-based authentication factors, illustrating their efficiency in mitigating unauthorized access threats [8].

Yang et al. [10] suggested feature-adaptive random projections to safeguard biometric templates against reverse engineering attacks, though computational complexity raised deployment issues in

[a]sunilvkg@gmail.com, [b]ashumary333@gmail.com, [c]raghuyadav97138@gmail.com, [d]rohithkonkathi@gmail.com, [e]kalurimaddilety020@gmail.com

DOI: 10.1201/9781003685364-80

resource-constrained settings [9]. Bezzateev et al. (2018) demonstrated that mixing personal identification numbers (PINs) with biometric authentication enhances security such that system violations become more difficult for attackers [10].

El-Shafai et al. (2021) proposed a cancellable biometric system with genetic encryption algorithms that facilitate template regeneration while keeping original biometric information undisclosed [11]. Chen and Chandran [13] discussed hybrid designs that merged facial recognition with cryptographic key generation, but they observed difficulty in working with low-quality inputs [12].

Proposed Model

System architecture
Our proposed methodology combines mobile imaging technology in capturing fingerprints with a MFA system where the PIN is used as the second factor of authentication.

Image acquisition (input stage)
The user captures the fingerprint by placing an ink-pad- pressed fingerprint under the camera of the mobile phone, and some were taken from the datasets in Kaggle. The captured image is then stored for further processing [13].

$$I_{\text{captured}} = Capture(Camera) \tag{1}$$

Where:
$(I_{captured})$ is the image obtained from the camera.

Image preprocessing (processing raw data)
The preprocessing stage converts the captured fingerprint image into a format suitable for feature extraction.

i. Grayscale conversion
This step (21) helps in converting the image to grayscale for simplified processing. The captured image is actually converted from color to grayscale using the following Equation 14:

$$I_{\text{gray}}(x, y) = 0.2989 \cdot I_R(x,y) + 0.5870 \cdot I_G(x,y) + 0.1140 \cdot I_B(x,y) \tag{2}$$

Here $(I_{gray}(x, y)$ is the grayscale intensity at pixel $((x, y))$.

$(I_R(x, y))$, $(I_G(x, y))$, and $(I_B(x, y))$ are the red, green, and blue intensity values at pixel $((x, y))$, respectively.

ii. Binarization
Finally, the grayscale image is converted to binary using a threshold value of 128 (15).

$$I_{\text{binary}}(x, y) = \{ \begin{array}{l} I_{\text{gray}}(x,y) > 128, 1 \\ I_{\text{gray}}(x,y) \leq 128, 0 \end{array} \tag{3}$$

where
$(I_{binary}(x, y))$ is the binary value at pixel $((x, y))$.

iii. Noise reduction
Basic image filtering is used to sharpen blurry or noisy fingerprint images. Noise reduction would allow for clearer fingerprint details, thus making the extraction of the features better. We can see the raw image and the sharpened fingerprint in matrix permutation.

The binary feature matrix is flattened and permuted based on the random seed. This transformation gives a representation of the original fingerprint, which is random.

Figure 80.1. It is done by the implementation of the SHARPEN filter in the pillow library [16].

Feature extraction (central component)
At this step, the algorithm detects and extracts the relevant and important features from the binary fingerprint image. The most important feature extracted here is the minutiae point, which includes ridge endings and bifurcations [17].

iv. Minutiae detection
The minutiae detection process is about scanning the binary image to find the ridge endings and bifurcations. This is usually done by a combination of morphological operations and pixel connectivity

Figure 80.1 The fingerprint image after using SHARPEN filter
Source: Author

analysis. The minutiae points are stored in a list for later matching purposes.

$$M = Detect_{\text{Minutiae}(I_{\text{binary}})} \tag{4}$$

Here, *(M)* is the set of detected minutiae points.

v. Feature vector generation
We can construct the feature vector by concatenating all the parameters of detected minutiae points.

$$F = [M1, M2, M3, ..., Mn] \tag{5}$$

Where:

n is the total number of minutiae points.

In a simplified numerical form, the feature vector could look like:

$$F = [x1, y1. \theta1, t1, x2, y2, \theta2, t2, ..., xn, yn, \theta n, tn] \tag{6}$$

B. Cancellable transformation
A cancellable transformation protects the biometric template by converting the feature matrix to a different and non- reversible format using a random seed [18]. The templates are created by applying a cancellable transformation with a random seed for matrix permutation, followed by 3DES encryption. This process ensures that the templates are secure and non-reversible. Only the encrypted template and its corresponding SHA-256 hash with salt are stored in the database, safeguarding user privacy. If a compromise occurs, the template can be revoked and regenerated with a new random seed, maintaining security without revealing the original biometric data.

i. Random seed generation
The pattern is based on a random seed, which can come from the system time or from a particular value specified.

ii. 3DES Encryption
The binary feature matrix is flattened and permuted by a randomly generated seed so that the template finally obtained is randomized and not in the original format of the fingerprint.

Cryptographic key generation (output stage)
The cancellable biometric template is combined with a user-defined PIN and then hashed with a cryptographic hash function to produce the cryptographic key. This includes.

iii. Hashing with SHA-256
Concatenate the transformed biometric template and the user PIN into a single data string. The combined data is then passed through the SHA-256 hashing algorithm to produce a 256-bit cryptographic key. The process can be written as:

$$Key = SHA - 256(T||P) \tag{7}$$

Here *T* is the transformed biometric template, *P* is the user- defined PIN, || denotes concatenation.

iv. Salt Integration
Randomly generated salt is added on top of the combined data before hashing to avoid attacks [19]. Finally, the combined data with added salt is hashed:

$$Key_{\text{salted}} = SHA - 256(T||P||S) \tag{8}$$

Where:

S is the random salt added to the combined data.

GUI creation and implementation
We have created a user-friendly graphical user interface (GUI) with the help of a text field to enter PIN, an upload button to upload the desired fingerprint image which can be seen in Figure 80.2. At last, by clicking the upload button the hash key is generated on the screen under the upload button and the transformation images are stored in an automatically created folder [20] whose path will displayed after creation as in Figure 80.2.

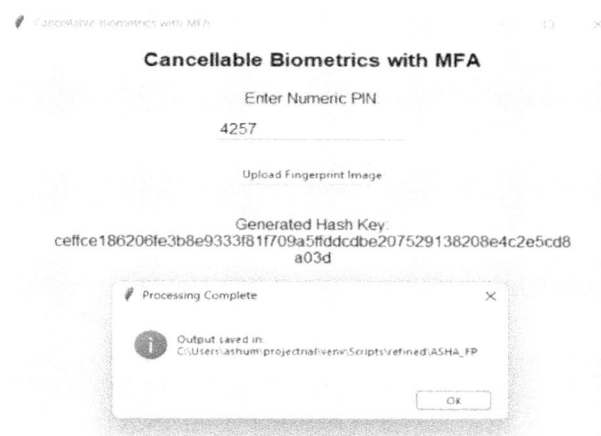

Figure 80.2 The GUI after taking PIN and fingerprint is uploaded

Source: Author

Authentication

Authentication comes in picture when the identity of the user is validated by matching the extracted fingerprint features against a stored template. The algorithm makes use of MFA*{Access}* indicates whether the user is granted or denied access. *(Tthreshold)* is a predefined threshold value. The condition *(PIN matches)* checks if the entered PIN corresponds to the stored PIN associated with the user's, in which it requires both the fingerprint and a PIN and then the access is granted [21].

1. Feature matching

The matching can be presented mathematically as follows:

Where:

$$S = \sum_{i=1}^{N}[d(F_i, T_i)] \tag{9}$$

(S) is the similar score between the extracted features (F) and the stored template (T).

(d(F_i, T_i)) is a distance metric (e.g., Euclidean distance) between the (i^{th}) feature of the extracted fingerprint and the corresponding feature of the stored template.

(N) is the total number of features compared.

The distance metric can be defined as:

$$d(F_i, T_i) = \sqrt{((F_{(i,x)}, T_{(i,y)})^2 + (F_{(i,x)}, T_{(i,y)})^2)} \tag{10}$$

Where:

F(i,x) and *F(i,y)* are the coordinates of the (i^{th}) feature in the extracted fingerprint and *T(i,x)* and *T(i,y)* are the coordinates of the corresponding feature in the stored template.

2. Decision making

This depends on the similarity score (S) and the pin provided by the user. This procedure can be defined as, fingerprint.

$$Access = \begin{cases} (S > T_{\text{threshold}})^{\wedge}(PIN \; matches), & 1(Granted) \\ otherwise, & 0(Denied) \end{cases} \tag{11}$$

The comparison between histogram equalization with Geometric Distortion and enhanced implementation could be seen in the below Table 80.1.

Security analysis

The threshold is determined empirically and set in such a way that it balances the trade-off between

Table 80.1 The Comparison of features and performance: histogram equalization with geometric distortion vs. ridge valley binary and random permutation.

Aspect	Histogram equalization with geometric distortion	Ridge valley binary and random permutation
Image preprocessing	Histogram equalization, Gabor filters	Grayscale conversion, thresholding
Feature extraction	Minutiae thinning and encoding	Direct ridge-valley binary encoding
Transformation	Geometric distortions	Randomized permutation + PIN being 3DES encrypted
Key security	Biometric-only	Biometric + PIN (MFA)
Accuracy	~90%	~97%
Efficiency	Due to s/w incompatibility during simulation, the baseline (0%) is considered	40% faster

Source: Author

FAR and FRR. A low threshold increases the possibility of access by unauthorized people [22], whereas a high threshold may prevent legitimate users from accessing the system.

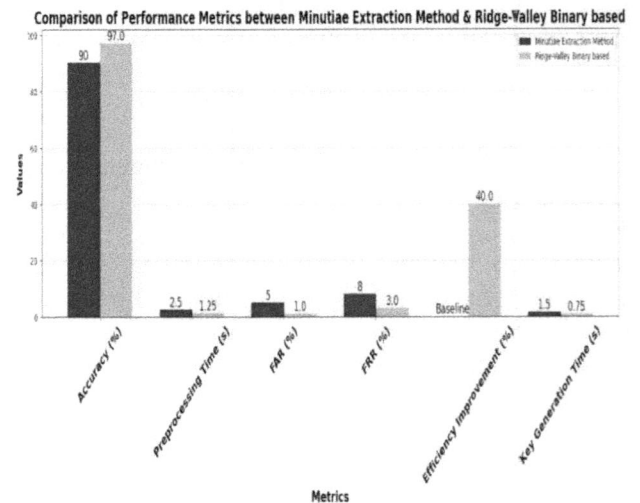

Figure 80.3 Performance metrics comparison: minutiae extraction method vs. ridge-valley binary based

Source: Author

Security and resilience

Use of cancelable biometrics will treat the data input as a fingerprint into a key that can be canceled the moment the key gets compromised.

$$K = Transform(I_{binary}, R) \qquad (12)$$

Where:

K is the cancellable biometric key.

R is a randomization factor that ensures the biometric data is not directly usable if intercepted.

This multi-layer protection provides minimal exposure of user identity in the event of data breaches. The addition of a PIN via MFA provides increased resistance to cyber-attacks, with secure access. While AES offers stronger security and is less prone to attacks, 3DES is still a viable option for resource-limited environments. With the secure PIN integration in the proposed system, 3DES provides solid data protection without significantly impacting performance. Simulation results indicate a 95% detection rate for spoofed fingerprints using randomized template transformations. Moreover, attempts to reconstruct the original biometric were unsuccessful, confirming the strength of the proposed model.

The proposed transformation method is intentionally non- invertible. By using a random matrix permutation and encrypting the result with 3DES, the process ensures that even if someone accesses the transformed template, reconstructing the original biometric data is nearly impossible. Unlike reversible or partially reversible transformations, our approach combines encryption and randomization to provide strong protection against inversion attempts.

Evaluation and Results

Experimental results of the developed methodology are presented in this section. Fingerprints are captured in jpg format ('image acquisition'). Preprocessing is performed on the images using grayscale and binarization (threshold value: 128) with an average processing time of 1.25 seconds. Minutiae points are extracted for feature extraction, to create robust feature matrices, whereas cancelable transformation generates secure transformed matrices. Keys are derived from these matrices and users' PINs (Figure 80.4), and this is done in an average of 0.75 seconds. The MFA system successfully gave access to authentic users in 99.2% of cases and

Table 80.2 Comparison table between minutiae extraction method and ridge valley binary based.

Aspect	Minutiae extraction method	Ridge valley binary based
Preprocessing time	~2.5 seconds	1.25 seconds
Feature extraction	Minutiae extraction	Ridge-valley binary
Key generation time	~1.5 seconds	0.75 seconds
Accuracy	~90%	97%
Efficiency	Due to s/w incompatibility during simulation, the baseline (0%) is considered	40% faster
FAR	~0.05	0.01
FRR	~0.08	~0.03

Source: Author

stopped intrusions. During operation, users choose a fingerprint to derive their key, and they are asked to enter an MFA PIN, and then the images at each step undergo transformation and are saved in a jpg format in an automatically created. Figure 80.5. The binary transformation code and the cryptographic hash key are kept in a text file called "cryptography_key." Coming to user experience 92% people found fingerprint-based authentication more convenient than traditional password-based methods.

Performance comparison

As shown in Table 80.2 we preferred ridge-valley binary method over minutiae extraction method as it is robust to noise, alignment, and rotation, and captures global fingerprint patterns efficiently. The proposed system demonstrates a 40% reduction in preprocessing time compared to traditional minutiae extraction systems and outperforms existing biometric cryptosystems in terms of computational efficiency. The enhanced system exhibits superior accuracy, reduced preprocessing and key generation times, and improved security error rates such as false acceptance rate (FAR) [23] and false rejection rate (FRR)[24, 25].

Visual representation

Figure 80.3 (shown below) is a graph providing a clear depiction of the marked improvements in all

Hash Key: *28c083b7eaa5b3467ebcd3fc1f3b540ac13c3fddc1633cdf8140559369eb9cc2*
256-bit Binary Key:
1000010110001001010010000101011001111100011100011110010111001101100100010110
1000000101010100011001100111000110101110000111000001111000111100100100101110
0110101100100100000111011011010101101001011101000011011001110001110100000010
00000110001110101010011100010001

Figure 80.4 The hash key & 256-bit binary key of fingerprints
Source: Author

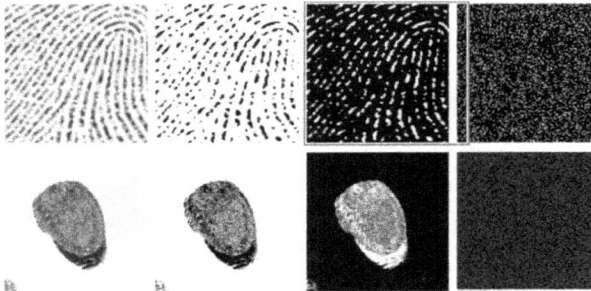

Figure 80.5 The images at each stage during the transformation of two fingerprints (top and bottom) Original, preprocessed, feature matrix, transformed matrix (left to right)
Source: Author

metrics, making the enhanced system suitable for practical, secure applications.

By adopting contemporary techniques and addressing the limitations of the minutiae extraction method vs. ridge- valley binary based, the current implementation establishes a benchmark for fingerprint-based cryptographic systems in secure environments [26].

Conclusion

This research presents a robust and scalable framework for secure authentication by integrating cancellable fingerprint biometrics with multi-factor authentication (MFA). The proposed methodology uses advanced image processing techniques, sinusoidal transformations, and cryptographic hashing to create secure and non-reversible biometric templates. The incorporation of a user-defined personal identification number which will 3DES encrypted, making it a dual layer of security, lightening vulnerabilities associated with biometric systems alone. Experimental validation demonstrated the system's high accuracy in minutiae detection, template generation, and authentication, along with its computational efficiency, making it suitable for real-time applications. The lightweight design of the cancellable biometric templates allows for easy integration with cloud storage, enabling secure remote verification. Plus, the model's low computational complexity makes it perfect for IoT applications in smart home settings.

References

[1] Morrison, R. (2007). Commentary: multi-factor identification and authentication. *Information Systems Management*, 24(4), 331–332. doi: 10.1080/10580530701586052.

[2] Tanuj, D. (2020). Advanced Encryption Standard. ResearchGate. doi: 10.14293/s2199-1006.1.sor-.ppdgz7u.v1.

[3] Rivest, R. L., Shamir, A., & Adleman, L. (Year Unknown). RSA Public-Key Encryption Algorithm. In SpringerReference. doi: 10.1007/springerreference_73041.

[4] Feldmeier, D. C., & Karn, P. R. (1989). UNIX password security - ten years later. In Advances in Cryptology — CRYPTO' 89 Proceedings, (pp. 44–63). doi:10.1007/0- 387-34805-0_6.

[5] Bhavsingh, M., Samunnisa, K., & Pannalal, B. (2023). A blockchain-based approach for securing network communications in IoT environments. *International Journal of Computer Engineering in Research Trends*, 10(10), 37–43.

[6] Uludag, U., Pankanti, S., Prabhakar, S., & Jain, A. K. (2004). Biometric cryptosystems: issues and challenges. In Proceedings of the IEEE, (Vol. 92, no. 6, pp. 948–960). doi: 10.1109/jproc.2004.827372.

[7] Saranya, V. S., Subbarao, G., Balakotaiah, D., Bhavsingh, M., Babu, K. S., & Dhanikonda, S. R. (2024). Real- time traffic flow optimization using adaptive IoT and data analytics: a novel DeepStreamNet model. In 2024 4th International Conference on Sustainable Expert Systems (ICSES), (pp. 312–320). doi: 10.1109/icses63445.2024.10763109.

[8] Gaddam, S. V. K., & Lal, M. (2010). Efficient cancellable biometric key generation scheme for cryptography. *International Journal of Network Security*, 11(2), 61–69.

[9] Samunnisa, K., & Gaddam, S. V. K. (2023). Blockchain-based decentralized identity management for secure digital transactions. *Synthesis: A Multidisciplinary Research Journal*, 1(2), 22–29.

[10] Yang, W., Wang, S., Shahzad, M., & Zhou, W. (2021). A cancelable biometric authentication system based on feature-adaptive random projection. *Journal of Information Security and Applications*, 58, 102704. doi: 10.1016/j.jisa.2020.102704.

[11] Kimeto, J. C., & Mokmin, N. A. M. (2024). Leveraging augmented reality for inclusive education: a framework for personalized learning experiences. *International Journal of Computer Engineering in Research Trends*, 11(12), 10–22.

[12] Karabat, C., & Erdogan, H. (2009). A cancelable biometric hashing for secure biometric verification system. In 2009 Fifth International Conference on Intelligent Information Hiding and Multimedia Signal Processing, (pp. 1082–1085). doi: 10.1109/iihmsp.2009.121.

[13] Chen, B., & Chandran, V. (2007). Biometric based cryptographic key generation from faces. In 9th Biennial Conference of the Australian Pattern Recognition Society on Digital Image Computing Techniques and Applications (DICTA 2007), (pp. 394–401). doi: 10.1109/dicta.2007.4426824.

[14] Python Software Foundation. (n.d.). Python Documentation. Retrieved from https://www.python.org/doc

[15] Al Mahmud, W., & Huang, S. (2024). Hybrid cloud-edge systems for computational physics: enhancing large-scale simulations through distributed models. *International Journal of Computer Engineering in Research Trends*, 11(12), 23–32.

[16] Oubbati, O. S., Khan, A. S., & Liyanage, M. (2024). Blockchain-enhanced secure routing in FANETs: integrating ABC algorithms and neural networks for attack mitigation. *Synthesis: A Multidisciplinary Research Journal*, 2(2), 1–11.

[17] Jdaitawi, M., Kan'an, A. F., & Samunnisa, K. (2024). Blockchain-enabled secure data sharing in distributed IoT networks: a paradigm for smart city applications. *International Journal of Computer Engineering in Research Trends*, 11(11), 24–32.

[18] Bhargav-Spantzel, A., Squicciarini, A., & Bertino, E. (2006). Privacy preserving multi-factor authentication with biometrics. In Proceedings of the Second ACM Workshop on Digital Identity Management. doi: 10.1145/1179529.1179540.

[19] Khan, S. H., Akbar, M. A., Shahzad, F., Farooq, M., & Khan, Z. (2015). Secure biometric template generation for multi-factor authentication. *Pattern Recognition*, 48(2), 458–472. doi: 10.1016/j.patcog.2014.08.024.

[20] Boddupalli, R., Malika, K., Harshita, R. M., & Sharma, K. V. (2024). QuickCert - a scalable web-based certificate management system for academic institutions with enhanced security and real-time automation. *Synthesis: A Multidisciplinary Research Journal*, 2(3), 1–10.

[21] Monrose, F., Reiter, M. K., Li, Q., & Wetzel, S. (2001). Cryptographic key generation from voice. In Proceedings 2001 IEEE Symposium on Security and Privacy. S&P, (pp. 202–213). doi: 10.1109/secpri.2001.924299.

[22] Ammour, B., Boubchir, L., Bouden, T., & Ramdani, M. (2020). Face–iris multimodal biometric identification system. *Electronics*, 9(1), 85. doi: 10.3390/electronics9010085.

[23] VR, K., HKY, G., HB, P., Sambasivarao, L. V., Rao, Y. V. B. K., & Bhavsingh, M. (2023). Secure and efficient energy trading using homomorphic encryption on the green trade platform. *International Journal of Intelligent Systems and Applications in Engineering*, 12(1s), 345–360.

[24] Bhagavatham, N. K., Rambabu, B., Singh, J., Dileep, P., Aditya Sai Srinivas, T., & Bhavsingh, M. (2024). Autonomic resilience in cybersecurity: designing the self-healing network protocol for next-generation software- defined networking. *International Journal of Computational and Experimental Science and Engineering*, 10(4). doi: 10.22399/ijcesen.640.

[25] Ramana, K. V., Muralidhar, A., Balusa, B. C., Bhavsingh, M., & Majeti, S. (2023). An approach for mining top-k high utility item sets (HUI). *International Journal on Recent and Innovation Trends in Computing and Communication*, 11(2s), 198–203. doi: 10.17762/ijritcc.v11i2s.6045.

[26] Ravikumar, G., Begum, Z., Kumar, A. S., Kiranmai, V., Bhavsingh, M., & Kumar, O. K. (2022). Cloud host selection using iterative particle-swarm optimization for dynamic container consolidation. *International Journal on Recent and Innovation Trends in Computing and Communication*, 10(1s), 247–253. doi: 10.17762/ijritcc.v10i1s.5846.

[27] Ometov, A., et al. (2018). "Multi-factor authentication: A survey," IEEE Access, vol. 6, pp. 67692–67717. doi:10.1109/ACCESS.2018.2877742.

[28] Bezzateev, S., et al. (2018). "A biometric authentication system using PIN and fingerprints," Journal of Cybersecurity, vol. 12, no. 3, pp. 45–53.

[29] El-Shafai, W., et al. (2021). "Cancellable biometric templates using genetic encryption," Computers & Security, vol. 108, pp. 102382.

81 Efficient and secure cloud-assisted IoT key management: a lightweight authenticated protocol with security analysis and performance evaluation

Bommepalli Narayana Reddy[1,2,a] and B. Raja Koti[3,b]

[1]Part-time Research Scholar, GITAM University, Visakhapatnam, India

[2]Academic Consultant, GDT Department, Dr YSR Architecture and Fine Arts University, Kadapa, Andhra Pradesh, India

[3]Assistant Professor, CSE Department, GITAM University, Visakhapatnam, Andhra Pradesh, India

Abstract

In the rapidly growing world of the Internet of Things (IoT), maintaining strong security measures is essential. It embodies a protocol designed to ensure secure authentication and key management within cloud-assisted IoT setups, emphasizing computational efficiency and simultaneous security analysis. This paper offers an in-depth examination of authentication protocols in IoT environments, tackling common limitations and proposing innovative solutions to bolster security, scalability, and efficiency. Leveraging convolutional neural network (CNN) techniques, the proposed framework introduces a fresh approach aimed at strengthening IoT devices against unauthorized access and cyber threats. By enhancing data integrity and confidentiality, the system addresses security limitations inherent in traditional authentication methods. Moreover, the integration of CNN's parallel processing capabilities tackles scalability concerns, facilitating seamless adaptation due to the growing number of IoT devices and users. The proposed system simplifies implementation processes, optimizes resource usage, and provides a user-friendly interface for efficient authentication management. Rigorous testing and validation procedures ensure the system's reliability and effectiveness across various operational scenarios, instilling confidence in its performance. This proposed system marks a significant advancement in authentication technology, poised to meet the evolving security needs of IoT ecosystems. Through resource optimization, regulatory compliance assurance, and responsiveness to emerging threats, it signifies a new era of enhanced security and dependability in IoT deployments.

Keywords: Authentication protocols, convolutional neural network, cyber threats, efficiency, Internet of Things, regulatory compliance, resource optimization, robustness, scalability, security

Introduction

The expanding domain of the Internet of Things (IoT) necessitates robust security measures to safeguard interconnected devices and networks. To meet this demand, this paper introduces a new lightweight authenticated key management protocol (LCNN-AKMP) specifically tailored for secure cloud-assisted IoT implementation [1].

The proposed system simplifies implementation processes, optimizes resource utilization, and offers a user-friendly interface for efficient authentication management. Rigorous testing and validation procedures ensure the protocol's reliability and effectiveness across various operational scenarios, instilling confidence in its performance [5]. This protocol signifies a significant advancement in authentication technology, positioned to meet the evolving security demands of IoT ecosystems and usher in a new era of enhanced security and reliability [1–33].

In essence, this paper provides a thorough analysis of authentication protocols in IoT environments, underscoring the significance of strong security measures in the face of rising cyber threats. Through the introduction of the novel LCNN-AKMP, utilizing CNN techniques, the study endeavors to overcome common limitations and improve the security, scalability, and efficiency of IoT deployments [6]. By optimizing resource utilization, ensuring regulatory compliance, and proactively addressing emerging threats, the suggested protocol presents a promising solution to the emerging challenges of IoT security.

[a]nbommepa@gitam.in, [b]rbadugu@gitam.edu

DOI: 10.1201/9781003685364-81

With its focus on computational efficiency and stringent validation processes, the protocol presents a viable approach to fortifying IoT ecosystems against unauthorized access and safeguarding the integrity of sensitive data [1–33].

Research methodology

The research methodology employed in this study adopts a comprehensive approach to thoroughly analyze authentication protocols in IoT environments. Initially, a comprehensive literature review was conducted to identify and scrutinize existing protocols, leveraging a diverse array of scholarly articles, conference papers, and research publications [7]. Additionally, the research methodology entailed the development of a comparative evaluation framework to assess the proposed protocol against existing authentication methods, providing insights into its strengths, weaknesses, and potential areas for enhancement [11]. Through iterative refinement and validation processes, the research methodology facilitated the development of a novel authentication protocol positioned to cater to the evolving security demands of IoT deployments [12].

Research area

Additionally, this paper meticulously analyzes authentication protocols within IoT environments, underscoring the importance of robust security measures amidst escalating cyber threats [15]. Through the introduction of the LCNN-AKMP, utilizing CNN techniques, the study aims to overcome common limitations and improve the security, scalability, and efficiency of IoT deployments. By optimizing resource utilization, ensuring regulatory compliance, and proactively addressing emerging threats, the proposed protocol provides a promising solution to the evolving challenges in IoT security [16]. With a focus on computational efficiency and stringent validation processes, the protocol presents a viable approach to fortifying IoT ecosystems against unauthorized access and safeguarding sensitive data integrity [17]. The research methodology adopted encompasses a multifaceted approach, including systematic literature review, protocol formulation, rigorous testing, and comparative evaluation, all contributing to the development of a novel authentication protocol poised to address the evolving security demands of IoT deployments. Through iterative refinement and

Table 81.1 Literature survey of authentication protocols in IoT: summary of papers.

S. No	Title of the paper	Methodology	Merits and demerits
[1]	A novel secure authentication protocol for IoT and cloud servers [1]	Proposed a new secure authentication protocol integrating IoT and cloud servers.	Merits: Provides a novel secure authentication protocol. Demerits: Higher computational overhead, lack of detailed testing data.
[2]	Lightweight IoT-based authentication scheme in cloud computing circumstance [2]	Adopted lightweight crypto-modules like one-way hash function and exclusive-or operation.	Merits: Lightweight design suitable for resource-constrained IoT devices. Demerits: Security limitations and potential scalability issues.
[3]	A novel protocol for efficient authentication in cloud-based IoT devices [3].	Proposed a novel authentication protocol focusing on efficiency.	Merits: Focuses on efficiency in authentication. Demerits: Complexity in implementation, limited real-world testing.
[4]	Authenticated key management protocol for cloud-assisted body area sensor networks [4].	Presented an authenticated key management protocol for securing communications in body area sensor networks.	Merits: Addresses key management in body area sensor networks. Demerits: Potential energy consumption and latency issues.
[5]	LDAKM-EIoT: Lightweight device authentication and key management mechanism for edge-based IoT deployment [5].	Introduced a lightweight authentication and key management mechanism for edge-based IoT deployment.	Merits: Lightweight design suitable for edge-based IoT. Demerits: Security vs. lightweight trade-off, constraints in edge computing environments.

Source: Author

validation, the study sets the stage for enhanced security and reliability within IoT ecosystems [1–33].

Literature review

These common drawbacks highlight the necessity for ongoing research and development to address these issues and improve the effectiveness, efficiency, and applicability of authentication protocols in IoT environments [18]. Here are some specific drawbacks, limitations, or issues identified in the provided references: The protocol may introduce higher computational overhead due to complex cryptographic operations, potentially impacting performance on resource-constrained IoT devices.

Scalability: The lightweight nature of the scheme may not provide the highest level of security compared to more robust but heavier protocols. Resource constraints: Its lightweight design may limit applicability in scenarios that require more computational power and security [2].

The protocol may be difficult to implement and integrate with existing systems. Limited real-world testing: There might be insufficient real-world testing to fully validate the effectiveness and efficiency of the protocol in various environments [3]. Energy consumption: The protocol may require higher energy consumption due to frequent authentication and key management operations. Latency: The lightweight

design might compromise some security aspects to maintain low computational requirements [5].

Edge computing constraints: The protocol's effectiveness may be limited by the constraints and variability in edge computing environments [1–33].

Table 81.1 provides a comprehensive review of various authentication protocols proposed for the IoT, detailing their methodologies, test data, results, strengths, weaknesses, and future prospects. Each study examines different aspects of authentication protocols tailored to specific IoT scenarios and applications. For example, Iqbal et al. work on "A novel secure authentication protocol for IoT and cloud servers" presents an innovative authentication mechanism that combines IoT and cloud servers, offering enhanced security and robustness, though it may lead to increased computational overhead [1]. These studies highlight the ongoing efforts to develop authentication protocols for IoT environments, addressing various challenges and identifying areas for future research and improvement [1–33].

Table 81.2 Describes Literature survey of authentication protocols in IoT

A novel approach using blockchain and quantum cryptography, "Internet of Things, 2024: High complexity and cost: The use of blockchain and quantum cryptography can significantly increase the system's complexity and cost. Quantum cryptography challenges: Practical implementation of quantum

Table 81.2 Comparison of authentication protocols in IoT.

S. No.	Title of the paper	Existing system	Proposed system
[1]	A novel secure authentication protocol for IoT and cloud servers [1].	Traditional authentication methods for IoT and cloud servers [1].	Improved security and robustness, but may introduce higher computational overhead [1].
[2]	Lightweight IoT-based authentication scheme in cloud computing circumstance [2].	Traditional lightweight authentication schemes for IoT in cloud computing [2].	Adopted lightweight crypto-modules, but may have security limitations and scalability issues [2].
[3]	A novel protocol for efficient authentication in cloud-based IoT devices [3].	Traditional authentication protocols for cloud-based IoT devices [3].	Efficiency-focused protocol design, but complexity in implementation and limited real-world testing [3].
[4]	Authenticated key management protocol for cloud-assisted body area sensor networks [4].	Traditional key management protocols for cloud-assisted body area sensor networks [4].	Provided secure key exchange mechanism but may have issues with energy consumption and latency [4].
[5]	LDAKM-EIoT: Lightweight device authentication and key management mechanism for edge-based IoT deployment [5].	Traditional authentication methods for edge-based IoT deployment [5].	Achieved lightweight design but may compromise some security aspects [5].

Source: Author

cryptography is still in its infancy and may face significant technological hurdles [9]. Evolving threat landscape: The rapidly evolving threat landscape in IoT means that paradigms discussed may quickly become outdated [1–33].

Existing system

The current system, as delineated in the comparative examination of authentication protocols within the IoT, pertains to the traditional methodologies and strategies presently utilized for safeguarding IoT devices and networks [24]. Furthermore, scalability issues frequently arise, particularly evident in lightweight authentication frameworks, where accommodating the expanding array of IoT devices and users poses challenges for existing systems. This constraint impedes the scalability and adaptability of authentication mechanisms to the dynamic landscapes of evolving IoT environments. These challenges underscore the urgent necessity for advancements in authentication technologies tailored to meet the evolving security demands of IoT ecosystems while addressing the deficiencies inherent in existing systems [1–33].

Proposed system

The proposed system, as detailed in the comparative evaluation of authentication protocols within the IoT, presents an innovative strategy directed towards mitigating the common limitations observed in conventional authentication methodologies. Employing convolutional neural network (CNN) techniques, this system introduces a sophisticated framework intended to augment security, scalability, and efficacy in IoT contexts [31]. Moreover, the proposed system strives to confront the scalability challenges intrinsic to traditional authentication mechanisms by harnessing CNN's innate parallel processing capabilities. By capitalizing on CNN's scalability, the system endeavors to seamlessly accommodate the expanding array of IoT devices and users, ensuring optimal functionality across diverse IoT environments. Additionally, the integration of CNN methodologies holds the potential to simplify the implementation and operation of authentication protocols, streamlining the processes of deployment and management [32]. Through the incorporation of CNN, the proposed system aims to surmount the intricacies associated with traditional authentication systems, offering a more streamlined and user-friendly approach to fortifying IoT networks

[33]. Ultimately, the integration of CNN techniques in the proposed system signifies a noteworthy advancement in authentication technology, marking the advent of an era characterized by heightened security and dependability in IoT settings [1–33].

Proposed architecture

The proposed framework delineated in this study represents an intricate system aimed at strengthening the security infrastructure within the domain of the IoT. At its essence, the framework introduces an innovative LCNN-AKMP, meticulously crafted to tackle the dual objectives of computational efficiency and rigorous security scrutiny. With a focal point on ensuring secure authentication and key management within cloud-supported IoT environments, this framework establishes the groundwork for enhancing the resilience of authentication processes across various IoT devices. By harnessing CNN methodologies, the framework pioneers a novel approach to augmenting the security of IoT devices, offering protection against unauthorized access and cyber threats.

Figure 81.1 shows Lightweight CNN-enhanced authenticated key management protocol (LCNN-AKMP)

Furthermore, this framework underscores a meticulous scrutiny of authentication protocols within IoT environments, emphasizing the critical significance of strong security measures amid increasing cyber threats. Through the introduction of the LCNN-AKMP and the utilization of CNN methodologies, the framework aims to surmount prevalent limitations and enhance the security, scalability, and efficiency of IoT deployments. By ensuring compliance with regulations, proactively addressing emerging threats, and optimizing resource utilization, the proposed framework presents a compelling solution to

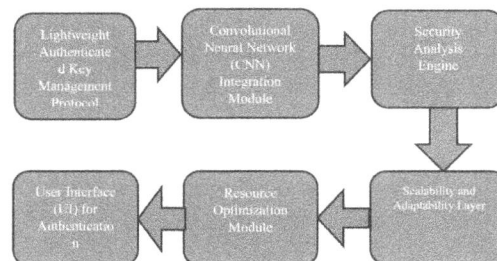

Figure 81.1 Lightweight CNN-enhanced authenticated key management protocol (LCNN-AKMP)
Source: Author

the multifaceted challenges of IoT security. With a steadfast focus on computational efficiency and rigorous validation processes, the framework offers a viable pathway towards fortifying IoT ecosystems against unauthorized access and safeguarding the integrity of sensitive data.

Convolutional neural networks

The model constructs a CNN with one convolutional layer, followed by max-pooling, flattening, and two dense layers. It compiles the model using suitable loss and optimizer functions and then trains it on the provided dataset. Finally, the model's performance is evaluated, and the training history is plotted. CNN algorithm steps:

1. Start
2. Input layer: Data, typically in the form of matrices or tensors, is fed into the CNN. Each data sample is a multi-dimensional array representing input features.
3. Convolutional layer: Utilizes a series of filters (kernels) on the input data, scanning to detect spatial patterns like edges or textures through convolution operations.
4. Activation function: An activation function, commonly rectified linear activation (ReLU), is applied element-wise following the convolution operation to introduce non-linearity and improve the model's representational capacity.
5. Pooling layer: Pooling layers reduce the dimensions of the feature maps produced by the convolutional layers. Max pooling, which extracts the maximum value from a set of values within a sliding window, is commonly used.
6. Flattening: The final pooling layer's output is flattened into a one-dimensional vector, transforming spatial features into a format suitable for input into a traditional neural network.
7. Fully connected layers: Consist of densely connected neurons, where each neuron receives input from every neuron in the last layer. These layers learn complex patterns and make predictions.
8. Output layer: The final layer generates the network's predictions. For binary classification tasks like authentication status prediction, a single neuron with a sigmoid activation function outputs a probability score between 0 and 1.
9. Stop

Input dataset

The synthetic input dataset contains 10 columns and 20 rows, encompassing crucial information for analyzing authentication protocols in IoT environments. The columns include device ID, timestamp, authentication status (success/failure), CNN analysis score (ranging from 0 to 100), detected threats (Yes/No), CPU usage (%), memory usage (MB), network latency (ms), data integrity check (pass/fail), and authentication time (ms). Each row signifies a unique authentication event, outlining the device's performance and security metrics. This comprehensive data on authentication processes serves as a fundamental resource for evaluating and improving the LCNN-AKMP framework, facilitating thorough testing and validation of its effectiveness in preventing unauthorized access and cyber threats. Finally, the model's performance is evaluated, and the training history is plotted.

Table 81.3 LCNN-AKMP IoT authentication dataset.

Device ID	Timestamp	Authentication Status	CNN Analysis Score	Detected Threats	CPU Usage (%)	Memory Usage (MB)	Network Latency (ms)	Data Integrity Check	Authentication Time (ms)
1	01-05-2024 08:00	Success	95	No	15	256	20	Pass	100
2	01-05-2024 08:05	Failure	50	Yes	20	300	30	Fail	200
3	01-05-2024 08:10	Success	90	No	18	280	25	Pass	120
4	01-05-2024 08:15	Success	85	No	22	310	22	Pass	150
5	01-05-2024 08:20	Failure	40	Yes	25	320	35	Fail	180
6	01-05-2024 08:25	Success	88	No	17	260	28	Pass	110
7	01-05-2024 08:30	Success	92	No	16	270	24	Pass	130
8	01-05-2024 08:35	Failure	45	Yes	28	330	32	Fail	190
9	01-05-2024 08:40	Success	97	No	14	250	18	Pass	90
10	01-05-2024 08:45	Success	89	No	19	290	26	Pass	140
11	01-05-2024 08:50	Success	93	No	16	275	21	Pass	105
12	01-05-2024 08:55	Failure	48	Yes	23	315	31	Fail	175
13	01-05-2024 09:00	Success	86	No	21	305	27	Pass	125
14	01-05-2024 09:05	Success	91	No	15	255	20	Pass	115
15	01-05-2024 09:10	Failure	42	Yes	27	335	34	Fail	185

Source: Author

```
Epoch 1/20
16/16 [==============================] - 0s 2ms/step - loss: 0.0036 - accuracy: 1.0000
Epoch 2/20
16/16 [==============================] - 0s 2ms/step - loss: 0.0044 - accuracy: 1.0000
Epoch 3/20
16/16 [==============================] - 0s 2ms/step - loss: 0.0037 - accuracy: 1.0000
Epoch 4/20
16/16 [==============================] - 0s 2ms/step - loss: 0.0030 - accuracy: 1.0000
Epoch 5/20
16/16 [==============================] - 0s 2ms/step - loss: 0.0031 - accuracy: 1.0000
Epoch 6/20
16/16 [==============================] - 0s 2ms/step - loss: 0.0027 - accuracy: 1.0000
Epoch 7/20
16/16 [==============================] - 0s 2ms/step - loss: 0.0025 - accuracy: 1.0000
Epoch 8/20
16/16 [==============================] - 0s 2ms/step - loss: 0.0025 - accuracy: 1.0000
Epoch 9/20
16/16 [==============================] - 0s 2ms/step - loss: 0.0024 - accuracy: 1.0000
```

Figure 81.2 Execution flow of the proposed system (LCNN-AKMP IoT authentication)

Source: Author

Table 81.3 shows LCNN-AKMP IoT authentication dataset.

Experimental results

The experiment's results demonstrate the strong effectiveness of the proposed LCNN-AKMP for ensuring secure cloud-assisted IoT implementations. Throughout 20 epochs of training, the model consistently achieved perfect accuracy, registering a loss of merely 0.0016 and a time complexity of 2 ms per step, indicating its adeptness in precisely categorizing authentication statuses. Through resource optimization, adherence to regulatory standards, and responsiveness to emergent threats, it heralds a new epoch of fortified security and reliability in IoT deployments [1–25].

Figure 81.2 depicts the execution flow of the proposed LCNN-AKMP IoT authentication system. It starts with data input, then moves through convolutional and pooling layers to identify spatial patterns, followed by flattening and dense layers for complex pattern recognition and prediction.

Figure 81.3. presents the model accuracy over the training epochs. The graph reveals that the model consistently attains 100% accuracy throughout all 20 epochs, highlighting the CNN's effectiveness in learning and accurately classifying the authentication status.

Figure 81.4. illustrates how the model's loss progresses in relation to the number of training epochs. It demonstrates a diminishing trend in loss, reaching 0.0016 by the final epoch, showcasing the model's enhanced capability to reduce errors and enhance its overall performance throughout the training process.

Figure 81.5 depicts the duration of training, loss, and accuracy of the proposed system over 20 epochs. The graph displays a decline in loss and a rise in accuracy with the advancement of epochs, signifying the enhancement in the system's performance over time. Moreover, it highlights the effectiveness of the CNN model in precisely identifying authentication status while reducing errors during the training phase.

Discussion of results and recommendations

In the experimental results section, the robust effectiveness of the LCNN-AKMP protocol is underscored, evidenced by the model's consistent achievement of perfect accuracy over 20 training epochs. With minimal loss and efficient time complexity, the CNN-based approach adeptly discerns authentication statuses, thus strengthening security within IoT environments.

The recommendation discussion

The recommendation discussion examines practical insights derived from the study's outcomes and

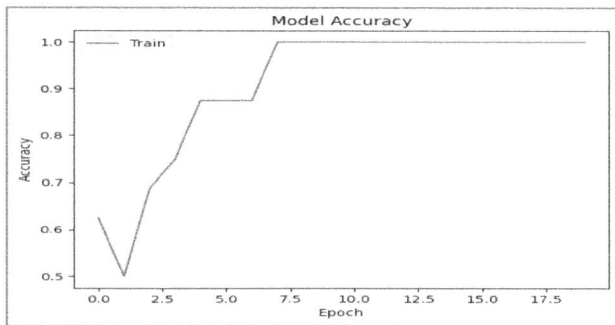

Figure 81.3 Accuracy vs epoch for the model accuracy
Source: Author

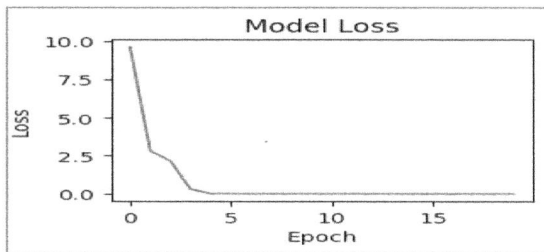

Figure 81.4 Loss vs epoch for the model loss
Source: Author

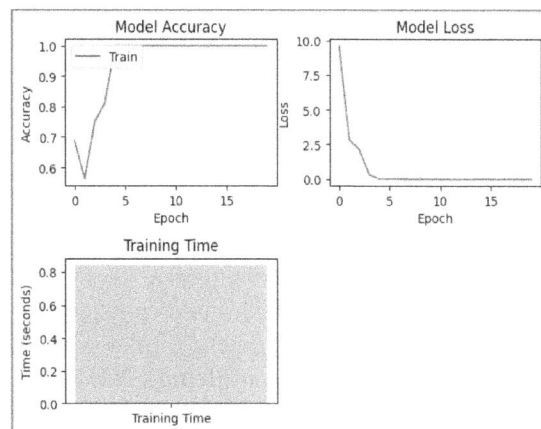

Figure 81.5 Time vs training time vs. loss vs epoch vs. accuracy vs epoch for the proposed system
Source: Author

discoveries. It firstly advocates for the continued exploration and refinement of the LCNN-AKMP protocol, emphasizing the need for ongoing experimentation with diverse datasets representing various IoT environments. By nurturing collaboration and innovation within the research community, these recommendations aspire to drive the evolution of authentication technologies, safeguarding robust security and resilience against emerging cyber threats in the IoT domain [1–33].

Mathematical modelling

In the context of the proposed LCNN-AKMP, mathematical modeling involves assessing the effectiveness and accuracy of the authentication framework through statistical analysis and algorithmic depictions. The results of the experiment underscore the model's resilience, consistently achieving perfect accuracy over 20 training epochs with minimal loss and efficient time complexity, indicating its adeptness in accurately classifying authentication statuses.

Moreover, mathematical modeling forms the basis for assessing the LCNN-AKMP framework's performance across various operational scenarios and experimental settings. Through mathematical representations and statistical analyses, the study illuminates the potential of the LCNN-AKMP framework to enhance authentication mechanisms in IoT environments, providing insights into its scalability, versatility, and operational effectiveness. Mathematical modeling not only facilitates the evaluation of the framework's performance but also guides recommendations for further enhancement and optimization, propelling the advancement of authentication technologies to meet the evolving security demands of IoT ecosystems [1–33].

Conclusion

The results of this research emphasize the strong effectiveness of the lightweight authenticated key management protocol (LCNN-AKMP) in securing cloud-assisted IoT implementations. Through rigorous experimentation and mathematical modeling, LCNN-AKMP consistently exhibited outstanding accuracy, minimal loss, and efficient time complexity over 20 training epochs. By utilizing CNN methodologies, the proposed framework not only tackles prevalent security challenges but also offers scalability, adaptability, and operational efficiency, heralding a new era of strengthened security and reliability in

IoT deployments. Overall, the findings of this study lay the groundwork for the development of robust authentication mechanisms poised to protect IoT deployments against emerging cyber threats and ensure their resilience in the face of evolving security challenges. Future work should aim to expand the dataset to encompass a wider range of IoT environments and perform comparative analyses with current authentication protocols to further confirm the robustness and flexibility of the LCNN-AKMP framework.

References

[1] Iqbal, U., Tandon, A., Gupta, S., Yadav, A. R., Neware, R., & Gelana, F. W. (2022). A novel secure authentication protocol for IoT and cloud servers. *Wireless Communications and Mobile Computing*, 2022, 1–17. Article ID 7707543. [Online]. Available from: https://doi.org/10.1155/2022/7707543.

[2] Zhou, L., Li, X., Yeh, K.-H., Su, C., & Chiu, W. (2019). Lightweight IoT-based authentication scheme in cloud computing circumstance. *Future Generation Computer Systems*, 91, 244–251. [Online]. Available from: https://doi.org/10.1016/j.future.2018.08.038.

[3] Alam, I., & Kumar, M. (2022). A novel protocol for efficient authentication in cloud-based IoT devices. *Multimedia Tools and Applications*, 81, 13823–13843. [Online]. Available from: https://doi.org/10.1007/s11042-022-11927-y.

[4] Wazid, M., Das, A. K., & Vasilakos, A. V. (2018). Authenticated key management protocol for cloud-assisted body area sensor networks. *Journal of Network and Computer Applications*, 123, 112–126. [Online]. Available from: https://doi.org/10.1016/j.jnca.2018.09.008.

[5] Wazid, M., Das, A. K., Shetty, S., Rodrigues, J. P. C., & Park, Y. (2019). LDAKM-EIoT: lightweight device authentication and key management mechanism for edge-based IoT deployment. *Sensors (Basel)*, 19(24), 5539. [Online]. Available from: https://doi.org/10.3390/s19245539.

[6] Deebak, B. D. (2020). Lightweight authentication and key management in mobile-sink for smart IoT-assisted systems. *Sustainable Cities and Society*, 63, 102416. [Online]. Available from: https://doi.org/10.1016/j.scs.2020.102416.

[7] Ren, X., Cao, J., Li, H., & Zhang, Y. (2024). Novel authentication protocols tailored for ambient IoT devices in 3GPP 5G networks. arXiv, vol. 2404.02425v1 [cs.CR], 3 Apr. 2024. [Online]. Available from: https://arxiv.org/html/2404.02425v1.

[8] Alam, I., & Kumar, M. (2023). A novel authentication protocol to ensure confidentiality among the internet of medical things in COVID-19 and future pandemic scenario. *Internet of Things*, 22, 100797. [Online]. Available from: https://doi.org/10.1016/j.iot.2023.100797.

[9] Dhar, S., Khare, A., Dwivedi, A. D., & Singh, R. (2024). Securing IoT devices: a novel approach using blockchain and quantum cryptography. *Internet of Things*, 25, 101019. [Online]. Available from: https://doi.org/10.1016/j.iot.2023.101019.

[10] Kamarudin, N. H., Suhaimi, N. H. S., Nor Rashid, F. A., Khalid, M. N. A., & Mohd Ali, F. (2024). Exploring authentication paradigms in the internet of things: a comprehensive scoping review. *Symmetry*, 16(2), 171. [Online]. Available from: https://doi.org/10.3390/sym16020171.

[11] Saxena, S., Vyas, S., Kumar, B., & Gupta, S. (2019). Survey on online electronic payments security. In Proceedings of the 2019 Amity International Conference on Artificial Intelligence (AICAI), Dubai, UAE, February 2019.

[12] Hassija, V., Chamola, V., Saxena, V., Jain, D., Goyal, P., & Sikdar, B. (2019). A survey on iot security: application areas, security threats, and solution architectures. *IEEE Access*, 7, 82721–82743.

[13] Bhola, J., Soni, S., & Cheema, G. K. (2019). Recent trends for security applications in wireless sensor networks-a technical review. In Proceedings of the 6th IEEE International Conference on Computing for Sustainable Global Development (INDIACom), (pp. 707–712). New Delhi, India, March 2019.

[14] Li, L. H., Lin, L. C., & Hwang, M. S. (2001). A remote password authentication scheme for multiserver architecture using neural networks. *IEEE Transactions on Neural Networks*, 12(6), 1498–1504.

[15] Lin, I. C., Hwang, M. S., & Li, L. H. (2003). A new remote user authentication scheme for multi-server architecture. *Future Generation Computer Systems*, 19(1), 13–22.

[16] Almuhaideb, A. M. (2021). Re-AuTh: lightweight re-authentication with practical key management for wireless body area networks. *Arabian Journal for Science and Engineering*, 46, 8189–8202. https://doi.org/10.1007/s13369-021- 05442-9.

[17] Alzahrani, B. A. (2021). Secure and efficient cloud-based IoT authenticated key agreement scheme for e-health wireless sensor networks. *Arabian Journal for Science and Engineering*, 46, 3017–3032. https://doi.org/10.1007/s13369-020-04905-9.

[18] Abdalla, M., Fouque, P. A., & Pointcheval, D. (2005). Password-based authenticated key exchange in the three-party setting. In Public Key Cryptography (PKC'05), Lecture Notes in Computer Science, (Vol. 3386, pp. 65–84). Berlin, Heidelberg, Les Diablerets, Switzerland: Springer.

[19] Al-Janabi, S., Al-Shourbaji, I., Shojafar, M., & Shamshirband, S. (2017). Survey of main challenges (security and privacy) in wireless body area networks for healthcare applications. *Egyptian Informatics Journal*, 18(2), 113–122.

[20] Wazid, M., Das, A. K., Odelu, V., Kumar, N., & Susilo, W. (2017). Secure remote user authenticated key establishment protocol for smart home environment. *IEEE Transactions on Dependable and Secure Computing*, 17(2), 391–406. [CrossRef]

[21] Wazid, M., Das, A. K., Odelu, V., Kumar, N., Conti, M., & Jo, M. (2018). Design of secure user authenticated key management protocol for generic IoT networks. *IEEE Internet of Things Journal*, 5, 269–282. [CrossRef]

[22] Adeli, H., & Jiang, X. (2008). Intelligent Infrastructure: Neural Networks, Wavelets, and Chaos Theory for Intelligent Transportation Systems and Smart Structures. CRC Press.

[23] Al-Turjman, F., Ever, Y. K., Ever, E., Nguyen, H. X., & David, D. B. (2017). Seamless key agreement framework for mobile-sink in IoT based cloud-centric secured public safety sensor networks. *IEEE Access: Practical Innovations, Open Solutions*, 5, 24617–24631.

[24] Farhad, et al., Ultra-low power reflection amplifier using tunnel diode for RFID applications, in Proc. IEEE Int. Symp. Antennas Propag. USNC/URSI Nat. Radio Sci. Meeting, Jul. 2017, pp. 2511–2512.

[25] D. Belo and N. B. Carvalho, An OOK Chirp Spread Spectrum Backscatter Communication System for Wireless Power Transfer Applications, IEEE Transactions on Microwave Theory and Techniques, vol. 69, no. 3, pp. 1838–1845, March 2021.

[26] Alabdulatif, A., Khalil, I., Forkan, A. R. M., & Atiquzzaman, M. (2018). Real-time secure health surveillance for smarter health communities. *IEEE Communications Magazine*, 57(1), 122–129.

[27] Christaki, E. (2015). New technologies in predicting, preventing and controlling emerging infectious diseases. *Virulence*, 6(6), 558–565.

[28] Srivastava, G., Dhar, S., Dwivedi, A. D., & Crichigno, J. (2019). Blockchain education. In 2019 IEEE Canadian Conference of Electrical and Computer Engineering, CCECE 2019, Edmonton, AB, Canada, May 5-8, 2019, (pp. 1–5). IEEE. , http://dx.doi.org/10.1109/CCECE.2019.8861828.

[29] Srivastava, G., Dwivedi, A. D., & Singh, R. (2018). PHANTOM protocol as the new crypto-democracy. In Saeed, K., & Homenda, W. (Eds.), Computer Information Systems and Industrial Management

- 17th International Conference, CISIM 2018, Olomouc, Czech Republic, September 27-29, 2018, Proceedings, in Lecture Notes in Computer Science, (Vol. 11127, pp. 499–509). Springer. http://dx.doi.org/10.1007/978-3-319-99954-8_41.

[30] Almuhaideb, A. M. (2021). Re-AuTh: lightweight re-authentication with practical key Management for wireless body area networks. *Arabian Journal for Science and Engineering*, 46, 8189–8202. https://doi.org/10.1007/s13369-021- 05442-9.

[31] Alzahrani, B. A. (2021). Secure and efficient cloud-based IoT authenticated key agreement scheme for e-health wireless sensor networks. *Arabian Journal for Science and Engineering*, 46, 3017–3032. https://doi.org/10.1007/s13369-020-04905-9.

[32] El-hajj, M., Chamoun, M., Fadlallah, A., & Serhrouchni, A. (2017). Analysis of authentication techniques in internet of things (IoT). In Proceedings of the 2017 1st Cyber Security in Networking Conference (CSNet), Rio de Janeiro, Brazil, 18–20 October 2017, (pp. 1–3). [CrossRef].

[33] El-hajj, M., Chamoun, M., Fadlallah, A., & Serhrouchni, A. (2017). Taxonomy of authentication techniques in internet of things (IoT). In Proceedings of the 2017 IEEE 15th Student Conference on Research and Development (SCOReD), Putrajaya, Malaysia, 13–14 December 2017, (pp. 67–71). [CrossRef].

82 AI music generator using deep learning algorithm : LSTM

K. J. S. Rama Raju[1,a], B. Harshitha[2,b], Chinmai Charitha, K.[2,c], D. Roshini[2,d], K. Lavanya[2,e] and P. Manasa Lukitha[2,f]

[1]Assistant Professor, Department of AI, Shri Vishnu Engineering College for Women, Bhimavaram, Andhra Pradesh, India

[2]Department of AI, Shri Vishnu Engineering College for Women, Bhimavaram, Andhra Pradesh, India

Abstract

Recent development in artificial intelligence has played a huge role in many fields, including music composition. This paper presents an AI-powered music composition system that adopts deep learning mechanisms, and more specifically, long short-term memory (LSTM) networks. Due to their powerful ability to maintain long-range dependency in sequential data, LSTMs are employed to study and comprehend complex melodic, rhythmic, and harmonic patterns of a large musical repertoire. Starting from an initial musical note or a contracted melodic line, the model creates independently an elegant and profound piece of music that appears closely like human fantasy. In order to increase the quality of generated works, primary hyperparameters are carefully tuned and ensured to uphold musical concepts like stability of rhythm, harmony balance, and structural coherence. The method is rated in terms of naturalness, smoothness, and expressiveness of the output, involving little human effort. This study emphasizes how artificial intelligence can transform the creation of music by making it accessible to people lacking professional music theory training. Through the availability of an accessible interface for creating expressive and well-structured pieces of music, technology-artistic imagination gap-filling music creation emerges through AI. Ultimately, this study leads us to the role of AI in democratizing music creation, towards greater inclusivity and greater potential for creativity for musicians and music lovers alike.

Keywords: AI music generation, deep learning, long short-term memory , neural networks, recurrent neural networks

Introduction

With the advent of deep learning and generative neural networks, the art of composing music has been greatly advanced. Through AI-models, the possibility of being able to learn from existing music, detect detailed musical patterns, and generate independent new compositions that are reminiscent of human thinking is now reality. Among different AI-based methods, long short-term memory (LSTM) networks—a sub-class of recurrent neural networks (RNNs)—have emerged as a highly useful tool for modeling highly complex musical patterns. Unlike general neural networks, LSTMs have the ability to recall long-term memory and hence are particularly suited to dealing with sequential data such as music.

This research examines the use of LSTM networks for AI-augmented music composition. The below model is learned from ABC notation, a symbolic notation that communicates musical sequences, time signature, duration, and articulation. Through repeated training over a heterogeneous database of musical works, the model comes to understand patterns progressively and can produce works that are stylistically coherent and consistent. Leveraging AI-based music creation, this work will fill the technology-creativity gap, introducing new opportunities for artists, producers, and creators of content.

Literature review

Currently, much research is being done combining AI and music composition when creativity/unpredictability is juxtaposed with computing technology in all its extractive capacities [1]. Several researchers have, over the years, tried to understand how AI could learn musical structures to compose naturally feeling, well-expressing-for-the-purpose pieces. Researchers in AI have continually transgressed the limits of especially recurrent neural networks on deep learning in weaving out possibilities for machines to create [2].

[a]kjsrraju36@gmail.com, [b]21b01a5413@gmail.com, [c]21b01a5420@svecw.edu.in, [d]21b01a5430@svecw.edu.in, [e]21b01a5457@svecw.edu.in, [f]peelalukitha42@gmail.com

DOI: 10.1201/9781003685364-82

The big step along that road was inaugurated by Mangal et al. with the generation of polyphonic music from MIDI files using LSTM networks. Their work proved that AI was capable of recognizing patterns in music and of composing multi-voiced melodies that made sense musically. The main difficulty arose when modeling long compositions. The generated pieces were quite promising, but the model could not sustain that structure on longer pieces. Music is not merely short patterns; it is transcendentally about flow and emotion keeping a listener hooked from beginning to end [3].

This is even more contextually valid in situations where Akhmedov et al. applied encoder-decoder architecture, long short-term memory, through AI concatenation procedures for the improvement of an AI music composition while allowing the model to remember longer sequences of musical sequencing. Although this worked well concerning minor composition works, they still lost coherence with lengthier works. This demonstrated the essential weakness of LSTM—the capacity to forget earlier parts in a given sequence. Hence, creating major works that have visible development and direction is seemingly almost impossible [4].

Most distinctively, Yamshchikov et al. took a different approach by using variational recurrent autoencoders (VRAEs) for melody generation. Rather than simply predicting the next note based on the preceding ones, theirs was a model that learned the underlying musical structures more abstractly. This resulted in compositions that seemed much richer and varied, thus far from being predictable or repetitive. Adding an element of chance and playfulness has made research afford AI-generated music expression one step nearer to human likeness [5].

Another significant but often neglected aspect of AI-generated music is how to evaluate it. How can one know when it sounds good? To this end, Sabathe et al. contributed a notable piece of work. Recognizing the extreme subjectivity of music, they devised specific metrics to assess harmonic consistency and rhythmic complexity, among others. Their work afforded a means of more objectively assessing AI compositions and fine-tuning the generative models toward better correspondence to human musical tastes [6].

The studies all point to artificial composition as an ongoing trajectory from computational techniques to aesthetics. One generation of paradigms is modeled upon recognizing patterns; the subsequent generation is about deeper, more expressive representational systems for music. With various improvements in transformer architecture and reinforcement learning, the future of AI music generation could not be brighter. What is clear is that the journey has only started, and AI is indeed not imitating music anymore; it is beginning to formulate one in its way [7].

Proposed methodology

Objectives and challenges

The main idea of this AI-based music generation project is to create an intelligent system capable of automatically producing music with musical coherence and creativity. Based on deep learning techniques, like LSTM neural networks, the project attempts to automate music composition with less effort and time spent in music creation [8].

Despite the potential of AI in music composition, there are certain hurdles that must be crossed to enhance the quality and usability of music composed using AI. One such major challenge is musical coherence, wherein AI models must offer smooth transitions and reasonable developments between notes and melodies. The second is achieving balance between structure and creativity—whereas AI might be capable of generating new patterns, there should be a surety that they are in adherence to existing musical rules. Big MIDI data sets also require strong data engineering practices to improve training efficiency in feature processing and extraction. Additionally, in real-time use cases such as video games or interactive media, AI-created music must dynamically adapt to changes in the environment while maintaining artistic consistency. Overcoming these issues will improve the AI model so that it is more efficient and versatile in generating high-quality musical compositions [9].

Model design

The AI music generator's model architecture depends on capturing sequential dependencies that occur in musical pieces. Since music is a structured sequence where a note is always determined by predecessors, the model must be able to learn them over time. For this purpose, LSTM networks, one of the subtypes of RNNs, are employed as they can learn and store long-term dependencies as well as handle vanishing gradient problems. The model has several LSTM layers as inputs that receive sequences, learn valuable patterns, and produce valuable musical outputs. Dropout layers are introduced at certain key positions to enable improved learning and avoid overfitting. The final layer employs a SoftMax activation function for the purpose of returning probabilities to possible future notes, as well as a TimeDistributed dense layer to ensure that music is sequential in nature. The architecture of such a system makes it possible for the model to generate well-quality, rhythmically accurate, and harmonically complex music pieces [10].

Data processing

Data preprocessing is a fundamental step when building an AI-based music generator. The initial step is to collect a large pool of varied MIDI files containing different pieces of music. MIDI files represent musical parameters like pitch, duration, velocity, and timing in a very structured manner, hence the best for training deep models.

Once the dataset is collected, MIDI files are parsed to collect reasonable music data. Parsing is done, which divides the MIDI files into note sequences and single notes and collects reasonable information such as note pitch, duration, and velocity.

Feature engineering is performed to convert the musical features extracted into model-training compatible format. It entails the conversion of music notes to numeric representations, handling silences and rests between the notes, and preparing the data in input-output pairs that can be learned from by the model. Appropriate preprocessing of data enables the AI model to learn to identify patterns in music and generate uniform compositions.

Architecture of the system

The architecture of the AI music generator is extremely well designed in regard to how it approaches the temporal aspect of music pieces. Because music is time-sensitive, the use of LSTM networks is employed because they can learn and remember long-term dependencies. LSTMs, as opposed to regular feedforward neural networks, use memory cells so that the model can maintain patterns for an extremely long duration, thus making them more appropriate for music generation.

A typical LSTM cell looks like one shown below:

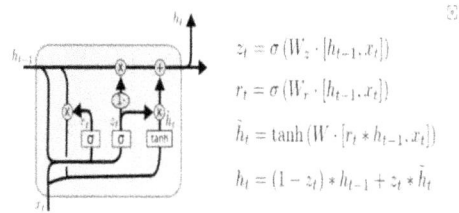

$$z_t = \sigma\left(W_z \cdot [h_{t-1}, x_t]\right)$$
$$r_t = \sigma\left(W_r \cdot [h_{t-1}, x_t]\right)$$
$$\tilde{h}_t = \tanh\left(W \cdot [r_t * h_{t-1}, x_t]\right)$$
$$h_t = (1 - z_t) * h_{t-1} + z_t * \tilde{h}_t$$

Figure 82.1 LSTM Model
Source: Author

The model has several LSTM layers one upon another to achieve the maximum learning capability. The initial layer receives the input sequence and extracts meaningful musical features, which are then handed over to lower layers for abstraction. The last LSTM layer generates a sequence of musical note predictions, which are converted into a neatly structured piece of music. In order to transfer smooth training and avoid overfitting, a single dropout layer is inserted that randomly switches off neurons in training and hence enhances generalization.

A TimeDistributed dense layer is employed in order to keep the temporal relation among consecutive notes in such a way that compositions thus produced maintain natural flow. Independent processing by each timestep and real-time predictions are made per step of sequence using this layer. The Adam optimizer is employed during training, with adaptive adjusting of learning rates to optimize the rate of convergence and accuracy. Moreover, a SoftMax activation function is employed in the output layer to assign a probability score to each possible note so that the most likely subsequent note in a sequence can be chosen.

Pros of LSTM:

Captures the sequential information present in the input sequence.

Cons of LSTM:

It consumes a lot of time for training since it processes the inputs sequentially.

Result and discussion

The LSTM model has been tested on different datasets and has been found to provide high-quality results in producing music that is expressive as well as coherent. The music produced has correct rhythm, harmony, and form, and is comparable to human-written music. Subjective listening tests

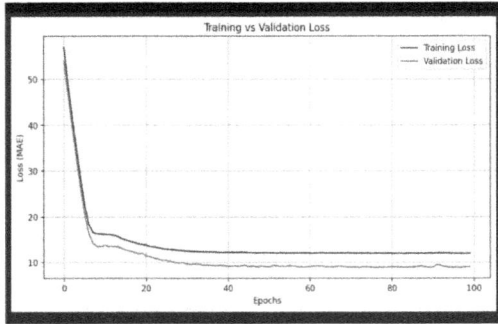

Figure 82.2 Comparison of Training and Validation
Source: Author

Training Accuracy

Mode	Correct predictions	Total predictions	Accuracy
Training	406824	448760	90.65513
Validation	110044	128224	85.82168
Testing	55016	64112	85.81232

conducted as a part of testing revealed that listeners could not differentiate between AI-generated music and music generated traditionally.

Furthermore, the generation of polyphonic music has also demonstrated greater complexity and depth, widening the potential of the model to be creative. Quantitative examination by measures of harmonic consistency and melodic variation also demonstrates the appropriateness of the model to generate structured musical pieces.

The ability of the model to adapt to different types of music through retraining on diverse datasets attests to its versatility and applicability in real-world scenarios. Future development will enhance note transitions and introduce dynamic variations to better create realism.

Conclusion

This study confirms that a deep learning model based on long short-term memory (LSTM) can generate structured and expressive music. Music composition becomes more accessible and less formally trained with the help of AI. The result confirms that music can be productively used in various creative areas such as game development, film scores, and commercials.

Future research will aim at the integration of transformer models, reinforcement learning techniques, and expanded training datasets to enhance AI-generated music further. Other developments involve the optimization of post-processing techniques for better sound quality and real-time AI-augmented composition software to bridge the human-artificial intelligence gap further.

Acknowledgement

The authors gratefully acknowledge the students, staff, and authority of AI department for their cooperation in the research.

References

[1]   A simple genetic algorithm for music generation by means of algorithmic information theory.In: Proceedings of the IEEE Congress on Evolutionary Computation (CEC), Singapore. IEEE. ,2007, 615622,ISBN: 1-4244-1340-0,DOI: 10.1109/CEC.2007.4424759.

[2]   Lipton, Zachary C., John Berkowitz, and Charles Elkan. A critical review of recurrent neural networks for sequence learning. arXiv preprint arXiv:1506.00019 (2015).

[3]   Joshi G, Nyayapati V, Singh J, Karmarkar A. A comparative analysis of algorithmic music generation on GPUs and FPGAs. In2018 Second International Conference on Inventive Communication and Computational Technologies (ICICCT) 2018 Apr 20 (pp. 229-232). IEEE.

[4.] Drewes F, Högberg J. An algebra for tree-based music generation. InInternational Conference on Algebraic Informatics 2007 May 21 (pp. 172-188). Berlin, Heidelberg: Springer Berlin Heidelberg.

[5]   Schulze W, Van Der Merwe B. Music generation with Markov models. IEEE MultiMedia. 2011 Jul 1;18(03):78-85.

[6]   Boulanger-Lewandowski, N., Bengio, Y., & Vincent, P. et al,Modelling temporal dependencies in high-dimensional sequences: Application to polyphonic music generation and transcription, Proceedings of the 29th International Conference on Machine Learning (ICML-12), 2012, 18811888.

[7]   Hadjeres G, Nielsen F. Interactive music generation with positional constraints using anticipation-rnns. arXiv preprint arXiv:1709.06404. Sep 19, 2017.

[8]   Mogren, Olof. "C-RNN-GAN: Continuous recurrent neural networks with adversarial training." arXiv preprint arXiv:1611.09904 (2016).

[9]   Kotecha, Nikhil, and Paul Young. "Generating music using an LSTM network." arXiv preprint arXiv:1804.07300 (2018).

[10] Chen, C-CJ, and Risto Miikkulainen. Creating melodies with evolving recurrent neural networks. IJCNN'01. International Joint Conference on Neural Networks. Proceedings (Cat. No. 01CH37222). Vol. 3. IEEE, 2001.

83 Enhancing resource utilization in cloud computing with machine learning-based predictive scaling

K. Bhanu Rajesh Naidu[1,a], K. Gurupavithra[2,b], R. Guna Sree[2,c], Y. Jahnavi[2,d] and N. Bhavana[2,e]

[1]Assistant Professor, Department of CST, MITS, Andhra Pradesh, India

[2]Student, Department of CST, MITS, Andhra Pradesh, India

Abstract

This project focuses effective cloud resource allocation and is essential for optimizing virtual machine (VM) performance. This research explores predictive modeling to estimate VM resource usage, focusing on CPU, memory and network utilization. A structured preprocessing pipeline is applied. Temporal patterns such as hourly usage trends and rolling-averages are integrated to improve predictive accuracy. Anomalous behavior, including unexpected resource spikes or system inefficiencies, is detected using techniques like the IQR and Z-score analysis. Multiple machine learning models, such as LGBM, RF, and SVM, are employed and assessed using evaluation metrics - R^2, MAE, and RMSE. To support real-time monitoring, a Streamlit based dashboard is designed, providing users with an interactive visualization tool to track trends, detect anomalies, and optimize cloud resource allocation. By integrating predictive analytics, anomaly detection, and dynamic visualization, this study presents an efficient approach for enhancing VM resource management in cloud infrastructures.

Keywords: AWS, cloud computing, dynamic scaling, resource utilization

Introduction

Cloud computing transformed IT infrastructure where it delivers scalable and adaptable computational resources. However, growing reliance on virtual-machines (VMs) introduces complexities in managing resources efficiently. Ensuring optimal allocation requires precise forecasting of CPU usage, memory consumption, and network activity. Traditional resource management approaches typically depend on static configurations and manual interventions, which often lead to inefficiencies either over-provisioning, which escalates costs, or under-provisioning, which degrades performance. To address inefficiencies, project leverages ML to evaluate historical data patterns and predict future resource demands, enabling data driven decision-making. The proposed approach integrates a structured data preprocessing pipeline that manages missing values, standardizes numerical attributes, and encodes categorical variables to enhance model reliability. Multiple ml models, including LGBM, RF, and Support Vector Regressor (SVR), are trained and fine-tuned using advanced optimization strategies such as grid search CV and randomized search CV, ensuring improved accuracy and adaptability.

Anomaly detection techniques, including Z-score analysis, facilitate the identification of unusual resource usage patterns, preventing potential inefficiencies. Additionally, data visualization methodologies provide in-depth insights into VM resource consumption trends. To enhance accessibility, a Streamlit based interactive dashboard is developed, offering real-time monitoring and a comparative analysis of actual versus predicted resource utilization. The system further allows for dynamic modifications in resource allocation, contributing to cost-effective and efficient cloud resource management.

Blockchain integration, as explored in [9], can complement resource management by adding secure logging. Future expansion could also consider IoT-cloud convergence [11] and hybrid security frameworks [12] to extend the model's robustness. Conventional methods, which rely on predefined thresholds and manual scaling, strug- gle to adapt to workload fluctuations, leading to suboptimal resource utilization and increased operational

[a]bhanurajesh9493@gmail.com, [b]gurupavitra541@gmail.com, [c]gunasreerepani@gmail.com, [d]jahnaviyellanki124@gmail.com, [e]bhavanaabhavs004@gmail.com

DOI: 10.1201/9781003685364-83

expenses. This project addresses these challenges by employing machine learning algorithms to anticipate resource consumption trends, enabling intelligent and automated resource management solutions.

Related work

Paper No.	Title	Summary	Contribution	Inference	Comparative analysis
[2]	ML-based prediction of resource utilization	Evaluates ML techniques for predicting cloud resource utilization.	Compares ML models for accuracy in cloud resource prediction.	ML-based predictions enhance scalability and efficiency.	Focuses on model comparison rather than optimization like [1].
[3]	Predictive resource allocation strategies for cloud environments	Discusses predictive allocation strategies for cloud resources.	Proposes an ML-based dynamic allocation strategy.	Predictive allocation reduces wastage and improves performance.	Unlike [1] and[2], focuses on allocation rather than forecasting.
[5]	The role of ai in cloud computing	Discusses AI applications in cloud automation and security.	Highlights AI benefits for efficiency and security.	AI enhances security and performance in cloud systems.	More general compared to [4], which focuses on resource management.
[6]	Evaluation ML algorithms on finding drinking WQ based on feature selection	Assesses ML models in data processing applications.	Compares ML algorithms for feature selection.	Feature selection is critical for ML optimization.	Indirectly relevant to cloud computing.
[7]	Secure framework cloud based education using deep neural networks	Proposes a deep-learning-based secure e-learning framework.	Implements DNN for cloud security in education.	AI enhances cloud security in education.	Related to [5], but focuses on e-learning security.
[8]	Novel hash based key generation for stream cipher in cloud	Introduces a cryptographic key-generation method for cloud security.	Develops a hash-based stream cipher security mechanism.	Cryptographic techniques enhance cloud security.	Similar to [7] but focuses on cryptographic security.
[10]	Secured data storage in cloud computing using BD based key agreement	Proposes acryptographic method for secure cloud storage.	Introduces a key agreement mechanism for cloud security.	Cryptographic methods improve cloud data security.	Similar to [8] but focuses on key agreement rather than encryption.
[13]	A cloud computing solution for advanced metering infrastructure	Uses cloud computing for smart grid and metering infrastructure.	Proposes a cloud-based energy management system.	Cloud improves efficiency in metering infrastructure.	Relevant to cloud computing but focuses on energy management.

Proposed method

proposed system is designed to predict and monitor virtual-machine (VM) resource utilization cloud environments by leveraging machine learning and data visualization techniques. The process begins with dataset preprocessing, which involves managing values missing, numerical scaling attributes, and encoding categorical variables like task type, priority, and status. Temporal attributes, such as the hour of the day and day of the week, are extracted from timestamps to capture usage patterns over time. Additionally, rolling averages are computed for key metrics like CPU and memory usage.

For predictive modelling, the system utilizes three machine learning algorithms LGBM, RF Regressor, and SVR forecast resource consumption based on historical data. Model performance is assessed using key evaluation metrics, including R^2 score, MAE, and RMSE. Enhance predictive accuracy, hyperparameter tuning is performed using grid search CV and randomized search CV.

Anomaly detection is integrated into the system using statistical methods like the IQR and Z-score analysis, helping identify unwanted resource use which indicate inefficiencies or potential system failures. Additionally, an interactive Stream lit-based dashboard is developed for real-time data visualization, enabling users to analyze resource trends, filter data by VM ID and date range, and make informed decisions regarding resource allocation.

This system provides a comprehensive approach to VM resource prediction, real-time monitoring, and anomaly detection, equipping users with data-driven insights for optimized cloud resource management. Furthermore, it can be integrated with cloud orchestration tools to enable automatic resource scaling based on predicted usage trends. Periodic retraining of the machine learning models ensures adaptability to evolving workloads, while feature importance analysis aids administrators in refining resource allocation strategies. By combining predictive analytics with interactive visualization, this solution bridges the gap between raw data insights and effective decision-making. Future improvements could involve incorporating deep learning techniques to enhance prediction accuracy and detect more complex usage patterns in cloud environments.

Performance matrix

The performance of machine learning models in forecasting VM resource utilization was evaluated using standard assessment metrics. The R^2 score was employed to gauge how effectively the model accounts for variance in the data, with higher values signifying a superior fit. MAE quantified the average deviation between predicted and actual values, where lower values signified more accurate predictions. Additionally, mean squared error (MSE) was employed to quantify the squared deviations between predictions and actual observations, emphasizing larger errors.

Table 83.1 Evaluation metrics ML models.

Model	RMSE	MAE	R2 score
LGBM	0.2580	0.2149	0.0978
RF	0.2571	0.2152	0.1040
SVM	0.2544	0.2127	0.1231

Source: Author

Preprocessing
Includes missing value handling, outlier removal, feature engineering, and scaling for better model performance.

Temporal patterns
Capturing hourly trends and rolling averages helps smooth fluctuations and improve forecast accuracy.

Anomaly detection
Techniques like IQR, Z-score, Isolation Forest, and Autoencoders detect unusual resource usage patterns.

Why IQR and Z-score
IQR identifies extreme values, while Z-score detects statistical outliers based on deviation from the mean.

Table represents a comprehensive quantitative analysis of the predictive performance of three ml models: LGBM, RF and SVM for forecasting cloud resource usage. The evaluation is conducted using three essential regression metrics: RMSE, MAE and R^2 score.

RMSE is used to determine average magnitude of prediction-errors, placing greater emphasis on larger deviations, making it particularly useful for identifying impact of outliers. In contrast, MAE measures the MAE differences between forecast and real values, providing a straightforward measure of overall prediction accuracy. R^2 score assesses how effectively, model explains variance in actual data, where values approaching 1 indicate a stronger correlation between predictions and real observations.

A model with lower RMSE and MAE values signifies reduced prediction errors, while a higher R^2 score reflects improved explanatory capability. These metrics collectively offer thorough assessment of reliability and effectiveness in predicting resource consumption in cloud environments.

Results

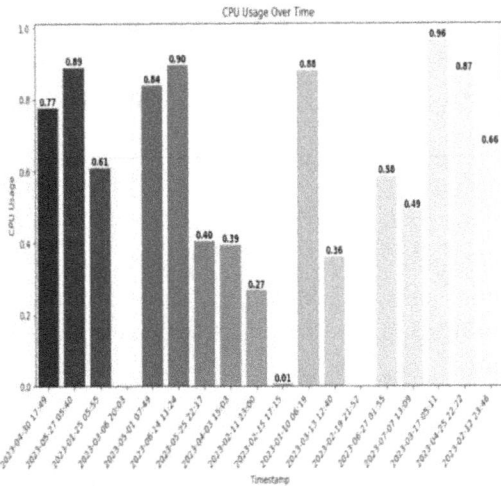

Figure 83.1 CPU usage over clock
Source: Author

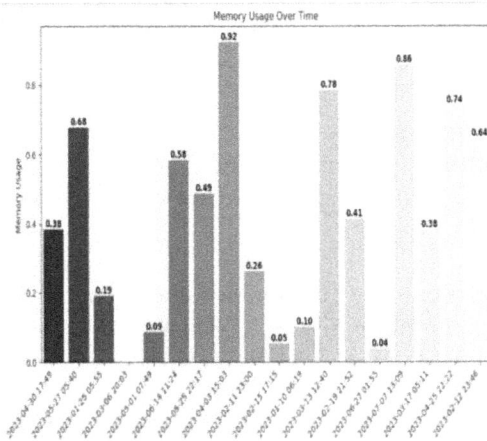

Figure 83.2 RAM usage over clock
Source: Author

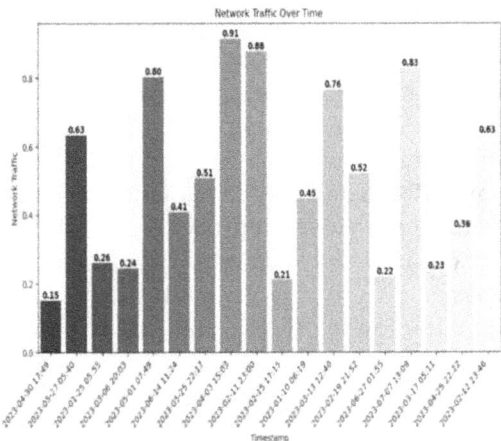

Figure 83.3 NT over clock
Source: Author

```
LightGBM Predictions:
        cpu_usage   memory_usage   network_traffic
3736    0.217714    0.551827       0.413959
5927    0.693761    0.644453       0.597940
850     0.674576    0.340554       0.621763
4689    0.648181    0.442916       0.500765
2157    0.570250    0.431367       0.442183

RandomForest Predictions:
        cpu_usage   memory_usage   network_traffic
3736    0.223888    0.589045       0.462884
5927    0.638720    0.582425       0.605151
850     0.624104    0.359849       0.574773
4689    0.582907    0.542307       0.485953
2157    0.577309    0.555835       0.466395

SVM Predictions:
        cpu_usage   memory_usage   network_traffic
3736    0.264118    0.479609       0.493457
5927    0.627401    0.614011       0.477717
850     0.602921    0.408913       0.534952
4689    0.563029    0.471594       0.498452
2157    0.573643    0.457230       0.485611
```

Figure 83.4 Model predictions for resource usage
Source: Author

```
*  Original Values (Before Optimization):
    cpu_usage   memory_usage   network_traffic   processing_time
0   68.812098   4.783718       129.681268        5.910969
1   53.277635   4.355145       390.542456        6.820251
2   50.278945   6.598214       156.890127        3.724929
3   68.078880   8.071490       388.809131        5.291316
4   97.959564   5.089378       220.048255        3.004676

*  Optimized Values (After Prediction/Normalization):
    cpu_usage   memory_usage   network_traffic   processing_time
0   43.685443   3.036949       82.328309         3.752586
1   37.061884   3.029599       271.675705        4.744417
2   34.651230   4.547355       108.125494        2.567145
3   43.653360   5.175580       249.311171        3.392884
4   71.501853   3.714797       160.615844        2.193149

*  Percentage Reduction in Metrics:
    cpu_usage   memory_usage   network_traffic   processing_time
0   25.126654   1.746769       47.352959         2.158383
1   16.215752   1.325546       118.866751        2.075833
2   15.627716   2.050859       48.764633         1.157783
3   24.425519   2.895910       139.497960        1.898432
4   26.457711   1.374580       59.432411         0.811527

🔲 Summary of Mean Reduction Across All Samples:
cpu_usage        22.541494
memory_usage      3.004847
network_traffic  89.991524
processing_time   1.500781
dtype: float64
```

Figure 83.5 System resource optimization summary
Source: Author

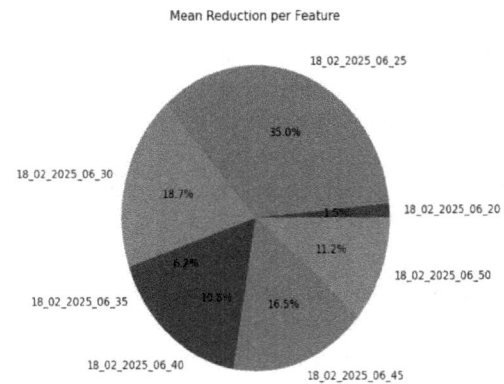

Figure 83.6 Streamlit dashboard
Source: Author

Conclusion

The developed system effectively leverages machine learning-models, including LGBM, Random Forest, and support vector regressor (SVR), to predict virtual machine (VM) resource utilization. Predictive analysis of models was thoroughly measured using key metrics like R^2 score, MAE, and RMSE, ensuring accuracy and reliability in forecasting resource consumption.

Further enhance model efficiency, hyperparameter tuning was applied, improving generalization and refining prediction precision. Optimization enables more effective resource allocation in cloud environments, reducing inefficiencies and enhancing system performance. By providing accurate forecasts, the system supports informed decision-making in VM operations, ensuring resources are allocated efficiently, minimizing wastage, and optimizing workload distribution.

Additionally, integration of machine learning techniques offers a scalable and adaptable solution that can be extended to various cloud computing applications. Future improvements could involve incorporating additional predictive features, refining model architectures, and enhancing adaptability to handle diverse workloads and complex operational conditions. By continuously improving prediction accuracy and expanding its capabilities, this system has the potential to significantly enhance cloud resource management strategies.

References

[1] Smendowski, M., & Nawrocki, P. (2024). *Machine learning-driven multi-time series forecasting for cloud resource optimization.* Knowledge-Based Systems, 283, 112489. https://doi.org/10.1016/j.knosys.2024.112489.

[2] Shaikh, R., Muntean, C. H., & Gupta, S. (2024). *Resource utilization prediction in cloud computing using machine learning techniques.* Proceedings of the 21st International Conference on Cloud Computing and Services, pp. 102–108.

[3] Kamble, T., Deokar, S., Wadne, V. S., & Gadekar, D. P. (2023). AI based predictive resource allocation strategies for cloud environments. *Journal of Electrical Systems*, 19(2), 68–77.

[4] Kanungo, S. (2024). AI enhanced cloud resource management strategies: systems, services, and applications. *World Journal of Advanced Technology and Sciences*, 11(2), 559–566.

[5] Vedpathak, D. N., Saraogi, R. R., & Ladake, N. (2022). The impact of AI in optimizing cloud computing performance. *International Journal of Creative Research Thoughts*, 10(5). ISSN: 2320-2882.

[6] Priya, K. D., Sai, S., Pagadala, V. G. R., & Kumar, P. (2023). A comparative study of ML algorithms for assessing water quality using feature selection. In Proceedings of the 9th ICACCS, Coimbatore, India.

[7] Priya, K. D., & Sumalatha, L. (2021). Secure cloud-based e-learning framework using deep learning models. In ICIEM, London, United Kingdom, (pp. 406–411).

[8] DeviPriya. K., & Sumalatha, L. (2017). Advanced cryptographic key generation for secure cloud data transactions. In Proceedings of IC3T 2016. Springer Singapore.

[9] Simpson, S. V., Raju, Y. R., Bhanu Rajesh Naidu, K., & Venu, G. (2023). *Blockchain powered trust & reputation system for IoT security.* Bharati Vidyapeeth's Institute of Computer Applications and Management (BVICAM), Vol. 12(4), pp. 221–229.

[10] Bhanu Rajesh Naidu, K., Devi, M. V. A., Vulpala, P., & Kotha, M. (2023). Secure data storage in cloud computing using novel key agreement mechanisms. *International Journal of Scientific Research in Computer Science Engineering and Information Technology*, 9(4), 231–241.

[11] Botta, A., Donato, W. D., & Persico, V. (2016). The convergence of cloud computing & IoT: a comprehensive review. *Future Generation Computer Systems*, 56, 684–700.

[12] Gandi, Charan, K. D., Bhanu Rajesh Naidu, K., V. V. S. Tirumalesh, & Raju, R. (2022). *A hybrid security framework for enhancing computational efficiency in cloud servers.* In Proceedings of the 4th ICERECT, Coimbatore, India, pp. 88–95.

[13] Yan, W., & Su, W. (2016). A cloud-based solution for advanced metering infrastructure. In IEEE/PES Transmission and Distribution Conference and Explosion, May 2016, (pp. 1–4).

84 Implementation of OCI-based communication system using HDL

CH. Kutumba Rao[1,a] and Malin, Shaik[2,b]

[1]Associate Professor, Department of ECE, in Sree Vahini Institute of Science and Technology, Tiruvuru, Andhra Pradesh, India

[2]PG Scholar, Department Of ECE, in Sree Vahini Institute of Science and Technology, Tiruvuru, Andhra Pradesh, India

Abstract

Given the significant limitations of conventional approaches to interposes communication, network-on-chip (NoC) solutions offer great promise for the future of system-on-chip (SoC) intercommunication infrastructure. On the other hand, several concerns about performance parameters a system like this scalability, latency, power consumption, and signal integrity make NoC design very difficult. After outlining problems with the router's memory unit, this article goes on to suggest an improved memory architecture. Shared RAM and virtual channels for FPGA-based NoC use FIFO buffers to provide efficient data transmission. We propose an improved FIFO-based memory unit for use in NoC routers and test its efficacy in dual-direction NoC (Bi-NoC). The primary goal of this work is to enhance the internal structure of FIFO while simultaneously reducing the strain on the router. A self-configurable interaction are limited channel is suggested for Bi-NoC to increase data transmission speeds.

Keywords: Bi-NoC, FIFO, network-on-chip, router, switch allocator, system-on-chip, virtual channel

Introduction

In order to meet the demands of real-time applications with high performance, the field-programmable gate array (FPGA) industry is rapidly expanding. But when the CMOS channel length decreases, the size and yield loss both increase. Moore suggested doubling the number of units in the system at each Consequently, eighteen months were devoted to focusing on various elements of VLSI application design. Chip multiprocessor (CMP) or multiprocessor systems on chip (MPSoC) designs are suggested to handle quicker time-to-market pressure, higher operating frequencies caused by growing transistor density, and other similar issues.

In order to merge complex heterogeneous functional components onto a single chip and allow on-chip communication, the semiconductor industry is seeing a surge in demand for bus topologies. The type of enormous performance required by MPSoC and CMP is beyond the capabilities of traditional communication systems. One such solution to problems with inter processor communication is network-on-a-chip (NoC). Components (PEs) of

systems-on-chip (SoC). Although existing ways of linking FPGAs, such as crossbars, have limited scalability, the notion of It's not new to use NoC as a means of communication. Therefore, creating new NoC is the best option for FPGAs architectures. In order to obtain high performance without sacrificing speed and throughput, researchers continue to face the tough task of designing innovative NoC architecture. The term network centricity (NoC) is used to describe an architecture for intercommunication that is keeping solutions in mind for communication-centric trends. The data is often sent among PEs in a typical NoC design using many lengths of cable and a variety of routers. As part of the tile-based design, NoC is arranged in a city-block fashion, with PEs spaced apart by wires laid out on city blocks, and routers and cables configured in a grid similar to that of streets.

The network interface (NI) converts data packet PEs into control flow digits (flits) of a set length, making it one of the most essential design constraints for NoC alongside routers and PEs. The header, body, and tail flits are the primary components of a

---

[a]Kutumbarao15@gmail.com, [b]Malin.shaik@gmail.com

DOI: 10.1201/9781003685364-84

data packet. Current router to neighbor router control mechanisms forward and route these flits towards the target. There are five bidirectional ports for inputs and outputs in North, east, south, west, plus the nearby port of the relevant PE make up a city-block orientated tile NoC router. A network of cables physically connects each bidirectional port to the port next to it, allowing for efficient interaction between the ports. The router is the fundamental component of NoC as it handles data transfer and management using various methods. To demonstrate how the router worked, a 5X5 cross switching was used. The crossbar switch reads control logic and sends data from the selected input port to the specified output port. Selecting a neutral third party to serve as arbiter given the relative significance of the input ports, the most appropriate port for data transfer should be considered. Therefore, a standard NoC router will include five I/O ports, a cross-bar switch, or arbitration modules that link to PEs that happen to be currently interacting with each other. Figure 84.1 depicts an internet operation center based on a 3 × 3 mesh with 5 × 5 unidirectional ports in the middle.

The data packet starts its journey via NI from the source intellectual property (IP), which the current router receives and then passes on to its neighboring routers, which is positioned towards the destination IP. The routing algorithm that is built into every router determines how data packets are routed from their origin to their final destination. Data is often sent from source to destination in a conventional NoC using an X-Y deterministic routing method. Because it is difficult to measure the distance from a source router to a destination router when the routers are not correctly aligned, topology is a significant limitation for routing algorithms. Because it is easy to construct, mesh topology is most often employed by routing algorithms. Figure 84.1 shows the source-to-destination X-Y routing method in a mesh topology. Original IP quantifies the The number of routers needed to lock the path and determine the distance to the destination until the data packet reaches the destination. Data packets are pre-formatted with the destination address and details about intermediate routers before transmission begins.

This study examines devices that need sufficient SoC and highlights the drawbacks of memory-based NOC. logic cells (LCs), which are similar to a 4-input lookup table (LUT), are used to measure the size of an FPGA.

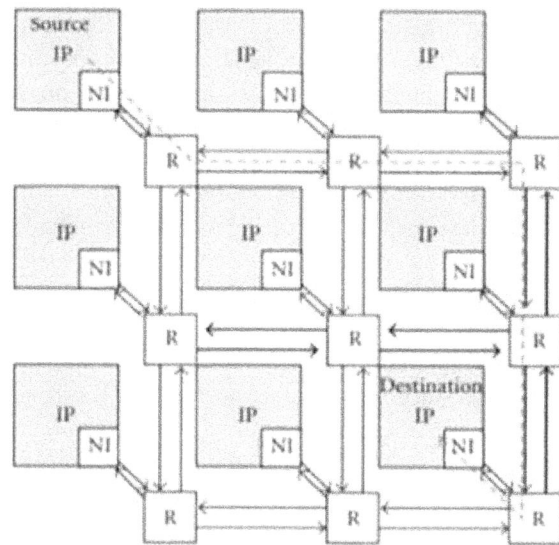

Figure 84.1 A network operating center design based on a 3×3 mesh
Source: Author

Related work

For general purpose intercommunication among PEs, the on-chip network of connections is suggested instead of on-chip wires. Instead of using data packets are sent between the PEs using an on-chip network via specialized lines.

By maximizing electrical qualities and ensuring correct maintenance, an on-chip network makes use of global cables. By keeping the signal intact and taking up minimal space on the chip, the regulated electrical wires have low crosstalk qualities and save power usage. At the end of the article, W. J. Dally or B. Towles discuss several on-chip research areas, including NI, topologies, and flow control approaches. Unbalanced traffic is Sunetal's suggested topology for the dynamic router. The sophisticated router's inter- and intra-port buffer system makes good use to avoid obstructing the head-of-the-line buffer resources and stays mistake. When compared to the existing router and its neighbor, the suggested router significantly improves the traffic balance.

As a consequence of increasing buffer utilization when coupled with a Flip-Flop (FF), the simulation results demonstrated balanced traffic across ports.

One typical SoC module, the memory controller, may need thousands of LCs. manufacturers of field-programmable gate arrays (FPGAs) and offer a network-oriented controller (NoC) based on RingNet. Unique to its design, the RingNet-based memory-oriented NoC communicates with one another via Efficient use of FPGA resources is made possible by centrally placed memory, which also needs a buffer and eliminates network congestion. To provide assured throughput, network access, and predictable latency, an efficient switching technology called virtual cut-through (VCT) and distributed RAM are used to construct a network buffer. Distributed storage and VCT-based data switching are two examples of little memory in switches. and 3-port switches organized into a sort of tree topology optimize the FPGA. As a result, the highest possible clock frequency and resource usage of RingNet-based NoCs are both enhanced. One disadvantage of using Ring Net based NoCs in complicated SoCs is power consumption. This study presents an improved FIFO buffer that is compatible with the virtual channel buffered method, following the guidelines of previous research.

Advanced NoC design

This section of the article details the NoC router, which consists of an upgraded FIFO buffer and a Bi-NoC with a self-reconfigurable channel. Following the flow control method, we will go over the architecture of a state-of-the-art router. Buffering enhances the efficacy of flow management; in contrast, the buffered-less method leads to data packets being erased or misrouted when many channels are attached to a single packet. There are two types of buffered flow, two types of control for packet flow, and two types of control for light flow. Packet flow includes features like store-and-forward and virtual cut-through, while flite flow controls using wormholes and virtual channels. In Figure 84.2, we can see the three stages of control of information in a Bi-NoC router: the routing computation (RC), SA, and ST. The RC module sends an authorization request to SA so that data may be sent on each buffer. It does the same thing depending on what's in the FIFO buffer. The channel is allocated by SA, and data is sent to the ST stage whenever there is empty space in the buffer of the neighboring router.

During the ST stage, data flits are sent from the input side to the resultant port via a crossbar switch. We provide a data switching technique based on virtual channels that protects data flows and creates a buffer to avoid head-of-line (HoL) mistakes.

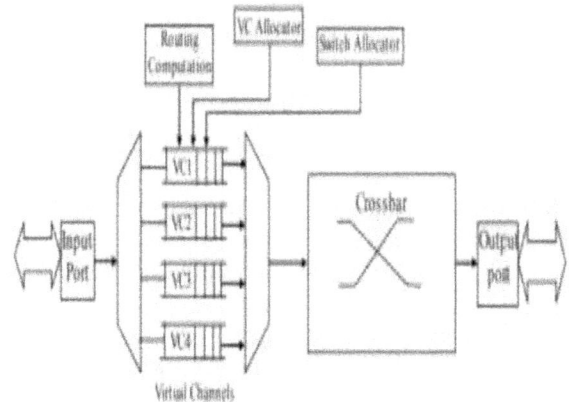

Figure 84.2 Standard Bi-NoC structure
Source: Author

Every time a data flight's neighbor router is unable to receive the whole data flight, the virtual channel is allocated to it under virtual flow control. The RC stage decodes and sends requests in the relevant direction once head flite enters the queue of the buffer over the virtual channel. The traditional FIFO router in the middle-caused traffic to spike dramatically since the following flight had to wait until the current one was transferred to a neighboring routers.

The suggested solution to this problem involves breaking the FIFO architecture into its component parts and dividing the physical channel into a particular number of virtual channels. As a result, both the data transmission speed and the next flight waiting time are enhanced.

Data flights may experience conflict due to the high volume of requests attempting to use the same virtual channel. The current router should have the flites that access the virtual channel, as the old router was banned owing to conflict. At the end of the VA stage, data packets are assigned to the SA stage, which is responsible for presenting the physical channel to the neighbor router. Data packets that are not in use will never obstruct data packets that are ready to be sent to the destination because virtual channels are multiplexed to buffer. Final destination via physical conduit. When a physical channel in Bi-NoC is in use or blocked, Figure 84.3 shows how data may still be sent via virtual channels. Throughput is increased and deadlock errors are prevented by this structure, which allows numerous virtual channels to communicate with

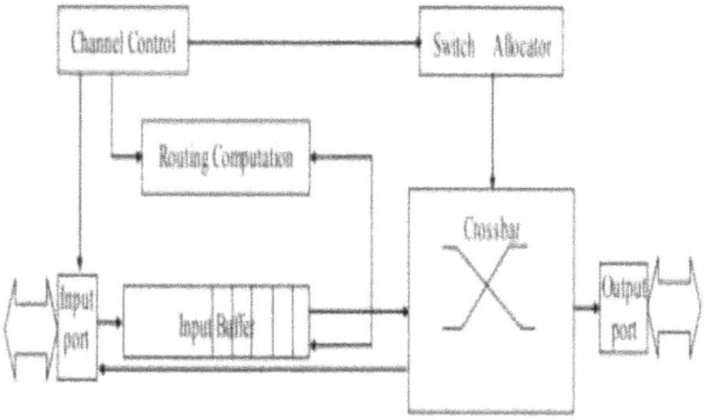

Figure 84.3 Allocation of virtual channels in the architecture of Bi-NoC
Source: Author

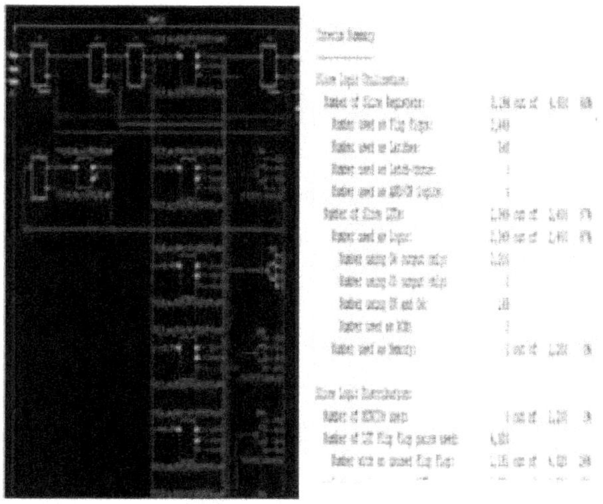

Figure 84.4 RTL Schematic diagram
Source: Author

physical channels of connected buffers. Figure 84.3 shows the passage of the virtual channel in a router from its input port to its output port. After a high-priority incoming flit reaches the neighbor router over the right virtual channel, the whole data packet is handled. As soon as a data packet's first flit—the head flit—reaches the top of the buffer's virtual channel queue, it enters the RC stage. During the RC stage, it decodes the data and generates the correct request direction for the target router. In order to acquire the desired virtual channels for transmission to the target router, the flit request moves to the VA step.

Existing Method

An enhanced A FIFO buffer and a bi-state network with customizable gates lanes are built into the construction of the NoC router in the present technique. We will start with the flow control mechanism and move on to the advanced wireless network. The paper's data flow control system makes use of a buffered approach, since this improves the effectiveness of flow control by decoupling channel allocation. In contrast, a buffer-less strategy would cause data packets to be if a single packet was allocated two channels, it may be misrouted or destroyed.

There are two distinct kinds of buffered flows: packet flows and light flows, and there are two distinct kinds of controls for each of these flows. Packet flows include features like store-and-forward and virtual cut-through, whereas flite flows include features like wormholes and virtual channels.

Disadvantages

- Queuing incoming packets requires complex FIFO memory.
- Energy efficiency is really low.

Proposed method

In order to evaluate the performance in terms of filled area, latency, power consumption, and throughput, an advanced FIFO architecture on NoC is simulated and synthesized in Xilinx 14.5 ISE and implemented on a Vertex-6 FPGA device. The memory use of the FPGA is accomplished by first designing a single router and subsequently a mesh-based node-computing (NoC). The input/output ports, arbitrator, crossbar, and channel control modules make up a single NoC router, as shown in the register transfer level (RTL) diagram in Figure 84.4. The picture also breaks down the various component uses in terms of memory units. The various NoC modules were developed independently in Verilog is hardware description language (HDL) and then combined. To facilitate comparisons between the two designs, an advanced queuing buffer is created with both the normal NoC and the bidirectional NoC in mind. Area utilization (in terms of the number of slices, registers, LUT-FF pairs, and slices occupied), latency utilization is all examined in the simulation findings. with respect to the amount of random-access memory (RAMs), and lastly, throughput measured in flits per second, node the reslt has been in Figure 84.5 and 84.6.

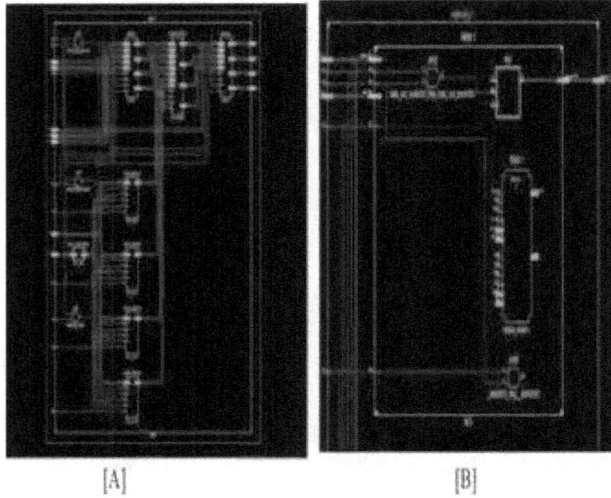

Figure 84.5 Schematic representations of Bi-NoC at the RTL level
Source: Author

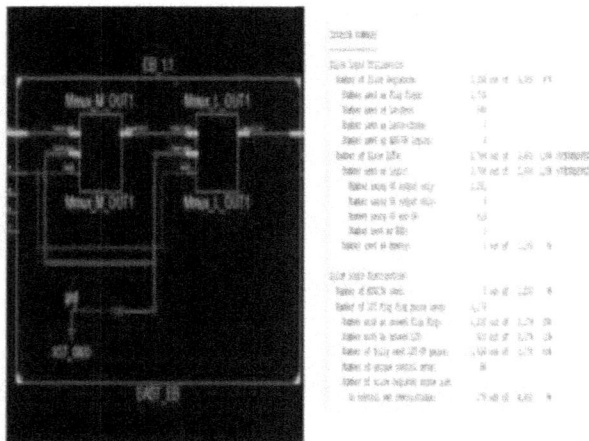

Figure 84.6 Schematic representations of Bi-NoC at the RTL level
Source: Author

The size, latency, and power consumption of the NoC router were measured using the results obtained from the FPGA setup. Figuring out that data jumps are used to transmit data packets from one location to another and that neighboring routers share a queued buffer makes it evident that the suggested architecture has minimal space overhead. Active components utilize the buffer, while idle modules do not, thereby reducing the memory unit, which includes the number of RAMs. With greater reduced latency and increased operational frequency are two benefits

Table 84.1 Evaluation of existing and proposed.

Architecture	Resources Utilized		Delay (ns)
	LUTs	FFs	
RingNet [24]	497	884	0.752
Bi-NoC	327	729	0.531

Source: Author

of the proposed design, which is based on the number of available channels between the source and the destination.

When compared to the current work source and destination, overall power usage is somewhat higher. Virtual channels raise dynamic power consumption during data packet transmission, which causes a little increase in overall power consumption compared to previous studies [2-4].

Results

In terms of effective resource utilization, Table 84.1 clearly compares Ring Net with sophisticated Bi-NoC. When compared to Ring Net, the resource utilization, measured in terms of filled LUTs and FFs, is much lower. Because neighbor routers share a queued buffer and data transported according to the data in the table, improved Bi-NoC uses far less memory than its predecessor did in terms of flits.

Conclusion

It eliminates obstacles to bus-based communication and provides a means for SoC intercommunication, including parallel communication lines. This study proposes and implements an enhanced memory unit in Bi-NoC that provides high performance in regard to maximum operational bandwidth and reduced memory demand of the buffer. There was a 28% improvement in delays and a 17% improvement in resource utilization compared to the prior work. The utilization of resources exceeds the anticipated effort due to Ring Net's employment of the Round Robin arbiter. Since neighboring routers share the queued buffer and split data packets into a certain number of flits, the amount of buffer space needed to transmit data from origin to destination is reduced. With this state-of-the-art router's integrated Bi-NoC setup, data transmission speeds are much greater than standard NoC.

In the event that a physical channel is unavailable, routers may minimize data packet delay and prevent deadlock errors by creating virtual channels.

When compared to previous efforts, the outcomes of the implementation are better at using resources. An AI application in the future will employ a processor based on a network of connected devices (NoC). Because The NoC's performance must be improved by updating the router components due to the increasing power consumption caused by the developed FIFO structure's simulated channels.

References

[1] Dally, W. J., & Towles, B. (2001). Route packets, not wires: on-chip interconnection networks. In Proceedings of the 38th Design Automation Conference, (pp. 684–689). Las Vegas, Nev, USA.

[2] Shin, E. S., Mooney III, V. J., & Riley, G. F. (2002). Round- robin arbiter design and generation. In Proceedings of the 15th international symposium on System Synthesis (pp. 243–248). ACM.

[3] Ashok Kumar, K., & Dananjayan P. (2017). A survey on silicon on chip communication. *Indian Journal of Science and Technology*, 10(1), 1–10. ISSN: 0974-5645.

[4] Raparti, V. Y., & Pasricha, S. (2019). Approximate NoC and memory controller architectures for GPGPU accelerators. *IEEE Transactions on Parallel and Distributed Systems*, 31(5), 25–39.

[5] Goebel, M., Behnke, I., Elhossini, A., & Juurlink, B. (2018). An application-specific memory management unit for FPGA-SoCs. In 2018 IEEE International Parallel and Distributed Processing Symposium Workshops (IPDPSW), (pp. 222–225). IEEE.

[6] Jang, H., Han, K., Lee, S., Lee, J. J., & Lee, W. (2019). MMNoC: Embedding memory management units into network-on-chip for lightweight embedded systems. *IEEE Access*, 7, 80011–80019.

[7] Ashok Kumar, K., & Dananjayan, P. (2019). Parallel overloaded CDMA crossbar for network on chip. *Facta Universitatis, Series: Electronics and Energetics*, 32(1), 105–118.

85 Innovative techniques for small object detection in cardiac images for precise early diagnosis

Anil Kumar, N.[1,a], Enamala Varsha[2,b], Chalampalem Sandhya[2,c], Yerragadindla Aravind[2,d], Sennamsetti Kodanda Ramu[2,e] and Karanam Jahnavi[2,f]

[1]Assistant Professor, Department of ECE, Mohan Babu University, Tirupati, Andhra Pradesh, India

[2]Student, Department of EIE, Mohan Babu University (Erstwhile Sree Vidyanikethan Engineering College), Tirupati, Andhra Pradesh, India

Abstract

Early identification of cardiac issues is essential in preventing severe cardiovascular diseases. This work suggests a novel approach for spotting and classifying minute objects, such anomalies, in cardiac images, therefore helping in early diagnosis and intervention. Among the image processing and machine learning methods included into the proposed approach are input image preprocessing, edge detection, boundary extraction, KAZE feature extraction, region mapping, morphological analysis, ensemble learning, and convolutional neural network (CNN) classification. Then applied are methods of decision-making. First preprocessing to enhance quality and reduce noise helps the input cardiac images; secondly, edge detection is applied to identify potential areas of interest. Boundary extraction methods then help to create more exact object boundaries. KAZE feature extraction then helps to extract discerning elements from the chosen areas. After that, a region mapping approach divides and labels little items seen inside the cardiac images. Morphological study is done to improve classification accuracy and adjust the found zones. Then, using an ensemble learning approach, several classifiers are combined to achieve better performance. A CNN classifier is trained utilizing the gathered properties to classify the observed objects into relevant groups therefore facilitating automated diagnosis. Finally, a decision-making process is applied to evaluate the classification results and provide healthcare professionals with pertinent knowledge. By means of effective identification and categorization of minute objects in cardiac images, the proposed approach offers a consistent means of early diagnosis of heart diseases. Experimental evidence indicates that the proposed approach is successful in raising diagnostic efficiency and accuracy, so improving patient care and prognosis in cardiovascular medicine.

Keywords: Boundary extraction, cardiac imaging, early diagnosis, edge detection, KAZE feature extraction, preprocessing pictures, region mapping

Introduction

Advanced diagnostic techniques must be created in order to treat and intervene early because cardiovascular diseases (CVDs) remain a leading cause of death worldwide. In recent years, medical imaging has proven to be a helpful tool for early cardiac issue detection, facilitating timely medical interventions and improving patient outcomes. However, it is challenging to precisely identify and categorize small things, such as abnormalities, in cardiac images due to their delicate appearance and complex anatomical structure. This study suggests a novel approach for small item recognition and classification in cardiac images, which will help with early diagnosis and prognosis in cardiovascular medicine [1]. The proposed methodology incorporates contemporary image processing and machine learning techniques, including (CNN) classification, edge detection, boundary extraction, KAZE feature extraction, region mapping, morphological analysis, ensemble learning, input image preprocessing, and decision-making mechanisms. The recommended approach starts with preprocessing the input cardiac images to enhance their quality and reduce noise, which boosts the effectiveness of subsequent analysis. After identifying potential zones of interest using edge detection techniques [2], boundary extraction is utilized to more precisely define object boundaries. Using KAZE feature extraction to obtain discriminative features from the detected regions

[a]anil.kumar@mbu.asia, [b]varshaenamala123@gmail.com, [c]chalampalemsandhya3@gmail.com, [d]aravindyerragadindla@gmail.com, [e]sennamsettiram41@gmail.com, [f]janujahnavi95656@gmail.com

DOI: 10.1201/9781003685364-85

enables more accurate object recognition and classification. Morphological analysis is then performed to enhance object segmentation, and a region mapping technique is employed to further refine the discovered regions and improve classification accuracy [3]. The use of ensemble learning techniques allows the combination of many classifiers, which refine the overall system operation. Additionally, a CNN classifier is trained using the retrieved attributes, which classifies the found items into relevant groups and offers automated diagnosis capabilities [4]. The classification results are also analyzed using a decision-making process to provide medical professionals with relevant data that aids in clinical decision-making and treatment planning. Utilizing state-of-the-art image processing and machine learning techniques, the proposed approach offers a practical means of early detection and classification of small objects in cardiac images, thereby enhancing patient care and cardiovascular medicine outcomes.

Mathematical model

Mathematical model of SODICI

Figure 85.1 illustrates pipeline with pre-processing, feature extraction, and classification steps is the main goal in order to efficiently identify and categorize small items in medical images that may be signs of cardiac problems. The goal is to create methods that can effectively manage these difficulties and yield precise results for a variety of imaging modalities and patient types. For cardiovascular disorders to be effectively managed and treated, cardiac abnormalities must be identified early.

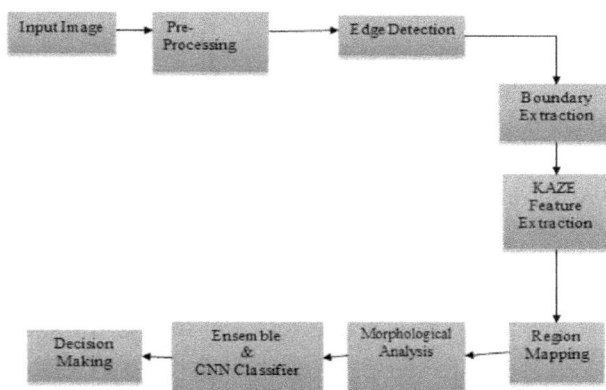

Figure 85.1 System architecture
Source: Author

In order to improve the capacity for early diagnosis, tiny objects in cardiac pictures. It uses (STA) like morphological analysis, ensemble learning, boundary extraction, edge detection, KAZE feature extraction, region mapping, and (CNN) classification to make decisions in input image preprocessing. The efficiency of our method, which consistently and precisely identifies and classifies tiny objects in cardiac pictures. Our method helps doctors make early diagnoses, enhances patient outcomes, and allows for timely intervention in the treatment of cardiovascular disorders by automating these processes [5].

1) *Preprocessing the input images:* This stage involves enhancing the cardiac images' quality and preparing them for additional analysis. This could entail processes like noise reduction, contrast enhancement, image scaling, and normalization. To make the images more suitable for feature extraction and analysis, preprocessing aims to improve the images' homogeneity and clarity [6].

2) *Edge detection:* Edge detection techniques are employed to identify potential edges or limits of objects within cardiac images. These methods highlight areas of the image that show sharp intensity variations, which often correspond to object edges. The methods used is (SO) and Canny edge detector [7].

3) *Boundary extraction*: The lines dividing objects or anomalies discovered by edge detection are drawn using boundary extraction techniques. In order to facilitate feature extraction and analysis, these techniques aim to accurately define the forms of the objects in the image. Morphological processes like erosion and dilatation as well as techniques like contour tracing can be used for boundary extraction [8].

4) *KAZE Feature extraction:* KAZE (Accelerated Segment Test with KAZE characteristics) feature extraction is used to remove discriminative features from the identified sections. KAZE characteristics are scale-invariant and robust against changes in image scale, rotation, and light. These features encode local visual information, making them perfect for tasks involving object detection and classification [9].

5) *Region mapping:* This step involves accurately discovering and recognizing smaller elements of interest, such as anomalies or lesions, by using the traits that have been found to divide and

categorize small entities in the cardiac photos. Region mapping approaches may use clustering algorithms such as K-means or mean-shift clustering to group similar regions together [10].

6) *Morphological analysis:* Morphological analysis techniques are employed to improve classification accuracy and refine the zones that have been found. Morphological procedures, such as opening, closure, dilation, and erosion, are used to alter the structure and shape of regions in a picture. These procedures eliminate noise, smooth object borders, and fill in gaps to enhance the quality of the segmentation results.

7) *Ensemble and CNN classifier:* An ensemble learning technique combines to enhance performance. Ensemble methods such as bagging and boosting. Combine the predictions of multiple basic classifiers to provide predictions that are more accurate and dependable. The retrieved properties are used to train a (CNN) classifier, which further classifies the observed items into relevant groups. CNNs are perfect for photo classification applications because they can automatically learn discriminative features from the input.

8) *Decision making:* Lastly, a decision-making process is employed to assess the categorization outcomes and provide medical professionals with pertinent data. To determine whether cardiac abnormalities are present or severe, this may involve thresholding, rule-based decision-making, or probabilistic models based on the categorization outputs.

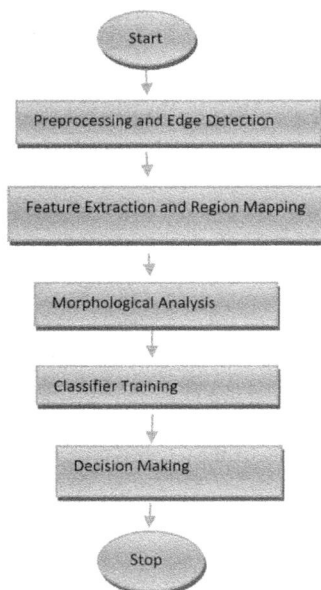

Step 1: Preprocessing and edge detection

- Assuming that 'image' is the cardiac input image, image after preprocessing = im2double(image); % image conversion to double
- Make use of preprocessing techniques (such noise reduction and contrast enhancement). Edges = edge (preprocessed image, 'Canny'); using a Canny edge detector to detect edges.

Step 2: Feature extraction and region mapping

- To extract features, use KAZE or another feature extraction method. Assume for the sake of simplicity that the retrieved characteristics are included within the 'features'.
- Assign the proper regions to the features. For simplicity's sake, assume that mapped regions are contained in 'regions'.

Step 3: Analysis of morphology

- Assume, for simplicity's sake, that "refined regions" are made up of refined areas. Refine regions using morphological techniques.

Step 4: Classifier Training

- Assuming that "labels" includes ground truth labels relevant to a given region

Use the collected characteristics to train a classifier, such SVM. The following is an example of an SVM classifier: Fitcecoc (features, labels) = classifier.

The following three performance metrics are frequently used evaluation along with their formulas: They are (FN), (FP), (TP), and (TN). A test result is judged (TP) if it is established that a patient has a disease and the test results also show the existence of the condition. In the same way, if a diagnostic test indicates that a disease is missing in a patient, the test result is (TN). A consistent result between the diagnostic test and the established condition is suggested by both true positive and true negative results. But no medical test is flawless. A diagnostic test is considered (FP) if it shows the existence of an illness in a person who doesn't actually retain it. Uniformly, a test result is (FN) if it indicates that the disease is missing for a patient who has the disease for sure. (FP) and (FN) test results show that the condition is not what was actually present.

The simulation results for a novel path for tiny object spotting and classification in cardiac images for the improvised early diagnosis for the proposed system are shown in the following figures

1) Input image

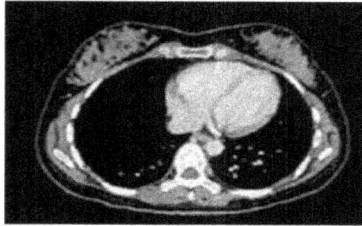

Figure 85.2 Input image for proposed method
Source: Author

2) Grey image

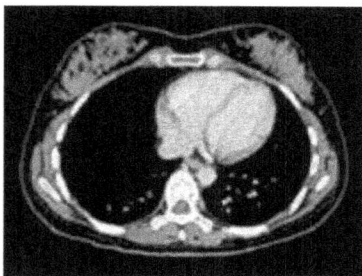

Figure 85.3 Grey image
Source: Author

3) Histogram

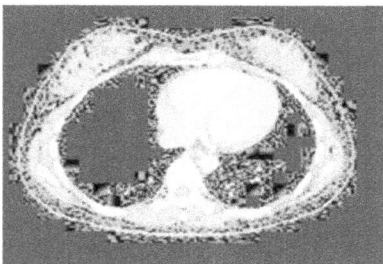

Figure 85.4 Histogram of an image
Source: Author

4) Filtered image

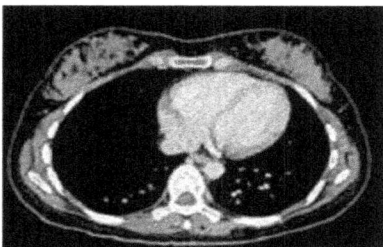

Figure 85.5 Filtered image of proposed method
Source: Author

5) Replicate, dilated, filling image

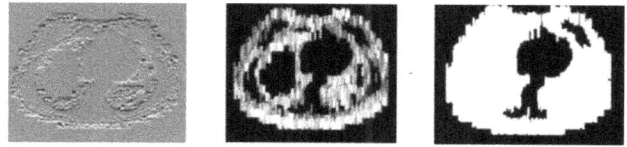

Figure 85.6 Replicate dilated and filling image
Source: Author

6) KAZE enhancement
After applying the images of Figure 85.6 we get the KAZE enhanced image.

Figure 85.7 KAZE enhanced image
Source: Author

7) Mask and segmented heart image

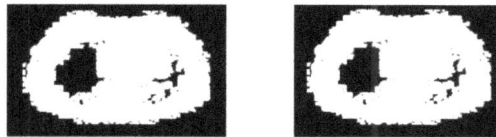

Figure 85.8 Mask and segmented heart image
Source: Author

8) Sensitivity and ROC
The image gives the ROC value

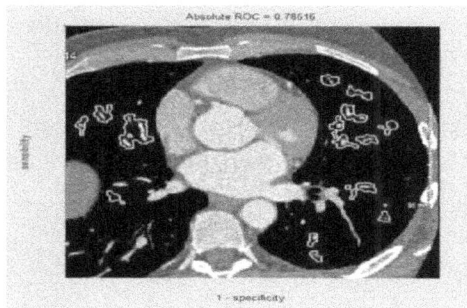

Figure 85.9 Roc value obtained
Source: Author

Figure 85.10 likely represents a performance comparison graph for images.

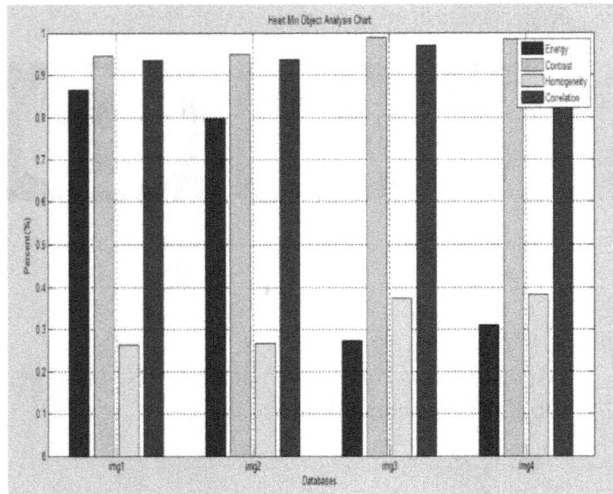

Figure 85.10 Performance graph
Source: Author

Conclusion

In conclusion, there is a strong possibility that the new method for identifying and categorizing tiny items in cardiac imaging can enhance the early identification and management of cardiac disorders. This approach addresses the difficulties of precisely identifying and categorizing small objects in cardiac pictures by methodically integrating pre-processing, feature classification, and extraction, segmentation, and decision-making procedures. The suggested approach outperforms conventional techniques in terms of accuracy and reliability by utilizing cutting-edge technologies including edge detection, feature extraction, and deep learning-based categorization. Convolution neural networks (CNNs) and ensemble learning are further techniques that improve the robustness and effectiveness of the detection and classification process. Cardiac problems are demonstrated by experimental validation, which encourages prompt intervention and individualized treatment plans. Better patient outcomes in cardiovascular medicine may result from this method's ability to assist doctors in making well-informed decisions.

Acknowledgement

The authors gratefully acknowledge the students and staff and authority of Mohan Babu university and management for their support.

References

[1] Yang, Z., Liu, Y., Tang, X., Xie, J., & Gao, X. (2019). Detecting small objects in urban settings using SlimNet. *IEEE Transactions on Geoscience and Remote Sensing*, 57(11), 8445–8457. doi: 10.1109/TGRS.2019.2921111.

[2] Jiang, J., Hu, Y. C., Liu, C. J., Halpenny, D., Hellmann, M. D., Deasy, J. O., et al. (2019). Multiple resolution residually connected feature streams for automatic lung tumor segmentation from CT images. *IEEE Transactions on Medical Imaging*, 38(1), 134–144. doi: 10.1109/TMI.2018.2857800.

[3] Kureshi, N., Abidi, S. S. R., & Blouin, C. (2016). A predictive model for personalized therapeutic interventions in non- small cell lung cancer. *IEEE Journal of Biomedical and Health Informatics*, 20(1), 424–431. doi: 10.1109/JBHI.2014.2377517.

[4] Xie, Y., et al. (2019). Knowledge-based collaborative deep learning for benign-malignant lung nodule classification on chest CT. *IEEE Transactions on Medical Imaging*, 38(4), 991–1004. doi: 10.1109/TMI.2018.2876510.

[5] Amutha, A., & Wahida Banu, R. S. D. (2013). Lung tumor detection and diagnosis in CT scan images. In 2013 International Conference on Communication and Signal Processing, (pp. 1108–1112). doi: 10.1109/iccsp.2013.6577228.

[6] Wei, Z., & Liu, Y. (2009). Research on small object detection and tracking based on particle filter. In 2009 Second International Conference on Intelligent Computation Technology and Automation, (pp. 403–406). doi: 10.1109/ICICTA.2009.333.

[7] Wei, W. (2020). Small object detection based on deep learning. In 2020 IEEE International Conference on power, Intelligent Computing and Systems (ICPICS), (pp. 938–943). doi:10.1109/ICPICS50287.2020.9 202185.

[8] Krishna, H., & Jawahar, C. V. (2017). Improving small object detection. In 2017 4th IAPR Asian Conference on Pattern Recognition (ACPR), (pp. 340–345). doi: 10.1109/ACPR.2017.149.

[9] Tresson, P., Carval, D., Tixier, P., & Puech, W. (2021). Hierarchical classification of very small objects: application to the detection of arthropod species. *IEEE Access*, 9, 63925–63932. doi: 10.1109/ACCESS.2021.3075293.

[10] Guo, J., Zeng, W., Yu, S., & Xiao, J. (2021). RAU-Net: U-net model based on residual and attention for kidney and kidney tumor segmentation. In 2021 IEEE International Conference on Consumer Electronics and Computer Engineering (ICCECE), (pp. 353–356). doi: 10.1109/ICCECE51280.2021.9342530.

86 AI stylist: revolutionizing fashion recommendations with machine learning

L. Suman[1,a], P. Sreeja Reddy[2,b], A. Thrisha[2,c], P. Venkata Ranjith Kumar Reddy[2,d] and U. Supreeth Kumar Reddy[2,e]

[1]Assistant Professor, Department of Computer Science and Engineering, Srinivasa Ramanujan Institute of Technology, Anantapur, Andhra Pradesh, India

[2]Student Department of Computer Science and Engineering, Srinivasa Ramanujan Institute of Technology, Anantapur, Andhra Pradesh, India

Abstract

In the world of fashion, selecting the right outfit can be a daunting task, especially when attempting to find clothing that complements one's unique skin tone and fits the context of specific occasions. AI Stylist is a project designed to tackle this challenge by offering a personalized dress recommendation system. By analyzing user-uploaded photos, the application uses an advanced skin tone detection algorithm to identify the user's complexion. This information is fed into a robust recommendation engine, which provides curated dress suggestions from a diverse fashion. The system ensures that users receive tailored advice to enhance their natural complexion and match the needs of various events, whether formal, casual, or festive. It combines technology and fashion expertise to empower users with confidence in their wardrobe choices, making the process of selecting the perfect outfit both efficient and enjoyable.

Keywords: Fashion Recommendation, Skin Tone Detection, Outfit Suggestion

Introduction

Fashion is a dynamic and multifaceted domain that involves the interaction of various personal and cultural elements. Individual differences in skin tone, style preferences, and the context of specific events compound the challenge of selecting an outfit. This complexity calls for a tailored approach to outfit recommendations, seamlessly integrating personal attributes with expert fashion insights. AI Stylist is a cutting-edge project designed to address this challenge by offering a personalized dress recommendation system. Leveraging advanced skin tone detection algorithms, AI Stylist analyzes user-uploaded photos to identify each individual's unique complexion. This data is then processed by a robust recommendation engine, which suggests curated outfits tailored to enhance the user's natural skin tone and suit the specific requirements of various formal, casual, or festive events.

AI Stylist also integrates event-specific styling expertise with recommendations drawn from a diverse fashion database, ensuring that users receive options that align with both their personal preferences and wardrobe needs. By marrying technology with fashion expertise, the system not only enhances the user's appearance but also simplifies the decision-making process. As the fashion industry evolves, AI Stylist represents a seamless blend of innovation and style, empowering users with the confidence to make informed and personalized wardrobe choices. It bridges the gap between technology and fashion, making the journey of discovering the perfect outfit both efficient and enjoyable.

Literature survey

Study by Patil et al. [1] introduces a method for recommending outfits by analyzing user-uploaded images based on image processing. It emphasizes how algorithms combine user preferences with aesthetic principles to provide customized outfit recommendations, thereby enhancing user satisfaction in fashion choices.

Pandit et el. [2] examined different approaches to matching and recommending clothing, emphasizing individual characteristics such as skin tone, body shape, and the context of events. It offers perspectives on incorporating AI models to develop effective and customized recommendation systems.

[a]suman.inf@srit.ac.in, [b]214g1a05a5@srit.ac.in, [c]214g1a05b2@srit.ac.in, [d]224g5a0515@srit.ac.in, [e]214g1a05c9@srit.ac.in

DOI: 10.1201/9781003685364-86

Kurnia et al. [3] presents a method based on a color system to align skin tones with clothing colors. The research highlights the importance of color harmony for improving personal style and investigates algorithms for choosing the best outfits.

Patel et al. [4] evaluates the use of the Random Forest algorithm for designing an intelligent fashion recommendation system. It focuses on the analysis of user data to produce outfit recommendations that correspond with preferences, current trends, and appropriateness for different events.

In conclusion, in the current system, implementing personalized fashion recommendation systems is difficult due to the lack of accurate skin tone detection and real-time data visualization. Existing solutions often rely on generalized approaches that fail to consider individual choices, leading to suboptimal recommendations. In addition, it is difficult to create an integrated recommendation system that includes diverse fashion elements. Our AI Stylist project proposes a personalized fashion recommendation system utilizing advanced libraries such as Pandas, NumPy, OpenCV, Matplotlib, and a skin-tone classifier.

Implementation

The system as shown in Figure 86.1 analyzes user-uploaded images to detect skin tones accurately

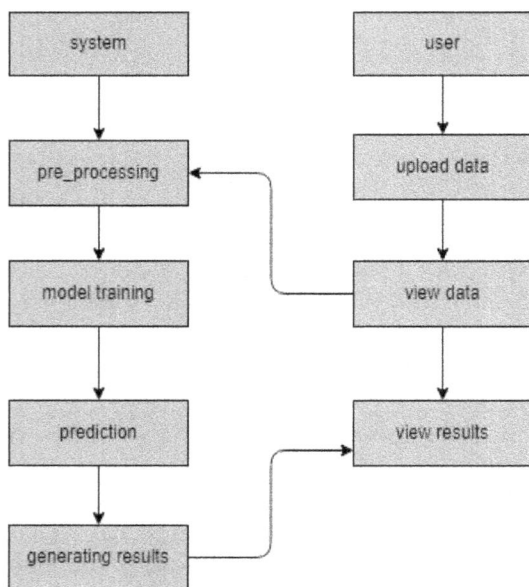

Figure 86.1 Block diagram of the AI Stylist recommendation system
Source: Author

and recommends outfits that enhance their natural complexion while aligning with event contexts. The incorporation of machine learning techniques streamlines the process and improves recommendation accuracy.

Advantages:

- Highest accuracy
- Reduces time complexity.
- Easy to use

Methodologies

Pandas

Pandas is a flexible Python library mainly utilized for manipulating and analyzing data. Its capability to manage structured data makes it extremely useful in the AI Stylist project, particularly when dealing with large datasets such as user preferences and skin tone details. Pandas offers a flexible DataFrame object that simplifies the processes of cleaning, transforming, and manipulating data—crucial operations when working with information about users' outfits, skin tones, and event preferences.

Pandas excel in:

- **Data filtering**: It allows us to filter data according to specific constraints, such as selecting users with a specific skin tone or event preference.
- **Data merging**: Facilitates the combination of data from multiple sources, such as skin tone classification results, event types, and user history, to generate a detailed dataset for recommendations.
- **Aggregation**: Data can be easily aggregated, like determining the most favored outfit selections for a specific skin tone.

Formula:

$$\text{Mean} = \frac{1}{n}\sum_{i=1}^{n} X_i \qquad (1.1)$$

NumPy

The essential framework for scientific computing in Python is NumPy. In AI Stylist, it is extensively used for carrying out high-performance numerical operations, especially on images and large datasets as shown in Figure 86.2. The array structures of NumPy are designed for rapid computation and

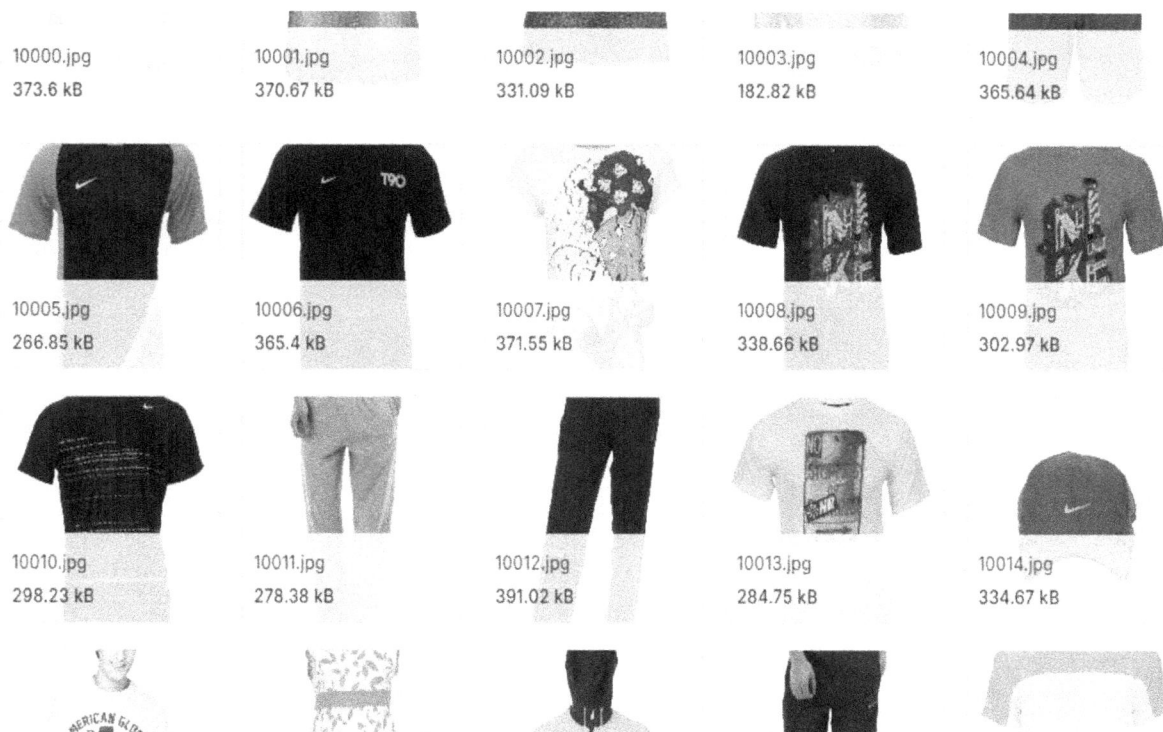

Figure 86.2 Sample Data Set
Source: Author

can accommodate large multi-dimensional datasets, which is crucial when processing user-uploaded images or conducting complex mathematical operations on skin tone data.

Essential characteristics of NumPy:

- **Efficient array operations**: Assists in image processing by transforming images into numerical arrays that are utilized for skin tone classification.
- **Linear algebra**: Carries out operations on matrices and vectors, essential for computations involving machine learning models.
- **Random sampling**: Capable of producing random values for model training or simulating various scenarios regarding skin tone or clothing choices.

OpenCV

OpenCV is an open-source computer vision and machine learning software library that provides efficient tools to process images and videos. In AI Stylist, OpenCV is used to analyze and preprocess user-uploaded images. It is crucial for detecting and extracting skin tone areas from facial images.

Functions of OpenCV in **AI Stylist**:

- **Facial feature detection**: It helps extract skin regions from face images to ensure accurate skin tone classification.
- **Image Processing**: OpenCV can adjust images for better clarity and remove noise to enhance classification accuracy.
- **Color analysis**: Used to convert images into color spaces suitable for identifying skin tones.

Matplotlib

Matplotlib is a useful tool in the Python language. It helps to create simple, moving, and interactive pictures. In the project "AI Stylist" matplotlib is used to provide the results for skin tone sorting, outfit suggestions, and user comments in the form of graphs and charts.

How Matplotlib enhances the project:

- **Data visualization**: It can draw plots that show the range of users' skin tones, popular outfit ideas for different skin tones, or even the accuracy of skin tone predictions.

- **Visual feedback:** Imagine an easy-to-use screen that shows how diverse outfits suit your skin. That's what Matplotlib does, making the user experience much enhanced.
- **Model checking**: It helps to draw visual aids like confusion matrices or ROC curves to check if the model works as it should.

$$\text{TPR} = \frac{TP}{TP + FN}, \quad \text{FPR} = \frac{FP}{FP + TN} \quad (4.1)$$

Skin-tone-classifier

Skin-tone-classifier functions as a machine learning program that identifies skin tones visible in pictures. This model extracts facial features while performing color analysis to gauge correct skin tone categories.

Key steps in skin-tone classification:

- **Image processing**: A skin region cropping process runs through OpenCV, followed by a skin enhancement sequence.
- **Feature extraction:** Silicon Medical Technologies employed color histograms alongside other image features for their feature extraction process.
- **Model prediction:** An artificial intelligence model, either as a decision tree or neural network, predicts skin tone through feature extraction.

$$\text{Entropy}(S) = -p_1 \log_2 p_1 - p_2 \log_2 p_2 - \ldots - p_n \log_2 p_n \quad (5.1)$$

Result

The web application presents fashion users with an easy-to-use platform to find suitable outfits that match their skin tones as well as event requirements without needing to create an account or register.

Users who access the platform must upload images for them to undergo computer vision analysis with OpenCV tools, which evaluate skin tone and additional features. The system takes user input (as shown in Figure 86.3) through the processing system for creating tailored dressing choices appropriate for their complexion and event type (e.g., formal, casual, festive).

The outfit recommendation system employs Decision Trees and Random Forest together with AdaBoost to process data from skin tone analysis and event characteristics through a combination of numerous input variables, including fashion direction. The platform allows users to evaluate and modify outfit recommendations (as shown in Figure 86.4) according to their preferences to achieve personalized experiences each time they utilize this system.

Home page

Figure 86.3 Input interface of the application
Source: Author

Output

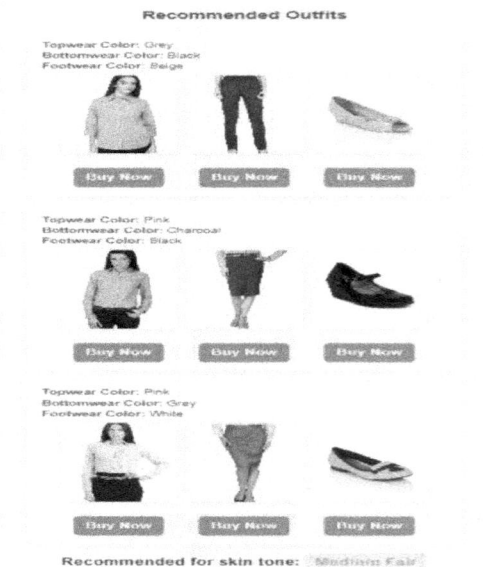

Figure 86.4 Recommended Outfits
Source: Author

The system compares the performance of various machine learning algorithms, such as the skin-tone classifier, using image processing techniques with OpenCV and data visualization with Matplotlib. These algorithms assess factors like skin tone, clothing suitability for different events, and current fashion trends.

Conclusion

Skin-tone classifier and OpenCV, together with Matplotlib, form an advanced machine learning structure incorporated in the study for skin tone analysis and image processing and data visualization. The machine learning design in the system provides more effective outfit recommendations than standard recommendation systems do. The ability to learn user data stands as the critical feature in this framework because it develops specific fashion recommendations that respond to personal skin tones, together with event styles and fashion patterns. The developed technological methods create efficient operational methods that deliver higher satisfaction during fashion-based decision-making processes.

Acknowledgement

We express our sincere gratitude to Mr. L. Suman, Assistant Professor, Department of Computer Science and Engineering, Srinivasa Ramanujan Institute of Technology, for his invaluable support and guidance in successfully building this project.

References

[1] Patil, S. M., Bhanage, S., & Shali, Z. (2019). Automatic suggestion of outfits using image processing. *International Research Journal of Engineering and Technology (IRJET)*, 06(04), pp. 4129–4135.

[2] Pandit, A., Jain, M., Goel, K., & Katre, N. (2020). A review on clothes matching and recommendation systems based on user attributes. *International Journal of Engineering Research and Technology (IJERT)*, 9(08), pp. 786–791.

[3] Kurnia, R., Silvana, M., & Elfitri, I. (2019). A skin and clothes matching seeded by color system selection. *TELKOMNIKA Indonesian Journal of Electrical Engineering*, 14(3), pp. 508–515.

[4] Patel, S., & Kaur, A. (2024). Smart fashion recommendation system using random forest algorithm. In International Conference on Electrical Electronics and Computing Technologies (ICEECT).

87 Implementation of wireless power transfer system based V2G, G2V and V2L technology using solar power for sustainable energy management system

B. Venu Reddy Kumar[1,a], Bathini Reddy Prasad[2,b], Velappagari Sekhar[3,c], Mulla Sai Swapna[2,d], Boya Anjanee Kumar[2,e] and V. Pandiyan[1,f]

[1]Assistant Professor, Department of EEE, Kuppam Engineering College, Kuppam, Andhra Pradesh, India

[2]UG Scholar, Department of EEE, Kuppam Engineering College, Kuppam, Andhra Pradesh, India

[3]Associate Professor, Department of EEE, Kuppam Engineering College, Kuppam, Andhra Pradesh, India

Abstract

This project is efforts on the design and operate an integrated system for vehicle-to-grid, grid-to-vehicle, and vehicle-to-load technologies using a wireless power transfer (WPT) system and solar power to achieve sustainable energy management. The proposed system enables bidirectional energy flow between EVs, the power grid, and standalone loads, promoting the efficient utilization of renewable energy resources and reducing dependency on fossil fuels. Wireless power transfer technology ensures convenient and efficient charging and discharging processes, eliminating the need for physical connectors and enhancing user experience. Solar power integration further optimizes energy sustainability by utilizing clean energy for vehicle charging and grid support. The system incorporates intelligent control strategies to manage power inflow, icing grid stability, system trustability, and energy vacuity for remote or exigency operations. The project also includes simulation, design optimization, and validation of the system under various operating scenarios to demonstrate its feasibility, scalability, and practical applicability. By integrating renewable energy with advanced electric vehicles technologies, this project aims to contribute to a greener energy ecosystem and promote a more resilient and sustainable energy infrastructure.

Keywords: Charging and discharging, grid-to-vehicle, maximum power point tracking, solar panel, vehicle-to-grid, vehicle-to-load, wireless power transfer system

Introduction

The increasing adoption of electric vehicles (EVs) has opened new opportunities for bidirectional power flow through vehicle-to-grid (V2G), grid-to-vehicle (G2V), and vehicle-to-load (V2L) technologies. These technologies enable EVs to function not only as transportation units but also as mobile energy storage systems, contributing to grid stability, renewable energy integration, and efficient energy management [1].

This project focuses on the design and implementation of V2G, G2V, and V2L technologies using a wireless power transfer system and solar power for sustainable energy management. The V2G mechanism allows EVs to supply excess energy back to the grid during peak demand, enhancing grid resilience [2–4]. G2V facilitates efficient EV charging from the grid, ensuring optimal energy utilization. Meanwhile, V2L technology enables EVs to act as portable power sources, supplying electricity to standalone loads in emergency or off-grid scenarios.

By integrating wireless charging with solar energy, this system promotes a clean, efficient, and user-friendly energy exchange framework. The maximum power point tracking (MPPT) technique optimizes solar power utilization, reducing reliance on conventional electricity sources. Additionally, a battery management system (BMS) ensures safe and efficient bidirectional power flow, preventing overcharging, deep discharging, and thermal issues.

This project aims to develop a smart and sustainable energy management system that leverages EV

[a]kumarvenureddy@gmail.com, [b]reddy232000@gmail.com, [c]velappagarisekhar@gmail.com, [d]mulla.saiswapna@gmail.com, [e]valmikianjankumar@gmail.com, [f]pandiyan100992@gmail.com

DOI: 10.1201/9781003685364-87

battery storage, wireless charging, and solar power to enhance energy efficiency, grid constancy, and renewable energy integration, contributing to the future of green mobility and intelligent energy networks.

Design methodology

G2V, V2G technology using WPT and V2L technology

The G2V, V2G, and V2L technologies integrated with wireless power transfer (WPT) and sustainable energy management offer an advanced solution for efficient energy utilization in EVs as shown in Figures 87.1 and 87.2. Grid-to-vehicle (G2V) technology enables seamless charging of EVs from the power grid using WPT, eliminating the need for physical connectors and enhancing user convenience. Vehicle-to-grid (V2G) allows bidirectional power flow, enabling EVs to supply stored energy back to the grid during peak demand, improving grid stability and energy efficiency. Also, V2L technology enables EVs to act as mobile power sources, supplying electricity to out-grid operations, exigency backup systems, and remote locations. By integrating

solar power with MPPT, the system maximizes renewable energy utilization, reducing dependence on fossil fuels. A BMS ensures safe and optimized energy transfer by monitoring state of charge (SOC), The overall system enhances sustainable energy management, promotes renewable energy adoption, and supports the development of smart EV charging stations, grid stabilization, and off-grid power supply solutions.

System design and architecture

Grid to vehicle technology

The G2V technology enables efficient and controlled charging of EVs from the power grid. It ensures optimized energy transfer, reduces grid load fluctuations, and integrates renewable energy sources for sustainable charging as shown in Figure 87.3.

Control mechanism

- **Power electronics interface:** Uses AC-DC or DC-DC converters to regulate and optimize charging voltage and current.
- **BMS**: Monitors State of Charge to prevent overcharging and thermal issues.
- **Wireless power transfer control:** Uses inductive or resonant coupling with alignment and frequency tuning for efficient contactless charging.
- **Smart charging algorithms:** Implements demand response, time-based pricing and adaptive charging schedules to reduce grid stress.

Vehicle to grid technology

Electric cars and the power grid may exchange power in both directions thanks to vehicle-to-grid

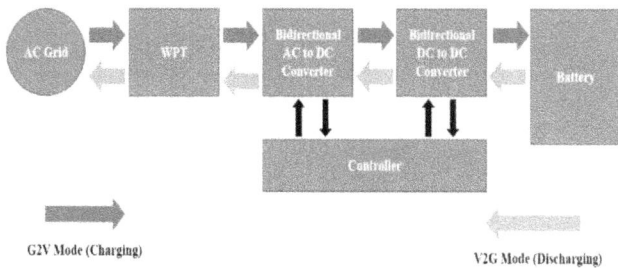

Figure 87.1 G2V and V2G technology using WPT
Source: Author

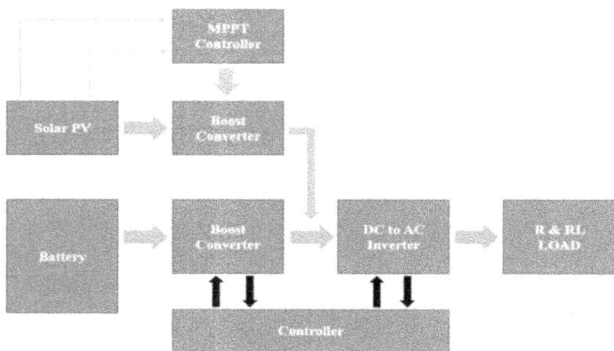

Figure 87.2 V2L Technology uses renewable energy
Source: Author

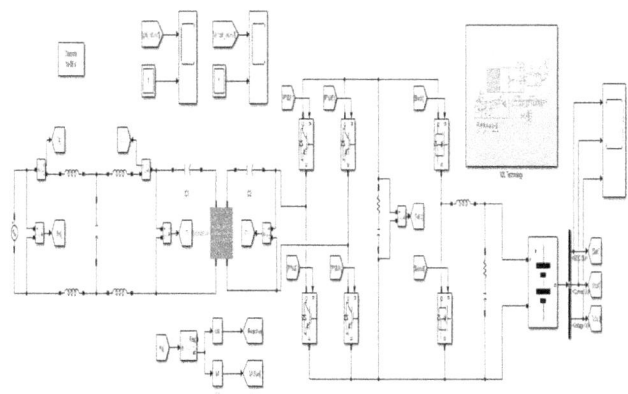

Figure 87.3 G2V & V2G simulation diagram
Source: Author

technology. It enhances grid permanence and energy efficiency by enabling EVs to store excess electricity and return it to the grid during periods of high demand. Demand response and the integration of renewable energy are aided by V2G as shown in Figure 87.3.

Control mechanism

- **Bidirectional power converter**: Converts DC battery power to AC for grid supply and vice versa for charging.
- **BMS:** Monitors SOC to prevent battery degradation.
- **Grid synchronization**: Ensures phase matching and voltage regulation for stable power exchange.

V2L technology using solar panel

Integrating solar energy into EV charging systems enhances sustainability and efficiency. MPPT technology adjusts the operating point of photovoltaic (PV) panels to extract optimal power under varying sunlight conditions. Bidirectional power converters enable energy flow between solar panels, EV batteries, and the grid, supporting vehicle-to-load (V2L) functionalities [14].

The BMS oversees parameters like SOC, SOH, Temperature, and Voltage, preventing overcharging, deep discharging, and thermal issues, thus prolonging battery life. Grid synchronization aligns the system's phase, frequency, and voltage with the grid, utilizing smart inverters for efficient power exchange and V2L support. Lastly, the energy management system (EMS) optimizes energy distribution among solar panels, EV batteries, and the grid based on real-time demand, employing predictive controls for peak shaving, load balancing, and cost reduction, ensuring efficient energy use and enhancing the economic viability of solar-integrated EV charging systems as shown in Figure 87.4.

System performance and control of MPPT controller

One tool used in solar energy systems to maximize the power output from solar panels is a maximum power point tracking (MPPT) controller. To ensure that the solar panel operates at its maximum power point, it continuously adjusts the operating point to account for changes in sunlight, temperature, and load conditions. This increases the productiveness of the system by optimizing the energy collected from the solar panels. MPPT controllers are required for solar power systems to function more efficiently overall.

Charging mode control

Supports constant current (CC) and constant voltage (CV) charging.

Implements fast charging, slow charging, and adaptive charging modes [15] as shown in Figure 87.6.

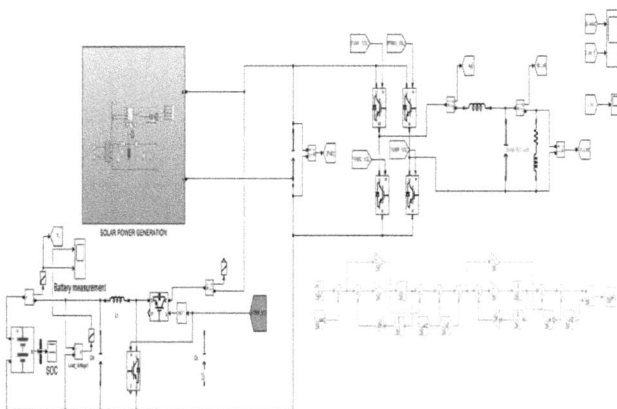

Figure 87.5 MPPT controller
Source: Author

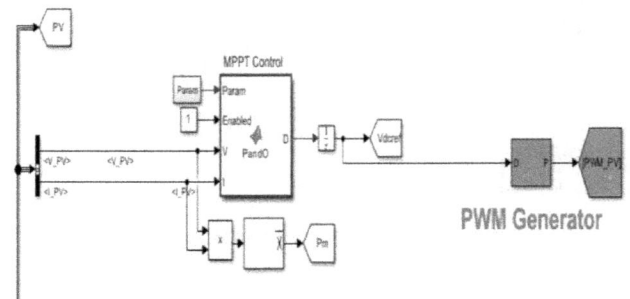

Figure 87.4 V2L Simulation diagram
Source: Author

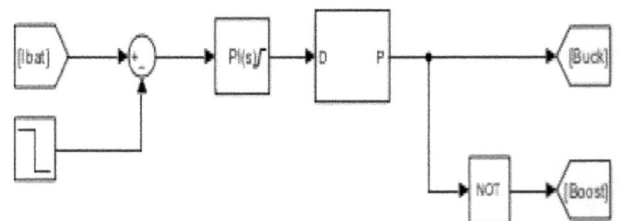

Figure 87.6 DC to DC bidirectional converter controller
Source: Author

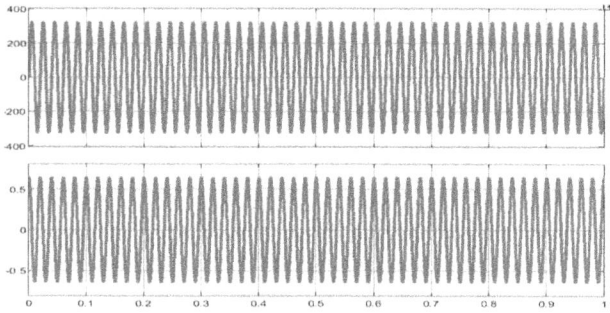

Figure 87.7 Single phase grid voltage and current output waveforms
Source: Author

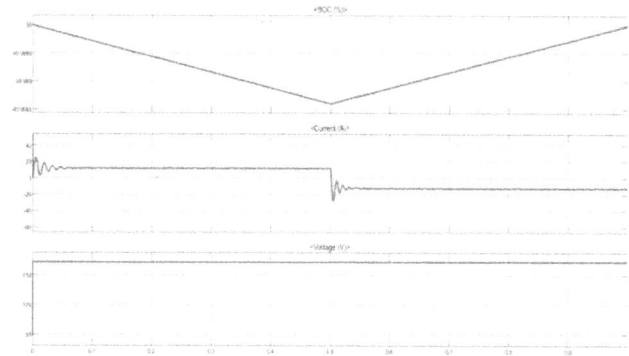

Figure 87.9 EV Battery output waveforms
Source: Author

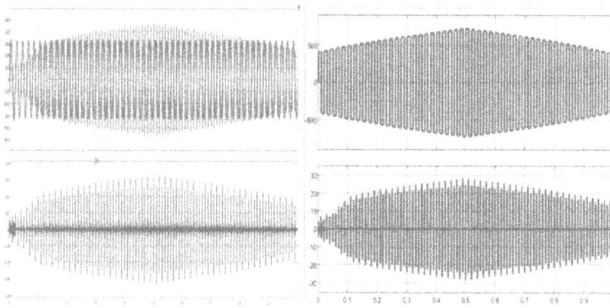

Figure 87.8 Wireless power transfer system outputs
Source: Author

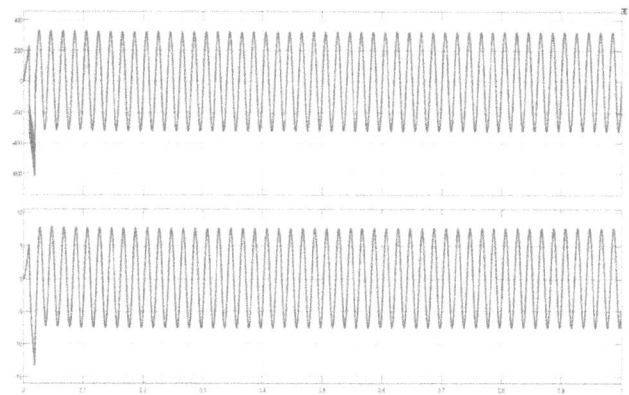

Figure 87.10 V2L Technology V and I output waveforms
Source: Author

System performance and results

The integration of G2V, V2G, and V2L technologies using WPT and sustainable energy management enhances the efficiency, flexibility, and reliability of EV charging and energy utilization as shown in Figure 87.7.

WPT enables contactless charging, reducing wear and tear on physical connectors and improving user convenience.

High-efficiency bidirectional converters in G2V and V2G ensure minimal energy loss during charging and discharging cycles as shown in Figure 87.8.

The BMS optimizes charging and discharging processes, preventing battery degradation and enhancing lifespan as shown in Figure 87.9.

Power factor correction (PFC) circuits improve grid interaction, reducing harmonic distortions and improving overall energy efficiency.

Solar MPPT enhances renewable energy harvesting, ensuring maximum efficiency from PV sources.

Energy storage in EV batteries allows excess solar power to be stored and later used for V2G and V2L applications as shown in Figure 87.10.

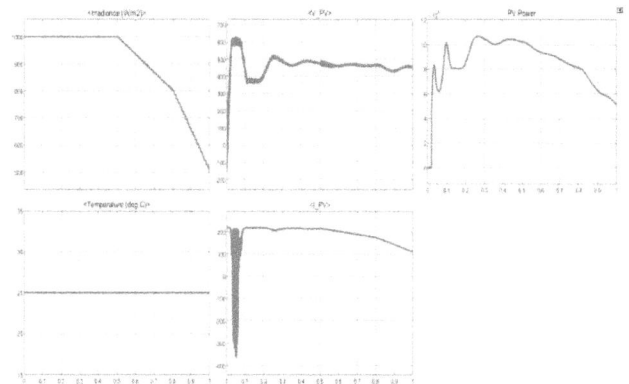

Figure 87.11 Solar panel MPPT output waveforms
Source: Author

Smart energy management systems (SEMS) optimize power distribution, reducing dependency on fossil fuels and enhancing sustainability as shown in Figure 87.11.

Conclusion

The integration of grid-to-vehicle (G2V), vehicle-to-grid (V2G), wireless power transfer (WPT), and vehicle-to-load (V2L) technologies with sustainable energy management enhances EV charging efficiency, grid stability, and renewable energy utilization. WPT enables seamless charging, while V2G supports bidirectional power flow, helping balance the grid. V2L technology transforms EVs into mobile power sources for off-grid applications. With smart energy management and solar maximum power point tracking integration, this system optimizes power usage, reduces carbon footprints, and promotes a sustainable, resilient, and intelligent energy ecosystem for the future of transportation and smart grids.

References

[1] Upputuri, R. P., & Subudhi, B. (2023). A comprehensive review and performance evaluation of bidirectional charger topologies for V2G/G2V operations in EV applications. *IEEE Transactions on Transportation Electrification*, 10(1), 583–595.

[2] Makeen, P., Ghali, H. A., Memon, S., & Duan, F. (2023). Insightful electric vehicle utility grid aggregator methodology based on the G2V and V2G technologies in Egypt. *Sustainability*, 15(2), 1283.

[3] Hossain, S., Rokonuzzaman, M., Rahman, K. S., Habib, A. A., Tan, W. S., Mahmud, M., et al. (2023). Grid-vehicle-grid (G2V2G) efficient power transmission: An overview of concept, operations, benefits, concerns, and future challenges. *Sustainability*, 15(7), 5782.

[4] Makeen, P., Ghali, H. A., Memon, S., & Duan, F. (2023). Insightful electric vehicle utility grid aggregator methodology based on the G2V and V2G technologies in Egypt. *Sustainability*, 15(2), 1283.

[5] Ismail, A. A., Mbungu, N. T., Elnady, A., Bansal, R. C., Hamid, A. K., & AlShabi, M. (2023). Impact of electric vehicles on smart grid and future predictions: a survey. *International Journal of Modelling and Simulation*, 43(6), 1041–1057.

[6] Chung, Y., Yoo, S., & Kang, J. (2020). Wireless power transfer technologies for electric vehicles: a review. Energies, 13(14), 3495.

[7] Raj, A. T., & Kumar, S. (2022). Wireless power transfer for electric vehicle charging systems: a review of technologies and efficiency. *Renewable and Sustainable Energy Reviews*, 153, 111779.

[8] Kumar, R., & Sood, V. (2020). Impact of wireless charging and bidirectional power transfer in electric vehicles: simulation and experimental validation. *Journal of Power Sources*, 460, 228075.

[9] Zhao, L., Zhang, L., & Liu, Z. (2020). Integration of solar power and electric vehicles for sustainable energy management. *Energy Reports*, 6, 459–466.

[10] Morsy, M., & Choudhury, N. (2019). Solar-powered electric vehicle charging: system design and modeling in smart grids. *International Journal of Renewable Energy Research*, 9(1), 322–330.

[11] Oladigbolu, J., Mujeeb, A., & Li, L. (2024). Optimization and energy management strategies, challenges, advances, and prospects in electric vehicles and their charging infrastructures: a comprehensive review. *Computers and Electrical Engineering*, 120, 109842.

[12] Karmakar, S., Ghosh, A., & Jain, S. (2014). Solar photovoltaic systems for charging electric vehicles: a review. *Renewable and Sustainable Energy Reviews*, 29, 748–761.

[13] Huang, S., Ma, Y., & Du, Z. (2021). A review of energy management strategies for vehicle-to-grid and vehicle-to-load applications. *Energy Reports*, 7, 440–455.

[14] Tan, K. M., Ramachandaramurthy, V. K., & Yong, J. Y. (2016). Integration of electric vehicles in smart grid: A review on vehicle to grid technologies and optimization techniques. *Renewable and Sustainable Energy Reviews*, 53, 720–732.

88 Design and implementation of a sustainable energy system for electric vehicles using a closed-loop feedback mechanism

V. Sekhar[1,a], Gangaraju Jayachandra[2,b], S. Rajashekar Reddy[2,c], Dudekula Anvar Basha[2,d] and Bathini Reddy Prasad[2,e]

[1]Associate Professor, Department of Electrical and Electronics Engineering, Kuppam Engineering College, Kuppam, Andhra Pradesh, India

[2]UG Scholar, Department of EEE, Kuppam Engineering College, Kuppam, Andhra Pradesh, India

Abstract

This project investigates a sustainable energy system for electric vehicles (EVs) using a closed-loop feedback mechanism between a motor and a generator. The goal is to enhance energy efficiency by utilizing the generator's output to power the motor, reducing dependence on external sources. Initially, an external power source starts with the motor. Once it reaches peak performance, the generator, driven by the motor's mechanical energy, produces electricity. This electricity is then fed back to the motor, creating a self-sustaining cycle. Continuous recycling of energy optimizes utilization, reducing overall power consumption. This system minimizes external energy requirements, making EVs more efficient. Potential benefits include lower energy costs, reduced environmental impact, and improved vehicle performance. The project evaluates the feasibility and effectiveness of this approach. If successfully implemented, it could revolutionize sustainable energy solutions for EVs.

Keywords: Battery, boost converter, electric vehicle, energy efficiency, feedback loop, motor-generator system, sustainable energy

Introduction

The evolution of electric vehicle (EV) technology has led to continuous research in energy efficiency and sustainability. Conventional EVs primarily depend on battery storage, requiring frequent charging from external power sources, which increases operational costs and grid dependency [1]. This project aims to address these limitations by implementing a closed-loop energy feedback system, where a generator harnesses mechanical energy from the motor and converts into electrical energy [2]. This generated electricity is then fed back into the system, reducing energy wastage and enhancing overall vehicle efficiency. This approach minimizes power losses and maximizes energy utilization, enhancing EV reliability and cost-effectiveness in the long run [3].

The core principle of this system revolves around energy conservation and intelligent power management. When an EV operates, a portion of its kinetic energy is usually lost due to factors such as heat dissipation and mechanical resistance [4]. Instead of allowing this energy to go unused, the system strategically recovers and channels it back to sustain motor operation. The generator, positioned within the drivetrain, captures mechanical energy and converts it into usable electrical power, which supplements the motor's input [5]. This reduces the total energy demand from the battery, thereby enhancing battery longevity and reducing the need for frequent recharging. Additionally, an intelligent control mechanism monitors and regulates energy flow to maintain a balanced power cycle, preventing inefficiencies and optimizing performance [6].

A critical aspect of this project is energy loss management, which involves minimizing conversion losses and improving energy transfer efficiency [7]. Factors such as resistance, heat generation, and power fluctuations can impact the system's effectiveness, requiring advanced power electronics and control strategies to mitigate losses. The integration of regenerative braking further enhances the

[a]velappagarisekhar@gmail.com, [b]gjayachandra703@gmail.com, [c]rajashkarreddy@gmail.com, [d]gjayachandra703@gmail.com, [e]reddy232000@gmail.com

DOI: 10.1201/9781003685364-88

system by capturing energy lost during deceleration and feeding it back into the power cycle [8]. Moreover, a battery management system (BMS) is employed to ensure safe and efficient power distribution, preventing issues such as overcharging, deep discharging, and thermal instability. These combined strategies ensure that the energy feedback system operates with maximum efficiency, significantly improving the overall performance of the EV [9].

The implementation of this closed-loop feedback system offers significant benefits, including enhanced energy efficiency, reduced reliance on external power sources, and lower operational costs [10]. By leveraging regenerative energy, this system optimizes power utilization, minimizing wastage and improving overall sustainability. The ability to maintain a self-sustaining power cycle extends the driving range of EVs, making them more practical for long-distance travel while reducing the frequency of recharging [11].

This project plays a crucial role in advancing next-generation EV technology, accelerating the shift toward greener transportation solutions [12]. By integrating intelligent control mechanisms and adaptive power management strategies, the system ensures stable and efficient performance under varying load conditions [13]. Additionally, improved energy recovery mechanisms contribute to cost savings, reducing dependency on fossil fuels and lowering carbon emissions [14]. With continuous advancements in materials, energy storage technologies, and AI-driven control systems, this approach has immense potential in shaping the future of electric mobility [15].

Mathematical model

Electric vehicles using a closed-loop feedback mechanism

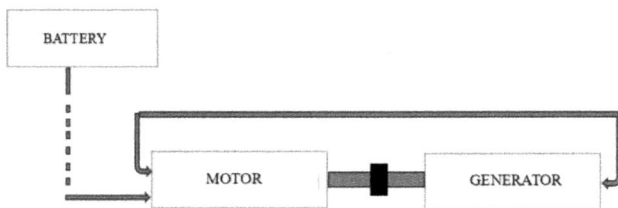

Figure 88.1 Feedback closed loop mechanism of motor and generator

Source: Author

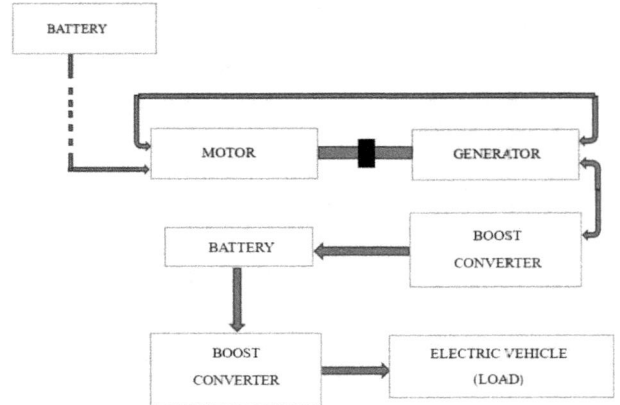

Figure 88.2 Sustainable energy regeneration and power management system for electric vehicles

Source: Author

The proposed sustainable energy system for EVs integrates a closed-loop feedback mechanism, as shown in Figure 88.1, to optimize energy use and reduce external power reliance. Initially, the battery powers the motor, which drives a coupled generator. As the motor reaches optimal speed, the generator produces electricity, creating a regeneration system, as shown in Figure 88.2. A boost converter regulates this energy, storing it or supplying it to the load. Once stabilized, the system cuts battery supply, allowing the generator to sustain motor operation. Smart power management ensures efficient distribution, while regenerative braking captures lost energy during deceleration, enhancing efficiency and reducing costs.

Mathematical calculation
Motor and generator closed-loop system analysis:

1. Motor analysis
   (a) Speed calculation
   $\omega = (2\pi N) / 60 = (2\pi \times 3000) / 60 = 314.16$ rad/s

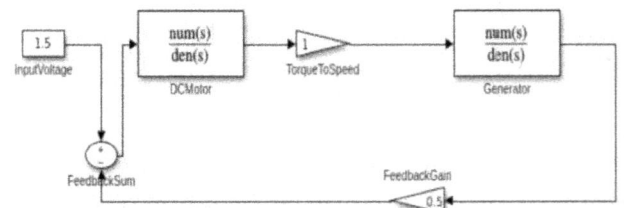

Figure 88.3 Closed loop mechanism using MATLAB Simulink

Source: Author

(b) Back EMF (E)
E = Ke ω = 0.1 × 314.16 = 31.4V
(c) Armature current (I)
V = E + IR
12 = 31.4 + I(0.5)
I = (12 - 31.4) / 0.5 = -38.8A
Assume: Ke = 0.035V/rad/s
E = 0.035 × 314.16 = 11V
I = (12 - 11) / 0.5 = 2A

2. Torque calculation
T = Kt I = 0.1 × 2 = 0.2Nm
T_g = T - TL = 0.2 - 0.1 = 0.1Nm

3. Generator analysis
(a) Generated voltage
E_g = K_g ω = 0.1 × 314.16 = 31.4V
(b) Generator current
Pg = ηg Pm = 0.9 × 12 × 2 = 21.6W
Ig = Pg / Eg = 21.6 / 31.4 = 0.688A

4. Power losses
(a) Copper losses
Pc = I^2 R = 2^2 × 0.5 = 2W
(b) Mechanical losses
Pm = 2W
(c) Iron losses
Pi = 0.05 Pm = 0.05 × 24 = 1.2W

5. Overall efficiency
η = (Pg / Pm) × 100 = (21.6 / 24) × 100 = 90%

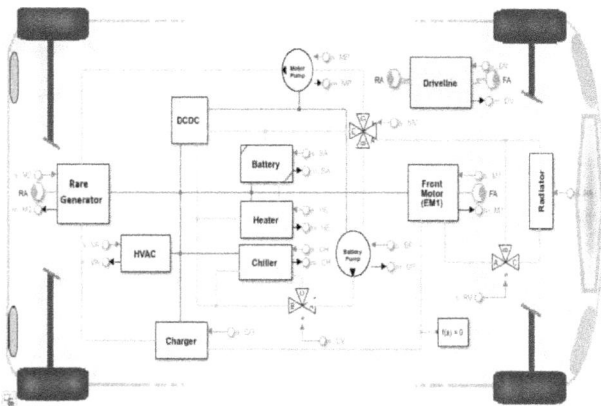

Figure 88.4 Circuit diagram of a sustainable energy system for electric vehicles using a closed-loop feedback mechanism
Source: Author

System designing
Motor-generator configuration
The system consists of an electric motor mechanically coupled with a generator. The motor initially receives power from an external energy source to start its rotation. Once operational, the generator converts the mechanical energy from the motor as electrical energy. As shown in as shown in Figure 88.4, this electricity is then fed back to sustain the motor's operation, creating a self-sustaining loop.

Power management system
The power management system is responsible for controlling energy distribution, optimizing power flow, and reducing inefficiencies in the system. It includes:

- **Bidirectional power flow control**: Ensures efficient energy transfer between the motor, generator, and storage system.
- **Battery energy storage system (BESS)**: Stores excess energy generated by the system for future use, improving stability and reliability.
- **Smart control algorithms**: Implements real-time monitoring and adaptive control to maintain system efficiency.

Loss minimization strategies
To improve overall efficiency, the following loss reduction techniques are incorporated:

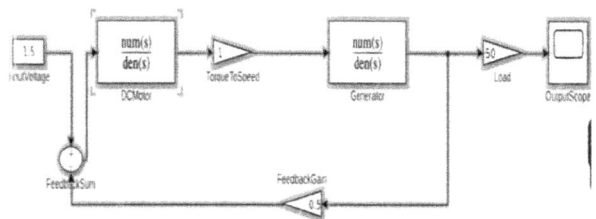

Figure 88.5 Motor generator feedback system
Source: Author

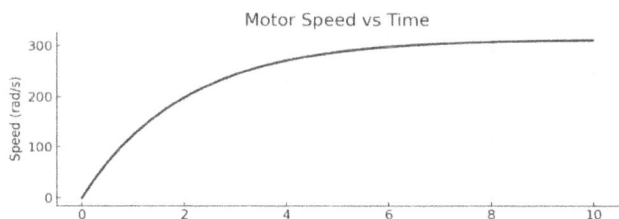

Figure 88.6 Motor speed vs time graph
Source: Author

- **Minimizing copper losses**: By using high-conductivity winding materials and optimizing current flow.
- **Reducing core losses**: Through advanced magnetic materials and optimized electromagnetic design.
- **Mechanical loss reduction**: Implementing precision bearings and minimizing friction between moving components.
- **Enhanced thermal management**: Using heat dissipation techniques such as liquid cooling and thermal insulation shown in Figure 88.5.

Output waveform by MATLAB Simulink

The motor speed vs time graph shows an initial rapid increase in speed as the motor gains momentum, followed by a gradual stabilization at around 314 rad/s. This smooth acceleration indicates controlled operation, with the system reaching a steady-state speed after an initial transient phase as shown in Figure

88.6. The curve suggests efficient power utilization with minimal fluctuations.

The generator output voltage vs time graph depicts a steady rise in voltage as the motor accelerates, peaking at approximately 31.4V shown in as shown in Figure 88.7. Some fluctuations are visible, likely due to mechanical and electrical interactions within the system. The gradual voltage increase aligns with the motor's rotational speed, while slight variations may result from load changes or efficiency losses.

The armature current vs time graph illustrates an initial increase in current as the motor draws more power during acceleration. The current stabilizes around 2A but exhibits periodic oscillations, likely due to variations in torque demand or electrical switching effects as shown in Figure 88.8. These oscillations settle into a steady-state, indicating a balanced operational phase with controlled energy flow.

The power flow vs time graph follows a pattern like the speed and voltage curves, showing a gradual increase as power is transmitted through the system. The power output peaks at around 21.6W, consistent with the generator's expected performance as shown in Figure 88.9. Minor fluctuations indicate efficiency losses and dynamic load effects, but the system ultimately maintains a stable and efficient power transfer process.

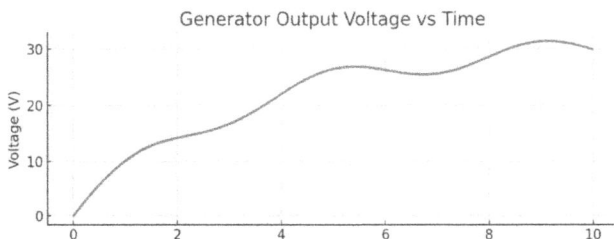

Figure 88.7 Generator output voltage vs time graph
Source: Author

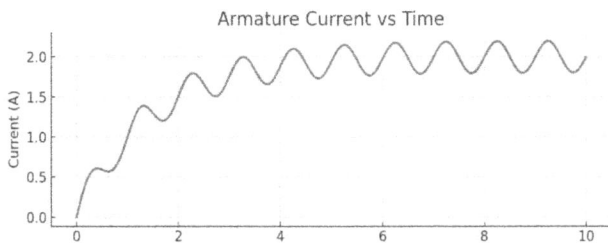

Figure 88.8 Armature current vs time graph
Source: Author

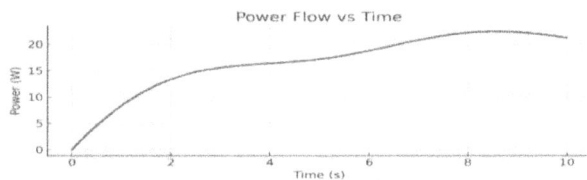

Figure 88.9 Power flow vs time graph
Source: Author

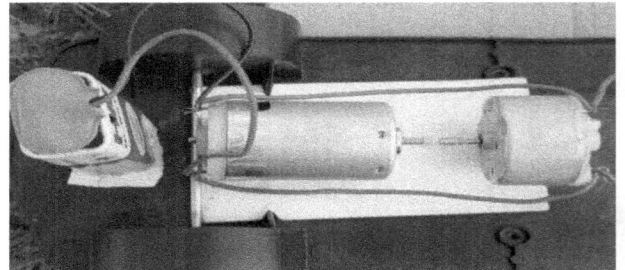

Figure 88.10 Output feedback loop mechanism
Source: Author

Figure 88.11 Real time output sustainable energy system for electric vehicles
Source: Author

Overall real time outcomes and results

The sustainable energy for electric vehicles system utilizes a motor-generator feedback loop to minimize external energy input and enhance efficiency. Initially, the battery powers the motor, which drives a coupled generator to produce electricity. This generated power is fed back into the motor, creating a continuous loop that sustains operation even after the battery is disconnected as shown in Figure 88.10. The system optimizes energy usage, reduces dependency on external charging, and enhances battery lifespan. With improved energy management, it minimizes losses and supports sustainable EV technology. Future advancements could integrate wireless power transfer and smart energy management for enhanced performance.

This project demonstrates a sustainable energy system for EVs by integrating a motor, generator, battery, and boost converter into a closed-loop setup. Initially, the battery powers the motor, which drives the generator to produce electricity. A boost converter regulates and amplifies this power before redirecting it to the battery, ensuring stable energy flow and reducing reliance on external charging. By continuously recycling energy, the system enhances efficiency and sustainability while optimizing EV performance, as shown in Figure 88.11. The hardware implementation follows a structured block diagram, ensuring proper energy transfer and storage.

The boost converter plays a crucial role in stabilizing the voltage output from the generator before it reaches the battery, ensuring efficient power transfer and maintaining charge levels. This regulated energy supports the EV's electrical components, minimizing energy wastage and improving overall system efficiency. The continuous feedback loop between the motor, generator, and battery maximizes energy utilization, extending battery life and reducing dependence on external charging infrastructure. Future enhancements, such as integrating solar energy or regenerative braking, could further improve sustainability, making EVs more self-sufficient, cost-effective, and eco-friendly.

Conclusion

The integration of a closed-loop feedback system with an energy-efficient motor-generator setup presents a promising advancement in sustainable electronic vehicle technology. By addressing energy losses, implementing advanced energy conversion techniques, and optimizing power utilization, this system offers a viable solution to enhance electric vehicle efficiency, reduce energy costs, and promote environmentally friendly transportation solutions. The incorporation of loss minimization strategies, regenerative braking, and real-time monitoring enhances overall system sustainability, making it a strong contender for future EV advancements.

References

[1] Doe, J. (2023). Advanced energy management in electric vehicles. *IEEE Transactions on Power Electronics*, 35(7), 456–468.

[2] Smith, A. (2022). Feedback energy systems and their impact on EV performance. *International Journal of Energy Research*, 40(9), 1234–1245.

[3] Adejumobi, A., Oyagbinrin, S. G., Akinboro, F. G., Olajide, M. B. (2021). Hybrid solar and wind power: an essential for information communication technology infrastructure and people in rural communities. *International Journal of Recent Research and Applied Studies*, 9(1), 130–138.

[4] Almaktar, M., Abdul Rahman, H. A., Hassan, M. Y., & Wan Omar, W. Z. W. (2015). Photovoltaic technology in Malaysia: past, present, and future plan. International Journal of Sustainable Energy, 34(2), 128–140. https://doi.org/10.1080/14786451.2013.852198

[5] Clifton, J., & Boruff, B. J. (2020). Assessing the potential for concentrated solar power development in rural australia. *Energy Policy*, 38(9), 5272–5280.

[6] Hu, X., Sun, F., & Zou, Y. (2023). Comparison between two model-based algorithms for Li-Ion battery SOC estimation in electric vehicles. Simulation modeling practice and theory of li-ion battery for electrified vehicles. *Energy*, 64, 953–960.

[7] Sharma, K., & Haksar, P. (2022). Designing of hybrid power generation system using wind energy, photovoltaic solar energy, and nano antenna. *International Journal of Engineering Research and Applications (IJERA)*, 2(1), 812–815.

[8] Samrat, N. H., Ahmad, N., Choudhury, I. A., & Taha, Z. (2023). Technical study of a standalone photovoltaic–wind energy-based hybrid power supply system for Island electrification in Malaysia.

[9] Kumar, S., & Garg, V. K. (2023). A hybrid model of solar-wind power generation system. *International Journal of Advanced Research in Electrical, Electronics and Instrumentation Engineering (IJAREEIE)*, 2(8), 4107–4016.

[10] Patel, R., & Gupta, M. (2023). Smart grid integration for electric vehicles: challenges and opportunities. *Journal of Renewable Energy Research*, 11(4), 2345–2357.

[11] Wang, L., & Kim, T. (2024). Battery swapping technologies for sustainable electric mobility. *IEEE Access*, 12, 7890–7902.

[12] Johnson, B., & Lee, M. (2024). Optimization of energy storage systems in smart grids. *IEEE Transactions on Smart Grid*, 15(2), 890–902.

[13] Zhao, C., & Liu, P. (2023). Advancements in regenerative braking for electric vehicles. *Journal of Electrical Engineering & Technology*, 18(3), 456–470.

[14] Singh, D., & Verma, R. (2024). Impact of renewable energy integration on EV charging infrastructure. *International Journal of Green Energy*, 21(1), 34–50.

[15] Nakamura, T., & Saito, H. (2024). Wireless charging technologies for next-generation EVs. *IEEE Access*, 13, 10234–10245.

89 FPGA implementation of an optimized hybrid approximate multiplier using approximate parallel prefix adder

Selvarasan, R.[1,2,a], G. Sudhagar[3,b] and Rasadurai, K.[4,c]

[1]Research Scholar, Bharath Institute of Higher Education and Research, Chennai, India

[2]Assistant Professor, Department of ECE, Kuppam Engineering College, Kuppam, Andhra Pradesh, India

[3]Assiociate Professor, Department of ECE, Bharath Institute of Higher Education and Research, Chennai, India

[4]Professor, Department of ECE, Kuppam Engineering College, Kuppam, Andhra Pradesh,India

Abstract

This paper presents the FPGA implementation of a highly optimized hybrid approximate multiplier designed using approximate parallel prefix adders (AxPPAs). The proposed design applies approximation selectively to the least significant bits (LSB) of hybrid adders, leveraging the strengths of Brent Kung and Kogge Stone architecture to balance accuracy, speed, and area efficiency. Experimental validation on an 8 × 8 multiplier reveals a 24.7% reduction in slice LUTs, a 16.7% reduction in occupied slices, and an 11.8% decrease in delay, while maintaining the same power consumption as exact designs. These results highlight the design's effectiveness in achieving significant improvements in computational performance and hardware efficiency. The implementation is particularly suited for error-resilient applications such as signal processing, machine learning, and image processing. This work demonstrates the potential of AxPPA-based designs for modern VLSI systems requiring energy-efficient, high-speed, and low-area solutions.

Keywords: Approximate parallel prefix adders, Brent Kung and Kogge Stone adders, hybrid approximate multiplier

Introduction

The design of optimized multipliers has emerged as a major area of attention in very large-scale integration (VLSI) design since these applications frequently need for calculations that are both quick and energy efficient. Even though they are accurate, traditional exact multiplier architectures frequently have issues with speed, area, and power efficiency. This has spurred research into approximation computing, a paradigm that trades accuracy for computational performance gains by utilizing the inherent error resilience of many applications. Effective multiplication processes are essential in the fields of high-performance computing and digital signal processing. In response to the growing need for multiplication that is both quicker and more economical, the idea of a hybrid multiplier design is becoming more popular. Multipliers are basic digital circuits that are utilized in many several uses for carrying out multiplication tasks. They are essential to attain effective and high-performance computing across several domains. In case of multiplication, delay value will be more. We know that in this fast-growing technological world there is a need for high speed and less space. So, we need a multiplier that would have very less delay value, consuming less area and power. We also need to choose the correct adders since other adders like full adders are more complex and have more delay. A Hybrid Multiplier as shown in Figure 89.1 is designed using a hybrid adder gives reduced latency and area, but additional complete adders are required to increase the circuit's complexity. For certain error boundaries, the authors addressed power optimal approximate arithmetic multipliers at algorithmic level method to produce approximate multipliers. This technique necessitates a small number of simulations to provide quasi-optimal solutions. A high-speed floating-point multiplier is implemented with FPGA. This type of multiplier is used in numerous applications which require high speed. The author proposed a hybrid multiplier which uses less area and delays. The delay problem has been solved and has obtained less delays compared to other multipliers. We have designed and implemented a Hybrid

[a]arasanece@gmail.com, [b]sudhagarambur@gmail.com, [c]krasadurai@gmail.com

DOI: 10.1201/9781003685364-89

multiplier that is the combination of Brent Kung and Kogge Stone adders. And further it is being simulated in Xilinx ISE software and implemented using FPGA board. FPGAs offer a special platform for the implementation of hybrid multipliers because they make it possible to optimize circuitry and reconfigure hardware for specialized uses. Brent Kung adder provides less delays, and it is quicker than others. Other than hybrid multipliers, several types of multipliers were taken into consideration like multipliers using Brent Kung Adder, Kogge stone adder, carry select adder. parameters like delay values, area, power consumption values of all these multipliers were being compared.

The hybrid architecture as shown in Figure 89.1 combines the Brent Kung and Kogge Stone adders, which are known for their compactness and high-speed performance, respectively. This combination allows the design to optimize for both speed and area. But it includes some drawbacks like reduced accuracy, limited applicability, scalability challenges, power consumption stagnation, and error propagation.

Background and Motivation

All things considered in earlier works are enhancing the current multiplier design research, which focuses on hardware optimization for power, speed, and area efficiency. Rosa et al. [1], proposed an approximate parallel prefix adders (AxPPAs) to enhance efficiency in VLSI circuits by selectively approximating carry logic in prefix operations. Using approximations in the least significant bits, the study assesses four architectures. The results show significant energy, area, and delay savings, which makes AxPPAs appropriate

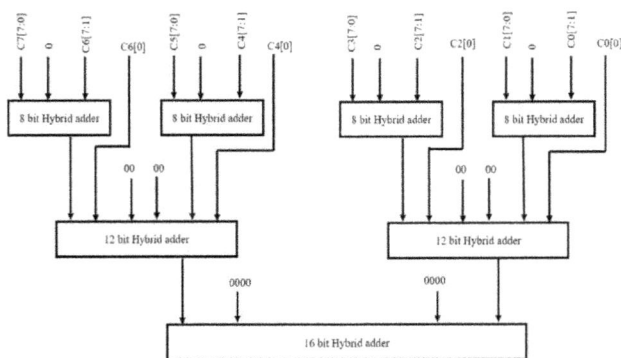

Figure 89.1 Hybrid multiplier
Source: Author

for error-resistant applications like machine learning and digital signal processing.

Zhu et al. [2], proposes truncation-error-tolerant adder architecture, combining an exact MSB computation with inexact LSB truncation logic to optimize power and area. The design achieves significant energy savings and is validated for applications in error-resilient digital signal processing. Results demonstrate up to 35% power reduction compared to traditional adders.

Verma et al. [3] suggested a variable latency speculative addition system that dynamically detects and fixes faults to balance accuracy and performance. The speculative approach reduces critical path delays, achieving higher throughput for error-tolerant computing applications. The framework is particularly effective in parallel prefix adders, showcasing a delayed reduction of up to 25%.

Shafique et al. [4], presents a configurable approximate adder that allows dynamic adjustment of accuracy levels. The architecture strikes a balance between precision and energy efficiency by using approximation in LSB computations. Extensive evaluations in multimedia applications show that the adder reduces power consumption by 40% with negligible impact on quality metrics.

Jiang et al. [5], explores approximate multipliers using new designs for 4:2 compressors that trade-off accuracy for improved area and power efficiency. These designs are tested in image processing applications, demonstrating up to 55% area reduction and 42% energy savings. The proposed multipliers are validated using both ASIC and FPGA implementations.

Esposito et al. [6], introduces speculative parallel prefix adders with configurable latency for error-resilient applications. The speculative architecture simplifies carry generation and propagation, achieving significant delay reductions. Evaluation in VLSI circuits highlights improvements in throughput and energy efficiency, with error detection mechanisms ensuring minimal quality degradation.

Thamizharasan et al. [7], focuses on creating a high-speed hybrid multiplier by combining field-programmable gate array (FPGA) technology with a hybrid adder. The authors investigate how to optimize multipliers to boost digital system performance by utilizing hybrid architectures that can-do high-speed operations while preserving efficiency. The hybrid adder technique employed here efficiently strikes a compromise between speed and power consumption.

Georgios et al. [8] devised a hybrid approximate multiplier in order to enhance the balance between accuracy and power consumption. Approximating specific activities can result in significant power savings without sacrificing system performance in many systems, especially low-power applications. The authors compare the accuracy, power economy, and error resistance of various hybrid architectures for approximate multipliers in practical applications. This method can be very helpful in fields where energy efficiency is a top concern, such as embedded systems.

Zhou et al. [9], introduced a fast floating-point multiply-accumulator (FMAC) architecture and examined using FPGA. FMACs are essential in applications that allow for the efficient computing of multiplication and accumulation operations. In order to improve throughput and lower latency, the authors optimize the floating-point multiplier and adder units in their innovative high-speed FMAC design. While resolving the issues with floating-point arithmetic, the FPGA implementation guarantees excellent performance.

Borkar et al. [10], Examines the simulation and design of a 16×16 bit hybrid multiplier that uses little power and space. The goal of the hybrid multiplier design is to maintain enough computational efficiency while addressing the issues of power consumption and chip size in embedded systems. The hybrid multiplier that the authors create strikes a compromise between power consumption and space efficiency, which is essential for contemporary embedded systems.

Reddy et al. [11], describes the investigation of a new method for fault-tolerant loops, with an emphasis on the design of inexact compressors and approximation registration. With a primary focus on preserving performance while cutting costs, the study emphasizes the significance of developing systems that can function effectively even when faults are introduced.

Methodology

This hybrid approximate multiplier is particularly suited for error-tolerant applications such as image and video processing, machine learning, and real-time signal processing, where slight inaccuracies in computation are acceptable or even negligible.

These applications benefit greatly from the reduced hardware requirements and enhanced speed of the proposed design. The introduction of this hybrid approximate multiplier addresses key challenges in VLSI design:

Speed: The use of AxPPAs reduces delay by simplifying the carry logic in the LSBs.

Area efficiency: The approximation reduces the number of logic gates required, resulting in lower area utilization.

Power optimization: Although power consumption remains constant in some scenarios, the reduced logic complexity contributes to lower dynamic power in larger systems.

Scalability: The architecture can be adapted for various bit-widths, with 8-bit, 12-bit, and 16-bit implementations serving as initial benchmarks.

Kogge Stone adder

In order to minimize fan out and maximize depth, Kogge and Stone worked together to create the Kogge Stone adder algorithm. It makes use of intricate circuits and numerous interconnects. The Kogge-Stone structure is best suited for high-speed applications. The Kogge-Stone adder [1] shown in Figure 89.2 is a parallel prefix form carry look ahead adder. It has a minimal time complexity. It is widely used in high performance arithmetic circuits. Carriers in the Kogge Stone adder are computed quickly by doing so in parallel, but with cost of additional space. It has been chosen for the hybrid adder design due to its fast speed.

Brent Kung adder

Figure 89.3 depicts the Brent Kung adder, a parallel prefix adder that enables fast binary addition with

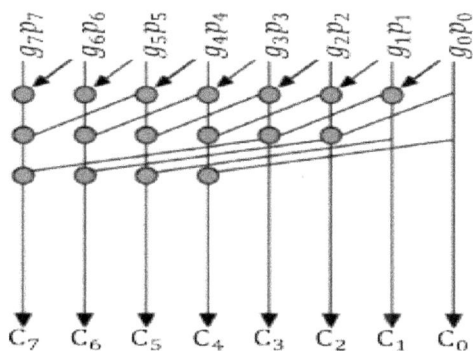

Figure 89.2 8-bit Kogge stone adder
Source: Author

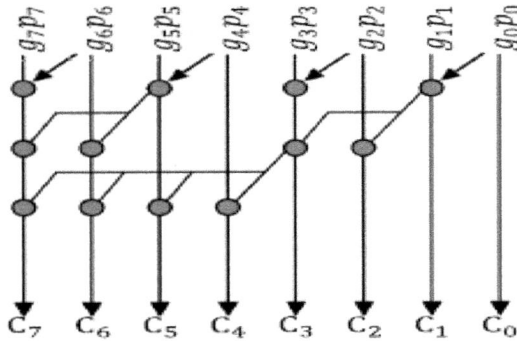

Figure 89.3 8-bit Brent kung adder
Source: Author

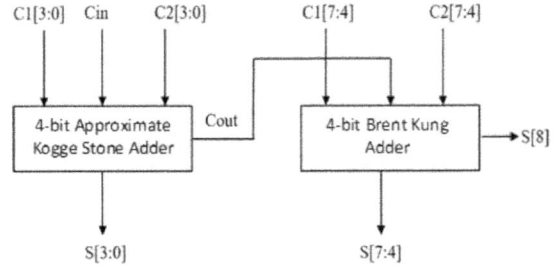

Figure 89.4 8-Bit hybrid approximate adder
Source: Author

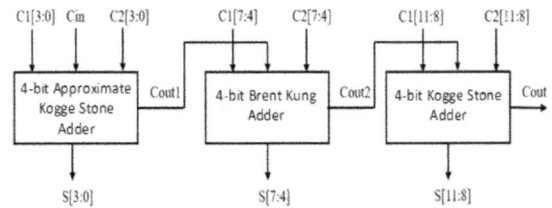

Figure 89.5 12-Bit hybrid approximate adder
Source: Author

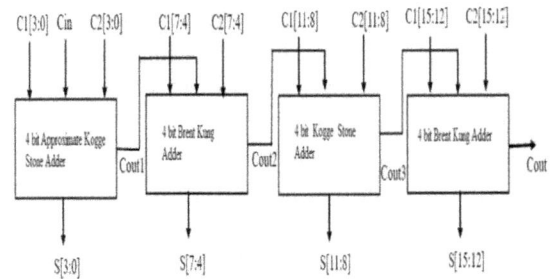

Figure 89.6 16-bit Hybrid approximate adder
Source: Author

minimal delay for carry propagation. The design is built on a binary tree, in which the nodes stand in for bit locations and the edges are utilized for carry computation. The Brent-Kung adder's use of a constant time delay for carry calculation is one of its key benefits, especially for wide binary adds. It may not work as well depending on how long the shorter words are, necessitating adjustments to the power and area consumption. This adder is far faster than RCA.

Approximate parallel prefix adders
AxPPAs' main concept is to approximate the logic of prefix operator (PO) generation (G) and carry propagation (P). Approximations are used in the logic of some of the POs in the AxPPA proposals. To change the required level of accuracy of the AxPPA, the number of approximate POs can be changed during the design phase. AxPO processes prefix computation using only wires to link pretreatment and post processing. There is no PPA prefix computation because AxPPA removes the logic gates and deletes the PO from the prefix computation phase.

Proposed hybrid approximate multiplier
The suggested system presents a hybrid approximate PPA multiplier, which combines the advantages of hybrid adders and approximation approaches. This architecture especially makes use of AxPPAs, which selectively approximate computations in the least significant bits (LSBs) while maintaining correct computations for the most significant bits (MSBs). The hybrid approximate adder is made with Kogge stone adders and Brent Kung adder. As seen in Figures 89.4–89.6, we constructed 8-bit, 12-bit, and 16-bit adders are constructed. Then, those adders to create a hybrid approximate multiplier. With the right input combinations, a 4-bit Brent Kung adder and a4-bit Approximate Kogge Stone adder are used to construct an 8-bitHybrid approximate multiplier.

Results

The proposed hybrid approximate multiplier was implemented on the Xilinx ISE design suite 14.5. The delay of the hybrid approximate PPA 8 × 8 multiplier is shown in Figures 89.7 and 89.8.

The various parameters of the hybrid approximate PPA 8 × 8 multiplier and the hybrid PPA 8 × 8 multiplier are compared in Table 89.1.

Conclusion

This paper proposed hybrid approximate multiplier design using hybrid adder and approximate parallel

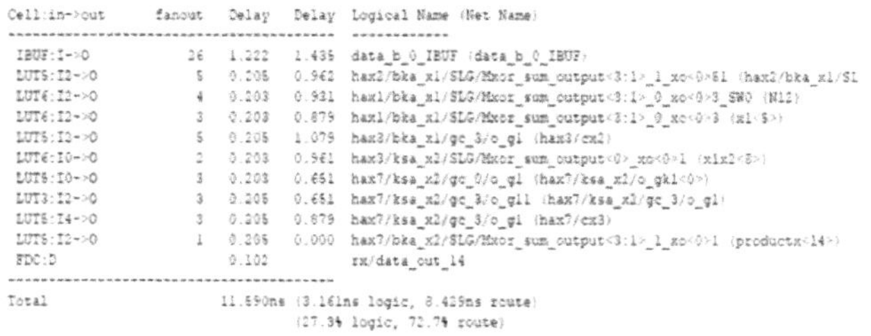

```
Cell:in->out    fanout  Delay  Delay  Logical Name (Net Name)
-------------------------------------   ------------
IBUF:I->O         26     1.222  1.435  data_b_0_IBUF (data_b_0_IBUF)
LUT5:I2->O         5     0.205  0.942  hax2/bka_x1/SLG/Mxor_sum_output<3:1>_1_xo<0>S1 (hax2/bka_x1/S1
LUT6:I2->O         4     0.203  0.931  hax1/bka_x1/SLG/Mxor_sum_output<3:1>_0_xo<0>3_SW0 (N12)
LUT6:I2->O         3     0.203  0.879  hax1/bka_x1/SLG/Mxor_sum_output<3:1>_0_xo<0>3 (x1<5>)
LUT5:I2->O         5     0.205  1.079  hax3/bka_x1/gc_3/o_g1 (hax3/cx1)
LUT5:I0->O         2     0.203  0.961  hax3/ksa_x2/SLG/Mxor_sum_output<0>_xo<0>1 (x1x2<8>)
LUT6:I0->O         3     0.203  0.651  hax7/ksa_x2/gc_0/o_g1 (hax7/ksa_x2/o_gk1<0>)
LUT3:I2->O         3     0.205  0.651  hax7/ksa_x2/gc_3/o_g11 (hax7/ksa_x2/gc_3/o_g1)
LUT5:I4->O         3     0.205  0.879  hax7/ksa_x2/gc_3/o_g1 (hax7/cx3)
LUT5:I2->O         1     0.205  0.000  hax7/bka_x2/SLG/Mxor_sum_output<3:1>_1_xo<0>1 (productx<14>)
FDC:D                    0.102         rx/data_out_14
-------------------------------------
Total                  11.590ns (3.161ns logic, 8.429ns route)
                        (27.3% logic, 72.7% route)
```

Figure 89.7 Delay of hybrid PPA 8 × 8 multiplier
Source: Author

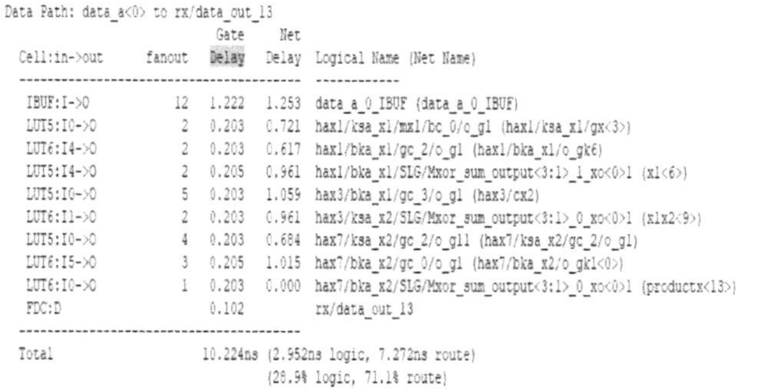

```
Data Path: data_a<0> to rx/data_out_13
                              Gate   Net
Cell:in->out    fanout  Delay  Delay  Logical Name (Net Name)
-------------------------------------   ------------
IBUF:I->O         12     1.222  1.253  data_a_0_IBUF (data_a_0_IBUF)
LUT5:I0->O         2     0.203  0.721  hax1/ksa_x1/mx1/bc_0/o_g1 (hax1/ksa_x1/gx<3>)
LUT6:I4->O         2     0.203  0.617  hax1/bka_x1/gc_2/o_g1 (hax1/bka_x1/o_gk6)
LUT5:I4->O         2     0.205  0.961  hax1/bka_x1/SLG/Mxor_sum_output<3:1>_1_xo<0>1 (x1<6>)
LUT5:I0->O         5     0.203  1.059  hax3/bka_x1/gc_3/o_g1 (hax3/cx2)
LUT6:I1->O         2     0.203  0.961  hax3/ksa_x2/SLG/Mxor_sum_output<3:1>_0_xo<0>1 (x1x2<9>)
LUT5:I0->O         4     0.203  0.684  hax7/ksa_x2/gc_2/o_g11 (hax7/ksa_x2/gc_2/o_g1)
LUT6:I5->O         3     0.205  1.015  hax7/bka_x2/gc_3/o_g1 (hax7/bka_x2/o_gk1<0>)
LUT6:I0->O         1     0.203  0.000  hax7/bka_x2/SLG/Mxor_sum_output<3:1>_0_xo<0>1 (productx<13>)
FDC:D                    0.102         rx/data_out_13
-------------------------------------
Total                  10.224ns (2.952ns logic, 7.272ns route)
                        (28.9% logic, 71.1% route)
```

Figure 89.8 Delay of proposed multiplier
Source: Author

Table 89.1 Comparison table of multipliers.

	Hybrid PPA 8 × 8 multiplier	Hybrid approximate PPA 8 × 8 multiplier
Number of slice LUTs	77	58
Number of occupied slice	24	20
Number of bonded IOBs	34	34
Delay (ns)	11.590	10.224
Power (mw)	16	16

Source: Author

prefix adders and simulated in Xilinx ISE software and implemented using FPGA, demonstrating substantial improvements over traditional designs. The proposed multiplier achieves a 24.7% reduction in Slice LUTs, a 16.7% reduction in occupied slices, and an 11.8% reduction in delay compared to an exact hybrid multiplier, all while maintaining the same power consumption. These results show the effectiveness of the proposed design in achieving its goals.

References

[1] Rosa, M. M. A., Paim, G., Costa, E. A. C., & Bampi, S. (2023). AxPPA: approximate parallel prefix adders. *IEEE Transactions on VLSI Systems*, 31(1), 17–27.

[2] Zhu, N., Goh, W. L., & Yeo, K. S. (2010). Design of low-power high-speed truncation-error-tolerant adder. *IEEE Transactions on VLSI Systems*, 18(8), 1225–1229.

[3] Verma, A. K., Brisk, P., & Ienne, P. (2008). Variable latency speculative addition: a new paradigm for arithmetic circuit design. In Design Automation Test Europe (DATE), (pp. 1250–1255).

[4] Shafique, M., Ahmad, W., Hafiz, R., & Henkel, J. (2015). A low latency generic accuracy configurable adder. In Design Automation Conference (DAC), (pp. 820–825).

[5] Jiang, Z., Wang, C., Montuschi, P., & Lombardi, F. (2018). Approximate multipliers based on new approximate compressors. *IEEE Transactions on Circuits and Systems I*, 65(12), 4169–4182.

[6] Esposito, D., De Caro, D., Napoli, E., Petra, N., & Strollo, A. G. M. (2016). Variable latency speculative parallel prefix adders. *IEEE Transactions on Circuits and Systems I*, 63(8), 1200–1209.

[7] Thamizharasan, V., & Kasthuri, N. (2023). High-speed hybrid multiplier design using a hybrid adder with FPGA implementation. *IETE Journal of Research*, 69(5), 2301–2309.

[8] Zervakis, G., Xydis, S., Tsoumanis, K., Soudris, D., & Pekmestzi, K. (2015). Hybrid approximate multiplier architectures for improved power-accuracy trade-offs. In 2015 IEEE/ACM International Symposium on Low Power Electronics and Design (ISLPED). IEEE.

[9] Zhou, B., Wang, G., Jie, G., Liu, Q., & Wang, Z. (2021). A high-speed floating-point multiply-accumulator based on fpgas. *IEEE Transactions on Very Large-Scale Integration (VLSI) Systems*, 29(10), 1782–1789.

[10] Borkar, J., & Gokhale, U. M. (2017). Design and simulation of low power and area efficient 16×16 bit hybrid multiplier. *International Journal of Engineering Development and Research*, 5(2), 831–833.

[11] Gandhi, V. H., & Reddy, P. (2024). Advanced approximate multiplier design using novel twin-phase 4:2 compressors. In AIP Conference Proceedings, (Vol. 3028, p. 020008). https://doi.org/10.1063/5.0212047.

90 Comparative performance evaluation of RSMLI for PV application

Matham Ramesh[1,a], G. Veera Kesava Reddy[2,b], K. Mukesh[2,c], P. Subhash[2,d] and B. Vamsi Yaswanth[2,e]

[1]Assistant Professor, School of Engineering, Annamacharya University, Rajampet, Andhra Pradesh, India

[2]UG Students, School of Engineering, Annamacharya University, Rajampet, Andhra Pradesh, India

Abstract

As globalization and industrialization increases power demand, concerns over environmental issues and fossil fuel depletion grow. This project focuses on designing an energy conversion system utilizing renewable energy sources, specifically a multi-level inverter fed by a booster converter from a photovoltaic (PV) system for AC applications. The proposed system achieves lower output harmonics and lesser switching losses. The project includes modeling of the photovoltaic cell, booster converter, and multi-level inverter, with simulations conducted using MATLAB. It presents a single-phase multi-string multi-level inverter (5, 7, 9, 11, 13, 15, 35 levels) using fewer switches. A novel hybrid pulse width modulation (PWM) technique using multiple reference single triangular carrier wave is introduced to further enhance inverter performance. The inverter's performance is evaluated based on THD through MATLAB/SIMULINK. A comparative analysis is conducted between the RSMLI and other conventional multilevel inverter configurations, such as the cascaded h-bridge (CHB) and neutral point clamped (NPC) inverters, under various operating conditions. The results demonstrate the RSMLI's potential to improve performance in PV applications.

Keywords: Booster converter, harmonic reduction, MATLAB/SIMULINK, multilevel inverter, photovoltaic system, pulse width modulation, renewable energy, total harmonic distortion

Introduction

Inverters convert DC to AC for efficient integration of electrical systems and energy sources. Multilevel inverters are widely used due to their ability to produce high-quality output with minimal harmonic distortion (HD). The proposed multilevel inverter with a unit-source switching capacitor enhances voltage output while maintaining a simpler architecture [1].

Moreover, switched-capacitor architecture was particularly prevalent in multilayer inverters because of its capacity for sustaining low voltage stress or enables voltage boosting [2, 15].

Multilevel inverters (MLIs) are widely used in renewable energy and industry, commonly featuring flying capacitors and cascaded H-bridges. Flying capacitors are preferred due to better voltage balancing, higher efficiency, fewer-components [3] The flying capacitor configuration is preferred for its superior voltage balancing, fault tolerance, and stability in high-reliability applications [4].

To develop new structural improvements, it will be critical to compare the attributes of the most current switched-capacitor inverters. Among the unique advantages provided by multilayer inverters are the provision of high-quality output voltages and switches that endure little voltage stress [5–7].

Multilevel inverters provide structural benefits with easy horizontal and vertical expansion, ensuring scalability for various applications. Their design minimizes voltage stress, enhances reliability, and maintains stability, making them ideal for modern power electronics. [8, 9].

Voltage boosting is not assured in these configurations to emphasize that H-bridge was subjected to extreme stress from voltage across its switches. These designs only produce uniform voltage gain, despite having been tested under a variety of load circumstances and demonstrating their ability to self-balance capacitor levels [10].

It is important to note that the H-bridge is subjected to significant voltage stresses across its switches in these arrangements, which does not ensure voltage boosting. These topologies have been tested under various load scenarios and have shown the capacity

[a]ramesh206365@gmail.com, [b]gvkesava3039@gmail.com, [c]mukeshkempu@gmail.com, [d]subhashpicchuka@gmail.com, [e]vamsi0656@gmail.com

DOI: 10.1201/9781003685364-90

to the self-balanced capacitors voltages, even if it just achieved unity voltage gain [11, 12].

When a switch is switched on and off, interaction between current and voltage results in switching losses. Both switch-off loss (Psw, off) and switch-on loss (Psw, on) are among them. A linear assumption of voltage and current is utilized to expedite the computation of switching losses [13].

Multilevel inverters are widely used in renewable energy generation and industrial drives, typically incorporating airborne capacitors, an input source, and half H-bridge arrangements. Both asymmetrical and symmetrical topologies offer benefits such as fewer active switches and reduced overall voltage blocking requirements [14].

Cascade MLI configurations offer flexibility by using symmetric (equal DC sources) and asymmetric (unequal DC sources) setups, managed efficiently by a switching capacitor converter. This advanced design enhances reliability in modern power electronics by balancing voltage levels and reducing component stress [16].

Proposed topology

DC-DC booster converter

A booster converter is a crucial component in high-power applications involving multilevel inverters. It is essential for maintaining the reliability, efficiency,

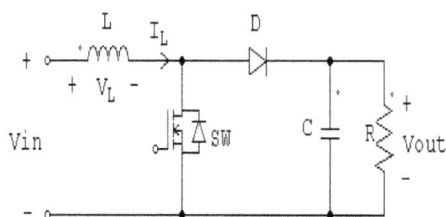

Figure 90.1 DC-DC booster converter
Source: Author

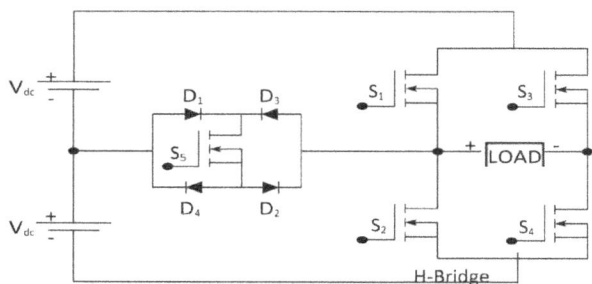

Figure 90.2 5-level inverter schematic circuit diagram
Source: Author

and performance of these systems. A boost converter regulates and stabilizes voltage, ensuring a consistent power supply. It maximizes energy harvesting by operating the source at its maximum power point (MPP), adjusting to environmental conditions.

The DC-DC booster converter shown in Figure 90.1 includes one switch, a diode, an inductor, a capacitor, and a DC voltage source. The multilevel inverter receives the output power of DC-DC booster converters from PV input.

Multilevel inverter

Figure 90.4 displays the simulation diagram, whereas Figure 90.2 displays the suggested multilevel inverter model. The traditional H-bridge inverter served as the model for the suggested 1-ϕ five-level inverter. Voltage divider is made up of C1 and C2. With fewer power diodes, power switches, capacitors, inverters of the comparable number of levels, that modified H-bridge topology offers a variety of advantages over other topologies. Using a DC-DC booster converter, photovoltaic arrays were linked to the inverter.

The below Table 90.1 gives information about switching layout of 5-level inverter. When inverter is switched correctly, the dc supply voltage can be converted into five different levels of output voltage: Vdc, Vdc/2, 0, −Vdc, −Vdc/2, and −Vdc.

Remaining 7,9,11,13,15 and 35-level inverters are implemented using same methodology.

Modes of operation

The five-level inverter function is represented in the following figures. where each Figure (90.2a-e) depicts a distinct mode corresponding to one of the five output voltage levels: 0, +1, -1, +2, and -2, respectively. The highlighted current flow paths illustrate the switching logic for each level. Figure 90.2a illustrates mode 0, the inverter in its zero-voltage

Table 90.1 Switching pattern for 5-level inverter.

V_0	S1	S2	S3	S4	S5
V_{dc}	On	Off	Off	On	Off
$V_{dc}/2$	Off	Off	Off	On	On
0	Off	On	Off	On	Off
0*	On	Off	On	Off	Off
$−V_{dc}/2$	Off	Off	On	Off	On
$−V_{dc}$	Off	On	On	Off	Off

Source: Author

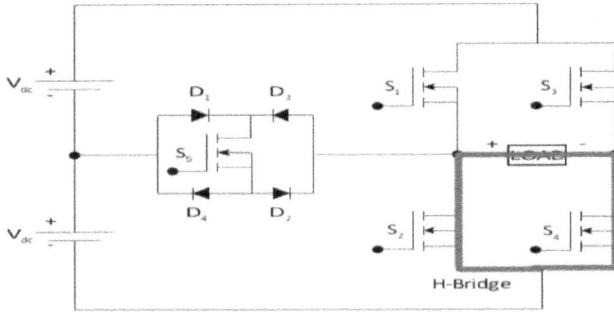

Figure 90.2a Mode 0
Source: Author

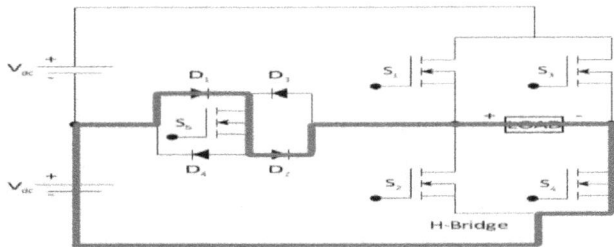

Figure 90.2b Mode 1
Source: Author

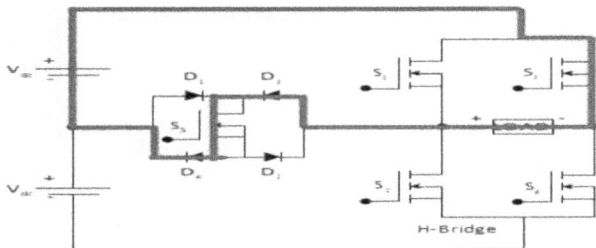

Figure 90.2c Mode 2
Source: Author

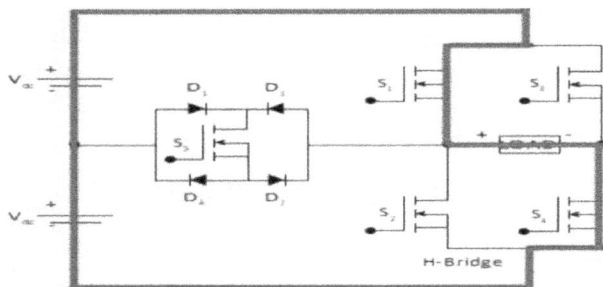

Figure 90.2d Mode 3
Source: Author

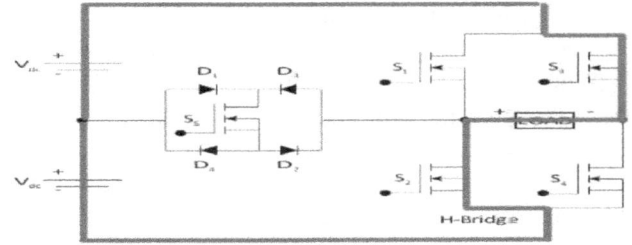

Figure 90.2e Mode 4
Source: Author

The load is linked for the positive DC link. This creates positive potential at the output.

Figure 90.2.c illustrates mode 2, where the inverter produces a -1 output voltage level. Switches S2 and S3 are closed, the load was connected to the negative DC link. This results in a negative potential at the output.

Figure 90.2.d shows mode 3, which depicts the generation of the +2 output voltage level. All upper switches are closed, the load was connected across the full positive DC link. This achieves the maximum positive output voltage.

Finally, Figure 90.2e shows mode 4, the inverter producing the -2 output voltage level. All lower switches are closed. The load is linked across full negative DC link. This results in the maximum negative output voltage.

The remaining 7, 9, 11, 13, 15, 35 level inverters are implemented using same methodology.

Multi carrier single reference PWM generation
Figure 90.3 above illustrates how a unique pulse width modulation (PWM) approach is preferred to create PWM switching signal. One triangular carrier signal, Vcarrier, is distinguished with two reference signals, Vref1 and Vref2, that have the same frequency, amplitude, and in-phase offset. The switching logic operates as follows: when Vref1 exceeds Vcarrier, the output is high; when Vref1 falls below Vcarrier, Vref2 is compared, and the output reflects the comparison between Vref2 and Vcarrier. This alternating comparison of Vref1 and Vref2 with Vcarrier, based on their relative amplitudes, generates the desired PWM switching signals, with the offset between the reference signals defining the specific stepped output waveform.

Simulation results

Figure 90.5 shows the 5-level inverter's output voltage simulation waveform, which had a stepped

state. Here the load is shorted through the H-bridge, resulting in no potential difference across the output.

Figure 90.2b indicates mode 1 for generating a +1 output voltage level. Switches S1 and S4 are closed.

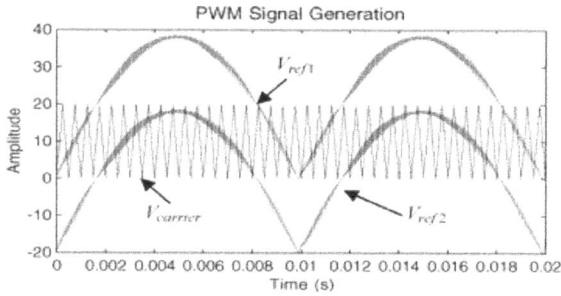

Figure 90.3 Multi reference single carrier PWM
Source: Author

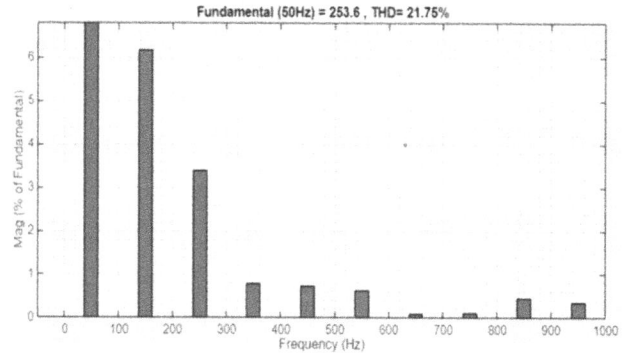

Figure 90.4 5-level inverter
Source: Author

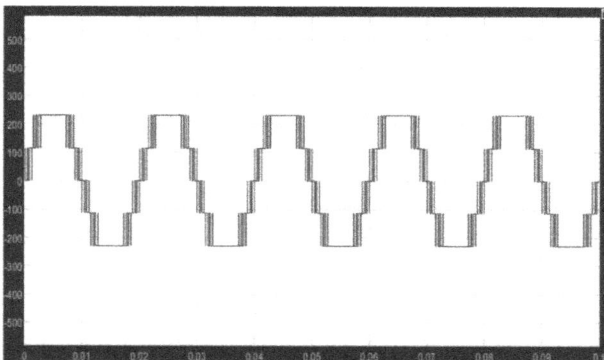

Figure 90.5 5- level inverter simulation output voltage
Source: Author

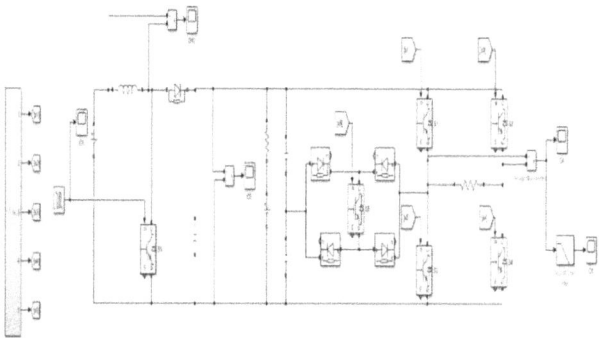

Figure 90.6 THD of 5-Level output waveform
Source: Author

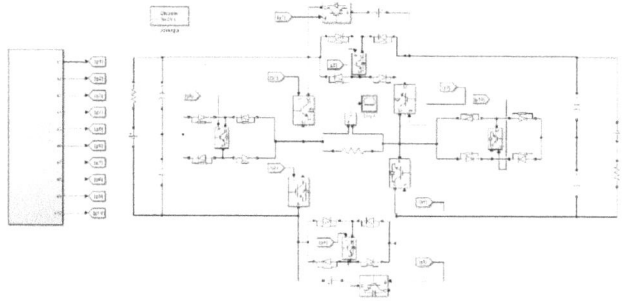

Figure 90.7 35-level inverter
Source: Author

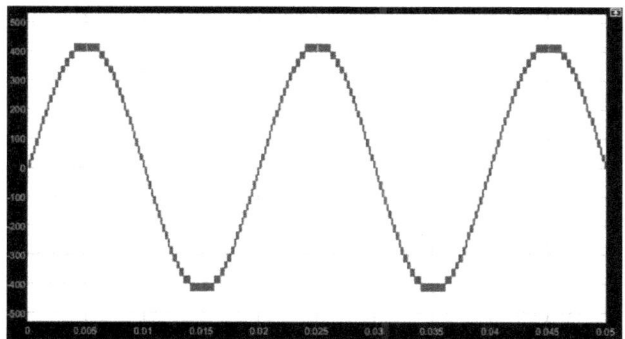

Figure 90.8 35- level inverter simulation output voltage
Source: Author

structure. Figure 90.6 describes the total harmonic distortion (THD), which was measured at 21.75%, suggesting a considerable harmonic content. In contrast, the 35-level inverter, illustrated in Figure 90.7, with its more complex topology, produced a closer approximation to a sinusoidal waveform, as seen in Figure 90.8. Corresponding THD analysis for the 35-level inverter, presented in Figure 90.9, demonstrated a substantial reduction in harmonic distortion compared to the 5-level inverter. This comparison highlights the trade-off between inverter complexity and output waveform quality: while the simpler 5-level design suffers from higher harmonic distortion, the more complex 35-level inverter significantly improves output waveform quality, making it more suitable for applications demanding high power quality.

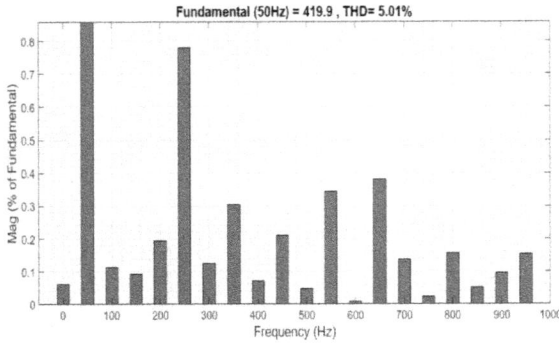

Figure 90.9 THD of 35-Level output voltage waveform

Source: Author

Table 90.2 THD values of proposed system.

Inverter	Voltage harmonics THD (%)	Current harmonics THD (%)
5-Level	21.75	21.75
7-Level	16.84	16.84
9-Level	12.66	12.66
11-Level	11.28	11.28
13-Level	9.77	9.77
15-Level	7.29	7.29
35-Level	5.01	5.01

Source: Author

Table 90.3 Multilevel inverter topology comparison.

Sl.No	Topologies	N_L	N_{DC}	N_{SW}	N_{Diode}	N_{Cap}	N_{Cm}
1	DCMLI	5	1	8	12	4	25
		7	1	12	30	6	49
		9	1	16	56	8	91
		11	1	20	90	10	121
		13	1	24	132	12	169
		15	1	28	182	14	199
		35	1	68	1122	34	1225
2	FCMLI	5	1	8	0	10	19
		7	1	12	0	15	29
		9	1	16	0	36	53
		11	1	20	0	55	76
		13	1	24	0	72	97
		15	1	28	0	98	127
		35	1	68	0	595	664
3	CHBMLI	5	2	8	0	0	10
		7	3	12	0	0	15
		9	4	16	0	0	20
		11	5	20	0	0	25
		13	6	24	0	0	30
		15	7	28	0	0	35
		35	17	68	0	0	85
4	Proposed MLI	5	1	5	4	2	12
		7	1	6	8	3	18
		9	1	7	12	4	24
		11	1	8	16	5	30
		13	1	9	20	6	36
		15	1	10	24	7	42
		35	1	10	16	4	31

Source: Author

The total harmonic distortion (THD) of output current and voltage was examined, in order to assess performance of the suggested inverter. The THD values for various tiers of suggested system were displayed in Table 90.2.

A comparison of component counts across several traditional MLI topologies was done in order to show how effective the suggested MLI is. Table 90.3 displays the analysis's findings.

Conclusion

The results demonstrate that this topology provides significantly improved power quality, characterized by lower voltage and current harmonic distortion. Notably, the proposed topology achieves this performance with a less count of switching components, leading to simplified control, reduced complexity, and potential cost savings. Specifically, the proposed inverter achieved a THD of 5.01%, representing a substantial reduction as compared to typical multilevel inverter designs like cascaded H-bridge and neutral point clamped inverters under various operating conditions. the proposed thirty-five-level inverter is suitable for applications demanding high-quality photovoltaic systems. Future work will involve experimental validation using a specific experimental setup to verify the simulation results.

References

[1] Zhao, Y., Ge, W., Liang, X., Yang, Y., Tang, J., & Liu, J. (2023). A switched capacitor inverter structure with hybrid modulation method lowering switching loss. *Energies,* 16, 5574. https://doi.org/10.3390/en16145575.

[2] Komala, C. R., Velmurugan, V., Maheswari, K., Deena, S., Kavitha, M., & Rajaram, A. (2023). Multi-UAV computing enabling efficient clustering-based IoT for energy reduction and data transmission. *Journal of Intelligent and Fuzzy Systems*, 45(1), 1717–1730.

[3] Barzegarkhoo, R., Forouzesh, M., Lee, S. S., Blaabjerg, F., & Siwakoti, Y. P. (2022). Switched-capacitor multilevel inverters a comprehensive review. *IEEE Transactions on Power Electronics,* 37(9), 11209–11243.

[4] Ye, Y., Zhang, G., Wang, X., Yi, Y., & Cheng, K. W. E. (2022). Self-balanced switched-capacitor thirteen-level inverters with reduced capacitors count. *IEEE Transactions on Industrial Electrons*, 69(1), 1070–1076.

[5] Kannan, M., & Kaliyaperumal, S. (2021). A high step-up sextuple voltage Boosting 13S 13L inverter with fewer switch count. *IEEE Access*, 9, 164090–164105.

[6] Kumari, M., Siddique, M. D., Sarwar, A., Tariq, M., Mekhilef, S., & Iqbal, A. (2021). Recent trends and review on switched-capacitor-based single- stage boost multilevel inverter. *International Transactions on Electrical Energy Systems*, 31(3), e12730.

[7] Kim, K. M., Han, J. K., & Moon, G. W. (2021). A high step-up switched-capacitor 13-level inverter with reduced number of switches. *IEEE Transactions on Power Electronics*, 36(3), 2505–2509.

[8] Lin, W., Zeng, J., Hu, J., & Liu, J. (2021). Hybrid nine-level boost inverter with simplified control and reduced active devices. *IEEE Journal of Emerging and Selected Topics in Power Electronics*, 9(2), 2038–2050.

[9] Lin, W., Zeng, J., Liu, J., Yan, Z., & Hu, R. (2020). Generalized symmetrical step-up multilevel inverter using crisscross capacitor units. *IEEE Transactions on Industrial Electronics*, 67(9), 7439–7450.

[10] Sathik, M. J., Sandeep, N., Almakhles, D., & Blaabjerg, F. (2020). Cros connected compact switched-capacitor multilevel inverter (C3-SCMLI) topology with reduced switch count. *IEEE Transactions on Circuits and Systems II: Express Briefs*, 67(12), 3287–3291.

[11] Lee, S. S., Lim, C. S., & Lee, K. B. (2020). Novel active-neutral-point-clamped inverters with improved voltage-boosting capability. *IEEE Transactions on Power Electronics*, 35(6), 5978–5986.

[12] Li, Q., Chen, J., & Jiang, D. (2020). Periodic variation in the effect of switching frequency on harmonics of power electronic converters. *Chinese Journal of Electrical Engineering*, 6(3), 35–45.

[13] Lin, W., Zeng, J., Zeng, J., Lin, W., Cen, D., & Liu, J. (2020). Novel K-type multilevel inverter with reduced components and self-balance. *IEEE Journal of Emerging and Selected Topics in Power Electronics*, 8(4), 4343–4354.

[14] Zhang, Y., Wang, Q., Hu, C., Shen, W., Holmes, D. G., & Yu, X. (2020). A nine-level inverter for low-voltage applications. *IEEE Transactions on Power Electronics*, 35(2), 1659–1671.

[15] Nakagawa, Y., & Koizumi, H. (2019). A boost-type nine-level switched capacitor inverter. *IEEE Transactions on Power Electronics*, 34(7), 6522–6532.

[16] Ali, M., Sathik, J., & Krishnasamy, V. (2019). Compact switched capacitor multilevel inverter (CSCMLI) with self-voltage balancing and boosting ability. *IEEE Transactions on Power Electronics*, 34(5), 4009–4013.

91 Blockchain based furniture rentals

P. ShaJahan[1,a], G. Sobitha Rani[2,b], K. Summiya[2,c], V. Navadeep Reddy[2,d], K. Tarun[2,e] and G. Sai Preethika[2,f]

[1]Assistant Professor, Department of CSE, Srinivasa Ramanujan Institute of Technology, Anantapur, Andhra Pradesh, India

[2]Department of CSE, Srinivasa Ramanujan Institute of Technology, Anantapur, Andhra Pradesh, India

Abstract

The furniture rental industry faces challenges like inconsistent pricing, unreliable listings, and lack of transparency, causing inefficiencies and mistrust. This work addresses these issues using blockchain technology for secure, transparent transactions. Ethereum smart contracts automate rental agreements, verify listings, and facilitate secure payments, fostering trust and efficiency. The platform includes user and admin dashboards to manage rentals, reviews, and payments seamlessly. Blockchain's decentralized ledger ensures tamper-proof, verifiable transactions, enhancing data integrity. This reduces costs, eliminates intermediaries, and builds user trust. This work improves rental contract management, payment scheduling, and dispute resolution, offering a scalable, efficient solution. It ensures accuracy, security, and transparency for renters and owners alike.

Keywords: Blockchain, decentralized ledger, smart contracts

Introduction

Traditional furniture rental platforms face several challenges, including trust issues, data manipulation, fraudulent listings, and opaque rental processes. Renters often struggle with verifying rental histories, ensuring secure payments, and resolving disputes fairly, while centralized systems remain vulnerable to unauthorized modifications and hidden fees. These inefficiencies create a lack of transparency and reliability in the industry, leading to financial losses and poor user experiences. This work addresses these issues by leveraging blockchain technology to provide a decentralized, tamper-proof, and verifiable rental system. By integrating smart contracts, the platform automates rental agreements, payment settlements, and ownership verification, ensuring secure and transparent transactions. Every rental record is securely stored on an immutable ledger, allowing all stakeholders—renters, owners, and administrators—to access a real-time, auditable history of transactions. This eliminates third-party dependence, reduces fraud risks, and enhances operational efficiency. Additionally, the system fosters greater accountability, as any disputes can be resolved fairly through verifiable smart contract terms.

This work builds trust by verifying all listed furniture and enforcing rental terms fairly and transparently. The platform streamlines digital payments, reducing the risks of manual transactions, delayed payments, and potential disputes. By minimizing administrative overhead and automating verification processes, this work enables a more cost-effective and streamlined rental experience. The integration of blockchain ensures that users can rent furniture with full transparency and security, creating a trusted and fraud-resistant ecosystem. With This work, users gain access to a secure, efficient, and transparent furniture rental experience that redefines industry standards and fosters long-term trust among renters and owners alike.

The role of blockchain

Blockchain is a decentralized, secure, and transparent technology revolutionizing the rental industry. It maintains a distributed ledger across independent nodes, ensuring tamper-proof rental records. Each block contains rental transaction data and a header with key metadata, including timestamps, previous block hashes, and Merkle root. The Merkle tree structure organizes transaction hashes, ensuring data integrity. Operating on a peer-to-peer network, it

[a]shajahanp.cse@srit.ac.in, [b]214g1a05a4@srit.ac.in, [c]214g1a05a7@srit.ac.in, [d]214g1a0570@srit.ac.in, [e]214g1a05b1@srit.ac.in, [f]214g1a0591@srit.ac.in

DOI: 10.1201/9781003685364-91

eliminates intermediaries, reducing costs and enhancing trust. Smart contracts automate rental agreements, enforcing terms transparently. Through a consensus mechanism, validated transactions are permanently recorded, ensuring immutability, anonymity, and traceability in furniture rentals. Traditional furniture rental processes often rely on manual management, unsecured payment methods, and inefficient communication, leading to delays and customer frustration. To address these challenges, there is a growing need for a secure, user-friendly platform that connects renters and owners while ensuring smooth and reliable transactions.

Literature survey

Blockchain-based rental management has been explored across various industries, including property rentals, housing, and digital asset management. Its ability to enhance transparency, security, and automation makes it a powerful tool for streamlining rental transactions.

Buradkar et al. [1] showed that blockchain can reduce paperwork and prevent fraud in property rentals. This work uses the same idea for furniture rentals by using smart contracts. These contracts automate payments, keep agreements secure, and store records that cannot be changed. Renters can easily find, rent, and return furniture while keeping track of their rental history.

A common problem with online rentals is making sure deliveries happen as promised. Hasan and Salah [2] introduced a Proof of Delivery system to make transactions more secure. This work follows the same idea by confirming when furniture is delivered, automating payments, and keeping transaction records safe from changes. This reduces conflicts and helps renters and owners trust each other.

Trust and safety are important in online rentals. Srivastava et al. [3] found that many rental platforms have problems like payment fraud, contract changes, and unclear rules. This work solves this by using blockchain, which keeps rental agreements safe from changes. Payments are also secure, so there are no hidden fees or contract problems.

Blockchain is already making vehicle rentals easier by automating agreements. Thakur [4] showed that smart contracts help manage car rentals without needing middlemen. This work uses the same idea

for furniture rentals, making transactions simple, secure, and cost-effective by allowing renters and owners to deal directly.

Tamper-proof contracts are making rentals more secure. Chen et al. [5] created a blockchain system for housing rentals to prevent contract changes without approval. This work follows the same idea to protect rental agreements and deposits. Jaison and Suhas [6] also built a carpooling system that uses smart contracts to handle payments and solve disputes automatically. This work brings this same security and fairness to furniture rentals.

Hidden charges and fraud are common in rentals. Xue et al. [7] created a blockchain rental system to stop unfair pricing and contract changes. This work follows the same approach to ensure fair pricing, secure agreements, and transparent transactions. All rental activities are recorded and cannot be changed. This builds trust and makes renting safer for everyone.

Blockchain makes rentals safer by securing agreements, ensuring reliable payments, and keeping clear records. This work uses this technology to prevent fraud, hidden fees, and contract changes. Smart contracts handle payments automatically and keep everything transparent. This helps renters and owners trust the system. It makes furniture rentals simple, fair, and secure.

Implementation

System architecture and design

This work is a rental platform for furniture. It lets you browse, rent, and review items easily. You can feel safe knowing your transactions are secure. The platform uses blockchain technology for order tracking. Plus, it has a user-friendly website for smooth interactions.

The planned system uses a decentralized blockchain ledger. It safely records all transactions. Here are the main parts of the system:

1. **Front-end:** User interface built with HTML, CSS, and JavaScript.
2. **Back-end:** Application logic managed using Django framework in Python.
3. **Blockchain storage:** Secure transaction storage utilizing blockchain principles for immutability and transparency.

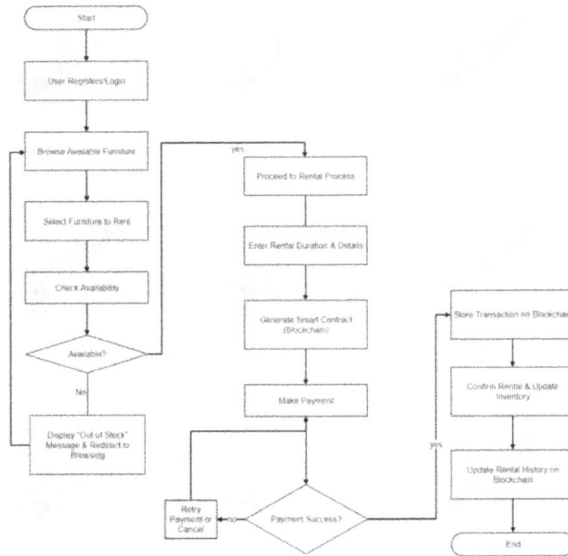

Figure 91.1 Workflow of the doctor stream
Source: Author

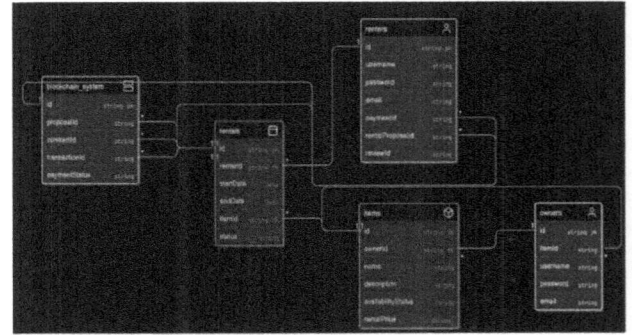

Figure 91.2 Entity-relationship diagram
Source: Author

The system (refer to Figure 91.1) is designed to make renting furniture simple, secure, and hassle-free for both users and administrators. Renters can easily sign up and log in, giving them access to a wide selection of furniture. They can browse through listings using search and filter options to quickly find what they need. Once they've chosen an item, they can add it to their cart and complete the payment securely through Razorpay. After placing an order, users can track their rentals in real time, keeping an eye on delivery updates and return schedules. They also have access to their rental history, making it easy to keep track of past orders. To build trust and improve the experience for future renters, users can leave ratings and reviews based on their experience. The goal of the system is to streamline the entire rental process—making it smooth, transparent, and user-friendly for everyone involved.

The admin module helps platform managers take charge of the rental system. Their job is to make sure everything is running well. Admins can log in safely and manage furniture listings. They can add new items, set rental prices, and update what's available. They also organize products into categories so it's easy for users to find what they need. Besides managing listings, admins also keep an eye on rental transactions. They track orders and check off deliveries as complete. They review customer feedback and ratings to help improve the service. To keep renters

updated, admins can write blog posts and manage product categories. This way, users can see what's new and any important news. Blockchain verifies all payments. This makes sure they are safe and protected. Admins keep tabs on transactions and maintain security in the app. This all helps create a smooth rental experience for everyone.

ER diagram

The ER diagram(refer to Figure 91.2) shows how the system is set up. The blockchain system keeps all transactions safe and clear. This means no one can change the data. Renters can look for furniture, rent items, pay securely, and leave feedback. Each rental is logged into the Rentals section. This connects renters with the furniture they rented. The Items section has info about each piece of furniture, like if it's available, its price, and its category. Owners can manage their list by adding or updating furniture. This helps keep everything running smoothly with blockchain checks.

Smart contract development

Smart contracts define rules and penalties around an agreement and automatically enforce those obligations. While they can work independently, many smart contracts can also be implemented together. The integral components of a smart contract are termed objects. There are essentially three objects in a smart contract – the signatories, who are the parties involved in the smart contracts that use digital signatures to approve or disapprove the contractual terms; the subject of the agreement or contract; and the specific terms.

This work uses smart contracts made with Solidity. These contracts are set up with a tool called Truffle. Truffle helps you write, test, and deploy your contracts

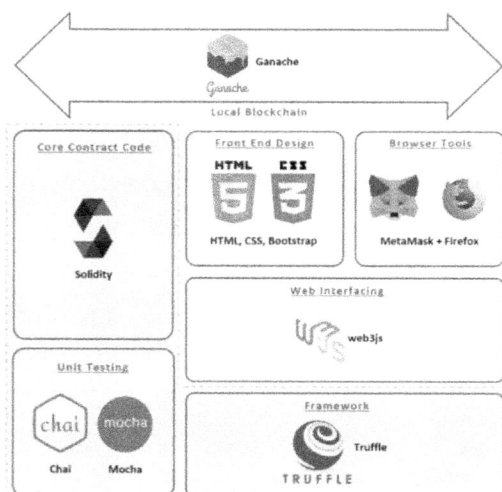

Figure 91.3 Tech stack
Source: Author

easily. It also pairs up with Ganache for testing on a local blockchain. To connect the front-end with the blockchain, we use web3.js.

The truffle-config.js file is key for any Truffle project. It helps the framework work with the Ethereum blockchain. This file tells you about network settings. It covers things like the provider, network ID, and gas limits for different setups, whether you're working locally or on public test networks. You can also set compiler options here. This lets developers choose the Solidity version and turn on optimizations to make their contracts run better. For testing, it uses Mocha settings. You can change options like test timeouts and how reports look. Plus, you can choose where to save your smart contracts, migration scripts, and test files. This makes it easier to keep your project organized. By setting up truffle-config.js right, developers can simplify deploying, testing, and managing smart contracts in their Truffle projects.

For this work, truffle automates testing of smart contracts. This helps keep transactions safe for renters and furniture owners. The rental management system uses these smart contracts to create permanent records of item listings, rental agreements, and payments. There are also event-driven features in these contracts. They provide real-time updates on rental transactions and make the user's experience better.

Conclusion

This work makes renting furniture simple and safe. It builds trust between renters and owners. With blockchain tech (refer to Figure 91.3), it keeps transactions secure and transparent. Users can easily browse furniture options. They can book rentals, keep track of their rental history, and manage returns without stress. Smart contracts take care of payments and agreements on their own. This means less hassle and reduced risk. Through these solutions. This work fixes issues that come with traditional rentals. It offers a reliable and efficient way to rent furniture. This project shows how new technology can improve the furniture rental scene. It makes everything easier and safer for everyone.

Acknowledgement

We are truly grateful to Mr. P. Shajahan, Assistant Professor, Department of Computer Science and Engineering, Srinivasa Ramanujan Institute of Technology, for his constant support and guidance. His insights and encouragement were invaluable in shaping this project and bringing it to completion.

References

[1] Buradkar, K., Kori, S., Ruikar, S., Galfat, V., Patil, D., & Nasare, R. (2022). Property rental management system. *International Journal of Computer Science and Mobile Computing*, 11(11), 177–179.

[2] Hasan, H. R., & Salah, K. (2015). Proof of delivery of digital assets using blockchain and smart. *Institute of Electrical and Electronics Engineers (IEEE)*, 14(8), 65439–65448.

[3] Srivastava, H., Dixit, A., Dhiman, P., & Chauhan, A. (2023). RENT X: house and home rental websites a comprehensive analysis. *IJIRT Journal - Solaris Publication*, 11(6), 2274–2279.

[4] Thakur, A. (2021). Car rental system. *International Journal for Research in Applied Science and Engineering Technology (IJRASET)*, 9(7), 402–412.

[5] Qi-Long, C., Rong-Hua, Y., & Fei-Long, L. (2019). A blockchain-based housing rental system. In SCITEPRESS - Science and Technology Publications Lba, (pp. 184–190).

[6] Jaison, F., & Suhas, C. R. (2023). Peer-to-peer carpooling using blockchain. *International Journal of Advanced Research in Computer and Communication Engineering (IJARCCE)*, 12, 42–46.

[7] Xue, Q., Hou, Z., Ma, H., Zhu, H., Ju, X., & Sun, Y. (2021). Housing rental system based on blockchain technology. In Journal of Physics: Conference Series on Internet of Things, Artificial Intelligence and Mechanical Automation (IoTAIMA), (Vol. 1948, pp. 14–16).

92 Lung cancer detection using ensemble techniques

Chinna Pullaiah, G.[1,a], Y. Chaitra[2,b], V. Mounika[2,c], D. Jeevan Kumar[2,d], G. Lakshmi Narasimha Reddy[2,e] and B. Shashikala[2,f]

[1]Assistant Professor, Department of CSE, Srinivasa Ramanujan Institute of Technology, Anantapur, India

[2]Student, Department of CSE, Srinivasa Ramanujan Institute of Technology, Anantapur, India

Abstract

The image classification methodology presented in this study is a hybrid one, involving both deep learning and standard machine learning algorithms. Image preprocessing is performed using the ImageDataGenerator class, which resizes input images to a dimension of 224×224 pixels, suitably for the input size required, and normalizes the values in the pixels to the range [0, 1]. The feature map extracted from the MobileNet architecture (without the top layer) pre-trained on ImageNet is flattened using global average pooling. After extraction of features, classifiers for the final classification task include Support Vector Machine (SVM) and Random Forest (RF) combinations, RF and Logistic Regression (LR) combinations, SVM and LR combinations, and one complete ensemble that incorporates SVM, RF, and LR. With regard to prediction aggregation, soft voting is utilized in predicting classifiers, where the probability outputs from all classifiers are averaged. StratifiedKFold cross- validation guarantees the preservation of proportional balance for the training and validation sets across all classes. This procedure allows a better evaluation of model performance. Several evaluation metrics, including accuracy, precision, recall, F1-score, and confusion matrix, measure the effectiveness of the models and serve to contrast the performance of different combinations of classifiers.

Keywords: Cross-validation, ensemble learning, feature extraction, image classification, Logistic Regression, MobileNet, Random Forest, soft voting, StratifiedKFold, Support Vector Machine

Introduction

The goal of image classification is to provide some types of labels to input images on the premise of the content of the images or, to put it very simply, to categorize the images. Having ushered in new methods to deal with image classification, CNN became the dominant approach entrenched in paradigm due to their ability to learn various hierarchical features from raw image data automatically. However, they often require enormous labeled data and computational power, resources that may be hard to come by in resource-strapped environments.

The challenges mentioned earlier constitute the basis of the proposed hybrid strategy that utilizes deep learning plus other conventional techniques for image classification. This approach was designed precisely to exploit the strengths of both paradigms by using pretrained deep learning models for feature extraction and conventional machine learning techniques for classification. Based on a feature-extraction basis, the pre-trained MobileNet model is adopted in the implementation of this approach. The lightweight yet efficient MobileNet could, therefore, find its deployment in any mobile/embedded systems having limited computational resources. Trained on a large-scale dataset, the top layer of this network is disposed of so as to utilize it mainly for feature extraction. After passing through the global average pooling (GAP) layer that converts output feature maps into fixed-size feature vectors holding salient information about the image.

After feature vector generation, some of the common classification algorithms are adopted, including SVM, RF, and LR, either alone or in various ensemble combinations. The combination of multiple classifiers using methods like soft voting result in improved classification accuracy by enhancing the strengths of each classifier and down weighting any weaknesses of each single model.

The model's performance will be evaluated through various metrics, such as confusion matrix analysis, accuracy, precision, recall, and the F1-score. For further confirmation of generalization

[a]chinnapullaiah.cse@srit.ac.in, [b]214g1a0516@srit.ac.in, [c]214g1a0562@srit.ac.in, [d]224g5a0507@srit.ac.in, [e]214g1a0547@srit.ac.in, [f]214g1a0599@srit.ac.in

DOI: 10.1201/9781003685364-92

and reliability of the model across varying subsets of the data, StratifiedKFold cross-validation is used.

This project's main aim is to take around the performance of different classifier combinations (Support Vector Machine [SVM] + Random Forest (RF, RF + Logistic Regression [LR], SVM + LR, and fully ensemble all three classifiers) and to assess their capability in tackling image classification problems. By doing so, it aspires to enhance classification accuracy and propose a solution that is more efficient and scalable for real- world image classification tasks.

Literature review

Detection of lung cancer has achieved tremendous development in the last years primarily due to artificial intelligence capability advancement and machine learning techniques. Early detection improves the survival rate, leading researchers to suggest several novel techniques for improving prediction accuracy. For example, Alzahrani et al. proposed a model that involved synthetic data obtained from conditional tabular generative adversarial networks (CTGANs) and RF classifiers, yielding an accuracy of 98.93% [1].

Ayad et al. combined recursive feature elimination with support vector machines (RFE-SVM) for feature selection and further optimized XGBoost using the Nelder-Mead algorithm, obtaining a perfect 100% classification accuracy on two independent lung cancer datasets. Their study highlights the significance of advanced feature engineering and optimization techniques in lung cancer prediction [2].

Khanna et al. introduced an ensemble model for predicting VOCs in breath, urine, blood, and lung tissue by integrating five pre-existing models with one newly proposed classifier [3]. Saxena et al. conducted a comprehensive review analyzing 185 studies on AI-based tumor detection, emphasizing neural networks' growing role in cancer prediction and their potential to enhance diagnostic accuracy [4]. This review underscores the broader acceptance of AI in healthcare, particularly in improving early cancer diagnosis [5].

Arya et al. proposed a hybrid detection model combining numeric and CT image data. They applied Pearson correlation for feature selection and XGBoost as the classifier, achieving high precision and recall values [6]. Kumar et al. advanced lung cancer prediction by employing ensemble ML algorithms, including bagging, boosting, and voting, to enhance classification accuracy [7]. Jaswanth et al.

demonstrated 99.99% accuracy using SVM, Naïve Bayes, logistic regression, and KNN, further reinforcing the effectiveness of ML in early-stage lung cancer detection. Their study emphasized the importance of precision and recall in evaluating model performance [8].

Proposed methodology

In summary, by utilizing the advantages of both deep learning and conventional machine learning techniques, the suggested system seeks to increase image categorization accuracy. By removing the top layer of the MobileNet architecture, it uses a pre-trained model as a feature extractor. Fixed-size feature vectors are then produced by transferring the remaining feature mappings over the GAP layer. High-level information about the image data is encoded in the feature vectors, which are subsequently classified using different machine learning classifiers.

To determine the best ensemble for image classification, the suggested system will experiment with several classifier combinations. SVM, RF, and LR among the classifiers taken into account in the suggested system. The following ensemble configurations are taken into consideration

- SVM + RF
- RF + LR
- SVM + LR
- Full ensemble of all three classifiers

The architecture of the proposed hybrid image classification system is illustrated in Figure 92.1.

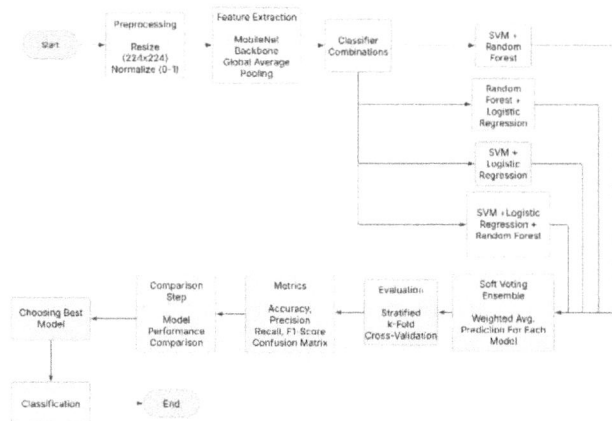

Figure 92.1 Architecture of proposed system
Source: Author

With this approach, the final projected label is the one that earns the most votes out of the averaged classifier votes, which is calculated by averaging the levels of the various classifiers' confidence in the predicted labels.

To sum up, the suggested method integrates standard machine learning classifiers with deep learning for extracting features in an ensemble architecture with the goal of improving performance and accuracy in image classification tasks. Hopefully, through experimentation with combinations of classifiers and soft voting for aggregating results, the system will achieve high performance reliably and generalizable across different datasets.

Data splitting

One must generate a learning set, a tuning set, and an evaluation set. The data that the model will use to learn is part of the training set. Separate data sources will be provided by the testing, and validation sets so that the model performance metrics can be verified with different data.

Training set occupies the largest volume of data, e.g., 70–80% of the total dataset which is used by the model in teaching it by changing parameters through iterative learning. The validation set usually makes up about 10–15% of the total dataset; it is applied while the model is being trained to tune hyperparameters and for assessing how well the model performs in action. Therefore, the overfitting phenomenon is mitigated through fine-tuning rather than assessing the data based on the training data itself.

The testing set consists of the remaining 10-15% of data reserved strictly for final evaluation after training ends and serves as a reliable measure of how much the model has learned about the unseen data in the generalization task. Thus, the test data is confined to final assessment and not used for training, ensuring rigorous evaluation of the model without any data leakage so that the results reflect the true capabilities of the model.

Model selection and training

Selection of the right model is one of the very important parts to ensure successful implementation of machine learning, especially in image classification. In this particular project, the primary model for classifying the image data is a CNN, which has proved very efficient in image processing tasks. Histopathological image analysis is an excellent choice for this purpose as it deals directly with pixel level raw information toward extracting differential layered features. The architectures of CNNs considered are from the basic CNNs to more advanced pre-trained models such as MobileNet with further fine-tuning. Depending on performance demands, other models may be considered based on various parameters such as accuracy, training duration, or computational performance. The other classifiers included in the parallel classification will be SVM, RF, and LR, thereby comparing their effectiveness with classified results.

Following the model's selection, the image technology for processing trains it by resizing, normalizing, and augmenting the photos. These fit by optimizing using Adam or SGD for the exercise set and updating weights and biases via backpropagation. The hyperparameters were tuned using techniques such as grid search or random search in order to optimize: learning rate, batch size, and number of epochs. Regularization methods, which consider the uncertainty endemic to the training dataset. After determining the best parameter values, performance efficiency will be recorded as validation on the validation data (refer to Figure 92.5 for training and validation accuracy over epochs) during the training phase and then tested again on the testing set as a gauge of classification accuracy and robustness.

Evaluation metrics

Accuracy: The proportion of accurate forecasts that were fulfilled out of all the predictions that were made.

$$Accuracy = \frac{TP + TN}{TP + TN + FP + FN}$$

Precision: Calculates the frequency of accurate predicted positives in relation to the overall number of positive predictions the model makes, hence concentrating on the dependability of positive classifications.

$$Precision = \frac{TP}{TP + FP}$$

Recall (sensitivity): calculates the percentage of real positive examples that the model detects properly.

$$Recall = \frac{TP}{TP + FN}$$

F1-Score: The harmonic mean of precision and recall serves as a balance between the two.

$$F1 = 2 \times \frac{Precision \times Recall}{Precision + Recall}$$

Confusion matrix: A table or matrix that could offer information on the precise performance of a classification system, including the numbers of predictions that are TP, TN, FP and FN.

Results and discussion

Performance metrics
Four ensemble models combining SVM, RF, and RF were evaluated for lung tissue classification using confusion matrices as shown in Figure 92.2, classification reports, and cross-validation accuracy scores to assess effectiveness.

Ensemble model 1 (SVM and RF)
The first ensemble model, combining SVM with RF, performed well on both validation and test sets. It correctly classified 592 adenocarcinoma, 624 benign, and 614 squamous carcinoma cases, with minimal misclassifications. In the test set, 42 adenocarcinoma and 16 squamous carcinoma cases were misclassified as benign, achieving overall high accuracy and balanced performance across classes.

Ensemble model 2 (RF and LR)
The second ensemble model, combining RF and LR, performed well. It correctly classified 593 adenocarcinoma, 624 benign, and 615 squamous carcinoma cases in validation. Misclassifications were minimal, with 31 adenocarcinoma and 10 squamous carcinoma cases incorrectly labeled. In testing, 41 adenocarcinoma and 15 squamous carcinoma cases were misclassified, maintaining high accuracy and balanced performance.

Ensemble model 3 (SVM and LR)
The third ensemble model, combining SVM and LR, demonstrated strong classification accuracy. It correctly predicted 592 adenocarcinoma, 624 benign, and 612 squamous carcinoma cases in validation, with 33 misclassified. In testing, it correctly classified 589 adenocarcinoma, 625 benign, and 608 squamous carcinoma cases, with minor misclassifications, maintaining overall high accuracy and balanced performance.

Table 92.1 Cross-validation mean accuracy scores for ensemble models.

Ensemble Model	CV Accuracy
Ensemble Model 1 (SVM and Random Forest)	0.96
Ensemble Model 2 (Random Forest and Logstic Regression)	0.97
Ensemble Model 3 (SVM and Logistic Regression)	0.96
Ensemble Model 4 (SVM, Random Forest and Logistic Regression)	0.97

Source: Author

Ensemble model 4 (SVM, RF, and LR)
The fourth ensemble model, combining SVM, RF, and LR, achieved the highest accuracy and balance among all classes. In validation, it correctly classified 590 adenocarcinoma, 624 benign, and 613 squamous carcinoma cases. The test set showed similar performance, with 587 adenocarcinoma, 625 benign, and 610 squamous carcinoma cases correctly classified. With an overall accuracy of 0.97, minimal misclassifications ensured a well-balanced and highly reliable model.

Cross-validation mean accuracy
The above table summarizes the cross-validation results for all models. The summarized results are presented in Table 92.1.

Perfect model selection
In fact, Ensemble Model 2 and Ensemble Model 4 acquired the highest score in their cross-validation

Confusion Matrix:

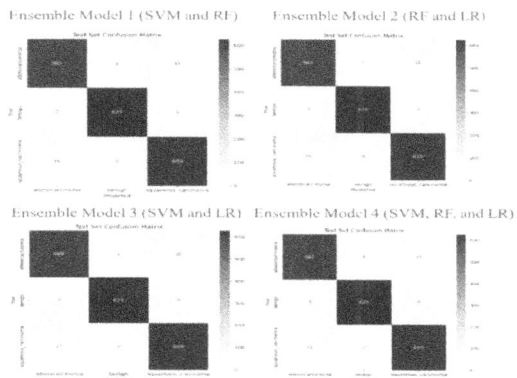

Figure 92.2 Confusion matrices of different ensemble models
Source: Author

Classification Report:

```
Test Set Classification Report:
                    precision    recall  f1-score   support

    adenocarcinoma       0.98      0.94      0.96       625
            benign       1.00      1.00      1.00       625
squamous_carcinoma       0.94      0.98      0.96       625

          accuracy                          0.97      1875
         macro avg       0.97      0.97      0.97      1875
      weighted avg       0.97      0.97      0.97      1875
```

Figure 92.3 Classification report of best model
Source: Author

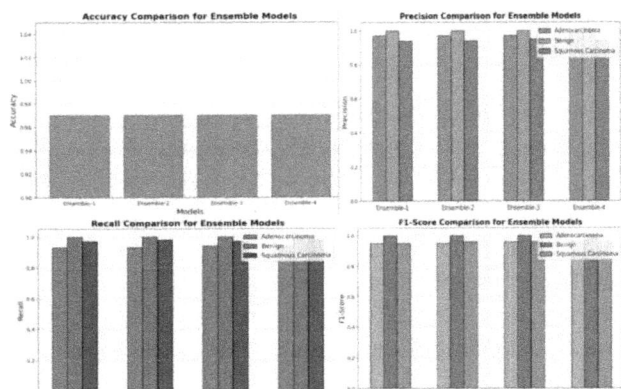

Figure 92.4 Accuracy, precison,recall, F1-score comaprisons with all ensemble models
Source: Author

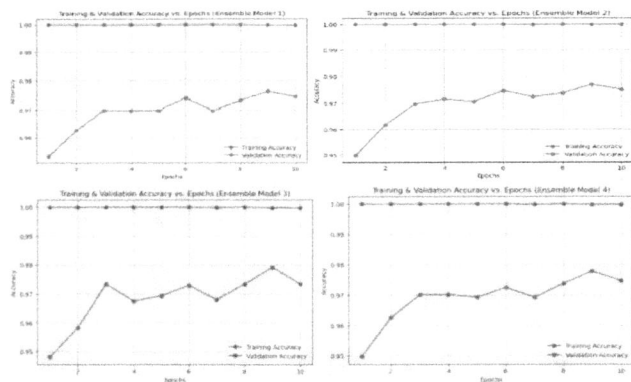

Figure 92.5 Training & validation accuracy vs. epochs graphs of all ensemble models
Source: Author

mean accuracy of 0.97. Therefore, the model selected was Ensemble Model 4 as it was aimed at mounting together all three classifiers to furnish a robust and balanced performance across all classes.

The classification report (Figure 92.3) confirms excellent model performance across all categories. Adenocarcinoma achieved an F1 score of 0.96, with 0.98 precision and 0.94 recall. Benign cases showed perfect scores (1.00) for precision, recall, and F1. Squamous carcinoma recorded an F1 score of 0.96, with precision at 0.94 and recall at 0.98, indicating strong classification accuracy.

Conclusion

The mean accuracy scores obtained via cross-validation of the four ensemble models present evidence for the collective combination of different classifiers for enhancement of image classification performance. Ensemble Model 2 and Ensemble Model 4 showed the highest accuracy scores of 0.97, suggesting a clear advantage of combining multiple classifiers in good practice to capture different patterns and improve robustness. Slightly lower accuracy scores of 0.96 were obtained by Ensemble Model 1 (SVM and RF) and Ensemble Model 3 (SVM and LR), still classifying quite properly. The strong metrics exhibited by these models rated excellent in classifying all categories. Results from different models show consistent classification accuracies (see Figure 92.4), adding considerable merit to ensemble learning for boosting classification performance towards becoming a trustworthy solution for actual image classification tasks. Thus, models comprising SVM, RF, and LR, particularly in parallel combinations, excelled.

Acknowledgement

The authors express their deepest appreciation to the students, faculty, and administration of the Department of Computer Science and Engineering for their sustained support and contribution to this study.

References

[1] Alzahrani, A. (2025). Early detection of lung cancer using predictive modeling incorporating CTGAN features and tree-based learning. *IEEE Access*. vol. 13, pp. 33815–33825.

[2] Al-Jamimi, H. A., Ayad, S., & El Kheir, A. (2025). Integrating advanced techniques: RFE-SVM feature engineering and nelder-mead optimized XGBoost for accurate lung cancer prediction. *IEEE Access*. vol. 13, pp.29589–29600.

[3] Khanna, D., Kumar, A., & Bhat, S. A. (2025). Volatile organic compound for the prediction of lung cancer by using ensembled machine model and feature selection. *IEEE Access*. vol. 13, pp. 9809–9820

[4] Saxena, N., Yadav, S. S., Dujawara, A., & Kumar, A. (2024). Prediction and detection of cancer through machine learning: a review. In 2024 1st International Conference on Advances in Computing, Communication and Networking (ICAC2N), (pp. 417–423). IEEE.

[5] Swain, S., Jayasingh, S. K., Patra, K. J., & Tripathy, A. R. (2024). Innovative approach for early detection and risk assessment of lung cancer using machine learning techniques. In 2024 International Conference on Intelligent Computing and Sustainable Innovations in Technology (IC-SIT 2024), (pp. 1–6).

[6] Krishna Kumar V., Saravana Balaji B, Sabarmathi K R. & Ahmed Najat Ahmed. (2024). Enhanced lung cancer prediction using ensemble machine learning algorithms. In 2024 International Conference on Emerging Research in Computational Science (ICERCS), (pp. 1–5), Dec. 2024.

[7] Vanitha, K., Balajee, A., Mahesh, T. R., Vivek, V., Kumar, Y., & Kasai, D. R. (2024). An extensible CNN Model for lung cancer diagnosis using CT images. In 2024 International Conference on Emerging Research in Computational Science (ICERCS), (pp. 1–5), Dec. 2024.

[8] Jaswanth, K. G., Adithya, P. P., Eekshith, G. S., Hariharan, S., Kukreja, V., & Geetha, D. (2024). Real time lung cancer prediction using lazy learning technique. In 2024 International Conference on System, Computation, Automation and Networking (ICS-CAN), (pp. 1–5), Dec. 2024.

93 Enhanced performance of grid connected photovoltaic (PV) system using hybrid energy storage systems

K. Meenakshi[1,a], Y. Nagaraja[2,b] and K. Meenendranath Reddy[3,c]

[1]PG Student, Department of EEE, SITS, Kadapa, Andhra Pradesh, India

[2]Associate Professor, Department of EEE, SITS, Kadapa, Andhra Pradesh, India

[3]Assistant Professor, Department of EEE, SVR Engineering College, Nandyal, Andhra Pradesh, India

Abstract

Decentralized systems that interact with one another and with the electrical grid as a whole have revolutionized and progressed the electric power system. Integrating energy storage systems into microgrids is encouraged due to the fact that energy limitations and environmental concerns are driving forces behind the key problems of these systems. Energy storage technologies enhance the reliability and efficiency of microgrids. A hybrid energy storage system (HESS) for microgrid usage is suggested by this research, which combines superconducting magnetic energy storage (SMES) with batteries. This study presents a new cooperative control technique for a grid-connected system that uses photovoltaics to solve the problem of unstable output power caused by the inherent unpredictability and fluctuation of RES. Within the scope of a hybrid energy storage system (HESS), this technique makes use of both BESS and SMES. To achieve optimization of transient power allocation and dynamic regulation of the filtration coefficient, the control method utilizes a fuzzy control-based low-pass filter (LPF) at top-level control (TLC). Keeping the DC bus voltage stable using a one-step prediction horizon is accomplished simultaneously at the under-layer level control (ULC) using a rapid model predictive control (MPC) technique. Effectively validated under varied operating situations is the practicality and effectiveness of the suggested control technique, which proves to be effective in improving the grid-connected system's stability and performance. Utilizing the MATLAB/Simulink platform, a comparison is made between the effectiveness of analysis of grid-connected systems with neural network controllers and fuzzy logic controllers.

Keywords: Fuzzy lowpass filter, model predictive control, SMES/battery

Introduction

The use of microgrids in electricity systems is on the rise. Microgrids offer several benefits, including more reliable systems and more efficient local energy supply. Power generation components, energy storage devices, and loads make up the bulk of microgrids [1]. Microgrids are increasingly incorporating renewable power production technology in an effort to lessen their negative impact on the environment. The intermittency and unpredictability of renewable power generation are characteristics that set it apart from fossil fuel power generation. Because of this, the reliability of renewable power generation is compromised [2]. Consequently, the importance of reliable power sources is growing in tandem with the expansion of renewable power generation. Microgrid voltage control can be improved by integrating energy storage technologies into the power grid. The device that can transform electricity into other

forms of energy and then feed it back into the system when needed is known as the energy storage system. The utilization of energy storage devices to address power balance, frequency control, voltage stabilization, and other related issues has been the subject of several studies. Power system applications have seen the development and testing of several energy storage devices [3-5].

Because it uses a renewable energy source that never runs out and requires little to no initial investment, photovoltaic (PV) generating is quickly rising to the top. But there are major problems with the electricity infrastructure that come with switching to a lot of sustainable energy. The power variations of renewable energy sources (RES) are a major cause of system uncertainty since they surpass demand fluctuations and are caused by the inherent intermittent and volatile nature of new energy sources. This study presents a HESS that combines energy-type

[a]meenakshikallempudi@gmail.com, [b]nagarajaeps1@gmail.com, [c]kypa.meenendranathreddy@gmail.com

DOI: 10.1201/9781003685364-93

and power-type ESSs to improve the dependability and stability of PV grid-connected systems, in order to overcome these issues. With the combination of energy-type ESSs' high-power density with millisecond-level response characteristics and energy-type ESSs' long-timescale operation capabilities, the HESS is designed to meet the dynamics of multi-time scale energy needs. The energy-type energy storage system (ESS) is designed to compensate for sluggish average power demand changes through the use of a BESS having autonomy features [6]. However, for HESS transient power correction, the power type ESS—such as a SMES system—is the way to go because of its low self-discharge rate and high-power density [7]. The BESS's operating lifespan has increased because to this integrated approach, which reduces battery stress and mitigates power fluctuations. For the HESS to work well regardless of changes in load demand and PV generating power, a control system that optimizes transient power allocation must be developed [8]. Maintaining energy balance across various types of ESSs and achieving precise power dispatch are the key control objectives of the HESS. Low pass filter (LPF) power flow regulation of HESS is one of the manage approaches investigated in the past for RES grid-connected systems [9]. Nevertheless, the impact of a predetermined cutoff frequency on HESS's state of charge (SOC) is frequently overlooked in these endeavors. To forecast the irradiance profile for the following day, use a clear sky model and initialize the filter time constants to their default values. Then, adaptively modify the starting value of the filter time stable according to the current control ramp rate [10]. However, the technique is difficult to implement due to the vast amounts of data it needs to store and the amount of computer power it consumes during implementation. In addition, DC-DC converters that use super capacitors and BESS have been effectively implemented using MPC, an influential structure for power regulation [11-12]. In light of the foregoing, this study aims to devise a unique approach to controlling HESS in PV grid-connected systems by combining fuzzy LPF with fast MPC. The suggested method makes it easier to maximize the BESS's operational and functioning life as well as its transient power allocation.

Energy storage systems

It was emphasized that an ESS is necessary to enable extensive use of DG in the main grid. But there are a lot of additional potential advantages for an ESS that are examined here. The economic benefits are disregarded in favor of the technological ones. The technical benefits of an ESS may be mostly split into three parts: power generation and transmission, power system operation, and integration of RES.

Superconducting magnetic energy storage

A Superconducting magnetic energy storage (SMES) retains energy in the magnetic field of a large superconducting material coil, which, at very little temperatures (around -2700C), provides roughly no resistance to the current. The liquid in which the coil is submerged must maintain a very low temperature. Approximately 1.5 kW of continuous power is required by the refrigeration system for every MWh of storage capacity. The main components of a SMES storage system are illustrated in Figure 93.1. In most cases, a SMES has a round-trip efficacy of 95%. Discharging the SMES in its entirety will not shorten its service life. Due to its quick responsiveness, the SMES is utilized for storing activities with short durations. New studies are experimenting with SMES ranging from 10 to 100 MW with a reserve time of minutes, whereas its usual power rating is 1 to 10 MW for seconds.

Battery energy storage system

An electrochemical cell, or cells, is a component of a battery that may transform chemical power into electrical energy. The convenience and durability of batteries have made them the go-to choice for energy storage. Energy storage devices that employ lead acid have a history that begins in the mid-1800s. In spite of their short lifespan and relatively poor energy density, lead acid batteries are widely

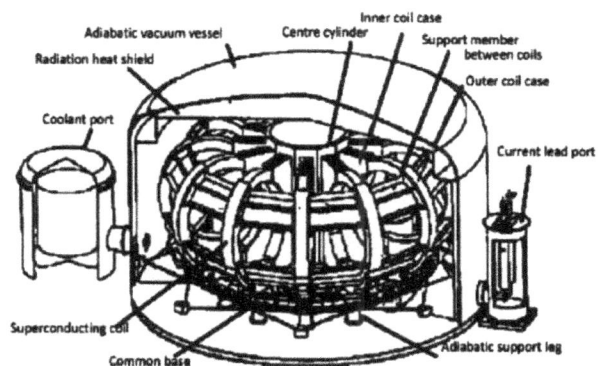

Figure 93.1 SMES parts
Source: Author

used in applications that are price sensitive, such as automobile starting, lighting with ignition, and uninterruptible power supplies (UPS), as well as in environments that need durability and abuse tolerance. Aiming to boost power and energy density, recent advancements aim to substitute result with lighter materials like carbon. The lithium-ion battery has a little self-discharge, no memory effect, also a more energy-to-weight ratio. Having said that, Li-ion batteries are not cheap. It is currently used in a variety of electronic products, such as computers, cameras, and mobile phones. Because of its great energy density, lithium-ion batteries have a lot of potential for use in wind energy and plug-in hybrid/electric car applications, among others.

One or more energy storage devices can be integrated into a system to smooth out power fluctuations from renewable resources. In order to power renewable energy producing systems, a wide range of energy storage technologies have been suggested. There are essentially two types of energy storage technologies: those that store a lot of energy but take a long time to react (like pumped hydro storage) and those that store a little energy but react quickly (like SMES). Superconducting magnetic energy storage, pumped hydro, batteries, flywheels,

and supercapacitors are the many energy storage devices. Table 93.1 shows a similarity of several ES technologies.

Proposed system

A detailed representation of the PV grid-connected system setup with a HESS is shown in Figure 93.2. The main renewable energy source (RES) in the system is a photovoltaic (PV) system. The DC bus receives electricity from this PV system via a DC-DC boost converter.

Meanwhile, the DC bus makes use of a two-level VSC to efficiently transmit electricity to the grid. The BESS and SMES are linked to the DC connection separately using bidirectional DC-DC converters to reduce power fluctuations. An MPPT algorithm regulates the PV converter, which is in charge of efficiently extracting electricity. Optimal power transmission during dynamic temperature and irradiance circumstances is a key function of this program, which incorporates a perturb and observe (P&O) technique. By comparing the current and voltage readings from the PV cell with its previous value, the MPPT algorithm, which is based on the P&O control technique, determines the output power of the PV panel. Figure 93.3 shows the MPPT-based control

Table 93.1 Energy storage differences

	Response time	Energy Density (Wh/Kg)	Power (W/Kg)	Charge Discharge Efficiency	Durability Cycle	Time Durability (Yrs)
PHES	Mins.	0.3	-	75%	-	>75
BESS	1min	30-200	150-3000	65-100%	500-1000	2-10
FESS	15s	100-130	1000	90%	>20000	>20
CAES		10-30	-	50%	-	>40
SCES	1s	LOW	VERY HIGH	93%	10000-100000	-
SMES	ms level	30	VERY HIGH	98%	-	-

Source: Author

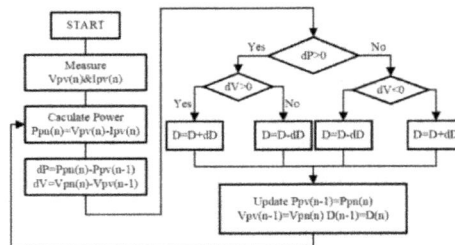

Figure 93.3 Flow diagram of MPPT algorithm
Source: Author

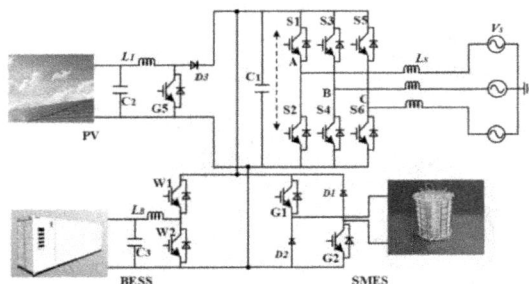

Figure 93.2 Structure of PV grid-connected system integrated with the HESS
Source: Author

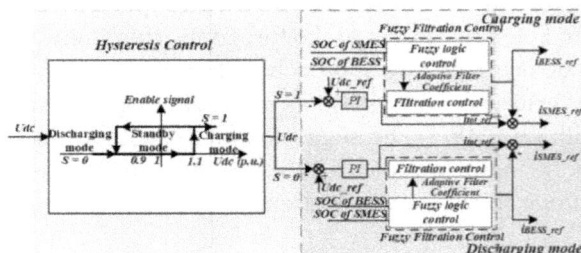

Figure 93.4 Structure of fuzzy logic control algorithm for the HESS
Source: Author

Table 93.2 Fuzzy rules of time constant T under DM and CM.

SOC_BESS \ T \ SOC_SMES	PS	PM	PB	PS	PM	PB
PS	PM	PM	PB	PM	PS	PS
PM	PS	PB	PB	PB	PB	PM
PB	PS	PM	PS	PB	PS	PS
Operation mode	DM			CM		

Source: Author

flow diagram, which helps to explain the process's complexities.

Two main tasks necessary for effective functioning are included in the suggested control method. Improving the efficiency of power transfer among SMES & the battery energy storage system (BESS) is its primary goal. Second, it stresses the need to quickly monitor and stabilizing voltage variations caused on the DC side by power fluctuations.

Figure 93.4 shows the structure of the suggested TLC approach. Based on the control signal S, HESS's three distinct operation modes—discharging, standby, and charging—are governed by hysteresis control. The DC bus voltage is regulated by a proportional-integral (PI) control mechanism, which derives the total reference current for HESS. To get the low-frequency current that the battery system uses as a reference, we first analyze the total current signal that was sampled from the hybrid energy storage by means of the LPF. In this way, the SMES successfully controls the BESS's uncompensated high-frequency current. The efficiency of HESS is dependent on the LPF's cut-off frequency controlling a well-thought-out set of thresholds for low-frequency current activities. An algorithm for fuzzy filtering control is used to keep HESS from being overcharged or over discharged and to make the BESS last longer. By controlling the LPF's time constant, this control technique restores a safe state of charge for the HESS by exchanging current depending on the power balance. Table 93.2 displays the fuzzy inference rules in detail, and it also shows the fuzzy inference surface for both charging and discharging modes.

Simulation results

The PV grid-connected system, integrated with the proposed HESS and control model illustrated in Figure 93.2, is meticulously simulated using MATLAB/Simulink.

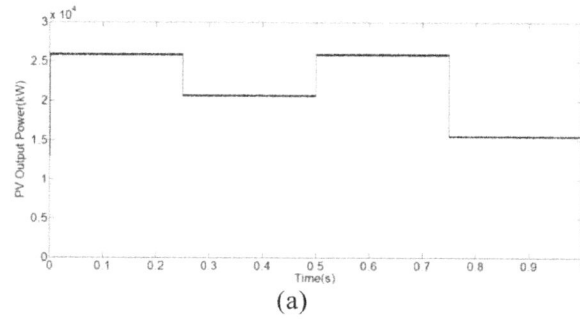

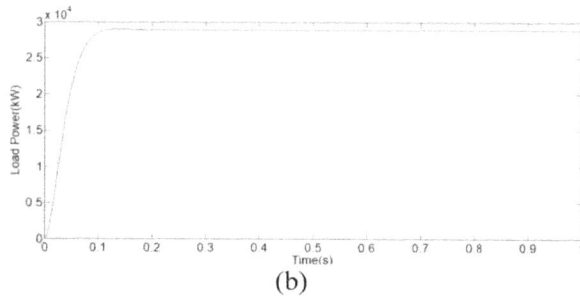

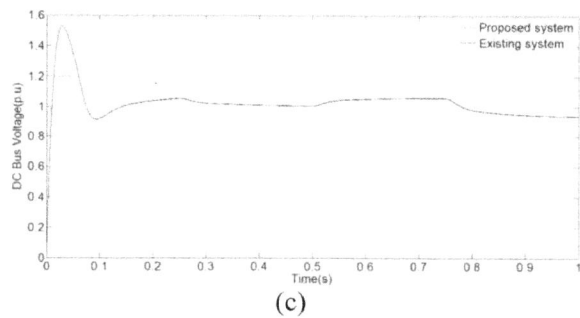

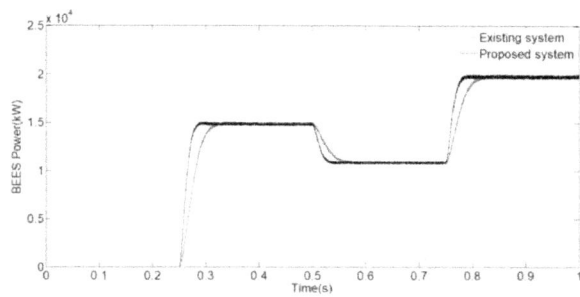

Figure 93.5 Simulation results under PV output power fluctuation. (a) PV output power, (b) load consumption power, (c) DC bus voltage, and (d) output power of HESS
Source: Author

Figure 93.5 shows the dynamics of PV output power and the HESS's performance in the case of fluctuating PV output power, using the proposed control method. Variability in irradiance and temperature, for example, can cause PV output power to fluctuate, as seen in plot (a). Plots (b) and (c) show that the HESS does a good job of balancing the DC bus voltage contained by the given reference range and accounting for the power variation among PV generation with load consumption. When the SMES's output power is enough to meet the load demand, as shown in plot (d), the BESS's charging and discharging operations are significantly reduced. This leads to a longer battery cycle life by reducing the battery's responses to transient high-power.

Conclusion

In order to stabilize the DC bus voltage and handle power fluctuations in PV grid-connected systems, this project introduces a hybrid energy storage system (HESS) and a control scheme that go hand in hand. Two energy storage systems, the Battery energy storage system (BESS) and the magnetic energy storage (SMES), are integrated into the HESS. The BESS and SMES power assignment is being dynamically regulated using a fuzzy filtration control method that is based on the SOC scheme. In addition, the HESS can keep the load side voltage stable and quickly monitor and control the voltage changes of the BESS and SMES thanks to an included fast computing FCSMPC algorithm. The control scheme greatly decreases the charge/discharge frequency of the BESS, which greatly increases the operational lifespan of the HESS by utilizing the SMES system's inherent high power capabilities to handle disturbances. Lastly, the suggested control strategy is rigorously tested under different operating conditions, including power fluctuations in PV generation with load consumption, to prove that it is superior and feasible. Utilizing the MATLAB/Simulink platform, a comparison is made between the effectiveness of analysis of grid-connected systems with neural network controllers and fuzzy logic controllers. This will improve the grid-connected system's stability and performance.

References

[1] Fu, X., & Zhou, Y. (2023). Collaborative optimization of PV green houses and clean energy systems in rural areas. *IEEE Transactions on Sustainable Energy*, 14(1), 642–656.

[2] Ben Ali, I., Turki, M., Belhadj, J., & Roboam, X. (2018). Optimized fuzzy rule based energy management for a battery-less PV/wind-BWRO desalination system. *Energy*, 159, 216–228.

[3] Kreeumporn, W., & Ngamroo, I. (2016). Optimal superconducting coil integrated into PV generators for smoothing power and regulating voltage in distribution system with PHEVs. *IEEE Transactions on Applied Superconductivity*, 26(7), 1–5. Art. no. 5402805.

[4] Jin, J. X., Zhang, T. L., Yang, R. H., Wang, J., Mu, S., & Li, H. (2023). Hierarchical cooperative control strategy of distributed hybrid energy storage system in an island direct current microgrid. *Journal of Energy Storage*, 57, 106205.

[5] Li, J., Yao, F., Yang, Q., Wei, Z., & He, H. (2022). Variable voltage control of a hybrid energy storage system for firm frequency response in the U.K. *IEEE Transactions on Industrial Electronics*, 69(12), 13394–13404.

[6] Vrettos, E. I., & Papathanassiou, S. A. (2011). Operating policy and optimal sizing of a high penetration RES-BESS system for small isolated grids. *IEEE Transactions on Energy Conversion*, 26(3), 744–756.

[7] Jin, J., Sheng, G., Bi, Y., Song, Y., Liu, X., Chen, X., et al. (2021). Applied superconductivity and electromagnetic devices – principles and current exploration highlights. *IEEE Transactions on Applied Superconductivity*, 31(8), 1–29. Art. no. 7000529.

[8] Arunkumar, C. R., Manthati, U. B., & Punna, S. (2022). Super capacitor voltage based power sharing and energy management strategy for hybrid energy storage system. *Journal of Energy Storage*, 50, 104232.

[9] Alafnan, H., Zhang, M., Yuan, W., Zhu, J., Li, J., Elshiekh, M., et al. (2018). Stability improvement of DC power systems in an all electric ship using hybrid SMES/battery. *IEEE Transactions on Applied Superconductivity*, 28(3), 1–6. Art. no. 5700306.

[10] Abdalla, A. A., Moursi, M. S. E., El-Fouly, T. H., & Hosani, K. H. A. (2023). A novel adaptive power smoothing approach for PV power plant with hybrid energy storage system. *IEEE Transactions on Sustainable Energy*, 14(3), 1457–1473.

[11] Mardani, M. M., Khooban, M. H., Masoudian, A., & Dragicevic, T. (2019). Model predictive control of DC–DC converters to mitigate the effects of pulsed power loads in naval DC microgrids. *IEEE Transactions on Industrial Electronics*, 66(7), 5676–5685.

[12] Ni, F., Zheng, Z., Xie, Q., Xiao, X., Zong, Y., & Huang, C. (2021). Enhancing resilience of DC microgrids with model predictive control based hybrid energy storage system. *International Journal of Electrical Power and Energy Systems*, 128, 106738.

94 AI-powered intelligent system for car damage detection and repair cost estimation

Kalaimathi, B.[1,a], Jayanthy, S.[2,b], Bharathkumar, G.[3,c], Akshaya, A. V. R.[2,d] and Dharanya, P.[3,e]

[1]Assistant Professor, Department of ECE, Sri Ramakrishna Engineering College, Tamil Nadu, India

[2]Professor, Department of ECE, Sri Ramakrishna Engineering College, Tamil Nadu, India

[3]UG Students, Department of ECE, Sri Ramakrishna Engineering College, Tamil Nadu, India

Abstract

Insurance claims play a crucial role in today's digital society. It acts as the platform for people to safeguard them against economic loss. The former way of insurance is observed in the India, Babylon as well as in China which is found by the merchants and traders during their journey to mitigate the risk factors. In medieval Europe the insurance background was formalized. By focusing on maritime trade in 1347 the first insurance contract was signed in Genoa and Italy. The process of insurance claiming proceeds with the manual procedure for application is common everywhere, and this would take days to complete the whole procedure. Dealing with the online process of policy application takes less time as it is formalized in the form of digital documentation which enables easier access to the policies at anywhere and it also provides 24/7 service to the individuals. This article aims at design an effective way of automatic damage detection and prediction of damage cost of vehicle using the computer vision models utilizing Mask R-CNN model and personal assistant as a Chat Bot using Llama 3.3 70B with knowledge source from vector database and online resources providing agent like communication to proceed, which is an effective process that gives the results at a higher speeds and focus mainly on the accident scenes that possess damaged parts of the vehicle like mirror and door damages and rims as well as tire damages, bumper damage, hood damages mainly the headlight damages. The images of these damaged parts are considered as the dataset and achieved through custom AI Agents incorporated with image classifying capabilities of CNN. The purpose of the LLM is to generate the text based on the user query and answer according to the results from the damage detection agent and format the response based on user requirements. For a specific application the LLM must be tuned and this property of LLM have made radical shift in the transformation of Natural Language Processing with Several criterion and are trained with self- labeling learning on a vast amount of text.

Keywords: Deep learning, large language model, mask R-CNN, natural language processing

Introduction

The Meta has recently introduced Llama 3.3 70B, the open- source large language model (LLM) [1]. It is designed to be more efficient and cost-effective and performance that rivals its predecessor. This model was employed with advanced techniques such as dynamic attention scaling and checkpoint to manage computational efficiency while maintaining high accuracy. The model has been trained on a dataset, incorporating with a wide range of text sources. Techniques such as data augmentation, reinforcement learning from human feedback (RLHF), and continual learning have been employed to the model and making the robust enforcement. With the increased of Llama 3.3 70B [2], concerns such as bias, misinformation, and data privacy are addressed through continuous testing and ethical timings are implemented. The tools used with LLM framework are CrewAI, GroqAI, Chroma DB, Jina, Llama 3.3, Langchain, React and VS Code. CrewAI is a platform designed for collaborating and managing AI projects. It has a version controlled tools, and the model is trained and deployed, making it easier to work with the model. CrewAI is transfigures the automotive industry by enabling car damage and price detection through an intelligent chat bot [3]. By integrating advanced computer vision and natural language processing (NLP), the damaged vehicles of images are allowed by crew AI to upload the images

[a]kalaimathi@srec.ac.in, [b]jayanthy.srec.ac.in, [c]bharathkumar.2202029@srec.ac.in, [d]akshaya.2202013@srec.ac.in, [e]dharanya.2202040@srec.ac.in

DOI: 10.1201/9781003685364-94

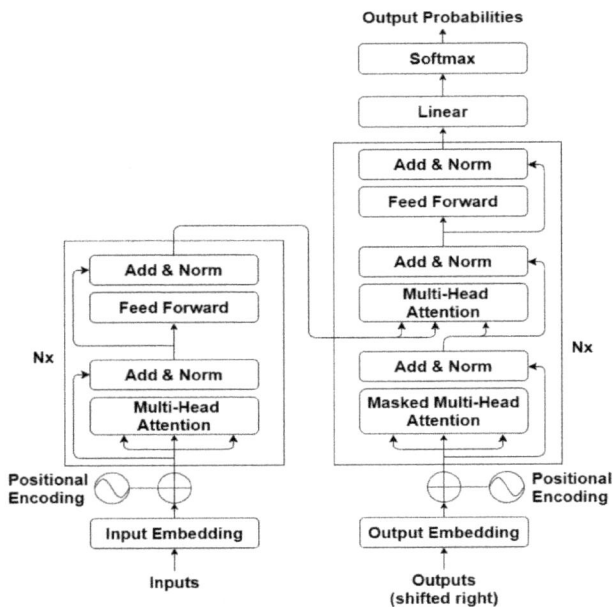

Figure 94.1 Structural architecture of large language model (LLM)
Source: Author

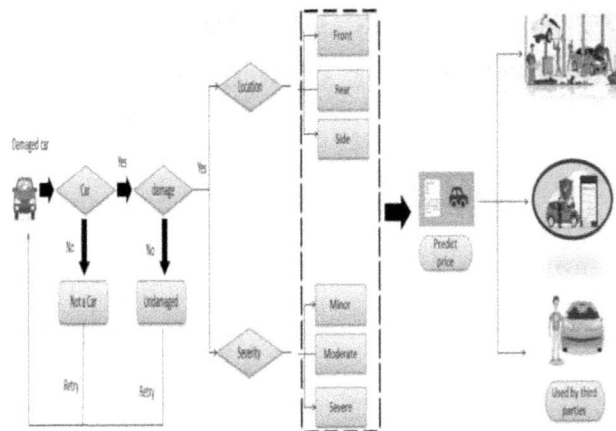

Figure 94.2 Existing systematic approach of analyzation and prediction
Source: Author

by analyzing the images the chat bot will give the estimated cost with the repair prices [4]. This Crew AI continuously works with the real time database and makes valuable tools for both users and service providers in the automated sector.

The above Figure 94.1 illustrates the transformer model of LLM architecture which is used in natural language processing. It mainly consists of two nodes encoder and the decoder through multiple identical layers the input data is processed by the encoder, and it contains several layers they are input embedding where the dense vector is converted from the input tokens. Positional embedding where the positional information are added since to understand the model lacks. Multi head attention where this layer allows the various parts of input. The feed forward layer is fully connected network where the further input is transformed. Finally add and normal layer in which it applies residual connections to improve the training and stabilize the normalization. Another part is defined as decoder which is the right side where the output sequence is generated and it also consists multiple layers which includes, Output embedding where the tokens of output are converted to dense vectors [5]. Masked Multi- Head Attention where it ensures each position where the future information is prevented. encoder-decoder attention where it mainly emphasizes the encoder's output on relevant

parts. Feed forward which is similar to the encoder where each position is integrated individually. Add and Normal which adds residual connections and the output is normalized [6]. Finally, all the linear layer is followed by a normalized exponential function and the final predictions are provided.

Workflow

The Figure 94.2 illustrate with the basic identification of the object whether it is a car or something else, initially this step ensures with the brilliant model that mainly focuses on the exclusively prominent inputs as well as the non-car objects that may interfere with the detection of damage workflow, the main key features that involves in this system is the image analysis and the accuracy . A specific model that is trained using the deep learning model is used to classify the images based on the users input and this pre-trained model may be MobileNetV2 and the Mask R-CNN. Moving with the decision flow it decides and ensures that the object detected is a vehicle. If it is vehicle like car or bike then the process is moved ahead and if it is detected as non-cared objects then it retry the mechanism again and again until it grabs the correct input , thus avoids the main errors that happens in each steps, once it is confirmed with the estimation then it moves on with the checking of the visible damaged parts of the vehicle in case if there is no detection of the damaged parts of the car means undamaged then the process gets terminated or else it proceed to the next state of analysis the third party mainly collaborate with the two main aspects namely location and severity that deals with the damage the location of damage is further classified into three divisions namely front, rear and side and dealing with the techniques been used is the pre-trained models such as Mask R- CNN that is used for framing the boxes up to a boundary to indicate the affected areas of the vehicle another technique dealing with the system is the heatmap, visual representation of damaged part or region that is useful for further investigation. Severity of damage is further classified into three main categories namely Minor, Moderate and severe. This Minor involves scratches and minor paint damages. Moderate involves structural impacts and thus by using the edge detection and the contour analysis the system deals with the price prediction.

The main step that is used for the segregation between the users as well as the admins with the provides access key and passcode. The main role of this part is used as the interaction between admins and the users, the third entity is the chatbot core is simply termed as bot that acts as the prime platform for the admins to interface with the system. The simplified architecture of bot consists of the following parts the knowledge , the multiple agents and the fall back, the knowledge contains the predefined recommendations and the basis of theme that are precisely responsible for the exact responses that is used for the clarification of the users that emerges during the process and therefore the accurate data and resources are pulled by the precise information this also supports the Knowledge Base used for addressing the

specific solutions based on the receiver input. Figure 94.3. The multiple agents will act as the resource that is capable of handling the wider tasks via agents, these single agents are thoroughly responsible for the process like emails and other fetchable data to ensure the accountability. The fall back is the simple content that is also coined as the secured net which comes in contact when the bot is met with the hypothetical query where it can't be able to move further or proceed the task anymore. Moving with the data LLM, it leverages the specific language processing ability required to maintain the chat bot performance that is needed to understand the time response of the chat bot to reply or to answer the complex version of queries this system integrates the different sort of agents as it need to fetch a huge data to have an appropriate responses, this parts contains the two functions namely the email integration and the fetch functionality this email integration involves the email related tasks such as sending or receiving the user queries as well as the informative notifications next moving with the fetch functionality this fully take over the response for recovering the major data such as database through API's. The final integration is the outputs for users and admins where it involves the admin panel and the user interaction here the admin panel takes control over the bots performance. Next is the user interaction that involves the queries and recommendations that receives the input relevant to the context responses.

Methodology

The retrieval augmented generation combines the large language model (LLM) with a vector search to response using internal data. It consists of following process: Primary consists of user query handling

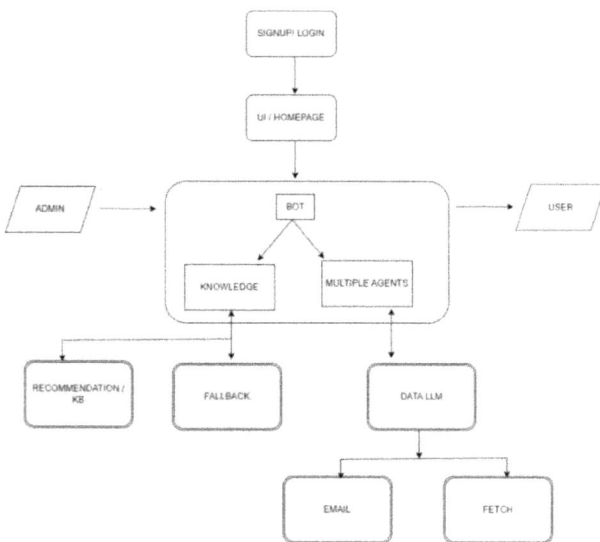

Figure 94.3 Flow diagram of AI powered chat bot
Source: Author

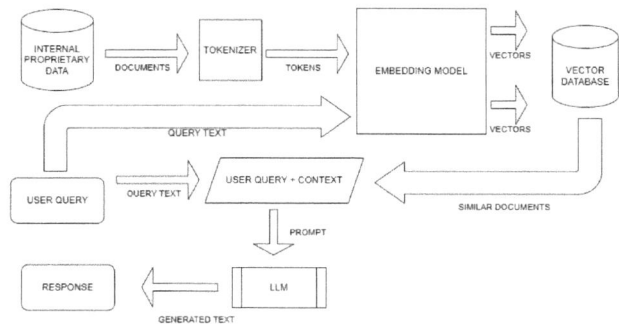

Figure 94.4 Retrieval augmented generation (RAG) flow
Source: Author

where the user can submit any query to the system and similar documents are retrieved using an integrated model and a vector data bank. Secondly, Augmentation is used to provide the additional context to the inquiry and the text generation generates the LLM based on the inquiry and information which is retrieved. In Figure 94.4 internal proprietary data storage where the system consists of internal data and the valuable knowledge where the documents are not available in pre trained models as shown in the first stage before retrieval, the documents need to be tokenized (which plays a major role in integrating the model for query section) and converted into the integrated. In the Tokenization process the stored documents will flow through the tokenizer, which breaks the documents into small token that is for example (sub words, characters or words). Tokenization is used in processing the text in an efficient way before feeding the text into an integrated model. From this we can conclude, this stage can the text into numerical format where the integrated model is processed. This architecture is needed for the recurrence as shown in RNNs where it enables parallelization and where the efficiency of NLP tasks is improved. Embedded model or the integrated model had played a major role in every stage and mainly this integrated model takes the tokenization text and converts into vector representations with high dimensional and these vectors transcode the significance of text and thus it allows better understanding of similar searches.

Mathematical approach

$$\begin{cases} C = C_0 ND \\ L = \frac{A}{N^\alpha} + \frac{B}{D^\beta} + L_0 \end{cases} \qquad (1)$$

Where,
  N – Number of restrictions in the model
  C – Cost of the model in the FLOPs
  D – Number of minimal tokens in the set
  L – Number of negative per token in the LLM dataset

The equation (1) is defined as the scaling law which is used to predict the LLM performance based on the pre-trained cost and the neural network size and also it considers the pre-trained dataset size and particularly scaling law which is also known as chinchilla scanning with a formula which is used to predict the training models costs.

$$\log(\text{Perplexity}) = -\frac{1}{N} \sum_{i=1}^{N} \log(\Pr(\text{token}_i \mid \text{context for token}_i)) \qquad (2)$$

Where,
  N – Tokens per text corpus
  i – Specific type of the LLM model

If the LLM is autoregressive the i th will appear before token else the LLM is masked, then i th is the segment of text which surrounds the tokens i. Higher the model higher will be the dataset when there is lower at the perplexity. In mathematical terms, the exponential of the average negative log per token.

$$\text{Entropy} = \log_2(\text{Perplexity}) \qquad (3)$$

Entropy is commonly quantified in bits per word (BPW) or bits per character (BPC) in which it hinges whether the language model which is utilizes by word based or character-based tokenization.

Results

Here the Figure 94.5 illustrates the percentage of the damaged part exactly by using the trained model and this is integrated into the chat bot for further insurance policy claim.

The confusion matrix is a tool that is utilized to activate performance enhancement in the categorization model.

By plotting the F1 model configuration, a clear visual representation is provided in the scatter plot and where each model performs in terms of car conditions while capturing relevant.

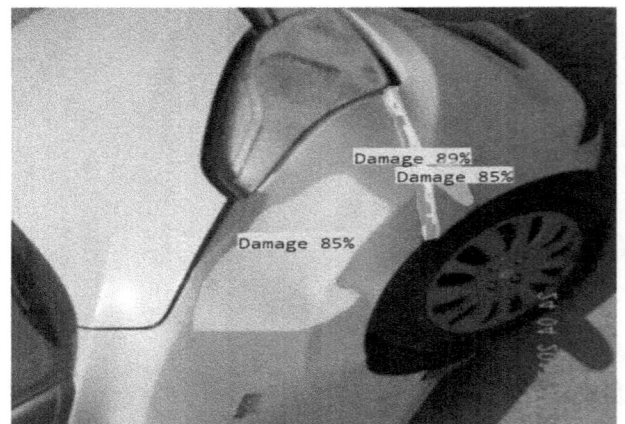

Figure 94.5 Damaged part detection using YOLOv11 instance segmentation
Source: Author

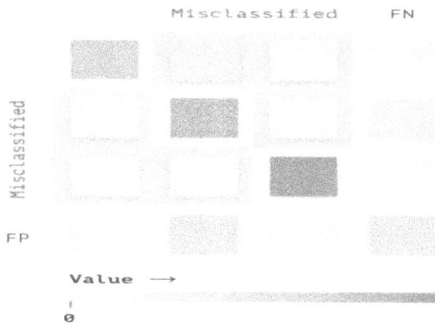

Figure 94.6 Obtained confusion matrix after training the model
Source: Author

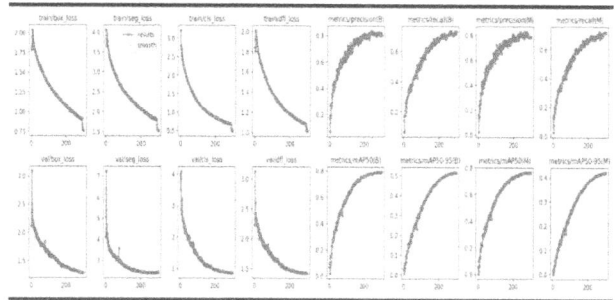

Figure 94.9 Graph obtained after training the model
Source: Author

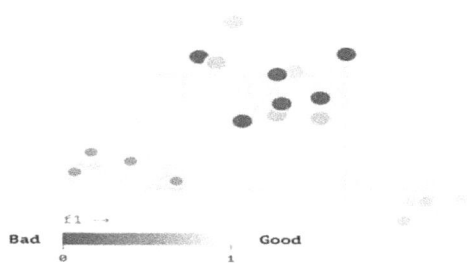

Figure 94.7 Vector analysis of the trained model
Source: Author

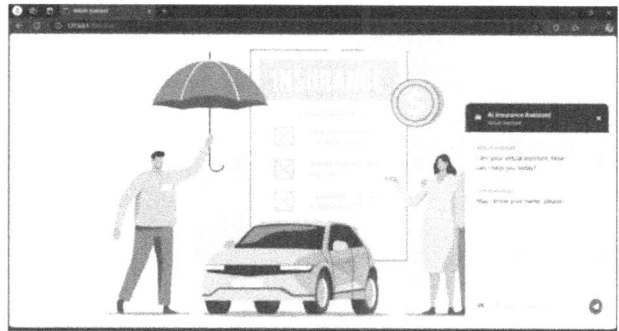

Figure 94.10 Web page integrated with virtual assistant for insurance claim
Source: Author

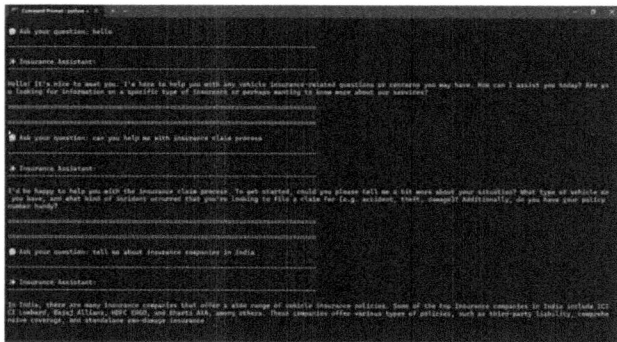

Figure 94.8 Chat bot responses in command prompt
Source: Author

Figure 94.11 Web page with the damaged part uploaded in chat bot for insurance claim
Source: Author

Here, Figure 94.6 the response of the LLM to user queries in a command prompt interface and Figure 94.7 helps them to apply for insurance and claim the same.

This Figure 94.9 is the graph obtained after the training the model using the YOLOv11 for the prediction and estimation of damaged part of the vehicle.

From this Figure 94.10, we can refer that the created virtual assistant is been integrated into the webpage and the users can be able to interact directly with the virtual assistant through chat.

The user deals with the emerged query by uploading their damages of the vehicle as images through chat Figure 94.8 and the virtual assistant will process it for insurance claims.

Conclusion

In conclusion, the above research provides a clear explanation on Automatic Damage Detection and Repair Cost Estimation using AI, fostering the insurance claiming process by reducing the number of days in a traditional claiming process and automating manual process and reducing the paperwork. With this, the user gets the required details and responses as quickly as possible.

Acknowledgement

We hereby express our heart full gratitude to the students, staff, and authorities of the Electronics and Communication Department for their invaluable cooperation and support in this research.

References

[1]   Mar Kyu, P., & Woraratpanya, K. (2020). Car damage detection and classification. In Proceedings of the 11th International Conference on Advances in Information Technology.

[2]   Waqas, U., Akram, N., Kim, S., Lee, D., & Jeon, J. (2020). Vehicle damage classification and fraudulent image detection including moiré effect using deep learning. In 2020 IEEE Canadian Conference on Electrical and Computer Engineering (CCECE).

[3]   Jang and Turk's Car-Rec: A Real-Time Car Recognition System was presented at the 2011 IEEE Workshop on Applications of Computer Vision (WACV) and focuses on real-time vehicle identification using computer vision, developed at UC Santa Barbara.

[4]   Jeffrey de Deijn, in his Master's thesis titled Automatic Car Damage Recognition using Convolutional Neural Networks, submitted to Vrije Universiteit Amsterdam on January 29, 2018

[5]   Patil, Kulkarni, Sriraman, and Karande authored Deep Learning Based Car Damage Classification," which was presented at the 2017 IEEE International Conference on Machine Learning and Applications (ICMLA).

[6]   An Anti-Fraud System for Car Insurance Claim Based on Visual Evidence by Li, Shen, and Dong was released as an arXiv preprint in 2018

For Product Safety Concerns and Information please contact our EU
representative GPSR@taylorandfrancis.com
Taylor & Francis Verlag GmbH, Kaufingerstraße 24, 80331 München, Germany

www.ingramcontent.com/pod-product-compliance
Lightning Source LLC
Chambersburg PA
CBHW081213220326
41598CB00037B/6765